ELEMENTS OF TRANSPORT PHENOMENA

LEIGHTON E. SISSOM, Ph.D., P.E.
Professor and Chairman
Department of Mechanical Engineering
Tennessee Technological University

DONALD R. PITTS, Ph.D.
Professor of Mechanical Engineering
Tennessee Technological University

McGRAW-HILL BOOK COMPANY
New York St. Louis San Francisco
Düsseldorf Johannesburg Kuala Lumpur
London Mexico Montreal
New Delhi Panama Rio de Janeiro
Singapore Sydney Toronto

ELEMENTS OF TRANSPORT PHENOMENA

Copyright © 1972 by McGraw-Hill, Inc. All rights reserved. Printed in the United States of America. No part of this publication may be reproduced, stored in a retrieval system, or transmitted, in any form or by any means, electronic, mechanical, photocopying, recording, or otherwise, without the prior written permission of the publisher.

Library of Congress Catalog Card Number 75-161670

07-057749-8

1234567890KPKP7987654321

This book was set in Times Roman, and was printed and bound by Kingsport Press, Inc. The designer was Nicholas Krenitsky; the drawings were done by John Cordes, J. & R. Technical Services, Inc. The editors were B. J. Clark and Laura Warner. Matt Martino supervised production.

CONTENTS

Preface   xi

List of Symbols   xiii

Part I. Philosophy and Fundamentals   1

   1. INTRODUCTION   3
      1-1   Transfer Phenomena   3
      1-2   Media   4
      1-3   Fluid Properties and the Continuum   5
      1-4   Equation of State   8
      1-5   The Perfect Gas   8
      1-6   Properties Derivable from Measurements   8
      1-7   Compressible and Incompressible Fluids   10
      1-8   Mass and Force   10
      1-9   Units and Dimensions   12
            Problems   13
            References   14

2. FUNDAMENTALS OF TRANSPORT PHENOMENA   15
   2-1 Fields   16
   2-2 Flux Density   16
   2-3 Field Intensity   17
   2-4 Rate Equations   17
   2-5 Unique Characteristics of the Rate Equations   26
   2-6 Conservation Laws   27
        Problems   29
        References   30

3. FUNDAMENTAL CONCEPTS FROM THERMODYNAMICS   31
   3-1 Equilibrium   32
   3-2 System and Control Volume   32
   3-3 Equality of Temperature   33
   3-4 The Zeroth Law of Thermodynamics   33
   3-5 Temperature Measurement and Thermometric Scales   34
   3-6 Properties and State of a Substance   35
   3-7 Work   38
   3-8 Heat   42
   3-9 The First Law of Thermodynamics   43
   3-10 The First Law for Processes in Closed Systems   45
   3-11 The Relationship between the First Law and Rate Processes   48
   3-12 Specific Heats   49
   3-13 The Second Law of Thermodynamics   50
   3-14 The First and Second Laws Combined   55
   3-15 The Third Law of Thermodynamics   56
        Problems   57
        References   61

Part II. Stationary Media   63

4. FLUID STATICS   65
   4-1 Static Equilibrium   65
   4-2 Hydrostatics   68
   4-3 Aerostatics   73
   4-4 Fluid Interfaces   76
   4-5 Forces on Submerged Surfaces   79
   4-6 Buoyancy   84
   4-7 Stability   86
        Problems   87
        References   102

5. STEADY-STATE CONDUCTIVE HEAT TRANSFER   103
   5-1 General Conductive Equation   103

CONTENTS

    5-2   One-dimensional Systems   107
    5-3   One-dimensional Systems with Heat Generation   111
    5-4   Convective Boundary Phenomena   115
*5-5   Heat Transfer from Fins   118
    5-6   Multidimensional Systems   127
    5-7   Analytical Solutions   127
    5-8   Conductive Shape-factor Solutions   134
    5-9   Numerical Solutions   138
          Problems   144
          References   151

6.   UNSTEADY-STATE CONDUCTIVE HEAT TRANSFER   152
    6-1   Physical and Mathematical Considerations   152
    6-2   Biot Modulus   153
    6-3   Lumped Systems   155
    6-4   One-dimensional Systems with Prescribed Surface Temperature   158
*6-5   One-dimensional Systems with Prescribed Convection Conditions   162
    6-6   Charts for One-dimensional Systems with Convective Boundary Conditions   164
    6-7   Extension to Multidimensional Systems   174
    6-8   Numerical Method   177
    6-9   Nonlinear Transient Conduction   180
          Problems   181
          References   185

7.   MOLECULAR MASS TRANSFER (DIFFUSION)   186
    7-1   The Diffusion Mode   187
    7-2   Fick's Law   188
    7-3   The Diffusion Coefficient   189
    7-4   Fluxes in Binary Mixtures   194
*7-5   Differential Form of the Mass-diffusion Equation   196
    7-6   Diffusion through a Stagnant Gas   200
    7-7   Diffusion through a Membrane   207
*7-8   Diffusion with Heterogeneous Chemical Reaction   209
*7-9   Diffusion with Homogeneous Chemical Reaction   213
*7-10 Transient Diffusion   215
          Problems   220
          References   225

8.   RADIATIVE HEAT TRANSFER   227
    8-1   Thermal Radiative Phenomena   227

* Sections marked with an asterisk can be omitted without loss of continuity.

- 8-2 Definitions and Properties 228
- 8-3 Blackbody Radiation 230
- 8-4 Radiant Exchange between Black Surfaces 237
- 8-5 Real Surfaces and Gray Surfaces 247
- 8-6 Radiant Exchange between Gray Surfaces 248
- 8-7 Radiant Exchange with Gases and Vapors 256
  - Problems 261
  - References 265

Part III. Moving Media 267

9. FLUID FIELDS 269
- 9-1 Description of a Flow Field 269
- 9-2 Fluid Motion 271
- 9-3 Types of Motion 274
- 9-4 Similitude 277
- 9-5 Relative Equilibrium 280
  - Problems 286
  - References 288

10. BASIC EQUATIONS 289

Conservation of Mass
- 10-1 Integral Form of the Continuity Equation 290
- 10-2 Differential Form of the Continuity Equation 295

Conservation of Momentum
- 10-3 Linear Momentum of a System 298
- 10-4 Integral Form of the Linear-momentum Equation 300
- 10-5 Differential Form of the Linear-momentum Equation 312
- 10-6 Moment of Momentum 318

Conservation of Energy
- 10-7 The First Law of Thermodynamics: Control-volume Analysis 322
- 10-8 Comparison of the First Law and Bernoulli's Equation 328
  - Problems 329
  - References 346

11. PERFECT FLUIDS 347
- 11-1 Rotation and Vorticity 348
- 11-2 Circulation 351
- *11-3 The Stream Function 354
- *11-4 The Velocity Potential 357

CONTENTS

    *11-5 Relationship between Stream Function and Velocity Potential   358
    *11-6 Boundary Conditions   360
    *11-7 Two-dimensional Simple Flows   361
    11-8 Closure   374
            Problems   374
            References   377

12. LAMINAR FLOW OF INCOMPRESSIBLE VISCOUS FLUIDS   378
    12-1 Isothermal Flow   378
    *12-2 Navier-Stokes Equations   380
    12-3 Flow in Tubes   386
    12-4 Channel Flow   393
    12-5 Flow over a Flat Plate   395
    12-6 Nonisothermal Flow   409
    *12-7 Energy Equation   411
    12-8 Convective Heat Transport on a Flat Plate   416
    12-9 Convective Heat Transport in a Tube   427
            Problems   431
            References   438

13. TURBULENT FLOW OF INCOMPRESSIBLE VISCOUS FLUIDS   440
    13-1 Isothermal Flow   440
    13-2 Equations of Motion for Turbulent Flow   442
    13-3 Fundamentals of Turbulent Flow in Pipes and Channels   448
    13-4 Empirical Relations for Turbulent Flow in Pipes   456
    13-5 Turbulent Flow over Surfaces   473
    13-6 Nonisothermal Flow   484
    13-7 Reynolds Analogy for Turbulent Flow   485
    13-8 Convective Heat Transport in Tubes   488
    13-9 Engineering Correlations for External Flow   494
    13-10 Heat Transfer to Liquid Metals   501
            Problems   504
            References   515

14. CONVECTIVE MASS TRANSFER   517
    14-1 Exact Analysis in a Laminar Boundary Layer   518
    14-2 Approximate Analysis of the Concentration Boundary Layer   525
    14-3 Analogy between Momentum, Heat, and Mass Transfer   529

14-4 Dimensional Analysis 535
14-5 Design Equations 536
14-6 Closure 540
     Problems 540
     References 544

15. ONE-DIMENSIONAL COMPRESSIBLE FLOW 546

15-1 The Speed of Sound 546
15-2 Mach Number and Mach Cone 548
15-3 Governing Equations 550
15-4 Isentropic Flow of a Perfect Gas 553
15-5 Isentropic Flow in a Varying-area Duct 558
15-6 Flow in Nozzles 561
15-7 Normal-shock Waves 566
15-8 Adiabatic Flow with Friction 573
15-9 Flow in Constant-area Frictionless Ducts with Heat Transfer (Rayleigh Flow) 582
15-10 Closure 585
     Problems 586
     References 592

16. FREE CONVECTIVE HEAT TRANSFER 593

16-1 Laminar Free Convection on a Vertical Plate 594
16-2 Solution of the Differential Equations 596
16-3 Integral Methods of Solution 602
16-4 Empirical Correlations for Natural Convection 605
16-5 Free Convection in Enclosed Layers 610
16-6 Combined Forced and Free Convection 612
     Problems 615
     References 617

17. MULTIPHASE PHENOMENA 619

17-1 Bubble Dynamics 619
17-2 Cavitation 624
17-3 Fundamentals of Two-phase Flow 628
17-4 Pressure Drop in Isothermal Flow 632
17-5 Fundamentals of Boiling Heat Transfer 636
17-6 Boiling-heat-transfer Correlations 642
17-7 Condensation Heat Transfer 648
17-8 Conduction with Phase Change 655
     Problems 660
     References 661

Part IV. Special Topics and Applications 665

18. THERMAL ANALYSIS OF HEAT EXCHANGERS 667
- 18-1 Introduction 667
- 18-2 Classification of Heat Exchangers 669
- 18-3 Heat-transfer Calculations 671
- 18-4 Effectiveness Method 681
- 18-5 Fouling Factors 689
- 18-6 Closure 691
  - Problems 692
  - References 694

19. OPEN-CHANNEL FLOW 695
- 19-1 The Speed of an Elementary Wave 696
- 19-2 Types of Motion 697
- 19-3 Steady, Uniform, Turbulent Flow 699
- 19-4 Optimum Shape of Cross Section 701
- 19-5 Transitions in Open Channels 702
- 19-6 The Analogy between Liquid and Gas Flow 705
  - Problems 711
  - References 715

20. FLOW THROUGH PERMEABLE MEDIA 717
- 20-1 Forces on Particles 719
- 20-2 One-dimensional Capillary-tube Model 722
- 20-3 Percolation: Fixed Bed 726
- 20-4 Fluidization: Expanded Bed 736
- 20-5 Solids Transport 741
  - Problems 746
  - References 748

Appendixes 749

- A. Tables of Properties and Functions 750
- B. Basic Equations for Selected Coordinate Systems 783
- C. Dimensional Analysis 786
  - C-1 Buckingham Pi Theorem 788
  - C-2 Momentum-transport Parameters 792
  - C-3 Heat-transport Parameters 795
  - C-4 Mass-transport Parameters 796
    - References 797

Answers to Odd-numbered Problems 799
Name Index 807
Subject Index 810

PREFACE

Reacting to the widespread desire to reduce the number of hours in the engineering curriculum while retaining its integrity, the authors and their colleagues have been working for 4 years to combine the elements of heat, mass, and momentum transfer into a single unified sequence. By eliminating redundant material courses could cover more topics in less time in the unified sequence, entitled Transport Phenomena, than had previously been possible in the separate transfer courses. More time was available for emphasizing fundamentals.

While it was possible to satisfy the needs of the Transport Phenomena sequence by carefully selecting material from a host of outstanding texts on its elements, no single text met the goals of the unified sequence. This book has evolved in response to that need. The authors claim little originality in its fundamental content but are pleased to introduce an approach they believe to be unique.

Relying upon the principle that physical phenomena are often so coupled that they can be described by common mathematical models, the text begins with the most elementary fluid condition—static. Analogous to the study of solid mechanics, the fundamental principles are then extended to moving media—forced and natural motion—and the student is carried through a process of increasing complexity as additional terms are introduced in the governing equations. The text concludes with carefully selected applications of interest to the chemical, civil, and mechanical engineer.

A salient feature of the text is its emphasis on the control-volume approach in both the overall and differential analyses. The book was written for the junior-level engineering student with a background in the fundamentals of solid mechanics and differential equations. It is presumed that the student has an elementary knowledge of vector analysis. Ideally, a student should have taken, or should take parallel with this study, a course in thermodynamics. Chapter 3, included as a review for those who have studied thermodynamics or as a bird's-eye view for those who have not taken a formal thermodynamics course, emphasizes the fundamental thermodynamic concepts basic to the study of the transport processes, the unifying definitions, and symbols.

The text was written for the chapters to be used in sequence as follows:

Quarter system: 1–7......8–13......14–20
Semester system: 1–10..............11–20

Each instructor is expected to be selective in the specialized materials, such as Chaps. 11 and 15, and in the applications. Sections marked with an asterisk in the Contents can be omitted without loss of continuity. The text may also be used, although perhaps less efficiently, for a first course in fluid mechanics and/or conduction and radiation heat transfer by studying the chapters in the following order:

Fluid mechanics: 1–3......4......9–11
Heat transfer: 1–3.....5–6........8

The authors are indebted to many teachers, students, and colleagues whose influence and encouragement have led to this work. Since much of the material came from notes and problems accumulated throughout the years, indebtedness is here expressed for material which cannot be specifically acknowledged due to loss of source or to such modification that it has come to seem original. A few classic problems will no doubt fall into this category. We are particularly grateful to Professors Griggs, Hewitt, Hribar, Purdy, Luckinbill, Miller, and Wallace for their classroom testing of the manuscript and for their invaluable critique.

Recognizing the limitations and compromises inherent in this integrated treatment, we hope that it will save time, emphasize the affinities among physical phenomena, and instill in the student an appreciation of the transfer processes.

Leighton E. Sissom
Donald R. Pitts

LIST OF SYMBOLS

a	acceleration
a_p	packing area
A	area
c	specific heat; velocity of propagation of radiation; speed of elementary surface wave; speed of sound
c_D	local drag coefficient
c_p	specific heat at constant pressure
c_v	specific heat at constant volume
C	capacitance
C_c	contraction coefficient
C_D	drag coefficient
$\mathbf{C}_p$	pressure coefficient, $p/\tfrac{1}{2}\rho V^2$
d	diameter
d_p	particle diameter
D	mass diffusivity; diameter
D_h	hydraulic diameter
D_p	average particle diameter
e	unit energy; roughness height

xiii

LIST OF SYMBOLS

E	electric potential; total energy; radiation energy emitted; total emissive power
E_T	isothermal bulk modulus of elasticity
f	friction factor
F	force; configuration factor
g	gravitational acceleration
g_c	gravitational constant, 32.2 lb_m-ft/lb_f-sec^2
G	shear modulus of elasticity; incident radiant energy
h	head; height (length in general), unit enthalpy, $u + pv$; convective-heat-transfer coefficient; Planck's constant, 6.625×10^{-27} erg-sec
h_f	head loss
H	total enthalpy, $U + p\mathscr{V}$; geometric flux
$\mathbf{i}$	unit vector in x direction
I	electric current; moment of inertia; radiation intensity
$\mathbf{j}$	unit vector in y direction
j	Colburn factor
J	mechanical equivalent of heat, 778.2 ft-lb_f/Btu; radiosity
J_A	mass flux of species A relative to the mass average velocity of the mixture
J_A^*	mass flux of species A relative to the axes moving at the molal average velocity
$\mathbf{k}$	unit vector in z direction
k	ratio of specific heats, c_p/c_v; thermal conductivity
k_c	convective-mass-transfer coefficient
k_L	loss coefficient
k_1	chemical rate constant
K	permeability
l	length
L	length
m	unit mass; parameter in fin equations, $\sqrt{hP/kA} = \sqrt{2h/kr}$
M	momentum; number of squares in a flow passage; parameter in finite-difference equations
n	polytropic exponent; Manning number
N	number of flow lanes
N_A	mass flux of species A relative to fixed axes
O	order of
p	pressure
p_A	partial pressure of species A
P	circumference; perimeter; wetted perimeter; property
q	heat flux; unit heat transfer
q'''	time rate of energy generation per unit volume
Q	total heat transfer; volumetric flow rate
r	radius
r_A'''	mass of species A produced per unit volume per unit time

LIST OF SYMBOLS

R	gas constant; radius; resistance
R_h	hydraulic radius, A/P
s	unit entropy; parameter in fin equations
S	total entropy; conduction shape factor, N/M; specific gravity; surface area; cross-sectional area
t	thickness; time
T	temperature; torque
T_b	bulk temperature
T_f	film temperature
u	unit internal energy; x component of velocity
u	unit vector
U	total internal energy
v	y component of velocity
v_*	shear velocity, y component
V	velocity
w	unit work; width; linear dimension; z component of velocity
W	total work; weight
$\dot{W}$	power (work rate)
x	quality
x, y, z	cartesian coordinates

Dimensionless Quantities

Bi	Biot modulus, hL/k
Eu	Euler number, $p/\rho V^2$
Fo	Fourier modulus, $\alpha t/L^2$
Fr	Froude number, $V/\sqrt{gl}$
Gr	Grashof number, $g\beta\rho^2 \Delta T_x^3/\mu^2$
Le	Lewis number, α/D
M	Mach number, V/c
Nu	Nusselt number, hx/k
Pe	Peclet number, **RePr**
Pr	Prandtl number, ν/α
Re	Reynolds number, Vl/ν
Sc	Schmidt number, ν/D
Sh	Sherwood number, $k_c x/D$
St	Stanton number, $h/c_p \rho V_\infty$
We	Weber number, $V^2 l \rho / \sigma$

Script

$\mathscr{D}$	flux density
$\mathscr{F}$	flux
$\mathscr{l}$	Prandtl mixing length

$\mathcal{M}$	molecular weight
$\mathcal{P}$	permeability
$\mathcal{R}$	universal gas constant, 1545.33 ft-lb$_f$/mole-°R
v	specific volume
$\mathcal{V}$	volume
$\mathcal{W}$	weight fraction

Greek Symbols

α	coefficient of linear expansion; absorptivity; thermal diffusivity, $k/\rho c_p$
β	coefficient of volume expansion
β_T	coefficient of compressibility
β_v	coefficient of tension
γ	angular deformation; specific weight
Γ	circulation
δ	boundary-layer thickness
δ_i	momentum thickness
δ_n	modified eigenvalue of transcendental equation
δ^*	displacement thickness
Δ	finite increment
ϵ	emissivity; eddy viscosity; effectiveness; void fraction
ζ	ratio of thermal to velocity boundary-layer thickness, δ_t/δ
η	efficiency; transformed y coordinate
θ	angle; reference temperature
κ_λ	monochromatic absorption coefficient
λ	second coefficient of viscosity; separation constant; lapse rate; mean free path; wavelength
λ_n	roots of transcendental equation
Λ	source (sink) flow rate, AV
μ	absolute viscosity
ν	kinematic viscosity; frequency of radiation
ξ	rate of normal strain; transformed x coordinate
ρ	mass density; reflectivity
σ	normal stress; surface tension; Stefan-Boltzmann constant, 0.1714×10^{-8} Btu/hr-ft^2-°R^4
Σ	summation
τ	shear stress; transmissivity
ϕ	angular coordinate; velocity potential
Φ	viscous-dissipation function
ψ	angular coordinate; stream function
ω	mass fraction, ρ_A/ρ; concentration; angular velocity; rotation; unit solid angle
Ω	vorticity, 2ω

LIST OF SYMBOLS

Subscripts

b	black body
c	center; centerline condition
e	exit; electric
f	saturated liquid; fin; film condition
fg	difference between saturated vapor and saturated liquid
g	saturated vapor
H	high-temperature reservoir (source)
i	initial; species i; inside
lm	log mean
L	low-temperature reservoir (sink); condition at linear distance equal to L
m	fixed value
M	model
n	integer
o	outside
p	profile; isobaric
P	prototype
r	ratio of model to prototype
s	surface condition
t	thermal
th	thermal
T	isothermal
v	isovolumetric
w	wall condition
x	upstream from normal shock
y	downstream from normal shock
yx	in a plane perpendicular to y and parallel to x
0	reference state; condition at zero linear distance; condition at zero time
∞	condition at infinity

Superscripts

$'$	fluctuating component; per unit length
$''$	per unit area
$'''$	per unit volume
$*$	sonic condition; critical state

Miscellaneous

$\cdot$	rate; per unit time
$-$	average of; distance to centroid or center of gravity
$\sim$	molar quantity

PART I
PHILOSOPHY AND FUNDAMENTALS

CHAPTER 1
INTRODUCTION

Historically the study of transfer phenomena has been categorized by processes. The study of some of these processes, which include the transfer of momentum, heat, mass, and electricity, has unfortunately been further restricted through alignment with specific branches of engineering. However, different physical phenomena can often be described by common mathematical models. This text relies upon this fact in seeking to unify the study of heat, mass, and momentum transfer.

1-1 TRANSFER PHENOMENA

The transfer process is characterized by the tendency toward equilibrium, a condition of no change. Common to a transfer process are the transport of some quantity, a driving force, and the move toward equilibrium. The characteristics of the mass of material through which the changes occur affect the *rate* of transport, and the geometry of the material affects the *direction*.

Consider what happens when a drop of dye is placed in water. The mass-transfer process causes the dye to diffuse throughout the water, reaching a state of equilibrium, which is easily detected visually. We can detect a similar change by smell when a small amount of perfume is sprayed into a room. The concentration becomes fainter at a point near the source as the perfume diffuses throughout the room. Anyone who has picked up a hot poker has felt the effects of heat transfer. The change in efflux of hot gases from a rocket engine can be noted by the sound. One can even sense the change by taste, as when a sugar cube dissolves and diffuses in the mouth. Hence, transfer processes are part of everyday experience.

In general, transfer processes occur simultaneously, and sometimes the individual fluxes interfere with one another. Heat and mass transfer occur simultaneously when a coolant is forced through a hot porous plate. In thermoelectric refrigeration an electric potential is used to extract heat from a storage chamber by causing a thermal potential to develop. In most cases, however, it is possible to separate the individual phenomena, recognizing that although they are coupled in fact, they obey common physical laws and can be described by common mathematical equations.

1-2 MEDIA

All matter is made up of *solid*, *liquid*, or *gas* or a combination of them. Since the transfer processes are affected by the medium through which the changes occur, it is imperative to understand the characteristics of each state.

A *solid* is generally thought of as a substance which offers resistance to change of shape (deformation), whereas, a *fluid* will deform continuously when subjected to a shearing stress, no matter how small. The mode of resistance distinguishes between a solid and a fluid.

A fluid may be either *liquid* or *gas* or a combination of the two. Fluids conform to the shape of their container. A liquid has a free surface, but a gas fills the entire container and has no free surface.

In order to understand the nature of a fluid, consider first the behavior of a solid. Assume that a solid block, bonded to a surface and a loading plate, is deformed by a force F as shown in Fig. 1-1. Within certain limits, depending upon the material used, the shear stress $\tau = F/A$ is proportional to the angular deformation γ, or

$$\tau = G\gamma \tag{1-1}$$

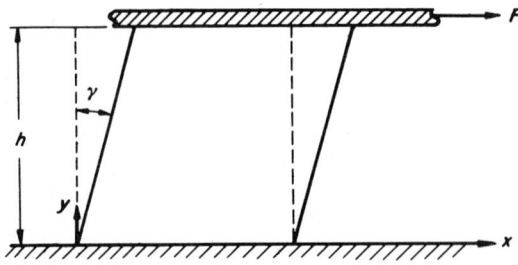

Fig. 1-1 Solid block subjected to shear deformation.

INTRODUCTION

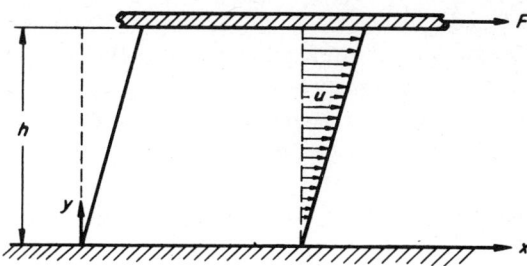

Fig. 1-2 Fluid element subjected to shear deformation.

where the constant of proportionality G is the shear modulus of elasticity of the material.

If the solid block is replaced by a fluid element of equal size, application of a force F will continue to deform the fluid element. The loading plate will move with a limiting velocity U proportional to the height h, and infinitesimal elements throughout the fluid will move with a velocity u proportional to their distance from the plate y, as shown in Fig. 1-2. For a large class of common fluids the shear stress is proportional to the velocity gradient, or

$$\tau = \mu \frac{du}{dy} \approx \mu \frac{\Delta u}{\Delta y} = \mu \frac{U}{h} \tag{1-2}$$

where the constant of proportionality μ is the absolute viscosity, one of the most important characteristics of a fluid.

We readily think of water and air as fluids, but many other substances which behave quite differently are also fluids, e.g., asphalt, glass, and Silly Putty. Blood is a fluid whose behavior varies widely, depending upon its content of hemocytes (blood cells), sugar, and plasma. "As slow as molasses in January" suggests the dependence of behavior on temperature. We ask then, what characteristics cause these substances—all fluids—to behave differently?

1-3 FLUID PROPERTIES AND THE CONTINUUM

To understand matter it is necessary to consider its molecules, which are in constant motion, colliding and rebounding not unlike billiard balls. To describe matter the history of each molecule must be known. This requires knowing each molecule's velocity and acceleration—quite impossible except statistically. In engineering applications, however, we are interested only in the manifestations of the molecular motion, i.e., what can be sensed and expressed in measurable terms? The answer is properties.

A property, an observable quantity, always has the same value when measured under the same conditions, regardless of how those conditions were reached. Consider a small closed container of gas. What happens as the number of molecules is reduced? The force per unit area on the wall of the container resulting from the collision of molecules, which is the *pressure p*, is decreased, since pressure is the effect of the average force resulting from repeated impacts of the molecules on the wall. There is a point, however, below which a reduction in one molecule produces a pressure which is discontinuous; hence, it is not reproducible when brought to the same conditions, and therefore it is not a property. This occurs when the *mean free path* of the molecules, the average distance traveled by the molecules between collisions, is of the same order of magnitude as the smallest significant length (the side of the container in the case under consideration). This point where behavior changes determines the lower bound of the *continuum*. The continuum results from a continuous distribution of matter [2].[1]

Figure 1-3 shows graphically how erratic the pressure might be if only a few molecules of gas were permitted in the container. As individual molecules or groups of them bombard the surface, an erratic pressure occurs. Measurement of pressure at time t_1 and t_2 yields two different values, violating the definition of a property. As defined then, a property has meaning only in a continuum. Noncontinuum behavior is treated in statistical mechanics and the kinetic theory of gases.

To further demonstrate the concept of a continuum, consider the property density ρ, defined as the mass per unit volume. Let $\delta \mathscr{V}$ be an arbitrary volume around a point P. A mass of material δm is contained within the volume. This quantity is meaningless as the volume $\delta \mathscr{V}$ approaches the size of the mean free path of the molecules. There is some volume $\delta \mathscr{V}_c$, however, that represents the lowest value at which this quantity will have meaning. This permits us to define the property density [3], i.e.,

$$\text{Density at point } p: \qquad \rho \equiv \lim_{\delta \mathscr{V} \to \delta \mathscr{V}_c} \frac{\delta m}{\delta \mathscr{V}} \tag{1-3}$$

Figure 1-4 illustrates that as molecules pass into and out of a volume less than $\delta \mathscr{V}_c$, $\delta m/\delta \mathscr{V}$ may be small (even zero) if few molecules occupy the space or large if many molecules occupy the space. For a property to have meaning it must be continuous. We may note that a spaceship passes from a continuum into a domain

[1] Numbered references appear at the end of each chapter.

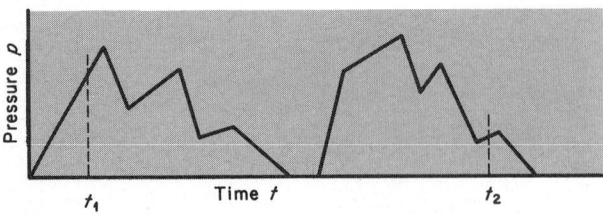

Fig. 1-3 Noncontinuum pressure distribution.

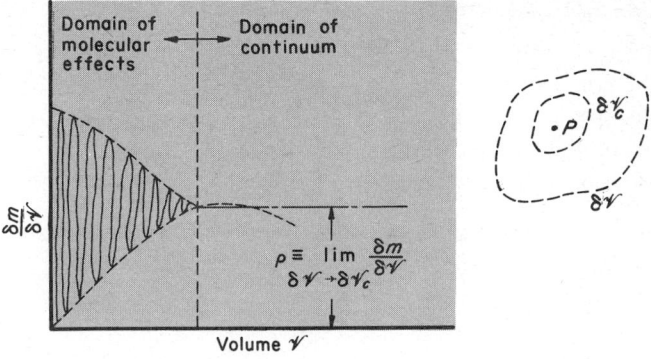

Fig. I-4 Definition of density.

of molecular effects as it leaves the atmosphere on a voyage into space. Noncontinuum effects are important, but they will not be treated in this text.

Specific volume v is the reciprocal of density; that is, $v \equiv 1/\rho$. *Specific gravity* S is the ratio of the density of a substance to that of pure water at 4°C and 76 cm Hg; thus

$$S \equiv \frac{\rho}{\rho_w} = \frac{\gamma}{\gamma_w}$$

where $\gamma = \rho g$ is the specific weight.

Temperature T is a property which enables us to determine whether two bodies or two adjacent fluid elements are in thermal equilibrium. It is a measure of the average translational kinetic energy of the molecules. We use the terms "hot" and "cold" in reference to high and low temperatures. Although temperature is a familiar property, it is difficult to define[1] because the definition must be indirect, through the concept of equality of temperature. The concept of temperature and its measurement is developed fully in Chap. 3.

The property *concentration* ω is of value when dealing with mixtures, such as dye in water or lemon in iced tea. In a diffusing mixture, the mass of individual species per unit volume, *mass concentration*, may be significant, as in the example of having enough sugar in one's coffee.

Properties are interrelated. For example, water and sulfuric acid, initially at the same temperature, will rise in temperature when they are mixed. The amount of temperature rise depends on the concentration. If they are contained in a rigid vessel at constant volume, the pressure will increase although the density remains constant. An unrestrained balloon filled with air increases in size when it is heated, as the pressure on the inside of the balloon remains equal to that on the outside.

---

[1] It is no more difficult to define temperature than to describe how cornbread tastes.

1-4 EQUATION OF STATE

The interrelation of fluid properties is in the domain of thermodynamics. During the conversion of energy within the fluid or between the fluid and its surroundings, the condition and motion of the fluid are affected. An *equation of state* relates the properties of the fluid as it undergoes a change. Fortunately, for most substances of engineering interest, the equation of state has a simple mathematical form, e.g.,

$$\rho = \phi(p,T) \tag{1-4}$$
$$p = f(\rho,T) \tag{1-5}$$
$$T = \xi(p,\rho) \tag{1-6}$$

These functional relations are always true for simple, compressible, pure substances although the equations which actually describe the relation may be quite formidable. But for most gases in a continuum, the equation is rather simple when pressures and temperatures are not extreme.

1-5 THE PERFECT GAS[1]

For a gas whose molecules collide perfectly elastically the equation of state is

$$p = \rho RT \tag{1-7}$$

R is a constant which depends solely on the molecular weight of the gas, T is the absolute temperature, discussed further in Chap. 3, and p is the absolute pressure. Values of the gas constant R are given for a number of common gases in Appendix Table A-10. Although strictly speaking the equation is applicable only to a hypothetical gas, sometimes referred to as a perfect *gas*, it very closely describes the behavior of many gases such as air, oxygen, and helium.

All gases approach perfect-gas behavior at low pressure, 1 atm or less, and at high temperatures, in excess of twice the critical temperature.[2] The perfect-gas equation yields good results when the density of the gas is of the order of one-thousandth of the density of the corresponding liquid. Real-gas effects are very pronounced near condensation conditions. *The student is cautioned against using the perfect-gas equation for water vapor* (*steam*).

1-6 PROPERTIES DERIVABLE FROM MEASUREMENTS

Since specific volume is the reciprocal of density, Eq. (1-4) can be rewritten

$$v = v(p,T) \tag{1-8}$$

---

[1] Often used synonymously with *ideal gas*. The more precise definition has been adopted here, however, reserving the term ideal gases for those perfect gases whose specific heats are functions of temperature only (Sec. 3-12). The term *perfect gas* will connote constant specific heats.

[2] The critical temperature is the temperature at which the vapor and liquid phases of a pure substance have identical properties when at the same pressure.

INTRODUCTION

Applying the chain rule of partial differentiation gives

$$dv = \left(\frac{\partial v}{\partial p}\right)_T dp + \left(\frac{\partial v}{\partial T}\right)_p dT \tag{1-9}$$

By measuring how the volume changes with pressure in an isothermal process (constant temperature) and with temperature in an isobaric process (constant pressure), it is possible to determine the variation in specific volume. More important, however, is that the partial derivatives have special physical significance [4], as discussed below.

The *coefficient of compressibility*, sometimes termed isothermal compressibility, is defined as

$$\beta_T \equiv -\frac{1}{v}\left(\frac{\partial v}{\partial p}\right)_T \tag{1-10}$$

and includes the first of the partial derivatives. It gives the change in volume with pressure when a process is carried out isothermally.

The second of the partial derivatives is a part of the *coefficient of volume expansion*:

$$\beta \equiv \frac{1}{v}\left(\frac{\partial v}{\partial T}\right)_p \tag{1-11}$$

Using these coefficients, Eq. (1-9) can be rewritten to give

$$dv = -\beta_T v\, dp + \beta v\, dT \tag{1-12}$$

or

$$\ln\frac{v_2}{v_1} = -\int \beta_T\, dp + \int \beta\, dT \tag{1-13}$$

Since β_T and β are defined in terms of properties, they are also properties. Equation (1-13) readily gives density changes, since

$$\ln\frac{\rho_1}{\rho_2} = \ln\frac{v_2}{v_1} \tag{1-14}$$

Analogous to Eq. (1-11), the *coefficient of linear expansion* α is defined by

$$\alpha \equiv \frac{1}{L}\left(\frac{\partial L}{\partial T}\right)_p \tag{1-15}$$

which is one-third of the coefficient of volume expansion in isotropic media, i.e., media that are the same in all directions.

Another property which arises physically is the *coefficient of tension*

$$\beta_v \equiv \frac{1}{\rho}\left(\frac{\partial p}{\partial T}\right)_v \tag{1-16}$$

which indicates the change in pressure with temperature in processes carried out at constant volume.

1-7 COMPRESSIBLE AND INCOMPRESSIBLE FLUIDS

The *isothermal bulk modulus of elasticity*, defined as the reciprocal of the coefficient of compressibility,

$$E_T = \frac{1}{\beta_T} \tag{1-17}$$

is often used as a measure of incompressibility. For liquid water at ordinary temperatures and pressures $E_T = 300,000$ psi. How compressible then is water? Consider the application of 1,000 psi pressure to 1 ft³ of water. Combining Eqs. (1-10) and (1-17) and writing in terms of average values,

$$\Delta v = -\frac{\bar{v}\,\Delta p}{\bar{E}_T} = -\frac{1,000}{300,000} = -\frac{1}{300} \text{ ft}^3$$

Therefore, the volume decreases (negative sign) by only 1 part in 300 under a pressure of 1,000 psi.

For a perfect gas $E_T = p$, a function of pressure alone. For processes that are not isothermal, the bulk modulus of elasticity is not a function of temperature but only of pressure. The implication is that *compressibility* is a function of pressure level. But more in keeping with common usage, the compressibility of a fluid is defined in terms of its density.

Fluids whose density changes are insignificant in a given process are said to be *incompressible*. Under normal engineering conditions liquids are considered incompressible; gases and vapors may be *compressible*, since their densities may change significantly.

Gas dynamics is the study of compressible flows and *hydrodynamics* that of incompressible flows.

1-8 MASS AND FORCE

In this age of rockets and space travel, it is important to distinguish carefully between mass and weight. Weight is the force of gravity on a body. Where there is no force of gravity, as in deep space, a body has no weight.

Mass is a quantity of matter. It consists of some number of molecules, depends upon molecular structure, and remains unaffected by external influences. We shall ignore Einstein's relation between mass and energy and consider mass as indestructible. Using a beam balance, we can measure mass comparatively. The system of Fig. 1-5 is balanced in the absence of external forces other than gravity, anywhere—on earth, in space, or on the moon. But the force exerted by a given mass depends upon the local gravitational field. A spring scale (Fig. 1-6) loaded with the same

INTRODUCTION

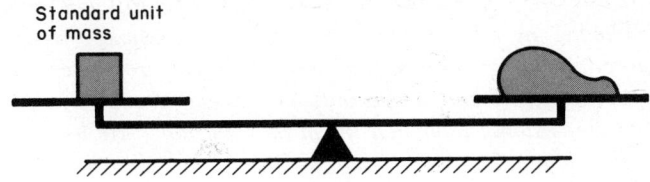

Fig. I-5 Beam balance for measuring mass.

mass deflects quite differently when it is located in different gravitational fields.[1] A mass which weighs 6 lb_f on earth would weigh approximately 1 lb_f on the moon and would weigh nothing in deep space.

In 1686 Sir Isaac Newton related force, mass, length, and time in his second law of inertia, which states that the force on a body of fixed mass is proportional to the product of its mass and acceleration.

$$\mathbf{F} \propto m\mathbf{a} \tag{1-18}$$

Choosing the proportionality constant to be $1/g_c$, we have

$$\mathbf{F} = \frac{m\mathbf{a}}{g_c} \tag{1-19}$$

The term g_c is a constant whose value depends only upon the units involved and not on the acceleration due to gravity g at a particular location.

In this text the use of the gravitational constant in an equation will depend upon its use in a bulk of the literature in a given area. In thermodynamics, for example, g_c is normally included in the equations, and the unit of mass is the pound mass (lb_m). In fluid mechanics, however, the slug mass unit is used, building upon that which is common in solid mechanics. Accordingly, the gravitational constant will usually be omitted from fluid-mechanics equations. Since the study of transport phenomena

[1] A pound is not a "pound the world around" unless, of course, it is a pound mass.

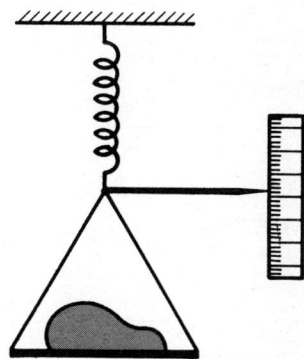

Fig. I-6 Spring scale for measuring weight.

crosses these fundamental areas, the gravitational constant g_c will be included or excluded in accordance with the most common treatment in the literature. The practicing engineer encounters tabulated properties in many different units. The gravitational constant g_c is only one of several quantities which are necessary in establishing dimensional homogeneity in an equation. Its use should be completely understood from the outset.

1-9 UNITS AND DIMENSIONS

A set of basic entities expressing our observations of the magnitudes of certain quantities is known as a *dimension*. Many units can be used to describe a dimension. For example,

$$36 \text{ inches} = 3 \text{ feet} = 1 \text{ yard} = 91.44 \text{ centimeters}$$

Inches, feet, yard, and centimeters are units, but they all represent a measure of length—dimension. In transport processes the *basic dimensions* are defined to be *force F, length L, time T, temperature θ*, and *mass M*.

Table 1-1 Dimensions and units

Dimensional system	Defined units	Derived unit	Gravitational constant g_c
Absolute			
Metric	Mass, g Length, cm Time, sec Temp., °K	Force: $\text{Dyne} = \dfrac{\text{g-cm}}{\text{sec}^2}$	$1 \dfrac{\text{g-cm}}{\text{dyne-sec}^2}$
English	Mass, lb_m Length, ft Time, sec Temp., °K	Force: $\text{Poundal} = \dfrac{lb_m\text{-ft}}{\text{sec}^2}$	$1 \dfrac{lb_m\text{-ft}}{\text{poundal-sec}^2}$
British			
Technical	Force, lb_f Length, ft Time, sec Temp., °R	Mass: $\text{Slug} = \dfrac{lb_f\text{-sec}^2}{\text{ft}}$	$1 \dfrac{\text{slug-ft}}{lb_f\text{-sec}^2}$
Engineering	Force, lb_f Mass, lb_m Length, ft Time, sec Temp., °R		$32.17 \dfrac{lb_m\text{-ft}}{lb_f\text{-sec}^2}$

INTRODUCTION

Table 1-1 summarizes the different dimensional systems, with units, which are used in engineering practice [1]. The derived units result from the application of Newton's second law with the defined units. Note that the British engineering system defines both mass and force, leaving no unit to be derived from Newton's second law. Relating mass units in the British system,

$$1 \text{ slug} \equiv 32.17 \text{ lb}_m \equiv 1 \text{ (lb}_f\text{-sec}^2/\text{ft)}$$

In general, the British engineering system will be used in this text; however, the student should develop facility in converting from one system to another.

Physical equivalence of units A scheme for converting units from one system to another with the least possible chance for error consists of establishing physical equivalence between quantities [2]. To apply this scheme, form a ratio of one unit and the proper number to another unit such that the ratio is unity. Consider the following statements of equivalence:

$$\frac{3 \text{ ft}}{1 \text{ yd}} \equiv 1 \qquad \frac{12 \text{ in}}{1 \text{ ft}} \equiv 1 \qquad \frac{2.54 \text{ cm}}{1 \text{ in}} \equiv 1$$

Since multiplying the right sides of the equivalences ($1 \times 1 \times 1 = 1$) does not change the measure of the quantity represented, it follows that the left sides can also be multiplied. Hence,

$$\frac{3 \text{ ft}}{1 \text{ yd}} \times \frac{12 \text{ in}}{1 \text{ ft}} \times \frac{2.54 \text{ cm}}{1 \text{ in}} = 91.44 \text{ cm/yd}$$

the conversion factor cited earlier. As another simple example, let us convert 60 mph to feet per second:

$$60 \frac{\text{miles}}{\text{hr}} \times \frac{5{,}280 \text{ ft}}{1 \text{ mile}} \times \frac{1 \text{ hr}}{3{,}600 \text{ sec}} = 88 \text{ ft/sec}$$

The inclusion of all units in a problem at the elementary level is the most important single thing a student can do to assure his success in solving the problem. The habit of including a unit analysis, when established early, becomes a part of the student's *modus operandi*.

PROBLEMS

1-1. A shear stress of 0.002 lb_f/in^2 causes a fluid to deform at the rate of 100 ft/sec-ft. What is the viscosity of the fluid in units of $\text{lb}_f\text{-sec}/\text{ft}^2$?

1-2. The specific gravity of kerosene is 0.82. What is its density if the density of water is 62.4 lb_m/ft^3?

1-3. What is the density of carbon dioxide stored at a pressure of 100 lb_f/in^2 and a temperature of 600°R? $R = 35.1$ ft-lb_f/lb_m-°R.

1-4. Standard gas bottles are pressurized to 2,000 psia. What is the density of hydrogen gas in such a container at 530°R if $R = 766.4$ ft-lb_f/lb_m-°R?

1-5. Air enters a tank 50 ft^3 in volume at 5 lb_m/sec while 2 lb_m/sec is discharged from the tank. If the temperature inside the tank remains constant at 500°R, find the rate of pressure rise inside the tank.

1-6. A truck tire has a constant volume of 6,000 in³ and contains 0.024 slug of air. What is the range of absolute pressures in the tire for a temperature range of 30 to 150°F?

1-7. (a) A weather balloon filled with 0.1 lb_m helium at a pressure of 14.7 psia has a volume of 10 ft³. What is the temperature of the helium?

(b) If the balloon rises to an elevation where the pressure is 10 psia, what is its volume, assuming constant temperature?

1-8. A 1-in-diam metal rod 20.000 ft long increases in temperature from 50 to 400°F. If the coefficient of linear expansion is 0.0000163 in/in-°F, what is its final length?

1-9. The coefficient of volume expansion for mercury is 0.0001819 in³/in³-°C. If the compressibility is 3.92×10^{-6} in³/in³-atm, what pressure will develop if liquid mercury at 1 atm is heated at constant volume from 82 to 83°C?

1-10. Find the equation of state of a substance which has a coefficient of compressibility of $1/p$ and a coefficient of volume expansion of $1/T$.

1-11. A test is conducted on the compression of rubber. If 0.016 lb_m of hydrogen is contained by a rubber balloon of 3 ft³ volume at 14.7 psia, what coefficient of compressibility can be expected if the balloon is placed in a vacuum chamber and the hydrogen pressure is raised by 1 psi?

1-12. A thermal-overload-protection device is being designed for use in an electrical system. The type of metal to be used is uncertain, but the choice has been narrowed to aluminum and nickel, with respective coefficients of linear expansion of 1.33×10^{-7} and 7.4×10^{-6}. If the metal strip is to be 9 in long and to operate over a range of 50°C, which metal would be better suited to ground the system by expanding over a 0.0059-in gap?

1-13. What is the isothermal bulk modulus of elasticity E_T of a perfect gas?

1-14. A piston of 1 ft² area exerts a pressure of 10 lb_f/in² on the moon, where there is negligible atmosphere and $g \approx g_c/6$. What mass will the piston have on the earth?

1-15. In a location where $g = 30$ ft/sec², what force is indicated from a 2-lb mass on (a) a beam balance and (b) a spring scale?

1-16. An inhabitant of outer space weighs 50 lb_f on a spring type of scale (calibrated where $\mathbf{g}_c = 32.17\ lb_m$-ft/$lb_f$-sec²) in the planet's atmosphere, where the local gravity acceleration is 6.0 ft/sec². This inhabitant appears in St. Michael, Alaska, where the local gravity acceleration is 32.22 ft/sec². Determine (a) his mass (our pounds mass) on his native planet; (b) his mass on earth; (c) his weight as indicated by the same spring scale in St. Michael.

1-17. What is the weight of 2 ft³ of hydrogen gas at 2,000 psia and 530°R where the gravitational acceleration is 29.8 ft/sec²?

1-18. (a) What airstream force is required to barely suspend a 0.05-lb_m ball above the exit of a vertical nozzle where $g = 30$ ft/sec²?

(b) Why would more force be necessary to suspend the ball farther from the nozzle?

1-19. A ball which weighs 2 lb_f at a point where the local acceleration is 30 ft/sec² is rolled down a plane inclined at a 30° angle with the horizontal at an acceleration of 1 ft/sec².

(a) What is its mass?

(b) How much would it weigh on the moon where $g_{moon} \approx g_{earth}/6$?

REFERENCES

1. Eskinazi, S.: "Principles of Fluid Mechanics," 2d ed., Allyn and Bacon, Boston, 1968.
2. Shames, I. H.: "Mechanics of Fluids," McGraw-Hill, New York, 1962.
3. Shapiro, A. H.: "The Dynamics and Thermodynamics of Compressible Fluid Flow," Ronald, New York, 1953.
4. Van Wylen, G. J., and R. E. Sonntag: "Fundamentals of Classical Thermodynamics," Wiley, New York, 1965.

CHAPTER 2
FUNDAMENTALS OF TRANSPORT PHENOMENA

We can study transport phenomena from two viewpoints, *lagrangian* or *eulerian*, and it is important to adopt the one which will yield accurate answers to our physical problems in the most straightforward manner.

In elementary solid mechanics the lagrangian method of analysis is used. It describes the behavior of discrete particles, or point masses, as they move in space. Fundamental laws, such as Newton's second law, apply directly to the discrete masses under consideration. The same viewpoint can also be used to study transport phenomena, but consider the complexity of describing the behavior of a particle of fluid as it flows through a region in space. Not only is it difficult to follow, but its shape may change continuously. Therefore, it is more advantageous to describe what happens *at a fixed point or in a fixed region* in space. This method, the eulerian method, allows us to observe phenomena at points of interest rather than trying to follow a particle throughout a region in space, e.g., the temperature at the nose of a rocket, the pressure at an elbow in a water main, the velocity at the tip of a compressor blade, or the concentration of perfume near the earlobe of a favorite girl. The eulerian method is used primarily in this book, but whenever results are easier to obtain by the lagrangian method, we shall not hesitate to switch.

2-1 FIELDS

A field is a region where things happen—observable things. We describe a thermal field in terms of temperatures at various locations, an electric field by point potentials, and a fluid field by velocities at different points. An acoustic field produced, say, by a band playing music may cause interactions in the form of dancing. We are a product of our environment, interacting with fields about us.

It is possible, and even probable, that several fields coexist in any given region. An airliner responds to the thrust of its jets (force field), required to overcome the effects of its gravitational field, while perturbing the ocean of air (aerodynamic field) through which it moves, at the same time being affected very slightly by the polar magnetic field. Interacting fluid, electric, magnetic, and thermal fields influence plasmas. While it is important to be able to predict phenomena resulting from interactions, it is necessary to segregate fields in order to understand their behavior.

In studying fields we encounter three types of quantities: scalars, vectors, and tensors. *A tensor is an ordered set of n quantities*, say $(m_1, m_2, \ldots, m_n)$. A second-order tensor involves nine components and arises in fields in such quantities as stress and strain. The components are represented by *scalars*, which *require only the specification of magnitude for a complete description*.

Many other physical phenomena, e.g., force, velocity, and acceleration, occur in ordered sets of three quantities. These phenomena can be represented by a first-order tensor, commonly called a *vector*. A vector is designated mathematically as

$$\mathbf{V} \equiv \mathbf{V}(x,y,z,t)$$

as in the case of velocity, or by the use of three scalar components each of which represents its magnitude in one of three orthogonal directions:

$$\begin{aligned} V_x &= f_1(x,y,z,t) \\ V_y &= f_2(x,y,z,t) \\ V_z &= f_3(x,y,z,t) \end{aligned} \tag{2-1}$$

Thus *a vector possesses both magnitude and direction*. Such quantities as temperature, concentration, volume, mass, and energy are scalars. Scalars are zero-order tensors.

A continuous distribution of these quantities—scalars, vectors, and tensors—described in terms of space coordinates and time constitutes a field.

2-2 FLUX DENSITY

Flux $\mathscr{F}$ is the transfer rate of some quantity. It may be gallons per minute, as in the case of fluid flow; Btu per hour, as in heat transfer; or, pounds mass per hour, as in the diffusion of water vapor. When expressed per unit surface area S, it is *flux density* $\mathscr{D}$

$$\mathscr{D} \equiv \lim_{\Delta S \to 0} \frac{\Delta \mathscr{F}}{\Delta S} \mathbf{n}$$

where $\mathbf{n}$ is a unit vector.

FUNDAMENTALS OF TRANSPORT PHENOMENA

While the flux of a liquid such as water is obvious, the flux associated with other transfer phenomena may be elusive to the inexperienced. The particular flux depends upon the field under consideration. It is characterized by flow (flux) lines common to the field (streamlines in the case of fluid flow). Flux is a scalar quantity; flux density is a vector.

2-3 FIELD INTENSITY

Field intensity ∇P refers to the strength of a field. It is expressed in terms of position. A high field intensity gives rise to a high flux density, and vice versa. Hence, flux density and field intensity are proportional. Expressing the field intensity as the change of a property P with respect to coordinate dimensions,

$$\mathscr{D} \propto \nabla P$$
$$\mathscr{D} = -C\,\nabla P \qquad (2\text{-}2)$$

where ∇P is the gradient of property P, the operator ∇ being defined by

$$\nabla \equiv \mathbf{i}\frac{\partial}{\partial x} + \mathbf{j}\frac{\partial}{\partial y} + \mathbf{k}\frac{\partial}{\partial z} \qquad (2\text{-}3)$$

in the rectangular coordinate system, and C is a constant of proportionality; $\mathbf{i}$, $\mathbf{j}$, and $\mathbf{k}$ are unit vectors in orthogonal directions. If C is positive, the negative sign indicates that flux is in the direction of decreasing intensity. The gradient of P is sometimes referred to as the *potential gradient* and represents a driving force.

Like the flux density and field intensity, the proportionality constant C depends upon the physical field. It is usually determined on a macroscopic basis by measurement of the flux density and field intensity. Its value may depend upon temperature, pressure, and position, except in homogeneous, isotropic media, where it is not a function of position.

2-4 RATE EQUATIONS

Equation (2-2) is a general equation which adequately describes the behavior in an electric field, a magnetic field, a fluid field, a thermal field, or a field of mass diffusion. It is based upon experience without regard for the microscopic details of the transfer mechanism.

The purpose of this section is to demonstrate the direct analogy between momentum, heat, and mass transfer. In one dimension, Eq. (2-2) reduces to

$$\mathscr{D}_y = -C\frac{\partial P}{\partial y} \qquad (2\text{-}4)$$

where $\mathscr{D}_y$ may represent the flux density of momentum, heat, or mass, depending upon the field under consideration. The partial derivative $\partial P/\partial y$ is the potential gradient, the intensity of the respective driving force. Rohsenow and Choi [6] and Artley [1] also present the rate equations in terms of general field properties.

Newton's viscosity equation   Consider a fluid confined between two parallel plates, the upper one being set in motion at a velocity U by a force F and the lower one being fixed (Fig. 2-1). Assume that the distance h between the plates is sufficiently small for the fluid particles to move in parallel paths. From experience we have observed that fluid particles adjacent to solid boundaries tend to adhere to the surface (easily observed when pouring motor oil or molasses). This same property generates an internal friction by adjacent fluid particles exerting a drag on each other and producing a shear stress $\tau_{yx} = F/A$ between adjacent fluid layers. The subscripts yx indicate that the stress is in a plane perpendicular to y and parallel to x, a nomenclature which is obviously necessary in three-dimensional systems.

Under steady-state conditions Newton observed that the shear stress is directly proportional to the velocity gradient,

$$\left(\frac{F}{A}\right)_{yx} = \tau_{yx} \propto \frac{\Delta u}{\Delta y} \propto \frac{U}{h}$$

His observation, repeatedly borne out by subsequent investigators, is equally valid at any position; i.e.,

$$\left(\frac{F}{A}\right)_{yx} = \mu \frac{\partial u}{\partial y} \tag{2-5}$$

where u is the fluid velocity in the x direction and μ is the absolute viscosity. This is of the same form as Eq. (2-4). Technically, this empirical relation, known as *Newton's equation of viscosity*, defines absolute, or dynamic, viscosity μ, which has dimensions

$$\mu = \left[\frac{M}{LT}\right] = \left[\frac{FT}{L^2}\right]$$

From this definition it is obvious that viscosity has meaning only when motion occurs. In engineering units $\mu = \{lb_m/\text{ft-sec}\}$ or $\{lb_f\text{-sec}/\text{ft}^2\}$.

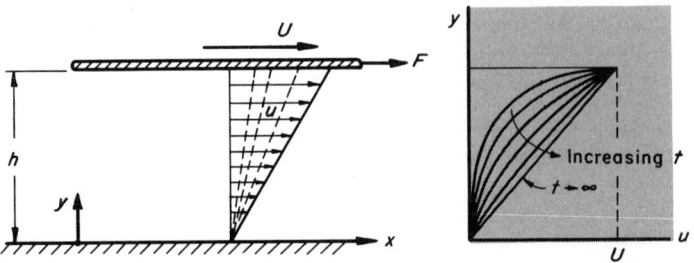

Fig. 2-1   Velocity distribution in fluid motion.

FUNDAMENTALS OF TRANSPORT PHENOMENA

Example What power (energy rate) is required for an ice skater to move on one blade at a speed of 30 mph? The skater's weight of 100 lb_f is evenly distributed over the area of the blade, which is 5.00 by 0.118 in. Assume the film to be 0.02 in thick.

Solution The blade glides on a film of water, close to 32°F, with an absolute viscosity of 3.5×10^{-5} lb_f-sec/ft² (from Fig. A-1). The power required is the product of the force required to move the skater and the velocity

$$\dot{W} = FV$$

For the small film thickness, it is reasonable to assume that a linear velocity distribution exists between the blade and the ice. Therefore, Newton's law of viscosity can be used to determine the force:

$$\dot{W} = \left(A\mu \frac{V}{h}\right)V = \frac{5.00 \times 0.118 \text{ in}}{0.020 \text{ in}} (3.5 \times 10^{-5} \text{ lb}_f\text{-sec/ft}^2)(44^2 \text{ ft}^2/\text{sec}^2) = 2.0 \text{ in-lb}_f/\text{sec}$$

Since 1 hp is 550 ft-lb_f/sec, the work rate in terms of horsepower is

$$\dot{W} = 2.0 \text{ in-lb}_f/\text{sec} \frac{\text{ft}}{12 \text{ in}} \frac{1 \text{ hp}}{550 \text{ ft-lb}_f} = 0.00303 \text{ hp}$$

Example Two horizontal coaxial disks 2 ft in diameter are spaced 0.050 in apart. The bottom one is fixed, and the upper one rotates at 5 rev/sec, requiring a torque of 8.62 ft-lb_f. Neglecting edge effects, what is the viscosity of the oil filling the space between the disks?

Solution Assuming a linear velocity distribution,

$$\frac{F}{A} = \mu \frac{du}{dy} = \mu \frac{u}{h}$$

where u is the local velocity, i.e., the velocity at radius r. The elemental force dF varies with

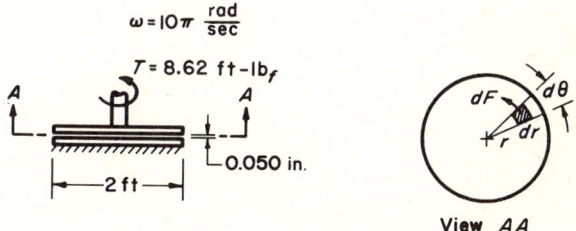

Fig. 2-2 Coaxial disks with oil interface.

the radius of the disk, since the shear area increases with radius. Therefore,

$$\frac{dF}{dA} = \mu \frac{r\omega}{h}$$

where the local velocity $u = r\omega$. But $dA = r \, dr \, d\theta$; therefore, integration with respect to θ yields

$$dF = \mu \frac{\omega}{h} r^2 \, dr \int_0^{2\pi} d\theta = \frac{2\pi\mu\omega}{h} r^2 \, dr$$

and torque $dT = r\,dF$, giving

$$T = \frac{2\pi\mu\omega}{h}\int_0^R r^3\,dr = \frac{\pi}{2}\frac{\mu\omega}{h}R^4$$

$$\mu = \frac{2hT}{\pi\omega R^4} = \frac{2(0.050\text{ in})(8.62\text{ ft-lb}_f)(1\text{ ft})}{\pi(10\pi/\text{sec})(1\text{ ft}^4)(12\text{ in})} = 7.27\times 10^{-4}\text{ lb}_f\text{-sec/ft}^2$$

Note that this is the approximate value for hydraulic fluid at 40°F given in Fig. A-1.

It is sometimes more advantageous to define *kinematic viscosity* ν

$$\nu \equiv \frac{\mu}{\rho}$$

which has dimensions of L^2/T with the corresponding units of square feet per hour, the name resulting because only kinematic quantities are involved.

The viscosity of fluids varies with temperature and pressure, being much more sensitive to temperature than pressure. Changes in temperature cause opposite variations in the viscosity of gases and liquids. An increase in the temperature of a liquid reduces its viscosity but increases the viscosity of a gas. This is intuitive for liquids but not readily apparent for gases. The variation of absolute and kinematic viscosity with temperature is given for a number of fluids in Figs. A-1 and A-2.

Although values for viscosity are obtained by macroscopic measurements, let us consider a gas from a microscopic standpoint in order to understand the basic mechanism. From observations we tend to think of viscosity as a property related to "stickiness." Basically, however, it arises because of momentum interchange between molecules. Molecules are constantly in motion, the motion being more pronounced at higher temperatures and lower pressures. As the gas moves, slow-moving molecules strike faster-moving ones, slowing them down. It is this momentum (the product of mass and velocity) interchange which gives rise to viscous shear, a measure of which is viscosity. The mechanism of momentum exchange in liquids is the same as in gases qualitatively, but the physical structure is much more complex since the molecules are closer and the molecular force fields have a greater effect on the momentum exchange in the collision process.

By analogy, suppose two trains loaded with coal are running on parallel tracks in the same direction. If workmen begin throwing coal from the slower train to the faster one, the train which "catches" the coal is slowed by the increased mass, because of the momentum component in the direction of motion of the train. Now imagine workmen on both trains, analogous to molecules in adjacent fluid layers, throwing coal back and forth from one train to the other. If the trains initially have unequal velocities and the mass-exchange rate is equal for both trains, the faster train is slowed. So it is with the momentum interchange between fluid layers.

Viscosity is often measured by observing the time required for a given amount of fluid to flow from a short small-bore tube. Viscosities of fuel oils are measured at 77 and 122°F, of lubricants at 100 and 210°F. Viscosity is often given in metric units which have special names:

μ : poise $\equiv$ 1 g/cm-sec $\equiv$ 100 centipoises

ν : stoke $\equiv$ 1 cm²/sec $\equiv$ 100 centistokes

The following conversion units facilitate direct conversion to engineering units:

μ : 1(lb$_f$-sec/ft²) = 479 poises

ν : 1(ft²/sec) = 30.48² stokes

The kinematic viscosity ν in square feet per second can be determined by the efflux time t in seconds from various viscometers by the following equations [5]:

Saybolt Universal: $\nu = \begin{cases} 0.00000245t - \dfrac{0.0021}{t} & 32 < t < 100 \\ 0.00000237t - \dfrac{0.00145}{t} & t > 100 \end{cases}$

Saybolt Furol: $\nu = \begin{cases} 0.0000241t - \dfrac{0.00198}{t} & 25 < t < 40 \\ 0.0000233t - \dfrac{0.000646}{t} & t > 40 \end{cases}$

Redwood no. 1 (English): $\nu = \begin{cases} 0.00000280t - \dfrac{0.00193}{t} & 34 < t < 100 \\ 0.00000266t - \dfrac{0.000538}{t} & t > 100 \end{cases}$

Redwood Admiralty (English): $\nu = 0.0000291t - \dfrac{0.02155}{t}$

Engler (German): $\nu = 0.00000158t - \dfrac{0.00403}{t}$

A kinematic-viscosity conversion table relating values obtained from the different types of viscometers is given as Appendix Table A-1.

Figure 2-3 illustrates schematically a concentric-cylinder device which can also be used to measure viscosity by application of Eq. (2-5). Direct measurement of torque and angular velocity are the only measurements necessary other than the physical dimensions. The drag on the end of the cylinder must be taken into account by calibration in using this technique.

As a corollary to Eq. (2-5), fluids which obey that equation are known as *newtonian fluids*. All gases and most liquids of engineering importance are newtonian. Fluids which do not behave in accordance with Eq. (2-5), nonnewtonian fluids, will

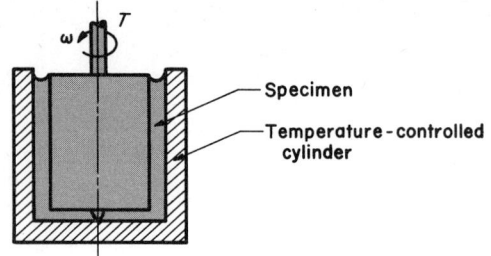

Fig. 2-3 Concentric-cylinder viscometer.

not be considered in this text. Nonnewtonian fluids, such as blood, pitch, molasses, molten rubber, and dough, are studied in a specialized field known as *rheology*.

Note again that Eq. (2-5) is the equivalent of Eq. (2-4) with the appropriate field properties replacing the general terms.

Fourier's heat-conduction equation Consider a homogeneous solid with parallel faces initially at uniform temperature T_i. While the lower face is maintained at a constant temperature T_i, let the upper face be suddenly changed to a constant temperature $T_1 > T_i$. The amount of heat transferred across an area in the y direction is directly proportional to the temperature gradient. Upon attaining the steady-state condition

$$\left(\frac{q}{A}\right)_y \propto \frac{\Delta T}{\Delta y} \propto \frac{T_1 - T_i}{h}$$

or, at any position,

$$\left(\frac{q}{A}\right)_y = -k\frac{\partial T}{\partial y} \qquad (2\text{-}6)$$

where the constant of proportionality k is the thermal conductivity of the solid. This relation, *Fourier's equation of heat conduction*, defines the thermal conductivity k,

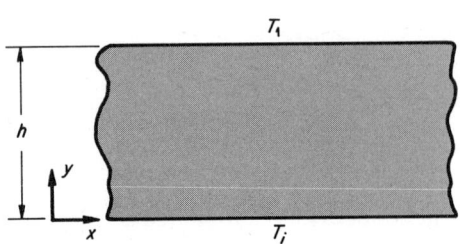
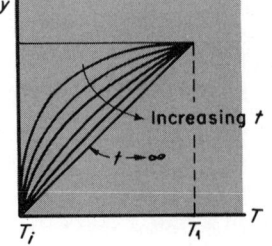

Fig. 2-4 Temperature distribution in a solid.

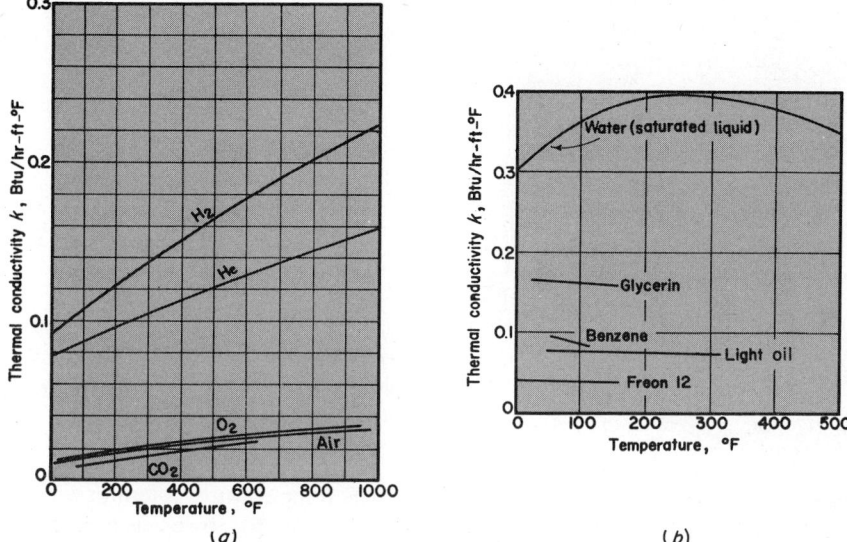

Fig. 2-5 Thermal conductivity of typical gases and liquids. (*From J. P. Holman, "Heat Transfer" 2d ed. Copyright 1968. McGraw-Hill Book Company. Used by permission.*)

which has dimensions

$$k = \left[\frac{\text{energy}}{\text{time} \times \text{length} \times \text{temperature}}\right]$$

It is commonly expressed in engineering units of Btu/hr-ft-°F. Values of thermal conductivity for common solids and fluids of engineering interest are given in Appendix A. The negative sign indicates that heat is transferred in the direction of decreasing temperature, in accordance with the second law of thermodynamics (discussed in Chap. 3).

Note that a fluid confined between two plates of constant temperatures will behave similarly except that the heat transfer gives rise to fluid motion, e.g., hot air rises, coupling heat and momentum transfer. A solid by itself is therefore more amenable to the study of the diffusion of heat.

The thermal conductivity is analogous to viscosity, since its value depends upon the energy exchange between molecules in motion. Faster-moving molecules impart some of their kinetic energy to slower-moving ones in the collision process. An increase in temperature increases molecular motion, transferring energy from regions of higher temperature to regions of lower temperature. Thermal conductivity varies with temperature and pressure, being much more sensitive to temperature than pressure. For engineering purposes it is independent of pressure in solids, liquids, and most gases below the critical pressure. Figures 2-5 and 2-6 show

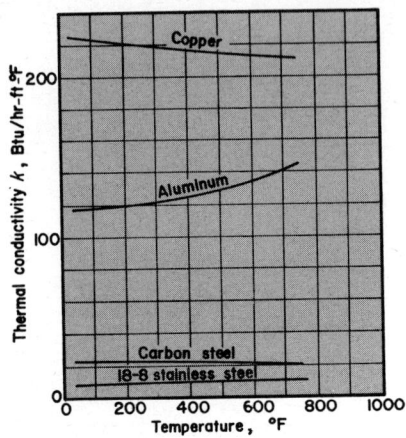

Fig. 2-6 Thermal conductivity of typical solids. (*From J. P. Holman, "Heat Transfer," 2d ed. Copyright 1968. McGraw-Hill Book Company. Used by permission.*)

the variation of thermal conductivity with temperature in common gases, liquids, and solids. It should be noted that all values of the given gases and liquids are less than those given for any solids. This would be expected upon considering the molecular spacing.

It is frequently convenient to use the ratio of a material's ability to transport energy to its capacity to store energy. This is the *thermal diffusivity* α, defined as

$$\alpha \equiv \frac{k}{\rho c} \tag{2-7}$$

where ρ is the mass density of the material and c is its specific heat.

Example To effect a bond between two metal plates, 1 and 6 in thick, respectively, heat is uniformly applied through the thinner plate by a radiant heat source. The bonding epoxy must be held at 120°F for a short time. When the heat source is adjusted to have a steady value of 96 Btu/hr-in², a thermocouple installed on the side of the thinner plate next to the source gives a temperature of 160°F. What is the thermal conductivity of the 1-in metal plate?

Solution The heat-conduction equation for this one-dimensional case may be written

$$\frac{q}{A} = k \frac{T_{hot} - T_{cold}}{\Delta x}$$

where Δx is the plate thickness. All quantities are known except k, which can be easily found:

$$k = \frac{q}{A} \frac{\Delta x}{\Delta T} = \left(96 \frac{\text{Btu}}{\text{hr-in}^2}\right) \frac{1 \text{ in}}{160 - 120°F} \frac{12 \text{ in}}{\text{ft}} = 28.8 \text{ Btu/hr-ft-°F}$$

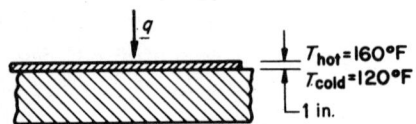

Fig. 2-7 Thermal bonding of metal plates.

FUNDAMENTALS OF TRANSPORT PHENOMENA

Judging from the order of magnitude of thermal conductivities of metals (Figure 2-6), the plate is probably an alloy, perhaps steel, rather than a pure metal.

The energy transfer in solids is by lattice vibration and by free-electron transport. Since in metals, free-electron transport is more prominent than lattice vibration, good heat conductors are also good electric conductors [3].

For many materials thermal conductivity varies linearly with temperature, i.e.,

$$k = k_0(1 + aT) \tag{2-8}$$

where k_0 is the value at zero temperature and a is a constant which depends upon the material. For such materials it is convenient to use an average value of thermal conductivity in making calculations of heat transfer.

Thermal conductivity can be measured in a variety of ways, all of which depend upon the observation of a temperature gradient across a specimen conducting a known amount of heat. Jakob [4], Schneider [7], Worthing and Halliday [9], and Wilkes [8] present the most popular methods.

Fick's diffusion equation For the third mechanism of transport, mass diffusion, consider two parallel plates with dry air between them. Let the top plate be made of a material which can be continuously saturated with water and let the bottom plate be covered with a desiccant, such as silica gel, which will continuously absorb all water that strikes it (Fig. 2-8). At time $t = 0$ imagine that the top plate is saturated with water and that it remains saturated with a mass concentration of ω_{W_h}. The water diffuses downward, producing a concentration gradient, which varies linearly with distance from the plate under steady-state conditions. Experience has shown that the gradient is directly proportional to the mass flux

$$J_A \equiv \left(\frac{\dot{m}}{A}\right)_y \propto \frac{\Delta \omega_W}{\Delta y} \propto \frac{\omega_{W_h}}{h}$$

for the steady state and

$$J_A \equiv \left(\frac{\dot{m}}{A}\right)_y = -\rho D_{WA} \frac{\partial \omega_W}{\partial y} \tag{2-9}$$

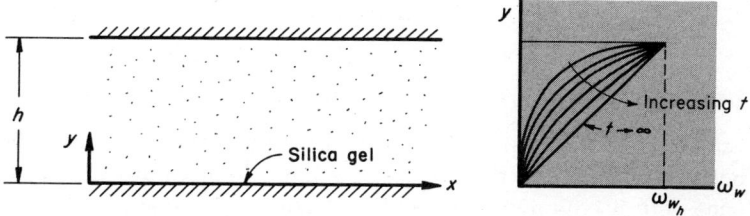

Fig. 2-8 Concentration distribution in a binary mixture.

at any position. Known as *Fick's equation of diffusion*, Eq. (2-9) defines the binary mass diffusivity which has dimensions and units

$$D_{WA} = \left[\frac{L^2}{T}\right] = \left\{\frac{\text{ft}^2}{\text{hr}}\right\}$$

analogous to the kinematic viscosity. This equation states that species W, the water, diffuses (moves relative to the mixture) in the direction of decreasing mass concentration, just as heat flows from a region of higher temperature to a region of lower temperature. Experimentally determined mass diffusivities of some typical gas, liquid, and solid binary mixtures are given in Appendix Tables A-5 to A-7. As would be expected by intuition $D_{\text{sol}} \ll D_{\text{liq}} < D_{\text{gas}}$. Unlike viscosity and thermal conductivity, mass diffusivity is a function of concentration as well as temperature and pressure.

Example As shown in Fig. 2-9, liquid B evaporates into air A from a large standpipe as a mixture of the fluid vapor and air is carried away at the top. The mass concentration of the fluid vapor decreases by 5 percent as it diffuses upward; that is, $\omega_B = \omega_{B_{z1}} - \omega_{B_{z2}} = 0.05$, while 0.2 lb$_m$/hr-ft^2 evaporates. Determine D_{BA}, the mass diffusivity of B with respect to A.

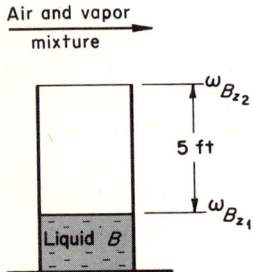

Fig. 2-9 Fluid evaporation into air.

Solution Fick's diffusion equation for a one-dimensional system is

$$\frac{\dot{m}}{A} = \rho D_{BA} \frac{\omega_{B_{z1}} - \omega_{B_{z2}}}{z_2 - z_1}$$

Therefore,

$$D_{BA} = \frac{\dot{m}}{\rho A} \frac{\Delta z}{\Delta \omega_B} = \frac{1}{\rho}\left(0.2 \frac{\text{lb}_m}{\text{hr-ft}^2}\right)\frac{5 \text{ ft}}{0.05} = \frac{20}{\rho} = \frac{20}{\rho_B}$$

2-5 UNIQUE CHARACTERISTICS OF THE RATE EQUATIONS

Equations (2-5), (2-6), and (2-9) are all of the form of Eq. (2-4); hence, these processes behave according to the same mathematical equation and may be treated in accordance with a common concept of diffusion. From these equations it is possible to describe how a property varies with position.

FUNDAMENTALS OF TRANSPORT PHENOMENA

Not only are the equations of the same form with equivalent transfer mechanisms, but the peculiar characteristic in each field (momentum diffusivity ν, commonly called kinematic viscosity, in momentum transfer; thermal diffusivity α in heat transfer; and mass diffusivity D in mass transfer) has the same dimensions, L^2/T. It is obvious then that dimensionless ratios can be formed with these properties. These dimensionless numbers are

$$\text{Prandtl number:} \quad \mathbf{Pr} \equiv \frac{\nu}{\alpha} = \frac{c\mu}{k} \tag{2-10}$$

$$\text{Schmidt number:} \quad \mathbf{Sc} \equiv \frac{\nu}{D} = \frac{\mu}{\rho D} \tag{2-11}$$

$$\text{Lewis number:} \quad \mathbf{Le} \equiv \frac{D}{\alpha} = \frac{\rho c D}{k} \tag{2-12}$$

and are significant physically when two or more of the transfer processes occur simultaneously.

The Prandtl number is important when both momentum and energy are propagated throughout the system, the Schmidt number for momentum and mass transfer, and the Lewis number when energy and mass transfer occur together. When the ratios are unity, the two simultaneous processes are said to be similar, i.e., to diffuse in the same ratio. This sometimes enables us to solve extremely complicated problems based upon a simple solution of a transfer process occurring alone.

For most gases the Prandtl number is close to unity, but for other fluids it varies widely, being very low for liquid metals (less than 0.05) and high for viscous oils (of the order of 1,000). The Schmidt number is approximately unity for gases but is large for liquids [6].

2-6 CONSERVATION LAWS

In addition to having rate equations of the same form, the *principle of conservation* holds for heat, mass, and momentum transfer, individually or coupled. Assuming that matter can neither be created nor destroyed and neglecting internal generation, experience has shown that a quantity entering an element of volume (later to be termed the control volume) is equal to that which leaves plus the quantity stored within the volume element. Writing this principle in a word equation gives

$$\text{Quantity entering volume element} = \text{quantity leaving volume element} + \text{quantity stored in volume element} \tag{2-13}$$

In order to bring the rate-process equations into this conservation principle, it can be expressed in terms of transfer rates, i.e.,

$$\text{Rate at which quantity enters} = \text{rate at which quantity leaves} + \text{rate of quantity accumulation} \tag{2-14}$$

This equation, sometimes referred to as the *equation of change*, applies to any transfer process.

To develop the analogy, consider a region of unit depth bounded by planes at y and $y + \Delta y$ (Fig. 2-10). Letting $\mathscr{D}$ be the flux density and P be the quantity per unit volume which changes with time, Eq. (2-14) becomes

$$\Delta x \, \mathscr{D}|_y = \Delta x \, \mathscr{D}|_{y+\Delta y} + \frac{\partial P}{\partial t} \Delta x \, \Delta y \tag{2-15}$$

Dividing by $\Delta x \, \Delta y$ and taking the limit as Δy approaches zero, i.e., describes the condition at a point, we get

$$\lim_{\Delta y \to 0} \frac{\mathscr{D}|_y - \mathscr{D}|_{y+\Delta y}}{\Delta y} = \frac{\partial P}{\partial t} \tag{2-16}$$

But the left side of the equation is the negative of the definition of the derivative. Hence,

$$-\frac{\partial \mathscr{D}}{\partial y} = \frac{\partial P}{\partial t} \tag{2-17}$$

Equation (2-17) is the equation of change which is valid for one-dimensional transfer processes. Replacing the flux density and the property which changes with time by their equivalents for the fields under consideration [refer to Eqs. (2-5), (2-6), and (2-9)], we get the conservation equations:

Momentum: $\quad \dfrac{\partial}{\partial y}\left(\mu \dfrac{\partial u}{\partial y}\right) = \dfrac{\partial u}{\partial t} \tag{2-18}$

Energy: $\quad \dfrac{\partial}{\partial y}\left(k \dfrac{\partial T}{\partial y}\right) = \rho c \dfrac{\partial T}{\partial t} \tag{2-19}$

Species: $\quad \dfrac{\partial}{\partial y}\left(\rho D_{ij} \dfrac{\partial \omega_i}{\partial y}\right) = \dfrac{\partial \omega_i}{\partial t} \tag{2-20}$

Equation (2-20) is written for species i, where i may be either air or water vapor in the the field depicted in Fig. 2-8.

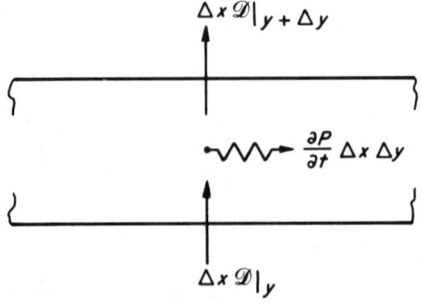

Fig. 2-10 Model for conservation laws.

FUNDAMENTALS OF TRANSPORT PHENOMENA

A similar analysis yields the one-dimensional conservation-of-mass equation (*continuity equation*).

Mass: $$\frac{\partial}{\partial y}\rho u = -\frac{\partial \rho}{\partial t} \tag{2-21}$$

These equations will be developed rigorously in Chap. 10 for three dimensions.

PROBLEMS

2-1. What is the density of kerosene at 50°F?

2-2. Determine the thermal diffusivity of benzene at 100°F and 1 atm pressure.

2-3. A 6-in-diam shaft revolves in a bearing at 900 rpm. Radial clearance between the shaft and the journal is 0.002 in. Determine the torque per foot of bearing required to overcome the resistance due to a lubricant having a dynamic viscosity of $\mu = 2 \times 10^{-4}$ lb_f-sec/ft².

2-4. The velocity profile for laminar flow in a round duct is

$$u = V_{max}\left[1 - \left(\frac{r}{R}\right)^2\right]$$

where V_{max} = const = velocity at centerline of duct

r = radial distance from pipe centerline

R = duct radius

(a) What is the maximum velocity?
(b) Show that the velocity gradient varies linearly with the radius.
(c) Determine the velocity gradient at the wall.
(d) What is the velocity gradient at the centerline?
(e) What is the wall shear stress when castor oil at 100°F flows in a 2-in-diam pipe with an average velocity of 10 ft/sec (V_{max} = 20 ft/sec)?

2-5. A 2-in-diam vertical rod is encased by a sleeve 3 ft long with a uniform radial clearance of 0.002 in. When the sleeve and rod assembly are immersed in water at 150°F, the effective weight of the sleeve is 4 lb_f. Neglecting the drag on the outer portion of the sleeve, how fast will the sleeve slide down the rod?

2-6. A 6-in-diam sleeve bearing has a radial clearance of 0.002 in which is filled with a light oil (ν = 50 ft²/sec, S = 0.80). What is the shear stress in the sleeve if the shaft which is supported by the bearing rotates at 300 rpm?

2-7. In a laboratory experiment a 6-in-long sleeve slides down a 2-in-diam rod with a radial clearance of 0.003 in. The assembly is immersed in a fluid, and the immersed sleeve weighs 1.0 lb_f. The measured steady (terminal) velocity of the sleeve is 8 ft/sec. Neglecting the drag on the outer surface of the sleeve, determine the dynamic viscosity of the fluid.

2-8. A 6-in-long cylindrical metal bar slides inside a tube filled with oil. The bar weighs 2 lb_f immersed in the oil. What is the viscosity of the oil, in lb_f-sec/ft², if the steady-state velocity of the bar is 0.318 ft/sec? The inside diameter (ID) of the tube, which may be used for computing areas, is 2.000 in and the clearance is 0.002 in.

2-9. What temperature must a stove burner maintain in order to bring 4 lb_m of water initially at 75°F to 212°F in 1 hr in a $\frac{1}{8}$-in-thick aluminum pan? The pan area exposed to the burner is 113 in².

2-10. The inside temperature of a house is to be maintained such that the inside surface temperature of the windows is 70°F. How much heat is transferred by conduction through a 4- by 8-ft picture window if the outside glass surface temperature is (a) 92°F; (b) 32°F? Assume 0.12 in glass.

2-11. A large slab of salt (initially dry) absorbs moisture from the air. What is the absorption rate if the mass concentration ω is 0.07 and 0.065 at 3 in above the surface and at the interface, respectively? $D_{WS} = 0.853$ ft²/hr.

2-12. Carbon dioxide at 100°F diffuses from a 50-ft² surface into air at a rate of 0.5 lb_m/hr. If the mass concentration ω at the surface is 0.0352 and is negligible 2 in from the surface, determine the mass diffusivity.

2-13. Water diffuses from a 100-ft² area at the rate of 4 lb_m/hr into dry air. How much must the mass concentration decrease over a vertical distance of 10 ft from the surface?

2-14. A fluid of density $\rho \neq \rho(t)$ flows with a velocity of $u = 4 - y^2$. Determine an expression for density when $y = 0$, $\rho = \rho_0$.

REFERENCES

1. Artley, J.: "Fields and Configurations," Holt, New York, 1965.
2. Bird, R. B., W. E. Stewart, and E. N. Lightfoot: "Transport Phenomena," Wiley, New York, 1960.
3. Holman, J. P.: "Heat Transfer," 2d ed., McGraw-Hill, New York, 1968.
4. Jakob, M.: "Heat Transfer," vol. I, Wiley, New York, 1949.
5. "Marks' Mechanical Engineers' Handbook," 6th ed., T. Baumeister (ed.), McGraw-Hill, New York, 1958.
6. Rohsenow, W. M., and H. Choi: "Heat, Mass, and Momentum Transfer," Prentice-Hall, Englewood Cliffs, N.J., 1961.
7. Schneider, P. J.: "Conduction Heat Transfer," Addison-Wesley, Reading, Mass., 1955.
8. Wilkes, G. B.: "Heat Insulation," Wiley, New York, 1950.
9. Worthing, A. G., and D. Halliday: "Heat," Wiley, New York, 1948.

CHAPTER 3
FUNDAMENTAL CONCEPTS FROM THERMODYNAMICS

In the transfer processes we seek the relationships between fluxes and field intensities in terms of field properties, physical properties of the transfer media, and the dimensions of space and time. *Thermodynamics deals with energy quantities which are transferred during the processes—work and heat.* Its principles and laws apply to all fields of engineering. This chapter sets forth some fundamental concepts necessary for subsequent study of the transfer processes, unifying the definitions and symbols of thermodynamics and the rate processes.

In its broadest sense the science of thermodynamics considers the conversion and transfer of energy. Classical, or macroscopic, thermodynamics is based upon man's observations. Its laws were developed inductively. No observable violations have occurred. Media are viewed from a continuum standpoint. Probabilistic, or microscopic, thermodynamics is based upon the interactions of molecules and the probability of their behaving in accordance with a set of laws which are identical to those developed in the classical approach. The two approaches are complementary in that the microscopic viewpoint describes fundamental behavior while the macroscopic viewpoint guarantees repeatability.

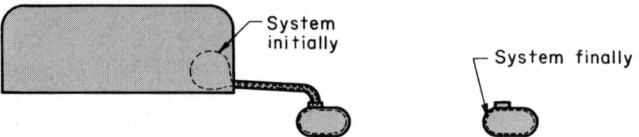

Fig. 3-1 Thermodynamic system.

3-1 EQUILIBRIUM

Thermodynamics is based upon an equilibrium condition or a series of equilibrium states. *Equilibrium is that state which is characterized by no change.* In the preceding chapter we noted that change occurs when the field intensity—any field intensity— varies throughout a region. Therefore, for equilibrium the intensity of all fields must be identical; no potential gradient can exist.

3-2 SYSTEM AND CONTROL VOLUME

A thermodynamic system is a fixed quantity of matter. It does not vary in mass or identity. Everything outside the system is termed the *surroundings*. The system and surroundings are separated by *boundaries*. Consider, for example, filling an automobile gasoline tank from a large tank truck. We may define the system as that amount of gasoline which will be transferred into the smaller tank, shown by the dashed lines in Fig. 3-1. The thermodynamics problem then becomes that of determining what happens to the gasoline between the initial equilibrium state and the final equilibrium state; it is a "bookkeeping process" of tabulating observable quantities initially and finally.

An alternative method of solving the same problem involves focusing attention on a fixed region in space, say the automobile tank. The fixed region is the *control volume*, and the thermodynamic problem can be solved by "standing" at the *control surface* (analogous to the system boundary) and observing the gasoline as it crosses.

All thermodynamic problems can be solved by using one of these two concepts, control volume or system. We shall use whichever is more convenient in any given problem. In some cases it will be more feasible to think in terms of a *deformable control volume*, typified by a balloon. At this point the student should ponder the

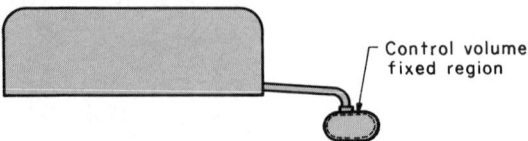

Fig. 3-2 Control volume.

FUNDAMENTAL CONCEPTS FROM THERMODYNAMICS

 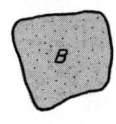

Fig. 3-3 Sketch illustrating equality of temperature.

analogy between the eulerian method of describing field properties and the thermodynamic concept of the control volume.

3-3 EQUALITY OF TEMPERATURE

We now seek to define the thermodynamic property temperature, for which we have no vocabulary. Difficult? But wait, we also wish to "invent" an instrument for measuring it.

Let two bodies (thermodynamic systems), A and B, one hot and one cold and completely isolated from all other bodies, be placed in contact with each other (Fig. 3-3). Let body A be hollow and contain a fluid whose volume changes as it becomes warmer or colder. In time both bodies undergo changes until all their properties cease to change. The *bodies* are then said to be *in thermal equilibrium*, i.e., to *have the same temperature*. Formally stated, two bodies have equal temperatures if there are no changes in their properties—electric resistance, expansion, chemical reaction, etc. But this definition must not be reversed. Two thermodynamic systems which are initially equal in temperature may change in temperature when brought in contact. For example, water and hydrochloric acid increase in temperature when mixed.

3-4 THE ZEROTH LAW OF THERMODYNAMICS

Now let us add a third body C to our example (Fig. 3-4). Further assuming no chemical or electrical interactions, if body A is brought into contact with the third body C with no observable changes, A and C are in thermal equilibrium, i.e., at the same temperature. If A and B remain in thermal equilibrium as before, then systems B and C are at the same temperature. A generalized axiom, which is perhaps obvious but which is not derivable from other observations in nature, can now be stated as the *zeroth law*: If two systems are equal in temperature to a third system, they are equal

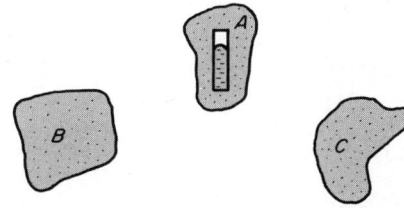

Fig. 3-4 Sketch illustrating Zeroth law.

in temperature to each other. This law is fundamental to all temperature measurement [2].

3-5 TEMPERATURE MEASUREMENT AND THERMOMETRIC SCALES

If a graduated scale were placed on body A of the preceding sections adjacent to its column of fluid, we could observe a difference in elevation of the fluid column as the temperature level changes. Thus, we have "invented" a *thermometer* which so far indicates only a difference in temperatures, relatively speaking, "hot" and "cold." Obviously, body A must be made of a material and contain a fluid which are conducive to observations. A thin glass tube filled with mercury (known as a *mercury-in-glass thermometer*) has the necessary thermometric properties.

It is now necessary to establish a numerical scale to which we can assign values of easily reproducible temperatures. In searching for a reproducible temperature for reference, Fahrenheit selected the temperature of the human body and assigned it the value 96. He assigned the value of 0 to a certain mixture of water, ice, and salt. On his scale the ice point was approximately 32. The scale was later revised in terms of (1) the *ice point*, the temperature of a mixture of ice and water in equilibrium with saturated air at a pressure of 1 atm, and (2) the *steam point*, the temperature of water and steam in equilibrium at a pressure of 1 atm. The values assigned these points were 32 and 212, respectively, on the *Fahrenheit scale*. On this slightly adjusted scale the normal body temperature is 98.6°F. The values assigned to the ice and steam points on the *Celsius* (formerly centigrade) *scale* are 0 and 100.

Suggested in this arbitrary choice of scales is the possibility of linear extrapolation as well as interpolation, indicated in Fig. 3-5. Observing the ratio of subdivisions between the boiling point and ice point on the two scales, $180/100 = 9/5$, we can easily establish a relationship between the Fahrenheit and Celsius temperatures:

$$T_F = \tfrac{9}{5}T_C + 32 \tag{3-1}$$

$$T_C = \tfrac{5}{9}(T_F - 32) \tag{3-2}$$

The concept of an absolute temperature scale, independent of any particular substance, arises in the second law of thermodynamics. It is necessary to use the

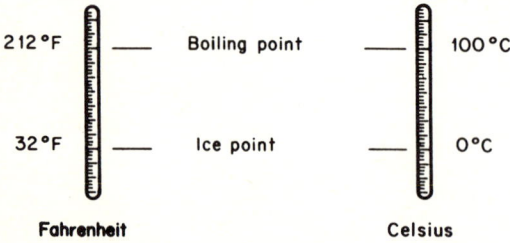

Fig. 3-5 Common thermometers.

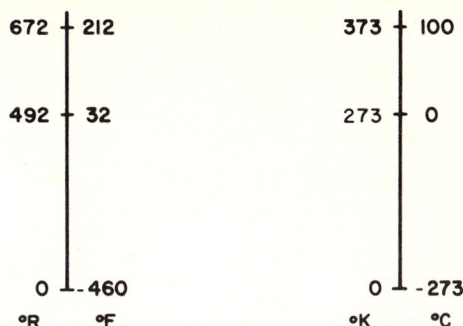

Fig. 3-6 Relation between common thermometric scales.

absolute temperature in all thermodynamic problems except those involving temperature differentials. Extending the thermometers of Fig. 3-5 to absolute zero, where all molecular activity ceases,[1] we get the absolute scales of Rankine (R), which corresponds to Fahrenheit (F), and Kelvin (K), which corresponds to Celsius (C). Figure 3-6 shows their relationship in round numbers. The exact relations are

$$°R = °F + 459.67 \tag{3-3}$$

$$°K = °C + 273.15 \tag{3-4}$$

In 1948 at the Ninth Conference on Weights and Measures an international temperature scale, based on a number of fixed and easily reproducible points, was revised from a centigrade scale adopted in 1927. As an example of its extent, two points on the scale are shown in Table 3-1. In 1954 the Tenth Conference on Weights and Measures redefined the centigrade scale in terms of a single fixed point, the triple point of water, and the ideal-gas temperature scale, which is described in any complete treatise on thermodynamics (see, for example, Lay [2] or Van Wylen and Sonntag [3]).

At these temperature extremes it is obvious that the common mercury-in-glass thermometer will not suffice because of the melting point of glass and the freezing point of mercury. More sophisticated devices must be employed. Thermocouples and platinum resistance thermometers cover the spectrum between these reference points. Their characteristics, as well as those of other temperature measuring instruments, are described in elementary instrumentation texts, e.g., Beckwith and Buck [1].

[1] This is one statement of the third law of thermodynamics.

Table 3-1

Temperature of equilibrium at 1 atm	°C
Oxygen point (liquid and vapor oxygen)	−182.970
Gold point (solid gold and liquid gold)	1063.000

3-6 PROPERTIES AND STATE OF A SUBSTANCE

A thermodynamic property is any measurement or quantity which serves to describe a system. Thermodynamic properties are either *intensive* or *extensive*. Intensive properties are independent of mass. Temperature, pressure, and density are intensive properties. Extensive properties vary directly with mass. Mass and total volume are extensive properties.

A property of a pure, simple, compressible substance can always be defined in terms of two independent intensive properties. For example, the pressure of a gas can be expressed in terms of its temperature and specific volume:

$$p = p(T,v) \tag{3-5}$$

A pure substance is also homogeneous and of fixed chemical composition. We sometimes speak of air as being pure; however, thermodynamically it is a *mixture* of several gases and vapors.

A *phase* is a quantity of matter which is homogeneous throughout. A substance may exist in any one or a combination of three phases—solid, liquid, and vapor. Two or more phases may coexist when in a common *state*, identified by two or more observable properties such as temperature and pressure. Change of phase and phase equilibrium can be understood by considering water. At a pressure[1] of 14.7 psia water is a solid (ice) when below 32°F, but at a temperature of 32°F solid, vapor, and liquid water can coexist. Further increases in temperature cause the liquid water to vaporize (turn to steam) until it is 100 percent water vapor above 212°F (Fig. 3-7). During this transition the *quality x*, the ratio of the mass of vapor to the total mass, changes from 0 to 1.00.

Figure 3-8 is a pressure-temperature diagram for a typical substance which expands upon freezing. The phase change of Fig. 3-7 can be represented by line AB since it is at constant pressure. As an interesting sidelight, consider what happens when a cold front sweeps across a space at constant barometric pressure, say from B to A in Fig. 3-8. As the temperature drops, rain occurs as the temperature crosses the vaporization curve: the water vapor (steam) condenses. It may snow or hail as the temperature decreases across the fusion line: the liquid water freezes. Snow may

---

[1] Pressure level is discussed in Chap. 4.

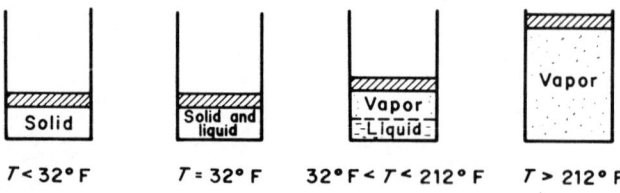

Fig. 3-7 Constant-pressure phase change of water.

FUNDAMENTAL CONCEPTS FROM THERMODYNAMICS

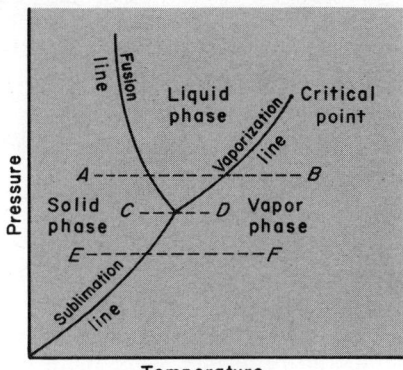

Fig. 3-8 Pressure-temperature diagram for a pure substance which expands upon freezing.

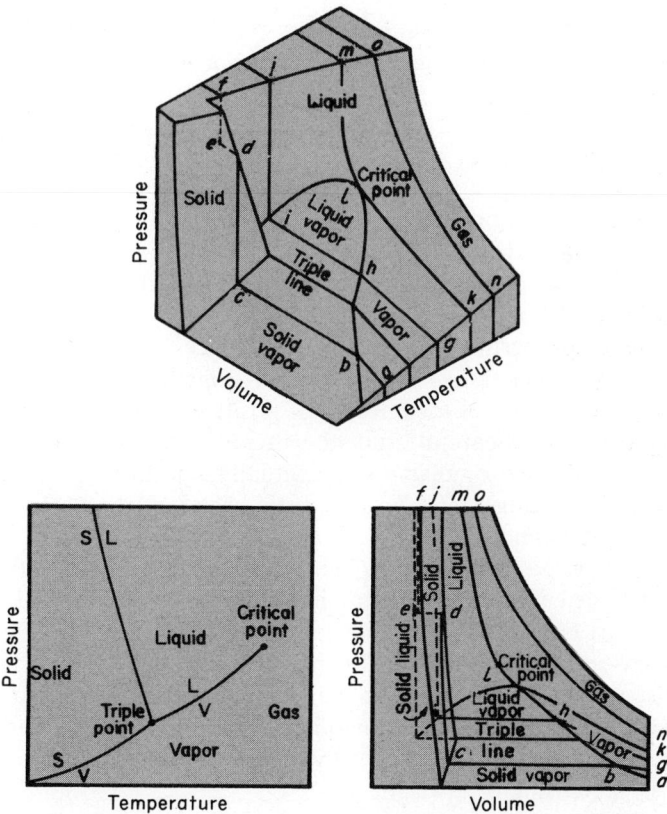

Fig. 3-9 pvT surface for a substance which expands upon freezing.

occur at a pressure below *CD* as the temperature drops, crossing the sublimation curve from *D* to *C* or from *F* to *E* without passing through a liquid phase. Curves which represent the relationship between pressure and temperature of pure substances are called *vapor-pressure curves*. The point common to the sublimation, fusion, and vaporization lines is the *triple point*, and a fluid can exist in all three phases at this point. At the *critical point* it is impossible to distinguish between the liquid and vapor phases.

Since pure substances can be described in terms of three measurable properties, pressure, specific volume, and temperature, as indicated by Eq. (3-5), their interrelation can be plotted on three mutually perpendicular coordinates. The resultant pvT surface graphically portrays this relationship. Figure 3-9 shows a pressure-volume-temperature surface, with its respective two-dimensional projections, for a substance which expands upon freezing (such as water).

3-7 WORK

Work, one of the basic quantities transferred during a thermodynamic process, is defined from elementary mechanics as a force **F** acting through a displacement x, where x is positive in the direction of the force; i.e.,

$$W \equiv \int_1^2 \mathbf{F} \cdot d\mathbf{x} \qquad (3\text{-}6)$$

This basic relation enables us to determine the work required to raise weights, propel missiles, or strum a banjo string. But this definition of work is too limited for thermodynamics, where the concern is with the interactions between a system and its surroundings. Therefore, we shall define work compatible with our concepts of systems, properties, and processes. Hence, *work is done by a system if the sole effect external to the system* (*on the surroundings*) *could be the raising of a weight*. Work done *by* a system is assumed to be positive, and work done *on* a system is considered negative. The careful student will note that this definition does not state that a weight is raised or that a force actually acts through a distance. This definition is necessary because of the need to distinguish between work and heat in the second law of thermodynamics.

It is possible to show that the definition of work from mechanics [Eq. (3-6)] is included in the thermodynamic definition. Consider the apparatus shown schematically in Fig. 3-10. The system of Fig. 3-10*a* interacts with its surroundings. No work occurs according to the mechanics definition. No force has moved through a distance. But the thermodynamist must ask if a weight "could" be made to rise with the given system. Obviously, by replacing the circuit by an electric motor and hoist depicted in Fig. 3-10*b*, there *is* thermodynamic work. The sole effect "could be" the raising of a weight.

The term *sole effect* in the definition of work implies that another effect might be external to the system. Consider the effect of pushing a block along a horizontal

FUNDAMENTAL CONCEPTS FROM THERMODYNAMICS

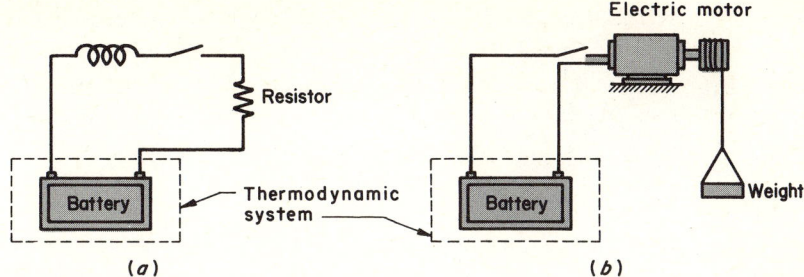

Fig. 3-10 Difference between thermodynamics and mechanics definitions of work

rough plane by inflating a balloon, the thermodynamic system shown in Fig. 3-11a. In this case mechanics work occurs, but another external effect, i.e., heat generated by friction on the rough plane, occurs. Now with the identical thermodynamic system, the balloon in Fig. 3-11b and a frictionless mechanism, thermodynamic work occurs. The raising of a weight is the *sole* external effect. Here we see that mechanics work fits into the definition of thermodynamic work [2].

The term *external* in the definition of work suggests that work is defined only with reference to a system boundary. If we take another system in Fig. 3-11b to include everything—weight, mechanism, and balloon—no work is done, since there are no external effects. Work has meaning only during transit. A system does not possess work. Furthermore, it depends upon the route of transfer. *It is a function of the path.* By a simple rearrangement of the mechanism of Fig. 3-11, interchanging the short and long arms, the value of work changes for an identical movement of the system. The work is more for this latter *path*. Because of this characteristic, an increment of work is designated as δW rather than dW, since dW implies an exact differential, which is valid only in the case of point functions (independent of path).

A large class of thermodynamic problems involves the expansion or compression of a gas, typified by the piston-cylinder arrangement shown in Fig. 3-12. Assuming that the system's external resistance is infinitesimally smaller than the internal pressure p, the work is given by

$$\delta W = pA\,dx = p\,d\mathcal{V} \tag{3-7}$$

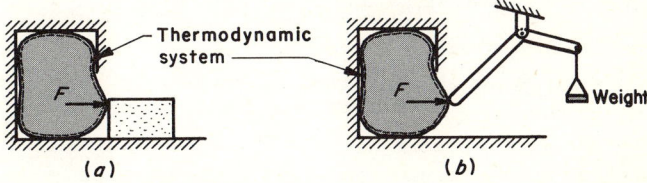

Fig. 3-11 Relation between the thermodynamics and mechanics definitions of work.

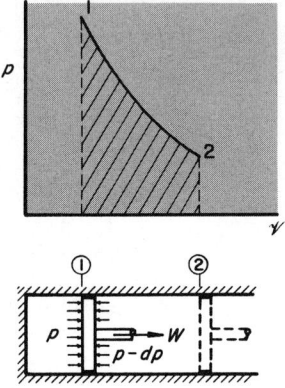
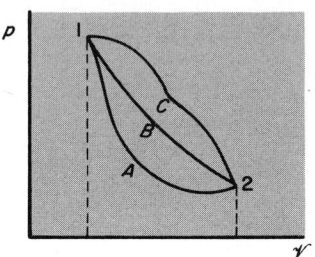

Fig. 3-12 Work done by an expanding gas in a piston-cylinder combination.

Fig. 3-13 The path function work.

where A is the piston area and $\mathscr{V}$ is the volume. This equation is valid only for a *quasi-steady process*, i.e., one in which *thermodynamic equilibrium exists at each infinitesimal step, a reversible process*. To be reversible (1) the system must be closed, (2) there must be no viscous effects within the system, (3) all properties (particularly pressure in this case) must be uniform on the system boundaries, and (4) there can be no other influencing fields—magnetic, gravitational, etc.

In order to integrate the right side of Eq. (3-7), the relationship between p and $\mathscr{V}$ must be known. The path of the process must be specified. The integral of the left side of the equation is never written as $W_2 - W_1$ since this would suggest a point function, an exact differential. It may be written as

$$_1W_2 = \int_1^2 \delta W = \int_1^2 p\, d\mathscr{V} \tag{3-8}$$

where $_1W_2$ is interpreted as the work done during the process. The $p\mathscr{V}$ curve of Fig. 3-12 shows that the work is represented by the area under the curve, and it may be graphically determined if it is not possible to define p in terms of $\mathscr{V}$.

Further emphasis can be given to work's dependence on path by considering other multiple paths which may be taken by a process. It is obvious from Fig. 3-13 that the work along path A is less than along path B or along path C, that is,

$$(_1W_2)_A < (_1W_2)_B < (_1W_2)_C$$

Thermodynamic properties are point functions, independent of path. For example,

$$\int_1^2 dp = p_2 - p_1 \tag{3-9}$$

and the change in pressure depends only upon the values of the pressure at point 1 and point 2.

FUNDAMENTAL CONCEPTS FROM THERMODYNAMICS

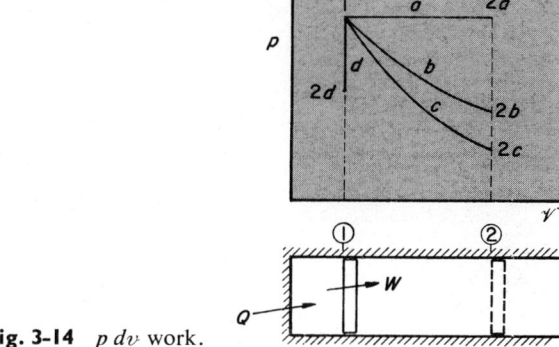

Fig. 3-14 $p\,dv$ work.

To illustrate the work done by a system with a moving boundary, let a gas be heated and expand against a piston of area A. The resistance to expansion is in accordance with the equation

$$p\mathscr{V}^n = \text{const} \tag{3-10}$$

where n is a constant known as a polytropic exponent (to be discussed later). Figure 3-14 illustrates some common processes encountered in thermodynamic problems. The work done by the expanding gas can be determined as follows:

$$p = \frac{\text{const}}{\mathscr{V}^n} = \frac{p_1 \mathscr{V}_1^n}{\mathscr{V}^n} = \frac{p_2 \mathscr{V}_2^n}{\mathscr{V}^n}$$

$$_1W_2 \equiv \int_1^2 p\,d\mathscr{V} = \text{const}\int_1^2 \frac{d\mathscr{V}}{\mathscr{V}^n} = \text{const}\left[\frac{\mathscr{V}^{-n+1}}{-n+1}\right]_1^2$$

$$= \frac{\text{const}}{1-n}(\mathscr{V}_2^{1-n} - \mathscr{V}_1^{1-n}) = \frac{p_2\mathscr{V}_2^n \mathscr{V}_2^{1-n} - p_1\mathscr{V}_1^n \mathscr{V}_1^{1-n}}{1-n}$$

$$= \frac{p_2\mathscr{V}_2 - p_1\mathscr{V}_1}{1-n} \tag{3-11}$$

It is now a simple matter to solve for the following special cases of the polytropic process illustrated in Fig. 3-14:

Isobaric: $\quad p_1 = p_{2a} = \text{const} \quad n = 0$

$$_1W_2 = p(\mathscr{V}_2 - \mathscr{V}_1) \tag{3-12}$$

Isothermal: $\quad n = 1 \quad p_1\mathscr{V}_1 = p_2\mathscr{V}_2 = \text{const}$

$$_1W_2 = \int_1^2 p\,d\mathscr{V} = p_1\mathscr{V}_1 \int_1^2 \frac{d\mathscr{V}}{\mathscr{V}} = p_1\mathscr{V}_1 \ln\frac{\mathscr{V}_2}{\mathscr{V}_1} \tag{3-13}$$

Isentropic: $n = k = \dfrac{c_p}{c_v} =$ ratio of specific heats[1]

$$_1W_2 = \frac{p_2 \mathscr{V}_2 - p_1 \mathscr{V}_1}{1 - k} \tag{3-14}$$

Isovolumetric: $\mathscr{V}_2 = \mathscr{V}_1$ piston does not move

$$_1W_2 = 0 \quad \text{since } d\mathscr{V} = 0 \tag{3-15}$$

Polytropic: $n = n$

This is the general case given by Eq. (3-11).

Example Describe the work done in a piston-cylinder arrangement filled with gas where the reversible process occurs in accordance with the equation $p\mathscr{V}^{4/3} = 20$ lb$_f$-ft^2. The initial volume is 1 ft^3, and the final volume is 2 ft^3.

Solution Since the process is reversible,

$$_1W_2 = \int_1^2 p\, d\mathscr{V}$$

But $p = 20/\mathscr{V}^{4/3}$; therefore,

$$_1W_2 = 20 \int_{\mathscr{V}_1=1}^{\mathscr{V}_2=2} \mathscr{V}^{-4/3}\, d\mathscr{V} = -20(3)[\mathscr{V}^{-1/3}]_{\mathscr{V}_1}^{\mathscr{V}_2} = -60\left(\frac{1}{2^{1/3}} - \frac{1}{1^{1/3}}\right)$$
$$= -60(0.794 - 1) = 60(0.206) = 12.36 \text{ ft-lb}_f$$

Work is done by the system (the gas) since the answer is positive. The student should verify the units of the answer.

3-8 HEAT

The other form of energy of significance in transfer processes, heat, is defined in terms of temperature. *Heat is the energy which is transferred across the boundaries of a system interacting with the surroundings by virtue of a temperature difference.*

Heat Q, like work, is not possessed by a system. It has meaning only during a process. For example, an oven does not contain heat (it contains thermal energy), but heat may be transferred from an oven to the kitchen. The heat may pass through a glass window or through a superinsulation (path-dependent); hence, it is a path function similar to work. Thus,

$$_1Q_2 \equiv \int_1^2 \delta Q \neq Q_2 - Q_1 \tag{3-16}$$

Just as a sign convention was established for work, we shall assume the heat transferred into a system to be positive and that transferred out to be negative.

---
[1] To be discussed later.

FUNDAMENTAL CONCEPTS FROM THERMODYNAMICS

3-9 THE FIRST LAW OF THERMODYNAMICS

Since the first law of thermodynamics is a relation between the fundamental quantities of heat and work, let us look further at their distinctions and similarities. The difference between heat and work can be illustrated by a slight modification of the example shown in Fig. 3-10. Let the energy generated by the flowing electric current in the resistor be used to heat a gas in a closed tank, as shown in Fig. 3-15. Clearly, work is transferred across the boundary of the system in Fig. 3-15a, since electricity crosses the system boundary. Note that a different system was chosen from that of Fig. 3-10. But this is the beauty of thermodynamics: we are at liberty to choose a system which will facilitate the solution we seek.

Now let us consider a system which includes only the resistor, shown in Fig. 3-15b. The heat generated by the flow of current in the resistor must leave the system. Hence, heat is transferred. In both cases the temperature of the gas increases. In Fig. 3-15a work crosses the boundary of the system chosen; in Fig. 3-15b heat crosses the boundary of the system chosen.

Neither heat nor work is a property of the system. They are boundary phenomena, path-dependent, inexact differentials. Both are forms of energy in transit and have meaning when a system undergoes a change of state. The sign convention adopted by choice is illustrated in Fig. 3-16.

The conventional units of work are foot-pounds force; of heat, the British thermal unit (Btu). The Btu was originally defined as that quantity of heat required to raise 1 lb_m of water from 59.5 to 60.5°F, which is referred to as the 60° Btu. There is a more precise definition in terms of electrical units, but it is not necessary to the basic understanding of the unit [3].

To understand the first law of thermodynamics we must understand a *cycle*, defined as *the passing of a system through a series of states but returning to its initial condition*. Consider an ice-cream freezer shown schematically in Fig. 3-17. The ingredients, milk, eggs, sugar, flavoring, etc., are contained in the system chosen. Work is transferred to the system by the paddle, causing the temperature of the system to rise, but the heat resulting from the increased temperature is transferred to the surrounding brine. Work goes in; heat comes out.

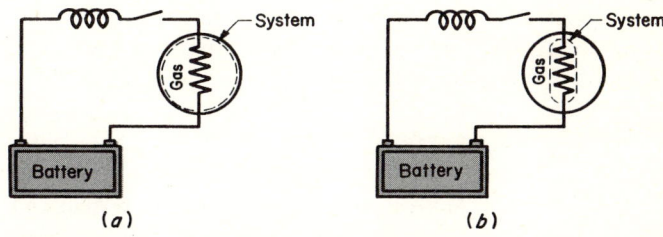

Fig. 3-15 An example of the difference between heat and work.

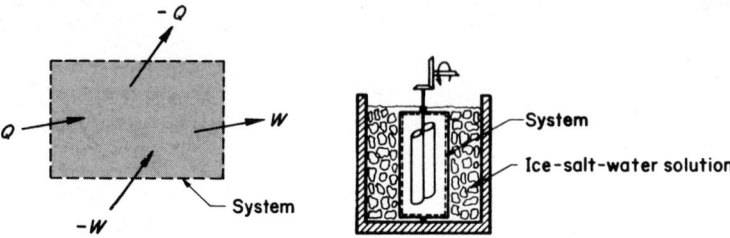

Fig. 3-16 Sign convention for heat and work.

Fig. 3-17 Ice-cream freezer illustrating a cycle and first-law principles.

What happens when all the energy added by work is extracted by the heat transfer? The system returns to its initial state, passing through a cycle. Note that for the system chosen the work is negative and the heat is negative. The careful student will note that the total work and heat transferred in the cycle is different from zero, i.e.,

$$\delta W \neq 0 \quad \delta Q \neq 0 \tag{3-17}$$

As a matter of fact, for the system in question

$$\delta W < 0 \quad \text{and} \quad \delta Q < 0 \tag{3-18}$$

With a little ingenuity we can measure the work and heat transferred. Equipping the input shaft with a pulley and weight will give the work, while the heat transfer can be measured by ice meltage. Before leaving this example, we should observe that more heat must be extracted than added by the work if we are to freeze the ice cream.

In 1843 a British scientist, Joule, carried out a number of experiments similar to the preceding example with various configurations. In all cases, he observed that the work done on the system was directly proportional to the quantity of heat removed from the system. Mathematically,

$$-\oint dW \propto -\oint \delta Q$$

or

$$\oint \delta W = J \oint \delta Q \quad \text{cycle} \tag{3-19}$$

where the proportionality constant J is the mechanical equivalent of heat, the value of which depends upon the units chosen. Equation (3-19) is the mathematical statement of the first law of thermodynamics. This law, which is the basic law of the conservation of energy, was deduced from observations. It is given the status of a law only because no contradiction to it has ever been found.

It is evident from Eq. (3-19) that work and heat can be expressed in equivalent units. Expressing work in foot-pounds force and heat in Btu,

$$J = 778 \text{ ft-lb}_f/\text{Btu}$$

FUNDAMENTAL CONCEPTS FROM THERMODYNAMICS

Fig. 3-18 Schematic representation of first-law quantities.

Equation (3-19) does not suggest that heat and work are the same thing, but it does establish the relationship between the two. While discussing units, recall that power is work rate, or work per unit time. Therefore, the following conversion factors will be useful:

$$1 \text{ hp} = 33{,}000 \text{ ft-lb}_f/\text{min} = 2545 \text{ Btu/hr}$$
$$1 \text{ kw} = 44{,}200 \text{ ft-lb}_f/\text{min} = 3412 \text{ Btu/hr}$$

3-10 THE FIRST LAW FOR PROCESSES IN CLOSED SYSTEMS

Most of our thermodynamic problems are concerned with processes rather than cycles. Systems rarely return to their initial state. Therefore, to be useful the first law should be formulated for easy application to processes. Remembering that δQ and δW are not exact differentials but in fact algebraic quantities whose sum is taken over the complete cycle to arrive at Eq. (3-19), we may rewrite it as

$$\oint (\delta Q - \delta W) = 0 \qquad (3\text{-}20)$$

where the proportionality constant J has been taken as unity, realizing that we must always use compatible units. Since the cyclic integral of the quantity $\delta Q - \delta W$ is zero, this is an exact differential, hence a property of the system. Denoting this new property by E,

■ $\delta Q - \delta W = dE$ process   (3-21)

The change in this property, called *energy E*, is equal to the difference between the heat supplied to the system and the work done on the system during any change of state. It does not matter that this property cannot be measured directly because it is always used as a deviation from some arbitrary reference level.

Unlike work and heat, energy is possessed by the system as represented in Fig. 3-18. It is an extensive property which is given the lowercase symbol e for a system of unit mass. Unlike work and heat, which are surface phenomena, energy E is a volume phenomenon which depends upon the collection of mass within the system. As a matter of fact, the equation is valid only when the same mass is present throughout the process—for a closed system.

Physically, the energy E represents all the energy of a system in a given state. It may be in a variety of forms. In thermodynamics it is common to consider this energy of three forms, kinetic, potential, and internal. Kinetic energy (KE) results from motion; potential energy (PE) results from mass position; and internal energy

U comes from internal molecular motion or mass composition. Other forms of energy, e.g., chemical or electric, are usually negligible in elementary thermodynamic problems but may be included when necessary. In symbolic form the energy E is

$$E = U + \text{KE} + \text{PE} \tag{3-22}$$

Recalling the relations for kinetic and potential energies from elementary mechanics, we have

$$E = U + \frac{mV^2}{2g_c} + \frac{mgz}{g_c} \tag{3-23}$$

or, dividing by mass, we can express the energy per unit mass of substance, i.e.,

$$e = u + \frac{V^2}{2g_c} + \frac{g}{g_c} z \tag{3-24}$$

We can "integrate" Eq. (3-21) to get

$$_1Q_2 - {_1W_2} = E_2 - E_1 \tag{3-25}$$

Substituting the equivalent for E from Eq. (3-23) gives

$$_1Q_2 - {_1W_2} = U_2 - U_1 + \frac{m(V_2^2 - V_1^2)}{2g_c} + \frac{mg}{g_c}(z_2 - z_1) \tag{3-26}$$

or, if more convenient on a per unit mass basis,

$$_1q_2 - {_1w_2} = u_2 - u_1 + \frac{V_2^2 - V_1^2}{2g_c} + \frac{g}{g_c}(z_2 - z_1) \tag{3-27}$$

These equations are in terms of energy changes and tell us nothing about absolute values of the respective energies. We can assign reference values to them. For example, the kinetic energy is normally taken as zero when a body has zero velocity with respect to the earth. Similarly, the potential energy may be assigned the value of zero at some arbitrary datum. From Joule's experiments we can observe that internal energy is a function of temperature. Hence, the selection of a reference temperature fixes the reference level of internal energy.

The possibility of selecting reference levels based on temperature makes the tabulation of thermodynamic properties convenient. Properties are normally tabulated for saturated states with the following subscripts designating the respective saturated states:

$(\cdot)_f \equiv$ saturated liquid

$(\cdot)_g \equiv$ saturated vapor

$(\cdot)_{fg} \equiv (\cdot)_g - (\cdot)_f$

For fluids which are a combination of vapor and liquid the quality x defined by

$$x \equiv \frac{\text{mass of vapor}}{\text{total mass}}$$

FUNDAMENTAL CONCEPTS FROM THERMODYNAMICS

makes it convenient to determine system properties. For example,

$$u = u_f + x u_{fg} \tag{3-28}$$

$$v = v_f + x v_{fg} \tag{3-29}$$

$$h = h_f + x h_{fg} \tag{3-30}$$

The last equation introduces a new thermodynamic property, enthalpy h, defined by

$$h \equiv u + pv \tag{3-31}$$

which is common in problems involving a transfer of mass across a control surface. Further consideration of this concept will be deferred until Chap. 10, when the energy equation (the first law) will be derived for a control volume. The student is cautioned to bear in mind that the relations of this section pertain to closed systems. It should also be obvious that the relations will differ when applied to stationary media and moving media.

Example Gas, initially at 50 psia, is contained in a frictionless piston-cylinder combination (Fig. 3-19). The spring exerts a force against the piston which is directly proportional to the gas volume. Atmospheric pressure of 14.7 psia acts on the spring side of the piston. (a) Determine the work done by the gas in expanding from 0.5 to 2.5 ft³. (b) If heat in the amount of 44.7 Btu is added during the process, what is the change in internal energy?

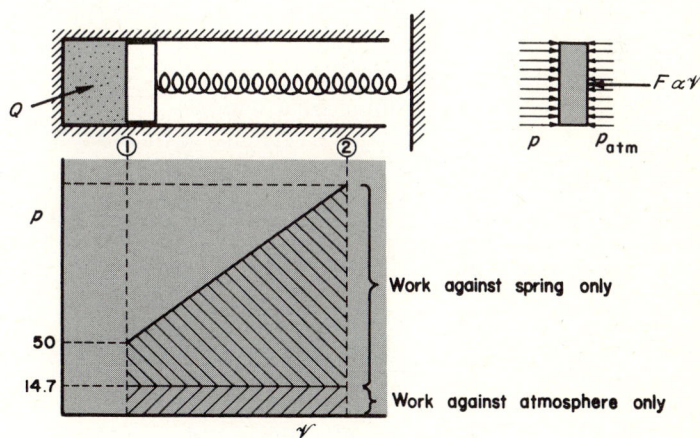

Fig. 3-19 Simultaneous work and heat transfer.

Solution (a) Since the spring force is proportional to the volume, the spring constant k ($F = k\mathscr{V}$) can be found by making a force balance on a free-body diagram of the piston, i.e.,

$$pA = p_{atm}A + k\mathscr{V}$$

or, at position 1,

$$k = \frac{(p_1 - p_\text{atm})A}{\mathscr{V}_1} = 70.6A$$

The pressure at position 2 can then be determined:

$$p_2 = p_\text{atm} + \frac{k\mathscr{V}}{A} = 14.7 + 70.6(2.5) = 191.2 \text{ psia}$$

Since the pressure variation is linear, the work can be found most easily by taking the area under the $p\mathscr{V}$ diagram:

$$_1W_2 = \frac{191.2 + 50}{2}(2.5 - 0.5)(144) = 34{,}800 \text{ ft-lb}_f$$

(b) Since there is no change in kinetic and potential energies, the first law for a process gives the change in internal energy:

$$U_2 - U_1 = {_1Q_2} - {_1W_2}$$
$$\Delta U = (44.7 \text{ Btu})(778 \text{ ft-lb}_f/\text{Btu}) - 34{,}800 \text{ ft-lb}_f = 0$$

We might erroneously conclude that this process is a cycle, since the heat transfer is equal to the work transfer. However, the volume did not return to its initial value.

3-11 THE RELATIONSHIP BETWEEN THE FIRST LAW AND RATE PROCESSES

Thermodynamics is fundamentally a science of static systems, systems in equilibrium. By observing a thermodynamic system at equilibrium states we can determine the amount of heat and work that were transferred between those states. In general, this tells us nothing about the rate of transfer based upon the field intensity. It tells us nothing about what goes on inside the system. That is the fundamental purpose of this text—describing what happens inside the system during the process.

Time is not a relevant parameter in classical thermodynamics, but it is often useful to write the first law for an interval of time in order to get average energy-transfer rates. Dividing Eq. (3-21) by δt and taking the limit as δt approaches zero gives

$$\frac{\delta Q}{\delta t} - \frac{\delta W}{\delta t} = \frac{dE}{\delta t} \tag{3-32}$$

$$\text{Power} = \dot{W} \equiv \lim_{\delta t \to 0} \frac{\delta W}{\delta t} \tag{3-33}$$

$$\text{Rate of heat transfer} = \dot{Q} \equiv \lim_{\delta t \to 0} \frac{\delta Q}{\delta t} \tag{3-34}$$

$$_1\dot{Q}_2 - {_1\dot{W}_2} = \frac{dE}{dt} \tag{3-35}$$

or

$$\dot{Q} - \dot{W} = \frac{dU}{dt} + \frac{d(\text{KE})}{dt} + \frac{d(\text{PE})}{dt} \tag{3-36}$$

FUNDAMENTAL CONCEPTS FROM THERMODYNAMICS 49

3-12 SPECIFIC HEATS

If a red hot iron ingot of 20 lb_m is quenched in a 20-lb_m pail of cold water, we know intuitively that the iron will cool and the water will become hot. Experience has shown that the temperature change of the iron is not equal to the temperature change of the water. Furthermore, this is the case for all materials. This characteristic is due to a property of the material known as specific heat c. It is the amount of heat required to change the temperature of a unit mass by 1° under certain conditions. Mathematically,

$$\delta q = c_n \, dT \tag{3-37}$$

where the subscript n denotes the process. In the absence of kinetic- and potential-energy changes Eq. (3-21) becomes

$$\delta q - \delta w = du \tag{3-38}$$

for a unit mass of substance.

In a constant-volume process no work can be done, since the boundary is fixed. Therefore,

$$\delta q|_v = du|_v \tag{3-39}$$

and

$$du = c_v \, dT \quad \text{or} \quad c_v \equiv \left.\frac{\partial u}{\partial T}\right|_v \tag{3-40}$$

when Eq. (3-39) is substituted into Eq. (3-37). In a constant-pressure process

$$\delta q|_p - p \, dv|_p = du|_p \tag{3-41}$$

$$\delta q|_p - d(pv)|_p = du|_p \tag{3-42}$$

or

$$\delta q|_p = d(u + pv)|_p \tag{3-43}$$

Introducing the enthalpy from Eq. (3-31) gives

$$dh = c_p \, dT \quad \text{or} \quad c_p \equiv \left.\frac{\partial h}{\partial T}\right|_p \tag{3-44}$$

Other processes may be used, but specific heats are normally tabulated only for constant-pressure and constant-volume processes.

For a perfect gas $pv = RT$. Substituting this into Eq. (3-31) and differentiating gives

$$h = u + RT \tag{3-45}$$

$$dh = du + R \, dT \tag{3-46}$$

$$\frac{dh}{dT} = \frac{du}{dT} + R \tag{3-47}$$

But for an *ideal gas* internal energy and enthalpy are functions of temperature only, i.e.,

$$u = f(T) \tag{3-48}$$
$$h = \phi(T) \tag{3-49}$$

Therefore, the definitions of specific heat from Eqs. (3-40) and (3-44) may be expressed in terms of total differentials. Thus,

$$c_{v_0} \equiv \frac{du}{dT} \tag{3-50}$$

$$c_{p_0} \equiv \frac{dh}{dT} \tag{3-51}$$

where the zero subscript denotes the specific heat of an ideal gas. Combining Eqs. (3-47), (3-50), and (3-51), we have

$$c_{p_0} = c_{v_0} + R \tag{3-52}$$

Example A 10-lb_m mixture of gases, which may be assumed perfect, is transported in a rigid container from Miami to Denver. During the trip a refrigerated car extracts 90 Btu from the system while 7,780 ft-lb_f of work is done on the system. If the equilibrium temperatures of the end states are equal to the local ambient temperatures (Miami, 100°F; Denver, 60°F), determine the equivalent specific heat for the mixture assuming that it does not vary with temperature within this temperature range. Assume 1 mile difference in elevation.

Solution Using the first law,

$$_1Q_2 - {_1W_2} = \Delta U + \Delta(\cancel{KE})^{\,0} + \Delta(PE)$$

$$W = (7{,}780 \text{ ft-}lb_f)\frac{\text{Btu}}{778 \text{ ft-}lb_f} = 10 \text{ Btu}$$

$$\Delta U = m\int_{T_1}^{T_2} c_v\, dT = mc_v(T_2 - T_1) = -400 c_v \text{ } lb_m\text{-}°R$$

$$\Delta(PE) = m\frac{g}{g_c}(z_2 - z_1) = (10 \text{ }lb_m)\frac{32.2 \text{ ft/sec}^2}{32.2 \text{ }lb_m\text{-ft/}lb_f\text{-sec}^2}\frac{5{,}280 \text{ ft}}{778 \text{ ft-}lb_f/\text{Btu}} = 68 \text{ Btu}$$

Observing the thermodynamic sign convention and substituting these values into the first law gives

$$-90 + 10 = -400 c_v + 68$$

$$c_v = \frac{80 + 68}{400} = 0.37 \text{ Btu/}lb_m\text{-}°R$$

3-13 THE SECOND LAW OF THERMODYNAMICS

The first law of thermodynamics establishes a relationship between heat and work but places no conditions on the direction of transfer. The second law of thermodynamics is the *directional* law.

FUNDAMENTAL CONCEPTS FROM THERMODYNAMICS

Limitations of the first law To illustrate the directional characteristic of the second law, let us return to the example of the ice cream freezer (Fig. 3-17). We added work to the system and extracted heat. Now let us reverse the process—add heat and get work out of the system. There is no conceivable way in which a weight might be returned to its original position by reversing the process. *It is impossible to fully convert all heat into work.* The process is *irreversible*.

Consider another example. A flywheel is stopped by a friction brake. In the process of stopping the flywheel the brake gets hot, and its internal energy is increased by an amount equal to the loss of kinetic energy of the flywheel. The first law would be satisfied if the hot brake gave up its energy to the flywheel, causing it to resume rotation. But there is no conceivable way in which this can happen. The process is irreversible.

Two bodies at different temperatures are placed in thermal contact in an insulated box. Heat is transferred from the high temperature body in accordance with the first law, causing the low temperature body to get warmer. The energy given up by the high temperature body is gained by the low temperature body in coming to thermal equilibrium. Letting the process be reversed would not violate the first law since it is concerned with the conservation of energy, but the same amount of energy cannot be transferred from the low temperature body to the high temperature body. Heat has never been observed to "flow uphill." The process is irreversible.

Some factors which cause irreversibility are (1) friction, (2) finite temperature difference, (3) unrestrained expansion, and (4) mixing of different substances. *In a cyclic process it is possible to convert all work into heat, but it is impossible to convert all the heat into work* [2].

Heat engine A heat engine is any device which operates cyclically and has as its primary purpose the conversion of heat into work. For example, a steam power plant has its working fluid, water, returning periodically to its initial state. As shown in Fig. 3-20, liquid water is pumped into the boiler, where it is vaporized and drives the turbine, producing work, some of which may be used to drive the condensate pump. Choosing the system as shown, only heat and work cross the boundary.

The system can be simplified (Fig. 3-21) as receiving heat from a high temperature reservoir (*source*) and rejecting heat to a low temperature reservoir (*sink*). *A thermal reservoir is a body which can receive or reject heat indefinitely without having its temperature affected.* The ocean and the atmosphere are examples of reservoirs.

Thermal efficiency η_{th} is defined as

$$\eta_{th} \equiv \frac{\text{energy effect sought}}{\text{energy input required}}$$

For the heat engine, the energy effect sought is the work output W, and the energy input required to produce it is the heat input Q_H; therefore,

$$\eta_{th} = \frac{W}{Q_H} \qquad (3\text{-}53)$$

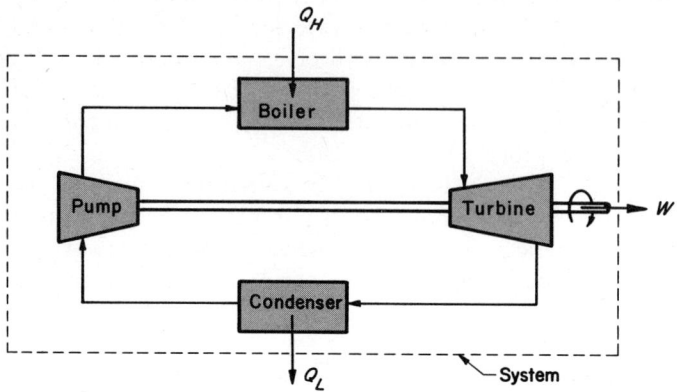

Fig. 3-20 The steam power plant as a heat engine.

For the cycle, from the first law,

$$W = \oint \delta W = \oint \delta Q = Q_H - Q_L \tag{3-54}$$

Therefore,

$$\eta_{\text{th}} = \frac{Q_H - Q_L}{Q_H} = 1 - \frac{Q_L}{Q_H} \tag{3-55}$$

The thermal efficiency can approach 100 percent only as the heat rejected approaches zero. From experience, $Q_L \geq \tfrac{2}{3} Q_H$. Hence, the thermal efficiency is normally less than one third of what the first law would permit if all the heat were converted to work.

There are two classic statements of the second law, both of which are negative statements and cannot be proved. However, since neither has ever been experimentally violated, we shall accept them as law. They are:

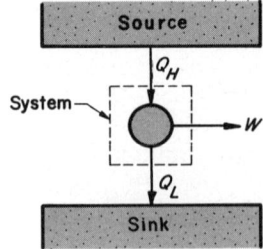

Fig. 3-21 Heat engine.

FUNDAMENTAL CONCEPTS FROM THERMODYNAMICS

Kelvin–Planck: It is impossible to construct a device which will operate in a cycle and produce no effect other than the raising of a weight and the exchange of heat with a single reservoir.

Clausius: Heat cannot pass spontaneously from a low temperature body to a high temperature body.

Proof of the equivalence of these two statements can be established by contradiction and is included in any complete treatise on thermodynamics [2, 3].

Since reservoir temperatures do not change, the thermal efficiency can be written as a function of the reservoir temperatures:

$$\eta_{th} \equiv 1 - \frac{Q_L}{Q_H} = f(T_L, T_H) \tag{3-56}$$

where T is the absolute, thermodynamic temperature. There are many functional relations which will satisfy this equation; however, let us assume the simplest (valid for a reversible engine)

$$\frac{Q_L}{Q_H} = \frac{T_L}{T_H} \tag{3-57}$$

Hence,

$$\eta_{th} = 1 - \frac{T_L}{T_H} \tag{3-58}$$

and the maximum efficiency occurs when the temperature of the sink is zero. This suggests the concept of absolute zero temperature and a *thermodynamic temperature scale* which is independent of the thermometric substance, since the determination of thermal efficiency requires only the measurement of heat and work and is independent of the working substance [4].

Measurement of the thermal efficiency of a reversible engine operating between the ice-point temperature T_i and the steam-point temperature T_s gives

$$\eta_{th} = 1 - \frac{Q_i}{Q_s} = 0.2680 = 1 - \frac{T_i}{T_s} \tag{3-59}$$

$$\frac{T_i}{T_s} = 0.7320 \tag{3-60}$$

The size of temperature increments may be chosen arbitrarily in accordance with Eq. (3-57); but choosing the difference between the ice point and the steam point on the Fahrenheit scale (as discussed in Sec. 3-5) gives

$$T_s - T_i = 180 \tag{3-61}$$

which can be solved simultaneously with Eq. (3-60) to get

$$T_s = 672°R \quad \text{and} \quad T_i = 492°R$$

Example To cool a building in the summer 100,000 Btu/hr must be extracted. The cooling system requires 19,800 Btu/hr of work. How much heat is rejected to the surroundings, and what is the thermal efficiency?

Solution From the first law the heat rejected to the surroundings is

$$-Q_H + 100{,}000 + 19{,}800 = 0$$

Comparing Fig. 3-22 with Fig. 3-21, the effect is reversed from that of the heat engine; hence, it is called a reversed heat engine or heat pump. For this case the energy ratio, called the *coefficient of performance* (COP), is given by

$$\text{COP} \equiv \frac{\text{energy effect sought}}{\text{energy input required}} = \frac{Q_L}{W} = \frac{100{,}000}{19{,}800} = 5.05$$

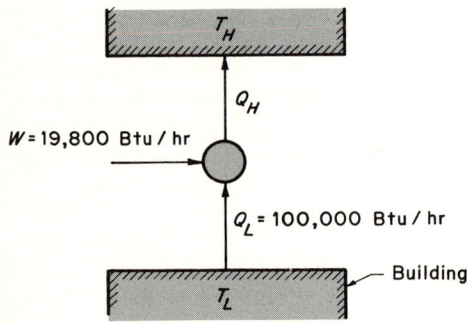

Fig. 3-22 Reversed heat engine (heat pump).

Clausius inequality We rewrite Eq. (3-57)

$$\frac{Q_H}{T_H} - \frac{Q_L}{T_L} = 0 \qquad (3\text{-}62)$$

The heat rejected is negative by our sign convention. Hence, the summation of the ratios of heat transfers to absolute temperatures for a reversible cycle operating between two reservoirs is zero

$$\sum_2 \frac{Q}{T} = 0 \qquad (3\text{-}63)$$

Extrapolating this to the general case of a reversible cycle involving any number of heat reservoirs at different temperatures, we have

$$\lim_{n \to \infty} \sum_n \frac{Q}{T} = \oint \frac{\delta Q}{T} = 0 \qquad (3\text{-}64)$$

But for an irreversible cycle, the cyclic integral is less than zero. Hence

$$\blacksquare \qquad \oint \frac{\delta Q}{T} \leq 0 \qquad \text{cycle} \qquad (3\text{-}65)$$

FUNDAMENTAL CONCEPTS FROM THERMODYNAMICS

The inequality holds for irreversible cycles, and the equality is valid for reversible cycles.

Following the same procedure as in the development of the first law, we now seek to state the second law for processes. For the reversible case the cyclic integral is zero, making $\delta Q/T$ an exact differential and hence a property of the system. Denoting this new property by S, we get

■ $$\left.\frac{\delta Q}{T}\right|_{\text{rev}} = dS \quad \text{process} \tag{3-66}$$

Analogous to internal energy of the first law, this property, called the *entropy S*, enables us to treat the second law quantitatively. Entropy has a unique value depending upon the state of the system, and its change between two states is independent of the process, or path, since it is an exact differential.

3-14 THE FIRST AND SECOND LAWS COMBINED

Dividing the first law equation (3-21) by T for a reversible process, we get

$$\left.\frac{\delta Q}{T}\right|_{\text{rev}} - \left.\frac{\delta W}{T}\right|_{\text{rev}} = \left.\frac{dE}{T}\right|_{\text{rev}} \tag{3-67}$$

which can be combined with Eq. (3-66) to give

$$(T\,dS = dE + \delta W)_{\text{rev}} \tag{3-68}$$

In the absence of motion, gravity effects, magnetization, etc., $dE = dU$, and $\delta W = p\,d\mathscr{V}$ for a quasi-steady process, as discussed in Sec. 3-7. Therefore,

$$T\,dS = dU + p\,d\mathscr{V} \tag{3-69}$$

Alternatively, since $H \equiv U + p\mathscr{V}$,

$$dH = dU + p\,d\mathscr{V} + \mathscr{V}\,dp \tag{3-70}$$

which gives the following equation when substituting Eq. (3-70) into Eq. (3-69):

$$T\,dS = dH - \mathscr{V}\,dp \tag{3-71}$$

Equations (3-69) and (3-71) are known as the classical $T\,dS$ equations. Note that they are in terms of properties only. Therefore, entropy can be calculated without measuring either heat or work, knowing the relationship between p, $\mathscr{V}$, T, and U for the system. Note also that since the equations are in terms of properties only, they are no longer restricted to reversible processes. The properties T, S, U, p, $\mathscr{V}$, and H change by the same amount in any process (reversible or irreversible) between any given pair of equilibrium states. They are frequently written per unit mass, i.e.,

■ $$T\,ds = du + p\,dv \tag{3-72}$$
■ $$T\,ds = dh - v\,dp \tag{3-73}$$

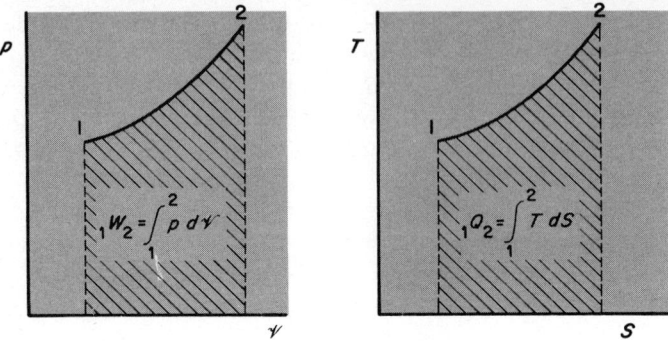

Fig. 3-23 Heat and work for reversible processes.

Just as the area under a reversible path on a $p\mathscr{V}$ diagram represents work, the area under a reversible path on a TS diagram represents heat, as shown in Fig. 3-23.

Example The constant-pressure specific heat of a hydrocarbon is

$$c_{p0} = 4 + 0.016T$$

where T is in degrees Rankine. Assuming it behaves as a perfect gas, what is its change in entropy from $p_1 = 20$ psia, $T_1 = 400°R$ to $p_2 = 100$ psia, $T_2 = 800°R$? The gas constant $R = 50.8$ ft-lb$_f$/lb$_m$-°R.

Solution Using the equation of state for a perfect gas and Eq. (3-73),

$$pv = RT$$
$$T\,ds = dh - v\,dp$$

but $dh = c_{p0}\,dT$, which can be combined to give

$$ds = c_{p0}\frac{dT}{T} - R\frac{dp}{p}$$

Integrating yields

$$s_2 - s_1 = 4\int_{T_1}^{T_2}\frac{dT}{T} + 0.016\int_{T_1}^{T_2}dT - R\int_{p_1}^{p_2}\frac{dp}{p}$$

$$= 4\ln\frac{T_2}{T_1} + 0.016(T_2 - T_1) - R\ln\frac{p_2}{p_1}$$

$$= 4\ln 2 + 0.016(400) - \frac{50.8\text{ ft-lb}_f/\text{lb}_m\text{-°R}}{778\text{ ft-lb}_f/\text{Btu}}\ln 5$$

$$= 9.067\text{ Btu/lb}_m\text{-°R}$$

3-15 THE THIRD LAW OF THERMODYNAMICS

The second-law relationship for entropy, Eq. (3-66), can account only for changes in entropy—one state relative to another. Although this is adequate for thermodynamic

FUNDAMENTAL CONCEPTS FROM THERMODYNAMICS

calculations, it is sometimes advantageous to speak in terms of absolute entropy, which requires the third law of thermodynamics. Simply stated, it is that *the entropy of a pure substance is zero at absolute zero.*

In a probabilistic sense, entropy is a measure of the disorder of a system. At absolute zero, there is no translational molecular activity, hence no disorder, or zero entropy.

PROBLEMS

3-1. Associate and discuss the eulerian and lagrangian points of view with the thermodynamic concepts of system and control volume.

3-2. Which of the following are intensive properties?

(a) Temperature (b) Volume
(c) Density   (d) Pressure
(e) Length   (f) Electric charge

3-3. On the pvT surface in Fig. 3-9 what is the quality between points: (a) i and L, (b) L and h, (c) h and g, (d) o and n, (e) j and i?

3-4. From Fig. 3-7 find the quality for each of the four cases if the container is rigid with a volume of 1 ft³ and the total mass is 1 lb$_m$. The percentage of vapor of the total mass in case (c) is 25 percent.

3-5. The quality of a 10-lb$_m$ mixture is 95 percent. If the mixture is in a 2-ft³ rigid container, what is the density of the vapor?

3-6. Two identical containers hold different mixtures. The first container holds a mixture which has a vapor density of 0.04 lb$_m$/ft³ with a quality of 20 percent. The second container holds a mixture which has a vapor density of 0.1 lb$_m$/ft³ with a quality of 50 percent. Compare the total mass in the two containers.

3-7. Sketch the following cycle on a PV diagram labeling the points:

a to *b* Isothermal expansion
b to *c* Constant volume: pressure increases
c to *d* Isentropic compression
d to *a* Isobaric compression

3-8. For the processes depicted by curves *a*, *b*, and *c* of Fig. 3-14, which process involves the most work? Why?

3-9. During a constant-temperature process air is expanded in a piston-cylinder arrangement until the final volume is twice the initial volume. Assuming that air is a perfect gas and that 1 lb$_m$ was at 100°F initially, how much work in foot-pounds force is done in the expansion?

3-10. A perfect gas is compressed from 5 to 1 ft³. During the compression process the pressure varies inversely with the volume. For an initial pressure of 100 psia, how much work (foot-pounds force and Btu) is done in compressing the gas?

3-11. The equation of state for a van der Waals gas is

$$\left(p + \frac{a}{v^2}\right)(v - b) = RT$$

where *a* and *b* are constants for a particular gas. Determine the work of a moving boundary per unit mass for an isothermal process.

3-12. A spherical balloon has a diameter of 10 in and contains air at a pressure of 20 psia. The diameter of the balloon increases to 12 in due to heating, and during this process the pressure is proportional to the diameter cubed. Calculate the work in foot-pounds force done by the air during this process.

3-13. A gas at 20 psia is expanded at constant pressure from 1 to 4 ft³. It is then further expanded at constant temperature to 6 ft³. Finally, heat in the amount of 20 Btu is added at constant volume. How much work in foot-pounds force is done in the total process?

3-14. A 6-in-long square rubber bar is secured by its end to a wall. How much work in foot-pounds force and Btu is done in stretching the bar to a length of 10 in if the forces exerted at different lengths are

Length, in	6	7	8	9	10
Force, lb_f	0	3	5	10	16

3-15. According to the definition of heat and work in the thermodynamic sense, determine whether the heat exchange is greater than zero, equal to zero, or less than zero in the systems undergoing the following changes. Do the same for work.

(a) A piece of copper heated to 1000°F is dropped into water at 60°F; consider the copper as the system.

(b) Gas in a large reservoir is used to charge a small rigid evacuated bottle; consider the gas being transferred as the system and neglect heat transfer between the reservoir and bottle, both of which are insulated from the surroundings.

(c) Heat is added to a gas in a rigid container such that the pressure and temperature increase; consider the gas as the system.

(d) In a combustion experiment, a mixture of hydrogen and oxygen in a soap bubble is ignited by a minute spark and the bubble expands; consider the gaseous mixture as the system and neglect conductivity of the wall of the bubble.

(e) An insulated wire is stretched; consider the wire as the system.

(f) Gas from a bottle is used to inflate a balloon which is originally flat; consider the gas as the system and assume that all components involved are nonconducting.

(g) A turbulent stream of liquid is settling in a large tank; consider the liquid as the system.

3-16. A process occurs in which a gas is expanded and then compressed to its original state.
(a) What is the change in its total energy E?
(b) How are heat and work related for the process?
(c) What is the relationship among the changes in potential, kinetic, and internal energies?

3-17. Which of the following statements is true? Why?

(a) $\oint du \geq 0$ (b) $\oint du = 0$ (c) $\oint du \leq 0$

3-18. A gas at 100 psia with an enthalpy of 50 Btu and a volume of 1 ft³ is expanded in a constant-pressure process to 5 ft³. What is its internal energy at the end of the expansion process?

3-19. During a period of 10 hr 3800 Btu is extracted from a system having an initial volume of 12 ft³ and an initial pressure of 80 psia.

(a) Assuming the pressure varies inversely with the volume, what is the change in internal energy if the final volume is 1 ft³?

(b) If the initial enthalpy is 1600 Btu, determine the final internal energy.

3-20. An insulated rigid tank of volume $\mathscr{V}$ contains m_1 lb_m of air at p_1 and T_1. A valve is opened, and the tank fills until its pressure is equal to that of the atmosphere p_a. Neglecting kinetic energy and assuming that the air acts like a perfect gas, i.e., it obeys the ideal-gas equation of state and internal energy and enthalpy are functions of temperature only, derive an expression for the amount of mass that flows into the tank; i.e., find $m_1 - m_2 = \phi(\mathscr{V}, T_a, p_a, p,$ appropriate constants).

FUNDAMENTAL CONCEPTS FROM THERMODYNAMICS

3-21. A piston-cylinder arrangement contains 1 lb_m of water. Initially a saturated liquid at 170 psia, it is expanded until the pressure is 20 psia and the specific volume is 0.0168 ft^3/lb_m. For the process, find the change in (*a*) temperature, (*b*) specific volume, (*c*) specific enthalpy, and (*d*) internal energy.

3-22. A piston-cylinder arrangement contains 1 lb_m of a specific gas initially at 20 psia with an enthalpy of 10 Btu/lb_m and with an initial volume of 1 ft^3. While the pressure is held constant, 10,000 Btu/lb_m of heat is added. If the final volume is 6 ft^3, what is the final internal energy?

3-23. Initially a saturated vapor at 100 psia, 2 lb_m of water is then compressed to a pressure of 1,000 psia at a temperature of 1400°F. Assuming that the pressure varies inversely with the volume, determine the heat transferred.[1]

3-24. Assume that 2 lb_m of water is initially at 70 psia and 302.93°F and occupies a volume of 10 ft^3 in a piston-cylinder arrangement. It is compressed, with the pressure varying inversely with the volume, to a pressure of 90 psia, a temperature of 320.28°F, and a volume of 7.8 ft^3. Determine (*a*) the energy added as heat and (*b*) the change in density of the vapor.

3-25. A home food freezer, which contains 200 lb_m of food with average specific heat $c = 0.6$ Btu/lb_m-°F, uses 1 kw (3412 Btu/hr) of electric power.
    (*a*) Will the food remain frozen if 2400 Btu/hr is transferred?
    (*b*) How many pounds mass can be preserved at 25°F?

3-26. If 1 lb_m of ice at 32°F requires 144 Btu (latent heat) to melt, what is its change in entropy during the melting process?

3-27. An inventor claims to have developed an engine which takes in 80,000 Btu at a temperature of 265°F, delivers 16 kwhr of work, and rejects the rest at $-90°F$. Is his claim valid? Defend your answer.

3-28. If a refrigerator requires 20 hp while operating between a reservoir at 0°F and the atmosphere at 70°F, at what rate must it transfer heat from the 0°F reservoir? Assume reversible.

3-29. If 8000 Btu of work is delivered by a reversible engine, which receives heat from a reservoir at 1000°F and rejects heat at 300°F, find the heat transferred to the engine.

3-30. A mixture of water is initially at 500 psia and 80 percent quality. It is expanded to a pressure of 300 psia and a quality of 70 percent. Is this process possible for an isolated system?

3-31. An isolated system is made up of saturated water vapor at 110°F. Will a mixture at 150°F with a quality of 95 percent result from the following compression and expansion processes?

	Compression	Expansion
Q	$-10{,}000$ Btu/lb_m	4000 Btu/lb_m
W	-5000 Btu/lb_m	1000 Btu/lb_m
Δv	33.62 ft^3/lb_m	
Δh		1075 Btu/lb_m

3-32. (*a*) For a constant-pressure process, the value of the exponent in $pV^n = C$ is (*i*) 0, (*ii*) ∞, (*iii*) 1.0, (*iv*) k, (*v*) not given.
    (*b*) For a reversible isothermal process from small to large volume, the heat transfer is (*i*) positive (*ii*) negative, (*iii*) 0, (*iv*) ∞, (*v*) not given.
    (*c*) When heat is reversibly added to a gas during a process, the work is (*i*) positive, (*ii*) negative, (*iii*) 0, (*iv*) ∞, (*v*) not given.

---

[1] Suggested reference: J. H. Keenan and F. G. Keyes, "Thermodynamic Properties of Steam," Wiley, New York, 1936.

(d) When heat is reversibly added to a gas while it expands and while its temperature drops, the change in entropy is (i) positive, (ii) negative, (iii) 0, (iv) ∞, (v) not given.

(e) For a polytropic expansion to a larger volume with decreasing temperature, the heat exchange is (i) positive, (ii) negative, (iii) 0, (iv) ∞, (v) not given.

(f) In a reversible nonflow process the pressure (i) can (ii) cannot rise if the volume is increasing.

3-33. It was noted that Eq. (3-72) is valid for any process. Since $T\,ds$ is equal to δq for a reversible process only, the effect of irreversibility is contained in either the term du or $p\,dv$. Do you agree? If so, which term contains this effect? Explain.

3-34. A well-insulated piston-cylinder combination contains air. A weight holds the piston at position 1, where the pressure is p_1 and the volume is $\mathscr{V}_1$. The weight is suddenly removed and the

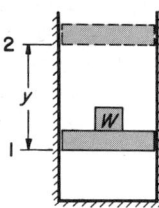

Fig. P3-34

piston moves *very rapidly* from position 1 and comes to rest, after bobbing up and down, at position 2, where the pressure is p_2.

$R = 53.3$ ft-lb$_f$/lb$_m$-°R $\quad p_0 =$ barometric pressure, 14.7 psia

$c_p = 0.24$ Btu/lb$_m$-°R $\quad \mathscr{M} = 28.97$, molecular weight

$c_v = 0.17$ Btu/lb$_m$-°R $\quad A =$ piston area

$k = 1.40$

(a) Derive an expression for the work done by the gas in terms of $p_1, p_2, \mathscr{V}_1$, and the appropriate constants.

(b) Get a numerical answer for the case where $p_1 = 20{,}000$ lb$_f$/ft^2, $p_2 = 10{,}000$ lb$_f$/ft^2, $\mathscr{V}_1 = 1.0$ ft^3.

(c) What is the volume at position 2?

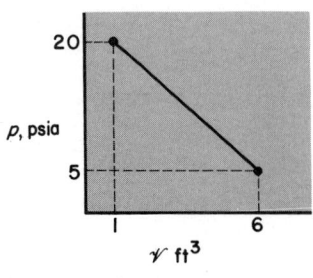
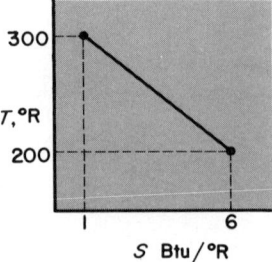

Fig. P3-35

FUNDAMENTAL CONCEPTS FROM THERMODYNAMICS

(d) If the gas is now recompressed by suddenly placing the same weight on the piston, what will be the final volume (say $\mathscr{V}_3$)?

(e) Discuss the significance of the answer found in part (d) as compared with the initial volume $\mathscr{V}_1$.

(f) Calculate the entropy change from state 1 to state 2.

(g) Calculate the entropy change from state 2 to state 3, the process indicated in part (d).

3-35. For the reversible process shown in Fig. P3-35 determine the change in (a) internal energy and (b) enthalpy.

3-36. A rigid container of constant volume contains 1 lb_m of air at 15 psia and 70°F. It is heated by means of an electric heating element to 500°F. What is the change in entropy of the air?

3-37. A container of water vapor is divided by a partition; each side has a volume of 1 ft^3. On side A the mass is 0.0035 lb_m, while on the opposite side B the mass is 0.5 lb_m. Side A is initially saturated at 150°F, and side B has a quality of 1 percent at 10 psia. Find the change in entropy for the process resulting in a final temperature of 170°F at 6 psia when the partition is removed.

3-38. A perfect gas in a rigid container undergoes an isothermal process in which the pressure is decreased to one-half its original value. If the volume is 1 ft^3 and the initial pressure is 20 psia, find the change in entropy for a temperature of 70°F.

3-39. Air changes states from $T_1 = 150°F$, $v_1 = 10$ ft^3/lb_m to $T_2 = 1500°F$, $v_2 = 1$ ft^3/lb_m. What is its change in entropy?

3-40. A perfect gas ($R = 53.3$ ft-lb_f/lb_m-°R, $c_p = 0.24$ Btu/lb_m-°R) changes states from $p_1 = 10$ psia, $T_1 = 70°F$ to $p_2 = 100$ psia, $T_2 = 1300°F$. Find its change in entropy.

REFERENCES

1. Beckwith, T. G., and N. L. Buck: "Mechanical Measurements," Addison-Wesley, Reading, Mass., 1961.
2. Lay, J. E.: "Thermodynamics," Merrill, Columbus, Ohio, 1963.
3. Van Wylen, G. J., and R. E. Sonntag: "Fundamentals of Classical Thermodynamics," Wiley, New York, 1965.
4. Wark, K.: "Thermodynamics," McGraw-Hill, New York, 1966.

PART II
STATIONARY MEDIA

CHAPTER 4
FLUID STATICS

In Chap. 1 it was pointed out that a fluid deforms continuously when subjected to shearing stresses. Therefore, for an element of fluid at rest, stress must always act normal to the surface of the element. This normal stress, in the absence of motion, is termed *pressure*. Fluids having constant velocities may be treated as static since their acceleration is zero.

Fluid statics is a special case of momentum transfer. *No momentum is transferred.* A study of fluid statics provides insight into the analysis of the more complicated fluid dynamics problems. Furthermore, most pressure measuring devices commonly used when fluid is in motion depend upon the transmission of a fluid pressure through static fluids; hence, an understanding of fluid statics and the attendant pressure distributions is fundamental to an understanding of rate processes.

4-1 STATIC EQUILIBRIUM

Forces which exist in fluid systems are either *body* or *surface forces*. Body force is proportional to the system's mass and is commonly termed weight.[1] Surface forces

---
[1] In more advanced cases other body forces, e.g., that produced by a magnetic field, may be present.

are commonly given in terms of orthogonal components tangent and normal to the surface in question, i.e., shear and normal forces. For static equilibrium only body forces and normal forces are present.

Pascal's law Before looking at the variation of pressure in a fluid element, let us first consider the relationship between the pressures on the surface of the element. We must resolve this in order to know "which" pressure is being varied. Consider the fluid wedge shown in Fig. 4-1. Assume that the element is small enough for the pressures to be uniformly distributed over the respective surfaces. This is exactly the case as the dimensions approach zero, i.e., at a point. The forces on the vertical xy planes are equal and opposite.

Noting that the body force, the weight, is the product of specific weight, $\gamma = \rho g$, and volume, the following equations result upon summing forces in the x and y directions, respectively:

$$p_x(\Delta y\, \Delta z) = p(\Delta x\, \Delta z)\frac{\sin \alpha}{\cos \alpha} \tag{4-1}$$

$$p_y(\Delta x\, \Delta z) = p(\Delta x\, \Delta z)\frac{\cos \alpha}{\cos \alpha} + \rho g\left(\Delta x\, \frac{\Delta y\, \Delta z}{2}\right) \tag{4-2}$$

Observing that $(\sin \alpha)/(\cos \alpha) = \tan \alpha = \Delta y/\Delta x$, we have

$$p_x = p \tag{4-3}$$

$$p_y - p = \rho g\, \frac{\Delta y}{2} \tag{4-4}$$

We let the element approach a point; i.e., as Δx, Δy, and Δz approach zero, Eq. (4-4) gives

$$p_y = p \tag{4-5}$$

The results of Eqs. (4-3) and (4-5) can be generalized by considering a tetrahedron rather than a wedge, giving

$$p = p_x = p_y = p_z \tag{4-6}$$

a scalar field, independent of direction.

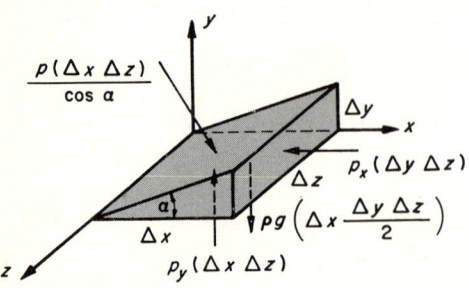

Fig. 4-1 Forces on fluid wedge.

FLUID STATICS

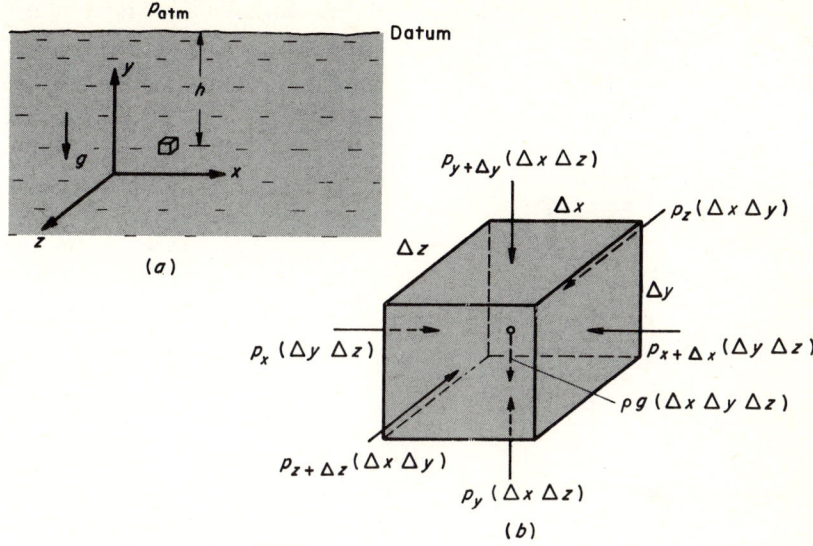

Fig. 4-2 Forces on fluid element.

Equation (4-6) is a statement of *Pascal's law: the pressure exerted at a point in a static or uniformly moving fluid is equal in all directions.* (It is also valid for an accelerating frictionless fluid.) This conclusion could have been deduced physically. Going back to the assumption of uniform pressure distribution on each surface, it is not reasonable to expect a discontinuity in pressures at the "corners" of the element, since we normally expect physical phenomena to be continuous. The physical phenomenon in this case (pressure) is continuous; hence, it is the same on all surfaces at a point.

Pressure variation In order to determine the variation of pressure, $p = f(x,y,z)$, consider the fluid element shown in Fig. 4-2. The forces acting on the element of fluid are the surrounding fluid stresses (pressures) and the force due to gravity (weight).

Applying Newton's first law, $\sum \mathbf{F} = 0$, gives the following scalar equations in the three mutually orthogonal directions:

$$\sum F_x = p_x(\Delta y\, \Delta z) - p_{x+\Delta x}(\Delta y\, \Delta z) = 0 \tag{4-7a}$$

$$\sum F_y = p_y(\Delta x\, \Delta z) - p_{y+\Delta y}(\Delta x\, \Delta z) - \rho g(\Delta x\, \Delta y\, \Delta z) = 0 \tag{4-7b}$$

$$\sum F_z = p_z(\Delta x\, \Delta y) - p_{z+\Delta z}(\Delta x\, \Delta y) = 0 \tag{4-7c}$$

Dividing each of the equations by $\Delta x \, \Delta y \, \Delta z$ and taking the limit as Δx, Δy, and Δz approach zero, we get

$$\sum f_x = -\frac{\partial p}{\partial x} = 0 \tag{4-8a}$$

$$\sum f_y = -\frac{\partial p}{\partial y} - \rho g = 0 \tag{4-8b}$$

$$\sum f_z = -\frac{\partial p}{\partial z} = 0 \tag{4-8c}$$

where f is the external force per unit volume or, in vector form,

$$\sum \mathbf{f} = -\nabla p + \rho \mathbf{g} = 0 \tag{4-9}$$

It should be observed that $\mathbf{g}$ may be taken in an arbitrary direction (coordinate axes may be oriented differently), in which case it may be envisioned as having three mutually orthogonal components g_x, g_y, and g_z with the respective pressure distributions being affected accordingly, e.g.,

$$-\frac{\partial p}{\partial x} + \rho g_x = 0 \tag{4-10}$$

It is obviously necessary to know the nature of ρ and g in order to integrate the equation. Two important cases will be considered.

4-2 HYDROSTATICS

For the case of g in the negative y direction, $p \neq f(x,z)$. Therefore, the equations may be written as ordinary derivatives. Equation (4-8b) becomes

$$\frac{dp}{dy} = -\rho g \tag{4-11}$$

Assuming an incompressible fluid, $\rho = \text{const}$, and a constant gravitational field, $g = \text{const}$, we have

$$\int_{p_0}^{p} dp = -\rho g \int_{y_0}^{y} dy$$

$$p - p_0 = -\rho g(y - y_0) \tag{4-12}$$

where the subscript zero designates a reference level. If the surface of the fluid body of Fig. 4-2a is taken as the datum,

$$p - p_{\text{atm}} = \rho g h = \gamma h \tag{4-13}$$

FLUID STATICS

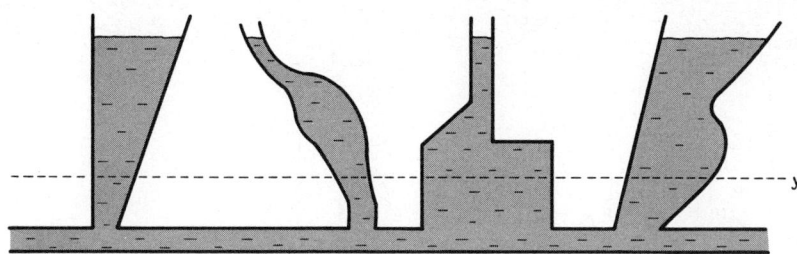

Fig. 4-3 Hydrostatic paradox.

since $y - y_0 = -h$. Equation (4-13), known as the *hydrostatic equation*, indicates that pressure depends only upon the given depth below the free surface. Figure 4-3 shows a fluid in containers of different shape and orientations, but the pressure is the same in every container at any point in the plane y. This is sometimes referred to as the *hydrostatic paradox*. The alert student will see how this concept can be used to level two corners of a building by using a garden hose filled with water.

Pressure level The pressure above that of the atmosphere, $p - p_{atm}$, is called *gauge pressure* p_g. Hence,

$$p_g = \gamma h \qquad (4\text{-}14)$$

Since many pressure-measuring devices indicate pressure with respect to the surroundings (the atmosphere in most cases), it is often more convenient to use gauge pressure. Figure 4-4 shows the relationship between gauge, absolute, and vacuum pressures.

Standard atmospheric pressure is the mean pressure at sea level. *Local atmospheric pressure*, at any elevation, is measured by a barometer. The most common is the mercury barometer, which consists of a glass tube filled with mercury, sealed at one end, and inverted with its open end submerged in a body of mercury. Figure 4-5

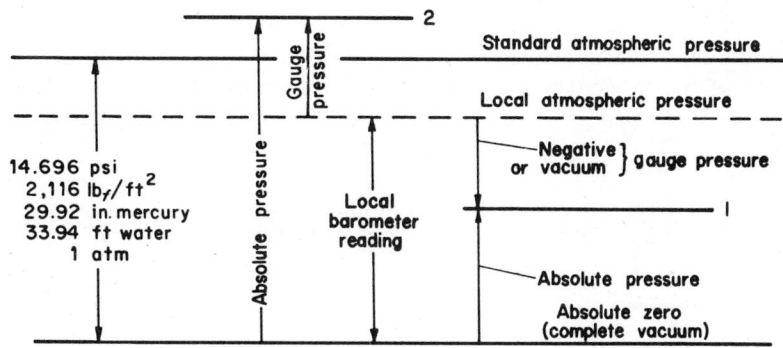

Fig. 4-4 Pressure relationship.

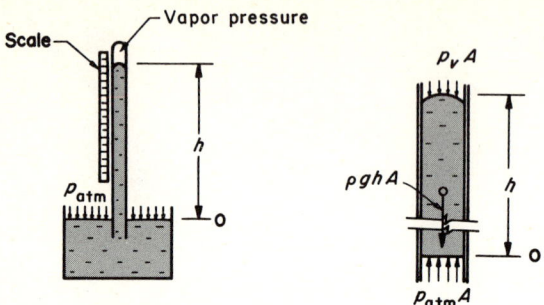

Fig. 4-5 Barometer.

shows a barometer schematically. A scale at the side of the fluid column gives the height, providing a measure of the atmospheric pressure.

A simple force balance on the fluid in the tube gives

$$p_{\text{vap}} + \rho g h = p_{\text{atm}} \tag{4-15}$$

where p_{vap} is the vapor pressure of the fluid, negligible in the case of mercury. Therefore for a mercury barometer

$$p_{\text{atm}} = \gamma h \tag{4-16}$$

Obviously, fluids of high densities permit shorter columns.

Pressures below the local atmospheric pressure are negative gauge or vacuum. For example, the pressure at station 1 of Fig. 4-4 might be

$$p_1 = 9 \text{ psia} = -5.7 \text{ psig} = 5.7 \text{ psi vacuum}$$

The reader will note, however, that in most cases involving force balances either absolute or gauge pressures may be used since the absolute pressure simply involves adding the barometric pressure to the gauge pressure, i.e.,

$$p_{\text{abs}} = p_g + p_{\text{bar}} \tag{4-17}$$

As an example, consider Fig. 4-2. The barometric pressure must be added to all sides of the element to give the absolute pressure; however, the barometric pressure on one side cancels that on the opposite side, making it immaterial which pressure is used.

Example What is the pressure *indicated* by pressure gauge C if the *indicated* pressures are $p_A = 45$ psi and $p_B = 20$ psi? The barometric pressure is 30.55 in Hg.

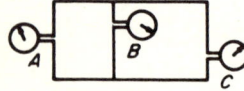

Fig. 4-6 Pressure tank.

FLUID STATICS

Solution The indicated pressure is gauge pressure, which is influenced by the gauge's environment. The environments of *A* and *B* are different. To get the absolute pressure at *A* the barometric pressure must be added to the indicated (gauge) pressure:

$$p_{atm} = 30.55 \text{ in Hg} \frac{14.7 \text{ psi}}{30 \text{ in Hg}} = 15 \text{ psia}$$

Therefore,

$$p_A = 45 + 15 = 60 \text{ psia}$$

But this is the absolute pressure in the left-hand tank. Gauge *B* also indicates the pressure in the left-hand tank as follows:

$$p_B = 60 \text{ psia} = p_B|_g + p_B|_{env}$$

$$p_B|_{env} = 60 - 20 = 40 \text{ psi}$$

which is the absolute pressure in the right-hand tank, but gauge *C* indicates the pressure in the right-hand tank also, i.e.,

$$p_C = 40 \text{ psia}$$

and the indicated pressure is the gauge pressure,

$$p_C = 25 \text{ psig}$$

Manometry One of the most convenient ways of measuring pressure is by determining how much it displaces a column of fluid (or fluids). For measuring large pressures mercury is commonly used as the manometer fluid. Water is used sometimes in measuring gas pressures.

Consider the simple U-tube of Fig. 4-7. The pressure on the meniscus[1] at *B* is equal to the pressure in the tank plus the pressure resulting from the weight of the column of fluid h_1. Since the meniscus *B* and point *C* are at the same elevation and joined by a common fluid, their pressures are identical. The pressure acting at *D* is that of the atmosphere and can be determined by a barometer. From this reasoning and the hydrostatic equation the pressure at any point in the manometer can be written:

$$p_B = p_A + \gamma h_1 = p_C \tag{4-18}$$

$$p_B = p_{atm} + \gamma_0 h_2 \tag{4-19}$$

$$p_A = p_{atm} + \gamma_0 h_2 - \gamma h_1 \tag{4-20}$$

No particular procedure is necessary in writing manometer equations. It is normally convenient to begin at a meniscus (any meniscus) and add pressures resulting from fluid columns when moving downward; subtract when moving upward.

Figure 4-8 shows a *differential manometer* between two tanks (pipes, or unknown pressure sources) which contain fluids of specific weights γ and γ_E, respectively.

---

[1] Fluid interfaces are discussed in Sec. 4-4.

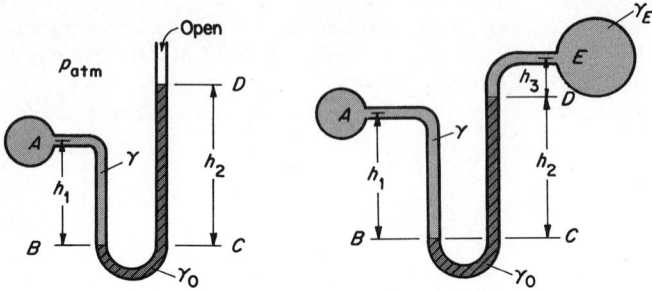

Fig. 4-7 U-tube manometer.

Fig. 4-8 Differential manometer.

Applying the hydrostatic equation gives

$$p_A + \gamma h_1 - \gamma_0 h_2 = p_D \tag{4-21}$$

$$p_E + \gamma_E h_3 = p_D \tag{4-22}$$

$$p_A - p_E = \gamma_0 h_2 - \gamma h_1 + \gamma_E h_3 \tag{4-23}$$

These simple principles can be extended to complex manometer arrangements where many fluids are involved. It is often convenient to use the specific gravity $S \equiv \gamma/\gamma_W$ in the manometer equations, where γ_W is the specific weight of water.

Example A 4- by 4-ft tank contains acetylene tetrabromide ($S = 2.96$). (*a*) With the manometer arrangement shown in Fig. 4-9, what pressures are indicated by the gauges at A and B? (*b*)

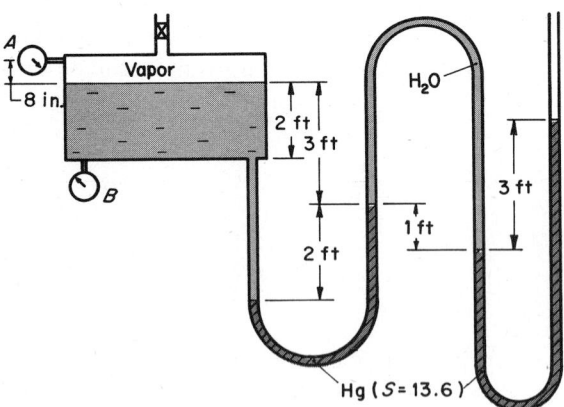

Fig. 4-9 Manometer tank.

What is the weight of the acetylene tetrabromide in the tank? (*c*) If the barometric pressure is 14.6 psia, is the valve open or closed on the tank?

FLUID STATICS

Solution (a) Neglecting the vapor head,

$$p_A + \gamma_w[5(2.96) - 2(13.6) + 1(1) - 3(13.6)] = 0$$

Note that p_A will be gauge pressure since the pressure at the open end of the tube was set equal to zero. Solving,

$$p_A = 22.5 \text{ psig}$$

Beginning with this value, p_B can be found:

$$p_B = p_A + \gamma_w S h_B = 22.5 + \frac{62.4}{144} 2.96(2) = 25.2 \text{ psig}$$

(b) Weight, $W = \gamma \mathscr{V} = \gamma_w S \mathscr{V}$

$$W = 62.4(2.96)[4(4)(2)] = 5{,}920 \text{ lb}_f$$

(c) If the valve were open, the gauge at A would read zero; therefore, the valve is closed.

4-3 AEROSTATICS

We have looked at the pressure variation in an incompressible fluid. Now let us consider a compressible fluid. The same equation

$$\frac{dp}{dy} = -\rho g \qquad (4\text{-}11)$$

must be solved. In this case, however, $\rho \neq$ const. Restricting our attention to a perfect gas, two cases will be considered in establishing a model atmosphere.

Isothermal atmosphere For a constant temperature the perfect-gas equation, $p = \rho RT$, becomes

$$\frac{p}{\rho} = C \qquad (4\text{-}24)$$

When substituted into Eq. (4-11), this relation permits the separation of variables and integration:

$$\frac{dp}{dy} = -\frac{p}{C} g \qquad (4\text{-}25)$$

$$\int_{p_0}^{p} \frac{dp}{p} = C' \int_{y_0}^{y} dy \qquad (4\text{-}26)$$

where $C' = -g/C$. Carrying out the integration,

$$\ln \frac{p}{p_0} = \frac{g}{C}(y_0 - y) \qquad (4\text{-}27)$$

or

$$p = p_0 \exp\left[\frac{g}{RT}(y_0 - y)\right] \qquad (4\text{-}28)$$

where the subscript zero is the reference level and of course T must be in absolute measure.

The assumption of constant temperature establishes a temperature-elevation gradient of zero. This gradient is defined as the lapse rate λ,

$$\lambda \equiv \frac{dT}{dy} \tag{4-29}$$

Iso-lapse-rate atmosphere For a constant lapse rate, integration of Eq. (4-29) gives a linear variation of temperature with elevation, i.e.,

$$T = T_0 + \lambda(y - y_0) \tag{4-30}$$

Combining Eqs. (4-11) and (4-29) gives

$$dp = -\rho \frac{g}{\lambda} dT \tag{4-31}$$

Introducing $\rho = p/RT$ from the perfect-gas equation,

$$\int_{p_0}^{p} \frac{dp}{p} = -\frac{g}{\lambda R} \int_{T_0}^{T} \frac{dT}{T} \tag{4-32}$$

$$\ln \frac{p}{p_0} = \frac{g}{\lambda R} \ln \frac{T_0}{T} \tag{4-33}$$

Replacing T by its equivalent from Eq. (4-30) and solving for pressure, we have

$$p = p_0 \left(\frac{T_0}{T_0 + \lambda y} \right)^{g/\lambda R} \tag{4-34}$$

The standard atmosphere There are several model atmospheres, but the standard is the International Civil Aviation Organization (ICAO) standard, which is defined up to an elevation of 65,800 ft, as shown in Fig. 4-10. The NACA standard atmosphere (1955) and the ARDC model atmosphere (1956) are consistent with the ICAO standard up to this elevation. Beyond this there are several extensions, including the U.S. standard atmosphere of 1962, which is shown. The hot and cold profiles represent United States military extreme-temperature criteria. The pictorial inserts represent steps in man's historical achievement in learning to exist at the altitudes shown.

Atmospheric regimes are noted on the high-altitude chart of Fig. 4-11. In the *troposphere*, the weather domain, the lapse rate of the model atmosphere is

FLUID STATICS

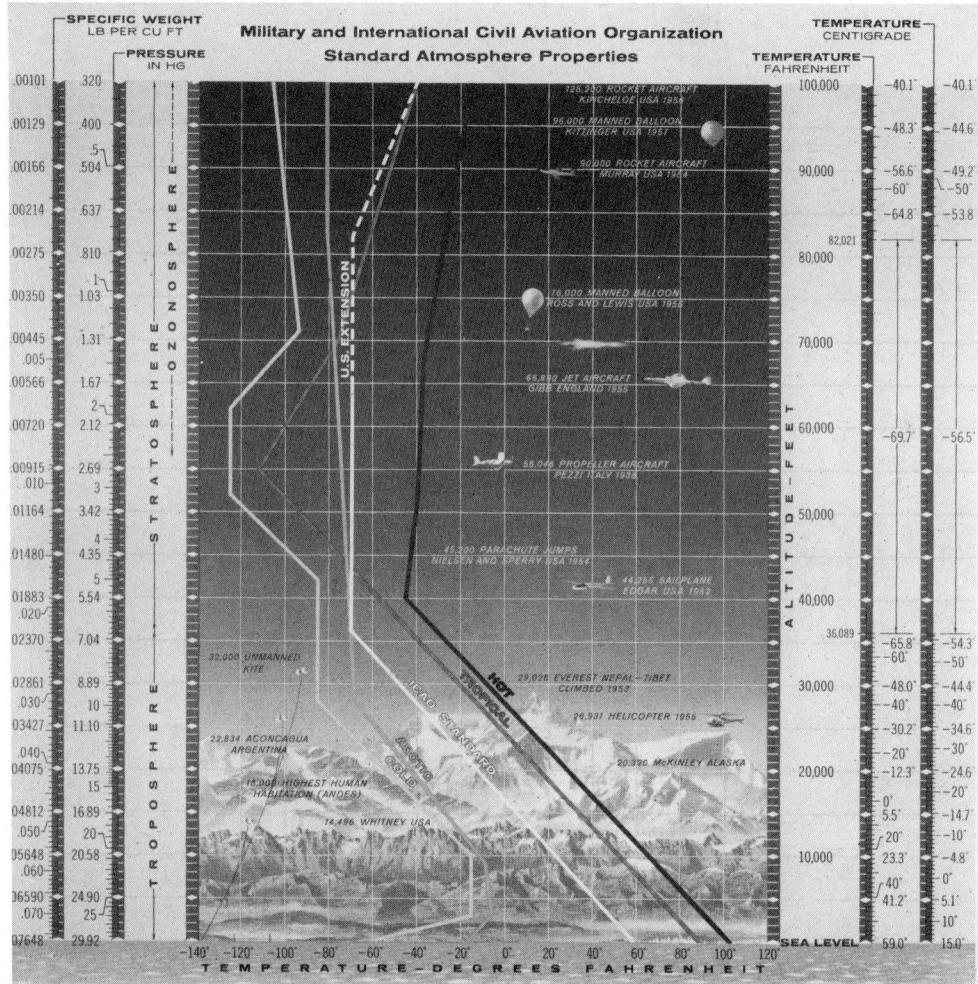

Fig. 4-10 Atmosphere chart. (*The Garrett Corporation*.)

3.56°F/1,000 ft [3], with a sea-level temperature of 59°F (518.7°R), giving

$$T = 519 - 0.00356y \quad °R \tag{4-35}$$

where y is the elevation above sea level in feet.

The geopotential altitude shown in Fig. 4-11 provides a measure of the energy required to accelerate a body at an altitude compared with that at sea level. For example, the energy required to lift a body 2,000,000 geometric feet is only 1,824,988 times that required to lift it 1 ft at sea level, because of the decrease in the acceleration of gravity. Atmospheric properties are given with respect to altitude in Figs. 4-10 and 4-11.

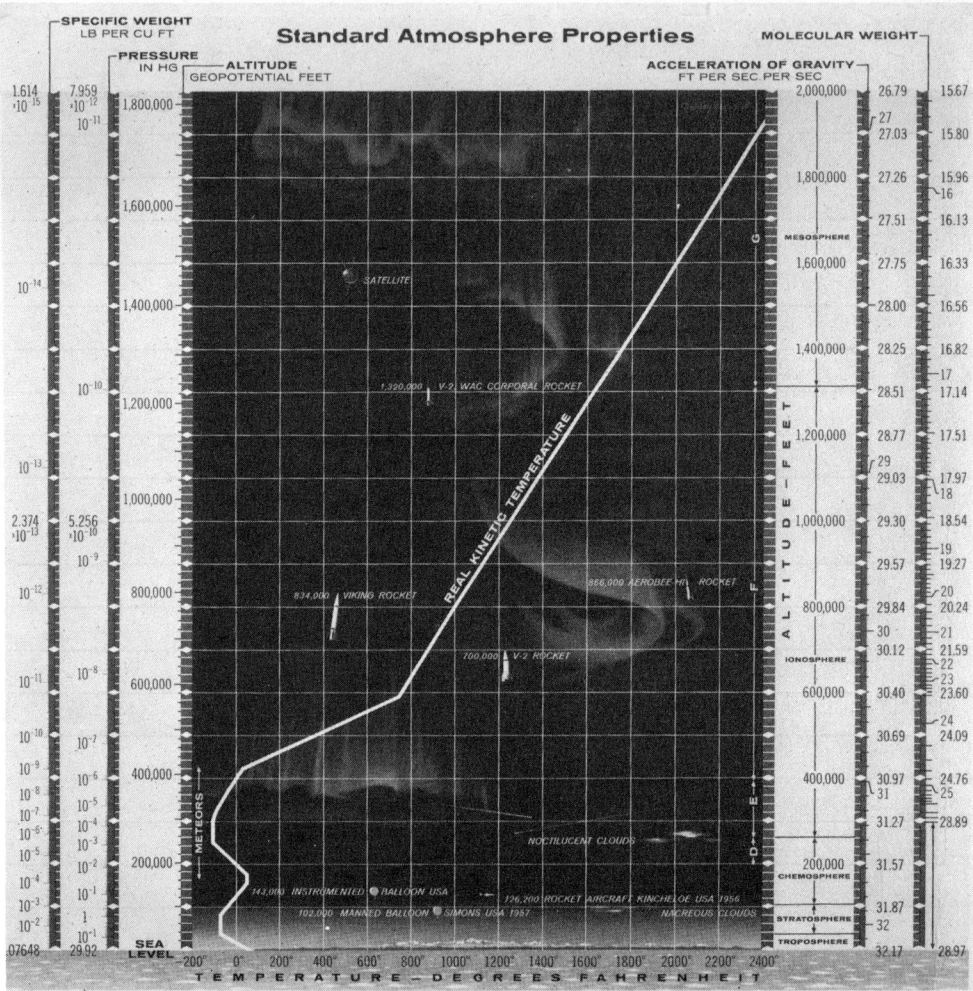

Fig. 4-11 High-altitude chart. (*The Garrett Corporation.*)

4-4 FLUID INTERFACES

Fluids, by their very nature, must interface with solids or other fluids. The behavior of a fluid at an interface depends upon the relative forces of *cohesion*, the attraction of fluid molecules for each other, and *adhesion*, the attraction of fluid molecules for the molecules of an adjacent surface. A fluid molecule on the interior of the fluid body is equally and oppositely attracted by adjacent molecules. However, a molecule on a fluid surface may not be attracted equally in all directions. It may be attracted more by a solid boundary, as water molecules are to glass, or it may be attracted more by adjacent fluid molecules, as mercury is in a glass tube.

FLUID STATICS

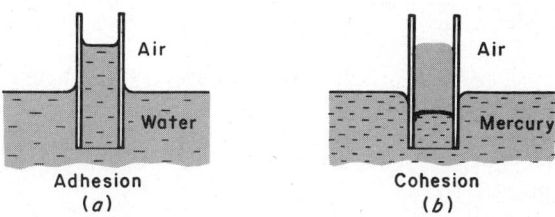

Fig. 4-12 Surface wetting of capillary tubes: (a) adhesion; (b) cohesion.

Fluid-solid interfaces depend upon the *wetting characteristics* of the fluid, as illustrated in Fig. 4-12 by small glass tubes in water and mercury. Water wets the surface, whereas mercury does not. A fluid-fluid interface, as between the air and the water or between the air and the mercury in Fig. 4-12, is known as a *meniscus*.

When a fluid body is large and surface interactions are negligible, as between its vapor or a gas, the surface is known as a *free surface*. A pond has a free surface enshrouded by an ocean of air. The energy associated with free surfaces is responsible for interesting and common phenomena. A steel needle which has been coated with an oil or wax will float on the surface of water. Fat drops in soup float as disks; candle wax flows up the wick; a soap film forms inside a wire loop dipped in a soap solution; a raindrop is spherical except for the drag forces on it; and everyone knows what happens to the water on a duck's back.

Surface tension Free surfaces behave similarly to flexible membranes, requiring energy to alter their shapes. This surface energy, erroneously called *surface tension* σ, is the force per unit length of any line on the free surface of a liquid which is necessary to hold that surface together at the line.

$$\sigma = \frac{dF}{dl} \qquad (4\text{-}36)$$

As a specific case consider a spherical bubble or liquid drop as shown in Fig. 4-13. The difference in pressure between the inside and outside of the bubble, $p_i - p_o$, must be offset by the surface tension in order to be in equilibrium. A simple force

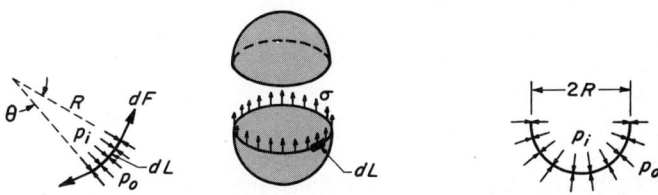

Fig. 4-13 Bubble or liquid drop.

balance gives

$$p_i - p_o = \frac{2\sigma}{R} \tag{4-37}$$

The general case for compound curvature (different radii in orthogonal directions) gives

$$p_i - p_o = \sigma\left(\frac{1}{R_1} + \frac{1}{R_2}\right) \tag{4-38}$$

where R_1 and R_2 are the respective radii [1]. If $R_1 = R_2$, as in the spherical bubble, Eq. (4-38) reduces to Eq. (4-37). For a cylindrical surface of radius R_1, $R_2 \to \infty$, giving

$$p_i - p_o = \frac{\sigma}{R_1} \tag{4-39}$$

The work required to change the character of the free surface can be determined by the thermodynamic equation for reversible work [Eq. (3-8)]. Therefore,

$$W = \int (p_i - p_o)\, d\mathscr{V} \tag{4-40}$$

where $\mathscr{V}$ is the volume of the fluid drop.

Capillarity The capillary effect is a sponge action—the rising of sap in trees, the flow of kerosene up the wick of a lantern, the movement of water in tightly packed soil.

Consider the rise of a fluid in a small-bore tube in an air environment, as shown in Fig. 4-14. The total upward surface-tension force must support the weight of the fluid column. The following equation is convenient for determining the surface tension of a fluid [1]:

$$2\pi R\sigma \cos \alpha = \pi R^2 h(\rho - \rho_a)g$$

$$\sigma = \frac{ghR(\rho - \rho_a)}{2 \cos \alpha} \tag{4-41}$$

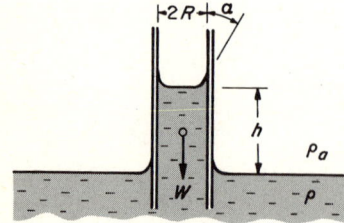

Fig. 4-14 Capillarity.

FLUID STATICS

The wetting characteristics of fluids and capillarity are obviously of importance in design of manometry equipment, tube sizing, etc.

4-5 FORCES ON SUBMERGED SURFACES

In the preceding sections pressure variations in fluids have been considered. The distributed forces resulting from the pressure-area product on finite areas can be replaced by a resultant force in determining external reactions to a system. This section deals with this resultant force, its line of action, and its point of application on a surface in an incompressible fluid. For this discussion the surface has no thickness, hence no weight; therefore, it will remain in static equilibrium, since the forces on opposite sides are equal and opposite.

Plane surfaces We seek the resultant hydrostatic force **F** on the inclined plane of Fig. 4-15. The force is equal and opposite on the top and bottom in Fig. 4-15a. Equivalent hydrostatic situations are shown in Fig. 4-15b and c, where the plane surface may be thought of as the gate of a dam with water on either side.

To analyze the problem, the plane is extended to the free surface and is oriented such that a surface element of width dy is at a uniform depth from the free surface. The elemental force exerted on this element is

$$dF = p\, dA$$
$$dF = \gamma h\, dA = \gamma y \sin\theta\, dA \tag{4-42}$$

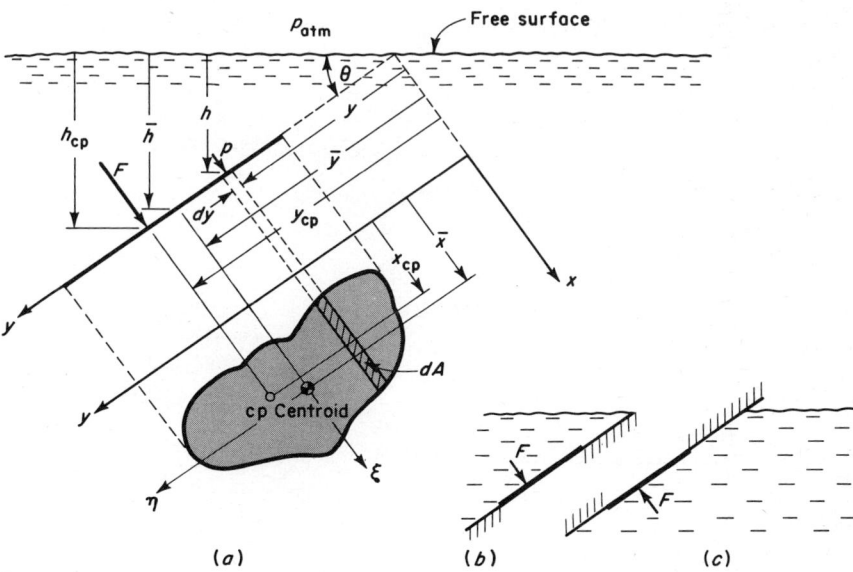

Fig. 4-15 Fluid forces on plane submerged surface.

Integrating over the plane area and noting that $\int_A y\, dA$ is the first moment of area about the x axis, we have

$$F = \gamma \sin\theta \int_A y\, dA = \gamma \sin\theta\, \bar{y} A = \gamma \bar{h} A \qquad (4\text{-}43)$$

where $\bar{h}$ is the depth from the free surface to the centroid. The force may then be written simply as

$$F = \bar{p} A \qquad (4\text{-}44)$$

from the hydrostatic equation, where $\bar{p}$ is the pressure at the centroid.

The result of Eq. (4-44) suggests that we may imagine the pressure at the centroid to extend over the entire area and compute accordingly. This can easily be seen by letting the plate be horizontal, as shown in Fig. 4-16a. The resulting diagram of pressure distribution, called a *pressure prism*, is of constant magnitude. If the same plane is rotated as shown in Fig. 4-16b, the prism becomes trapezoidal but the resultant force is the same as long as the centroid is held at the same depth. It is obvious that the volume of the pressure prism is the magnitude of the resultant force. It is equally obvious that the force acts at different points in the two cases shown in Fig. 4-16a and b. The resultant acts through the center of gravity of the pressure prism and is normal to the surface since no shear forces are present. The resultant acts at the centroid of the plane in Fig. 4-16a, but it acts below the centroid in Fig. 4-16b. We shall now determine where the resultant force acts for the general case of the inclined plane surface.

The point of application of the resultant force, designated by the subscript "cp" in Fig. 4-15, is called the *center of pressure*. To locate it in the y direction, equate the moment of the resultant force $y_{\text{cp}} F$ to the moment of the distributed forces about the x axis; thus

$$F y_{\text{cp}} = \int_A y p\, dA = \int_A \gamma y^2 \sin\theta\, dA$$

$$\gamma \bar{y} \sin\theta\, A y_{\text{cp}} = \gamma \sin\theta \int_A y^2\, dA$$

$$y_{\text{cp}} = \frac{I_{xx}}{\bar{y} A} \qquad (4\text{-}45)$$

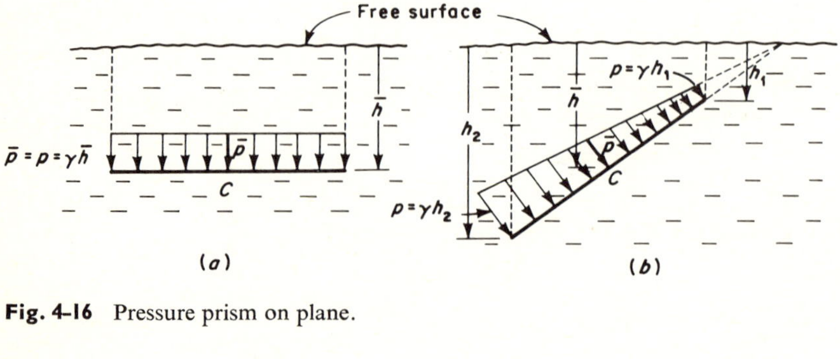

Fig. 4-16 Pressure prism on plane.

FLUID STATICS

where I_{xx} is the second moment of area, or moment of inertia, about the x axis. Applying the parallel-axis theorem, $I_{xx} = I_{\xi\xi} + A\bar{y}^2$,

$$y_{cp} = \bar{y} + \frac{I_{\xi\xi}}{A\bar{y}} \qquad (4\text{-}46)$$

where $I_{\xi\xi}$ is the moment of inertia about the plane's centroidal axis ξ. Since $I_{\xi\xi}$ is always positive, the center of pressure is always below the centroid by an amount equal to $I_{\xi\xi}/A\bar{y}$.

To locate the center of pressure in the x direction, we equate the moment about the y axis with the corresponding moment from the pressure distribution; thus

$$Fx_{cp} = \int_A xp\,dA = \gamma \sin\theta \int_A xy\,dA$$

$$\gamma\bar{y}\sin\theta\,Ax_{cp} = \gamma \sin\theta\,I_{xy}$$

$$x_{cp} = \frac{I_{xy}}{\bar{y}A} \qquad (4\text{-}47)$$

where I_{xy} is the product of inertia about the reference axes. Using the parallel-axis theorem, $I_{xy} = I_{\xi\eta} + \overline{xy}A$,

$$x_{cp} = \bar{x} + \frac{I_{\xi\eta}}{\bar{y}A} \qquad (4\text{-}48)$$

where $I_{\xi\eta}$ is the product of inertia about the plane's centroidal axes ξ and η. When either of the centroidal axes is an axis of symmetry of the surface, $I_{\xi\eta}$ vanishes and the center of pressure lies on $x = \bar{x}$. Since $I_{\xi\eta}$ may be either positive or negative, the center of pressure may lie on either side of the line $x = \bar{x}$.

An additional, extremely useful observation can be made by referring to the pressure prisms of Fig. 4-16. The force on any element dA is given by

$$F = \gamma \int h\,dA = \gamma \mathscr{V} \qquad (4\text{-}49)$$

where $\mathscr{V}$ is the volume of fluid above the surface, either real, as in Fig. 4-15b, or imaginary, as in Fig. 4-15c. The vertical force is then simply the weight of the fluid, real or imaginary, which lies above the surface plus the force due to atmospheric pressure at the fluid surface.

Example A rectangular flood gate (Fig. 4-17) opens by hydrostatic pressure. If the gate weighs 25 tons per foot of depth, what head H is required to open it?
Solution Assuming unit depth, the resultant force is

$$F = \gamma\bar{h}A = 0.707\gamma\bar{y}A$$

and it acts at the center of pressure located at

$$\frac{I_{\xi\xi}}{A\bar{y}} = \frac{100}{12\bar{y}}$$

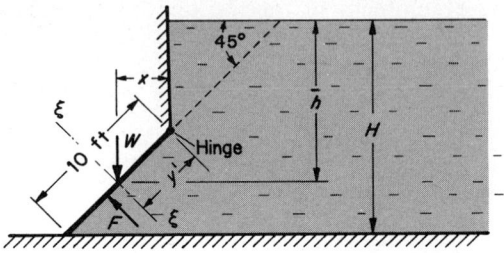

Fig. 4-17 Flood gate.

below the centroid. Taking moments about the hinge gives

$$Wx = F\left(y' + \frac{I_{\xi\xi}}{A\bar{y}}\right)$$

$$50{,}000(0.707)(5) = 0.707\gamma\bar{y}(10)\left(5 + \frac{100}{12\bar{y}}\right)$$

$$\frac{5{,}000(5)}{62.4\bar{y}} - \frac{100}{12\bar{y}} = 5$$

$$401 - 8.33 = 5\bar{y}$$

$$\bar{y} = \frac{392.67}{5} = 78.52 \text{ ft}$$

Therefore,

$$H = 0.707(\bar{y} + 5) = 59.1 \text{ ft}$$

Curved surfaces The concepts of plane surfaces will now be applied to curved surfaces, depicted generally in Fig. 4-18. The elemental area dA can be projected onto the three mutually orthogonal planes by taking the dot product of $d\mathbf{A}$ with the respective unit vectors, $\mathbf{i}, \mathbf{j}$, and $\mathbf{k}$. The projected areas are equal on any plane parallel to the respective principal planes.

Now consider the elemental fluid volume bounded by the yz plane, the submerged surface, and the projection elements formed by $d\mathbf{A} \cdot \mathbf{i}$ as shown in Fig. 4-18b. For equilibrium the horizontal force on the elemental area of the yz plane must equal that in the negative x direction at the surface. But $dA_x = d\mathbf{A} \cdot \mathbf{i}$; therefore,

$$dF_{xy}\big|_{z \text{ plane}} = dF_x\big|_{\text{surface}} \qquad (4\text{-}50)$$

Similarly, the horizontal force in any direction can be determined. Hence, *the horizontal force on any submerged curved surface is the vector sum of forces on two projected areas in orthogonal planes.*

The vertical force follows readily from the concept introduced with Eq. (4-49). The vertical force on the fluid element shown in Fig. 4-18c must be equal to the weight

FLUID STATICS

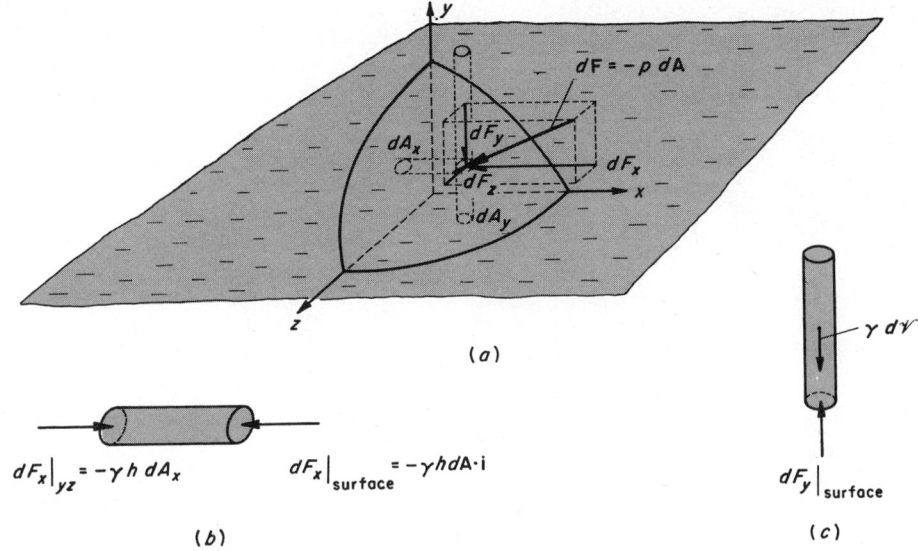

Fig. 4-18 Fluid forces on curved submerged surface.

of the fluid element above it. Therefore, *the vertical force on any submerged curved surface is equal to the weight of the fluid, real or imaginary, above the surface.*

Example A barrier holds water as shown in Fig. 4-19. Neglecting the weight of the barrier, what moment T_0 is produced on the hinge pin at 0 per unit length of the barrier?

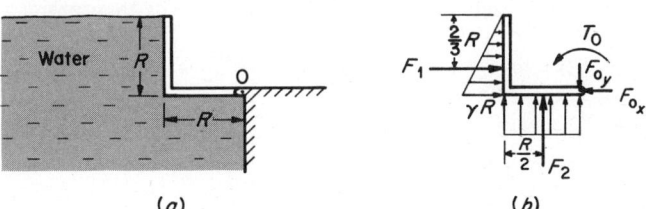

Fig. 4-19 Force distribution on simple water barrier.

Solution A free-body diagram is shown in Fig. 4-19b, with the pressure distribution resulting from the water. The resultant forces F_1 and F_2 act through the centroid of the pressure prisms. The magnitude of the pressure at the bottom of the barrier is γR; therefore,

$$F_{0_x} = F_1 = \tfrac{1}{2}\gamma R R = \tfrac{1}{2}\gamma R^2$$

$$F_{0_y} = F_2 = \gamma R R = \gamma R^2$$

$$T_0 = \tfrac{1}{3}RF_1 + \tfrac{1}{2}RF_2 = \tfrac{1}{6}\gamma R^3 + \tfrac{1}{2}\gamma R^3 = \tfrac{2}{3}\gamma R^3$$

Example Let the barrier of the above example be replaced by a cylinder of radius R (Fig. 4-20). How much must the cylinder weigh per foot if its contact with the wall is frictionless?

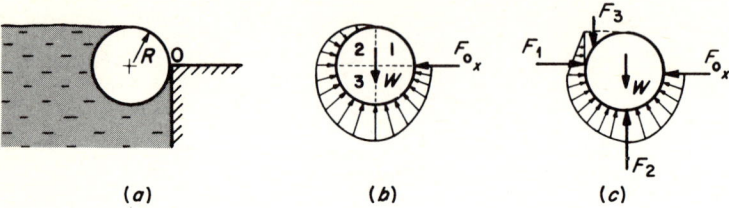

Fig. 4-20 Force distribution on cylindrical water barrier.

Solution The pressure distribution of the free-body diagram shown in Fig. 4-20b can be replaced by its equivalent in Fig. 4-20c since the horizontal force on quadrant 2 of the cylinder is equal to that on a projected vertical plane. Therefore,

$$F_1 = \tfrac{1}{2}\gamma R^2$$

The resultant force on the bottom half of the cylinder surface is equal to the weight of the "imaginary" fluid above it; hence

$$F_2 = \left(\frac{\pi R^2}{2} + 2R^2\right)\gamma$$

The resultant vertical force on quadrant 2 is due to the weight of the water above quadrant 2:

$$F_3 = \left(R^2 - \frac{\pi R^2}{4}\right)\gamma$$

For equilibrium, excluding the force due to atmospheric pressure,

$$W = F_2 - F_3 = \left(\frac{\pi R^2}{2} + 2R^2 - R^2 + \frac{\pi R^2}{4}\right)\gamma = \left(1 + \frac{3\pi}{4}\right)\gamma R^2$$

4-6 BUOYANCY

The *buoyant force* F_B on a body which is submerged or floating in a static fluid (or fluids) is the net vertical force resulting from the pressure distribution exerted by the fluid (or fluids) on the body. A body is in *flotation* when in contact with a fluid (or fluids) only. The techniques of the preceding section permit us to evaluate the buoyant force.

Consider a potato-shaped body immersed in a fluid of specific weight γ, as shown in Fig. 4-21. The dotted line indicates the outermost periphery of the body. The resultant force on the bottom surface BCD is the weight of the fluid above the surface, enclosed by $ABCDEA$. Similarly, the resultant force on the top surface is the weight of the fluid enclosed by $ABC'DEA$. Since the buoyant force is the net

FLUID STATICS

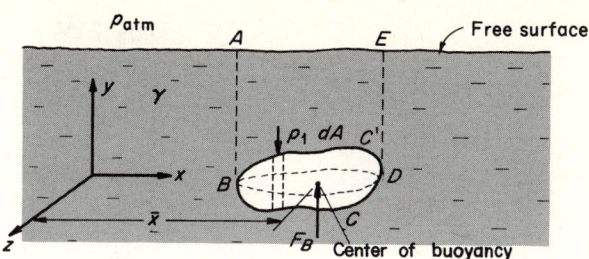

Fig. 4-21 Body in flotation.

vertical force,

$$F_B = W_{ABCDEA} - W_{ABC'DEA} \tag{4-51}$$

$$F_B = \gamma(\mathscr{V}_{ABCDEA} - \mathscr{V}_{ABC'DEA}) \tag{4-52}$$

But the volume difference in parentheses is identically equal to the volume of the body; hence,

$$F_B = \gamma \mathscr{V} \tag{4-53}$$

where γ is the specific weight of the fluid and $\mathscr{V}$ is the volume of the submerged body. Therefore, *the buoyant force is equal to the weight of fluid displaced.* This is known as *Archimedes' principle.* The same formulation applies to a floating body when the volume is only that portion below the free surface.

To find the line of action of the buoyant force, which acts through the center of buoyancy, moments are taken about a convenient axis z. The sum of the elemental weight-distance products is equal to the buoyant force times the distance through which it acts $\bar{x}$:

$$\gamma \int x \, d\mathscr{V} \equiv \gamma \mathscr{V} \bar{x} \tag{4-54}$$

$$\bar{x} = \frac{1}{\mathscr{V}} \int x \, d\mathscr{V} \tag{4-55}$$

But since this is the expression for the centroid of the volume displaced by the body, *the line of action of the buoyant force is through the centroid of the volume of fluid displaced.*

For a body in flotation in two or more immiscible fluids, the buoyant force is the sum of the partial weights;

$$F_B = \gamma_1 \mathscr{V}_1 + \gamma_2 \mathscr{V}_2 + \cdots \tag{4-56}$$

However, the line of action does not pass through the centroid of the total volume displaced by the body but must be determined by considering each displaced volume

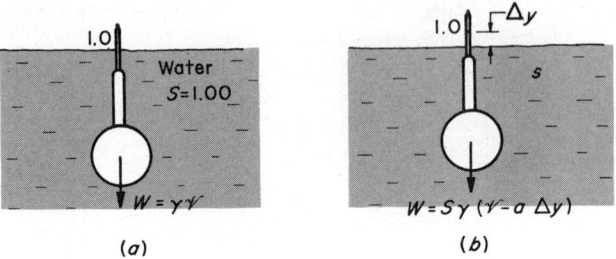

Fig. 4-22 Hydrometer in flotation.

segment, giving

$$\bar{x} = \frac{\gamma_1 \int x\, d\mathscr{V}_1 + \gamma_2 \int x\, d\mathscr{V}_2 + \cdots}{\gamma_1 \mathscr{V}_1 + \gamma_2 \mathscr{V}_2 + \cdots} \tag{4-57}$$

Example A *hydrometer* is used to determine the specific gravity of liquids. The principle is demonstrated in Fig. 4-22. The stem of the hydrometer has area a, and the hydrometer displaces a volume $\mathscr{V}_0$ when floating in water. When placed in another liquid, it is displaced by the amount Δy. Marking on the stem provides a measure of the specific gravity. Determine the displacement as a function of the specific gravity of the liquid.

Solution The weight of the hydrometer of Fig. 4-22a is equal to that in Fig. 4-22b; hence

$$\gamma \mathscr{V}_0 = S\gamma(\mathscr{V}_0 - a\, \Delta y)$$

$$\Delta y = \frac{\mathscr{V}_0}{a} \frac{S-1}{S}$$

4-7 STABILITY

Stability is the measure of a body's tendency to remain in equilibrium or to return to equilibrium upon being displaced. Figure 4-23 illustrates the types of equilibrium. In general, a submerged body is stable only when its center of gravity lies below its center of buoyancy, as illustrated by the balloon of Fig. 4-24. When displaced, the couple Wx tends to restore equilibrium.

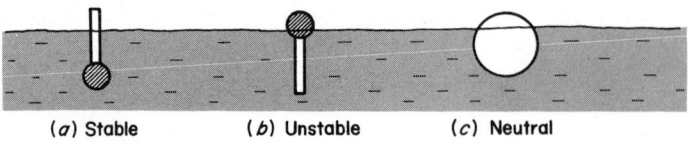

Fig. 4-23 Types of equilibrium.

FLUID STATICS

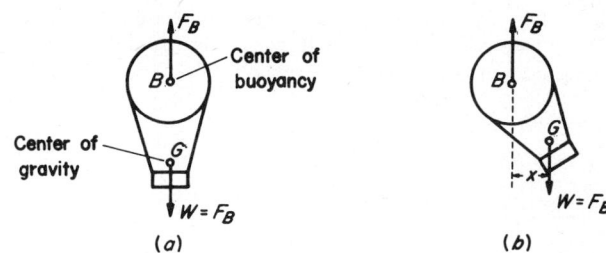

Fig. 4-24 Balloon in stable equilibrium.

For floating bodies it is not necessary that the center of gravity lie below the center of buoyancy in order to have stability. We shall consider the simple case of a prismatic body as shown in Fig. 4-25. The center of gravity is at G and the center of buoyancy is designated by B, with the subscript zero referring to the equilibrium position.

As the body is tipped, the center of buoyancy shifts and the buoyant force and weight develop a righting couple equal to $\overline{MG}W \sin \theta$, which will return the body to its stable condition as long as the point M is above G. If the body is tipped further, the center of buoyancy shifts, with M moving to G (neutral equilibrium) and if further tipped, below G to the unstable condition. It is obvious that if M shifts below G, the couple which is developed upsets the body rather than restores it. The distance $\overline{MG}$, known as the *metacentric height*, is a measure of the body's stability [2, 4].

PROBLEMS

4-1. The pressure distribution on a 2-in-diam disk is approximated by $p = 10e^{-r^2}$, where r is the radius in inches and p is the pressure in pounds force per square inch. What is the force on the disk caused by this pressure?

4-2. The barometer stands at 30 in Hg.

(a) What is the gauge pressure of the atmosphere? The absolute pressure?

(b) What is the gauge pressure at the bottom of a 10-ft swimming pool? The absolute pressure?

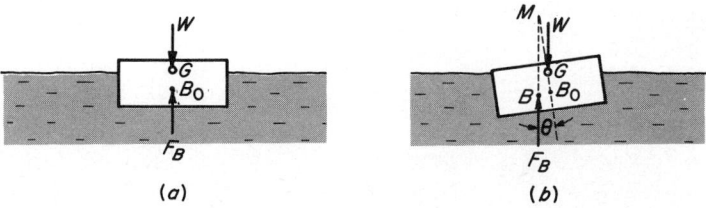

Fig. 4-25 Stability of prismatic body.

4-3. Two cylindrical vessels have the same base area and the same height of liquid. The pressure on the base is the same for both vessels by virtue of the hydrostatic paradox, and yet the weight of liquid in each vessel differs. Explain.

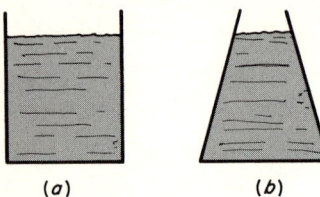

Fig. P4-3

4-4. On Jan. 23, 1960, the bathyscaphe *Trieste* descended to an ocean depth of 35,800 ft. Determine the pressure on its hull at that depth.

4-5. For the two types of water towers shown determine the pressure at the base of each tower when they are (*a*) full and (*b*) one-half full.

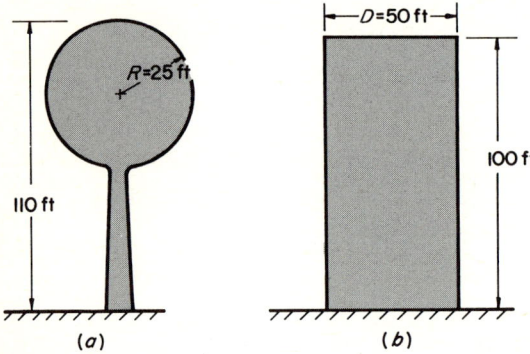

Fig. P4-5

4-6. Neglecting the weight of the vessel, determine (*a*) the force exerted on the bottom of the vessel by the water, (*b*) the force exerted on the annular area *ABCD*, (*c*) the tension in the tank wall, (*d*)

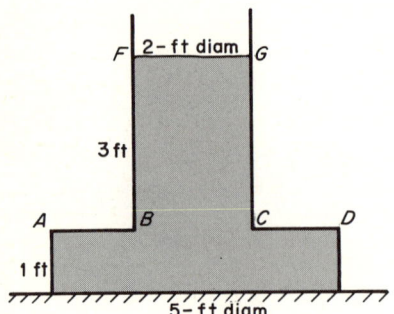

Fig. P4-6

FLUID STATICS

the force exerted against the surface supporting the vessel, and (e) the maximum stress in vertical section BCFG. Sketch the pressure distribution (to some scale) of the vessel on the fluid.

4-7. If the top 2 ft of water in Prob. 4-6 is replaced by oil with specific gravity 0.8, sketch the pressure distribution (to the same scale used in Prob. 4-6).

4-8. For the system shown in the figure, find x such that the bar is horizontal. What is p_1? p_2? Neglect the weights of the pistons and bar.

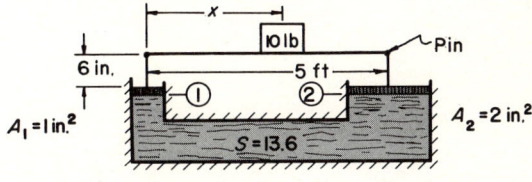

Fig. P4-8

4-9. A glass contains 5 in of water. A straw is in the glass and is at an angle of 75° with the horizon. The straw touches the bottom of the glass and is 10 in long. What pressure must one develop at the top of the straw in order to begin to drink the water? Express your answer in both absolute and gauge values. $P_{atm} = 14.7$ psi.

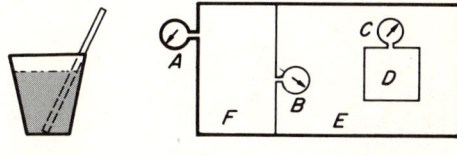

Fig. P4-9   **Fig. P4-10**

4-10. Consider the system shown in the figure. The pressures indicated by gauges B and C are 60 and 40 psig, respectively. The absolute pressure in container D is 60 psia. Find the pressure indicated by gauge A. The barometer stands at 30 in Hg.

4-11. To prevent the freezing of a water-level gauge in a river-measuring station, 6 ft of kerosene ($S = 0.80$) is put in the tube as shown in the figure. A float rides on the top of the column of kerosene. Since the float is supposed to record the water surface elevation, what correction C must be made in the records to give the correct water level?

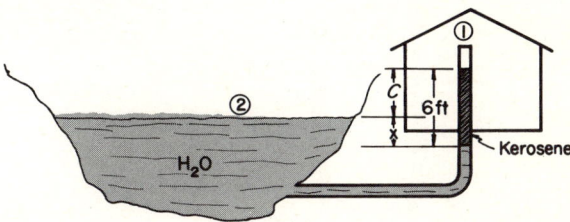

Fig. P4-11

4-12. A 20-in-diam vertical standpipe 30 ft tall is made of $\frac{1}{8}$-in-thick steel.
  (a) Find the maximum tensile stress in the pipe when it is full of water. Where does it occur?
  (b) If the standpipe of part (a) is standing in water 15 ft deep, where does the maximum stress occur? How does it compare with that in part (a)?

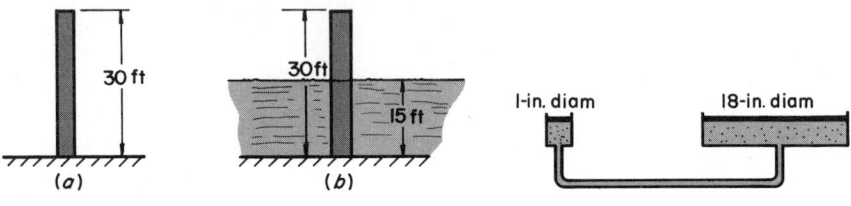

Fig. P4-12   **Fig. P4-13**

4-13. A small 1-in-diam cylinder is connected to a large 18-in-diam cylinder by a $\frac{1}{4}$-in copper tube connected to the bottom of both cylinders. Oil ($S = 0.80$) is in both cylinders and the connecting tube. Neglecting the weight of the pistons, how much force would have to be applied perpendicular to the 1-in piston to start raising a load of 2 tons on the large piston?

4-14. What is the maximum depth for which a vacuum pump can be used to pump water from a well? Explain.

4-15. Find the force F necessary to lift the large weight. Develop an expression for the displacement of the large block x_2 in terms of the displacement of the small block x_1.

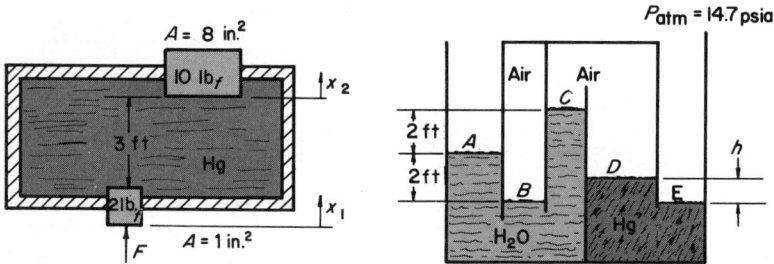

Fig. P4-15   **Fig. P4-16**

4-16. The container shown holds air, water, and mercury. Find h.

4-17. For the manometer shown determine h.

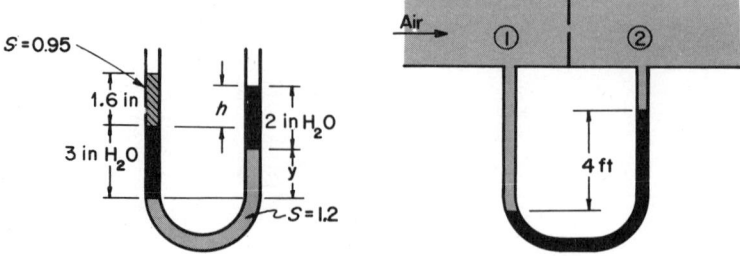

Fig. P4-17   **Fig. P4-18**

FLUID STATICS

4-18. A U-tube manometer is connected across an orifice plate.
  (a) For $p_1 = 45$ psig and $p_2 = 32$ psig what is the specific gravity of the manometer fluid?
  (b) If the manometer fluid is mercury and $p_1 = 60$ psig, determine the gauge pressure p_2.

4-19. For the draft gauge shown, what is the gauge pressure in the tube in inches of water?

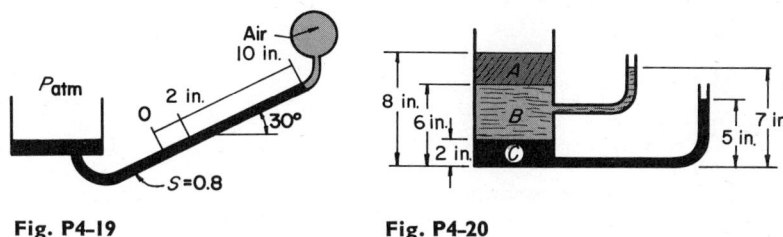

Fig. P4-19   Fig. P4-20

4-20. What is the specific weight γ of liquid B? The atmospheric pressure is 33.5 ft H_2O. $S_O = 5$.

4-21. A young engineer is asked to find p_A. He says that $p_A = 15$ psia since the manometer shows equal heights. Do you agree? If so, explain. If not, what is p_A?

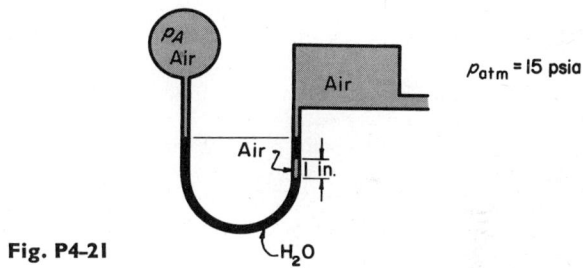

Fig. P4-21

4-22. For the system what does the gauge record as p_1? What is the absolute value of p_2? What is the absolute pressure of the air? Is the air compressed? $p_{atm} = 14.7$ psi.

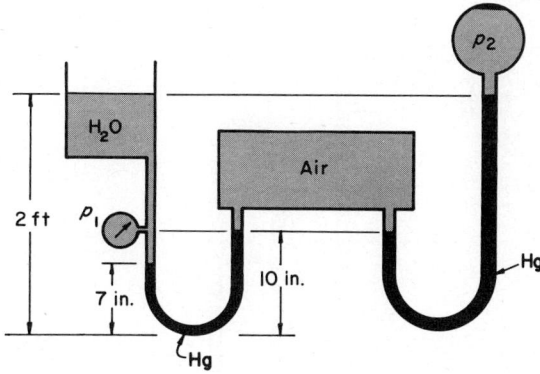

Fig. P4-22

4-23. The pressure loss across a restriction in a pipeline is frequently used in measuring a fluid flow rate. Consider the venturi meter with a manometer as shown.
(a) Find p_A if $p_B = 10$ psia.
(b) For $p_A = 20$ psia and a barometric pressure of 29.0 in Hg, find p_B in feet of water, gauge.

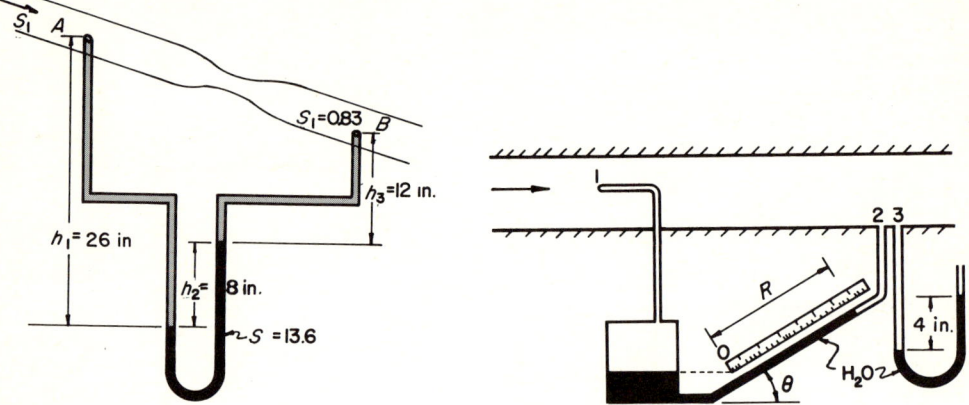

Fig. P4-23

Fig. P4-24

4-24. An inclined-tube manometer is used to measure the difference in dynamic pressure at 1 and static pressure at 2 in a flowing gas.
(a) With the zero point on the inclined scale adjusted to coincide with the water level in the manometer well as shown, derive an expression for the pressure differential in terms of the fluid displacement R and the angle of the tube θ.
(b) If the local barometric pressure is 30 in Hg, what is the dynamic pressure in psia? $R = 10$ in; $\theta = 30°$.

4-25. For the system shown, what is the absolute pressure in the tank?

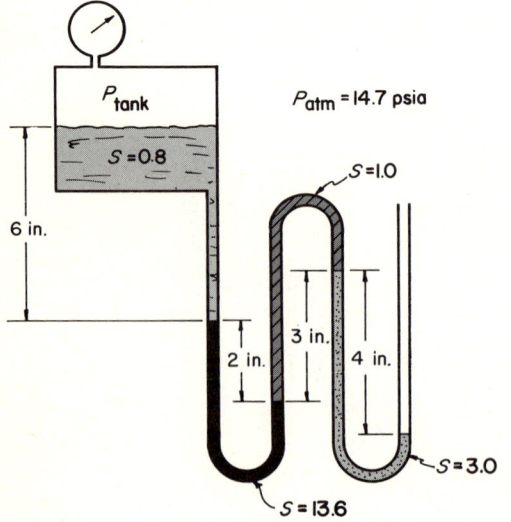

Fig. P4-25

FLUID STATICS

4-26. A 4- by 4-ft tank contains acetylene tetrabromide ($S = 2.96$). With the manometer arrangement shown:
  (a) Find the pressures indicated by the gauges at A and B.
  (b) If the barometer reading is 29 in Hg, is the valve open or closed?
  (c) What is the weight of the acetylene tetrabromide in the tank?

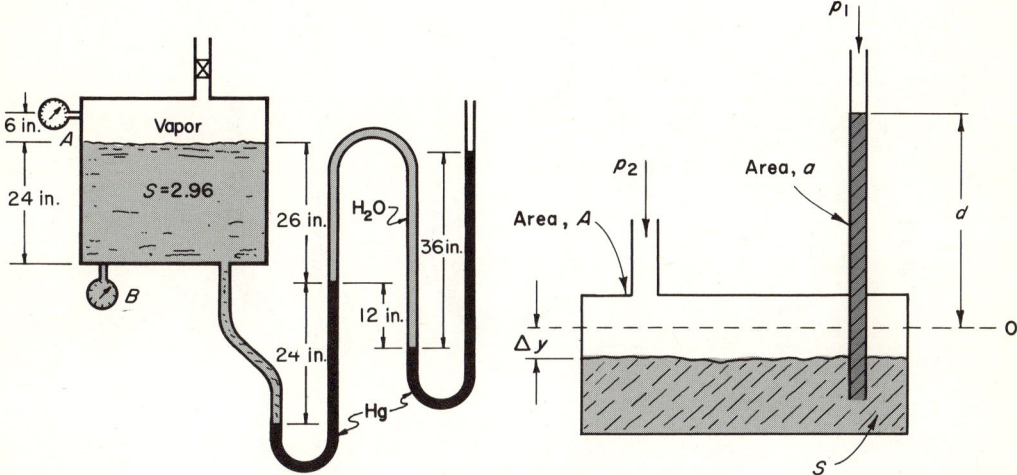

Fig. P4-26   Fig. P4-27

4-27. For the well manometer shown derive an expression for the scale graduation d (in terms of A, a, and S) for a pressure differential of 1 ft of water.

4-28. Is the air chamber shown connected to a vacuum pump or to a compressor? $p_{atm} = 14.7$ psia.

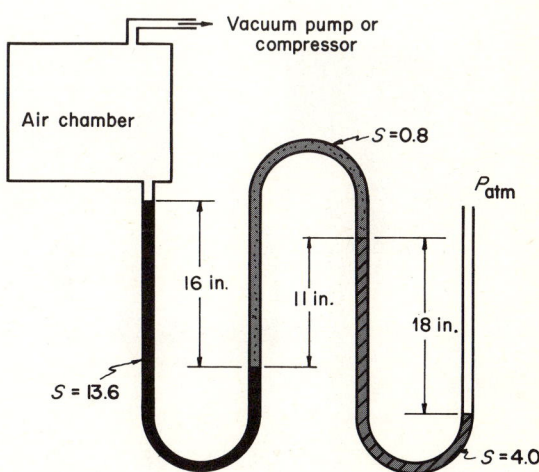

Fig. P4-28

4-29. A tank of water is connected to the manometer arrangement as shown. Atmospheric pressure is 14.7 psia. What is the absolute pressure at A?

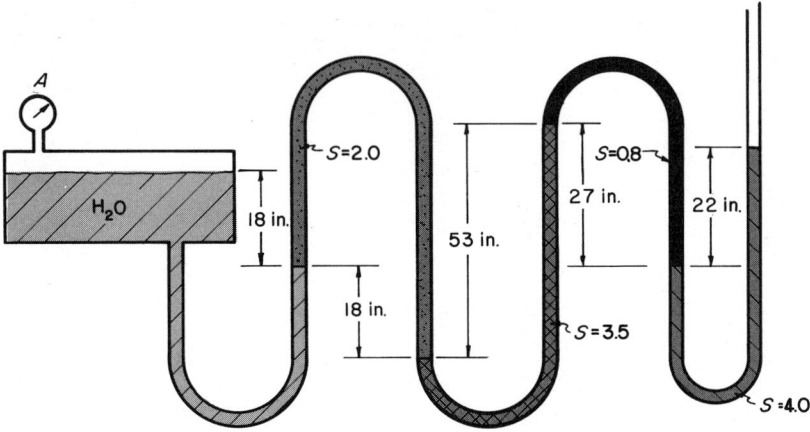

Fig. P4-29

4-30. A car lift rack in a garage uses an air-over-hydraulic system as shown schematically. If the supply pressure is as shown, what is the maximum capacity of the lift?

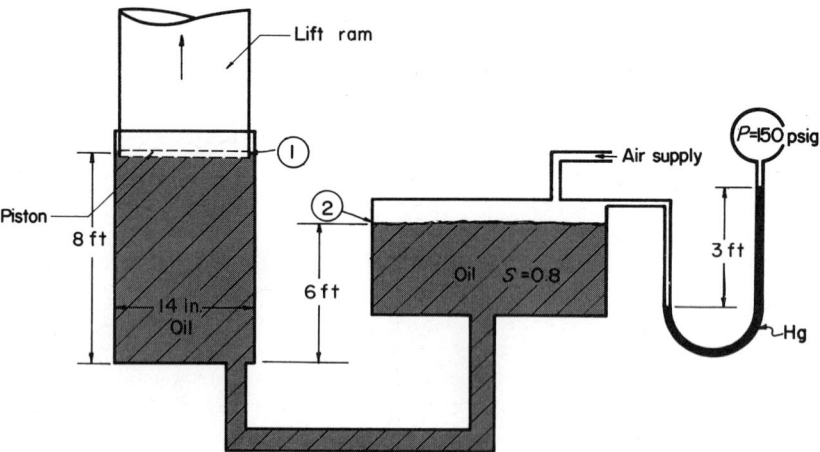

Fig. P4-30

4-31. Mercury fills a 0.05-in-diam glass tube with a meniscus as shown. Determine the surface tension.

FLUID STATICS

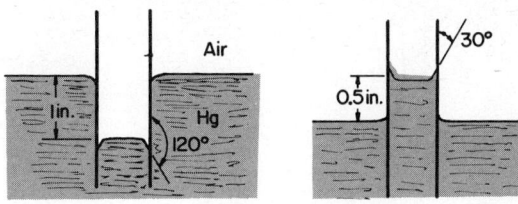

Fig. P4-31   Fig. P4-32

4-32. A 0.03-in-diam tube produces a meniscus in glycerin at 70°F as indicated. What is the surface tension?

4-33. Two bubbles are formed simultaneously from a soap-water solution. The first bubble is spherical and has a radius R. The second bubble is of compound curvature with orthogonal radii of R and $2R$. p_i is 15 psia, and p_o is 14.7 psia. Develop an expression for the surface tension of the spherical bubble in terms of that of the compound bubble. If the bubbles begin to rise, which will burst more quickly?

4-34. Develop an expression for the diameter of a spherical gas bubble at a depth h in fresh water if the diameter is 1.0 in at the water surface. The ambient pressure is 15 psi.

4-35. Find the total force on the bottom surface of the container. Compare this force with the total weight of the water. By performing a force balance on the fluid, resolve this apparent paradox.

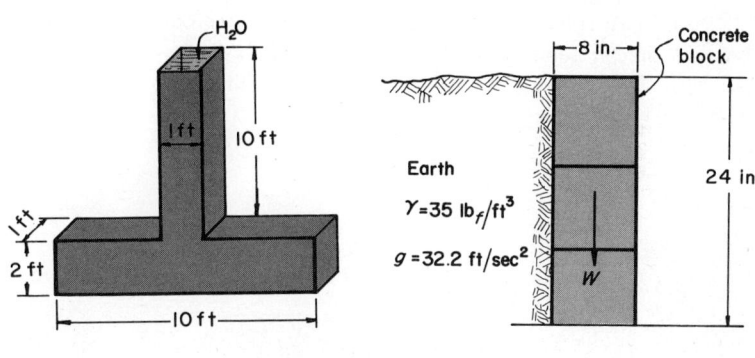

Fig. P4-35   Fig. P4-36

4-36. Many builders use concrete-block retaining walls for earth elevation changes. Unfortunately, they are often poorly designed and during spring rains are prone to failure by tipping over. A well-known engineering handbook suggests that such walls be designed as though the earth were a fluid with a specific weight of 35 lb_f/ft^3. If the average density of concrete blocks is 12 lb_m/ft^3 (this includes voids), is the 2-ft-high wall in the figure stable?

4-37. An 8-in-ID hydraulic cylinder has a 100,000-lb_f lifting capacity. For an allowable tensile stress of 12,000 psi, what should be the outside diameter of the cylinder?

4-38. A small reservoir dam is built of reinforced concrete having a specific weight 2.5 times that of water. For the indicated water level:
 (a) What force does the water exert on the dam?
 (b) Where is the resultant force applied?

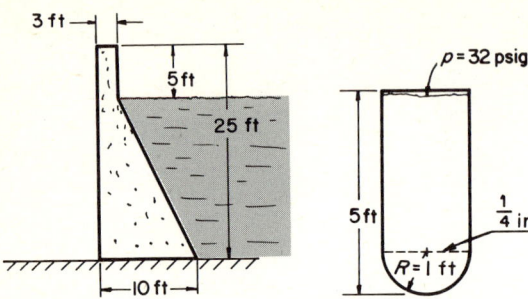

Fig. P4-38 Fig. P4-39

4-39. A hot-water heater has a hemispherical bottom which is welded to its cylindrical side. The effective cross-sectional area of the weld is $\frac{1}{4}$ in. Determine the axial tensile stress in the weld.

4-40. A 3- by 3-ft wall of sandbags is built on a river levee. What should be the average density of the sandbags (including voids) in order for the wall not to tip over assuming the bags to be tied together to form an integral unit?

4-41. The center of pressure in a drainage canal of rectangular cross section is located 18 in below the centroid of the vertical walls.
   (*a*) How deep is the water if the canal is flowing full?
   (*b*) What is the total force on a 1-ft section of the wall?

4-42. A circular flood gate is restrained as shown.
   (*a*) If the gate weighs 40,000 lb$_f$, what force *F* is required to hold the gate closed when the water is 30 ft deep?
   (*b*) If the force *F* were not applied, at what water depth would the gate begin to open?

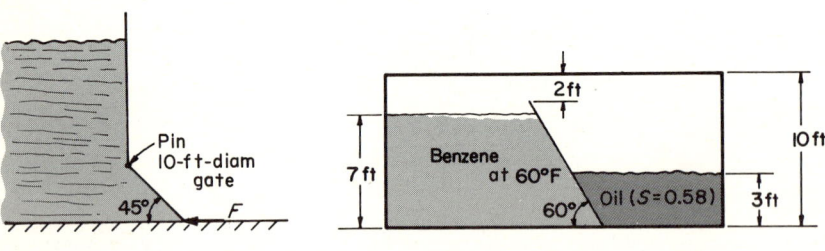

Fig. P4-42 Fig. P4-43

4-43. A storage tank for benzene and oil has an inclined divider welded to the bottom of the tank.
   (*a*) For the indicated levels, what moment must a 1-ft section of weld resist?
   (*b*) What combination of liquid levels would create the largest bending moment? What would it be?

4-44. A drum-type gate of cylindrical cross section is to be used for flood control. Neglecting the friction in the hinge, how much must the gate weigh per lineal foot if it is to open 5° when the water level is 10 ft above the hinge?

FLUID STATICS

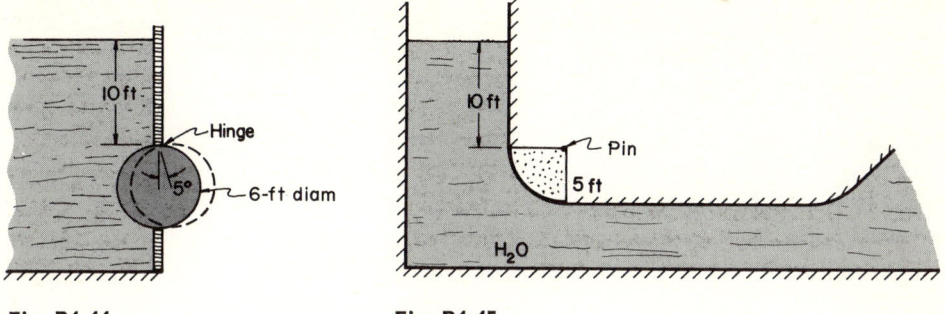

Fig. P4-44   Fig. P4-45

4-45. Determine the moment supported by the pin per foot of gate. Neglect the weight of the gate.

4-46. The rectangular flood gate shown is designed to begin opening when the water level is 17 ft higher than the hinge point. What is the weight per foot of the gate?

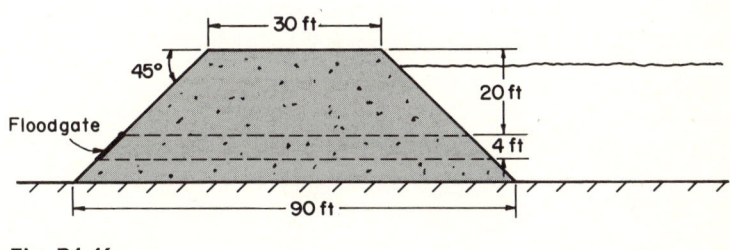

Fig. P4-46

4-47. The rectangular gate shown weighs 10,000 lb_f/ft. Find h. The pin cannot support a moment.

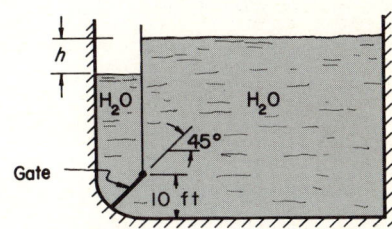

Fig. P4-47

4-48. Calculate the moment M required to hold the rectangular gate in the position shown. Neglect the weight of the gate and assume $\gamma_{H_2O} = 64.0$ lb_f/ft^3.

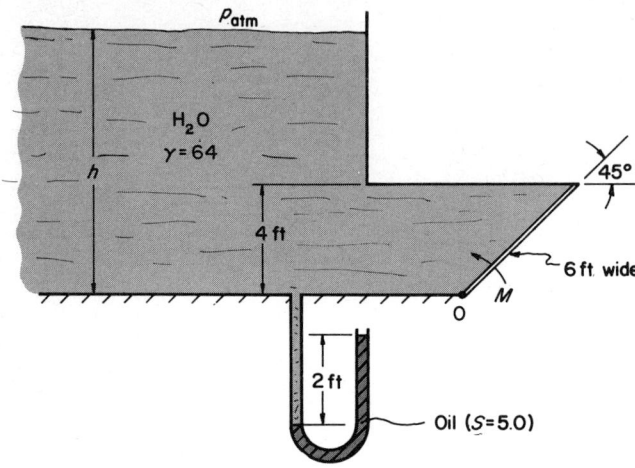

Fig. P4-48

4-49. Determine the torque required to hold the butterfly valve in the closed position.

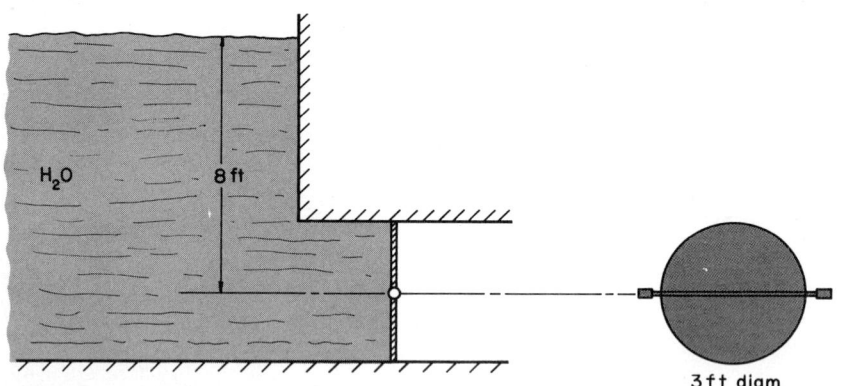

Fig. P4-49

4-50. Determine x such that the moment exerted by the pin is zero.
4-51. Determine the value of x so that the gates open simultaneously.
4-52. A water-wave buffer is constructed of a curved section and a right-angle section. The curved part weighs 2,000 lb_f/ft, and the right-angle part weighs 3,000 lb_f/ft; both sections are homogeneous solids. What moment M is required at the pin to keep it in the position shown?
4-53. The right-hand side of a water storage tank is filled at the rate of 5,000 gal/hr. The square

FLUID STATICS

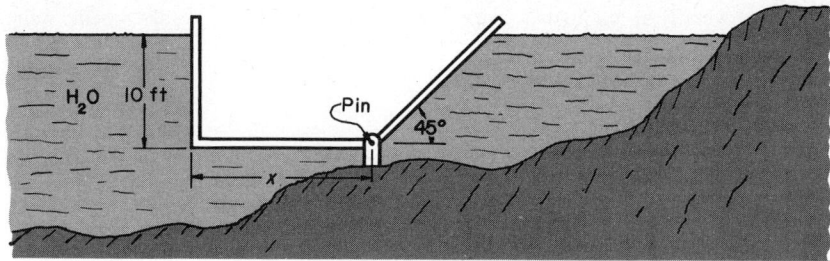

Fig. P4-50

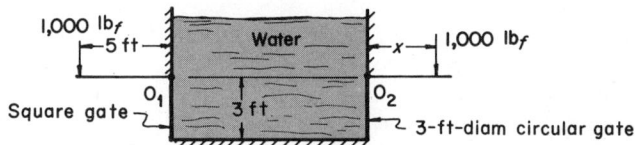

Fig. P4-51

Fig. P4-52

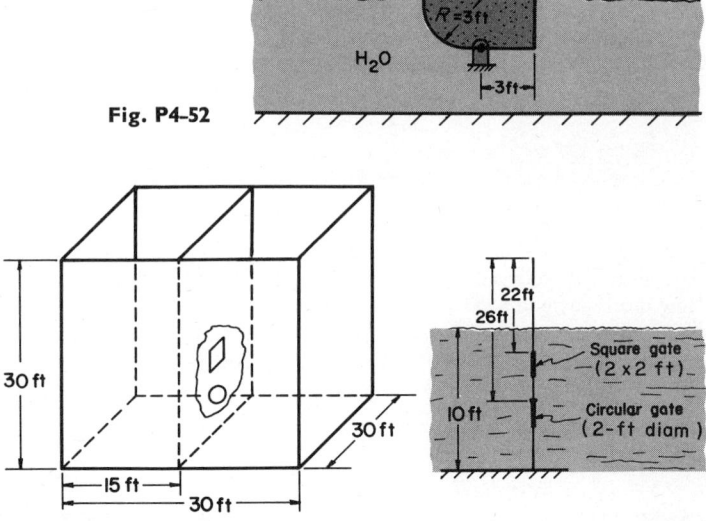

Fig. P4-53

gate has a 3,000-ft-lb$_f$ moment holding it closed, and the circular gate is held by a 4,000-ft-lb$_f$ moment. Both sides of the tank are initially at a depth of 10 ft.

  (a) Which gate will open first? At what depth?
  (b) How much time elapses until the gate opens?

4-54. A 180-lb$_f$ man stands on a 14-in-diam log ($\gamma = 36$ lb$_f$/ft^3) which floats in water. The weight of the man causes the log to sink just below the surface of the water. How long is the log?

4-55. A cylindrical log floats in water as shown. What is its average specific weight?

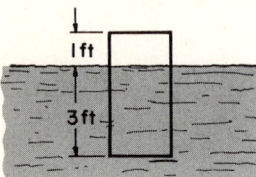

Fig. P4-55

4-56. A 1-ft cube of brass ($\gamma = 535$ lb$_f$/ft^3) floats in mercury. How much of the cube is above the surface?

4-57. An iceberg ($\gamma = 47$ lb$_f$/ft^3) has a volume of 10^5 ft^3 protruding out of seawater ($\gamma = 64$ lb$_f$/ft^3). How much of the iceberg is below the surface?

4-58. A high-altitude weather balloon weighing 120 lb$_f$ is filled with helium. At an atmospheric pressure of 14.7 psia the balloon is spherically shaped, with a diameter of 20 ft. For a given temperature $\gamma_{\text{air}} = 0.072$ lb$_f$/ft^3 and $\gamma_{\text{He}} = 0.008$ lb$_f$/ft^3. What is the magnitude of the buoyant force?

4-59. It is desired to use a 2-ft-diam beach ball to stop a small drain in a swimming pool. Obtain a relationship between the drain diameter d and the minimum water depth h for which the ball will remain in place.

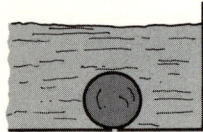

Fig. P4-59

4-60. A barge used in the intercoastal waterways of the Gulf of Mexico is 60 ft long, 20 ft wide, and 10 ft deep. It weighs 40 tons empty and requires 1.5 ft of freeboard for protection against waves. How much payload can it safely transport?

4-61. To save himself in a flood a small animal jumps onto a piece of wood which sinks uniformly a total of 1.3 in into the water. The wood is 1 ft square by 2 in thick and has a specific gravity of 0.52. How heavy is the animal?

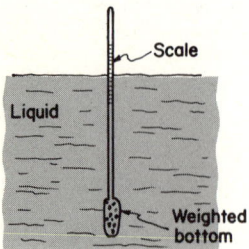
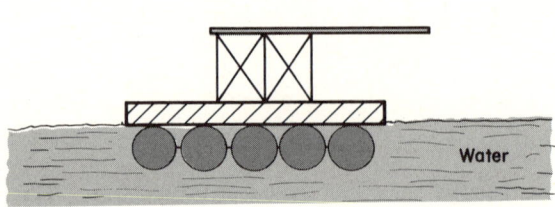

Fig. P4-62   Fig. P4-63

FLUID STATICS

4-62. If the hydrometer shown weighs 0.08 lb_f and has a stem diameter of 0.20 in, compute the distance between specific-gravity S markings 1.0 and 1.1.

4-63. A diving platform floats on five 50-gal drums which are completely submerged when the platform is loaded to capacity. The empty platform weighs 350 lb_f. How many 150-lb_f people can the platform support?

4-64. A small submarine is cruising in the Atlantic Ocean at a depth of 1,000 ft. To bring the ship to the surface, the captain releases the ship's ballast, after which the internal volume of the submarine (10^4 ft^3) is filled with air at 14.7 psia. The ship weighs 100 tons, and the specific weight of seawater is 64 lb_f/ft^3.

    (*a*) What is the force which accelerates the ship to the surface?
    (*b*) What is the normal stress on the hull of the submarine at the 1,000-ft cruising depth?

4-65. Determine the readings of scales A and B. Neglect the weight of the container.

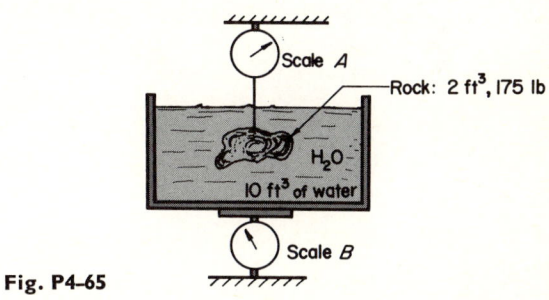

Fig. P4-65

4-66. What is the weight of the gauge plus the container? What pressure is indicated by the gauge?

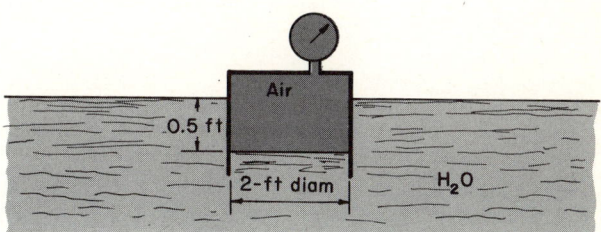

Fig. P4-66

4-67. A 200-lb_f man has a specific weight of 65 lb_f/ft^3 with his breath exhaled. After taking a deep breath and jumping into a swimming pool, he floats vertically upward until his head just breaks the surface of the water. How many cubic inches of air did he draw into his lungs?

4-68. A cube having a specific weight of 60 lb_f/ft^3 floats in a container filled with oil and water as shown. If the bottom surface of the cube remains parallel to the surface of the container, find x.

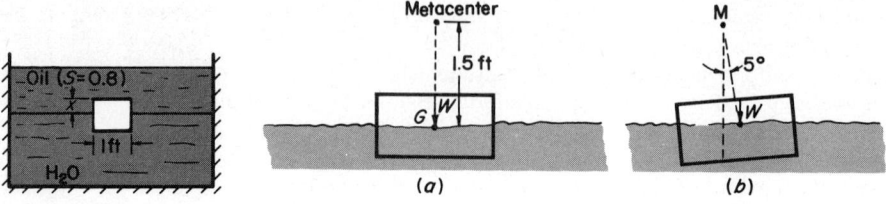

Fig. P4-68 **Fig. P4-69**

4-69. A 50-lb$_f$ rectangular block of wood ($S = 0.5$) floats in water as shown. When tipped 5°, what is the value of the restoring couple?

4-70. An 18-ft-diam balloon is filled with helium ($\gamma = 0.008$ lb$_f$/ft^3). For the angle of tilt and location of center of mass as shown, determine the restoring moment. Assume $\gamma_{\text{air}} = 0.072$ lb$_f$/ft^3.

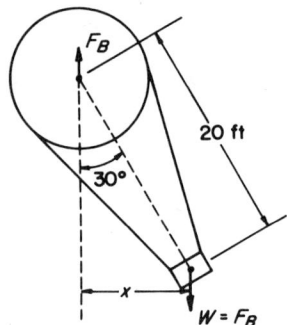

Fig. P4-70

4-71. A 40-ft-long barge rolls to the position shown. For a total weight of 40 tons, what is the value of the restoring moment?

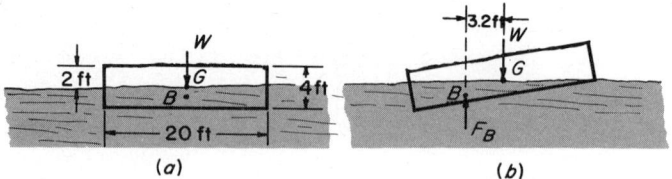

Fig. P4-71

REFERENCES

1. Eskinazi, S.: "Principles of Fluid Mechanics," 2d ed., Allyn and Bacon, Boston, 1968.
2. Shames, I. H.: "Mechanics of Fluids," McGraw-Hill, New York, 1962.
3. Shepherd, D. G.: "Elements of Fluid Mechanics," Harcourt, Brace & World, New York, 1965.
4. Streeter, V. L.: "Fluid Mechanics," 4th ed., McGraw-Hill, New York, 1966.

CHAPTER 5
STEADY-STATE CONDUCTIVE HEAT TRANSFER

From experimental evidence we know that whenever a thermal field gives rise to gradients within a material, heat transfer will occur by conduction. The basic rate equation describing this process was discussed in Chap. 2 and is known as Fourier's[1] heat-conduction equation. In the present chapter and the next we direct our attention to a detailed study of this phenomenon, invoking the principle of field segregation.

5-1 GENERAL CONDUCTIVE EQUATION

Consider any homogeneous body of material (*solid*, *liquid*, or *gas*) exposed to a thermal field. If it is a liquid or a gas, we impose the restriction that either it is not in motion or all parts of the system under consideration have the same velocity. Thus there are no *macroscopic* movements within the system under analysis. A physical example of a nonsolid material for which a conductive-heat-transport analysis can be performed is water freezing in a pipe under the no-flow condition. We shall develop the general equation for a *solid* element of material, but we should

[1] Named for the French mathematician Jean Fourier (1768–1830).

keep in mind that the equation holds equally well for liquid or gaseous materials with suitable restrictions.

In general, we are concerned with three-dimensional designs, and these sometimes give rise to thermal gradients in all directions. If we can specify the gradient in each of three mutually perpendicular directions, however, we can determine the heat flux in any direction and the temperature at any point in the material. Let us consider then a small element of material of a solid and arbitrarily impose a cartesian coordinate system. This could be a small cube from, say, a nuclear-reactor element, an electric wire carrying a current, or an insulation board on the wall of a house. The first law of thermodynamics [Eq. (3-36)] applied to this volume element with no change in potential and kinetic energy can be expressed as

Rate at which heat enters + rate at which work enters
= rate at which heat leaves + rate at which work leaves
+ rate at which internal energy increases   (5-1)

For a simple incompressible substance, the net work done on the system is converted to internal energy, and consequently we can write this equation as

$$q_x + q_y + q_z + q_{\text{conv}} = +q_{x+\Delta x} + q_{y+\Delta y} + q_{z+\Delta z} + \frac{\partial U}{\partial t} \qquad (5\text{-}2)$$

where the heat fluxes are

$$q_x = -k \left.\frac{\partial T}{\partial x}\right|_x \Delta y\, \Delta z \qquad q_{x+\Delta x} = -k \left.\frac{\partial T}{\partial x}\right|_{x+\Delta x} \Delta y\, \Delta z$$

$$q_y = -k \left.\frac{\partial T}{\partial y}\right|_y \Delta x\, \Delta z \qquad q_{y+\Delta y} = -k \left.\frac{\partial T}{\partial y}\right|_{y+\Delta y} \Delta x\, \Delta z$$

$$q_z = -k \left.\frac{\partial T}{\partial z}\right|_z \Delta x\, \Delta y \qquad q_{z+\Delta z} = -k \left.\frac{\partial T}{\partial z}\right|_{z+\Delta z} \Delta x\, \Delta y$$

The rate of conversion of thermodynamic work to thermal energy per unit volume will be designated by q'''. Thus,

$$q_{\text{conv}} = q''' \Delta x\, \Delta y\, \Delta z$$

and since ρc represents the internal-energy storage capacity per unit volume,

$$\frac{\partial U}{\partial t} = \rho c\, \Delta x\, \Delta y\, \Delta z\, \frac{\partial T}{\partial t}$$

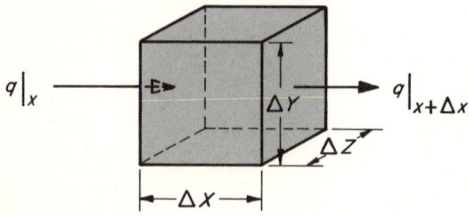

Fig. 5-1 Volume element for derivation of the general heat-conduction equation (heat flux in y and z directions omitted for clarity).

STEADY-STATE CONDUCTIVE HEAT TRANSFER

Substituting these into Eq. (5-2), grouping all conduction terms, and dividing by the volume $\Delta x\, \Delta y\, \Delta z$ yields

$$\frac{k\frac{\partial T}{\partial x}\big|_{x+\Delta x} - \frac{\partial T}{\partial x}\big|_{x}}{\Delta x} + \frac{k\frac{\partial T}{\partial y}\big|_{y+\Delta y} - \frac{\partial T}{\partial y}\big|_{y}}{\Delta y} + \frac{k\frac{\partial T}{\partial z}\big|_{z+\Delta z} - \frac{\partial T}{\partial z}\big|_{z}}{\Delta z} + q''' = \rho c \frac{\partial T}{\partial t} \quad (5\text{-}3)$$

In the limit as Δx, Δy, and Δz approach zero we have by definition of the derivative

$$\frac{\partial}{\partial x}\left(k\frac{\partial T}{\partial x}\right) + \frac{\partial}{\partial y}\left(k\frac{\partial T}{\partial y}\right) + \frac{\partial}{\partial z}\left(k\frac{\partial T}{\partial z}\right) + q''' = \rho c \frac{\partial T}{\partial t} \quad (5\text{-}4)$$

and for constant thermal conductivity, this further reduces to

$$\blacksquare \quad \frac{\partial^2 T}{\partial x^2} + \frac{\partial^2 T}{\partial y^2} + \frac{\partial^2 T}{\partial z^2} + \frac{q'''}{k} = \frac{1}{\alpha}\frac{\partial T}{\partial t} \quad (5\text{-}5)$$

where α is the thermal diffusivity, $k/\rho c$. This equation is referred to as the general heat-conduction equation for any stationary medium having constant thermal conductivity.

Even though the q''' term of this derivation originated as a result of net thermodynamic work done on a substance (an example of which is "I^2R heating" in an electric conductor), this can be considered as a thermal energy *generation* per unit volume. This type of problem is encountered in nuclear fission, with a conversion of matter to energy which is not accounted for in Eq. (3-36).

Special cases

1. For systems which contain no heat sources, Eq. (5-5) reduces to

$$\frac{\partial^2 T}{\partial x^2} + \frac{\partial^2 T}{\partial y^2} + \frac{\partial^2 T}{\partial z^2} = \frac{1}{\alpha}\frac{\partial T}{\partial t} \quad (5\text{-}6)$$

which is known as the *Fourier equation*.

2. For systems which are steady with respect to time but have heat sources, Eq. (5-5) becomes

$$\frac{\partial^2 T}{\partial x^2} + \frac{\partial^2 T}{\partial y^2} + \frac{\partial^2 T}{\partial z^2} + \frac{q'''}{k} = 0 \quad (5\text{-}7)$$

which is known as the *Poisson equation*.

3. Finally, if the system is both steady and free of heat sources, the heat-conduction equation can be expressed as

$$\frac{\partial^2 T}{\partial x^2} + \frac{\partial^2 T}{\partial y^2} + \frac{\partial^2 T}{\partial z^2} = 0 \quad (5\text{-}8)$$

which is known as the *Laplace equation*.

Other coordinate systems It is frequently advantageous to begin an analysis with the heat-conduction equation in other than cartesian coordinates. Recalling from Chap. 1 that temperature is simply a *conserved* property, we note that Eq. (5-5) can be written for any coordinate system as

■ $$\nabla^2 T + \frac{q'''}{k} = \frac{1}{\alpha}\frac{\partial T}{\partial t} \tag{5-9}$$

where the operator ∇^2 (del squared) is known as the *laplacian*. In the cartesian coordinate system this is

$$\nabla^2(\cdot) = \frac{\partial^2(\cdot)}{\partial x^2} + \frac{\partial^2(\cdot)}{\partial y^2} + \frac{\partial^2(\cdot)}{\partial z^2} \tag{5-10}$$

It is relatively simple to formulate $\nabla^2(\cdot)$ for the cylindrical and spherical coordinate systems shown in Fig. 5-2a and b [10]. The result in cylindrical coordinates is

$$\frac{\partial^2 T}{\partial r^2} + \frac{1}{r}\frac{\partial T}{\partial r} + \frac{1}{r^2}\frac{\partial^2 T}{\partial \phi^2} + \frac{\partial^2 T}{\partial z^2} + \frac{q'''}{k} = \frac{1}{\alpha}\frac{\partial T}{\partial t} \tag{5-11}$$

and in spherical coordinates

$$\frac{1}{r}\frac{\partial^2}{\partial r^2}rT + \frac{1}{r^2 \sin\psi}\frac{\partial}{\partial \psi}\left(\sin\psi\,\frac{\partial T}{\partial \psi}\right) + \frac{1}{r^2 \sin^2\psi}\frac{\partial^2 T}{\partial \phi^2} + \frac{q'''}{k} = \frac{1}{\alpha}\frac{\partial T}{\partial t} \tag{5-12}$$

It should be remembered that Eqs. (5-5) to (5-12) are applicable only for media having constant thermal conductivity. Although this condition is rarely if ever exactly true (the thermal gradients usually give rise to variations in properties), it is frequently *assumed* for the sake of the resulting mathematical simplifications, and the assumption is quite reasonable for moderate temperature gradients.

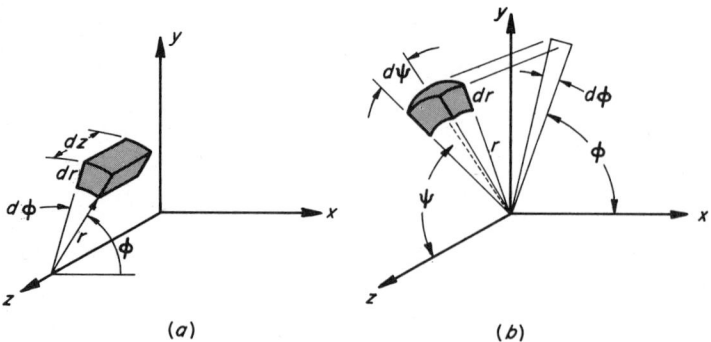

Fig. 5-2 Control volume for general heat-conduction equation: (*a*) cylindrical; (*b*) spherical.

STEADY-STATE CONDUCTIVE HEAT TRANSFER

5-2 ONE-DIMENSIONAL SYSTEMS

We now wish to consider the simplest physical configurations encountered in heat transfer by conduction. This class of problems results, for example, on very large plane walls subjected to different temperatures on their two faces or in long hollow tubes with a constant inside temperature different from that on the outside. Throughout this chapter we shall restrict our attention to steady-state applications; i.e., we shall not consider problems where temperature is a function of time.

Plane wall Consider the composite wall of Fig. 5-3 consisting of two different materials a and b. For a steady state we note that the heat flux q is constant at each plane of area A perpendicular to the direction of heat flow. If the surfaces 1 and 3 are held at temperatures T_1 and T_3, respectively, T_1 being greater than T_3, the heat flow through material a, obtained by integrating Fourier's law, is

$$q = -k_a A \frac{T_2 - T_1}{L_a} \tag{5-13}$$

and through material b is

$$q = -k_b A \frac{T_3 - T_2}{L_b} \tag{5-14}$$

We note in passing that the simplest case of thermal conduction is that of a plane wall of a single material with no internal heat sources, such as material a. Then Eq. (5-13) constitutes the complete problem formulation.

Returning to the present case of a composite wall, from the last two equations we have

$$T_1 - T_2 = q \frac{L_a}{k_a A} \qquad T_2 - T_3 = q \frac{L_b}{k_b A}$$

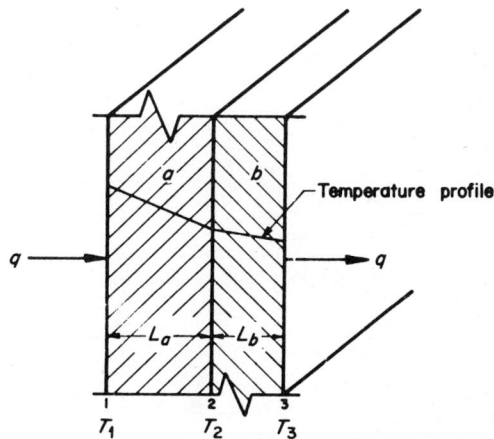

Fig. 5-3 One-dimensional heat flow through a plane wall.

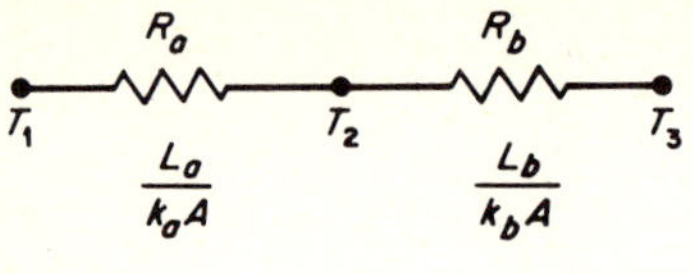

Fig. 5-4 Electrical analog for two-layer composite wall.

the heat flux q being the same through both walls. Thus,

$$T_1 - T_3 = (T_1 - T_2) + (T_2 - T_3) = q\left(\frac{L_a}{k_a A} + \frac{L_b}{k_b A}\right)$$

and hence,

$$q = \frac{T_1 - T_3}{L_a/k_a A + L_b/k_b A} \tag{5-15}$$

Now we note at this point that Eq. (5-15) can be written

$$\text{Heat flux} = \frac{\text{overall temperature difference}}{\Sigma \text{ thermal resistances}} \tag{5-16}$$

where each $L_i/k_i A$ is interpreted as the thermal resistance of a single thickness of conductive material. This form suggests the use of an electrical analogy. For the simple case of two materials shown in Fig. 5-3, the electric network would be as shown in Fig. 5-4. The electrical analogy permits extension of the preceding work to relatively complex problems involving simultaneous heat flux through series and parallel thermal resistances.

Example A load-bearing masonry wall consists of a 4-in brick outer face with $\frac{3}{8}$-in mortar joints, an 8-in concrete wall, and a $\frac{5}{8}$-in insulating board on the inside. The outer and inner temperatures are 10° and 70°F, respectively. Determine the heat flux.

Solution The resistances are for unit height and unit width of wall

Brick: $\quad R_b = \dfrac{\frac{4}{12}}{k_b(4)(2\frac{5}{8})/12} = \dfrac{1}{0.38(2.625)} = \dfrac{1.002}{\text{Btu/hr-°F}}$

Mortar: $\quad R_m = \dfrac{\frac{4}{12}}{k_m(4)(\frac{3}{8})(\frac{1}{12})} = \dfrac{1}{0.44(0.375)} = \dfrac{6.06}{\text{Btu/hr-°F}}$

Concrete: $\quad R_c = \dfrac{\frac{8}{12}}{k_c(1)} = \dfrac{0.666}{0.54} = \dfrac{1.234}{\text{Btu/hr-°F}}$

Insulating board: $\quad R_i = \dfrac{\frac{5}{8}(\frac{1}{12})}{k_i(1)} = \dfrac{0.625}{0.09(12)} = \dfrac{0.578}{\text{Btu/hr-°F}}$

where physical properties are taken from Appendix A except k_m, which is 0.44 Btu/hr-ft-°F. Then, by analogy with dc electric network theory

$$q = \frac{T_1 - T_4}{\Sigma R} = \frac{T_1 - T_4}{R_b R_m/(R_b + R_m) + R_c + R_i} = \frac{10 - 70}{[1.002(6.06)]/(1.002 + 6.06) + 1.234 + 0.578}$$

$$= \frac{10 - 70}{2.672} = -22.5 \text{ Btu/hr-ft}$$

STEADY-STATE CONDUCTIVE HEAT TRANSFER

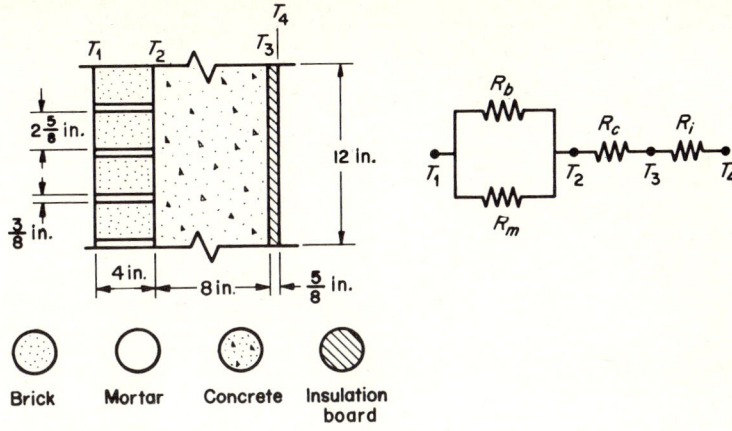

Fig. 5-5 Composite wall and electric circuit.

The minus sign indicates that the direction of the heat flux is from the inside to the outside, since we were solving for the heat flux from outer to inner surfaces. This simplified method of treating a composite wall assumes one-dimensional heat transfer, which is exactly true only in the simple case where all materials have equal thermal conductivities. Consequently, these problems are actually two-dimensional, but the simplified one-dimensional approach is frequently acceptable for engineering calculations.

From our discussion of thermal conductivity in Chap. 2 we recall that the thermal conductivity of most materials is dependent upon temperature. For many engineering materials, a reasonable approximation is given by Eq. (2-8)

$$k = k_0(1 + aT)$$

which is, of course, a linear function of temperature, the constant a being the slope and k_0 being the initial value of the thermal conductivity. Substituting this expression into Fourier's law for the plane wall and separating variables yields

$$\frac{q}{A} \int_{x_1}^{x_2} dx = - \int_{T_1}^{T_2} k_0(1 + aT)\, dT \tag{5-17}$$

Integrating, we have

$$q = \frac{k_0 A}{x_2 - x_1} \left[(T_1 - T_2) + \frac{a}{2}(T_1^2 - T_2^2) \right] \tag{5-18}$$

or

$$q = -\frac{A(T_2 - T_1)}{x_2 - x_1} k_0 \left(1 + a\frac{T_1 + T_2}{2}\right) \equiv -k_m A \frac{\Delta T}{\Delta x} \tag{5-19}$$

which indicates that the problem is reduced to evaluating the thermal conductivity at the mean temperature of the material.

Cylinders A second class of one-dimensional problems frequently encountered in practice is radial heat transfer through hollow cylinders. A typical problem would involve the heat loss from an insulated thick-walled pipe, as shown in Fig. 5-6. The inside pipe radius is r_1, the outside pipe radius (and inside insulation radius) is r_2, and the outside insulation radius is r_3. The temperatures corresponding to the three radii are T_1, T_2, and T_3, respectively. For a pipe of length L, the area for radial heat flow in the system is

$$A = 2\pi r L \tag{5-20}$$

and substitution into Fourier's law yields

$$q = -(2\pi r L) k \frac{dT}{dr} \tag{5-21}$$

For the pipe section (material a) only, the boundary conditions are

$$\begin{aligned} \text{At } r = r_1: & \quad T = T_1 \\ \text{At } r = r_2: & \quad T = T_2 \end{aligned} \tag{5-22}$$

Solving Eq. (5-21) for material a yields

$$q = \frac{2\pi k_a L (T_1 - T_2)}{\ln (r_2/r_1)} \tag{5-23}$$

and by comparison with Eq. (5-16) we see that the *thermal resistance* of material a is

$$R_a = \frac{\ln (r_2/r_1)}{2\pi k_a L} \tag{5-24}$$

In a completely analogous manner, the thermal resistance of the insulation material is

$$R_b = \frac{\ln (r_3/r_2)}{2\pi k_b L} \tag{5-25}$$

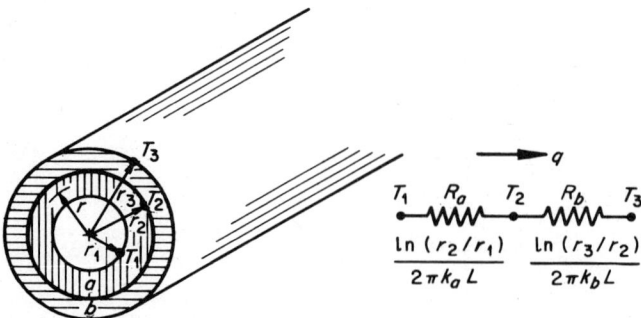

Fig. 5-6 Composite cylindrical wall and electrical analog.

and thus the complete solution is

$$q = \frac{2\pi L(T_1 - T_3)}{[\ln(r_2/r_1)]/k_a + [\ln(r_3/r_2)]/k_b} \tag{5-26}$$

Obviously this approach can be extended to any number of layers of material by application of Eq. (5-16), which holds for radial as well as plane wall systems.

Throughout the treatment of multilayered systems, we have tacitly assumed no contact resistance between the layers of different materials. Frequently this is not the case, and contact resistance can become a significant part of the total. A good introductory treatment of this complicated phenomenon is given by Holman [6].

Spheres If the temperature is a function of the radius only, the steady-state heat flux by conduction in a sphere is readily determined. In this case the area at any given radius is

$$A_r = 4\pi r^2 \tag{5-27}$$

and substituting this into Fourier's law and then integrating yields

$$q = \frac{4\pi k(T_1 - T_2)}{1/r_1 - 1/r_2} \tag{5-28}$$

where T_1 and T_2 are the inner and outer temperatures corresponding to r_1 and r_2 respectively. From this, the thermal resistance of a single spherical layer is

$$R = \frac{1/r_1 - 1/r_2}{4\pi k} \tag{5-29}$$

and the extension to a multilayer spherical system is obvious.

5-3 ONE-DIMENSIONAL SYSTEMS WITH HEAT GENERATION

A number of practical applications of conductive heat transfer involve systems with internal heat generation, e.g., electric coils, nuclear-fission elements, electric resistance elements, and chemically reacting systems. In many cases, the heat generation per unit volume is essentially constant, and if the temperature is a function of only one space variable, the problem is amenable to simple mathematical analysis. Some simple cases follow.

Plane wall with heat generation: Case I Consider a plane wall of thickness $2L$ with the other dimensions (height and length) much greater than the thickness (Fig. 5-7). The two faces of the wall are maintained at T_1 and T_2, which are constant. A problem of this type may arise in an electric conductor (bus bar). The appropriate differential equation can be written by application of (5-7) and is

$$\frac{d^2T}{dx^2} + \frac{q'''}{k} = 0 \tag{5-30}$$

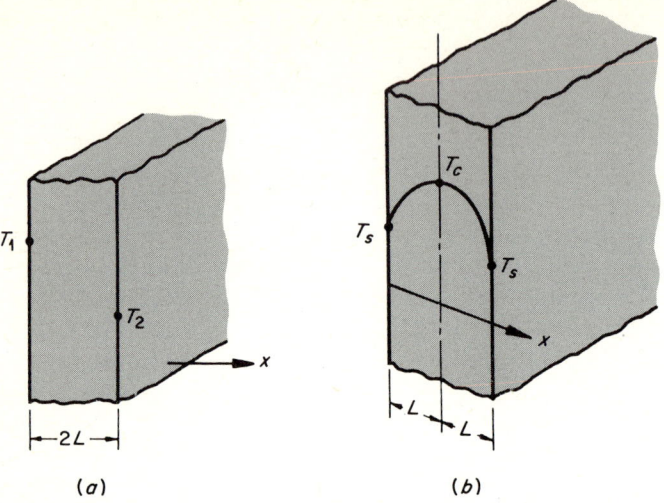

Fig. 5-7 Plane wall with internal heat generation: (*a*) asymmetrical and (*b*) symmetrical boundary conditions.

with boundary conditions:

$$T = \begin{cases} T_1 & \text{at } x = 0 \\ T_2 & \text{at } x = 2L \end{cases} \tag{5-31}$$

The general solution to Eq. (5-30) is

$$T = -\frac{q'''}{2k} x^2 + C_1 x + C_2 \tag{5-32}$$

Applying the first boundary condition yields

$$C_2 = T_1$$

and applying the second boundary condition yields

$$C_1 = \frac{T_2 - T_1}{2L} + \frac{q''' L}{k}$$

Thus, the temperature distribution is given by

$$T = \left[\frac{T_2 - T_1}{2L} + \frac{q'''}{2k} (2L - x) \right] x + T_1 \tag{5-33}$$

and the heat flux can readily be obtained by differentiating Eq. (5-33) with respect to x and applying Fourier's equation.

STEADY-STATE CONDUCTIVE HEAT TRANSFER

Plane wall with heat generation: Case 2 From Eq. (5-33) it is apparent that for the simpler case where $T_1 = T_2 = T_s$, as shown in Fig. 5-7b, we have

$$T = \frac{q'''}{2k}(2L - x)x - T_s \tag{5-34}$$

Denoting the midplane temperature at $x = L$ by T_c, we find

$$T_c - T_s = \frac{q'''}{2k}L^2 \tag{5-35}$$

That T_c is the maximum temperature is easily verified by applying the condition for an extremal. Differentiation of Eq. (5-34) yields

$$\frac{dT}{dx} = \frac{q'''}{k}(L - x) \tag{5-36}$$

and the temperature gradient is zero for $x = L$.

Applying Eq. (5-36) to evaluate the heat flux yields

$$-kA\frac{dT}{dx}\bigg|_{x=2L} = -kA\frac{q'''}{k}(-L) = q'''AL \tag{5-37}$$

It is informative to reconsider this last expression with regard to the physical problem. The right side of this equation is the product of the volume AL (where A is the area perpendicular to the x direction) and the volumetric heat-generation rate; i.e., this is the rate of energy generated in one-half of the plate. Since this is equal to the heat flux at the surface by conduction, we conclude that the heat generated in the right half of the plate is conducted out the right face.

A similar treatment at $x = 0$ would yield a result identical with Eq. (5-37) except that the sign would be negative due to our choice of coordinate system. Finally, we note that by combining Eqs. (5-34) and (5-35) we obtain

$$\frac{T - T_s}{T_c - T_s} = 2\frac{x}{L} - \left(\frac{x}{L}\right)^2 \tag{5-38}$$

which clearly indicates the parabolic form of the temperature distribution. A nondimensional temperature formed with T and two known temperatures as used in Eq. (5-38) will be found convenient in many conduction problems.

Cylinder with heat generation: Another frequently encountered configuration with uniform heat generation is the long solid cylinder. In this case, the length precludes the existence of an axial temperature gradient, and for uniform surface temperature there is no azimuthal gradient. For steady state, the appropriate differential equation can be obtained from Eq. (5-11) and is

$$\frac{d^2T}{dr^2} + \frac{1}{r}\frac{dT}{dr} + \frac{q'''}{k} = 0 \tag{5-39}$$

Two boundary conditions may be stated by noting that there can be no thermal gradient at the centerline of the rod by virtue of symmetry, i.e.,

$$\left.\frac{dT}{dr}\right|_{r=0} = 0$$

and at the outer surface the temperature is usually known. Thus

At $r = r_s$: $T = T_s$

Equation (5-39) may be written

$$\frac{d}{dr}\left(r\frac{dT}{dr}\right) = \frac{-rq'''}{k} \tag{5-39a}$$

and a first integration is readily seen to yield

$$r\frac{dT}{dr} = -\frac{r^2}{2}\frac{q'''}{k} + C_1$$

Application of the boundary condition at the centerline indicates that C_1 is zero. A second integration results in

$$T = -\frac{r^2}{4}\frac{q'''}{k} + C_2$$

and by applying the remaining boundary condition we find

$$C_2 = T_s + \frac{r_s^2}{4}\frac{q'''}{k}$$

Thus the temperature distribution is

$$T - T_s = \frac{r_s^2 q'''}{4k}\left[1 - \left(\frac{r}{r_s}\right)^2\right] \tag{5-40}$$

Denoting the centerline temperature by T_c, we can write this result in a convenient nondimensional form as

$$\frac{T - T_s}{T_c - T_s} = 1 - \left(\frac{r}{r_s}\right)^2 \tag{5-40a}$$

It is informative to note that the minimum temperature in the rod is quite obviously the surface temperature, and by Eq. (5-40) it is readily apparent that the maximum temperature occurs at $r = 0$ and is T_c.

Example An AWG no. 10 stainless-steel wire, 1 ft long is used as an electric resistance heater in a laboratory experiment. The measured voltage drop across the wire is 20 volts. and the measured current is 40 amp. The wire surface temperature, measured with an attached thermocouple, is 600°F. Find the maximum temperature in the wire.

STEADY-STATE CONDUCTIVE HEAT TRANSFER

Solution The maximum temperature is the centerline temperature and can be determined with Eq. (5-40). Calculating q''' from the given data,

$$\text{Electric power input} = I^2R = q'''\pi r_s^2 L = q'''\pi(1)\frac{0.051^2}{12^2}$$

$$= 56.7 \times 10^{-6} q'''$$

or, by Ohm's law,

$$I^2R = I^2\frac{E}{I} = EI = 20(40) = 800 \text{ watts}$$

Hence

$$q''' = \frac{800}{\pi r_s^2 L} \text{ watts} = \left(\frac{800}{56.7 \times 10^{-6}} \text{ watts/ft}^3\right)(3.412 \text{ Btu/hr-watt})$$

$$= 48.1 \times 10^6 \text{ Btu/hr-ft}^3$$

Using $k = 10$ Btu/hr-ft-°F,

$$T_{\max} = T_c = T_s + \frac{r_s^2 q'''}{4k} = 600°F + \frac{(18.07 \times 10^{-6})(4.81 \times 10^7)}{4 \times 10} = 622°F$$

Having determined T_c, we note the advantage of a nondimensional temperature expression such as Eq. (5-40a) in obtaining the temperature at another radius, since

$$T = 600 + (622 - 600)\left[1 - \left(\frac{r}{0.051}\right)^2\right]$$

5-4 CONVECTIVE BOUNDARY PHENOMENA

To this point in our study of conductive heat transport, we have avoided discussing the mechanism of heat removal (or addition) external to the boundaries of the system. We have evaluated the surface heat flux by Fourier's equation applied to the conductive material without regard to the external phenomenon. Frequently, the primary heat flux at the surface is due to convective transport. *This implies a mass or bulk movement of the medium, with this motion transporting or convecting the energy flux.*

The basic equation for convective heat transport is Newton's law of cooling

$$q = hA \, \Delta T \tag{5-41}$$

where h = convective-heat-transfer coefficient, Btu/hr-ft²-°F
A = area perpendicular to direction of heat flux, ft²
ΔT = temperature difference between surface and convective fluid, °F

It is important to note a basic difference between this law and Fourier's equation. The k in the latter is a *property* of the medium, whereas the h in Newton's law is a function of many variables, including velocity, fluid properties, etc. Thus Eq. (5-41) is merely a definition of h.

A complete study of convective heat transport is deferred until Part III, Moving Media. At the present, however, we need a working knowledge suitable for formulating an energy balance of the type

$$-kA\left.\frac{dT}{dx}\right|_{\text{boundary}} = hA\,\Delta T$$

This equation enables us to use a boundary condition other than a known temperature in the solution of many types of problem, including the cases already treated in this chapter.

Overall heat-transfer coefficient In many applications the boundary temperatures of a problem are not specified, and it becomes advantageous to formulate the problem in terms of convective coefficients at the boundaries. Consider the plane wall of Fig. 5-8a, which could represent for example the wall of a cold-storage room. The outside temperature at a distance from the wall (undisturbed by the heat flow) is T_1; the inside undisturbed temperature is T_4. A fluid boundary layer exists along both vertical surfaces. The methods of obtaining numerical values for the outside and inside average convective coefficients $\bar{h}_o$ and $\bar{h}_i$, respectively, are presented in Part III, Moving Media; for the present we assume them to be specified.

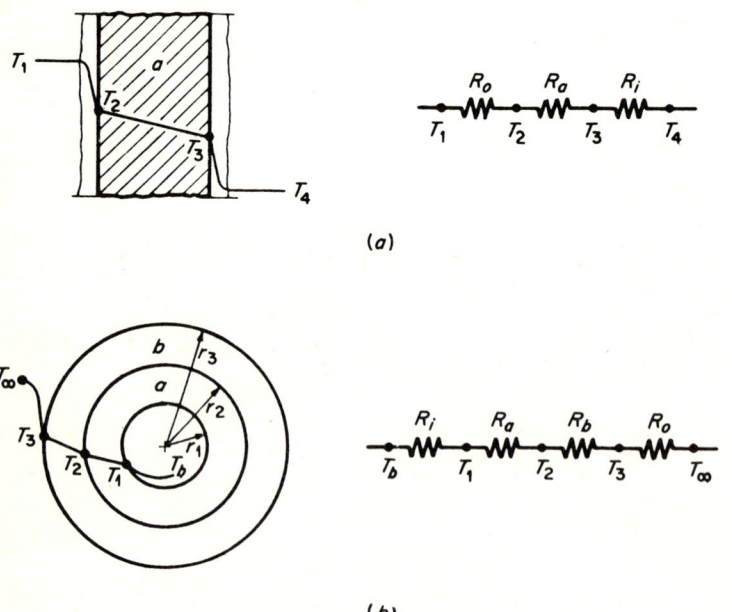

Fig. 5-8 One-dimensional heat flow with convective boundaries: (a) plane wall; (b) cylindrical wall.

STEADY-STATE CONDUCTIVE HEAT TRANSFER

By Eq. (5-41) we have

$$\frac{q}{A} = \bar{h}_o(T_1 - T_2) = \bar{h}_i(T_3 - T_4)$$

and thus

$$q = \frac{T_1 - T_2}{1/\bar{h}_o A} = \frac{T_3 - T_4}{1/\bar{h}_i A}$$

which indicates that each $1/\bar{h}A$ is interpreted as the thermal resistance of a single convective boundary. Coupling these with the conductive resistance within the wall, Eq. (5-16) can be applied to the present problem to yield

$$\frac{q}{A} = \frac{T_1 - T_4}{1/\bar{h}_o + L_a/k_a + 1/\bar{h}_i} \tag{5-42}$$

It is sometimes advantageous to express the heat flux per unit area in terms of the *overall heat-transfer coefficient U* defined by

$$U \equiv \frac{1}{A \Sigma R_{\text{th}}} \tag{5-43}$$

for any geometry, and thus

$$\frac{q}{A} = U(\Delta T)_{\text{overall}} \tag{5-42a}$$

For the plane wall of Fig. 5-8a we obtain

$$U = \frac{1}{1/\bar{h}_o + L_a/k_a + 1/\bar{h}_i} \tag{5-44}$$

by inspection of Eq. (5-42). For the general case of a multilayered plane wall with inner and outer convective boundaries

$$U = \frac{1}{1/\bar{h}_o + L_a/k_a + L_b/k_b + \cdots + 1/\bar{h}_i} \tag{5-44a}$$

For a cylindrical problem the overall coefficient depends upon the area chosen for use in Eq. (5-43) or (5-42a). In most cases, the outside area (see Fig. 5-8b) of a cylinder is selected as the base area for analysis, and the electrical analog for a two-layer cylinder with specified inner and outer heat-transfer coefficients and fluid temperatures is as shown in Fig. 5-8b. The overall coefficient for this problem is

$$U_o = \frac{1}{\dfrac{r_3}{r_1 \bar{h}_i} + \dfrac{r_3 \ln(r_2/r_1)}{k_a} + \dfrac{r_3 \ln(r_3/r_2)}{k_b} + \dfrac{1}{\bar{h}_o}} \tag{5-45}$$

where the subscript denotes that U_o is applicable with area based on the outer cylinder surface. The extension to a multilayered cylindrical system is

$$U_o = \frac{1}{\dfrac{r_n}{r_i \bar{h}_i} + \dfrac{r_n \ln(r_2/r_1)}{k_{1-2}} + \cdots + \dfrac{r_n \ln(r_n/r_{n-1})}{k_{(n-1)-n}} + \dfrac{1}{\bar{h}_o}} \qquad (5\text{-}45a)$$

where n is 1 greater than the number of layers and the subscripts on conductivity indicate the bounding radii of a material layer.

5-5 HEAT TRANSFER FROM FINS

In this section we shall consider heat transfer from various forms of surface extensions employed to increase the heat transfer to or from a wall. Such designs are quite common in engineering applications and are widely used in heat exchangers.

Uniform cross section Perhaps the simplest form of an extended surface is a straight rod or pin of uniform cross-sectional area attached to a plane wall, as illustrated in Fig. 5-9. Considering the element of length Δx removed from the rod, we can write an energy balance for steady-state conditions as follows:

Energy in at x = (energy out at $x + \Delta x$) + energy out along exposed surface

Now assuming the *radial* temperature gradients to be negligible, we can express the conductive energy fluxes in terms of the temperature gradient in the x direction, and for a uniform cross-sectional area A and circumference P we have

Energy in at $x = -kA \left.\dfrac{dT}{dx}\right|_x$

Energy out at $x + \Delta x = -kA \left.\dfrac{dt}{dx}\right|_{x+\Delta x}$

Energy out along exposed surface $= \bar{h}(P\Delta x)(T - T_\infty)$

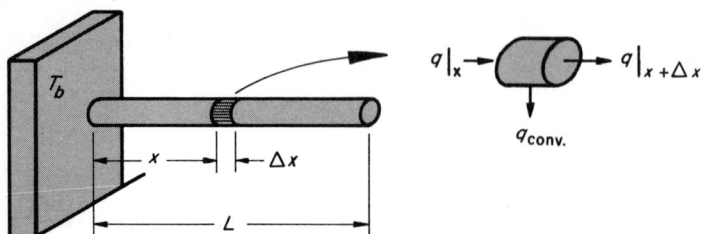

Fig. 5-9 Rod-type extended surface.

STEADY-STATE CONDUCTIVE HEAT TRANSFER

In the last expression, T_∞ is the surrounding-fluid temperature at a distance sufficiently removed to be unaffected by the rod temperature. Substituting these expressions into the energy balance, we obtain by taking the limit as $\Delta x \to 0$

$$\frac{d^2 T}{dx^2} - \frac{\bar{h}P}{kA}(T - T_\infty) = 0 \tag{5-46}$$

Letting $\theta = T - T_\infty$ and $m = \sqrt{\bar{h}P/kA} = \sqrt{2\bar{h}/kr}$, this can be written

$$\frac{d^2\theta}{dx^2} - m^2\theta = 0 \tag{5-46a}$$

which is the standard form of an ordinary second-order differential equation with constant coefficients. The general solution is

$$\theta = C_1 e^{mx} + C_2 e^{-mx} \tag{5-47}$$

An immediate boundary condition is the base or wall temperature; thus

$$\text{At } x = 0: \quad \theta = T_b - T_\infty = \theta_b \tag{5-48}$$

The second boundary condition depends upon the physical situation, and of the many possibilities, two will be considered here:

Case 1 A very long pin with the temperature at the end essentially the same as that of the fluid T_∞
Case 2 A pin of finite length losing heat at its end by convection

The first of these is not common but affords considerable mathematical simplicity in the solution which follows. For an infinitely long pin, the second boundary condition becomes

$$\text{At } x = \infty: \quad \theta = 0 \tag{5-49}$$

From the boundary condition of Eq. (5-49), we see that $C_1 = 0$. Then applying Eq. (5-48), we obtain $C_2 = \theta_b$, and the complete solution for case 1 is

$$\theta = \theta_b e^{-mx} \tag{5-50}$$

The heat transfer from the pin can be evaluated by either of two methods since the energy enters the pin by conduction at the base and leaves by convection along the surface. Thus, the total heat transfer from the pin can be found from

$$q = -kA \left.\frac{dT}{dx}\right|_{x=0} = -kA \left.\frac{d\theta}{dx}\right|_{x=0} = -kA(-m\theta_b e^{-m(0)}) = kAm\theta_b$$

But $m = \sqrt{\bar{h}P/kA}$; hence,

$$q = \sqrt{\bar{h}PkA}\,\theta_b \tag{5-51}$$

The same result can easily be found by integrating the convective heat transfer over the area of the pin.

Turning to the pin of finite length, case 2, the second boundary condition is

$$-k\frac{dT}{dx}\bigg|_{x=L} = \bar{h}_L \theta\big|_{x=L} \qquad (5\text{-}52)$$

where $\bar{h}_L$ is the convective coefficient applicable to the rod end. The determination of the constants C_1 and C_2 for this case is slightly more difficult algebraically than for case 1. This problem is treated in detail by Chapman [4], who shows that the complete solution for the temperature distribution is

$$\frac{\theta}{\theta_b} = \frac{\cosh m(L-x) + (\bar{h}_L/mk)\sinh m(L-x)}{\cosh mL + (\bar{h}_L/mk)\sinh mL} \qquad (5\text{-}53)$$

and the heat flux from the pin is

$$q = \sqrt{hPkA}\,\theta_b\,\frac{\sinh mL + (\bar{h}_L/mk)\cosh mL}{\cosh mL + (\bar{h}_L/mk)\sinh mL} \qquad (5\text{-}54)$$

It is important to note the difference between $\bar{h}$ and $\bar{h}_L$. For an extended surface of this configuration, it is obvious that the velocity which is normally perpendicular to the rod would not be uniform around the circumference. Thus, an average value for the heat-transfer coefficient along the rod would probably differ from the value for the end surface, which would be subjected to a uniform velocity.

Since this entire analysis is for a one-dimensional temperature gradient, we are restricted to pins of small diameter. (A large diameter would result in a significant radial temperature gradient.) This introduces a question regarding the significance of the end convective-heat-loss term in case 2 since the surface area A must be quite small. Frequently this term is ignored completely, and the resulting temperature distribution is

$$\frac{\theta}{\theta_b} = \frac{\cosh m(L-x)}{\cosh mL} \qquad (5\text{-}55)$$

with the corresponding heat flux

$$q = \sqrt{hPkA}\,\theta_b\,\tanh mL \qquad (5\text{-}56)$$

Although this is sometimes referred to as the case of an insulated end, it is a misnomer since insulation is not usually applied to any part of an extended surface.

Turning to the case of a *rectangular* fin of constant cross-sectional area, i.e., a *straight* rectangular fin, as depicted in Fig. 5-10, we can apply an approach similar to that used for the rod, and this yields Eq. (5-46). Again we restrict our consideration to the case of small thickness t, and the temperature distribution is obviously one-dimensional (a function of x only). Since the width w is then much larger than

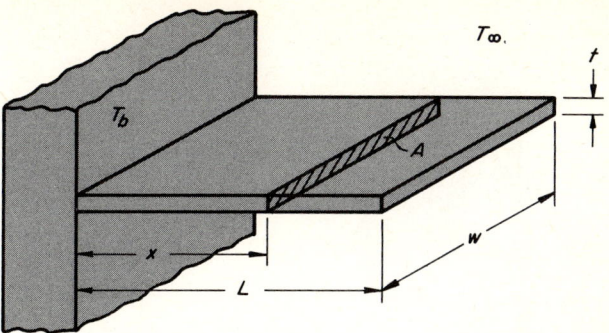

Fig. 5-10 Straight rectangular fin.

the thickness t, the fin perimeter is

$$P \simeq 2w$$

The cross-sectional area is

$$A = wt$$

and consequently

$$m \simeq \sqrt{\frac{2\bar{h}}{kt}}$$

Then the preceding equations for the temperature distribution and the heat flux, Eqs. (5-50) to (5-56), with the appropriate boundary conditions, apply to the rectangular fin as well as to the rod with the appropriate boundary conditions. The only difference is that $m^2 = 2\bar{h}/kt$ for the fin whereas $m^2 = 2\bar{h}/kr$ for the rod or pin.

Nonuniform cross section The mathematical analysis of the previous section was greatly simplified by the fact that both the area and the perimeter of the fin are independent of the x position. In many applications, this is not the case. Consider, for example, the annular fin of uniform thickness, as shown in Fig. 5-11. In this case, the cross-sectional area and the perimeter are functions of radial position. The result of an energy balance on an elemental control volume of radial dimension Δr is the differential equation

$$\frac{d^2\theta}{dr^2} + \frac{1}{r}\frac{d\theta}{dr} - \frac{2\bar{h}}{kt}\theta = 0 \tag{5-57}$$

where

$$\theta = T - T_\infty$$

and the major assumptions are (1) the temperature distribution is symmetrical with respect to the angular coordinate and (2) the thickness t is much smaller than the

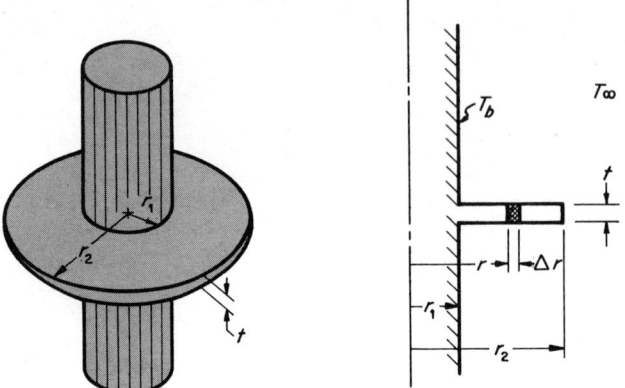

Fig. 5-11 Annular fin of uniform thickness.

effective length, $r_2 - r_1$. For details of the derivation the reader is referred to Chapman [4] or to Schneider [9]. At this point we note that Eq. (5-57) is slightly more complicated than that obtained for uniform area. This is Bessel's differential equation of order zero and has the solution

$$\theta = BI_0(mr) + CK_0(mr) \qquad (5\text{-}58)$$

where I_0 = modified Bessel function of first kind
K_0 = modified Bessel function of second kind
$m^2 = 2\bar{h}/kt$
B, C = constants (to be determined by boundary conditions)

It is important to keep in mind that the product of m and r is dimensionless and that tabulated values of the Bessel functions are available for a wide range of numerical arguments.

The two constants B and C are determined from the boundary conditions, one of which is

$$\text{At } r = r_1: \qquad \theta = T_b - T_\infty = \theta_b \qquad (5\text{-}59)$$

A second practical boundary condition is obtained by noting that the heat loss from the end may usually be neglected, and thus

$$\text{At } r = r_2: \qquad \frac{d\theta}{dr} = 0 \qquad (5\text{-}60)$$

This is more frequently true for the annular fin than for the rectangular surface since the surface area is increasing with r, resulting in a higher dissipation per unit of radial distance than the heat loss per unit of length for the rectangular fin.

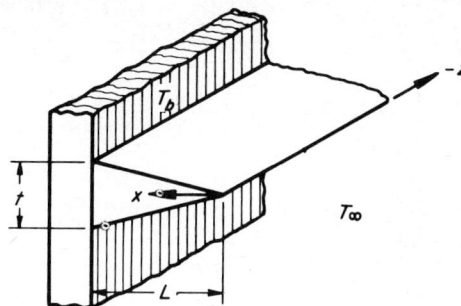

Fig. 5-12 Straight fin with rectangular profile.

Determining the constants and substituting them into Eq. (5-58) yields

$$\frac{\theta}{\theta_b} = \frac{I_0(mr)K_1(mr_2) + K_0(mr)I_1(mr_2)}{I_0(mr_1)K_1(mr_2) + K_0(mr_1)I_1(mr_2)} \tag{5-61}$$

Differentiating θ with respect to r and using this evaluated at r_1 with Fourier's equation to determine the heat flux into the fin, we find

$$q = 2\pi kmt\theta_b \frac{K_1(mr_1)I_1(mr_2) - I_1(mr_1)K_1(mr_2)}{K_0(mr_1)I_1(mr_2) + I_0(mr_1)K_1(mr_2)} \tag{5-62}$$

A short table of values of Bessel functions is included in Appendix A.

Another fin design of nonuniform cross-sectional area less common than the annular fin is the *triangular* or *tapered* fin attached to a plane or straight wall, as shown in Fig. 5-12. The solution for this case as given by Chapman [4] is

$$\frac{\theta}{\theta_b} = \frac{I_0(2sx^{\frac{1}{2}})}{I_0(2sL^{\frac{1}{2}})} \tag{5-63}$$

where

$$s = \sqrt{\frac{2fhL}{kt}} \quad \text{and} \quad f = \sqrt{1 + \left(\frac{t}{2L}\right)^2}$$

The heat transfer per unit width of the fin (in the z direction) is found, by application of Fourier's equation, to be

$$q = \frac{-kt\theta_b s I_1(2sL^{\frac{1}{2}})}{L^{\frac{1}{2}} I_0(2sL^{\frac{1}{2}})} \tag{5-64}$$

There are numerous other fin configurations, including straight and curved fins of parabolic and trapezoidal profile, and spines. For a more detailed treatment the reader should consult Ref. 9.

Fin efficiency A highly simplified method of calculating the heat transfer from a fin is afforded by the concept of fin efficiency. As an indication of the effectiveness of a fin in transferring heat, the fin efficiency, defined by

$$\eta_f = \frac{\text{actual heat transferred}}{\text{heat that would be transferred if entire fin were at base temperature}}$$

is introduced. This is a rather obvious choice, since in numerical calculations it permits the designer to compute the heat transfer from a finned surface as

$$q = \bar{h}(A_w + \eta_f A_f)\Delta T \tag{5-65}$$

where A_w is the area of wall, tube, etc., between fins and A_f is the total surface area of the fins.

This is illustrated in Fig. 5-13. A_w is the outer surface area of the tube exposed to the surrounding fluid; i.e., per unit length this is the circumference minus nt, where n is the number of radial fins of thickness t. The A_f is the actual fin surface per unit length (of n fins).

For the case of a rod or rectangular fin with negligible heat loss at the end, the efficiency becomes

$$\eta_f = \frac{\sqrt{\bar{h}PkA}\,\theta_b \tanh mL}{\bar{h}PL\theta_b} = \frac{\tanh mL}{mL} \tag{5-66}$$

For width w (in the z direction),

$$mL = \sqrt{\frac{\bar{h}(2w + 2t)}{kwt}}\,L$$

and for most practical applications

$$2w + 2t \simeq 2w$$

Thus,

$$mL = L\sqrt{\frac{2\bar{h}}{kt}} = L^{\frac{3}{2}}\sqrt{\frac{2\bar{h}}{kLt}}$$

but Lt is the *profile* area A_p of the fin, and consequently

$$mL \simeq L^{\frac{3}{2}}\left(\frac{2\bar{h}}{kA_p}\right)^{\frac{1}{2}} \tag{5-67}$$

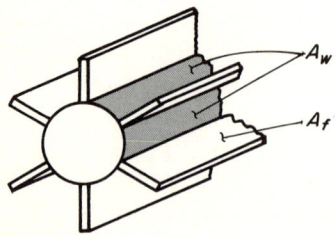

Fig. 5-13 Areas for use in Eq. (5-65).

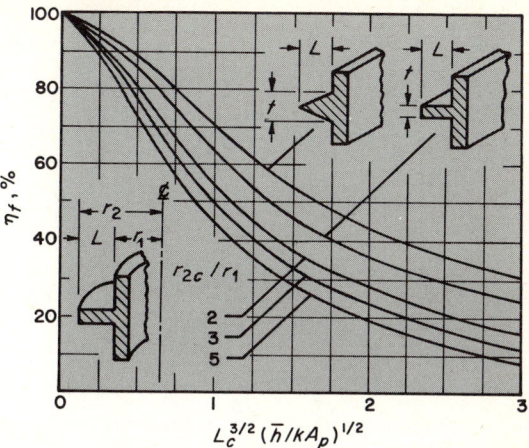

Fig. 5-14 Efficiencies of fins.

We could use this with Eq. (5-66) to calculate the fin efficiency with negligible end heat loss. It would be more useful, however, to have an expression for fin efficiency which accounts for heat loss at the tip. Harper and Brown [5] have shown that a corrected length

$$L_c = L + \frac{t}{2} \tag{5-68}$$

can be used to modify the equations developed for no heat loss at the tip with reasonable accuracy. For limitations on this, the reader should consult Ref. 5. Intuitively, one would suspect that a very thick fin with a large tip area or an unusually high heat-transfer coefficient at the end would invalidate this simplified approach.

Using the *corrected* length, efficiencies have been plotted for the triangular, the rectangular, and the annular fins in Fig. 5-14.

Note that the profile area A_p is the product of $L_c t$ for the rectangular and annular fins and is the product of $L_c t/2$ for the triangular fin. Also L_c is given by Eq. (5-68) for the rectangular and annular fins, whereas there is no correction for the triangular case; that is, $L_c = L$.

Example A rectangular fin 0.05 in thick and 2 in long extends from a plane wall. The fin material is mild steel ($k = 26$ Btu/hr-ft-°F), and the external heat-transfer coefficient $\bar{h}$ may be taken as 10 Btu/hr-ft²-°F. The surrounding air temperature is 80°F, and the wall temperature is 300°F. Calculate the heat loss from the fin.

Solution For this finite length, the situation corresponds to case 2 for the boundary conditions, and the heat flux can be found by Eq. (5-54). For this problem, however, we note that the tip area per unit width is 0.05 in², whereas the area of the sides (top and bottom) is 4 in² per inch of depth. Thus, the heat transfer from the tip may be ignored, and the heat flux can be calculated using the simpler solution of Eq. (5-56). Hence,

$$q = \sqrt{hPkA}\, \theta_b \tanh mL$$

For a 1-ft depth,

$$A = \frac{0.05}{12}(1) = 0.00416 \text{ ft}^2$$

$$P = 2\left(1 + \frac{0.05}{12}\right) \text{ ft} \simeq 2 \text{ ft}$$

$$m \simeq \sqrt{\frac{2h}{kt}} = \sqrt{\frac{2(10)(12)}{26(0.05)}} = \sqrt{185} = 13.6$$

Thus,

$$q = \sqrt{10(2)(26 \times 0.00416)}(300 - 80) \tanh\left[13.6\left(\tfrac{2}{12}\right)\right]$$

$$= 316.3 \text{ Btu/hr-ft}$$

If we use Eq. (5-68) to account for heat loss at the tip, we obtain

$$q = \sqrt{20(26)(0.00416)}(300 - 80) \tanh\left(13.6 \,\frac{2.025}{12}\right)$$

$$= 317.0 \text{ Btu/hr-ft}$$

which is slightly more accurate.

Example Annular copper 3-in-OD fins are placed on a 1-in-OD tube. The fins are 0.10 in thick. The tube surface is at 450°F, and the surrounding fluid temperature is 70°F. Find the heat loss from each fin if $\bar{h} = 240$ Btu/hr-ft²-°F.

Solution The easiest way to determine the heat flux from an annular fin is to use Fig. 5-14. The necessary parameters are

$$L_c = 1.50 - 0.5 + \frac{0.10}{2} = 1.05 \text{ in} = 0.0875 \text{ ft}$$

$$\frac{r_{2c}}{r_1} = \frac{1.50 + 0.10/2}{0.5} = \frac{1.55}{0.5} = 3.1$$

$$A_p = t(r_{2c} - r_1) = \frac{0.10(1.55 - 0.5)}{144} = 7.3 \times 10^{-4} \text{ ft}^2$$

$$L_c^{3/2} \sqrt{\frac{h}{kA_p}} = 0.0248 \sqrt{\frac{240 \times 10^4}{215 \times 7.3}} = 1.01$$

From Fig. 5-14, $n_f = 49$ percent. If the entire fin were at the base temperature, 450°F, the heat flux would be

$$q_{450} = 2\bar{h}\pi(r_{2c}^2 - r_1^2)\,\Delta T$$

$$= 2(240)(\pi)\,\frac{2.4 - 0.25}{144}(450 - 70) = 8560 \text{ Btu/hr}$$

whereas, because the temperature diminishes with radius, the actual heat flux is $n_f q_{450}$, or $q = 0.49(8560) = 4194$ Btu/hr.

STEADY-STATE CONDUCTIVE HEAT TRANSFER

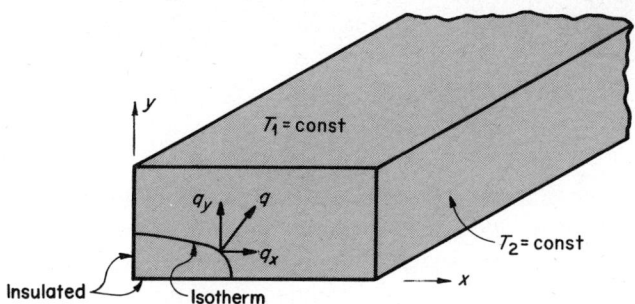

Fig. 5-15 Two-dimensional steady-state conductive heat transfer.

5-6 MULTIDIMENSIONAL SYSTEMS

So far we have been concerned with conductive problems in which there was a significant temperature gradient in only one spatial direction. We now focus our attention on the more general case, which may involve thermal gradients in two or three directions. For steady state and constant thermal conductivity, the appropriate differential equation describing the thermal field is Laplace's equation, which in cartesian coordinates is

$$\frac{\partial^2 T}{\partial x^2} + \frac{\partial^2 T}{\partial y^2} + \frac{\partial^2 T}{\partial z^2} = 0 \tag{5-8}$$

The increased mathematical difficulty in solving for the temperature gradients is immediately obvious: we must obtain solutions to a second-order *partial* differential equation, whereas in one-dimensional, steady-state systems an *ordinary* differential equation is involved.

It is important to keep our primary purpose in mind. We require the temperature gradients to determine the flux by Fourier's equation. As an example, consider a very long rectangular bar subjected to the surface conditions shown in Fig. 5-15. Clearly, the heat flux can be calculated vectorially if we can obtain $\partial T/\partial x$ and $\partial T/\partial y$.

There are several methods of determining the temperature distribution in a two- or three-dimensional system. These include (1) analytical, (2) graphical, (3) analogical, and (4) numerical. None of these constitutes a satisfactory approach for all engineering problems; but since each is appropriate for a certain class of problems, these methods may be said to *complement* each other. The remainder of this chapter will be devoted to a study of these four methods.

5-7 ANALYTICAL SOLUTIONS

Consider a long rectangular bar of homogeneous conductive material with surfaces held at constant temperatures, similar to that shown in Fig. 5-15. For a very long bar (z direction), we may neglect heat transfer along the length, and the problem reduces to the two-dimensional case, where the temperature must satisfy

$$\frac{\partial^2 T}{\partial x^2} + \frac{\partial^2 T}{\partial y^2} = 0 \tag{5-69}$$

if the material has uniform thermal conductivity. We shall use the classical *separation-of-variables technique* in solving this problem. Equation (5-69) is a linear second-order partial differential equation. The solution requires two boundary conditions for each independent variable (direction), i.e., the same number as the order of the highest derivative in each independent variable. Let us arbitrarily impose the boundary conditions shown in Fig. 5-16.

Substituting

$$\theta = T - T_{\text{ref}} \tag{5-70}$$

into Eq. (5-69), we obtain

$$\frac{\partial^2 \theta}{\partial x^2} + \frac{\partial^2 \theta}{\partial y^2} = 0 \tag{5-71}$$

We first consider the case where the boundary condition along the right edge is an imposed sine-wave temperature distribution. The boundary conditions then are

$$\theta = \begin{cases} 0 & \text{at } y = 0 \\ 0 & \text{at } x = 0 \\ 0 & \text{at } y = b \\ \theta_m \sin \dfrac{\pi y}{b} & \text{at } x = w \end{cases} \tag{5-72}$$

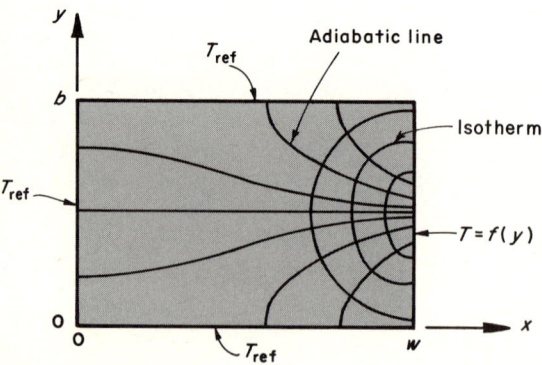

Fig. 5-16 Section of rectangular bar with isotherms.

STEADY-STATE CONDUCTIVE HEAT TRANSFER

and the problem is completely specified. The solution to Eq. (5-71) is obtained by assuming a product solution

$$\theta = XY \tag{5-73}$$

where $X = X(x)$ and $Y = Y(y)$. Substituting the assumed form of θ into Eq. (5-71) yields

$$-\frac{1}{X}\frac{d^2X}{dx^2} = \frac{1}{Y}\frac{d^2Y}{dy^2} \tag{5-74}$$

and the variables are separated. Since the right side is a function of y alone and the left side is a function of x alone, each side must be equal to some constant in order for the equality of this last expression to hold for all changes of x and y (which are independent of each other). Letting the separation constant be $-\lambda^2$, we have

$$\frac{d^2X}{dx^2} - \lambda^2 X = 0 \tag{5-75}$$

$$\frac{d^2Y}{dy^2} + \lambda^2 Y = 0 \tag{5-76}$$

The solutions to these two equations depend upon the value of λ^2 (zero, negative, or positive). It can easily be shown that $\lambda^2 = 0$ or $\lambda^2 < 0$ will yield a product solution that cannot satisfy the imposed sine-function boundary condition. The details of this are left as an exercise. The remaining possibility is $\lambda^2 > 0$, and for this the solutions to Eqs. (5-75) and (5-76) are

$$X = C_1 e^{-\lambda x} + C_2 e^{\lambda x} \tag{5-77}$$

$$Y = C_3 \cos \lambda y + C_4 \sin \lambda y$$

Consequently, the general solution becomes

$$\theta = (C_1 e^{-\lambda x} + C_2 e^{\lambda x})(C_3 \cos \lambda y + C_4 \sin \lambda y) \tag{5-78}$$

The constants are determined by application of the boundary conditions of (5-72). From these we have

Condition 1: $\quad 0 = (C_1 e^{-\lambda x} + C_2 e^{\lambda x})C_3$

Condition 2: $\quad 0 = (C_1 + C_2)(C_3 \cos \lambda y + C_4 \sin \lambda y)$

Condition 3: $\quad 0 = (C_1 e^{-\lambda x} + C_2 e^{\lambda x})(C_3 \cos \lambda b + C_4 \sin \lambda b)$

Condition 4: $\quad \theta_m \sin \dfrac{\pi y}{b} = (C_1 e^{-\lambda w} + C_2 e^{\lambda w})(C_3 \cos \lambda y + C_4 \sin \lambda y)$

From the first of these,

$$C_3 = 0$$

from the second

$$C_1 = -C_2$$

and using these, together with the third, yields

$$0 = C_2(e^{\lambda x} - e^{-\lambda x})C_4 \sin \lambda b = C \frac{e^{\lambda x} - e^{-\lambda x}}{2} \sin \lambda b = C \sinh \lambda x \sin \lambda b$$

For this to be true for all values of x,

$$\sin \lambda b = 0 \tag{5-79}$$

which holds if

$$\lambda = \frac{n\pi}{b} \tag{5-80}$$

where n is any integer. Recalling that the original differential equation (5-71) is linear, the principle of *superposition* holds; i.e., the sum of any number of solutions constitutes a solution, and consequently we can express the solution as the sum of an infinite series

$$\theta = \sum_{n=1}^{\infty} C_n \sinh \frac{n\pi x}{b} \sin \frac{n\pi y}{b} \tag{5-81}$$

since a different solution exists for each integer n and each solution has its own separate constant of integration C_n. Then, by the fourth boundary condition,

$$\theta_m \sin \frac{\pi y}{b} = \sum_{n=1}^{\infty} C_n \sinh \frac{n\pi w}{b} \sin \frac{n\pi y}{b}$$

By examination of the sine terms on each side of the equation, we see that $n = 1$ is the only integer permitted, and

$$C_n = \theta_1 = \frac{\theta_m}{\sinh (\pi w/b)} \tag{5-82}$$

The final solution is

$$\theta = \theta_m \frac{\sinh (\pi x/b)}{\sinh (\pi w/b)} \sin \frac{\pi y}{b} \tag{5-83}$$

and this temperature distribution is sketched in Fig. 5-16.

At this point it is informative to consider the choice of the boundary condition at $x = w$. The sine-wave form may have appeared to be rather arbitrary at first, but it is clear that this choice permitted a final solution without the use of an infinite series. Consider what would appear to be the simplest nontrivial boundary condition at $x = w$, namely, a constant temperature θ_m. Then the determination of C_n requires the expansion of θ_m in a Fourier series over the interval $0 < y < b$. The

STEADY-STATE CONDUCTIVE HEAT TRANSFER

details will be found in Schneider [9], and the resulting temperature distribution is

$$\frac{\theta}{\theta_m} = \frac{2}{\pi} \sum_{n=1}^{\infty} \frac{(-1)^{n+1}+1}{n} \sin\frac{n\pi y}{b} \frac{\sinh(n\pi x/b)}{\sinh(n\pi w/b)} \qquad (5\text{-}84)$$

More generally, if the temperature at the surface $x = w$ is a function of y, $\theta = \theta(y,w) = F(y)$, then the solution as shown by Schneider is

$$\theta = \frac{2}{b} \sum_{n=1}^{\infty} \frac{\sinh(n\pi x/b)}{\sinh(n\pi w/b)} \sin\frac{n\pi y}{b} \int_0^b F(y) \sin\frac{n\pi y}{b} \, dy \qquad (5\text{-}85)$$

Example Consider a long square bar (Fig. 5-17) of homogeneous composition (uniform properties) with the boundary conditions $\theta = 100°F$ on the top surface and $\theta = 0$ on all other xz and yz surfaces. Find the temperature along the centerline of the bar.

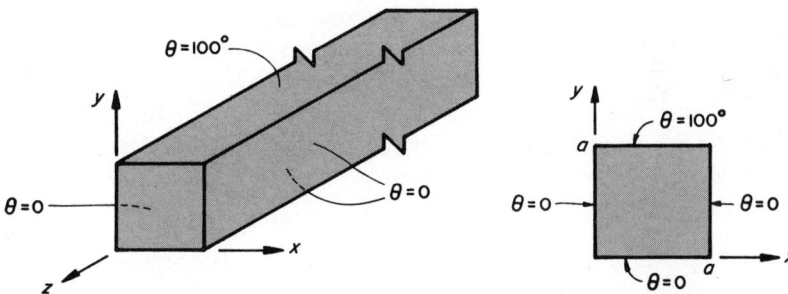

Fig. 5-17 Long square bar.　　**Fig. 5-18** Boundary conditions.

Solution Since the bar is very long, we may assume there is no z-direction thermal gradient, and the problem is two-dimensional. A vertical slice of the bar with the appropriate differential equation and boundary conditions (Fig. 5-18) is

$$\frac{\partial^2 \theta}{\partial x^2} + \frac{\partial^2 \theta}{\partial y^2} = 0$$

and the solution can be obtained with Eq. (5-84) modified to account for the relocation of the constant-temperature boundary. This is

$$\frac{\theta}{\theta_m} = \frac{2}{\pi} \sum_{n=1}^{\infty} \frac{(-1)^{n+1}+1}{n} \sin\frac{n\pi x}{w} \frac{\sinh(n\pi y/w)}{\sinh(n\pi b/w)}$$

But $b = w = a$; therefore at $y = x = a/2$

$$\frac{\theta}{\theta_m} = \frac{2}{\pi} \sum_{n=1}^{\infty} \frac{(-1)^{n+1}+1}{n} \sin\frac{n\pi}{2} \frac{\sinh(n\pi/2)}{\sinh n\pi}$$

$n = 1$: $\quad \sin\dfrac{\pi}{2} \dfrac{\sinh(\pi/2)}{\sinh \pi} = \dfrac{1(2.3013)}{11.5487} = 0.199$

$n = 2$: $\quad \sin \pi = 0$

$n = 3$: $\quad \sin\dfrac{3\pi}{2} \dfrac{\sinh(3\pi/2)}{\sinh 3\pi} \simeq \dfrac{-1(56)}{6{,}254} = -0.00895$

Truncating the series at $n = 3$,

$$\frac{\theta}{\theta_m} \simeq \frac{2}{\pi}[2(0.199) - \tfrac{2}{3}(0.00895)]$$

$$\simeq 0.647(0.398 - 0.00597) = 0.2498$$

While the preceding method illustrates the general approach in solving two-dimensional problems, a real insight into the principle of superposition is afforded by considering the temperature at the center of the plane if $\theta = 100°$ along each edge. Then clearly the center temperature would be $\theta = 100°F$. Since the sides are of the same length, it is equally clear that one-fourth of this center temperature is due to the single heated boundary at $y = a$, and for this problem the center temperature due to one heated boundary is

$$\theta(x,y)\big|_{a/2,a/2} = 25°F$$

From the preceding example we conclude that the *principle of superposition* is an important tool for the solution of two-dimensional steady-state conduction problems. This permits the division of a problem into two or more simpler problems having known solutions such as those given by Eqs. (5-83) to (5-85). While a rigorous proof of the validity of the principle of superposition and of the conditions under which it may be used is not our purpose, it is in order to enunciate some practical rules and suggestions.

1. Superposition may be applied to any linear, homogeneous differential equation such as the Laplace equation.
2. The problem is to be divided into a number of simpler subproblems each having only one *nonhomogeneous* boundary condition. (For our purposes we define a nonhomogeneous boundary condition as one for which the dependent variable, namely θ, is not zero.)
3. The number of such simpler subproblems is the same as the number of nonzero boundary conditions in the original complete problem.
4. The existence of a nonhomogeneous boundary condition can be eliminated by choosing an appropriate reference temperature in the definition of θ.
5. The physical geometry of all of the subproblems must be the same as that of the original problem.

The following example will clarify some of these concepts.

Example Determine the centerline temperature of the long bar of the last example if the upper surface is at 200°F, the right-hand vertical surface is at 300°F, and the other two surfaces are at 100°F.

Solution To remove two of the nonhomogeneities, define

$$\theta = T - 100°F$$

and then the complete problem is as shown in Fig. 5-20. Now this problem can be divided into two simpler problems (Fig. 5-21), each having two homogeneous boundary conditions in one direction and one homogeneous and one nonhomogeneous boundary condition in the other direction. For this simplified geometry, the centerline temperatures resulting from

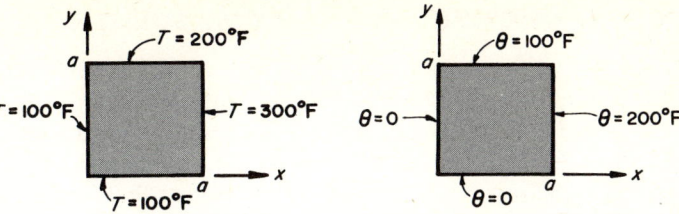

Fig. 5-19 Boundary conditions in T.

Fig. 5-20 Boundary conditions in θ.

subproblems 1 and 2 are, respectively,

$\theta_1 \big|_{a/2,a/2} = 50°F$ and $\theta_2 \big|_{a/2,a/2} = 25°F$

Then, by superposition,

$\theta = \theta_1 + \theta_2 = 75°F$

and

$T = 75°F + 100°F = 175°F$

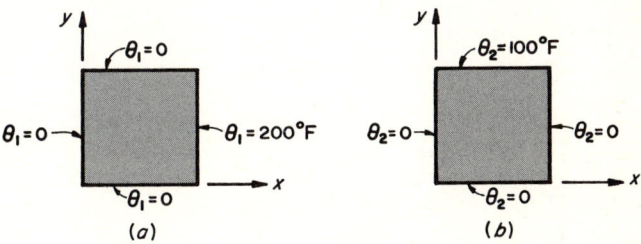

Fig. 5-21 (a) Subproblem 1; (b) subproblem 2.

The extreme simplicity of this last example permits a solution by inspection for each of the subproblems, but this is generally not possible. If, for example, we required the temperature at any point other than the center, it would be necessary to use a series solution such as that given by Eq. (5-84). The application of such analytical solutions is generally quite laborious, and this suggests that other methods may be advantageous in multidimensional conduction problems. The remaining sections of this chapter are devoted to graphical, analogical, and numerical techniques.

As a final comment concerning analytical solutions, it is possible to apply the separation-of-variables technique to three-dimensional problems by the assumption that

$$\theta(x,y,z) = X(x)Y(y)Z(z) \tag{5-86}$$

The general solution results in six constants which must be determined by application of the boundary conditions, this being considerably more complicated than the two-dimensional solution. A number of such problems are treated in Ref. 1.

5-8 CONDUCTIVE SHAPE-FACTOR SOLUTIONS

Consider the two-dimensional system (Fig. 5-22) which results from a heated pipe being encased in a long rectangular insulation jacket. The inside insulation temperature is at T_1, which is higher than T_2, resulting in an outward conductive heat flux. If *uniformly spaced* lines are constructed everywhere perpendicular to the *uniformly spaced* isotherms, the result is a group of heat-flow *lanes*. In Fig. 5-22, there are six such lanes in each quadrant, and if we can determine the heat flow for each lane, we can calculate the total heat transfer. Applying Fourier's law to an element *a-b-c-d* of one of these passages yields for unit depth

$$q = \frac{ky_2(1)(T_{ad} - T_{bc})}{y_1} \qquad (5\text{-}87)$$

If $y_1 = y_2$, the individual elements such as *a-b-c-d* become *curvilinear squares*. Note that y_1 is $(ab + cd)/2$ and y_2 is $(ad + bc)/2$. Since the isotherms are uniformly spaced, the temperature potential across each square becomes

$$\Delta T = \frac{T_1 - T_2}{M} \qquad (5\text{-}88)$$

where M is the number of squares in each flow passage. Then if there are N such flow lanes for the entire configuration, we can express the total heat flux as

$$q = NK\frac{T_1 - T_2}{M} = Sk(T_1 - T_2) \qquad (5\text{-}89)$$

where

$$S = \frac{N}{M} \qquad (5\text{-}90)$$

is known as the *conduction shape factor*. The solution of a two-dimensional heat-conduction problem is reduced to the determination of S. After discussing two

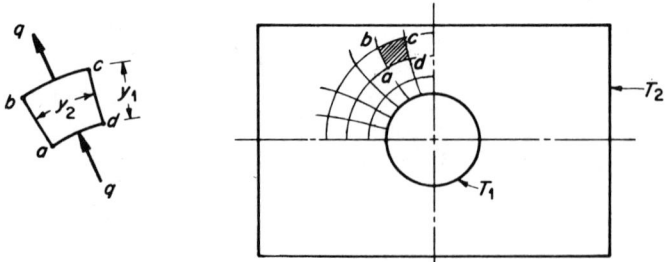

Fig. 5-22 Isotherms and heat-flow lines for conduction-shape-factor analysis.

STEADY-STATE CONDUCTIVE HEAT TRANSFER

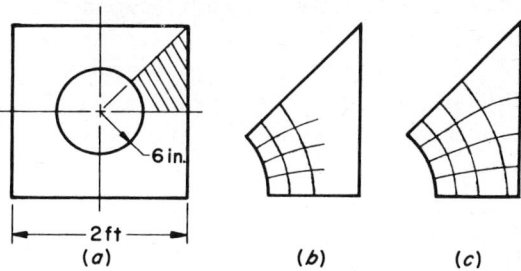

Fig. 5-23 Freehand plotting.

common methods of obtaining this factor, we give a tabulation of some common configurations.

Freehand plotting A practical method of obtaining the conduction shape factor involves the technique of freehand plotting, which is something between an art and a scientific technique. From the preceding introduction it is clear that a plot of equally spaced isotherms and flow lines is sufficient to determine the heat flux. The isotherms may be quite difficult to determine analytically but can frequently be drawn in freehand fashion together with the heat-flow (adiabatic) lines to yield reasonably accurate heat-transfer results.

To illustrate the technique, again consider the problem of a heated pipe in a block of insulation. For simplicity, let the block be of square cross section. This means that the vertical and horizontal centerlines, as well as the diagonals, constitute lines of symmetry. Since a line of physical and thermal symmetry is an adiabatic, or heat-flow, line, we can take advantage of this fact to reduce the work. In the present case, we shall use only one-eighth of the insulation and multiply the resulting S_i by 8. Figure 5-23b illustrates the beginning of the plot. Here, we have arbitrarily chosen to fix the number of heat-flux passages, $N = 4$, and thus we shall determine M.

While there is no complete list of infallible rules for drawing the plot, a number of suggestions are given by Brewley [2], which include:

1. Apply symmetry to divide the field into compartments.
2. Identify all known isotherms.
3. Flow lines bisect corners of isothermal boundaries.
4. If possible, start isotherms in a region where flow lines are uniformly spaced.
5. Begin with a crude network to find approximate orientation of isotherms and adiabatics.
6. Continuously modify to keep the flow lines orthogonal to the isotherms while forming a network of curvilinear squares.

Note that in Fig. 5-23b we began the plot at the inner surface, where flow lines are uniformly spaced. In Fig. 5-23c the network has been completed, and $M \approx 3.5$.

As a check, we could sketch diagonals in the curvilinear squares, and they should also form an orthogonal grid.

With a little practice, this method becomes an important tool. It requires a little patience, a soft lead pencil, a good eraser, and very slight drafting ability. As a final word of caution, *never* use a straightedge in sketching the isotherms or flow lines.

Electrical analog The analogy between thermal and electric flux transport for one-dimensional systems was discussed in Sec. 5-2. The analogy holds equally well in multidimensional systems; the electric potential E satisfies Laplace's equation, which for a two-dimensional problem is

$$\frac{\partial^2 E}{\partial x^2} + \frac{\partial^2 E}{\partial y^2} = 0 \tag{5-91}$$

in the cartesian coordinate system. Since the differential equation is *analogous*, the lines of constant electric potential would be identical with lines of constant thermal potential for similar boundary conditions.

This affords a simple method of experimentally determining the isotherms. Using a geometric model made from a high-resistance conductor, usually an electrically conductive paper such as Teledeltos or one of the conductive papers used in strip chart recorders, an electric field is established by applying suitably attached low voltage dc electrodes. The lines of constant voltage can be determined by means of a null detector, shown in Fig. 5-24, and small pinholes can be punched in the paper to permit filling in the lines. The heat-flow lines can be drawn by constructing them orthogonal to the lines of constant potential or by reversing the analog to apply the potential along the lines of symmetry in Fig. 5-24. The system should be mounted on a rigid insulating board, and equally good results can be obtained with a sensitive dc voltmeter replacing the null detection device.

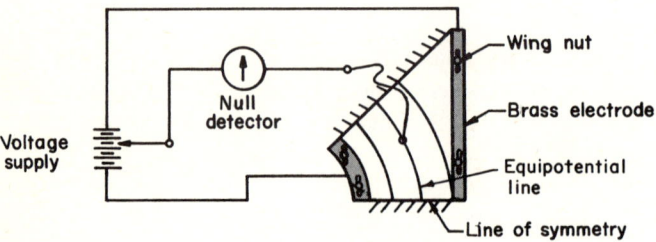

Fig. 5-24 Schematic of electrical analog for two-dimensional conduction.

Table 5-1 Conductive shape factors†

Physical description	Sketch	Shape factor
Conduction through a material of uniform k from a horizontal isothermal cylinder to an isothermal surface	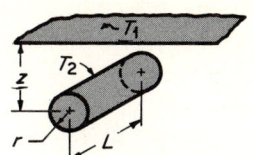	a. Finite length $$S = \frac{2\pi L}{\cosh^{-1}(z/r)} \quad \frac{z}{L} \ll 1$$ b. Infinite length (per unit length) $$\frac{S}{L} = \frac{2\pi}{\cosh^{-1}(z/r)}$$
Conduction in a medium of uniform k from a cylinder of length L to two parallel planes of infinite width and length L	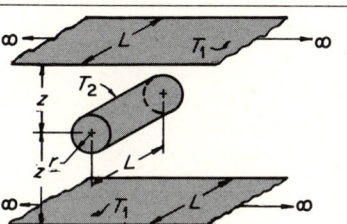	$$S = \frac{2\pi L}{\ln(4z/r)}$$
Conduction from an isothermal sphere through a material of uniform k to an isothermal surface	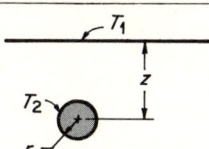	$$S = \frac{4\pi r}{1 - r/2z}$$
Conduction between two long isothermal parallel cylinders in an infinite medium of constant k	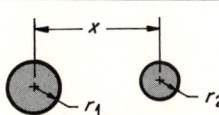	$$\frac{S}{L} = \frac{2\pi}{\cosh^{-1}[(x^2 - r_1^2 - r_2^2)/2r_1 r_2]}$$ $L \gg r$ $L \gg x$
Conduction between a vertical isothermal cylinder in a medium of uniform k and a horizontal isothermal surface	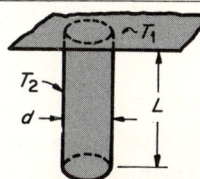	$$S = \frac{2\pi L}{\ln(4L/d)}$$ $L \gg d$
Conduction through an edge formed by intersection of two plane walls with inner wall temperature T_1 and outer wall temperature T_2 as shown‡	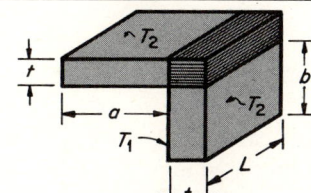	$S = 0.54L$ $a > t/5$ $b > t/5$
Conduction through a corner at intersection of three plane walls, each of thickness t, with uniform inner temperature T_1 and outer temperature T_2	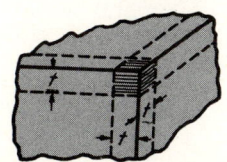	$S = 0.15t$ inside dimensions $> t/5$

† Summarized from J. E. Sunderland and K. R. Johnson, *Trans. ASHRAE*, **10**: 238–239 (1964).
‡ S for the plane wall is simply A/t, where A for the top wall shown is $A = aL$, for side wall, $A = bL$.

The objective of the electrical analog is the determination of the conductive shape factor S. The choice between freehand plotting and this method is usually dictated by accuracy and time requirements. With existing conducting-sheet analog equipment, results are generally obtained faster and are slightly more accurate than those obtained by the freehand method.

Extensions of the electrical-analog method to account for finite resistances simulating convection along boundaries as well as nonuniform thermal conductivity are given by Kayan [7, 8].

Shape factors for some geometrical configurations and thermal conditions of practical importance have been determined and are presented in Table 5-1.

Example A 4-in-diam steam line and a 2-in-diam chilled-water line for air conditioning are located horizontally 6 in apart in a large service trough packed with rock wool insulation, $k = 0.025$ Btu/hr-ft-°F. The steam-line surface temperature is 280°F, and the chilled-water-line surface temperature is 40°F. Calculate the heat transfer to the water for 40 lineal feet of piping.

Solution From Table 5-1, the shape factor is

$$\frac{S}{L} = \frac{2\pi}{\cosh^{-1}[(x^2 - r_1^2 - r_2^2)/2r_1r_2]}$$

$$x = \tfrac{6}{12} \text{ ft} \qquad r_1 = \tfrac{2}{12} \text{ ft} \qquad r_2 = \tfrac{1}{12} \text{ ft}$$

$$\frac{x^2 - r_1^2 - r_2^2}{2r_1r_2} = \frac{\tfrac{36}{144} - \tfrac{4}{144} - \tfrac{1}{144}}{2[\tfrac{2}{12}(\tfrac{1}{12})]} = \frac{31}{6} = 5.16$$

$$\frac{S}{L} = \frac{2\pi}{\cosh^{-1} 5.16} = 2.695$$

Thus,

$$q = \frac{S}{L} k \,\Delta T\, L = 2.695(0.025)(240)(40) = 64.7 \text{ Btu/hr}$$

5-9 NUMERICAL SOLUTIONS

The fourth general method of solving conduction problems is the application of a numerical technique to the solution of the temperature differential equation. The availability of high-speed digital computers throughout industry allows the engineer to solve many conduction problems previously considered impossible. Even without the high-speed computer, the numerical approach known as *relaxation* is applicable to many problems for which we are unable to obtain a closed-form mathematical solution. The fundamental numerical approach is essentially the same for the relaxation method as for a digital-computer solution, since the objective of either is the simultaneous solution of a number of resulting *algebraic* equations.

Let us focus our attention on a general two-dimensional conduction problem depicted in Fig. 5-25. This is a section of material subjected to thermal gradients in the x and y directions but for the sake of simplicity having no z-direction temperature gradient. Further, we assume constant properties and no heat generation.

STEADY-STATE CONDUCTIVE HEAT TRANSFER

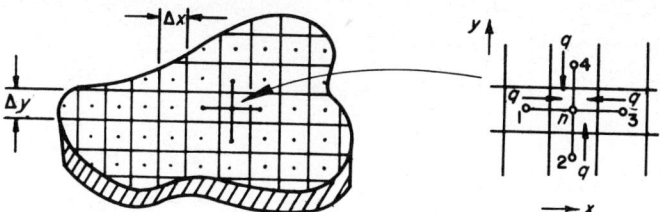

Fig. 5-25 Nomenclature for two-dimensional numerical technique.

convenient treatment is to subdivide the body with a grid network with $\Delta x = \Delta y$ and to identify the center of each square of the grid as a *nodal point*. Then we consider the heat transfer to occur between nodal points through fictitious rods having a conductance for depth b (z direction)

$$kA = k\,\Delta x\,b = k\,\Delta y\,b \tag{5-92}$$

Forming a steady-state energy balance on the interior nodal point n, we have

$$q_{1n} + q_{2n} - q_{3n} - q_{4n} = 0 \tag{5-93}$$

where q_{1n} denotes the conductive flux from nodal point 1 to nodal point n, etc.; thus

$$q_{1n} = k\,\Delta y\,b\,\frac{T_1 - T_n}{\Delta x}$$

Substitution of equations of this form into Eq. (5-93) yields

$$k\,\Delta y\,b\,\frac{T_1 - T_n}{\Delta x} + k\,\Delta x\,b\,\frac{T_2 - T_n}{\Delta y} + k\,\Delta y\,b\,\frac{T_3 - T_n}{\Delta x} + k\,\Delta x\,b\,\frac{T_4 - T_n}{\Delta y} = 0 \tag{5-94}$$

but $\Delta x = \Delta y$, and hence

$$T_1 + T_2 + T_3 + T_4 - 4T_n = 0 \tag{5-95}$$

or

$$T_n = \frac{T_1 + T_2 + T_3 + T_4}{4} \tag{5-96}$$

An equation of this form can be written for each interior nodal point of a body. Simultaneous solution of the resulting algebraic equations yields the temperature distribution. By Fourier's law we can then calculate the heat flux. Obviously, the accuracy of the solution will depend to a large degree upon the grid size selected; the finer the subdivision, the higher the accuracy. With a high-speed computer, the limitation on the grid size is usually the keypunch time required. For the relaxation technique, a larger grid is usually selected. This method will be explained by the following example.

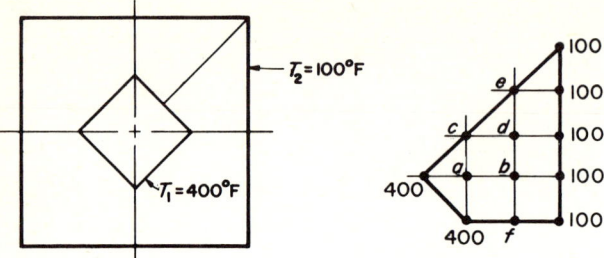

Fig. 5-26 Section of insulation block.

Example Consider the block of insulation material shown in Fig. 5-26 with a square inside duct at 400°F and an outside temperature of 100°F. Find the temperature distribution by the relaxation method.

Solution
1. Subdivide the body by means of a square grid, and identify the interior nodal points as a, b, c, etc. (make use of symmetry).
2. Write Eq. (5-95) for each interior nodal point. Thus,

Node a: $\quad 400 + 400 + T_b + T_c - 4T_a = 0$ (1)
Node b: $\quad T_a + T_f + 100 + T_d - 4T_b = 0$ (2)
Node c:[1] $\quad T_a + T_d - 2T_c = 0$ (3)
Node d: $\quad T_c + T_b + 100 + T_e - 4T_d = 0$ (4)
Node e:[1] $\quad T_d + 100 - 2T_e = 0$ (5)
Node f:[1] $\quad 200 + T_b + 50 - 2T_f = 0$ (6)

Notice that at this point we have six algebraic equations with six unknowns. The relaxation technique for solving these simultaneously proceeds as follows.

3. Assume values for the temperatures at the nodal points. These should be intelligent guesses in order to minimize the work.
4. Since the assumed temperatures will not be highly accurate, the right side of each of the nodal equations will not be zero, but will be a *residual* number due to inaccuracies. Thus we replace Eqs. (1) to (6) with equations of the form

$$400 + 400 + T_b + T_c - 4T_a = R_a \quad (1a)$$

etc., i.e., simply replace the zero with R_i, where i is the node identification. Calculate initial residuals for the assumed temperatures.

5. Set up a table of the form of Table 5-2, which eliminates the necessity of using the residual equations and affords an efficient technique for constructing the relaxation table. Notice that the effect of a unit change in all temperatures has been identified as a "block" effect. This is useful as there are two shortcuts:(1) *over relaxation* and (2) *block relaxation*. The first of these is merely relaxation of a nodal temperature beyond that which would at first appear to be necessary. The block-relaxation technique is advantageous when all (or most) residuals are of the same sign.

---

[1] Since these nodes are on lines of symmetry, only one-half of the heat flow through them can be attributed to the temperature value in the section of body shown.

STEADY-STATE CONDUCTIVE HEAT TRANSFER

Table 5-2 Effect of unit temperature change upon residuals

	R_a	R_b	R_c	R_d	R_e	R_f
$\Delta T_a = +1$	−4	+1	+1	0	0	0
$\Delta T_b = +1$	+1	−4	0	+1	0	+1
$\Delta T_c = +1$	+1	0	−2	+1	0	0
$\Delta T_d = +1$	0	+1	+1	−4	+1	0
$\Delta T_e = +1$	0	0	0	+1	−2	0
$\Delta T_f = +1$	0	+1	0	0	0	−2
Block = +1	−2	−1	0	−1	−1	−1

6. Using the first guessed or assumed nodal temperatures together with the residual equations, calculate the residuals at each nodal point due to the inaccuracies of the assumed values. Begin the relaxation table (see Table 5-3). Then change either the largest residual or apply a block change, whichever is more appropriate. After a number of changes, it is advisable to check the arithmetic by substituting the revised temperatures into the residual equations and thus obtain a check on the residuals. If an error is found, simply write the correct

Table 5-3 Relaxation table

	T_a	R_a	T_b	R_b	T_c	R_c	T_d	R_d	T_e	R_e	T_f	R_f
	330	−60	210	0	250	10	180	−10	150	−20	230	0
$\Delta T_a = -16$	314	4	210	−16	250	−6	180	−10	150	−20	230	0
$\Delta T_e = -11$	314	4	210	−16	250	−6	180	−21	139	2	230	0
$\Delta T_d = -6$	314	4	210	−22	250	−12	174	3	139	−4	230	0
$\Delta T_b = -6$	314	−2	204	2	250	−12	174	−3	139	−4	230	−6
$\Delta T_c = -6$	314	−8	204	2	244	0	174	−9	139	−4	230	−6
$\Delta T_f = -3$	314	−8	204	−1	244	0	174	−9	139	−4	227	0
Block = −1	313	−6	203	0	243	0	173	−8	138	−3	226	1
Check by eqs.			√	√		√		√		√		√
			−6	0		0		−8		−3		1
$\Delta T_d = -2$	313	−6	203	−2	243	−2	171	0	138	−5	226	1
$\Delta T_a = -2$	311	2	203	−4	243	−4	171	0	138	−5	226	1
$\Delta T_c = -3$	311	2	203	−4	243	−4	171	−3	135	1	226	1
$\Delta T_b = -1$	311	1	202	0	243	−4	171	−4	135	1	226	0
$\Delta T_c = -2$	311	−1	202	0	241	0	171	−6	135	1	226	0
$\Delta T_d = -1.5$	311	−1	202	−1.5	241	−1.5	169.5	0	135	−0.5	226	0
$\Delta T_c = -1.0$	311	−2	202	−1.5	240	0.5	169.5	−1	135	−0.5	226	0
Block = −1	310	0	201	−0.5	239	0.5	168.5	0	134	0.5	225	1
$\Delta T_f = +0.5$	310	0	201	0	239	0.5	168.5	0	134	0.5	225.5	0
Check by eqs.			√	√		√		√		√		√
			0	0		0.5		0		0.5		0
Solution	310		201		239		168.5		134		225.5	

residual in the appropriate column and proceed with the calculations. It may be noted from Table 5-3 that the first guess resulted in zero residuals at nodal points b and f. This does not mean that the correct temperatures were assumed but merely that the residuals at these points calculated using the erroneous values at the surrounding nodes were zero and is of no consequence. The computation in this example was stopped when all residuals were reduced to an absolute value of 1.0 or less. This corresponds to a maximum error of $\frac{1}{4}$°F for an interior point and $\frac{1}{2}$°F for a point on a line of symmetry.

The preceding analysis resulting in Eq. (5-95) does not hold for curved boundaries or boundaries with convective conditions. The former is treated by Schneider [9], and the general approach for a convective boundary condition is as follows. An energy balance on nodal point n of Fig. 5-27 is

$$k \, \Delta y \, b \frac{T_1 - T_n}{\Delta x} + k \frac{\Delta x}{2} b \frac{T_2 - T_n}{\Delta y} + k \frac{\Delta x}{2} b \frac{T_3 - T_n}{\Delta y} + h \, \Delta y \, b(T_\infty - T_n) = 0$$

Notice that the area for heat flow from nodal point 2 to node n is $b(\Delta x/2)$; this same area applies between nodes 3 and 0. Again, for $\Delta x = \Delta y$, this reduces to

$$\tfrac{1}{2}(2T_1 + T_2 + T_3) + \frac{h \, \Delta x}{k} T_\infty - \left(\frac{h \, \Delta x}{k} + 2 \right) T_n = 0 \tag{5-97}$$

The solution to the problem would be obtained by using an equation of this type for each nodal point on a convective boundary and one of the type given by Eq. (5-95) for each interior point. Each of these would be written in residual form with the right side replaced by R_i. A summary of useful residual equations is given in Table 5-4. It is rather easy to obtain nodal residual equations for other more specialized cases using the simple energy-balance approach.

The technique of numerical relaxation can be readily extended to three-dimensional problems. The choice between a relaxation and a computer approach is generally one of economics. For nonrepetitive problems it is hardly worthwhile to program the set of residual equations unless very high accuracy is required.

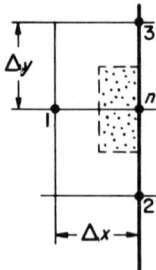

Fig. 5-27 Nomenclature for numerical solution of a problem with a convective boundary.

STEADY-STATE CONDUCTIVE HEAT TRANSFER

Table 5-4 Residual equations for numerical calculations, square grid

Physical description	Nodal residual equation
Convection boundary: 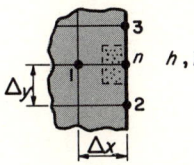	$\frac{1}{2}(2T_1 + T_2 + T_3) + \frac{h\,\Delta x}{k}T_\infty - \left(\frac{h\,\Delta x}{k} + 2\right)T_n = R_n$
Convection boundary at exterior corner: 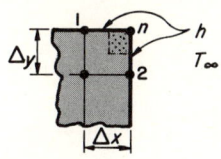	$\frac{1}{2}(T_1 + T_2) + \frac{h\,\Delta x}{k}T_\infty - \left(\frac{h\,\Delta X}{k} + 1\right)T_n = R_n$
Convection boundary at internal corner: 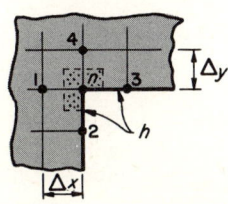	$T_1 + T_4 + \frac{1}{2}(T_2 + T_3) + \frac{h\,\Delta x}{k}T_\infty - \left(\frac{h\,\Delta x}{k} + 3\right)T_n = R_n$
Nodal point along a line of symmetry (insulated boundary): 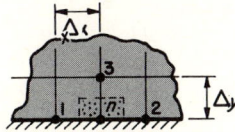	$\frac{1}{2}(T_1 + T_2) + T_3 - 2T_n = R_n$

PROBLEMS

5-1. For steady one-dimensional conduction, find the temperature at the interface T_2 for the composite wall ($k_A = 10$ Btu/hr-ft-°F, $k_B = 1$ Btu/hr-ft-°F) shown when L_B is (a) 1 in, (b) 2 in, (c) $\frac{1}{2}$ in.

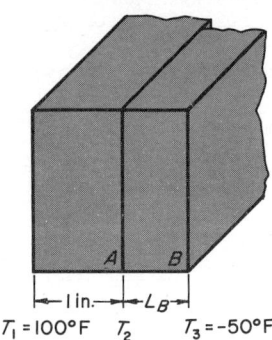

$T_1 = 100°F \quad T_2 \quad T_3 = -50°F$

Fig. P5-1

5-2. Determine the steady-state heat transfer q from a 20-ft-long cylinder having inside and outside radii of 9 and 10 ft, respectively. The thermal conductivity is 1.0 Btu/hr-ft-°F, and the inside and outside temperatures are, respectively, 400 and 100°F.

5-3. An industrial oven wall is made up of 9 in of fireclay brick (inside), 4 in of kaolin insulating brick, and 8 in of masonry brick (outside). The inner and outer surface temperatures, T_1 and T_4, are 400 and 70°F, respectively. Neglecting the resistance of the mortar joints, determine the temperatures T_2 and T_3 at the intermediate surfaces.

5-4. The inside surface of a spherical iron shell having an inner radius of 5 in and an outer radius of 6 in is at a uniform temperature of 100°F. The entire sphere is immersed in boiling water at 212°F. Assuming that the outer surface is at the same temperature as the water, what is the heat transfer?

5-5. A thick-walled tube of type 347 stainless steel, 1 in ID and 2 in OD is covered with a 2-in layer of molded pipe covering. What is the heat loss per foot of tube if the inside wall temperature of the pipe is maintained at 900°F and the outside of the insulation is at 100°F?

5-6. (a) A 4-in-OD mild-steel pipe carrying chilled water is to be covered with a 1-in thickness of asbestos and a 1-in thickness of rock wool. The pipe surface is at 35°F, and the outer insulation surface is at 80°F. To achieve the optimum insulating effect which insulation should be placed next to the pipe?

(b) For a single layer of insulation l in thick, derive an expression for the diameter D of the pipe in terms of l, k, q, and ΔT.

5-7. What thickness of rock-wool insulation is needed to guarantee that the temperature of the outer surface of a kitchen oven will not exceed 120°F? The maximum oven temperature, maintained by thermostatic control, is 500°F; the maximum steady-state electric energy input is 4,400 watts, and the oven is 2 by 2 by 2 ft.

5-8. A 10-in-OD mild-steel pipe of $\frac{1}{2}$-in wall thickness is covered by two layers of insulating materials (Fig. P5-8). This pipe is used to convey heated air in a test arrangement and is not a permanent installation.

(a) What is the heat transfer per lineal foot of pipe?
(b) What is T_3?

(c) If the insulating materials could be interchanged, i.e., if the 85 percent magnesia could be placed next to the pipe, what would be the heat transfer per lineal foot? (The magnesia is not suitable for long-term use above 600°F.)

(d) Compare and discuss the results of parts (a) and (c).

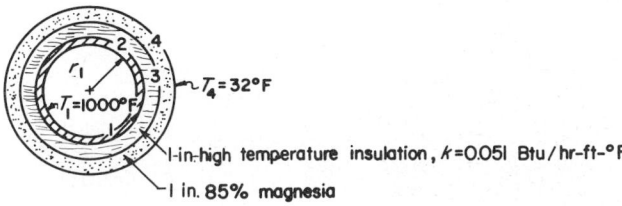

Fig. P5-8

5-9. Derive an expression for the total heat transfer from the tank of thermal conductivity k as shown.

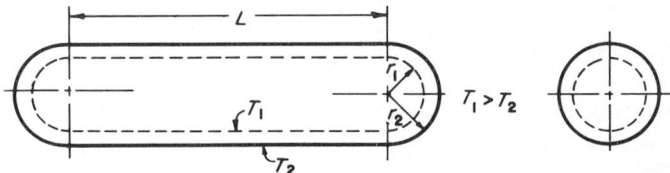

Fig. P5-9

5-10. The annular space between two thin concentric spherical shells having radii of 4 and 6 in is filled with bulk powdered insulating material. What wattage is required from an electric resistance heater located in the center of the smaller sphere in order to maintain a temperature difference between the two spherical shells equal to 40°F? Assume that the average thermal conductivity of the insulating material is 0.04 Btu/hr-ft-°F.

5-11. Two materials are in perfect thermal contact. The steady-state temperature distributions are as shown. If the thermal conductivity of the 3-in-thick material is $k_{12} = 10$ Btu/hr-ft-°F, what is the thermal conductivity of the 5-in-thick material k_{23}?

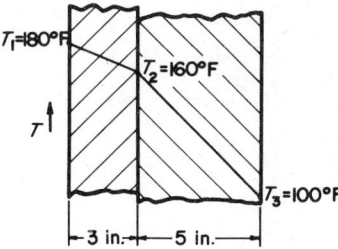

Fig. P5-11

5-12. The wall of a refrigeration truck body consists of $\frac{1}{16}$ in of sheet steel outside and $\frac{1}{8}$-in plywood ($k = 0.06$ Btu/hr-ft-°F) inside with a layer of glass-wool insulation between. The body is being designed to maintain an inside temperature of 10°F while the temperature on the outside is 70°F. What thickness of packed glass-wool insulation is needed to limit the heat transfer to 2 Btu/hr-ft²?

5-13. Find the heat transfer through the composite wall shown in Fig. P5-13. Assume one-dimensional flow. $A_B = A_D = \frac{1}{4}A_C$; $A_F = A_G$.

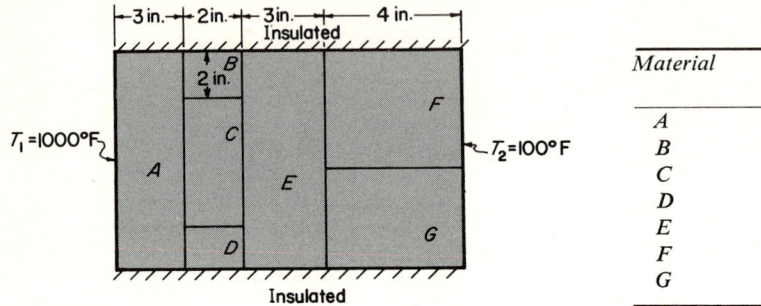

Material	Thermal conductivity k, Btu/hr-ft-°F
A	100
B	40
C	10
D	60
E	30
F	40
G	20

Fig. P5-13

5-14. Determine the thermal resistance of a wall which allows a heat transfer of 280 Btu/hr-ft² with a temperature difference of 100°F.

5-15. A spherical shell of radii r_1 and r_2 is made of a material with a thermal conductivity of $k = k_0 T^2$. Derive an expression for the heat transfer if the surfaces are held at temperatures T_1 and T_2, respectively.

5-16. A deep-sea research probe is constructed of a two-layer spherical shell. The inner layer is mild steel with an inside radius of 10 in, and the outer layer is type 304 stainless steel. Each layer is 1 in thick, and the two layers are in perfect thermal contact. The electronic gear inside the probe will give off energy resulting in a heat transfer of approximately 5000 Btu/hr-ft² based on outside surface area when the unit is surrounded by 40°F seawater. The inner surface wall should be less than 125°F for safe operation of the electronic equipment. Estimate the inner surface temperature under these conditions assuming the outer surface to be at the water temperature.

5-17. If the spaces of the rubber insulating wall shown in Fig. P5-17 are filled with dry air at an average temperature of 32°F, determine the thermal resistance of the wall and compare it with that of a solid wall.

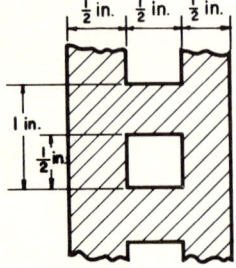

Fig. P5-17

STEADY-STATE CONDUCTIVE HEAT TRANSFER

5-18. A 200-amp electric current flows through a stainless-steel wire 0.1 in in diameter. The resistivity of the wire is 7×10^{-7} ohm-cm, and its length is 10 ft. If the outer surface temperature of the wire is 500°F, what is its center temperature?

5-19. A fluid of low electric conductivity at 200°F is heated by an immersed long iron bar 1 in thick by 5 in wide. The 1-in surfaces are insulated so that fluid is in contact with the 5-in-wide surfaces only. Heat is generated uniformly in the bar at a rate of 100,000 Btu/hr-ft³ by passing an electric current through it. Determine the unit surface conductance required to maintain the temperature of the bar below 400°F.

5-20. A food-storage freezer is operating in a 60°F environment and maintaining a $-20°F$ inside temperature. The outer and inner walls are made of 26-gauge (0.0184-in) sheet steel. Between the two layers of sheet steel are 2 in of fine unpacked glass-wool insulation. The average convective-heat-transfer coefficients are $\bar{h}_o = 4$ Btu/hr-ft²-°F and $\bar{h}_i = 2$ Btu/hr-ft²-°F.

(a) Determine the overall heat-transfer coefficient U.
(b) What is the heat transfer per unit area?

5-21. To reduce the energy loss from a $\frac{1}{2}$-in-OD metal hot-water line, a plumber decides to insulate the line with 0.47-in-thick insulation having a thermal conductivity of 0.09 Btu/hr-ft-°F. As a result of the high convective coefficient on the inside of the line, the surface of the metal tube can be considered to remain at a uniform temperature of 180°F. The line is surrounded by air at 80°F and for which $\bar{h}_o = 1.5$ Btu/hr-ft²-°F. In terms of percentage reduction, how successful was the plumber in reducing the heat loss?

5-22. A plane wall 5 in thick generates heat internally at the rate of 10 Btu/hr-ft³. One side of the wall is insulated, and the other side is exposed to an environment at 120°F. The convective-heat-transfer coefficient between the wall and the environment is 100 Btu/hr-ft²-°F. The thermal conductivity of the wall is 10 Btu/hr-ft-°F. Find the two surface temperatures.

5-23. What is the heat loss per lineal foot from a 4-in-OD, 3.5-in-ID cast-iron pipe insulated with 1 in of 85 percent magnesia? The pipe transports a fluid at 500°F with an inner heat-transfer coefficient $\bar{h}_i = 50$ Btu/hr-ft²-°F. The insulation surface is exposed to ambient air at 70°F with an outer heat-transfer coefficient $\bar{h}_o = 5$ Btu/hr-ft²-°F.

5-24. A constant-temperature cold-water reservoir for a laboratory experiment consists of a 4-in-ID type 304 stainless-steel pipe with $\frac{1}{2}$-in walls and a 2-in layer of rock wool covered with a thin cotton cloth on the outside. The inner pipe surface in contact with an ice-water mixture is maintained at 32°F. The outside convective coefficient is 2.0 Btu/hr-ft²-°F where the ambient temperature is 70°F. Draw an electrical analog and determine the heat transfer per lineal foot from the room to the pipe.

5-25. A $\frac{1}{8}$-in-diam copper heater wire 2 ft long is submerged in a 320°F fluid. The voltage drop across the wire is 60 volts, and the current measured is 120 amp. Calculate the center temperature of the wire if $\bar{h}_o = 1000$ Btu/hr-ft²-°F.

5-26. A vapor-to-liquid heat-exchanger surface of 1,000-in² face area is constructed of $\frac{3}{8}$-in nickel with a $\frac{3}{64}$-in plating of copper on the vapor side. The resistivity of the scale deposit on the vapor side is estimated to be 0.03 hr-ft²-°F/Btu, and the vapor and liquid-side convective coefficients are known to be 850 and 110 Btu/hr-ft²-°F, respectively. The heated vapor is at 233°F, and the liquid is at 155°F. Find the (a) overall heat-transfer coefficient U, (b) overall heat exchange, (c) temperature drop across the scale deposit, and (d) temperature at the copper-nickel interface.

5-27. The outside walls of a house consist of 4-in masonry brick, $\frac{1}{2}$-in pine sheathing, 2- by 4-in pine studs 16 in on center, and $\frac{1}{2}$-in sheetrock ($k = 0.09$ Btu/hr-ft-°F). Assume the thermal conductivity of the mortar joints to be equal to that of the brick.

(a) Find the overall heat-transfer coefficient U if the outside heat-transfer coefficient $\bar{h}_o$ is 3.5 Btu/hr-ft²-°F, the inside coefficient is 1.5 Btu/hr-ft²-°F, and air space between the studs has a heat-transfer coefficient of 1.2 Btu/hr-ft²-°F (for each internal surface). Calculate the heat transfer for a 1-ft depth per unit section (16 in) of wall.

(b) Repeat part (a) if the space between the studs is filled with fine glass-wool insulation.

5-28. A simplified method of determining thermal conductivity is to use the temperature gradient in a small-diameter long rod with one end attached to a high- (or low-) temperature source. What is the thermal conductivity of a $\frac{1}{2}$-in-diam rod extending from an oven into a 70°F environment with an external heat-transfer coefficient of 5.0 Btu/hr-ft^2-°F if the temperatures detected by two thermocouples located 6 in apart along the rod are 342 and 227°F?

5-29. A long 1-in-diam mild steel rod protrudes from a furnace wall which is maintained at 1000°F. The rod is surrounded by a fluid at 200°F. The heat-transfer coefficient between the rod and the fluid is 10 Btu/hr-ft^2-°F. Calculate the heat loss from the rod.

5-30. The end of a very long cylindrical stainless-steel rod is attached to a heated wall, and its surface is in contact with a cold fluid.

(*a*) If the rod diameter were doubled, by what percentage would the rate of heat removal increase?

(*b*) If the rod were made of aluminum, by what percentage would the heat-transfer rate change from that of the stainless steel?

5-31. Two long pieces of $\frac{1}{4}$-in square copper bar are to be silver-soldered together end to end. The surrounding air temperature is 80°F, and the melting point of the solder is 1200°F. If the heat-transfer coefficient between the copper and the air is 3 Btu/hr-ft^2-°F, find the minimum energy input required in watts to hold the soldered surface at 1200°F.

5-32. A long 1-in-diam brass rod ($k = 60$ Btu/hr-ft-°F) is heated by inserting part of it in a laboratory furnace. The major portion of the rod projects into ambient 80°F air. During steady state, two temperatures 4 in apart along the length of the rod are 312 and 215°F. What is the effective external heat-transfer coefficient h?

5-33. A heat exchanger intended for cooling an electronics package in a space vehicle is to utilize rectangular aluminum fins ($k = 119$ Btu/hr-ft-°F). The physical design dictates a fin length of 1.6 in, and weight and manufacturing considerations limit the choice to either (*a*) 0.032-in material spaced 0.25 in apart or (*b*) 0.064-in material spaced 0.50 in apart. The design temperatures are 195°F for the heat-exchanger wall and 68°F for the ambient temperature. Assuming an external heat-transfer coefficient $\bar{h}$ of 2.0 Btu/hr-ft^2-°F, select the better fin design based on the calculated heat loss per unit surface area of the heat exchanger.

5-34. Annular fins of mild steel are used on a 2-in-OD steam line for heating highly viscous fluids in a railroad tank car. The fins are 6 in OD, 0.102 in thick, and $\frac{3}{4}$ in apart. Assuming saturated steam at 50 psia in the pipe, that the pipe wall temperature is equal to that of the saturated steam, and that the external heat-transfer coefficient is 50 Btu/hr-ft^2-°F, calculate the heat transfer per linear foot of piping for a fluid temperature of 60°F.

5-35. To increase the heat dissipation from an air-cooled cylinder wall the installation of fins is under consideration. The wall temperature is 1200°F, and the heat-transfer coefficient between the solid surface and the 120°F ambient air is 15 Btu/hr-ft^2-°F. The two types under consideration are the straight rectangular fin and the tapered fin. Each fin is to be 1 in thick at the base, 4 in long, and made from aluminum. Compare the effectiveness of the two fins based on the heat flow per unit weight.

5-36. Determine the steady-state temperature distribution in an infinitely long two-dimensional fin of thickness l. The base temperature of the fin is $F(y)$, the ambient temperature is T_∞, and the convective-heat-transfer coefficient is large.

5-37. Find the temperature distribution in the semi-infinite rod in Fig. P5-37. The base temperature is θ_b, the ambient temperature is 0, and the convective-heat-transfer coefficient is large.

5-38. Analytically develop a shape factor for a hollow sphere with inner and outer radii r_1 and r_2 and thermal conductivity k. The inner and outer temperatures, T_1 and T_2, respectively, may be assumed constant.

STEADY-STATE CONDUCTIVE HEAT TRANSFER 149

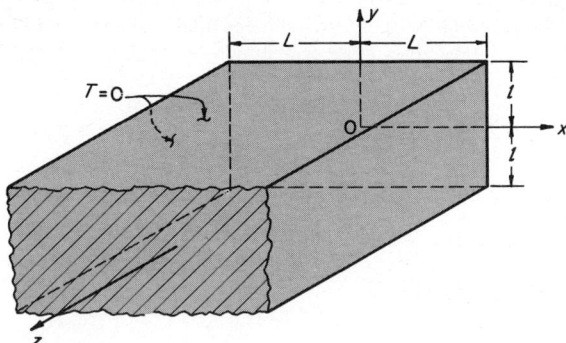

Fig. P5-37

5-39. Part of a laboratory system consists of a long $\frac{3}{4}$-in-OD copper tube embedded in the center of a square asbestos insulator with 3-in sides. The purpose of the pipe is to deliver hot water at 30 psia to the system, but the system requires that the water must not be boiling. The energy loss per foot of pipe is measured to be 35 Btu/hr-ft, and the outside surface temperature of the asbestos is at 150°F. Assuming the water and the tube are at the same temperature, does the water boil?

5-40. An underground concrete tunnel of 3- by 3-ft cross section has a 10-in-OD steam main passing through its geometrical center. Around the outside of the pipe and completely filling the tunnel is a fiber type of insulating material ($k = 0.05$ Btu/hr-ft-°F). What is the heat loss per foot from the pipe ($T_1 = 200°F$) to the ground ($T_2 = 55°F$)?

5-41. A long structural wedge-shaped beam in a processing plant has a cross section 4 in high, 2 in across the top, and 4 in on the bottom. It is symmetrical about a vertical centerline, and the sloping sides are insulated. The top of the wedge is subjected to a temperature of 1000°F while the base is maintained at 400°F.

(a) Determine the conduction shape factor S by the potential-plotting technique.

(b) If the thermal conductivity of the beam is 25 Btu/hr-ft-°F, what is the heat-transfer rate per lineal foot?

5-42. A circular cylinder 1 in in diameter and 1 ft long is used in a chemical experiment to measure the amount of heat given off by the reactants. After the reactants are mixed, the cylinder is placed vertically in a large block of material whose thermal conductivity is 25 Btu/hr-ft-°F; the top of the cylinder is flush with the upper surface of the block. Upon reaching steady-state conditions, the temperature of the cylinder is 850°F, and the temperature of the top of the block is 100°F. How much heat is given off by the reaction?

5-43. A heater is placed in a 1- by 1- by 1-ft aluminum container having $\frac{1}{2}$-in walls. The temperature of the walls inside the container is 400°F, and the temperature of the outside walls is 100°F. How much heat is conducted (a) through the corners of the container and (b) through the edges?

5-44. A long 3-in-OD sewer line is buried 3 ft below the surface of the earth ($k = 0.2$ Btu/hr-ft-°F). If 10 Btu/hr-ft of heat is given off from the line, how cold must the surface of the earth be for water to begin to freeze in the line, assuming negligible resistance between the water and the pipe?

5-45. The cavity of a volcano considered to be at a uniform temperature can be represented as having a spherical shape with a 60-ft diameter. From sonar measurements, the depth of the cavity is estimated to be about 220 ft below the surface of the earth ($k = 0.2$ Btu/hr-ft-°F). The heat transfer to the earth's surface is approximately 1 million Btu/hr when the temperature of the earth's surface is at 70°F. Estimate the temperature of the volcanic cavity.

5-46. The surfaces of a two-dimensional body are held at the temperatures shown in Fig. P5-46.

(a) For the square grid shown, write the appropriate residual equations for the three interior nodes.

(b) Set up a relaxation-pattern effect of unit change on the residuals.

(c) Using a relaxation table, find T_1, T_2, and T_3. Reduce residuals to ± 10.

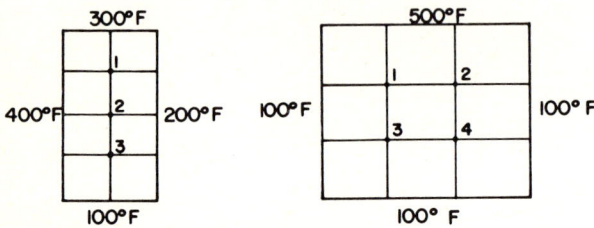

Fig. P5-46   Fig. P5-47

5-47. For Fig. P5-47 (square grid):

(a) Write the residual equations for points 1, 2, 3, and 4.

(b) Prepare an operation table for $+1$ change in temperature.

(c) Assuming $T_1 = T_2 = 300°F$ and $T_3 = T_4 = 200°F$, determine T_1, T_2, T_3, and T_4 by relaxation.

(d) With the assumed temperatures of part (c), use block relaxation alone to solve for the nodal temperatures.

5-48. An extruded stainless-steel beam of the cross section shown in Fig. P5-48 is used as a structural member in a furnace where the outer surface temperature is 2000°F. To prevent creep, cooling water flows through the beam. To determine the water flow rate necessary, the heat flow must be estimated. The inside surface temperature is 70°F.

(a) Estimate the heat transfer by sketching the isotherms and heat-flow lines in the left section.

(b) Assuming $T_d = 1700°F$, $T_e = 1400°F$, and $T_f = 1300°F$, determine the residual for point e.

(c) Write an expression for the heat flow per unit length into the beam using the 1- by 1-in grid as shown for the first approximation of temperatures in the relaxation process.

(d) Determine the heat transfer by the relaxation process.

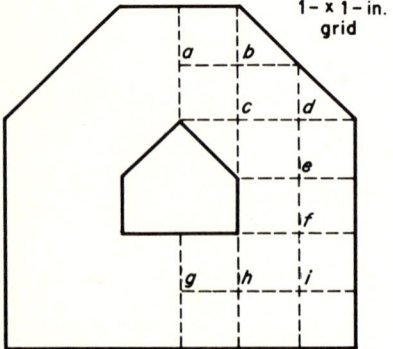

Fig. P5-48

STEADY-STATE CONDUCTIVE HEAT TRANSFER

5-49. A cylindrical pin fin is attached to a 300°F wall while its surface is exposed to a gas at 100°F. The convective-heat-transfer coefficient is 20 Btu/hr-ft^2-°F. The fin is made of stainless steel with a thermal conductivity of 10 Btu/hr-ft-°F. Determine the nodal temperatures by relaxation.

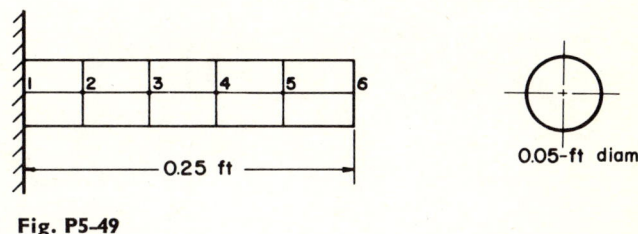

Fig. P5-49

REFERENCES

1. Arpaci, V. S.: "Conduction Heat Transfer," Addison-Wesley, Reading, Mass., 1966.
2. Brewley, L. V.: "Two Dimensional Fields in Electrical Engineering." Macmillan, New York, 1948.
3. Carslaw, H. S., and J. C. Jaeger: "Conduction of Heat in Solids," Oxford University Press, London, 1959.
4. Chapman, A. J.: "Heat Transfer," 2d ed., Macmillan, New York, 1967.
5. Harper, W. B., and D. R. Brown: Mathematical Equations for Heat Conduction in the Fins of Air-cooled Engines, *NACA Rept.* 158, 1922.
6. Holman, J. P.: "Heat Transfer," 2d ed., McGraw-Hill, New York, 1968.
7. Kayan, C. F.: An Electrical Geometrical Analog for Complex Heat Flow, *Trans. ASME*, **67**: 713 (1945).
8. Kayan, C. F.: Heat Transfer Temperature Patterns of a Multicomponent Structure by Comparative Methods, *Trans. ASME*, **71**: 9 (1949).
9. Schneider, P. J.: "Conduction Heat Transfer," Addison-Wesley, Reading, Mass., 1955.
10. Spiegel, M. R.: "Vector Analysis," Schaum Outline Series, McGraw-Hill, New York, 1959.

CHAPTER 6
UNSTEADY-STATE CONDUCTIVE HEAT TRANSFER

6-1 PHYSICAL AND MATHEMATICAL CONSIDERATIONS

In the last chapter, we considered the general class of conductive problems where none of the temperatures are time-dependent. There are numerous problems, however, where changes in conditions result in transient temperature distributions which may be quite significant. Such conditions are encountered in the manufacture of ceramics, brick, glass, and automobile tires, in food processing (cooking and freezing), and in metal forming and heat-treating. As a trivial but familiar specific example, consider the process of heating water to make instant coffee. To minimize the time, we turn the burner of the electric range to its maximum current setting and place a thin aluminum pan containing the water on it. The water will reach the desired boiling condition at just about the time the electric element reaches its maximum temperature, at which point we turn the current off.

Another example of transient conduction is afforded by the cooling of a metallic forging after the forming operation; no heat is added, and the forging cools by surface convection and radiation, the internal temperature at any given point being a transient function of time. The heat lost at the surface is *conducted* from the interior by the transient thermal gradient in the body.

UNSTEADY-STATE CONDUCTIVE HEAT TRANSFER

To analyze such problems, the general conduction Eq. (5-5) is used. For constant thermal conductivity this is

$$\frac{\partial^2 T}{\partial x^2} + \frac{\partial^2 T}{\partial y^2} + \frac{\partial^2 T}{\partial z^2} + \frac{q'''}{k} = \frac{1}{\alpha}\frac{\partial T}{\partial t}$$

and for a two-dimensional system with no internal heat generation it reduces to

$$\frac{\partial^2 T}{\partial x^2} + \frac{\partial^2 T}{\partial y^2} = \frac{1}{\alpha}\frac{\partial T}{\partial t} \tag{6-1}$$

Equation (6-1) is a second-order partial differential equation in three independent variables, x, y, and t. The solution requires for determination of all constants *one condition for each order of the highest-order derivative in each independent variable.* Thus, we need (1) two boundary conditions in the x direction, (2) two boundary conditions in the y direction, and (3) one time condition. The term boundary condition may be somewhat misleading, since this may include an internal condition; e.g., the centerline temperature gradient in a symmetrical body is usually zero and often serves as a boundary condition. The time condition is frequently an *initial condition*, and is often referred to as such.

In this chapter we shall restrict our analytical efforts to the solution of one-dimensional transient problems, though the solution of a two-dimensional problem can be obtained by means of the separation-of-variables technique employed in the steady-state square-bar problem of Chap. 5. Arpaci [1] analytically treats multi-dimensional transient problems in detail.

6-2 BIOT MODULUS

The approach to any transient conductive-heat-transfer problem should begin with a determination of the necessity for treatment of internal temperature gradients. To illustrate, consider the immersion of an initially heated cylinder whose length is large compared with the radius into a moving fluid stream at $T_\infty < T_c$ (Fig. 6-1). The fluid motion over the body results in a convective heat transfer between the stream and the

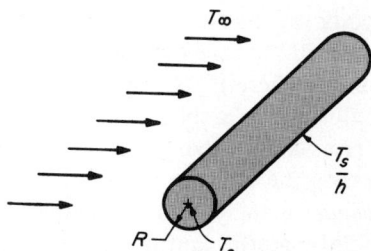

Fig. 6-1 Nomenclature for Biot modulus.

cylinder which can be expressed by Newton's law of cooling. This defines the average convective-heat-transfer coefficient $\bar{h}$, where

$$q = \bar{h} A_s (T_s - T_\infty) \tag{6-2}$$

This energy must be conducted to the surface, and thus

$$q = -k A_s \left. \frac{dT}{dr} \right|_{r=R} \tag{6-3}$$

Combining Eqs. (6-2) and (6-3) gives

$$\bar{h}(T_s - T_\infty) = -k \left. \frac{dT}{dr} \right|_R \simeq -k \frac{T_c - T_s}{R} \tag{6-4}$$

and by rearrangement we have

$$\frac{T_c - T_s}{T_s - T_\infty} \simeq \frac{\bar{h} R}{k} \equiv \text{Biot number} \tag{6-5}$$

From this, it is apparent that the dimensionless Biot number is indicative of the existence or lack of a significant internal temperature gradient. If this number is very small, the internal gradient must be small compared with the external temperature difference, $T_s - T_\infty$, and the heat transfer is then primarily controlled by convective transport at the surface. To further our understanding of the physical problem, let us now assume a very-small-radius cylinder of high thermal conductivity, say a small silver wire. If a high-velocity stream is blowing over it, we would expect the entire wire to rapidly assume the stream temperature. On the other hand, if the cylinder is a 6-in-diam baked clay model being cooled in atmospheric air with no forced velocity (this results in low $\bar{h}$), we would expect several hours to be required for the centerline temperature to reach the ambient value.

Generally, the Biot modulus may be thought of as the ratio of internal heat-flow resistance to external heat-flow resistance and is written as

$$\text{Dimensionless relative resistance:} \quad \mathbf{Bi} = \frac{\bar{h} L}{k} \tag{6-6}$$

where L is an appropriate linear distance obtained by dividing the volume of the body by its surface area (unless otherwise defined). (Note that this results in $L = R/2$ for the cylinder just discussed, but R is customarily used for a long cylinder or a sphere.) Using this method to determine L, bodies resembling a plate, a cylinder, or a sphere may be assumed to have *uniform* temperature with a resulting error of less than 5 percent whenever the Biot number is less than 0.1. In such cases, the body has negligible internal resistance to heat flow, and a *lumped-thermal-capacity* analysis is applicable. Throughout the preceding discussion, k is for the solid material and not the fluid.

UNSTEADY-STATE CONDUCTIVE HEAT TRANSFER

Example Find the Biot modulus for a rectangular parallepiped of steel with x, y, and z dimensions of 1, 2, and 3 in, respectively, in an annealing operation where $\bar{h} = 10$ Btu/hr-ft^2-°F and $k = 25$ Btu/hr-ft-°F.

Solution First determine L:

Volume $= 1(2)(3) = 6$ in^3

Surface area $= 2[1(2)] + 2[1(3)] + 2[2(3)] = 4 + 6 + 12 = 22$ in^2

$$L = \frac{\mathscr{V}}{A_s} = \frac{6}{22} \text{ in} = 0.0227 \text{ ft}$$

$$\text{Bi} = \frac{\bar{h}L}{k} = \frac{10(0.0227)}{25} = 0.00908$$

and obviously this problem can be treated by a lumped analysis.

6-3 LUMPED SYSTEMS

For problems with a suitably small Biot modulus, the transient conduction problem may be reasonably approximated by the assumption of uniform internal temperature, and the analysis involves the use of a lumped thermal capacity. A typical problem in this category is the quenching of a small metal forging in a heat-treatment operation. The first law of thermodynamics applied to this problem yields

$$\text{Heat flow out of forging during } dt = -\begin{pmatrix}\text{change of internal thermal} \\ \text{energy of forging during } dt\end{pmatrix}$$

Thus,

$$\bar{h}A_s(T - T_\infty)\, dt = -\rho c \mathscr{V}\, dT \tag{6-7}$$

where A_s is the surface area of the forging over which the average convective-heat-transfer coefficient is $\bar{h}$. Rearrangement of Eq. (6-7) yields

$$\frac{\bar{h}A_s}{\rho c \mathscr{V}}\, dt = -\frac{dT}{T - T_\infty} \tag{6-8}$$

which is an ordinary first-order differential equation in $T(t)$. To obtain the solution we require one time condition, namely, the initial condition

$$T = T_i \quad \text{at } t = 0$$

The solution is

$$\frac{T - T_\infty}{T_i - T_\infty} = \exp\left[-\left(\frac{\bar{h}A_s}{\rho c \mathscr{V}}\right)t\right] \tag{6-9}$$

and the temperature history is seen to be an exponential decay analogous to the voltage decay in a discharging dc electric capacitor. The system and form of the temperature decay are shown in Fig. 6-2.

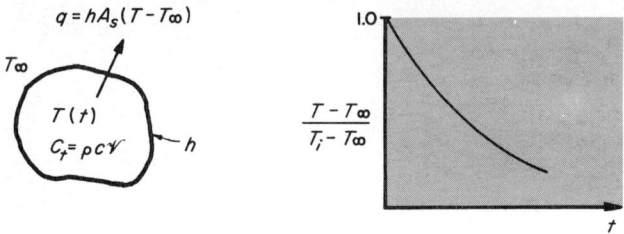

Fig. 6-2 Cooling of a lumped-thermal-capacity system.

In the analogous electric capacitor-discharge problem, the voltage decays according to

$$\frac{E}{E_i} = e^{-t/(RC)_e} \tag{6-10}$$

where the product $(RC)_e$ is the time constant and has units of time. For the thermal problem, the thermal capacitance is

$$C_t = \rho c \mathcal{V} \tag{6-11}$$

and the thermal resistance is

$$R_t = \frac{1}{\bar{h}A_s} \tag{6-12}$$

Thus, Eq. (6-9) can be written as

$$\frac{T - T_\infty}{T_i - T_\infty} = e^{-t/(RC)_t} \tag{6-13}$$

where $(RC)_t$ is the thermal time constant of the system and is the time required for the initial temperature difference, $T_i - T_\infty$, to decay by 63.2 percent.

If we introduce the dimensionless *Fourier modulus* defined by

Dimensionless time: $\quad \mathbf{Fo} \equiv \dfrac{\alpha t}{L^2} \tag{6-14}$

where L is the volume divided by the surface area (as defined for the Biot modulus), we have

$$\mathbf{BiFo} = \frac{\bar{h}L}{k}\left(\frac{k}{\rho c}\frac{t}{L^2}\right) = \frac{\bar{h}A_s}{\rho c \mathcal{V}} t$$

and Eq. (6-9) may be written

$$\frac{T - T_\infty}{T_i - T_\infty} = \exp(-\mathbf{BiFo}) \tag{6-15}$$

UNSTEADY-STATE CONDUCTIVE HEAT TRANSFER

It is significant that the solution of a lumped-thermal-capacity problem is independent of the thermal conductivity of the system.

The preceding treatment was for a system with cooling at the surface, frequently referred to as *newtonian cooling*. The opposite problem, the heating of a lumped-capacity system, is formulated exactly like the cooling problem, and the resulting equations are identical with those for cooling. (The sign convention and temperature differences result in an algebraically correct solution for either case.)

Example Determine the time constant for a copper-constantan thermocouple with the configuration shown in Fig. 6-3 when exposed to an airstream with $\bar{h} = 5$ Btu/hr-ft²-°F. What is the time required for the thermocouple to reach 249.5°F if the initial thermocouple temperature is

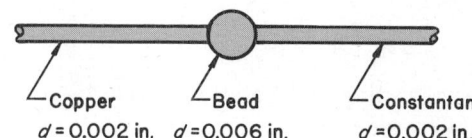

Fig. 6-3 Thermocouple. Copper $d = 0.002$ in. Bead $d = 0.006$ in. Constantan $d = 0.002$ in.

70°F and the air temperature is 250°F? (Assume the thermal conductivity of the junction to be 100 Btu/hr-ft-°F.)

Solution The thermal response is dictated by the bead temperature. Neglecting conduction through the wires, the bead temperature is controlled by the convective heat transfer at the surface.

The appropriate significant length is (neglecting area reduction due to wires)

$$L = \frac{\mathscr{V}}{A_s} = \frac{\frac{4}{3}\pi r^3}{4\pi r^2} = \frac{r}{3} = \frac{0.006}{6} \text{ in} = \frac{0.001}{12} \text{ ft} = 8.33 \times 10^{-5} \text{ ft}$$

The Biot modulus is

$$\text{Bi} = \frac{\bar{h}L}{k} = \frac{(5 \text{ Btu-hr-ft}^2\text{-}°F)(8.33 \times 10^{-5} \text{ ft})}{100 \text{ Btu/hr-ft-}°F} = 4.17 \times 10^{-6}$$

and obviously the problem may be treated by a lumped thermal analysis. The time constant is RC, where

$$R = \frac{1}{\bar{h}A_s} = \frac{1}{5(4\pi)(0.003/12)^2} = 255 \times 10^3 \text{ °F-hr/Btu}$$

$$C = c\rho\mathscr{V} = 0.1(557)(\tfrac{4}{3}\pi)((0.003/12))^3 = 3.64 \times 10^{-9} \text{ Btu/°F}$$

and

$$(RC)_t = (255 \times 10^3)(3.64 \times 10^{-9}) = 928 \times 10^{-6} \text{ hr} = 3.34 \text{ sec}$$

The time required for the thermocouple to reach 249.5°F is found from Eq. (6-13):

$$\ln \frac{T - T_\infty}{T_i - T_\infty} = \frac{-t}{(RC)_t}$$

Thus,

$$t = -(RC)_t \ln \frac{T - T_\infty}{T_i - T_\infty} = 3.34 \ln \frac{70 - 250}{249.5 - 250} = 19.67 \text{ sec}$$

This example is a very practical application of the lumped-capacity approach; the physical dimensions of the thermocouple chosen are very close to those currently available at modest cost. It also aids in answering an appropriate philosophical question regarding the exponential character of the solution given by Eqs. (6-9) or (6-13), which frequently is bothersome to students. This solution asserts that a body subjected to newtonian heating or cooling will never attain the temperature of the surrounding medium. While this assertion is mathematically correct, it is of little consequence as the difference in temperatures can be reduced to any desired level and certainly far below our present capability for measurement.

Frequently it is desirable to determine the rate of heat transfer from a body during newtonian cooling. The instantaneous heat flux can be obtained from Eq. (6-7) and is

$$q = \bar{h}A_s(T - T_\infty)$$

Substituting the solution given by Eq. (6-13) for the instantaneous temperature difference yields

$$q = \bar{h}A_s(T_i - T_\infty)e^{-t/(RC)_t} \tag{6-16}$$

The total heat flow from $t = 0$ to any specific time t_1 is

$$Q = \int_0^{t_1} q\,dt$$

and using Eq. (6-16) to perform this integration results in

$$Q = C_t(T_i - T_\infty)(1 - e^{-t_1/(RC)_t}) \tag{6-17}$$

The lumped-capacity method of analysis can be extended to the case of multiple systems with heat transfer between the *lumps* of the system, such as that encountered when two different objects are bolted together. This results in a separate linear differential equation for the temperature of each lump and requires the simultaneous solution of the resulting set of differential equations. A detailed treatment of this class of problem is given by Holman [6].

Another important class of problems is that resulting from periodic temperature changes in the medium surrounding a lumped system. This is encountered in chemical batch processing as well as thermal temperature control in buildings. Problems of this type are treated by Kreith [7].

6-4 ONE-DIMENSIONAL SYSTEMS WITH PRESCRIBED SURFACE TEMPERATURE

Perhaps the simplest case of transient conduction is that due to the sudden temperature change at the surface of a body of semi-infinite thickness, as shown in Fig. 6-4. The body is assumed to be infinite in the y and z directions and to extend to $+\infty$ in the x direction, the term semi-infinite arising because of the finite termination of the

UNSTEADY-STATE CONDUCTIVE HEAT TRANSFER

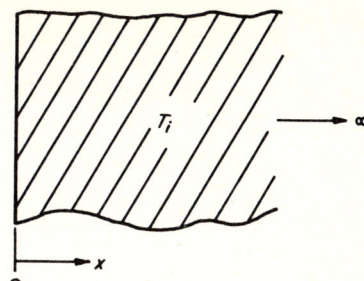

Fig. 6-4 Semi-infinite body.

body at $x = 0$. Such configurations are approximated in nature by the earth's surface, large bodies of water, etc.

Under the assumption of one-dimensional heat flow, the appropriate differential equation for the temperature distribution is

$$\frac{\partial^2 T}{\partial x^2} = \frac{1}{\alpha} \frac{\partial T}{\partial t} \qquad (6\text{-}18)$$

which requires two boundary conditions in the x direction and one time condition:

$$\begin{aligned}
&\text{Boundary condition 1:} \quad T(0,t) \simeq T_s \quad \text{for } t > 0 \\
&\text{Boundary condition 2:} \quad T(\infty,t) \simeq T_i \\
&\text{Initial condition:} \quad T(x,0) = T_i
\end{aligned} \qquad (6\text{-}19)$$

Complete details of the solution are given by Schneider [10]. For the sake of brevity, we note that

$$T = C t^{-\frac{1}{2}} e^{-x^2/4\alpha t} \qquad (6\text{-}20)$$

is a particular solution of Eq. (6-18). This can readily be verified by differentiation and direct substitution into the differential equation; the details are left as an exercise.

A second particular solution to Eq. (6-18) can be found by integrating the right side of Eq. (6-20) with respect to x. Thus,

$$T = C \int_0^x t^{-\frac{1}{2}} e^{-x^2/4\alpha t} \, dx \qquad (6\text{-}21)$$

and this can be shown to be a solution by the same technique as before. Defining

$$\xi \equiv \frac{x}{\sqrt{4\alpha t}}$$

Eq. (6-21) becomes

$$T = 2C\alpha^{\frac{1}{2}} \int_0^{x/\sqrt{4\alpha t}} \exp\left[-\left(\frac{x}{\sqrt{4\alpha t}}\right)^2\right] d\frac{x}{\sqrt{4\alpha t}}$$

or

$$T = 2C\alpha^{\frac{1}{2}} \int_0^{x/\sqrt{4\alpha t}} e^{-\xi^2}\, d\xi \qquad (6\text{-}22)$$

But

$$\operatorname{erf} x \equiv \frac{2}{\sqrt{\pi}} \int_0^x e^{-\xi^2} d\xi$$

is the standard notation for the Gauss error function which is tabulated in mathematical tables, and consequently we write

$$T = C_1 \operatorname{erf} \frac{x}{\sqrt{4\alpha t}} + C_2 \qquad (6\text{-}23)$$

as the general solution to the present problem, where C_1 and C_2 are arbitrary constants. Now applying the first boundary condition and noting that $\operatorname{erf} 0 = 0$, we find that

$$C_2 = T_s$$

Then applying the second boundary condition with $\operatorname{erf} \infty = 1$ yields

$$C_1 = T_i - T_s$$

and consequently

$$\frac{T - T_s}{T_i - T_s} = \operatorname{erf} \frac{x}{\sqrt{4\alpha t}} \qquad (6\text{-}24)$$

is the complete solution. It should be noted that T is a function of both x and time and that the solution satisfies the initial condition for all finite values of x. For convenience, the error function for arguments from 0 to 2 is plotted in Fig. 6-5, and this of course also represents the dimensionless temperature distribution.

The heat flow by conduction at any distance x in the body can be calculated by Fourier's equation. Expressing $T(x,t)$ in terms of the integral represented by erf x results in

$$T(x,t) = T_s + (T_i - T_s) \frac{2}{\sqrt{\pi}} \int_0^{x/\sqrt{4\alpha t}} e^{-\xi^2}\, d\xi$$

Performing the partial differentiation by Leibnitz' rule yields

$$\frac{\partial T}{\partial x} = \frac{T_i - T_s}{\sqrt{\pi \alpha t}} e^{-x^2/4\alpha t} \qquad (6\text{-}25)$$

and the heat flux at the surface ($x = 0$) can readily be evaluated by

$$q\big|_{x=0} = \frac{kA(T_s - T_i)}{\sqrt{\pi \alpha t}} \qquad (6\text{-}26)$$

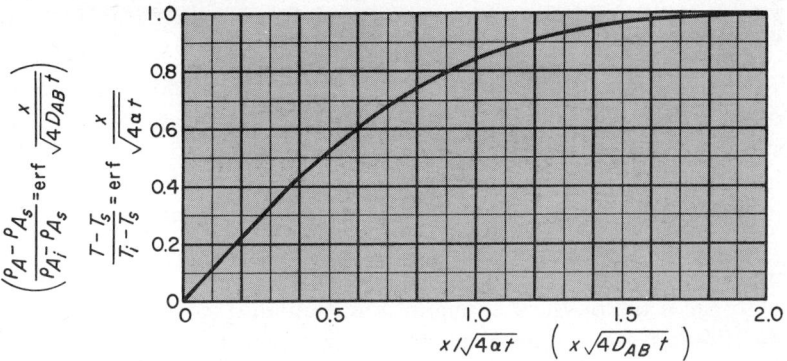

Fig. 6-5 Gauss error function. Expressions in parentheses are applicable to mass diffusion (Chap. 7).

This last result states that the heat flux entering the body diminishes with time, which is intuitively correct as the temperature gradient in the body must diminish with heat transfer.

Even though the preceding formulation was for a body of infinite extent in the positive x direction (semi-infinite body), the results are applicable under certain conditions to finite bodies with one-dimensional heat transfer. In particular if the temperature in the interior of the body is unaffected, the temperature distribution in the affected portion is identical with that in the semi-infinite body. As a general criterion, the solution for the semi-infinite body is applicable to bodies of finite thickness L[7] when

$$\frac{L}{\sqrt{4\alpha t}} \geq 0.5 \tag{6-27}$$

Example It is essential to locate water pipes below the depth to which freezing can occur in the soil. If the initial soil temperature is 50°F and the surface temperature drops rapidly to -5°F, to what depth will the freezing temperature penetrate during a 10-hr period? Assume dry soil with $\alpha = 0.01$ ft²/hr.

Solution The dimensionless temperature ratio is

$$\frac{32 - (-5)}{50 - (-5)} = \frac{37}{55} = 0.673 = \mathrm{erf}\,\frac{x}{\sqrt{4\alpha t}}$$

Then using Fig. 6-5 we find

$$\frac{x}{\sqrt{4\alpha t}} \simeq 0.7$$

which yields

$$x = 0.7\sqrt{4(0.01)(10)} = 0.433 \text{ ft}$$

for the depth to which freezing penetrates.

6-5 ONE-DIMENSIONAL SYSTEMS WITH PRESCRIBED CONVECTION CONDITIONS

Transient cooling or heating of a body due to a sudden change in the temperature of the surrounding fluid is considerably more important than the case of a prescribed surface temperature. Industrial applications include quenching of metals during heat treatment, drying operations, baking processes, etc. Generally such processes involve a constant convective-heat-transfer coefficient h. As a specific example we shall consider a very large plate of finite thickness $2L$ in the x direction with an initially uniform temperature T_i, as depicted in Fig. 6-6a.

Since the physical configuration, boundary conditions, and initial temperature distribution are symmetrical about the vertical centerline, the resulting transient temperature distribution will likewise be symmetrical in the body, and without loss of generality we may treat this problem as shown in Fig. 6-6b. Notice that the symmetry of the temperature distribution ensures no heat transfer across the centerline, and thus one boundary condition becomes $\partial T/\partial x = 0$ at $x = 0$.

Letting $\theta = T - T_\infty$, as in Sec. 5-4, the complete mathematical specification of the problem is

$$\frac{\partial^2 \theta}{\partial x^2} = \frac{1}{\alpha}\frac{\partial \theta}{\partial t} \tag{6-28}$$

with

Boundary conditions: $\quad \dfrac{\partial x}{\partial \theta} = \begin{cases} 0 & \text{at } x = 0 \\ -\dfrac{h}{k}\theta & \text{at } x = L \end{cases}$ (6-28a)

Initial condition: $\quad \theta = \theta_i \quad \text{at } t = 0$

Application of the classical separation-of-variables technique as used in Sec. 5.7 yields the following general solution of Eq. (6-28):

$$\theta = e^{-\lambda^2 \alpha t}(C_1 \sin \lambda x + C_2 \cos \lambda x) \tag{6-29}$$

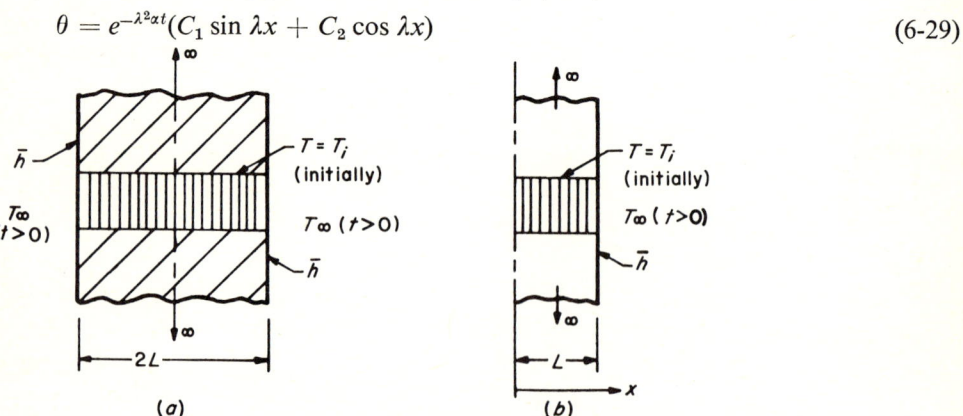

Fig. 6-6 One-dimensional system with a convective boundary condition.

UNSTEADY-STATE CONDUCTIVE HEAT TRANSFER

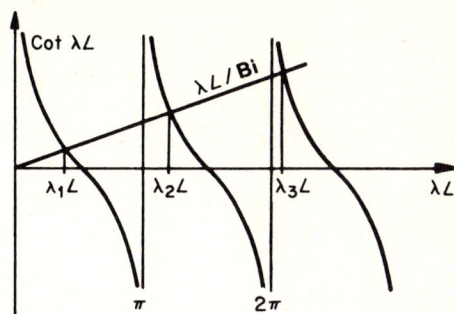

Fig. 6-7 Solution of transcendental equation for λ.

where $-\lambda^2$ is the separation parameter or constant. The details of obtaining Eq. (6-29) are left as an exercise; again the fact that this is a solution can readily be verified by differentiation and direct substitution into Eq. (6-28).

Application of the first boundary condition yields

$$\left.\frac{\partial T}{\partial x}\right|_{x=0} = e^{-\lambda^2 \alpha t} \lambda (C_1 \cos \lambda x - C_2 \sin \lambda x)\Big|_{x=0} = 0$$

Consequently,

$$C_1 = 0$$

and the solution simplifies to

$$\theta = C e^{-\lambda^2 \alpha t} \cos \lambda x \tag{6-30}$$

Application of the second boundary condition gives

$$C e^{-\lambda^2 \alpha t} \lambda (-\sin \lambda x)\Big|_{x=L} = -\frac{\bar{h}}{k}(C e^{-\lambda^2 \alpha t} \cos \lambda x)\Big|_{x=L}$$

and thus

$$\lambda \sin \lambda L = \frac{\bar{h}}{k} \cos \lambda L$$

This last expression can be rearranged to form

$$\cot \lambda L = \frac{\lambda L}{\bar{h}L/k} = \frac{\lambda L}{\mathrm{Bi}} \tag{6-31}$$

which is a transcendental equation in λ satisfied by an infinite number of values of the parameter λL. For the purpose of visualization this is illustrated in Fig. 6-7.

The values of λ satisfying Eq. (6-31) are known as *characteristic values* or *eigenvalues*, and the solution becomes

$$\theta = \sum_{n=1}^{\infty} C_n e^{-\lambda_n^2 \alpha t} \cos \lambda_n x \qquad (6\text{-}32)$$

where the λ_n's are roots of Eq. (6-31). A convenient method of determining these is to use an accurate graph of the type shown in Fig. 6-7. Fortunately, the series converges quite rapidly, and only a few of the roots are required to give a reasonably accurate answer.

In order to evaluate θ we need an expression for the coefficients C_n. Application of the initial condition results in

$$\theta_i = \sum_{n=1}^{\infty} C_n \cos \lambda_n x \qquad (6\text{-}33)$$

which requires that the C_n's be so chosen that θ_i can be represented by an infinite series of cosine terms. The Fourier cosine series, as treated in advanced engineering mathematics texts such as Kreysig [8], satisfies Eq. (6-33), where

$$C_n = \frac{2\theta_i \sin \lambda_n L}{\lambda_n L + \sin \lambda_n L \cos \lambda_n L} \qquad (6\text{-}34)$$

The final expression for the transient temperature then becomes

$$\frac{\theta}{\theta_i} = 2 \sum_{n=1}^{\infty} e^{-\lambda_n^2 \alpha t} \frac{\sin \lambda_n L}{\lambda_n L + \sin \lambda_n L \cos \lambda_n L} \cos \lambda_n x \qquad (6\text{-}35)$$

The energy transferred from the plate during a time interval from 0 to t_1 is

$$Q = \int_0^{t_1} q \, dt \qquad (6\text{-}36)$$

where

$$q = -kA \frac{\partial T}{\partial x}\bigg|_{x=L}$$

Differentiating Eq. (6-35) to obtain the temperature gradient at $x = L$ and performing the integration of Eq. (6-36) yields

$$\frac{Q}{Q_i} = 2 \sum_{n=1}^{\infty} \frac{1}{\lambda_n L} \frac{\sin^2 \lambda_n L}{\lambda_n L + \sin \lambda_n L \cos \lambda_n L} (1 - e^{-\lambda_n^2 \alpha t}) \qquad (3\text{-}37)$$

where the initial energy stored in the half-plate, $Q_i = AL\rho c \theta_i$, has been introduced.

6-6 CHARTS FOR ONE-DIMENSIONAL SYSTEMS WITH CONVECTIVE BOUNDARY CONDITIONS

Plate The series solutions for the temperature distribution and the heat removed from the plate of the preceding section are too cumbersome for rapid solution of practical problems. Fortunately it is possible to restate these in terms of the non-dimensional Biot and Fourier moduli of Eqs. (6-6) and (6-14). Introducing

$$\delta_n = \lambda_n L \tag{6-38}$$

we obtain

$$\frac{\theta}{\theta_i} = \frac{T - T_\infty}{T_i - T_\infty} = 2 \sum_{n=1}^{\infty} \exp(-\delta_n^2 \mathbf{Fo}) \frac{\sin \delta_n}{\delta_n + \sin \delta_n \cos \delta_n} \cos\left(\delta_n \frac{x}{L}\right) \tag{6-39}$$

and

$$\frac{Q}{Q_i} = 2 \sum_{n=1}^{\infty} \frac{1}{\delta_n} \frac{\sin^2 \delta_n}{\delta_n + \sin \delta_n \cos \delta_n} [1 - \exp(-\delta_n^2 \mathbf{Fo})] \tag{6-40}$$

where the δ_n's are a function of the Biot modulus.

These equations have been evaluated numerically by numerous investigators, and the graphical results of Heisler [5] are among the most widely used for the temperature distribution. Denoting the temperature at the centerline of the plate by subscript c, Eq. (6-39) simplifies to

$$\frac{\theta_c}{\theta_i} = \frac{T_c - T_\infty}{T_i - T_\infty} = 2 \sum_{n=1}^{\infty} \exp(-\delta_n^2 \mathbf{Fo}) \frac{\sin \delta_n}{\delta_n + \sin \delta_n \cos \delta_n} \tag{6-41}$$

and this is plotted for a wide range of Biot and Fourier moduli in Fig. 6-8. It is to be emphasized that this is for the *centerline* transient temperature in a symmetrical plate of *half-thickness L*, the symmetry being both geometrical and thermal.

The temperature at any position in the plate can be found by applying a position-correction factor to the centerline temperature. Figure 6-9 is a plot of θ/θ_c, that is, the ratio of $T - T_\infty$ at any position x/L to that at the center of the plate at a given dimensionless time (**Fo**).

A very convenient representation of the integrated heat transfer from the plate, Eq. (6-40), is provided in terms of the Biot and Fourier moduli in Fig. 6-10, where

$$Q_i = \rho c V \theta_i$$

is the initial energy difference. The application of these charts is illustrated in the following example.

Example A 4-in-thick steel slab very wide and long and initially at 400°F is suddenly exposed to a convective-fluid environment at 100°F. The average convective coefficient is $h = 50$ Btu/hr-ft²-°F, and the average value of α over the temperature range is 0.452 ft²/hr. Determine the center temperature and the temperature at a 1½-in depth after 10 min of exposure and the total heat removed during the first 10 min.

Solution The Heisler chart of Fig. 6-8 will be used to determine the center temperature of the slab, and then Fig. 6-9 will be used to obtain the temperature at a depth of 1½ in. From the problem statement

$\theta_i = T_i - T_\infty = 400 - 100 = 300$°F
$k = 25$ Btu/hr-ft-°F (assume constant)(from Table A-4)
$\alpha = 0.452$ ft²/hr (assume constant)
$L = 2$ in $= 0.167$ ft $\qquad h = 50$ Btu/hr-ft²-°F
$t = 10$ min $= 0.167$ hr $\qquad x = 0.5$ in $= 0.0417$ ft

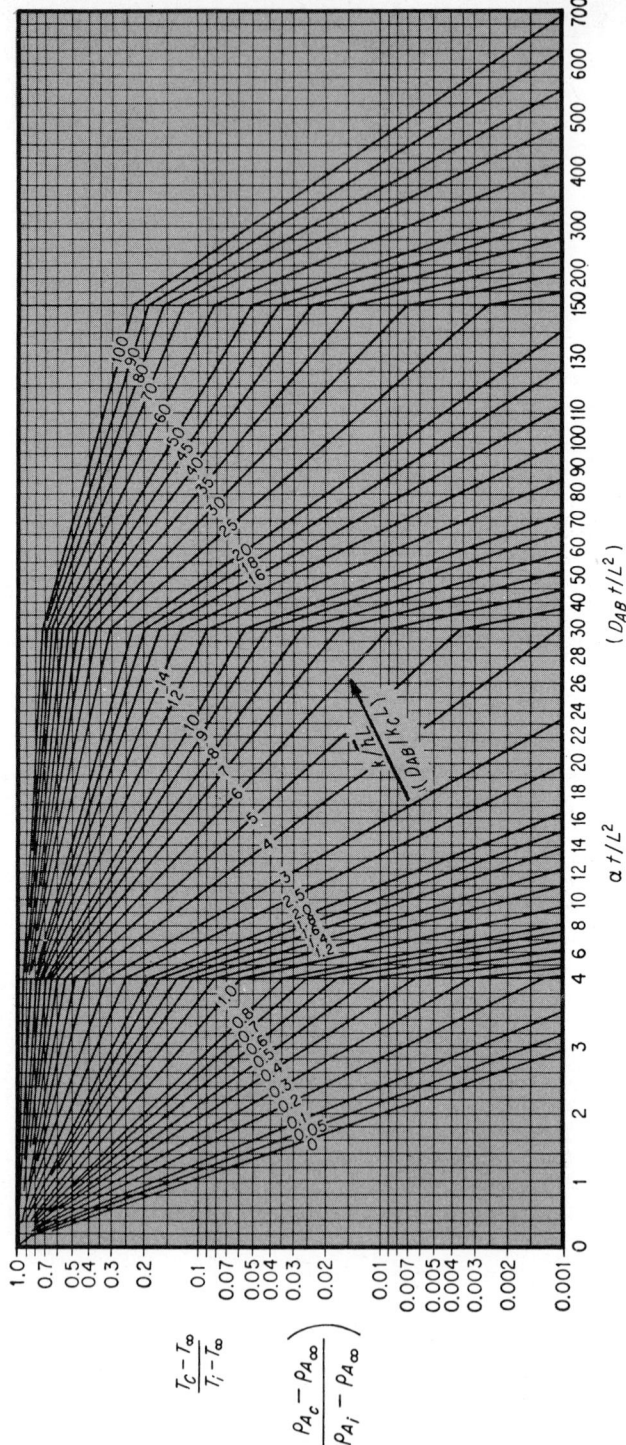

Fig. 6-8 Centerline transient temperature (and mass concentration) for an infinite plate of thickness 2L. [*From M. P. Heisler, Trans. ASME,* **69**:227 (1947).]

UNSTEADY-STATE CONDUCTIVE HEAT TRANSFER

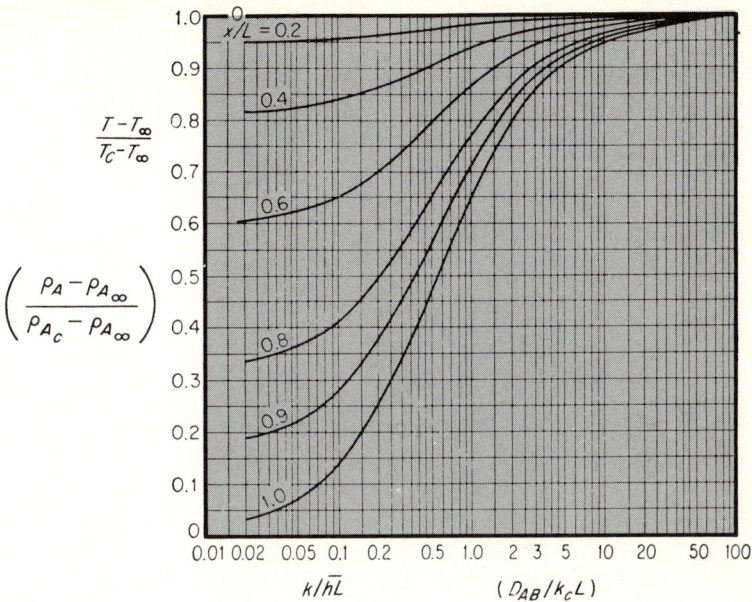

Fig. 6-9 Position-correction temperature (and mass-concentration) chart for an infinite plate of thickness $2L$. [*From M. P. Heisler, Trans. ASME,* **69**:227 (1947).]

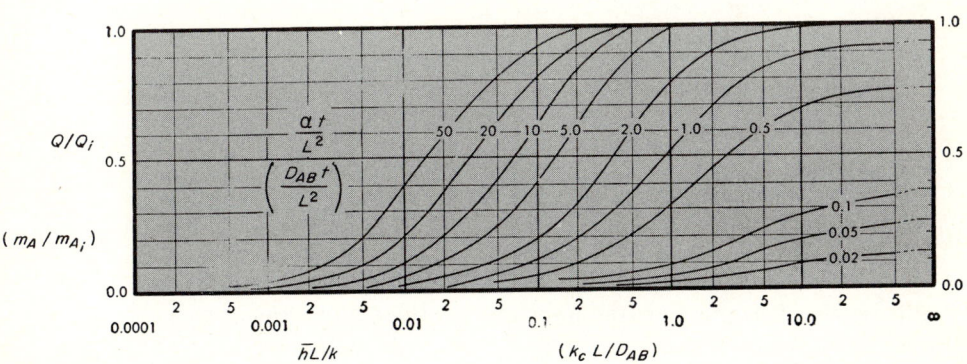

Fig. 6-10 Dimensionless heat flow (and mass flux) to or from an infinite plate of thickness $2L$. (*From H. Gröber, S. Erk, and U. Grigull, "Grundgesetze der Wärmeübertragung," 3d ed., Springer-Verlag, 1955, by permission.*)

Thus,

$$\text{Fo} = \frac{\alpha t}{L^2} = \frac{0.452(0.167)}{0.167^2} = 2.705$$

$$\frac{1}{\text{Bi}} = \frac{k}{hL} = \frac{25}{50(0.167)} = 2.995$$

From Fig. 6-8 we have

$$\frac{\theta_c}{\theta_i} = 0.47$$

$$\theta_c = T_c - T_\infty = 0.47(300) = 141$$

$$T_c = T_\infty + 141 = 241°\text{F}$$

From Fig. 6-9 at $x/L = 0.25$

$$\frac{\theta}{\theta_c} = 0.98$$

$$\theta = (0.98)(141) = 138.2°\text{F}$$

$$T = T_\infty + 138.2 = 238.2°\text{F}$$

To determine the total heat loss from the plate during the 10-min interval, we find from Fig. 6-10 at

$$\text{Fo} = 2.705$$

$$\text{Bi} = \frac{1}{2.995} = 0.334$$

$$\frac{Q}{Q_i} = 0.6$$

The initial energy capacity per unit area is (entire plate)

$$Q_i = \rho c 2 L \theta_i$$

where

$$\rho = 487 \text{ lb}_m/\text{ft}^3 \qquad c = 0.113 \text{ Btu/lb}_m\text{-}°\text{F}$$

and thus

$$Q_i = 487(0.113)(2)(0.167)(300°\text{F}) = 5515 \text{ Btu/ft}^2$$

so that

$$Q = 0.6(5515) = 3310 \text{ Btu/ft}^2$$

Long cylinder For a very long cylinder of radius R with axial symmetry and no lengthwise thermal gradient, the appropriate differential equation can be obtained directly from Eq. (5-11) and is

$$\frac{\partial^2 \theta}{\partial r^2} + \frac{1}{r}\frac{\partial \theta}{\partial r} = \frac{1}{\alpha}\frac{\partial \theta}{\partial t} \tag{6-42}$$

UNSTEADY-STATE CONDUCTIVE HEAT TRANSFER

with

Boundary conditions: $\quad \dfrac{\partial \theta}{\partial r} = \begin{cases} 0 & \text{at } r = 0 \\ -\dfrac{\bar{h}}{k}\theta & \text{at } r = R \end{cases}$ (6-43)

Initial condition: $\quad \theta = \theta_i \quad \text{at } t = 0$

The complete solution of Eq. (6-42) subject to these conditions is given by Arpaci [1], and the final result is

$$\frac{\theta}{\theta_i} = 2\sum_{n=1}^{\infty} \frac{1}{n^R} e^{-\lambda_n^2 \alpha t} \frac{J_0(\lambda_n r)J_1(\lambda_n R)}{J_0^2(\lambda_n R) + J_1^2(\lambda_n R)} \quad (6\text{-}44)$$

where the λ_n's are the roots of

$$\lambda_n R \frac{J_1(\lambda_n R)}{J_0(\lambda_n R)} - \text{Bi} = 0 \quad (6\text{-}45)$$

In these expressions the Biot number is $\bar{h}R/k$, J_0 and J_1 are the Bessel functions of the first kind of zero and first orders, respectively, and θ is $T(r,t) - T_\infty$. As in the case of the infinite flat plate, we can obtain series solutions for the centerline dimensionless temperature and the dimensionless accumulated heat flow from the cylinder in terms of the Biot and Fourier moduli. To facilitate use these have been plotted along with a position-correction chart in Figs. 6-11 to 6-13.

Example A very long 4-in-diam steel cylinder initially at 400°F is suddenly exposed to a convective-fluid environment at 70°F. The average convective coefficient is $\bar{h} = 100$ Btu/hr-ft²-°F, and $\alpha = 0.452$ ft²/hr (average value over the temperature range). Determine the temperature $\frac{1}{2}$ in from the surface after 5 min.

Solution The Heisler chart of Fig. 6-11 will be used to find the centerline temperature, and then the position-correction chart of Fig. 6-12 will be used to determine the temperature $\frac{1}{2}$ in from the outer surface. By the problem statement

$\theta_i = T_i - T_\infty = 400 - 70 = 330°F$

$k = 25$ Btu/hr-ft-°F (assume constant)(from Table A-4)

$\alpha = 0.452$ ft²/hr (assume constant)

$R = 2$ in $= 0.167$ ft $\qquad \bar{h} = 100$ Btu/hr-ft²-°F

$t = 5$ min $= 0.0833$ hr $\qquad r = 1.5$ in $= 0.125$ ft

Thus,

$$\text{Fo} = \frac{\alpha t}{R^2} = \frac{0.452(0.0833)}{0.167^2} = 1.35$$

$$\frac{1}{\text{Bi}} = \frac{k}{\bar{h}R} = \frac{25}{100(0.167)} = 1.50$$

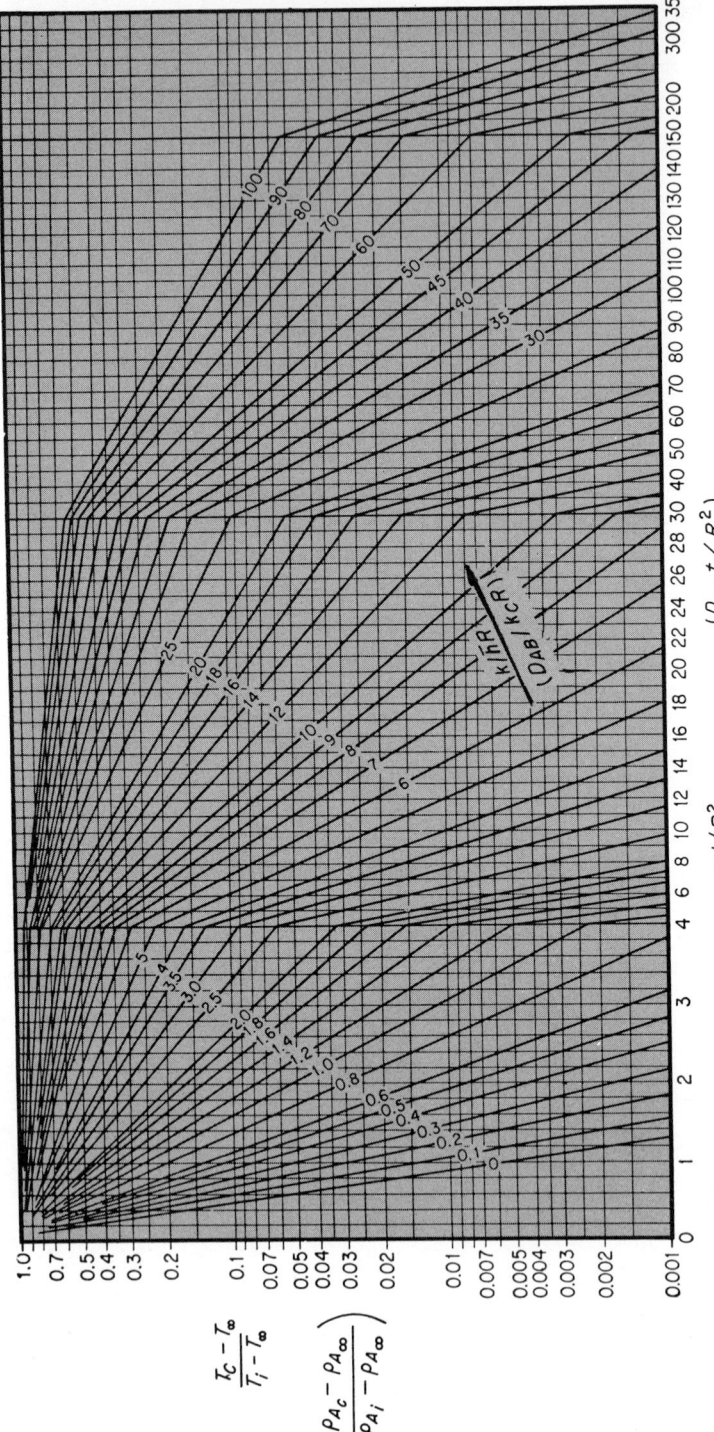

Fig. 6-11 Centerline transient temperature (and mass concentration) for an infinitely long cylinder of radius R. [*From M. P. Heisler, Trans. ASME,* **69**:227 (1947).]

UNSTEADY-STATE CONDUCTIVE HEAT TRANSFER

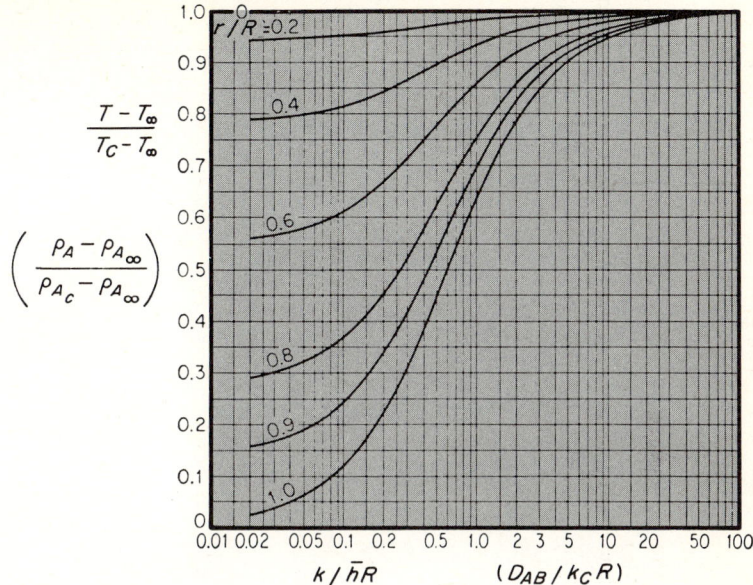

Fig. 6-12 Position-correction temperature (and mass-concentration) chart for an infinitely long cylinder. [*From M. P. Heisler, Trans. ASME,* **69**:227 (*1947*).]

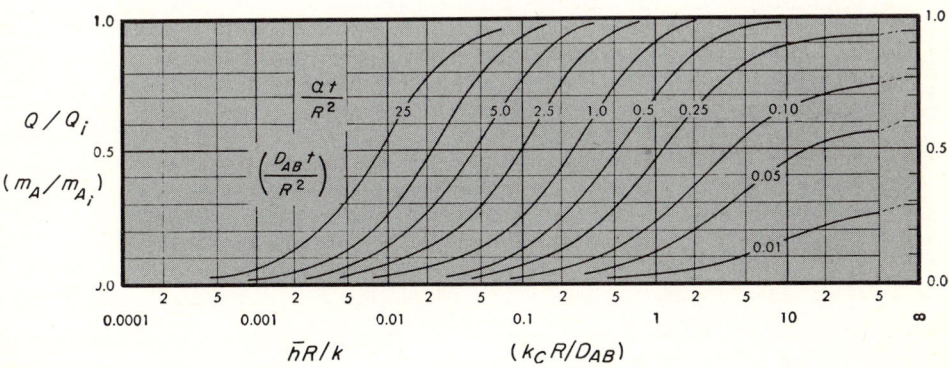

Fig. 6-13 Dimensionless heat flow (and mass flux) to or from an infinitely long cylinder. (*From H. Gröber, S. Erk, and U. Grigull, "Grundgesetze der Wärmeübertragung," 3d ed., Springer-Verlag, Berlin, 1955, by permission.*)

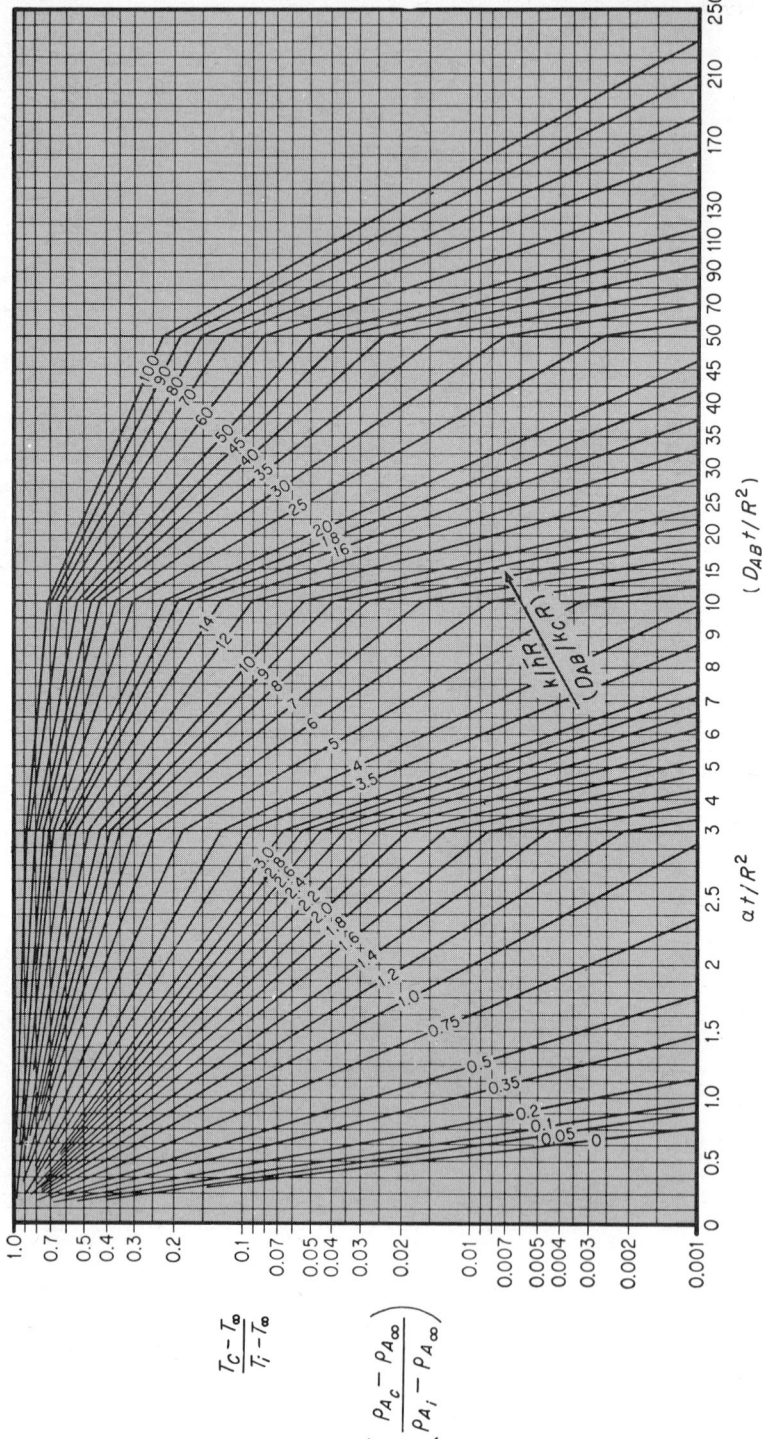

Fig. 6-14 Center transient temperature (and mass concentration) for a sphere of radius R. [From M. P. Heisler, *Trans. ASME*, **69**:227 (1947).]

UNSTEADY-STATE CONDUCTIVE HEAT TRANSFER

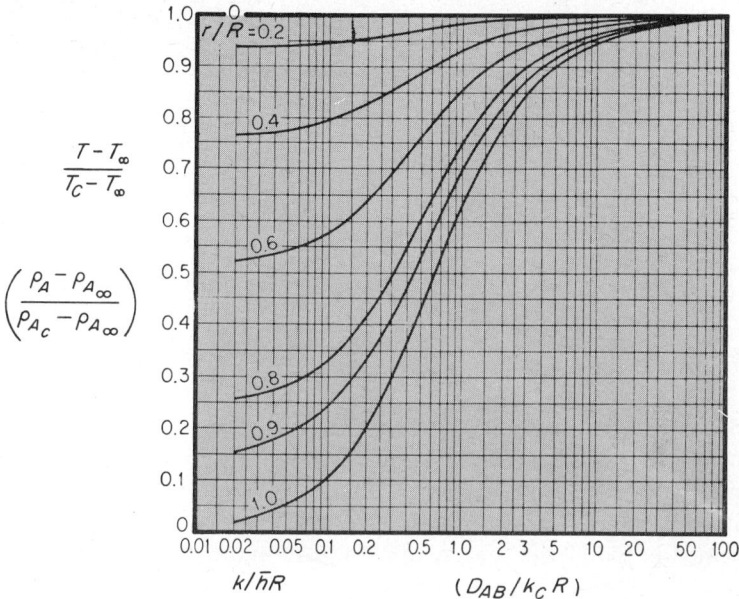

Fig. 6-15 Position-correction temperature (and mass-concentration) chart for a sphere of radius R. [*From M. P. Heisler, Trans. ASME,* **69**:227 *(1947)*.]

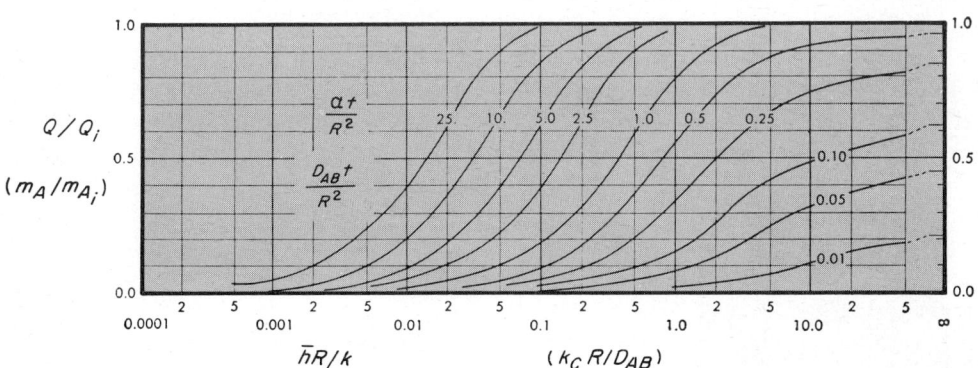

Fig. 6-16 Dimensionless heat flow (and mass flux) to or from a sphere of radius R. (*From H. Gröber, S. Erk, and U. Grigull, "Grundgesetze der Wärmeübertragung," 3d ed., Springer-Verlag, Berlin, 1955, by permission.*)

From Fig. 6-11 we have

$$\frac{\theta_c}{\theta_i} = 0.25$$

$$\theta_c = T_c - T_\infty = 0.25(330) = 82.5°F$$

$$T_c = T_\infty + \theta_c = 152.5°F$$

From Fig. 6-12 at $r/R = 0.125/0.167 = 0.75$

$$\frac{\theta}{\theta_c} = 0.84$$

$$\theta = 0.84(82.5°F) = 69.3°F$$

$$T = T_\infty + \theta = 70 + 69.3 = 139.3°F$$

Sphere The solution for the transient temperature in a sphere with a uniform convective surface heat flux is also given by Arpaci [1]. Again as for the infinite plate and the infinitely long cylinder, the solution has been numerically evaluated in terms of the dimensionless Biot and Fourier numbers, and these results are presented in the Heisler charts of Figs. 6-14 and 6-15 and the Gröber chart of Fig. 6-16.

6-7 EXTENSION TO MULTIDIMENSIONAL SYSTEMS

The solutions of the previous section for the plate aud cylinder are valid only when one-dimensional heat transfer is involved; e.g., for the cylinder there can be no axial temperature gradient. The more commonly encountered physical geometries are such that gradients in two or more directions exist, and the problem must be treated as multidimensional. To fix ideas, consider taking an ordinary brick from a kiln. The physical geometry of the brick is such that thermal gradients exist in the three mutually perpendicular directions, and the complete solution of the three-dimensional transient conduction equation in cartesian coordinates is required.

Fortunately, there is a mathematical technique which permits us to form the solution to a multidimensional problem from its one-dimensional components, for which we already have the solutions. Consider a two-dimensional transient problem resulting from the annealing of a very long rectangular bar of comparable width

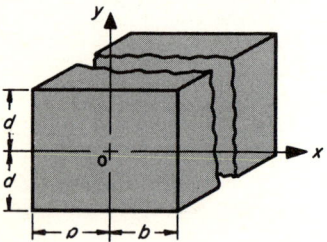

Fig. 6-17 Nomenclature for two-dimensional bar.

UNSTEADY-STATE CONDUCTIVE HEAT TRANSFER

and thickness in the x and y directions, respectively. Let us assume a uniform surface heat-transfer coefficient $\bar{h}$ and a uniform initial temperature T_i. The complete mathematical formulation of the problem is

$$\frac{\partial^2 \theta}{\partial x^2} + \frac{\partial^2 \theta}{\partial y^2} = \frac{1}{\alpha} \frac{\partial \theta}{\partial t} \tag{6-46}$$

with

Boundary conditions: $\quad \dfrac{\partial \theta}{\partial x} = \begin{cases} 0 & \text{at } x = 0 \\ -\dfrac{\bar{h}\theta}{k} & \text{at } x = b \end{cases}$

$$\tag{6-47}$$

$\dfrac{\partial \theta}{\partial y} = \begin{cases} 0 & \text{at } y = 0 \\ -\dfrac{\bar{h}\theta}{k} & \text{at } y = d \end{cases}$

Initial condition: $\quad \theta(x,y,0) = \theta_i$

where $\theta = T - T_\infty$. Here we note that there are two boundary conditions corresponding to each second-order derivative in different directions and one condition for the first-order time derivative.

To obtain a solution to Eq. (6-46), we again employ the separation-of-variables technique of Sec. 5-7. Assume a product solution of the form

$$\theta(x,y,t) = X(x,t) Y(y,t) \tag{6-48}$$

If this works, we can then obtain the solution of a two-dimensional transient conduction problem from the product of two one-dimensional transient solutions. Forming first and second partial derivatives from Eq. (6-48) and substituting into Eq. (6-46) yields

$$\frac{1}{X}\left(\frac{\partial^2 X}{\partial x^2} - \frac{1}{\alpha}\frac{\partial X}{\partial t}\right) = -\frac{1}{Y}\left(\frac{\partial^2 Y}{\partial y^2} - \frac{1}{\alpha}\frac{\partial Y}{\partial t}\right) \tag{6-49}$$

The left side of this equation is independent of y, and the right side is independent of x; thus each must be equal to some constant or function of time $\lambda(t)$. The geometry of the problem requires the solution in the y direction to be similar to that in the x direction, which can result only if λ is zero. This leads to two similar equations

$$\frac{\partial^2 X}{\partial x^2} = \frac{1}{\alpha}\frac{\partial X}{\partial t} \qquad \frac{\partial^2 Y}{\partial y^2} = \frac{1}{\alpha}\frac{\partial Y}{\partial t} \tag{6-50}$$

with the appropriate boundary and initial conditions of (6-47). Either of these, however, is identical with the previously solved one-dimensional problem of Eqs.

(6-28) and (6-28a), and thus we see, by Eq. (6-48), that the solution to the two-dimensional problem in cartesian coordinates is

$$\left(\frac{T-T_\infty}{T_i-T_\infty}\right)_{\substack{\text{long}\\\text{bar}}} = \left(\frac{T-T_\infty}{T_i-T_\infty}\right)_{\substack{2b\\\text{plate}}} \left(\frac{T-T_\infty}{T_i-T_\infty}\right)_{\substack{2d\\\text{plate}}} \tag{6-51}$$

where the plate solutions may be taken from Fig. 6-8. Similar treatments yield

$$\left(\frac{T-T_\infty}{T_i-T_\infty}\right)_{\substack{\text{parallelepiped}\\\text{(brick)}}} = \left(\frac{T-T_\infty}{T_i-T_\infty}\right)_{\substack{2b\\\text{plate}}} \left(\frac{T-T_\infty}{T_i-T_\infty}\right)_{\substack{2d\\\text{plate}}} \left(\frac{T-T_\infty}{T_i-T_\infty}\right)_{\substack{2L\\\text{plate}}} \tag{6-52}$$

and

$$\left(\frac{T-T_\infty}{T_i-T_\infty}\right)_{\substack{\text{short}\\\text{cyl}}} = \left(\frac{T-T_\infty}{T_i-T_\infty}\right)_{\substack{2L\\\text{plate}}} \left(\frac{T-T_\infty}{T_i-T_\infty}\right)_{\substack{\text{inf}\\\text{cyl}}} \tag{6-53}$$

where $2L$ is the length in the z direction for a brick-shaped object or the axial length of a short cylinder. These last two cases are of more general industrial importance than other geometrical shapes. A rather complete listing of solutions of this type for other configurations is given by Schneider [10].

Example A steel roller bearing at 1600°F is to be immersed in a liquid bath at 100°F during a heat-treating process. The bearing diameter is $\frac{1}{4}$ in, and its length is $\frac{1}{2}$ in. The heat-transfer coefficient $\bar{h}$ may be taken as 100 Btu/hr-ft²-°F. Determine the maximum temperature in the bearing 1 min after immersion; $\alpha = 0.425$ ft²/hr.

Solution The maximum temperature will obviously be at the center of the bearing, both radially and axially (at $r = 0$ and $\frac{1}{4}$ in from either end). The temperature at this point can be found from Eq. (6-53) with the two dimensionless temperatures obtained from Figs. 6-8 and 6-11.

$\theta_i = T_i - T_\infty = 1600 - 100 = 1500°F$

$k = 25$ Btu/hr-ft-°F (assume constant)(from Table A-4)

$\alpha = 0.425$ ft²/hr (assume constant)

$t = 1$ min $= 0.0168$ hr $\qquad \bar{h} = 100$ Btu/hr-ft²-°F

Plate solution

$$\text{Fo} = \frac{\alpha t}{L^2} = \frac{0.452(0.0167)}{(\frac{1}{48})^2} = 17.4$$

$$\frac{1}{\text{Bi}} = \frac{k}{\bar{h}L} = \frac{25}{100(\frac{1}{48})} = 12$$

From Fig. 6-8

$$\left(\frac{\theta_c}{\theta_i}\right)_{\substack{2L\\\text{plate}}} = \left(\frac{T_c-T_\infty}{T_i-T_\infty}\right)_{\substack{2L\\\text{plate}}} = 0.25$$

Cylinder solution

$$\text{Fo} = \frac{\alpha t}{R^2} = \frac{0.452(0.0167)}{(\frac{1}{96})^2} = 69.6$$

$$\frac{1}{\text{Bi}} = \frac{k}{hR} = \frac{25}{100(\frac{1}{96})} = 24$$

From Fig. 6-11

$$\left(\frac{\theta_c}{\theta_i}\right)_{\substack{\text{inf} \\ \text{cyl}}} = \left(\frac{T_c - T_\infty}{T_i - T_\infty}\right)_{\substack{\text{inf} \\ \text{cyl}}} = 0.0035$$

Then by Eq. (6-53),

$$\left(\frac{T - T_\infty}{T_i - T_\infty}\right)_{\text{max}} = 0.25(0.0035) = 0.000875$$

$$T_{\text{max}} = 100°\text{F} + 0.000875(1{,}500) = 101.3°\text{F}$$

It is interesting to note that if the bearing had been an infinitely long cylinder, the temperature at the centerline would have been $T_c = 100 + 0.0035(1{,}500) = 105.25°\text{F}$. Thus, the axial heat transfer has measurably lowered the temperature in the short-cylinder problem.

6-8 NUMERICAL METHOD

While the one-dimensional charts together with the extension to multidimensional problems offer ready solutions to transient problems of relatively simple physical geometry, there are numerous practical problems which do not lend themselves to exact mathematical solutions, because of the geometric shape or the nature of the boundary conditions. Such problems can be solved by numerical techniques.

To illustrate the basic technique, consider again the two-dimensional conduction problem of Fig. 5-25, where for the sake of simplicity we assume no z-direction temperature gradient, constant properties, and no internal heat generation. The appropriate differential equation for the temperature is

$$\frac{\partial^2 T}{\partial x^2} + \frac{\partial^2 T}{\partial y^2} = \frac{1}{\alpha}\frac{\partial T}{\partial t} \tag{6-54}$$

The problem is to cast this equation in finite-difference form. Forming an approximation of the first derivatives of T with respect to x at points c and a of Fig. 6-18 yields

$$\left.\frac{\partial T}{\partial x}\right|_c \simeq \frac{T_3 - T_n}{\Delta x} \qquad \left.\frac{\partial T}{\partial x}\right|_a \simeq \frac{T_n - T_1}{\Delta x}$$

The second derivative is then by definition

$$\frac{\partial^2 T}{\partial x^2} \simeq \frac{(\partial T/\partial x)_c - (\partial T/\partial x)_a}{\Delta x} = \frac{T_3 + T_1 - 2T_n}{(\Delta x)^2} \tag{6-55}$$

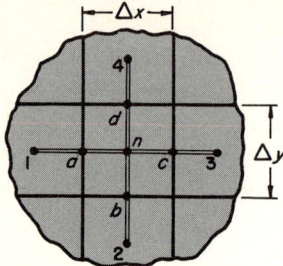

Fig. 6-18 Nomenclature for finite-difference equations.

and in a similar manner,

$$\frac{\partial^2 T}{\partial y^2} \simeq \frac{(\partial T/\partial y)_a - (\partial T/\partial y)_b}{\Delta y} = \frac{T_4 + T_2 - 2T_n}{(\Delta y)^2} \tag{6-56}$$

The time derivative can be approximated by

$$\frac{\partial T}{\partial t} \simeq \frac{T_n^{\tau+1} - T_n^{\tau}}{\Delta t} \tag{6-57}$$

and using Eqs. (6-55) to (6-57), Eq. (6-54) becomes

$$\frac{T_3^{\tau} + T_1^{\tau} - 2T_n^{\tau}}{(\Delta x)^2} + \frac{T_4^{\tau} + T_2^{\tau} - 2T_n^{\tau}}{(\Delta y)^2} = \frac{1}{\alpha} \frac{T_n^{\tau+1} - T_n^{\tau}}{\Delta t} \tag{6-58}$$

where T_n^{τ}, T_1^{τ}, T_2^{τ} ... designate the temperatures at nodes $n, 1, 2, \ldots$ at time τ and $T_n^{\tau+1}$ is the temperature at node n one time interval $\Delta \tau$ later. Now if we choose

$$\Delta x = \Delta y$$

Eq. (6-58) simplifies to

$$T_n^{\tau+1} = \frac{\alpha \, \Delta t}{(\Delta x)^2} (T_1^{\tau} + T_2^{\tau} + T_3^{\tau} + T_4^{\tau}) + \left[1 - \frac{4\alpha \, \Delta t}{(\Delta x)^2}\right] T_n^{\tau} \tag{6-59}$$

If the temperatures at the nodal points are known initially, the temperature at any node can be calculated at a time interval Δt later by Eq. (6-59). Obviously, this equation applies to any *interior* nodal point, and any number of calculations can be made corresponding to the desired number of time increments chosen to represent the total lapsed time. Applying this to every point in the grid network yields the temperature distribution at a selected number of time intervals Δt.

It is convenient to introduce the parameter

$$M = \frac{\Delta x^2}{\alpha \, \Delta t} \tag{6-60}$$

UNSTEADY-STATE CONDUCTIVE HEAT TRANSFER

and in terms of M, the one-, two-, and three-dimensional finite-difference equations are

$$T_n^{r+1} = \frac{1}{M}(T_1^r + T_3^r) + \left(1 - \frac{2}{M}\right)T_n^r \tag{6-61}$$

$$T_n^{r+1} = \frac{1}{M}(T_1^r + T_2^r + T_3^r + T_4^r) + \left(1 - \frac{4}{M}\right)T_n^r \tag{6-62}$$

$$T_n^{r+1} = \frac{1}{M}(T_1^r + T_2^r + T_3^r + T_4^r + T_5^r + T_6^r) + \left(1 - \frac{6}{M}\right)T_n^r \tag{6-63}$$

respectively. Note that in the one-dimensional case, Eq. (6-61) implies that the surrounding nodal points are T_1 and T_3 in the x direction, whereas in Eq. (6-63) it is clear that the surrounding nodal points in the z direction are 5 and 6.

In general, the smaller the time and space increments, the more accurate the solution and the larger these increments, the more rapidly the solution will be obtained. There is a restriction on the relationship of Δx and Δt for each physical problem, however, which simply stated is that the coefficient of T_n in the finite-difference equation cannot be negative. For example, if $1 - 2/M$ in Eq. (6-61) were negative, it would imply that the higher the temperature at node n, the lower it would be at a time interval later, a physically absurd situation. In order for all coefficients to be equal or greater than zero, we require

$M \geq 2$ for one-dimensional problems
$M \geq 4$ for two-dimensional problems
$M \geq 6$ for three-dimensional problems

and the selection of Δx then limits our choice of Δt. These requirements assure stability of the numerical solution. A detailed discussion of stability requirements and convergence of numerical solutions is given in Ref. 9.

While the preceding finite-difference equations can be used for any *interior* nodal point of a solid, they are not applicable to an exterior point on a convective boundary. An energy balance on the part of the one-dimensional system of Fig. 6-19 represented by the crosshatched area and unit depth is

$$\rho c \frac{\Delta x}{2} \Delta y \frac{T^{r+1} - T_n^r}{\Delta t} = k \Delta y \frac{T_1^r - T_n^r}{\Delta x} + \Delta y h(T_\infty - T_n^r) \tag{6-64}$$

which for $\Delta x = \Delta y$ becomes

$$T_n^{r+1} - T_n^r = \frac{2}{M}(T_1^r - T_n^r) + \frac{2}{M}\frac{\bar{h}\Delta x}{k}(T_\infty - T_n^r)$$

or

$$T^{r+1} = \frac{2}{M}\left(\frac{\bar{h}\Delta x}{k}T_\infty + T_1^r\right) + \left[1 - \frac{2}{M}\left(\frac{\bar{h}\Delta x}{k} + 1\right)\right]T_n^r \tag{6-65}$$

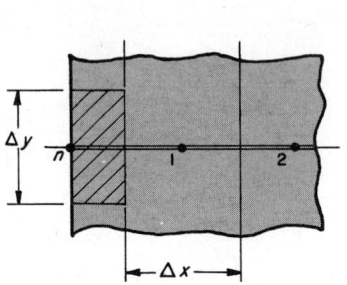
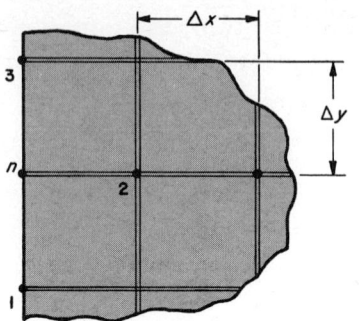

Fig. 6-19 Nomenclature for convective-boundary nodal point, one-dimensional.

Fig. 6-20 Nomenclature for convective-boundary nodal point, two-dimensional.

For a two-dimensional problem such as illustrated in Fig. 6-20, where $T_n \neq T_1 \neq T_3$, an approach analogous to the preceding yields

$$T_n^{\tau+1} = \frac{1}{M}\left(T_1^\tau + 2T_2^\tau + T_3^\tau + 2\frac{\bar{h}\Delta x}{k}T_\infty\right) + \left[1 - \frac{2}{M}\left(\frac{\bar{h}\Delta x}{k} + 2\right)\right]T_n^\tau \tag{6-66}$$

where $\Delta x = \Delta y$. To assure stability of these numerical solutions, we require

$$M \geq 2\left(\frac{\bar{h}\Delta x}{k} + 1\right) \quad \text{for one-dimensional problems}$$

$$M \geq 2\left(\frac{\bar{h}\Delta x}{k} + 2\right) \quad \text{for two-dimensional problems}$$

Clearly, Eq. (6-66) would require further modification if the nodal point were either an *interior* or *exterior* corner. For a detailed treatment of finite-difference techniques in transient conductive problems Dusinberre [3] should be consulted. The computational details for application of any of these nodal equations involve the knowledge of the initial temperature and the calculation of nodal temperatures at intervals of lapsed time. For hand calculation, it is advisable to set up a table to record the nodal temperatures throughout the field for each Δt interval.

The choice of a computer or manual program is largely one of economics; very complex and repetitive programs obviously should be computerized. In this regard, Schenck [9] is a helpful reference.

6-9 NONLINEAR TRANSIENT CONDUCTION

In many engineering problems the temperature differences are sufficiently large to cause appreciable error if constant conductivity is assumed. If the approximation of Sec. 5-2 for temperature dependence of the thermal conductivity

$$k = k_0(1 + aT)$$

is introduced, the transient differential equation becomes nonlinear. This form for conductivity variation was used by Dowty and Haworth [2], who employed numerical methods to obtain transient-temperature charts of the Heisler type for the case of a slab initially at a uniform temperature T_i and both surfaces suddenly changed to, and held at, an equal temperature T_b.

Zerkle and Sunderland [11] have presented time-temperature charts for a slab initially at T_i suddenly subjected to thermal radiant exchange with an environment at a constant temperature T_e. Their solution assumes a homogeneous slab with constant thermal properties and one face insulated. The nonlinearity arises due to the radiation at the other face which is proportional to the fourth power of the temperature. The reader interested in either of these nonlinear transient problems should consult the references cited for working charts and details of the solutions.

PROBLEMS

6-1. Consider a container full of frozen mercury that is suddenly exposed to an ambient temperature of 75°F. The container is cylindrical with a diameter of 2 in and a length of 6 in. Assume an average heat-transfer coefficient of 5 Btu/hr-ft²-°F. Could a lumped-system analysis be applied to this problem?

6-2. Consider a 14-lb$_m$ chunk of aluminum which is approximately spherical. It is at a uniform temperature of 600°F and is suddenly immersed in a fluid at 65°F. The convective-heat-transfer coefficient is 10 Btu/hr-ft²-°F. If you make the engineering approximation that the chunk is a sphere:

(a) Is it reasonable to neglect the internal resistance on making a thermal analysis?

(b) Regardless of your answer to part (a), determine the time required for the aluminum to cool to 100°F using the lumped-analysis approach.

6-3. An object in the shape of a right-circular cone has a diameter of 4 in and a height of 3 in. It has a thermal conductivity of 10 Btu/hr-ft-°F and a heat-transfer coefficient of 12 Btu/hr-ft²-°F. Could a lumped-system analysis be applied to this problem?

6-4. A triangular-shaped fin is 1 in thick and 6 in long at the base and is 2 in wide (perpendicular to the base surface). The fin is made of aluminum and has an average heat-transfer coefficient along the sides and ends of 15 Btu/hr-ft²-°F. There is no convective heat transfer at the base. Calculate the Biot modulus.

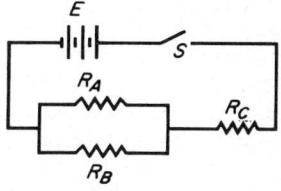

Fig. P6-5

6-5. Three resistors of equal resistance are in separate calorimeters, each of which contains 1.0 lb$_m$ of water. In a certain time t after closing the switch S the water in calorimeter A is 40°F hotter than before the switch was closed. How much has the temperature of the water in calorimeter C been increased?

6-6. Compare the temperature response of a 0.016-in-diam aluminum wire initially at 400°F when suddenly exposed to (*a*) forced air at 100°F ($\bar{h} = 12$ Btu/hr-ft^2-°F) and (*b*) still air at 100°F ($\bar{h} = 2$ Btu/hr-ft^2-°F). Sketch the temperature-time history of each case.

6-7. A stainless-steel $\tfrac{1}{2}$-in-diam ball bearing initially at a temperature of 100°F is suddenly exposed to a fluid whose temperature is 1000°F. The average heat-transfer coefficient between the bearing and the fluid is 12 Btu/hr-ft^2-°F.
    (*a*) Calculate the Biot modulus.
    (*b*) What is the temperature of the bearing after 1 min?

6-8. An iron ingot at a temperature of 2200°F is rolled into the shape of a rectangular parallelpiped with dimensions 1 ft by 6 in by 10 ft. The ingot is allowed to cool in a room where the temperature is 120°F. The average heat-transfer coefficient is 10 Btu/hr-ft^2-°F. Find the time required for the center temperature of the ingot to reach 400°F.

6-9. How deep must a water pipe be buried in the ground (assume dry, $\alpha = 0.016$ ft^2/hr) to maintain its temperature above 32°F during a 10 hr cold snap? The soil is initially at 60°F and in a very short time the surface temperature drops to 20°F.

6-10. A steel chisel, which may be approximated by a 1-in-diam cylinder 8 in long, at a temperature of 800°F is tempered by placing it in a liquid bath. The bath temperature is 75°F, and the average heat-transfer coefficient is 50 Btu/hr-ft^2-°F. What is the temperature of the chisel 30 sec after it is placed in the bath?

6-11. An epoxy which bonds at a temperature of 400°F is used to join two plates of brass face to face. The dimensions of the plates are $\tfrac{1}{4}$ by 2 by 4 in, and their initial temperature is 75°F. For bonding the plates will be placed in an oven, and for mass production it is necessary that the bonding time be less than 5 min. The average heat-transfer coefficient to the plates is 12 Btu/hr-ft^2-°F. What is the minimum temperature of the oven if the bonded faces are in complete thermal contact?

6-12. A home fireplace has 4-in walls made of fireclay brick. A fire is started which causes the surface of the brick to be heated to a temperature of 500°F. Determine the temperature on the opposite face of the brick after a period of 4 hr if the initial temperature was 60°F.

6-13. A very large and thick slab of steel is initially at a temperature of 1200°F when placed in an environment which changes its surface temperature to 90°F. What is the temperature of the slab at a depth of 4 in 20 min after its surface temperature is changed?

6-14. Water at 60°F is being used to reduce the temperature of a thick concrete wall which was initially at 180°F. Assuming the surface to be maintained at 60°F, how long would the water have to be applied to reduce the temperature 3 in below the surface of the wall to 100°F?

6-15. A large 2-ft-thick concrete isolation block for an air compressor is located in a paint factory which burns, subjecting the block to a uniform surface temperature of 2000°F. The ambient air was initially 70°F. How much time is required for the block to reach 1000°F at a depth of 1 ft? *Hint:* Consider the slab as a semi-infinite body.

6-16. A 4-in aluminum cube initially at a temperature of 80°F comes into contact with a surface at a temperature of 500°F. Assume all faces to be insulated except the one in contact. What is the center temperature of the cube after a period of 5 min?

6-17. A concrete wall 2 ft thick completely encloses a paint room in a factory. The wall is initially at 70°F when a fire erupts in the paint room and causes the inside of the wall to come in contact with hot gases at 1600°F. The outside of the wall is covered with a material that has a flash-point temperature of 400°F. The average heat-transfer coefficient on the hot side is 5 Btu/hr-ft^2-°F. Find the time required for the material on the outside of the wall to start burning.

UNSTEADY-STATE CONDUCTIVE HEAT TRANSFER 183

6-18. A long 2-in-diam aluminum cylinder initially at 500°F is suddenly exposed to an environment at 150°F. The average heat-transfer coefficient is 100 Btu/hr-ft²-°F. Find (a) the temperature at a radius of 0.9 in after 1 min and (b) the energy lost per unit length of the cylinder during the first minute after the cylinder is exposed to the environment.

6-19. A 4-in-diam iron ball is suddenly exposed to a stream of ice water. It is initially at 70°F, and $\bar{h} = 100$ Btu/hr-ft²-°F. What is its temperature $1\frac{1}{2}$ in from the surface 5 min later?

6-20. A solid steel 1.50 in-diam sphere is to be heat-treated in the following manner. The temperature of the whole sphere is to be raised uniformly to 1,400°F. Then the sphere is to be plunged into a large lead bath, where its surface is immediately brought to a temperature of 750°F and kept there until the center temperature of the sphere drops to 900°F. At this moment, the sphere is to be removed from the lead and quenched in a cold brine bath. The properties of the steel may be taken as follows: $\rho = 485$ lb$_m$/ft³, $c = 0.11$ Btu/lb$_m$-°F, $k = 20$ Btu/hr-ft-°F. How long should the sphere be kept in the lead bath if the heat-transfer coefficient is 200 Btu/hr-ft²-°F?

6-21. A 2-in-diam copper sphere initially at 300°F is cooled in a liquid bath at 100°F. The average heat-transfer coefficient is 100 Btu/hr-ft²-°F.
   (a) How long must the sphere remain in the bath for the center temperature to reach 120°F?
   (b) How much heat is removed from the sphere during this time?

6-22. Potatoes initially at 68°F are cooked by boiling at 212°F. Assume the potatoes to be approximately spherical with a $1\frac{1}{4}$-in radius. (Raw potatoes are normally 85 to 88 percent water by weight, and the external heat-transfer coefficient may be assumed to be greater than 100 Btu/hr-ft²-°F.)
   (a) Estimate the time required for the center to reach 200°F.
   (b) How much heat is added to each potato?

6-23. In preparation for a party, the host wishes to chill canned soft drinks from 75 to 35°F. The freezer temperature is 0°F, and an external heat-transfer coefficient of 5 Btu/hr-ft²-°F is appropriate. Assume the properties of the soft drink to be similar to those of water; the can dimensions are $2\frac{1}{2}$ by 6 in. By neglecting end effects approximate the time required for chilling.

6-24. In the winter citrus-fruit growers in Florida must concern themselves with frost damage to their crops. On a clear, windless night the average heat-transfer coefficient to spherically shaped fruit on trees is approximately 2 Btu/hr-ft²-°F. A part of this surface loss is actually due to radiation to the sky, which can be significantly reduced by the use of smudge pots. The ambient temperature is 52°F, and a cold front suddenly moves into the region, quickly lowering the temperature to 28°F, where it is expected to remain constant for 8 hr. Should smudge pots be used to protect 6-in-diam grapefruit? (Examine center temperature and surface temperature, assuming fruit properties to be the same as water.)

6-25. Tomatoes are processed in 8-in-diam cans 8 in long. After canning, the tomatoes are sterilized in a steam autoclave at 220°F. The initial temperature of the juice at the start of the sterilizing process is 120°F. Assume a very large heat-transfer coefficient, that the contents of the cans are at rest and have a density of 60 lb$_m$/ft³, specific heat of 0.9 Btu/lb$_m$-°F, and a thermal conductivity of 0.4 Btu/hr-ft-°F. Consider the temperature difference between the steam and the inner surfaces of the cans as negligible. How much time is required for the tomatoes to reach a sterilizing temperature of 200°F at the center of the can?

6-26. A steel ($k = 25$ Btu/hr-ft-°F) forging having a 2-in diameter and 4 in long is initially at 1600°F. It is heat-treated by immersion in a lead bath at 400°F. The convective-heat-transfer coefficient is 80 Btu/hr-ft²-°F. What is the temperature at the geometric center of the cylinder after 5 min?

6-27. At a picnic canned soft drinks which have been left in the sun are initially at a temperature of 100°F. They are then placed in a cooler containing ice water. The average heat-transfer coefficient is 50 Btu/hr-ft²-°F. The cans have a $2\frac{1}{2}$-in diameter and are 6 in long. Assume the properties of the soft drinks to be similar to those of water. Find the time required for the center temperature of the drinks to reach 40°F considering both radial and axial conduction.

6-28. A long piece of mild-steel stock 2 by 3 in initially at a temperature of 2200°F is heat-treated by quenching in an oil bath. The temperature of the oil bath is 200°F, and the average heat-transfer coefficient is assumed to be 80 Btu/hr-ft²-°F. What is the temperature at the geometric center after 2 min of quenching?

6-29. A solid 2-in glass cube initially at a temperature of 50°F is placed in boiling water. The average heat-transfer coefficient is 100 Btu/hr-ft²-°F. What is the temperature at the center of the cube after 10 min?

6-30. A masonry brick 2 by 4 by 8 in initially at 60°F is placed in a kiln at a temperature of 1800°F. Assume the average heat-transfer coefficient for the brick to be 10 Btu/hr-ft²-°F. What is the temperature at the geometric center after a period of 1 hr?

6-31. A home frozen-food chest has inner dimensions 3 by 3 by 8 ft. The walls are insulated with ½ in of fine glass wool. The chest is initially filled, and the stored items may be considered to have average properties two-thirds those tabulated for water. Assume the initial storage temperature is $-5°F$, the room temperature is 85°F, and there is a power failure. The average free-convection coefficient may be taken as 4 Btu/hr-ft²-°F. (Note that the resistance due to the insulation is significant.)

(a) How long can the power shortage last before the center temperature of the stored product reaches 20°F?

(b) Will this protect the stored items from spoiling; i.e., will the frozen-product temperature exceed 32°F at any location?

6-32. A 4-in-thick very long plate ($\alpha = 0.174$ ft²/hr) initially at 100°F is immersed in a fluid at a temperature of 800°F. By the numerical method calculate the temperature distribution across the plate at 0.5-in increments after 3 min.

6-33. A 1- by 1- by 2-in bar ($\alpha = 0.0868$ ft²/hr) is initially at a temperature of 80°F. Assume that the four rectangular surfaces are suddenly brought to temperatures of 100, 200, 300, and 400°F and the two square surfaces to 500 and 600°F. Using the numerical method, find the temperature at the geometric center of the bar after 15 sec.

6-34. A 2-in square steel bar initially at a temperature of 100°F is partially immersed in a fluid as shown. The fluid temperature is 500°F. The two upper faces of the bar are exposed to ambient air at a temperature of 80°F with an average heat-transfer coefficient of 10 Btu/hr-ft²-°F. (Assume $k = 25$ Btu/hr-ft-°F and $\alpha = 0.625$ ft²/hr.) Using the numerical method, calculate the temperature

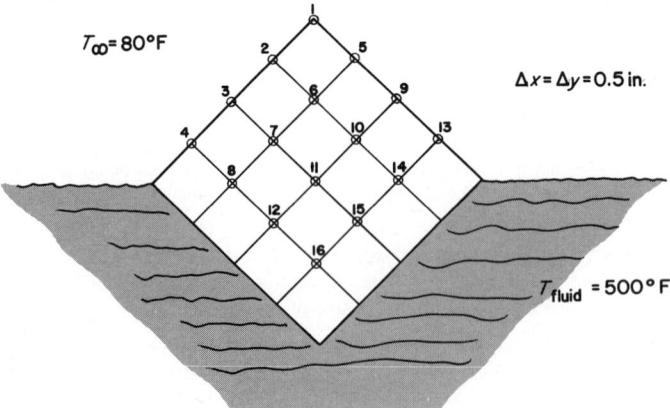

Fig. P6-34

distribution at the indicated nodal points after a period of (*a*) 1 min and (*b*) 2 min. The bar is very long; there is no heat flow along its length.

REFERENCES

1. Arpaci, V. S.: "Conduction Heat Transfer," Addison-Wesley, Reading, Mass., 1966.
2. Dowty, E. L., and D. R. Haworth: Solution Charts for Heat Conduction in Materials with Variable Thermal Conductivity, *ASME Paper* 65-WA/HT-29, 1965.
3. Dusinberre, G. M.: "Heat Transfer Calculations by Finite Differences," International Textbook, Scranton, Pa., 1961.
4. Gröber, H., S. Erk, and U. Grigull: "Grundgesetze der Wärmeübertragung," 3d ed., Springer-Verlag, Berlin, 1955.
5. Heisler, M. P.: Temperature Charts for Induction and Constant Temperature Heating, *Trans. ASME*, **69**: 227 (1947).
6. Holman, J. P.: "Heat Transfer," 2d ed., McGraw-Hill, New York, 1968.
7. Kreith, F.: "Principles of Heat Transfer," 2d ed., International Textbook, Scranton, Pa., 1965.
8. Kreysig, E.: "Advanced Engineering Mathematics," Wiley, New York, 1962.
9. Schenck, H.: "Fortran Methods in Heat Flow," Ronald, New York, 1963.
10. Schneider, P. J.: "Conduction Heat Transfer," Addison-Wesley, Reading, Mass., 1955.
11. Zerkle, R. D., and J. E. Sunderland: The Transient Temperature Distribution in a Slab Subject to Thermal Radiation, *J. Heat Transfer*, **87**: 117 (1965).

CHAPTER 7

MOLECULAR MASS TRANSFER (DIFFUSION)

In the preceding chapters we have dealt with single-component media whose behavior was characterized by local pressure and temperature gradients. In this chapter another driving force, concentration gradient, is introduced. This driving force causes the *transport of a component of a mixture from a region of high concentration to a region of low concentration.* The transport process is known as *mass transfer.*

The mechanisms of mass transfer are varied. Bird [2] classifies them into eight types:

1. Molecular (ordinary) diffusion, resulting from a concentration gradient
2. Thermal diffusion, arising from a temperature gradient
3. Pressure diffusion, which occurs by virtue of a pressure gradient
4. Forced diffusion, resulting from external forces other than gravity
5. Forced-convection mass transfer
6. Natural-convection mass transfer
7. Turbulent mass transfer, resulting from eddy currents in a fluid
8. Interphase mass transfer, occurring by virtue of nonequilibrium at an interface

MOLECULAR MASS TRANSFER (DIFFUSION)

These types divide naturally into two distinct modes of transport. The first four are molecular mass transfer; the last four are convective mass transfer. The process of molecular mass transfer, or diffusion, is presented in this chapter; convective mass transfer is covered in Chap. 14. Although the two modes often occur simultaneously, one mode usually dominates, and we can understand the mechanisms better by considering them separately.

Examples of mass transfer in everyday life are legion: the diffusion of sugar in a cup of coffee; vaporization of water in a teakettle; the movement of moisture-laden air over the ocean with its subsequent precipitation on dry land; combustion and air-conditioning processes; cloud formation; clothes drying. The chemical engineer is concerned with gas absorption, separation, crystallization, and extraction; the mechanical engineer confronts the mass-transfer process in humidification, drying, cutting and welding metals, ablation of heat shields in high-speed flight, deaeration of feedwater in steam boilers, and the production and heat treatment of metals; and civil engineers make use of mass transfer in waste treatment.

7-1 THE DIFFUSION MODE

This chapter will deal primarily with the molecular (ordinary) diffusion of binary (two-component) mixtures, typifying the diffusion process and being by far the most significant of the types of diffusion.

For the case of *thermal diffusion* in a binary mixture, the molecules of one component travel toward the hot region while the molecules of the other component tend to move toward the cold region; this is called the *Soret effect*. The inverse is the tendency to generate a thermal gradient with the development of a concentration gradient; this is called the *Dufour effect*. Thermal diffusion has been successfully used in the separation of isotopes.

Pressure diffusion results when a pressure gradient exists in a fluid mixture, e.g., in a closed deep well or in a closed tube which is rotated about an axis perpendicular to the tube's axis (centrifuge). The lighter component tends to move toward the low-pressure region.

An external force other than gravity in a mixture, when it acts in a different manner on the different components, results in *forced diffusion*. The diffusion of ions in an electrolyte in an electric field is a classic example of forced diffusion.

When thermal, pressure, and/or forced diffusion occur, a concentration gradient is developed, causing ordinary diffusion in the opposite direction. Upon reaching a steady state, the fluxes from the two (or more) types of diffusion sometimes offset each other, resulting in properties at a point being constant with time. The effects of thermal, pressure, and forced diffusion will be ignored in the introductory treatment of this chapter.

Mass transfer by diffusion is analogous to conduction heat transfer. Mass is transported by the movement of a species in the direction of its decreasing concentration, analogous to the energy exchange between molecules in the direction of decreasing temperature in conduction.

Ordinary diffusion may occur in gases, liquids, or solids. Because of the molecular spacing, the diffusion rate is much faster in gases than in liquids; it is faster in liquids than in solids.

7-2 FICK'S LAW

The fundamental equation (one-dimensional) of molecular diffusion, known as Fick's first law [9, 1, 19], derivable from the kinetic theory of gases and introduced as Eq. (2-9), can be written for a binary mixture as

$$J_A = -\rho D_{AB} \frac{\partial \omega_A}{\partial x} \tag{7-1}$$

where J_A = mass flux of molecular species A relative to mass average velocity of the mixture, $lb_m/hr\text{-}ft^2$

$\rho = \rho_A + \rho_B$, mass density (concentration) of the mixture, lb_m/ft^3

D_{AB} = mass diffusivity of species A with respect to species B, ft^2/hr

$\omega_A = \rho_A/\rho$, mass fraction, lb_m species A/lb_m mixture

Consider the two-compartment tank of Fig. 7-1 with one compartment containing gas A and the other compartment containing gas B. Both compartments are initially at a uniform pressure (see Ref. 8 for nonuniform pressures) and temperature throughout. When the partition between the compartments is removed, gas A moves to the right in accordance with Eq. (7-1) since its concentration in the right-hand compartment is initially zero; gas B moves to the left. This process continues until there is no difference in concentration throughout the vessel; i.e., molecules of gas A are uniformly interspersed with the molecules of gas B. The leftward diffusion of gas B is described by

$$J_B = -\rho D_{BA} \frac{\partial \omega_B}{\partial x} \tag{7-2}$$

Many alternate forms of the rate equation are given in the literature, written in terms of different potentials commonly encountered, e.g.,

$$J_A = -D_{AB} \frac{\partial \rho_A}{\partial x} \quad \text{constant density} \tag{7-3}$$

$$J_A = -\frac{D_{AB}}{R_A T} \frac{\partial p_A}{\partial x} \quad \text{ideal gas} \tag{7-4}$$

where ρ_A = mass concentration of species A, lb_m/ft^3

$p_A = p - p_B$, partial pressure of species A, lb_f/ft^2

R_A = gas constant for species A, $ft\text{-}lb_f/lb_m\text{-}°R$

T = absolute temperature of mixture, °R

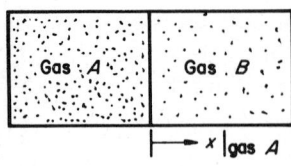

Fig. 7-1 Two-gas diffusion tank.

MOLECULAR MASS TRANSFER (DIFFUSION)

All the preceding forms are in terms of the pound mass. There are times, however, when it is more convenient for the engineer to work in terms of the pound mole. In this case the equations will be written with a tilde above quantities which are in terms of the pound mole. For example, the fundamental equation would become

$$\tilde{J}_A = -\tilde{\rho} \, D_{AB} \frac{\partial \tilde{\omega}_A}{\partial x} \tag{7-5}$$

where $\tilde{J}_A$ = molal flux of species A relative to molal avreage velocity of mixture, lb mole/hr-ft²
$\tilde{\rho} = \tilde{\rho}_A + \tilde{\rho}_B$, molal density of mixture
$\tilde{\omega}_A$ = mole fraction, lb mole species A/lb mole mixture

Also, for the other forms of the rate equation

$$\tilde{\rho}_A = \frac{\rho_A}{\mathcal{M}_A} \tag{7-6}$$

where $\tilde{\rho}_A$ = molar concentration of species A, lb mole/ft³
$\mathcal{M}_A$ = molecular weight of species A

Equation (7-6) is a convenient means of relating mass and molar concentrations. For simplicity, the pound mass will be primarily used throughout as the fundamental unit. Conversions are handily made by using Eq. (7-6).

7-3 THE DIFFUSION COEFFICIENT

The proportionality factor D_{AB} in Fick's law, known as the *mass diffusivity* or *coefficient of diffusion*, is a property of a specific system. Its value depends upon the system's pressure, temperature, and composition. Tables A-5, A-6, and A-7 give some experimental diffusivities for gases, liquids, and solids, respectively.

In the absence of experimental data, the following theoretical expressions give approximations, sometimes as valid as experimental values because of the difficulties encountered in their measurement.

Mass diffusivity in gases Using the kinetic theory of gases, Gilliland [10] evaluated the empirical coefficient required by the analysis of Sutherland [17], Chapman [6], and Jeans [14] to get the following equation for a binary gas system:

$$D_{AB} = 0.0069 \frac{T^{3/2}}{p(\tilde{\mathscr{V}}_A^{1/3} + \tilde{\mathscr{V}}_B^{1/3})^2} \sqrt{\frac{1}{\mathcal{M}_A} + \frac{1}{\mathcal{M}_B}} \tag{7-7}$$

where D_{AB} = mass diffusivity, ft²/hr
T = temperature of mixture, °R
p = pressure of mixture, atm
$\mathcal{M}$ = molecular weight
$\mathscr{V}$ = atomic or molecular volume (Table 7-1), ft³/lb mole

Table 7-1 Atomic and molecular volumes†

Substance	Volume, $ft^3/lb\ mole$‡	Substance	Volume, $ft^3/lb\ mole$‡
Air	29.9	Oxygen, molecule, O_2	7.4
Antimony	24.2	Coupled to two other elements:	
Arsenic	30.5	In aldehydes and ketones	7.4
Bismuth	48.0	In methyl esters	9.1
Bromine	27.0	In ethyl esters	9.9
Carbon	14.8	In higher esters and ethers	11.0
Chlorine, terminal as in R—Cl	21.6	In acids	12.0
Medial as in R—CHCl—R′	24.6	In union with S, P, N	8.3
Chromium	27.4	Phosphorus	27.0
Fluorine	8.7	Silicon	32.0
Germanium	34.5	Sulfur	25.6
Hydrogen, molecule, H_2	14.3	Tin	42.3
In compounds	3.7	Titanium	35.7
Iodine	37.0	Vanadium	32.0
Nitrogen, molecule, N_2	15.6	Zinc	20.4
In primary amines, N_2	10.5		
In secondary amines, N_2	12.0		

† Adapted from Warren M. Rohsenow and Harry Choi, Heat, Mass, and Momentum Transfer, © 1961. Reprinted by permission of Prentice-Hall, Inc., Englewood Cliffs, N.J.

‡ Except where noted, values given are for atomic volumes. For three-membered ring, e.g., ethylene oxide, deduct 6.0; for four-membered ring, e.g., cyclobutane, deduct 8.5; for five-membered ring, e.g., furan, deduct 11.5; for six-membered ring, e.g., benzene, deduct 15.0; for naphthalene ring, deduct 30.0; for anthracene ring, deduct 47.5

Kopp's law of additive atomic volumes applies in cases where compounds are involved [15], for example, $\tilde{\mathscr{V}}_{CH_4} = 14.8 + 4(3.7) = 29.6$ ft³/lb mole.

Of particular interest to the environmental engineer is the mass diffusivity of water vapor in air, given by the semiempirical relation

$$D_{wa} = \frac{0.000146}{p} \frac{T^{2.5}}{T + 441} \qquad (7\text{-}8)$$

where the units are the same as in Eq. (7-7) [5].

Example In order to produce a burning pattern of given characteristics in an asphalt heater, hydrogen, H_2, and methane, CH_4, are introduced at opposite ends of a 1-ft-long cylindrical manifold in an annular combustion chamber (Fig. 7-2a). The mixture, maintained at a temperature of 537°R and a total pressure of 1 atm, is continually injected into the combustion chamber, where it is burned with a stoichiometric mixture of oxygen. To maintain the desired burning pattern the partial pressures, in atmospheres, at the ends of the manifold are as shown in Table 7-2. Because of the geometry of the manifold, the diffusion of the hydrogen and methane may be assumed to be one-dimensional. Estimate the diffusion rates of both gases.

MOLECULAR MASS TRANSFER (DIFFUSION)

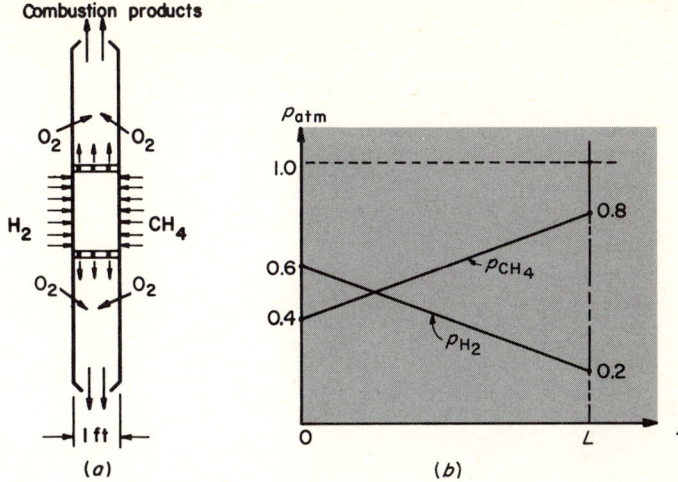

Fig. 7-2 Equimolal counterdiffusion of gases: (*a*) annular combustion chamber; (*b*) partial-pressure gradients in manifold.

Solution Assuming the gases to be perfect, Eq. (7-4) can be used since it is expressed in terms of partial pressures. It will first be necessary, however, to estimate the mass diffusivity $D_{H_2-CH_4}$. This can be done by using Eq. (7-7), where

$\tilde{V}_{H_2} = 14.3$ ft^3/lb-mole   and $\tilde{V}_{CH_4} = 29.6$ ft^3/lb-mole

using the atomic and molecular volumes from Table 7-1. From Table A-10

$\mathcal{M}_{H_2} = 2.02$ lb$_m$/lb mole $R_{H_2} = 766.53$ ft-lb$_f$/lb$_m$-°R

$\mathcal{M}_{CH_4} = 16.04$ lb$_m$/lb mole $R_{CH_4} = 96.40$ ft-lb$_f$/lb$_m$-°R

Therefore, using Eq. (7-7),

$$D_{H_2-CH_4} = D_{CH_4-H_2} = 0.0069 \frac{537^{\frac{3}{2}}}{1(14.3^{\frac{1}{3}} + 29.6^{\frac{1}{3}})^2} \sqrt{\frac{1}{2.02} + \frac{1}{16.04}}$$

$$= \frac{0.0069(12,500)}{(2.426 + 3.091)^2} \sqrt{0.495 + 0.0623} = \frac{0.0069(12,500)(0.747)}{32}$$

$$= 2.016 \text{ ft}^2/\text{hr} = 0.56 \times 10^{-3} \text{ ft}^2/\text{sec}$$

Note that this semiempirical value differs from the experimental value of 0.6730×10^{-3} ft^2/sec given in Table A-5. From the engineering point of view it is better to use the experimental

Table 7-2

	$x = 0$	$x = L$
p_{H_2}	0.6	0.2
p_{CH_4}	0.4	0.8

value; therefore,

$$D_{H_2-CH_4} = D_{CH_4-H_2} = 2.42 \text{ ft}^2/\text{hr}$$

Separating the variables in Eq. (7-4) and integrating,

$$J_A \int_0^L dx = -\frac{D_{AB}}{R_A T} \int_{p_A|_{x=0}}^{p_A|_{x=L}} dp_A$$

$$J_A = -\frac{D_{AB}}{R_A T} \frac{p_A(L) - p_A(0)}{L}$$

or for each gas

$$J_{H_2} = -\frac{D_{H_2-CH_4}}{R_{H_2} T} \frac{p_{H_2}(L) - p_{H_2}(0)}{L}$$

$$J_{CH_4} = -\frac{D_{CH_4-H_2}}{R_{CH_4} T} \frac{p_{CH_4}(L) - p_{CH_4}(0)}{L}$$

giving

$$J_{H_2} = -\frac{2.42 \text{ ft}^2/\text{hr}}{(766.53 \text{ ft-lb}_f/\text{lb}_m\text{-°R})(537°R)} \frac{0.2 - 0.6 \text{ atm}}{1 \text{ ft}} \frac{2{,}116 \text{ lb}_f/\text{ft}^2}{1 \text{ atm}}$$

$$= 0.0050 \text{ lb}_m/\text{hr-ft}^2$$

and

$$J_{CH_4} = -\frac{2.42 \text{ ft}^2/\text{hr}}{(96.4 \text{ ft-lb}_f/\text{lb}_m\text{-°R})(537°R)} \frac{0.8 - 0.4 \text{ atm}}{1 \text{ ft}} \frac{2{,}116 \text{ lb}_f/\text{ft}^2}{1 \text{ atm}}$$

$$= -0.0396 \text{ lb}_m/\text{hr-ft}^2$$

It is instructive to note the flux in terms of pound moles, namely,

$$\tilde{J}_{H_2} = \left(0.0050 \frac{\text{lb}_m}{\text{hr-ft}^2}\right) \frac{\text{lb mole}}{2.02 \text{ lb}_m} = 0.00248 \text{ lb mole/hr-ft}^2$$

$$\tilde{J}_{CH_4} = \left(-0.0396 \frac{\text{lb}_m}{\text{hr-ft}^2}\right) \frac{\text{lb mole}}{16.04 \text{ lb}_m} = -0.00248 \text{ lb mole/hr-ft}^2$$

or

$$\tilde{J}_{H_2} = -\tilde{J}_{CH_4}$$

commonly known as *equimolal counterdiffusion*. This might have been deduced by noting that

$$p_{H_2} + p_{CH_4} = p$$

giving

$$\frac{dp_{H_2}}{dx} + \frac{dp_{CH_4}}{dx} = 0$$

or

$$\frac{(JR)_{H_2} T}{D_{H_2-CH_4}} + \frac{(JR)_{CH_4} T}{D_{CH_4-H_2}} = 0$$

$$(JR)_{H_2} + (JR)_{CH_4} = 0$$

MOLECULAR MASS TRANSFER (DIFFUSION)

Introducing the universal gas constant $\mathscr{R} = RM$,

$$\left.\frac{J}{M}\right|_{H_2} \mathscr{R} + \left.\frac{J}{M}\right|_{CH_4} \mathscr{R} = 0$$

or

$$\tilde{J}_{H_2} + \tilde{J}_{CH_4} = 0$$

Mass diffusivity in liquids Because of the higher molecular density in liquids, their mass diffusivities are much smaller than those of gases. Little is known about the prediction of liquid mass diffusivities. For dilute solutions Wilke's [20] work led to the following semiempirical equation [15, 16]:

$$D_{AB} = \frac{4.0 \times 10^{-7} T}{\mu(\tilde{V}_A^{1/3} - \Phi_B)} \tag{7-9}$$

where D_{AB} = mass diffusivity, ft²/hr
 T = temperature of the solution, °R
 μ = dynamic viscosity of the solution (approximately equal to that of the solvent in dilute solutions), lb$_m$/ft-hr
 $\tilde{V}_A$ = atomic or molecular volume of solute (Table 7-1), ft³/lb mole
 Φ_B = 2.0 for water; 2.46 for ethyl alcohol (ethanol), C_2H_5OH; and 2.84 for benzene, C_6H_6

This equation gives approximate results for dilute solutions of nondissociating solutes. In concentrated solutions the viscosity changes with concentration. In the case of strong electrolytes dissolved in water, the diffusion rates are those of the individual ions [18].

Example Estimate the mass diffusivity of grain alcohol (ethanol) in a 95% water solution at 77°F and compare with the experimental data of Table A-6.
Solution From Table 7-1 the molecular volume for ethanol is

$$\tilde{V}_{C_2H_5OH} = 2\tilde{V}_C + 6\tilde{V}_H + \tilde{V}_O = 2(14.8) + 6(3.7) + 7.4 = 59.2 \text{ ft}^3/\text{lb mole}$$

From Table A-3 the dynamic viscosity is

$$\mu = (6.12 \times 10^{-4} \text{ lb}_m/\text{ft-sec})(3{,}600 \text{ sec/hr}) = 2.2 \text{ lb}_m/\text{ft-hr}$$

Using Eq. (7-9),

$$D_{ew} = \frac{(4.0 \times 10^{-7})(537)}{2.2(59.2^{1/3} - 2.0)} = (0.513 \times 10^{-4} \text{ ft}^2/\text{hr}) \frac{\text{hr}}{3{,}600 \text{ sec}} = 1.43 \times 10^{-8} \text{ ft}^2/\text{sec}$$

This is in good agreement with the value of 1.216×10^{-8} ft²/sec given in Table A-6. It emphasizes, however, the limitation of Eq. (7-9) to dilute solutions; values differ for more concentrated solutions.

Mass diffusivity in solids The mechanism of mass transfer in solids is very complex, involving the diffusion of gases or liquids throughout a heterogeneous network of pores and the diffusion of atoms within the solid material. Because of the interrelation of these mechanisms, theoretical prediction of mass diffusivities in solids has been unsuccessful. The first of these mechanisms is of great importance to the chemical engineer in catalysis, e.g., in reactors where the rate of reaction is critical or needs to be optimized. The atomic movement is important to the metallurgist in the production and heat treatment of metals. Solid-in-solid diffusion occurs, for example, when iron is heated in a bed of coke. In this case, the carbon diffuses into the iron, having the highest concentration near the surface.

The most accurate method of determining the mass diffusivity is by plating a radioactive isotope of the solute on one surface of the solvent. The rate of decay of surface radioactivity is then a direct measure of the rate of penetration of the solute into the solvent. Hutchison and Baird [13] cite some common mass diffusivities encountered in metallurgical processes. Experimental values for a few select solid systems are given in Table A-7.

7-4 FLUXES IN BINARY MIXTURES

In our study of the elementary physics and mechanics we learned to appreciate the concept of relative velocity of particles and single-component fluids (mixtures), realizing that one's viewpoint depends upon one's reference frame. In mass transfer the concept is somewhat complicated by the various species of a multicomponent mixture moving at different velocities. The fluxes encountered in the mass-transfer process are summarized in Table 7-3.

If an aggregate of particles which move retain the same position relative to each other, the velocity of any one particle is the velocity of the aggregate. When the individual particles move at different velocities, however, the velocity of the aggregate is not intuitively apparent. In this case we define the velocity of the aggregate as the mass flux of the component, say species A, divided by the mass concentration of A

$$u_A \equiv \frac{N_A}{\rho_A} \qquad (7\text{-}10)$$

But the mass flux of a binary mixture is

$$N = N_A + N_B \qquad (7\text{-}11a)$$

or

$$\rho u = \rho_A u_A + \rho_B u_B \qquad (7\text{-}11b)$$

Since $\omega_A \equiv \rho_A/\rho$, Eq. (7-11b) can be written

$$u = \omega_A u_A + \omega_B u_B \qquad (7\text{-}11c)$$

MOLECULAR MASS TRANSFER (DIFFUSION)

Table 7-3 Fluxes in binary systems (one-dimensional)

Mass flux, $lb_m/hr\text{-}ft^2$	Molal flux, $lb\ mole/hr\text{-}ft^2$	Reference frame
N	$\tilde{N}$	Fixed axes
J	$\tilde{J}$	Axes moving at the mass average velocity $u = \dfrac{1}{\rho}(\rho_A u_A + \rho_B u_B)$
J^*	$\tilde{J}^*$	Axes moving at the molal average velocity $\tilde{u} = \dfrac{1}{\tilde{\rho}}(\tilde{\rho}_A u_A + \tilde{\rho}_B u_B)$

The flux J_A is measured across a plane which moves with the mass average velocity of the mixture. For the interdiffusion of two fluids of equal density in a closed stationary tank, the plane is stationary. The flux J_A^* is that which crosses a plane moving with the molal average velocity of the mixture. For the case of equimolal counterdiffusion of two gases in a closed stationary vessel, illustrated by the first example of Sec. 7-3, the plane is fixed in space. In most other cases there is an additional flux because of *bulk flow* with respect to a fixed reference frame. It is important then, even in stationary media, to understand the various fluxes. We shall be concerned primarily, however, with relating the flux to a fixed reference, such as a fluid interface, rather than to average velocities. Hence, we shall primarily use N and $\tilde{N}$.

From their definitions, the fluxes are related as follows:

$$N_A = \rho_A u_A \tag{7-10}$$

$$J_A = \rho_A(u_A - u) = N_A - \rho_A u \tag{7-12}$$

$$J_A^* = \rho_A(u_A - \tilde{u}) = N_A - \rho_A \tilde{u} \tag{7-13}$$

where $u_A - u$ is the diffusion velocity of A with respect to u and $u_A - \tilde{u}$ is the diffusion velocity of A with respect to $\tilde{u}$. Eliminating N_A between Eqs. (7-12) and (7-13), we get

$$J_A + \rho_A u = J_A^* + \rho_A \tilde{u} \tag{7-14}$$

The bulk flow is given by the terms $\rho_A u$ and $\rho_A \tilde{u}$. Using Fick's law in Eq. (7-12), for example, illustrates how the bulk flow contributes to the absolute motion of species A; that is,

$$N_A = -\rho D_{AB} \frac{\partial \omega_A}{\partial x} + \rho_A u \tag{7-15}$$

Expressing u in terms of the mass fraction, using Eqs. (7-10) and (7-11c), and simplifying, we get

$$\blacksquare \quad N_A = -\rho D_{AB} \frac{\partial \omega_A}{\partial x} + \omega_A (N_A + N_B) \tag{7-16}$$

where N_A and N_B are the mass fluxes of species A and B, respectively, in the x direction. Extended to three dimensions, in vector form, the equation becomes

$$\mathbf{N}_A = -\rho D_{AB} \nabla \omega_A + \omega_A (\mathbf{N}_A + \mathbf{N}_B) \tag{7-17}$$

where the mass flux $\mathbf{N}_A$ is the vector sum of the flux resulting from the *concentration-gradient contribution*, $-\rho D_{AB} \nabla \omega_A$, and the flux resulting from the *bulk-motion contribution*, $\omega_A (\mathbf{N}_A + \mathbf{N}_B)$.

In order to solve either Eq. (7-16) or (7-17) something must be known about the relationship between $\mathbf{N}_A$ and $\mathbf{N}_B$, and the boundary conditions must be specified. We shall consider some simple cases commonly encountered in the practice of engineering, including (1) the diffusion of A into a stagnant B, $N_B = 0$, and (2) processes in which chemical reactions occur. For the latter, there are two types of chemical reactions: *homogeneous*, in which the chemical change occurs uniformly throughout the entire volume of the medium, and *heterogeneous*, in which the chemical change takes place only at a restricted location, such as the surface of a catalyst. The governing equations and boundary conditions for these two types of chemical reaction are quite different. The rate of production of a species in a homogeneous reaction appears in the differential equation as a production term, analogous to the heat-generation term of the heat conductive Eq. (5-5). For the heterogeneous type the rate of production appears in the boundary condition at the surface where the reaction occurs. Before considering specific cases, however, we shall develop the continuity equation for a given species, a conservation law requiring that a given species be neither created nor destroyed.

7-5 DIFFERENTIAL FORM OF THE MASS-DIFFUSION EQUATION

Consider the conservation of species A in the element of volume $\Delta x \, \Delta y \, \Delta z$ shown in Fig. 7-3. This could be a cube taken from a cup of coffee through which sugar, species A, is diffusing; it might represent an element of fuel burning. The flux of species A is shown only for the x direction for clarity.

Applying the conservation principle [Eq. (2-14)] to the volume element gives

Rate at which species A enters + rate at which species A is produced =

rate at which species A leaves + rate at which species A accumulates (7-18)

MOLECULAR MASS TRANSFER (DIFFUSION)

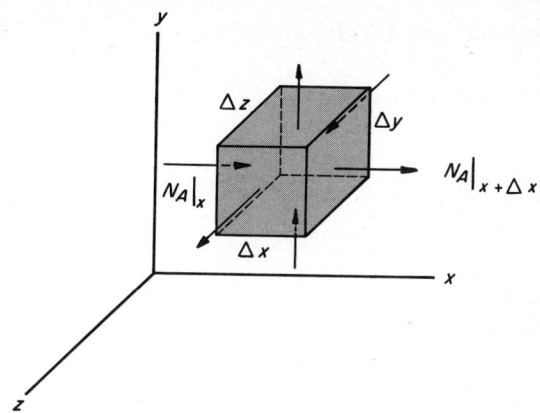

Fig. 7-3 Differential volume element.

This is directly analogous to the heat-conduction word equation (5-1). Species A accumulates at the rate of

$$\frac{\partial \rho_A}{\partial t} \Delta x \, \Delta y \, \Delta z$$

and the mass rate of A produced is

$$r_A''' \, \Delta x \, \Delta y \, \Delta z$$

where r_A''' is the mass of A produced per unit volume per unit time. Using these terms and the fluxes of Fig. 7-3 in Eq. (7-18), we get

$$N_A|_x \Delta y \, \Delta z + N_A|_y \Delta x \, \Delta z + N_A|_z \Delta x \, \Delta y + r_A''' \Delta x \, \Delta y \, \Delta z$$
$$= N_A|_{x+\Delta x} \Delta y \, \Delta z + N_A|_{y+\Delta y} \Delta x \, \Delta z + N_A|_{z+\Delta z} \Delta x \, \Delta y + \frac{\partial \rho_A}{\partial t} \Delta x \, \Delta y \, \Delta z$$

where the subscripts x, y, and z denote both the flux direction and the respective faces of the volume element. Dividing by $\Delta x \, \Delta y \, \Delta z$ and taking the limit as Δx, Δy, and Δz approach zero, we get

$$\frac{\partial N_{A_x}}{\partial x} + \frac{\partial N_{A_y}}{\partial y} + \frac{\partial N_{A_z}}{\partial z} + \frac{\partial \rho_A}{\partial t} - r_A''' = 0 \tag{7-19a}$$

or, in vector form,

$$\boldsymbol{\nabla} \cdot \mathbf{N}_A + \frac{\partial \rho_A}{\partial t} - r_A''' = 0 \tag{7-19b}$$

We can write the conservation equation for species B similarly, giving

$$\boldsymbol{\nabla} \cdot \mathbf{N}_B + \frac{\partial \rho_B}{\partial t} - r_B''' = 0 \tag{7-20}$$

Adding Eqs. (7-19b) and (7-20) gives

$$\nabla \cdot (\mathbf{N}_A + \mathbf{N}_B) + \frac{\partial(\rho_A + \rho_B)}{\partial t} - (r_A''' + r_B''') = 0 \tag{7-21}$$

but

$$\mathbf{N}_A + \mathbf{N}_B = \rho_A \mathbf{V}_A + \rho_B \mathbf{V}_B = \rho \mathbf{V}$$

$$\rho_A + \rho_B = \rho$$

and

$$r_A''' = -r_B'''$$

in order to conserve mass. Hence, Eq. (7-21) becomes

$$\nabla \cdot \rho \mathbf{V} + \frac{\partial \rho}{\partial t} = 0 \tag{7-22}$$

the *conservation equation for a mixture*. This is identical to the differential continuity equation of Chap. 10.

If we substitute the mass-flux equation (7-17) into the conservation equation (7-19b), the result is the *general binary diffusion equation for species A*:

$$-\nabla \cdot \rho D_{AB} \nabla \omega_A + \nabla \cdot \rho_A \mathbf{V} + \frac{\partial \rho_A}{\partial t} - r_A''' = 0 \tag{7-23}$$

where

$$\rho_A \mathbf{V} = \omega_A (\mathbf{N}_A + \mathbf{N}_B) \tag{7-24}$$

While this equation is too unwieldy to use in its present form, most engineering processes are such that its application to them simplify its form and facilitate its use. Table 7-4 presents some simplified forms of the equation with their restrictions.

The mass-diffusion equation (7-27) is analogous to the heat-conduction equation (5-5); Eq. (7-28) is analogous to Eq. (5-8). This similarity is the basis for the parallel treatment of heat conduction and mass diffusion. These equations have been expressed in vector form, making them applicable to any orthogonal coordinate system. With the aid of Appendix B the equations can be transformed to the desired coordinate system by expressing the substantial derivative $D\rho_A/Dt$ and the laplacian $\nabla^2 \rho_A$ in the appropriate forms.

For no chemical reaction, Eq. (7-27) reduces to

$$\frac{\partial \rho_A}{\partial t} = D_{AB} \nabla^2 \rho_A \tag{7-29}$$

which is known as *Fick's second law of diffusion*. Solutions to this equation for many commonly encountered boundary conditions of engineering interest may be found in the works of Carslaw and Jaeger [4] and Crank [7].

MOLECULAR MASS TRANSFER (DIFFUSION)

Table 7-4 Simplified forms of the mass-diffusion equation

Assumption	Conservation of species A equation		Used for
ρ = const D_{AB} = const	$\dfrac{\partial \rho_A}{\partial t} + \rho_A(\nabla \cdot \mathbf{V}) + (\mathbf{V} \cdot \nabla \rho_A) = D_{AB}\nabla^2 \rho_A + r_A'''$ $\dfrac{D\tilde{\rho}_A}{Dt} = D_{AB}\nabla^2 \tilde{\rho}_A + \tilde{r}_A'''$	(7-25a) (7-25b)	Dilute liquid solutions at constant pressure and temperature
$\tilde{\rho}$ = const D_{AB} = const	$\dfrac{\partial \tilde{\rho}_A}{\partial t} + (\tilde{\mathbf{V}} \cdot \nabla \tilde{\rho}_A) = D_{AB}\nabla^2 \tilde{\rho}_A + \tilde{r}_A'''$ $\qquad\qquad - \tilde{\omega}_A(\tilde{r}_A''' + \tilde{r}_B''')$	(7-26)	Low-density gases at constant temperature and pressure
$\tilde{\rho}$ = const D_{AB} = const $\mathbf{V}=0$ (no motion)	$\dfrac{\partial \tilde{\rho}_A}{\partial t} = D_{AB}\nabla^2 \tilde{\rho}_A + \tilde{r}_A'''$	(7-27)	Solids; stationary liquids; equimolal counterdiffusion of gases
ρ = const D_{AB} = const $\mathbf{V}=0$ $\partial \rho_A/\partial t = 0$ No chemical reaction	$\nabla^2 \rho_A = 0$	(7-28)	Any of the above media when the process is steady state

Boundary conditions The analogy between the diffusion of mass and heat is completed with the boundary conditions, which are also very similar. Some typical ones follow:

1. *The concentration at a surface* ($x=0$) *may be specified.* It may be in terms of
   a. Mass concentration $\rho_A = \rho_{A_s}$
   b. Molal concentration $\tilde{\rho}_A = \tilde{\rho}_{A_s}$
   c. Mass fraction $\omega_A = \omega_{A_s}$
   d. Mole fraction $\tilde{\omega}_A = \tilde{\omega}_{A_s}$
   e. Partial pressure (perfect gas) $p_A = p_{A_s}$
2. *The flux at a surface* ($x=0$) *can be given*, for example, $J_A = J_{A_s}$ or $N_A = N_{A_s}$. Cases of engineering interest are
   a. Specified mass flux

$$J_{A_s} = -D_{AB}\left.\dfrac{\partial \rho_A}{\partial x}\right|_{x=0} \qquad (7\text{-}30)$$

   b. Impermeable surface, e.g., vapor barrier,

$$J_{A_s} = 0 \qquad \left.\dfrac{\partial \rho_A}{\partial x}\right|_{x=0} = 0 \qquad (7\text{-}31)$$

c. *Mass flux to a surrounding fluid;* e.g., if the mass-diffusion equation is written for a solid through which diffusion occurs, mass may be lost at the surface to the surrounding fluid in accordance with the relation

$$N_{A_s} = k_c(\rho_{A_s} - \rho_{A_\infty}) \tag{7-32}$$

where k_c = *convective-mass-transfer coefficient*
ρ_{A_s} = surface mass concentration
ρ_{A_∞} = mass concentration in fluid stream

This equation is analogous to Newton's law of cooling, $q'' = h(T_s - T_\infty)$ [Eq. (5-41)]. Chapter 14 will deal primarily with this relation.

3. *The rate of chemical reaction may be specified.* For example, species A may disappear (or appear) at a surface in accordance with the first-order chemical reaction

$$N_{A_s} = k_1 \rho_A \tag{7-33}$$

where k_1 is the chemical rate constant.

The remainder of this chapter will be devoted to the application of the steady- and unsteady-state mass-diffusion equation to a variety of one-dimensional cases, each of which will be developed by making a *shell mass balance*. In each case the student should verify that the resulting differential equation could have been attained from the general equations of this section. Use of the various boundary conditions will also be illustrated.

7-6 DIFFUSION THROUGH A STAGNANT GAS

The Arnold diffusion cell of Fig. 7-4, which is often used to measure mass diffusivities experimentally, contains a pure liquid A which vaporizes and diffuses into the stagnant column of gas B. The fluid which diffuses upward is carried away by a gas stream

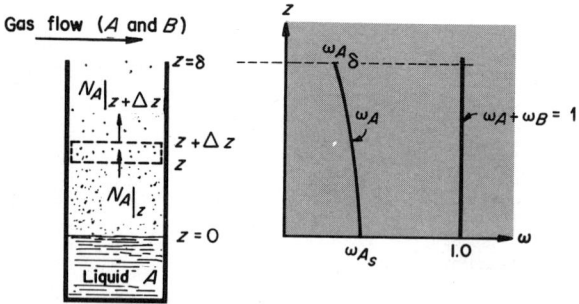

Fig. 7-4 Arnold diffusion cell.

MOLECULAR MASS TRANSFER (DIFFUSION)

at the top flowing normal to the cell. We shall assume that the liquid level is maintained at the position $z = 0$.

For steady-state diffusion, a shell mass balance over the control volume shown gives

$$SN_A|_z - SN_A|_{z+\Delta z} = 0 \tag{7-34}$$

where S is the cross-sectional area of the column. Dividing by $S \Delta z$ and taking the limit as Δz approaches zero, we get

$$-\lim_{z \to 0} \frac{N_A|_{z+\Delta z} - N_A|_z}{\Delta z} \equiv -\frac{dN_A}{dz} = 0 \tag{7-35}$$

The mass flux is given by Eq. (7-16), which for this case becomes

$$N_A = -\frac{\rho D_{AB}}{1 - \omega_A} \frac{d\omega_A}{dz} \tag{7-36}$$

since $N_B = 0$. Substitution of this mass flux into the diffusion equation (7-35) gives

$$\frac{d}{dz}\left(\frac{\rho D_{AB}}{1 - \omega_A} \frac{d\omega_A}{dz}\right) = 0 \tag{7-37}$$

which becomes

$$\frac{d}{dz}\left(\frac{1}{1 - \omega_A} \frac{d\omega_A}{dz}\right) = 0 \tag{7-38}$$

for a perfect-gas mixture at constant temperature and pressure with $\rho D_{AB} =$ constant. This second-order differential equation requires two boundary conditions which are shown in Fig. 7-4, i.e.,

(1) at $z = 0$: $\quad \omega_A = \omega_{As}$ \hfill (7-39a)

(2) at $z = \delta$: $\quad \omega_A = \omega_{A\delta}$ \hfill (7-39b)

Integrating Eq. (7-38) with respect to z gives

$$\frac{1}{1 - \omega_A} \frac{d\omega_A}{dz} = C_1 \tag{7-40}$$

and a second integration yields

$$-\ln(1 - \omega_A) = C_1 z + C_2 \tag{7-41}$$

where C_1 and C_2 are the constants of integration, which can be evaluated from the boundary conditions, giving

$$\blacksquare \quad \frac{1 - \omega_A}{1 - \omega_{As}} = \left(\frac{1 - \omega_{A\delta}}{1 - \omega_{As}}\right)^{z/\delta} \tag{7-42}$$

or, since $\omega_A + \omega_B = 1$,

$$\frac{\omega_B}{\omega_{Bs}} = \left(\frac{\omega_{B\delta}}{\omega_{Bs}}\right)^{z/\delta} \tag{7-43}$$

Equations (7-42) and (7-43) describe the mass fractions of A and B; their gradients are shown in Fig. 7-4. It is interesting to note that the mass fraction of B is not constant even though B is stagnant, being influenced by the diffusion of A. But for engineering calculations, we normally need the mass flux rather than the concentration gradient. To that end it is advantageous to determine the average mass fraction, defined by

$$\bar{\omega} = \frac{\int_{z_1}^{z_2} \omega \, dz}{\int_{z_1}^{z_2} dz} \tag{7-44}$$

For species B this becomes

$$\frac{\bar{\omega}_B}{\omega_{Bs}} = \frac{\int_0^\delta \frac{\omega_B}{\omega_{Bs}} dz}{\int_0^\delta dz} \tag{7-45}$$

where 0 and δ are taken as the locations between which the average mass fraction is needed. Substituting Eq. (7-43) into Eq. (7-45) and solving, we get

$$\frac{\bar{\omega}_B}{\omega_{Bs}} = \frac{\int_0^\delta \left(\frac{\omega_{B\delta}}{\omega_{Bs}}\right)^{z/\delta} dz}{\int_0^\delta dz} = \frac{1}{\delta} \left[\frac{\left(\frac{\omega_{B\delta}}{\omega_{Bs}}\right)^{z/\delta}}{\frac{1}{\delta} \ln\left(\frac{\omega_{B\delta}}{\omega_{Bs}}\right)} \right]_0^\delta \tag{7-46}$$

$$\bar{\omega}_B = \frac{\omega_{B\delta} - \omega_{Bs}}{\ln(\omega_{B\delta}/\omega_{Bs})} \equiv (\omega_B)_{lm} \tag{7-47}$$

which is defined as the log-mean mass fraction $(\omega_B)_{lm}$, being the difference between the terminal values divided by the natural logarithm of their ratio.

The mass flux is given by Eq. (7-36). Since it is constant throughout the stagnant gas, the equation may be written for any location in the gas column; choosing the liquid-gas interface, the *rate of evaporation* is

$$N_A\big|_{z=0} = -\frac{\rho D_{AB}}{1 - \omega_{A_0}} \frac{d\omega_A}{dz}\bigg|_{z=0} \tag{7-48}$$

or, since $\omega_B = 1 - \omega_A$, $N_B = 0$, and $N_A = \text{const}$

$$N_{A_s} = \frac{\rho D_{AB}}{\omega_{B_s}} \frac{d\omega_B}{dz}\bigg|_{z=0} \tag{7-49}$$

MOLECULAR MASS TRANSFER (DIFFUSION)

Differentiating Eq. (7-43) and evaluating the gradient at $z = 0$, we find the rate of evaporation to be

$$N_{A_s} = \frac{\rho D_{AB}}{\delta} \ln \frac{\omega_{B\delta}}{\omega_{B_s}} \tag{7-50}$$

or, in terms of the log-mean mass fraction,

$$N_{A_s} = \frac{\rho D_{AB}}{\delta(\omega_B)_{lm}} (\omega_{B\delta} - \omega_{B_s}) \tag{7-51a}$$

or

$$N_{A_s} = \frac{\rho D_{AB}}{\delta(\omega_B)_{lm}} (\omega_{A_s} - \omega_{A\delta}) \tag{7-51b}$$

If isothermal and isobaric, an alternative expression for the rate of evaporation, in terms of pressure, is

$$N_{A_s} = \frac{p D_{AB}}{RT\delta(p_B)_{lm}} (p_{A_s} - p_{A\delta}) \tag{7-52}$$

where

$$(p_B)_{lm} \equiv \frac{p_{B\delta} - p_{B_s}}{\ln(p_{B\delta}/p_{B_s})} \tag{7-53}$$

Example Although the assumption of ρD_{AB} = const made for the Arnold cell of Fig. 7-4 simplified the analysis, it was not necessary. The analysis can readily be made for a variable density by noting that

$$p = \rho RT = p_A + p_B = (\rho_A R_A + \rho_B R_B)T$$

for a perfect gas. The gas constant for the mixture is therefore given by

$$R = \frac{\rho_A}{\rho} R_A + \frac{\rho_B}{\rho} R_B = \omega_A R_A + \omega_B R_B$$

or, since $\omega_A + \omega_B = 1$,

$$R = (1 - \omega_B)R_A + \omega_B R_B = R_A + (R_B - R_A)\omega_B$$

In terms of ω_B Eq. (7-36) can be written

$$N_A = -\frac{p}{RT} \frac{D_{AB}}{\omega_B} \frac{d\omega_B}{dz}$$

for a perfect gas. Combining these results and integrating,

$$N_A \int_0^\delta dz = -\frac{p D_{AB}}{T} \int_{\omega_{B_s}}^{\omega_{B\delta}} \frac{d\omega_B}{\omega_B [R_A + (R_B - R_A)\omega_B]}$$

we get

$$N_{A_s} = \frac{p}{R_A T} \frac{D_{AB}}{\delta} \ln \frac{[R_A + (R_B - R_A)\omega_{B_s}]\omega_{B\delta}}{[R_A + (R_B - R_A)\omega_{B\delta}]\omega_{B_s}}$$

This result may be compared with Eq (7-50) to show the effect of assuming constant density.

Film theory A concept often useful in the modeling of mass-transfer processes follows from the foregoing analysis. Imagine that the wall of the Arnold diffusion cell is removed and that the same processes occur. In this case we can visualize a *fictive film* of stagnant (very slow-moving) gas adjacent to the liquid surface, as shown in Fig. 7-5.

If the "stagnant" film thickness δ is chosen such that the gas film offers the same resistance to diffusion as encountered in the combined process of molecular diffusion and diffusion by mixing of the moving fluid, the preceding analysis applies; Eqs. (7-51) and (7-52) give the mass flux. Expressing the boundary condition at the liquid-gas interface in the convective form of Eq. (7-32),

$$N_{A_s} = \frac{k_c}{RT}(p_{A_s} - p_{A_\delta}) \tag{7-54}$$

we see that the convective-mass-transfer coefficient k_c is

$$k_c = \frac{pD_{AB}}{\delta(p_B)_{lm}} \tag{7-55}$$

which can be readily evaluated for a given flow condition and film thickness.

In the example of equimolal counterdiffusion of gases (Sec. 7-3) the mass flux was found to be

$$N_{A_s} = J_{A_s} = \frac{D_{AB}}{RT\delta}(p_{A_s} - p_{A_\delta}) \tag{7-56}$$

expressed in the fictive-film terminology. In this case the convective-mass-transfer coefficient k_c would be

$$k_c = \frac{D_{AB}}{\delta} \tag{7-57}$$

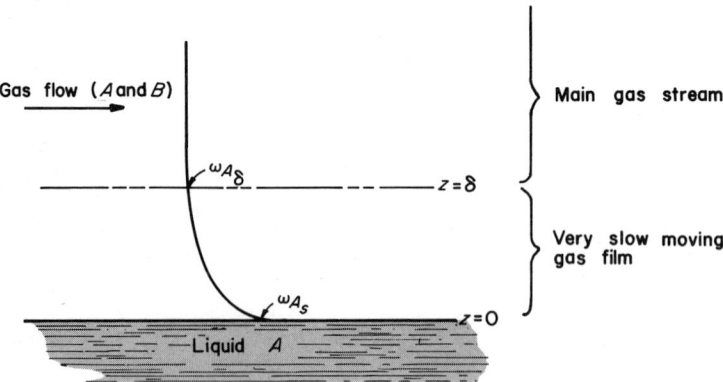

Fig. 7-5 Fictive-film model for mass transfer.

MOLECULAR MASS TRANSFER (DIFFUSION)

upon comparing Eqs. (7-54) and (7-56). The mass transfer in these two cases would therefore be equal only when the total pressure is equal to the log-mean pressure of species B. More attention will be given to the convective-mass-transfer coefficient in Chap. 14.

Example A service station attendant accidentally spills 5 gal of gasoline, which quickly spreads over a level concrete area of 25 ft². Estimate the time required for the gasoline to evaporate into quiescent dry air; $D_{ga} = 6.5$ ft²/hr. The temperature is 70°F, and it may be assumed that the evaporation takes place through a film 6 in thick. The vapor pressure of the gasoline is 2 psia.

Solution Since $p_g + p_a = 1$ atm (say 14.7 psia), Eq. (7-52) can be written

$$N_g = \frac{pD_{ga}}{RT\delta(p_a)_{lm}}(p_{g_\delta} - p_{g_s})$$

which combines with Eq. (7-53) and the perfect-gas equation to give

$$N_g = \frac{\rho D_{ga}}{\delta} \ln \frac{p_{a_\delta}}{p_{a_s}}$$

The density of the mixture $\rho = p/RT$ is approximately equal to the density of dry air (it could be calculated if the constituents of the gasoline were given). From Table A-3

$$\rho = 0.075 \text{ lb}_m/\text{ft}^3$$

The vapor pressure is the pressure at which a fluid and its vapor (gas) are in equilibrium, i.e.,

$$p_{g_s} = 2 \text{ psia}$$

Hence,

$$p_{a_s} = 14.7 - 2 = 12.7 \text{ psia} \quad \text{and} \quad p_{a_\delta} \simeq 14.7 \text{ psia}$$

Using these data, the mass flux is

$$N_g = \frac{(0.075 \text{ lb}_m/\text{ft}^3)(6.5 \text{ ft}^2/\text{hr})}{0.5 \text{ ft}} \ln \frac{14.7}{12.7} = 0.144 \text{ lb}_m/\text{hr-ft}^2$$

For an average gasoline density of 6 lb_m/gal the approximate evaporation time will be

$$t = \frac{5 \text{ gal}}{25 \text{ ft}^2} \frac{6 \text{ lb}_m}{\text{gal}} \frac{\text{hr-ft}^2}{0.144 \text{ lb}_m} = 8.35 \text{ hr}$$

Example Determine the convective-mass-transfer coefficient for a small droplet of species A vaporizing into an infinite atmosphere of species A and B. There is no relative motion between the droplet and the atmosphere other than that caused by the diffusion. Assume that the evaporation is quasi-steady.

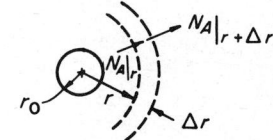

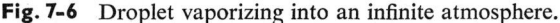

Fig. 7-6 Droplet vaporizing into an infinite atmosphere.

Solution This is the same process analyzed in this section except that in this case it is more convenient to work in the spherical coordinate system. For the quasi-steady process a mass balance on a spherical shell gives

$$4\pi(r^2 N_A)_r - 4\pi(r^2 N_A)_{r+\Delta r} = 0$$

Dividing by $4\pi \Delta r$ and taking the limit as Δr approaches zero, the differential equation is

$$\frac{d}{dr} r^2 N_A = 0$$

or

$$r^2 N_A = r_0^2 N_{A_s} = \text{const}$$

But N_A is given by Eq. (7-17) expressed in spherical coordinates, i.e.,

$$N_A = -\rho D_{AB} \frac{d\omega_A}{dr} + \omega_A (N_A + N_B)$$

and $N_B = 0$; hence,

$$N_A = -\frac{\rho D_{AB}}{1 - \omega_A} \frac{d\omega_A}{dr} = \frac{r_0^2}{r^2} N_{A_s}$$

Separating variables and integrating, when $\rho D_{AB} = \text{const}$,

$$-\rho D_{AB} \int_{\omega_{A_s}}^{\omega_{A\delta}} \frac{d\omega_A}{1 - \omega_A} = r_0^2 N_{A_s} \int_{r_0}^{\delta \to \infty} \frac{dr}{r^2}$$

$$\rho D_{AB} \ln \frac{1 - \omega_{A\delta}}{1 - \omega_{A_s}} = r_0 N_{A_s}$$

Since $\omega_A + \omega_B = 1$,

$$N_{A_s} = \frac{\rho D_{AB}}{r_0} \ln \frac{\omega_{B\delta}}{\omega_{B_s}}$$

or, in terms of the log-mean concentration,

$$N_{A_s} = \frac{\rho D_{AB}}{r_0} \frac{\omega_{B\delta} - \omega_{B_s}}{(\omega_B)_{lm}}$$

In terms of mass fraction the convective-mass-transfer coefficient given by Eq. (7-32) is

$$N_{A_s} = \rho k_c (\omega_{A_s} - \omega_{A\delta})$$

or

$$N_{A_s} = \rho k_c (\omega_{B\delta} - \omega_{B_s})$$

Equating the two expressions for N_{A_s} gives

$$k_c = \frac{D_{AB}}{r_0 (\omega_B)_{lm}}$$

MOLECULAR MASS TRANSFER (DIFFUSION)

7-7 DIFFUSION THROUGH A MEMBRANE

The mechanism of diffusion through a membrane parallels that of diffusion through a stagnant gas in that the governing differential equation is

$$\frac{dN_A}{dx} = 0 \tag{7-58}$$

however, the mass flux, given by Eq. (7-16), is

$$N_A = -\rho D_{AB} \frac{d\omega_A}{dx} \tag{7-59}$$

since $\omega_A \ll 1$ (implies $\rho = c$); hence,

$$\frac{d^2\omega_A}{dx^2} = 0 \tag{7-60}$$

which integrates to give the mass fraction

■ $\quad \omega_A = (\omega_{A_2} - \omega_{A_1})\dfrac{x}{L} + \omega_{A_1} \tag{7-61}$

using the boundary conditions specified in Fig. 7-7. The mass flux is then given by Eq. (7-59) evaluated at some convenient station; hence

$$N_{A_1} = \frac{\rho D_{AB}}{L}(\omega_{A_1} - \omega_{A_2}) \tag{7-62}$$

or if isothermal and isobaric,

■ $\quad N_{A_1} = \dfrac{D_{AB}}{RTL}(p_{A_1} - p_{A_2}) \tag{7-63}$

in terms of partial pressures.

In working with vapor barriers it is often convenient to define a parameter called *permeability* $\mathscr{P}$

$$\mathscr{P} \equiv \frac{N_{A_1}}{(p_{A_1} - p_{A_2})/L} = \frac{D_{AB}}{RT} \tag{7-64}$$

which is a measure of the membrane's ability to transmit vapor.

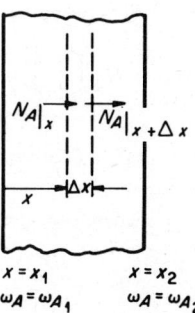

Fig. 7-7 Diffusion through a membrane.

Example A Pyrex tube of inside radius r_1, outside radius r_2, and length L contains a mixture of natural gas. Pyrex is permeable to helium but quite impermeable to all other gases, for example, $\mathcal{P}_{He} \simeq 25 \mathcal{P}_{H_2}$. Obtain an expression for the mass flux of helium through the tube in terms of the mass diffusivity of helium in Pyrex D_{hp}, the tube dimensions, and the mass fraction at the inner and outer surfaces of the tube ω_{h_1} and ω_{h_2}, respectively.

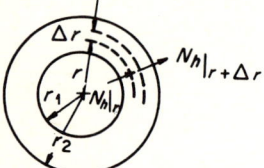

Fig. 7–8 Diffusion of helium through Pyrex.

Solution The mass flux is given by Eq. (7-17). Expressed in cylindrical coordinates,

$$N_h = -\rho D_{hp} \frac{d\omega_h}{dr}$$

since $\omega_h \ll 1$, making the bulk-flow term negligible. The conservation-of-species equation comes from a mass balance on a cylindrical shell, giving

$$2\pi L (rN_h)_r - 2\pi L (rN_h)_{r+\Delta r} = 0$$

Dividing by $2\pi L \, \Delta r$ and taking the limit as Δr approaches zero, we get the governing differential equation

$$\frac{d}{dr} rN_h = 0$$

Therefore,

$$rN_h = r_1 N_{h_1} = \text{const}$$

Combining the mass-flux equation and the continuity equation, we have

$$\frac{r_1}{r} N_{h_1} = -\rho D_{hp} \frac{d\omega_h}{dr}$$

Separating variables and integrating,

$$r_1 N_{h_1} \int_{r_1}^{r_2} \frac{dr}{r} = -\rho D_{hp} \int_{\omega_{h_1}}^{\omega_{h_2}} d\omega_h$$

$$r_1 N_{h_1} \ln \frac{r_2}{r_1} = \rho D_{hp} (\omega_{h_1} - \omega_{h_2})$$

Therefore, the mass flux is

$$N_{h_1} = \frac{\rho D_{hp}}{r_1} \frac{\omega_{h_1} - \omega_{h_2}}{\ln (r_2/r_1)}$$

MOLECULAR MASS TRANSFER (DIFFUSION)

7-8 DIFFUSION WITH HETEROGENEOUS CHEMICAL REACTION

In addition to the conservation equation and Fick's law of diffusion, the stoichiometry of a chemical reaction is significant in many processes. The reaction at a surface will be considered in this section. Such a reaction is approximately the case in a catalytic reactor, depicted schematically in Fig. 7-9. Species A diffuses steadily through a stagnant film to the catalytic surface, where it is instantaneously converted to A_3 in accordance with the reaction

$$3A \rightarrow A_3 \tag{7-65}$$

Following the reaction, A_3 diffuses outward from the catalytic surface. For every 3 moles of A which diffuse to the right 1 mole of A_3 diffuses to the left:

$$\tilde{N}_{A_3} = -\tfrac{1}{3}\tilde{N}_A \tag{7-66}$$

where the negative sign indicates countermotion of the two species.

We may choose to consider either species A or A_3. A mass balance on species A over a thin slab gives the familiar result

$$\frac{d\tilde{N}_A}{dx} = 0 \tag{7-67}$$

Similarly, for species A_3

$$\frac{d\tilde{N}_{A_3}}{dx} = 0 \tag{7-68}$$

Since Eq. (7-66) is valid for any location in the stagnant film, it may be used in Eq. (7-16), expressed in molal units, to give

$$\tilde{N}_A = -\frac{\tilde{\rho} D_{AA_3}}{1 - \tfrac{2}{3}\tilde{\omega}_A} \frac{d\tilde{\omega}_A}{dx} \tag{7-69}$$

Upon assuming D_{AA_3} to be constant, Eq. (7-69) can be substituted into Eq. (7-67), resulting in

$$\frac{d}{dx}\left(\frac{1}{1 - \tfrac{2}{3}\tilde{\omega}_A} \frac{d\tilde{\omega}_A}{dx}\right) = 0 \tag{7-70}$$

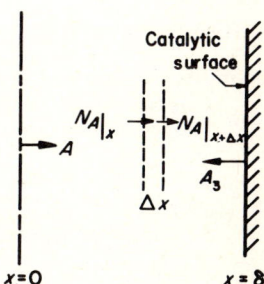

Fig. 7-9 Heterogeneous chemical reaction.

which integrates to

$$-\tfrac{3}{2}\ln(1 - \tfrac{2}{3}\tilde{\omega}_A) = C_1 x + C_2 \tag{7-71}$$

The boundary conditions are:

(1) at $x = 0$: $\quad \tilde{\omega}_A = \tilde{\omega}_{A_0}$ \hfill (7-72a)

(2) at $x = \delta$: $\quad \tilde{\omega}_A = 0$ \hfill (7-72b)

At first glance the second boundary condition may seem unusual, but it is resolved upon realizing that $\tilde{\omega}_{A_3} = 1$ at the catalytic surface. After applying the boundary conditions, the resulting concentration profile is described by

$$\blacksquare \quad 1 - \tfrac{2}{3}\tilde{\omega}_A = (1 - \tfrac{2}{3}\tilde{\omega}_{A_0})^{1-x/\delta} \tag{7-73}$$

Differentiating with respect to x, we get

$$\frac{d\tilde{\omega}_A}{dx} = \frac{3}{2\delta}(1 - \tfrac{2}{3}\tilde{\omega}_{A_0})^{1-x/\delta} \ln(1 - \tfrac{2}{3}\tilde{\omega}_{A_0}) \tag{7-74}$$

or

$$\left.\frac{d\tilde{\omega}_A}{dx}\right|_{x=0} = \frac{3}{2\delta}(1 - \tfrac{2}{3}\tilde{\omega}_{A_0}) \ln(1 - \tfrac{2}{3}\tilde{\omega}_{A_0}) \tag{7-75}$$

Using this result to evaluate the molar flux at $x = 0$, Eq. (7-69) gives

$$\blacksquare \quad \tilde{N}_{A_0} = -\frac{3\tilde{\rho} D_{AA_3}}{2\delta} \ln(1 - \tfrac{2}{3}\tilde{\omega}_{A_0}) \tag{7-76}$$

Finite reaction rates If the reaction rate in the preceding analysis is slow rather than instantaneous, described by the equation

$$\tilde{N}_A = k'' \tilde{\rho}_A \tag{7-77}$$

where k'' is the rate constant, the molal flux will be

$$\tilde{N}_{A_0} = -\frac{3\tilde{\rho} D_{AA_3}}{2\delta(1 + D_{AA_3}/k''\delta)} \ln(1 - \tfrac{2}{3}\tilde{\omega}_{A_0}) \tag{7-78}$$

The double prime on the rate constant is used to suggest that it pertains to an area (surface).

Example A pulverized spherical particle of coal burns in air at 2000°F. If the reaction $C + O_2 \to CO_2$ occurs very rapidly at the particle surface, estimate the time required for the particle to burn completely from an initial diameter of 0.010 in. Assume the coal to be pure carbon of 80 lb_m/ft^3 density; the mass diffusivity of oxygen in the mixture is 6.0 ft^2/hr.

Solution Illustrated in Fig. 7-10, air (21 percent O_2 and 79 percent N_2) diffuses inward through a stagnant film; the oxygen from the air reacts with the carbon, forming carbon dioxide, and the CO_2 diffuses outward. The reaction equation is more correctly written as

$$C + O_2 + \tfrac{79}{21}N_2 \to CO_2 + 3.76N$$

MOLECULAR MASS TRANSFER (DIFFUSION)

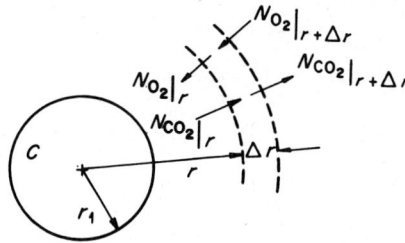

Fig. 7-10 Diffusion with heterogeneous chemical reaction.

however, the nitrogen, being inert, does not enter into the combustion process; i.e., the nitrogen which diffuses inward as a component of the air returns outward with the CO_2 and remains unchanged.

For a quasi-steady-state process a mass balance for oxygen over the spherical shell gives

$$4\pi(r^2\tilde{N}_{O_2})_r - 4\pi(r^2\tilde{N}_{O_2})_{r+\Delta r} = 0$$

Dividing by $4\pi \Delta r$, and taking the limit as Δr approaches zero, the governing differential equation is

$$\frac{d}{dr}r^2\tilde{N}_{O_2} = 0$$

A mass balance on either of the other constituents would have given similar results, i.e.,

$$\frac{d}{dr}r^2\tilde{N}_{CO_2} = 0 \quad \text{and} \quad \frac{d}{dr}r^2\tilde{N}_{N_2} = 0$$

The stoichiometry of the chemical reaction gives the relation between the fluxes, i.e.,

$$\tilde{N}_{CO_2} = -\tilde{N}_{O_2}$$

the negative sign indicating that diffusion of the two species is in opposite directions. The flux of oxygen is given by Eq. (7-16), modified to account for each component which might contribute to bulk motion,

$$\tilde{N}_{O_2} = -\tilde{\rho} D_{O_2-\text{mix}} \frac{d\tilde{\omega}_{O_2}}{dr} + \omega_{O_2}(\tilde{N}_{O_2} + \tilde{N}_{CO_2} + \tilde{N}_{N_2})$$

In this case, however, there is no net motion of the nitrogen, $\tilde{N}_{N_2} = 0$; therefore, the flux simplifies to

$$\tilde{N}_{O_2} = -\tilde{\rho} D_{O_2-\text{mix}} \frac{d\tilde{\omega}_{O_2}}{dr}$$

From the mass balance

$$r^2\tilde{N}_{O_2} = \text{const} = r_1^2\tilde{N}_{O_2}\big|_1$$

hence,

$$r^2\tilde{N}_{O_2} = r_1^2\tilde{N}_{O_2}\big|_1 = -r^2\tilde{\rho} D_{O_2-\text{mix}}\frac{d\tilde{\omega}_{O_2}}{dr}$$

The boundary conditions are

(1) at $r = r_1$: $\tilde{\omega}_{O_2} = 0$ instantaneous combustion
(2) at $r \to \infty$: $\tilde{\omega}_{O_2} = 0.21$

Separating variables and integrating with these boundary conditions, we get

$$r_1^2 \tilde{N}_{O_2}\big|_1 \int_{r_1}^{r \to \infty} \frac{dr}{r^2} = -\tilde{\rho} D_{O_2\text{—mix}} \int_0^{0.21} d\tilde{\omega}_{O_2}$$

$$r_1 \tilde{N}_{O_2}\big|_1 = -\tilde{\rho} D_{O_2\text{—mix}}(0.21)$$

The density of the mixture is given by

$$\tilde{\rho} = \frac{p}{\tilde{R}T}$$

To evaluate $\tilde{R}$, we shall assume standard conditions, where 1 lb mole occupies 359 ft³ at 1 atm pressure and 492°R, giving

$$\tilde{R} = \frac{p\mathscr{V}}{nT} = \frac{(1\text{ atm})(359\text{ ft}^3)}{(1\text{ lb mole mix})(492°\text{R})} \qquad \tilde{R} = 0.73 \frac{\text{atm-ft}^3}{\text{lb mole mix-°R}}$$

Therefore,

$$\tilde{\rho} = \frac{p}{\tilde{R}T} = \frac{1\text{ atm}}{(0.73\text{ atm-ft}^3/\text{lb mole-°R})(2460°\text{R})}$$

$$\tilde{\rho} = 0.000556 \text{ lb mole/ft}^3$$

The burning rate is

$$\tilde{N}_{O_2} = -\frac{\tilde{\rho} D_{O_2\text{—mix}}(0.21)}{r_1}$$

$$\tilde{N}_{O_2} = -\frac{\left(0.000556\,\frac{\text{lb mole mix}}{\text{ft}^3}\right)\left(6\,\frac{\text{ft}^2}{\text{hr}}\right)\left(0.21\,\frac{\text{lb mole O}_2}{\text{lb mole mix}}\right)}{0.010/12 \text{ ft}}$$

$$\tilde{N}_{O_2} = -0.84 \text{ lb mole O}_2/\text{hr-ft}^2$$

Hence,

$$\tilde{N}_C = 0.84 \text{ lb mole C/hr-ft}^2$$

and

$$\tilde{\rho}_C = \frac{\rho_C}{\mathscr{M}_C} = \frac{80\text{ lb}_m/\text{ft}^3}{12\text{ lb}_m/\text{lb mole}} = 6.67 \text{ lb mole/ft}^3$$

Therefore,

$$\dot{m} = 4\pi r_1^2 N_C = \left(0.84\,\frac{\text{lb mole}}{\text{hr-ft}^2}\right)\frac{\text{ft}^3}{6.67\text{ lb mole}}(4\pi)\left(\frac{0.01}{12}\right)^2 \text{ft}^2$$

$$= 1.1 \times 10^{-6} \text{ ft}^3/\text{hr}$$

The volume of the particle is

$$\mathscr{V} = \tfrac{4}{3}\pi r_1^3 = \tfrac{4}{3}\pi \left(\frac{0.01}{12}\right)^3 \text{ft}^3 = 2.42 \times 10^{-9} \text{ ft}^3$$

Therefore, the time required to burn is

$$t = \frac{\mathscr{V}}{\dot{m}} = \frac{2.42 \times 10^{-9} \text{ ft}^3}{1.1 \times 10^{-6} \text{ ft}^3/\text{hr}} \frac{3{,}600 \text{ sec}}{\text{hr}} = 7.9 \text{ sec}$$

7-9 DIFFUSION WITH HOMOGENEOUS CHEMICAL REACTION

In the preceding section we noted that the chemical reaction came into the results only in the boundary conditions; it was a boundary phenomenon. In this section we shall consider the case where the reaction occurs uniformly throughout the fluid—homogeneous chemical reaction—requiring an additional term in the governing differential equation. To illustrate the process, consider the model of Fig. 7-11. Gas A diffuses through a stagnant film to a catalytic surface where A is instantaneously converted into B (1 mole of A produces 1 mole of B). As B moves back through the film, it decomposes into A, $\tilde{r}_A''' = -\tilde{r}_B'''$, in accordance with the equation

$$\tilde{r}_B''' = -k''' \tilde{\rho}_B \tag{7-79}$$

where k''' is the reaction rate constant; the triple prime suggests that the rate constant pertains to the entire body of fluid (volume).

A mass balance for A on a thin slab of the fluid gives

$$S\tilde{N}_A|_x - S\tilde{N}_A|_{x+\Delta x} + \tilde{r}_A''' S \Delta x = 0 \tag{7-80}$$

where $\tilde{r}_A'''$ is the moles of A produced per unit volume per unit time. Dividing by the volume $S \Delta x$ and taking the limit as Δx approaches zero,

$$-\frac{d\tilde{N}_A}{dx} + \tilde{r}_A''' = 0 \tag{7-81}$$

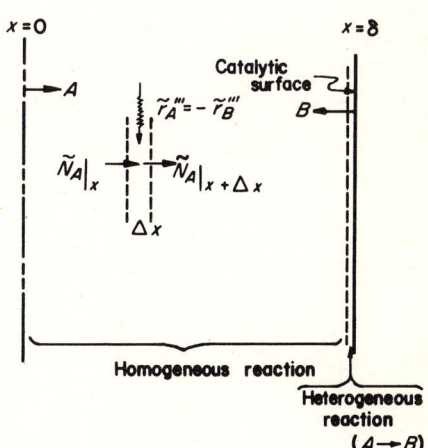

Fig. 7-11 Homogeneous chemical reaction in the presence of a catalytic surface.

But
$$\tilde{r}_A''' = -\tilde{r}_B''' = k'''\tilde{\rho}_B \tag{7-82}$$
and
$$\tilde{\rho}_B = \tilde{\rho} - \tilde{\rho}_A = \tilde{\rho}(1 - \tilde{\omega}_A) \tag{7-83}$$
Therefore,
$$\frac{d\tilde{N}_A}{dx} - k'''\tilde{\rho}(1 - \tilde{\omega}_A) = 0 \tag{7-84}$$

The diffusion Eq. (7-16), in terms of molal units, is
$$\tilde{N}_A = -\tilde{\rho}D_{AB}\frac{d\tilde{\omega}_A}{dx} + \tilde{\omega}_A(\tilde{N}_A + \tilde{N}_B) \tag{7-85}$$
but
$$\tilde{N}_A = -\tilde{N}_B \tag{7-86}$$
since $A \to B$ at the wall, giving
$$\tilde{N}_A = -\tilde{\rho}D_{AB}\frac{d\tilde{\omega}_A}{dx} \tag{7-87}$$
which can be solved simultaneously with Eq. (7-84), yielding
$$\frac{d^2\tilde{\omega}_A}{dx^2} + \frac{k'''}{D_{AB}}(1 - \tilde{\omega}_A) = 0 \tag{7-88}$$
when the product $\tilde{\rho}D_{AB}$ is constant. The relevant boundary conditions are

(1) at $x = 0$: $\quad \tilde{\omega}_A = \tilde{\omega}_{A_0}$ \hfill (7-89a)

(2) at $x = \delta$: $\quad \tilde{\omega}_A = 0 \quad$ instantaneous reaction \hfill (7-89b)

Assuming D_{AB} is constant, Eq. (7-88) is a linear second-order nonhomogeneous differential equation having constant coefficients; its solution is

$$\tilde{\omega}_{A_0} = (\tilde{\omega}_{A_0} - 1)\cosh\sqrt{\frac{k'''}{D_{AB}}}\,x - \frac{1 + (\tilde{\omega}_{A_0} - 1)\cosh\sqrt{k'''/D_{AB}}\,\delta}{\sinh(\sqrt{k'''/D_{AB}})\,\delta} \times \sinh\sqrt{\frac{k'''}{D_{AB}}}\,x \tag{7-90a}$$

or

$$\blacksquare \quad \tilde{\omega}_{A_0} = \frac{(\tilde{\omega}_{A_0} - 1)\sinh\left[\sqrt{\frac{k'''}{D_{AB}}}(\delta - x)\right] + \cosh\sqrt{\frac{k'''}{D_{AB}}}\,\delta - \sinh\sqrt{\frac{k'''}{D_{AB}}}\,\delta}{\sinh\sqrt{\frac{k'''}{D_{AB}}}\,\delta}$$
$$\tag{7-90b}$$

MOLECULAR MASS TRANSFER (DIFFUSION)

This form of the mole-fraction equation can be readily differentiated to substitute into Eq. (7-87) to give the molal flux:

$$\tilde{N}_A = \frac{\tilde{\rho}\sqrt{k''' D_{AB}}\,(\tilde{\omega}_{A_0} - 1)\cosh\sqrt{\frac{k'''}{D_{AB}}}(\delta - x)}{\sinh\sqrt{\frac{k'''}{D_{AB}}}\,\delta} \qquad (7\text{-}91)$$

7-10 TRANSIENT DIFFUSION

The preceding sections have considered only steady-state processes. Before a steady state can be reached, however, some time must lapse upon initiating a mass-transfer process before transients disappear, e.g., dropping a lump of sugar in a cup of coffee produces a concentration gradient which varies with time at the outset, as illustrated in Fig. 2-8.

Semi-infinite medium with prescribed surface concentration The simplest case of transient mass diffusion which is amenable to analytical solution is the one-directional mass transfer in a semi-infinite stationary medium with a prescribed surface mass concentration. The sudden flooding of a large portion of the earth's crust is an example. Many actual processes are adequately described during their transient period by the resulting equation. For no chemical reaction, Eq. (7-27) describes the condition depicted in Fig. 7-12; in the x direction

$$\frac{\partial \rho_A}{\partial t} = D_{AB}\frac{\partial^2 \rho_A}{\partial x^2} \qquad (7\text{-}92)$$

which requires two boundary conditions and one initial condition, namely,

Boundary condition 1: $\quad \rho_A(0,t) = \rho_{A_s}$

Boundary condition 2: $\quad \rho_A(\infty,t) = \rho_{A_i}$ $\qquad (7\text{-}93)$

Initial condition: $\quad \rho_A(x,0) = \rho_A$

The first boundary condition requires that the mass concentration ρ_A be held constant at the surface; the second requires the core of the body, at a large distance from the surface, to remain at its initial mass concentration ρ_{A_i}. The initial condition states that at any location in the medium at time zero the mass concentration is a constant, ρ_{A_i}.

At this point we note that the governing relation and conditions of Eqs. (7-92) and (7-93) are directly analogous to the transient thermal case described by Eqs. (6-18) and (6-19) in Sec. 6-4 when

$$\alpha \sim D_{AB} \quad \text{and} \quad T \sim \rho_A$$

i.e., when α and T are analogous to D_{AB} and ρ_A, respectively. Therefore, the solution

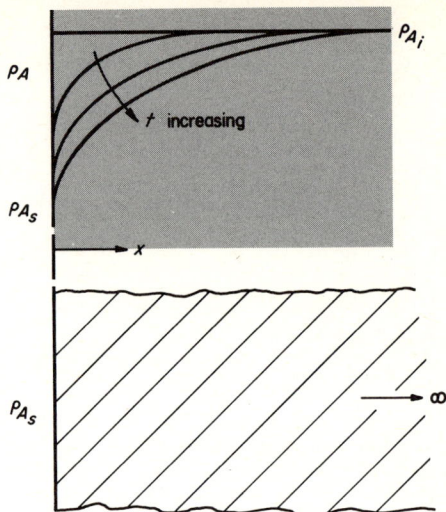

Fig. 7-12 Transient diffusion in a semi-infinite medium.

for the mass concentration and mass transfer readily follow from the thermal equations (6-24) and (6-26), giving

$$\frac{\rho_A - \rho_{A_s}}{\rho_{A_i} - \rho_{A_s}} = \text{erf} \frac{x}{\sqrt{4 D_{AB} t}} \qquad (7\text{-}94)$$

$$N_{A_s} = \frac{D_{AB}(\rho_{A_s} - \rho_{A_i})}{\sqrt{\pi D_{AB} t}} \qquad (7\text{-}95)$$

In the absence of tabulated values of the Gauss error function, Eq. (7-94) can be solved with the aid of the curve of Fig. 6-5.

One-dimensional systems with prescribed convection conditions Rather than knowing the surface condition, the more general case is when the mass concentration ρ_A of the surrounding fluid is known. This is the case in convective drying processes where a stream of inert gas is passed over the medium being dried. This requires introducing the convective-mass-transfer coefficient k_c defined by

$$N_A = k_c(\rho_A - \rho_{A_\infty}) \qquad (7\text{-}96)$$

or, for constant density,

$$N_A = \rho k_c(\omega_A - \omega_{A_\infty}) \qquad (7\text{-}97)$$

relating the mass transfer to the condition of the surrounding fluid.

For the drying of wet solids the drying rate is approximately constant as long as the surface remains wet. In this case the mass transfer is governed by the resistance at the liquid-gas interface. When dry spots begin to appear on the surface, the drying

MOLECULAR MASS TRANSFER (DIFFUSION)

rate begins to depend upon the internal resistance, decreasing as the body looses moisture. This decrease occurs during the *falling-rate period* and continues until the equilibrium concentration ρ_{eq} is reached. The equilibrium concentration ρ_{eq}, determined empirically for each solid, depends upon the temperature, pressure, and relative humidity of the inert gas. During the falling-rate period Eq. (7-92) is valid, and the convective boundary condition can be expressed by

$$N_{A_s} = k_c(\rho_{A_s} - \rho_{A\infty}) = K_c(\rho_{A_{l,s}} - \rho_{A_{l,eq}}) \tag{7-98}$$

where K_c is the coefficient which gives the mass-transfer rate when the liquid is at its equilibrium condition $\rho_{A_{l,eq}}$ [15]. It should be noted that ρ_A is the concentration of the vapor A in the gas, but ρ_{A_l} is the concentration of the liquid A in the solid.

Table 7-5 gives the three common geometrical configurations with the governing equations and boundary conditions. Since the equations are analogous to those of Chap. 6, the transient charts of Heisler [12] and Gröber [11] may be used to determine the mass concentration or mass fraction and the mass flux. Use of the charts is illustrated in the following example.

Example A florist transports a load of 2-in-diam open-cellular foam balls in an uncovered truck. The balls are completely dry at the outset, but a sudden, violent rainstorm occurs en route. What is the moisture content at the center of the balls 3 hr after the rain begins if $D_{AB} = 0.0005$ ft^2/hr?

Solution Since the mass density of the foam balls is not known, the mass concentration will be expressed in terms of the dry foam (dry basis), i.e.,

$\omega_{A\infty} = 1$ lb$_m$ water/lb$_m$ foam

$\omega_{A_i} = 0$

Assuming that the surface resistance is negligible, $k_c \to \infty$,

$$\frac{D_{AB}}{k_c R} \to 0$$

and

$$\frac{D_{AB} t}{R^2} = \frac{(0.0005 \text{ ft}^2/\text{hr})(3 \text{ hr})}{(\frac{1}{12} \text{ ft})^2} = 0.216$$

The mass concentration is then given by Fig. 6-14:

$$\frac{\omega_{A_c} - \omega_{A\infty}}{\omega_{A_i} - \omega_{A\infty}} \simeq 0.2$$

hence,

$$\frac{\omega_{A_c} - 1}{0 - 1} \simeq 0.2$$

$\omega_{A_c} \simeq 0.8$ lb$_m$ water/lb$_m$ foam

Multidimensional systems The separation-of-variables technique of Sec. 6-7 applied to finite bodies under transient thermal conditions is equally applicable to

Table 7-5 One-dimensional systems with prescribed convection conditions

	Large plate	Long cylinder	Sphere
Nomenclature			
Governing equation	$\dfrac{1}{D_{AB}}\dfrac{\partial \rho_A}{\partial t} = \dfrac{\partial^2 \rho_A}{\partial x^2}$	$\dfrac{1}{D_{AB}}\dfrac{\partial \rho_A}{\partial t} = \dfrac{1}{r}\dfrac{\partial}{\partial r}\left(r\dfrac{\partial \rho_A}{\partial r}\right)$	$\dfrac{1}{D_{AB}}\dfrac{\partial \rho_A}{\partial t} = \dfrac{1}{r^2}\dfrac{\partial}{\partial r}\left(r^2\dfrac{\partial \rho_A}{\partial r}\right)$
Boundary conditions	$\dfrac{\partial \rho_A}{\partial x} = \begin{cases} 0 & \text{at } x=0 \\ -\dfrac{k_c}{D_{AB}}(\rho_{A_s} - \rho_{A\infty}) & \text{at } x=L \end{cases}$	$\dfrac{\partial \rho_A}{\partial r} = \begin{cases} 0 & \text{at } r=0 \\ -\dfrac{k_c}{D_{AB}}(\rho_{A_s} - \rho_{A\infty}) & \text{at } r=R \end{cases}$	$\dfrac{\partial \rho_A}{\partial r} = \begin{cases} 0 & \text{at } r=0 \\ -\dfrac{k_c}{D_{AB}}(\rho_{A_s} - \rho_{A\infty}) & \text{at } r=R \end{cases}$
Initial condition	$\rho_A = \rho_{A_i}$ at $t=0$	$\rho_A = \rho_{A_i}$ at $t=0$	$\rho_A = \rho_{A_i}$ at $t=0$
Transient solution	Figs. 6-8 to 6-10	Figs. 6-11 to 6-13	Figs. 6-14 to 6-16

MOLECULAR MASS TRANSFER (DIFFUSION)

mass diffusion. Therefore, by direct analogy with Eqs. (6-61) to (6-63) we can get the mass fraction or mass concentration for finite bodies, i.e.,

$$\left(\frac{\rho_A - \rho_{A\infty}}{\rho_{A_i} - \rho_{A\infty}}\right)_{\substack{\text{long}\\\text{bar}}} = \left(\frac{\rho_A - \rho_{A\infty}}{\rho_{A_i} - \rho_{A\infty}}\right)_{\substack{2b\\\text{plate}}} \left(\frac{\rho_A - \rho_{A\infty}}{\rho_{A_i} - \rho_{A\infty}}\right)_{\substack{2d\\\text{plate}}} \tag{7-99}$$

$$\left(\frac{\rho_A - \rho_{A\infty}}{\rho_{A_i} - \rho_{A\infty}}\right)_{\substack{\text{parallele-}\\\text{piped}}} = \left(\frac{\rho_A - \rho_{A\infty}}{\rho_{A_i} - \rho_{A\infty}}\right)_{\substack{2b\\\text{plate}}} \left(\frac{\rho_A - \rho_{A\infty}}{\rho_{A_i} - \rho_{A\infty}}\right)_{\substack{2d\\\text{plate}}} \left(\frac{\rho_A - \rho_{A\infty}}{\rho_{A_i} - \rho_{A\infty}}\right)_{\substack{2L\\\text{plate}}} \tag{7-100}$$

$$\left(\frac{\rho_A - \rho_{A\infty}}{\rho_{A_i} - \rho_{A\infty}}\right)_{\substack{\text{short}\\\text{cyl}}} = \left(\frac{\rho_A - \rho_{A\infty}}{\rho_{A_i} - \rho_{A\infty}}\right)_{\substack{2L\\\text{plate}}} \left(\frac{\rho_A - \rho_{A\infty}}{\rho_{A_i} - \rho_{A\infty}}\right)_{\substack{\text{inf}\\\text{cyl}}} \tag{7-101}$$

The plate and infinite-cylinder solutions are taken from Figs. 6-8 and 6-11, respectively.

Example The maximum moisture content which can be tolerated in an oak roller, 1 ft diameter by 2 ft long, to be used in a textile machine, is 20 percent by weight. If the roller, initially having a moisture content of 35 percent (by weight), is placed in a kiln where the moisture content is maintained at 5 percent (by weight), how long must it be dried when (*a*) the ends of the roller are sealed with a vapor barrier, (*b*) the cylindrical surface is sealed with a vapor barrier, and (*c*) drying occurs on the entire surface? The surface resistance may be assumed negligible, and $D_{AB} = 0.0004$ ft²/hr.

Solution From the given conditions the concentration ratio, based on dry wood, is

$$\frac{\rho_{A_c} - \rho_{A\infty}}{\rho_{A_i} - \rho_{A\infty}} = \frac{0.2 - 0.05}{0.35 - 0.05} = 0.5$$

where the maximum moisture content is taken as that at the center of the roller. For negligible surface resistance

$$\frac{D_{AB}}{k_c R} \to 0 \quad \text{and} \quad \frac{D_{AB}}{k_c L} \to 0$$

(*a*) When the ends are sealed, Fig. 6-11 gives

$$\frac{D_{AB} t}{R^2} \simeq 0.2$$

hence,

$$t \simeq 0.2 \frac{(0.5 \text{ ft})^2}{0.0004 \text{ ft}^2/\text{hr}} \simeq 125 \text{ hr}$$

(*b*) When no mass is transferred at the cylindrical surface,

$$\frac{D_{AB} t}{L} \simeq 0.35$$

from Fig. 6-8, and

$$t \simeq 0.35 \frac{1 \text{ ft}}{0.0004 \text{ ft}^2/\text{hr}} \simeq 825 \text{ hr}$$

(c) Using the results of parts (a) and (b), we note that the mass-transfer rate at the cylindrical surface is $\frac{825}{125} = 6.6$ that which occurs at the ends of the cylinder. When both surfaces are exposed, an approximate time is given by

$$t' \simeq 125 - \frac{1}{6.6} 125 \simeq 106 \text{ hr}$$

This can be checked by using Eq. (7-101) in conjunction with Figs. 6-8 and 6-11. For this estimated time

$$\frac{D_{AB}t}{L^2} = \frac{(0.0004 \text{ ft}^2/\text{hr})(106 \text{ hr})}{1 \text{ ft}^2} = 0.0424$$

$$\frac{D_{AB}t}{R^2} = \frac{(0.0004 \text{ ft}^2/\text{hr})(106 \text{ hr})}{0.5^2 \text{ ft}^2} = 0.17$$

From Fig. 6-8

$$\left(\frac{\rho_{A_C} - \rho_{A_\infty}}{\rho_{A_i} - \rho_{A_\infty}}\right)_{2l \text{ plate}} \simeq 0.8$$

and from Fig. 6-11

$$\left(\frac{\rho_{A_C} - \rho_{A_\infty}}{\rho_{A_i} - \rho_{A_\infty}}\right)_{\text{inf cyl}} \simeq 0.7$$

Equation (7-101) gives

$$\left(\frac{\rho_{A_C} - \rho_{A_\infty}}{\rho_{A_i} - \rho_{A_\infty}}\right)_{\text{short cyl}} = 0.8(0.7) = 0.56$$

which is approximately equal to the ratio

$$\frac{\rho_{A_C} - \rho_{A_\infty}}{\rho_{A_i} - \rho_{A_\infty}} = 0.5$$

found at the outset from the given conditions. Since this approximation is within the accuracy of the transient charts, the time is

$$t \simeq 106 \text{ hr}$$

It should be noted that after this time the roller is overdried everywhere except at its center.

PROBLEMS

7-1. (a) Does a transport of electric charge also involve a mass transport?
(b) Does the transport of energy (heat) result in a mass transport?
(c) Does mass diffusion result in a transfer of heat?
(d) If the answer to (a), (b), or (c) is yes, what is the gradient that causes mass transport? Heat transport?
(e) Can heat be transferred into regions of higher temperature?
(f) Can mass be transferred into regions of higher concentration?

MOLECULAR MASS TRANSFER (DIFFUSION)

7-2. Helium at 70°F ($\rho = 5 \text{ lb}_m/\text{ft}_3$) diffuses through a large sheet of 0.12-in.-thick Pyrex glass ($\rho = 50 \text{ lb}_m/\text{ft}$). The mass fractions of helium on the two faces of the Pyrex sheet are 9×10^{-3} and $3 \times 10^{-3} \text{ lb}_m$ He/lb$_m$ He + Py. If the mass diffusivity is 4.5×10^{-11} cm^2/sec, plot the variation of helium mass fraction with thickness x.

7-3. If the density of water vapor is $1.58 \times 10^{-3} \text{ lb}_m/\text{ft}^3$ and the density of air is $6.95 \times 10^{-3} \text{ lb}_m/\text{ft}^3$, estimate the mass flux of the water vapor with the mass fraction gradient as shown. The mass diffusivity of water vapor in air is 0.908 ft^2/hr.

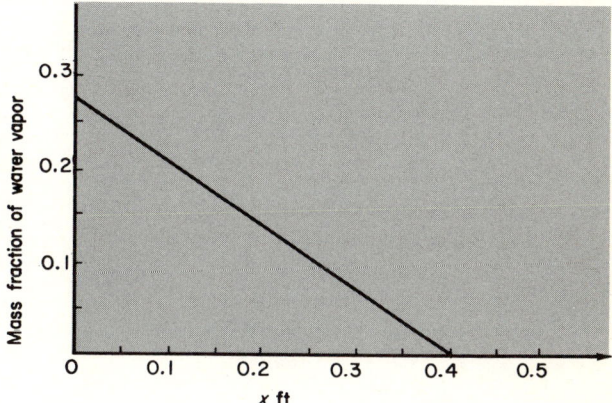

Fig. P7-3

7-4. A partition separates acetone, C_3H_6O, from air, both at 32°F. Initially the acetone is a pure substance. The partition is removed, and after 2 hr the mass flux is $J_A = 1.5 \times 10^{-1} \text{ lb}_m/\text{ft}^2$. Find the mass fraction of acetone 1 ft to the right of the initial position of the partition. The densities of the components are $\rho_{\text{acetone}} = 0.1 \text{ lb}_m/\text{ft}^3$ and $\rho_{\text{air}} = 0.0785 \text{ lb}_m/\text{ft}^3$.

7-5. Given $J_A + J_B = 0$, show that $D_{AB} = D_{BA}$ for a binary mixture.

7-6. Define solute and solvent. How can you tell the difference between them in a solution when they are not explicitly defined?

7-7. Estimate the coefficient of diffusivity of benzene in water.

7-8. Use the Gilliland equation to calculate the diffusivity of benzene, C_6H_6, in air at 77°F and 1 atm pressure. The molecular weights of benzene and air are 78 and 29, respectively. Note that benzene is a six-membered ring. Compare this with the experimental value of 0.341 ft^2/hr.

7-9. Calculate the diffusivity of carbon dioxide gas into nitrogen gas at 1 atm and 528°R by the Gilliland equation. Compare the value with that given in Table A-5.

7-10. Calculate the coefficient of diffusivity of carbon dioxide gas into carbon monoxide gas at a temperature of 32°F and a pressure of 1 atm. Compare this with the experimental value of Table A-5.

7-11. Calculate the coefficient of diffusivity of water vapor in air at 1 atm total pressure and at a temperature of 70°F using the Gilliland equation and the semiempirical equation for a water-vapor–air mixture.

7-12. Estimate the diffusivity of iodine, I_2, in a 98% benzene solution at 77°F. Would this method be useful if the solution were 60% benzene?

7-13. Estimate the diffusivity of methanol, CH_3OH, in water at standard atmospheric conditions. Compare with the value given in Table A-6.

7-14. Estimate the mass diffusivity of glycerol in a 99% solution of water. Compare with the experimental data given in Table A-6.

7-15. Consider one-dimensional mass transfer for a mixture of molecular oxygen and carbon dioxide at 70°F and a constant total pressure of 2 atm. The oxygen will be called gas A and the carbon dioxide will be called gas B. From the conditions $\tilde{\omega}_A = \frac{1}{4}$, $u_A = 20$ ft/hr, $u_B = -4$ ft/hr, calculate the following:

(a) $\tilde{\omega}_B$
(b) $\mathscr{M}$, $\mathscr{M}_A$, and $\mathscr{M}_B$ in lb_m/lb mole
(c) $\tilde{\rho}$, $\tilde{\rho}_A$, and $\tilde{\rho}_B$ in lb moles/ft³ *Hint:* Use the perfect-gas relationship.
(d) ω_A and ω_B
(e) ρ, ρ_A, and ρ_B in lb_m/ft³
(f) $u_A - \tilde{u}$ and $u_B - \tilde{u}$ in ft/hr
(g) $u_A - u$ and $u_B - u$ in ft/hr
(h) $\tilde{N}$, $\tilde{N}_A$, and $\tilde{N}_B$ in lb moles/hr-ft²
(i) N, N_A, and N_B in lb_m/hr-ft²

7-16. Define bulk flow and discuss its significance.

7-17. For the system shown, use a shell balance to show that $N_{B_z} = -N_{A_z}$ assuming steady state with constant pressure and temperature.

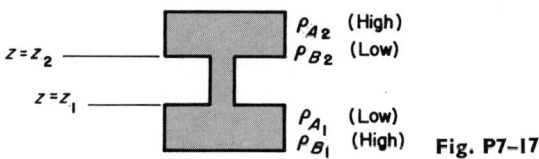

Fig. P7-17

7-18. A catalytic reactor is used to change water, H_2O, to heavy water, H_3O. If the process could be simplified to studying the change of H_2 to H_3 in the catalytic reactor shown, where the surface $z = \delta$ produces an irreversible and instantaneous reaction of the type $3H_2 \rightarrow 2H_3$, find an expression for the molar flux through the film.

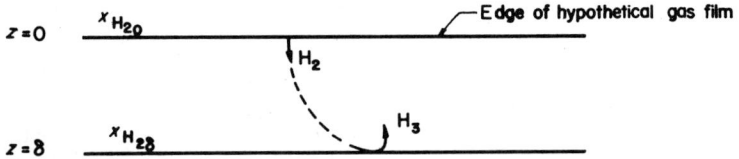

Fig. P7-18

7-19. The water in an outdoor swimming pool is maintained at 80°F. On a day when the temperature is 65°F and the relative humidity is 40 percent, estimate the rate of water loss per square foot by the fictive-film method assuming a 2-in stagnant air layer with relative humidity 100 and 40 percent at the two surfaces of the film.

7-20. A vapor B diffuses through a layer of stagnant air 0.25 in thick under steady-state conditions. Assuming the vapor density to be negligible at the upper surface of the stagnant film, the measured evaporation rate to be 2.0 lb_m/hr ft², the total pressure to be 14.7 psia, the temperature to be constant at 70°F, and the partial pressure of B to be 3.6 psia at the liquid interface, calculate ρD_{AB} of B into air.

MOLECULAR MASS TRANSFER (DIFFUSION)

7-21. Calculate the time required for 0.5 lb_m of carbon dioxide to diffuse through a stagnant air layer if the area normal to the direction of mass flux is 50 ft², the stagnant or fictive film is 2.0 in thick, the molal concentration at the carbon dioxide surface is $\tilde{\rho}_{CO_2} = 0.0008$ lb mole/ft³, and the concentration $\tilde{\rho}_{CO_2}$ at the upper fictive-film surface is assumed to be zero. The pressure is 1 atm, and the temperature is 70°F. (Use the Gilliland equation to calculate the diffusivity.)

7-22. An Arnold cell is used to determine the diffusivity of water into hydrogen at 32°F. If the results agree with values of Table A-5 and the cell is 8 in in diameter and 6 in from the interface to the surface, how much water must be supplied to the cell to maintain a constant level? Assume that the vapor pressure is 1.4 psia and that the hydrogen is at 14.7 psia; $\rho = 0.009$ lb_m/ft³.

7-23. If completely dry air at 70°F and 1 atm pressure blows gently over the cell shown and the partial pressure of the water vapor at the liquid–air interface is 2.0 psia, calculate the rate of evaporation in lb_m/hr-ft². Assume constant pressure and temperature and perfect-gas theory. $D_{WA} = 1.0$ ft²/hr.

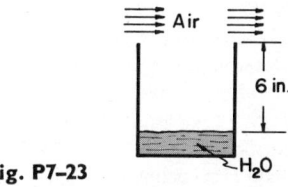

Fig. P7–23

7-24. Determine the mass-transfer rate from a drop of water at 70°F suspended in still air with a relative humidity of 70 percent. Assume the air at the interface has a 100 percent relative humidity and the droplet has a 0.1-in diameter. $D_{AB} \simeq 1$ ft²/hr.

7-25. A mixture of benzene vapor and carbon dioxide at 32°F has an average density of 0.138 lb_m/ft³ in the device shown. The concentration of benzene at the surface is 50 percent, and at a point 0.3 in above the surface the concentration of benzene is 20 percent. Estimate the convective-mass-transfer coefficient of benzene to the carbon dioxide under these conditions, assuming $D_{AB} = 0.136$ ft²/hr.

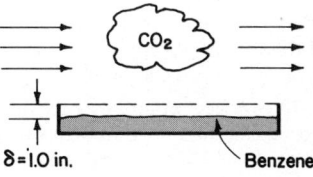

Fig. P7–25

7-26. If the vapor pressure of chloroform is 3 psia and a dish of it is placed in a cold (32°F) room for several hours, estimate the mass-transfer rate if the evaporation takes place through a film 0.25 in thick.

7-27. A scheme for reducing weight on large transport aircraft involves using helium gas for tire pressurization. Assume a tire to be toroidal, with a 24-in ID and a 60-in OD (18 in in cross-sectional diameter), an initial pressure of 150 psia, a tire wall thickness of 1.1 in, and $D_{He-rubber} = 3.0 \times 10^{-11}$ ft²/hr; estimate the helium loss during a 12-hr flight. The temperature may be assumed constant at 30°F, and the perfect-gas equation of state may be used.

7-28. Hydrogen at 1 atm and 932°F is held in a silicon dioxide container. The walls of the container are 0.1 in thick. Estimate the maximum mass-transfer rate from the container.

7-29. A fiber fire hose contains water at 55°F and 80 psia. If the diffusivity of water into the hose is 1.3 ft²/hr, estimate the mass-diffusion rate if the hose is 0.125 in thick with a 3-in ID.

7-30. Compare the permeability of silicon dioxide to helium at 68°F with that to hydrogen at 932°F.

7-31. In a wet season the soil beside the basement wall of a building accumulates a 20 percent moisture concentration. If the wall is 18 in thick and $\omega = 0.05$ at the center, determine the mass-transfer rate through the wall if the mass diffusivity is 3.94×10^{-4} ft²/hr.

7-32. How thin can the wall of a 2-in-diam Pyrex tube be if it is not to pass more than 2.365×10^{-10} lb_m/hr-ft² helium at 68°F to normal atmosphere?

7-33. Determine an expression for the mass flux from a square duct and compare it with the flux from a cylindrical duct if the radius of the cylindrical duct is equal to one-half the side of the square duct.

7-34. Carbon monoxide is burned by passing it through a square porous plate 3 ft on a side to a pure oxygen atmosphere, $2CO + O_2 \rightarrow 2CO_2$, where $\rho_{mix} = 0.93$ lb_m/ft³ and $\tilde{\rho}_{mix} = 0.037$ lb mole/ft³. Determine the flow rate of carbon monoxide to the plate to maintain a constant combustion rate. Assume a film from $x = 0.01$ ft to $x = 2$ ft.

7-35. Like other hydroxides, barium hydroxide releases oxygen when heated. The substance is formed, however, in accordance with the equation $BaO + H_2O \rightarrow Ba(OH)_2$. Suppose a sphere of BaO is suspended in a large beaker of water. Determine the initial rate of manufacture of $Ba(OH)_2$ if the reaction is instantaneous, assuming that the sphere has a 1-in diameter initially and

$$\tilde{\rho}_{mixture} = 4.3 \text{ lb mole/ft}^3 \qquad D_{H_2O-mix} = 3.4 \times 10^{-2} \text{ ft}^2/\text{hr}$$

7-36. For the reaction $2CO + O_2 \rightarrow 2CO_2$, determine the mass-flux rate of CO when

$$D_{AB} = 0.532 \text{ ft}^2/\text{hr} \qquad k''' = 4.2 \text{ hr}^{-1}$$

$$\tilde{\rho}_{mix} = 0.0047 \text{ lb mole/ft}^3 \qquad \delta = 2 \text{ ft}$$

At $x = 0$ $\quad \tilde{\omega}_A = 0.8$

7-37. A large slab of wet calcium chloride 4 in thick is allowed to dry in ambient air. The initial moisture content (wet basis) is 0.2 lb water per pound of calcium chloride. The ambient-air humidity ratio is 0.018. If the surface resistance to mass transfer is twice that of the internal resistance and the diffusivity of water in calcium chloride is 0.0009 ft²/hr, calculate the moisture content at 1-in intervals from the surface after 50 hr.

7-38. A new pond has very thick clay walls and bottom. If the initial moisture content of the clay (before filling the pond with water) is 0.08 lb water per pound of wet clay, the diffusivity of water into clay is 0.0003 ft²/hr, and the equilibrium moisture content of the clay in contact with water is 0.60 lb water per pound of clay, estimate the time required for the moisture content 1 ft below the surface to reach 25 percent (on a wet basis). The surface resistance is negligible.

7-39. The moisture content of the soil in a nursery is initially 0.5 lb_m water per pound mass of wet soil. If the diffusivity of water in the soil is 0.004 ft²/hr and the surface concentration is suddenly reduced to 0.05 lb_m water per pound mass of wet soil, estimate the time required for the moisture content 6 in below the surface to reach 25 percent on a wet basis; i.e., when will the plants need to be watered? The initial soil condition may be considered constant for a very large depth.

7-40. A piece of 1-in-thick plywood is used in a concrete form. After the concrete has set and the form is removed, it rains, and the plywood receives a moisture content of 23 lb_m water/ft³. (Plywood in this condition has a tendency to warp.) If the plywood is laid out on a grating and there is negligible surface resistance, how long is required for the moisture content at the center to reach 0.1 lb_m water/ft³ if the relative humidity is 70 percent at 60°F? Assume $D_{AB} = 0.0002$ ft²/hr.

7-41. A large tanker truck overturns and spills a corrosive fluid over a field. If the mass diffusivity of the fluid into the soil is 0.0038 ft²/hr and the fluid remains on the soil 30 min before evaporating to the air, to what depth is plant and insect life likely to be destroyed if a concentration of 0.1 percent by weight will destroy most life?

7-42. How deep should a bridge abutment be set in an area where water stands 180 days if the moisture concentration at the base is not to exceed 8 lb_m/ft³ and the diffusivity of water into the soil is 4.7×10^{-3} ft²/hr? $T_{av} = 60°F$.

7-43. A porous 2-in-diam rod which is very long is subjected to a cross flow of a certain fluid such that the coefficient of mass transfer is 5 ft/hr. The diffusivity of the fluid into the rod is 3.6 ft²/hr. Determine the mass fraction at the center of the rod after 3 min.

7-44. A charcoal briquet is formed with an initial moisture content of 35 lb_m/ft³. The briquet is approximately spherical with a radius of 0.1 ft. It is placed in a forced-air dryer with a moisture concentration of 0.02 lb_m/ft³. If the diffusivity of water in charcoal is 0.05 ft²/hr and the surface resistance is negligible, estimate the time required to dry the center of the briquet to a concentration of 3.5 lb_m water/ft³.

7-45. Estimate the concentration of hydrogen 0.5 in below the surface of a nickel sphere with a 2-in diameter after exposure to a pure hydrogen atmosphere at 330°F for 180 days. What is the concentration at the center?

7-46. If a 1- by 3- by 5-in dry sponge is submerged in a tank of 60°F water, how much time is required before the center reaches a concentration of 60 lb_m water/ft³, assuming $D_{AB} = 7$ ft²/hr and $k_c \to \infty$?

REFERENCES

1. Bennett, C. O., and J. E. Myers: "Momentum, Heat, and Mass Transfer," McGraw-Hill, New York, 1962.
2. Bird, R. B.: Theory of Diffusion, *Advan. Chem. Eng.*, **1**: 170 (1956).
3. Bird, R. B., W. E. Stewart, and E. N. Lightfoot: "Transport Phenomena," Wiley, New York, 1960.
4. Carslaw, H. S., and J. C. Jaeger: "Conduction of Heat in Solids," 2d ed., Oxford University Press, London, 1959.
5. Chaddock, J. B.: Mass Transfer, chap. 4 in "ASHRAE Guide and Data Book: Fundamentals," American Society of Heating, Refrigerating, and Air-conditioning Engineers, New York, 1967.
6. Chapman, W.: *Trans. Roy. Soc.*, **A217**: 165 (1917).
7. Crank, J.: "The Mathematics of Diffusion," Oxford University Press, London, 1957.
8. Dyer, D. F.: Bulk and Diffusional Transport of Gases at Non-uniform Pressures, *Trans. Faraday Soc.*, **63**(3): 531 (1967).
9. Fick, A.: *Ann. Physik.*, **94**: 59 (1855).
10. Gilliland, E. R.: *Ind. Eng. Chem.*, **26**: 681 (1934).
11. Gröber, H., S. Erk, and U. Grigull: "Grundgesetze der Wärmeübertragung," 3d ed., Springer-Verlag, Berlin, 1955.
12. Heisler, M. P.: Temperature Charts for Induction and Constant Temperature Heating, *Trans. ASME*, **69**: 227 (1947).
13. Hutchinson, T. S., and D. C. Baird: "The Physics of Engineering Solids," Wiley, New York, 1963.
14. Jeans, Sir James: "Dynamical Theory of Gases," Cambridge University Press, London, 1921.
15. Rohsenow, W. M., and H. Y. Choi: "Heat, Mass, and Momentum Transfer," Prentice-Hall, Englewood Cliffs, N.J., 1961.

16. Sherwood, T. K., and R. L. Pigford: "Absorption and Extraction," 2d ed., McGraw-Hill, New York, 1952.
17. Sutherland, W.: *Phil. Mag.*, **36:** 507 (1893).
18. Treybal, R. E.: "Mass Transfer Operations," 2d ed., McGraw-Hill, New York, 1968.
19. Welty, J. R., C. E. Wicks, and R. E. Wilson: "Fundamentals of Momentum, Heat and Mass Transfer," Wiley, New York, 1969.
20. Wilke, C. R.: *Chem. Engr. Progr.*, **46:** 95–104 (1950).

CHAPTER 8
RADIATIVE HEAT TRANSFER

8-1 THERMAL RADIATIVE PHENOMENA

In the preceding chapters the conductive mode of heat transfer has been emphasized, along with some consideration of convective transport in the boundary conditions. In the present chapter we introduce a totally different energy-transfer mechanism, *thermal electromagnetic radiation*, which is radiation emitted by virtue of the temperature of a body.

Rays of heat exhibit the wave and quantum characteristics typical of electromagnetic radiation. For example, they can be readily polarized, and they show interference when reunited after traveling different path lengths. As a result of their nature, we recognize that thermal radiation is a part of the electromagnetic spectrum as shown in Fig. 8-1. The velocity of propagation of radiation in a vacuum is

$$c = \lambda \nu = 3 \times 10^{10} \text{ cm/sec} \tag{8-1}$$

where λ is the wavelength and ν is the frequency of the radiation. Clearly the spectrum could be presented in terms of wavelength or frequency, wavelength being the arbitrary choice for Fig. 8-1.

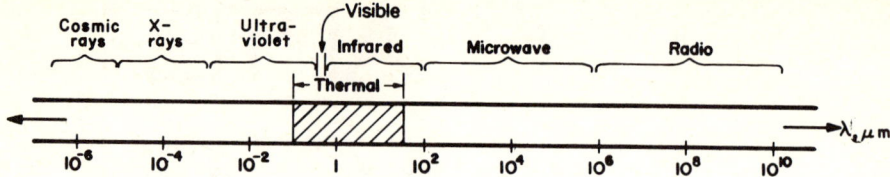

Fig. 8-1 Electromagnetic-radiation spectrum.

Wavelength is measured in any convenient linear dimension—meters, feet, centimeters, micrometers, or angstroms. Since 1 micrometer (μm) is one-millionth of a meter and 1 angstrom (Å) is 10^{-4} μm,

$1\ \mu\text{m} = 10^{-4}\ \text{cm} = 3.94 \times 10^{-5}\ \text{in}$

$1\ \text{Å} = 10^{-8}\ \text{cm} = 3.94 \times 10^{-9}\ \text{in}$

A convenient rule of thumb for obtaining a physical insight into typical radiation wavelengths is

$25\ \mu\text{m} \simeq 0.001\ \text{in}$

Figure 8-1 indicates that thermal radiation occurs within a relatively small part of the total spectrum, encompassing most of the infrared and part of the ultraviolet regions. It is interesting to note the extremely narrow range of the visible-light portion, which extends from approximately 0.4 to 0.7 μm.

Historically the study of thermal-radiative phenomena is particularly interesting. Application of the classical laws of physics gave erroneous results for radiated energy, and this eventually led to Planck's formulation of the quantum theory [9]—not only one of the most far-reaching contributions to the physical sciences but the origin of modern physics. As a result, we now recognize that radiant energy is transmitted in the form of discrete quanta, each of which has an energy

$$E = h\nu \tag{8-2}$$

where h is Planck's constant, 6.625×10^{-27} erg-sec. While this is highly important to fundamental research in radiative phenomena, very little use will be made of it in the present chapter, which is devoted to applied heat-transfer calculations.

8-2 DEFINITIONS AND PROPERTIES

Consider a *beam* or *bundle* of radiant energy incident upon a body as shown in Fig. 8-2. Of the total energy incident upon the body G part may be reflected, part may be absorbed, and part may be transmitted. With the definitions

$\alpha \equiv$ fraction of incident energy absorbed $=$ absorptivity

$\rho \equiv$ fraction of incident energy reflected $=$ reflectivity

$\tau \equiv$ fraction of incident energy transmitted $=$ transmissivity

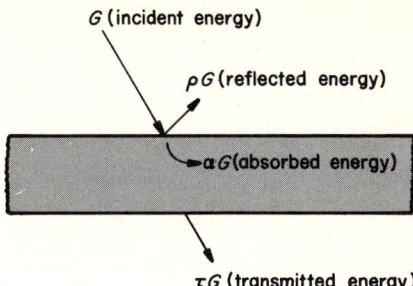

Fig. 8-2 Radiant-energy terminology.

an energy balance on the body shown in Fig. 8-2 becomes

$$G = \alpha G + \rho G + \tau G \tag{8-3}$$

Dividing Eq. (8-3) by G yields

$$1 = \alpha + \rho + \tau \tag{8-4}$$

Generally speaking, the transmissivity of most solids is zero, all thermal radiation entering the body being absorbed within approximately 0.1 in of the body for electric nonconductors according to Eckert [3] and in a much thinner layer for materials which are conductors. Thus, for most solids used in engineering applications we may consider

$$\alpha + \rho = 1.0 \tag{8-5}$$

This equation is also frequently applicable for liquids, though the transmissivity of a liquid is a strong function of its thickness.

Most gases reflect very little of the radiant energy incident upon their interface, and Eq. (8-4) reduces to

$$\alpha + \tau = 1.0 \tag{8-6}$$

The reflection of radiant energy from a surface is described in terms of two ideal models, *diffuse* and *specular reflectors*. Figure 8-3a depicts perfectly specular reflection, which is approximated by a mirror surface. Note specifically that the angle of incidence ψ is equal to the angle made by the reflected energy θ. Figure 8-3b illustrates a diffuse reflector; in this case the magnitude of the energy reflected in a direction θ is proportional to $\cos \theta$, where again θ is measured from the normal to the surface.

Neither model can accurately represent a real surface. If surface roughness elements are very small compared with the wavelength λ of the radiation, the surface is nearly specular; when roughness elements are large, the surface reflects diffusely. This concept introduces a very significant fact in practical radiation calculations, specifically that the *surface roughness has a very pronounced influence upon most thermal-radiation properties of materials*. Indeed, the field of radiative heat transfer

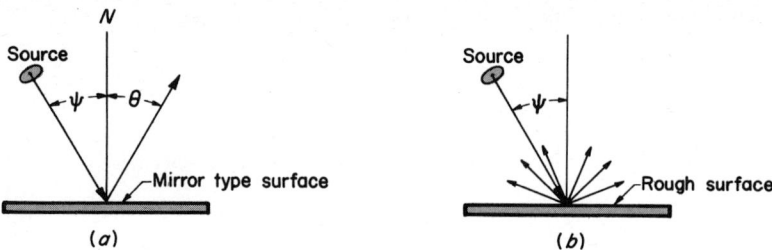

Fig. 8-3 Ideal-reflection models: (*a*) perfectly specular ($\psi = \theta$); (*b*) perfectly diffuse.

is in serious need of a method of adequately defining surface conditions and accurately predicting radiation properties from such descriptions.

8-3 BLACKBODY RADIATION

The ideal surface for a study of thermal radiation is the *blackbody*, defined as a body which absorbs all incident thermal radiation, i.e.,

$$\alpha_b = 1.0 \tag{8-7}$$

for all wavelengths of radiant energy. Thus, the reflectivity of a blackbody is zero. Historically the term blackbody arose because a black object typically absorbs more incident thermal energy than any other color.

Blackbody surfaces do not exist; this concept serves as a model only. A blackbody can be approximated by a cavity in a material of absorptivity less than unity in the manner illustrated in Fig. 8-4. Here a single bundle (ray) of energy is shown entering a small opening into a cavity (*Hohlraum*) with resulting specular reflections from the inner surfaces. If the entering energy is G_0, the amount leaving the first surface by reflection is then

$$G_1 = (1 - \alpha)G_0 \tag{8-8}$$

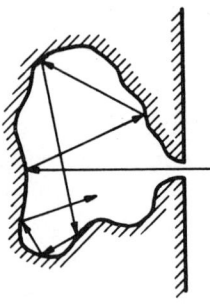

Fig. 8-4 Approximation to a blackbody.

RADIATIVE HEAT TRANSFER

The amount reflected from the second surface is

$$G_2 = (1 - \alpha)G_1 = (1 - \alpha)[(1 - \alpha)G_0]$$
$$G_2 = (1 - \alpha)^2 G_0 \tag{8-9}$$

and from the nth reflecting surface by induction is

$$G_n = (1 - \alpha)^n G_0 \tag{8-10}$$

This approach to a thermal blackbody is analogous to looking through a keyhole into a darkened room. The light passing through the hole is completely absorbed by the surfaces of the room, and nothing is visible.

The total energy emitted per unit area and per unit time by a body is termed the *emissive power E*. For a blackbody, Stefan determined experimentally and Boltzmann derived analytically the following expression for emissive power:

■ $$E_b = \sigma T^4 \tag{8-11}$$

where $\sigma = 0.1714 \times 10^{-8}$ Btu/hr-ft²-°R⁴, the Stefan-Boltzmann constant.

Any surface which is nonblack will have an emissive power E less than that of a blackbody at the same temperature. The ratio of these two terms at the same temperature is known as the total *emissivity*,

$$\epsilon \equiv \frac{E}{E_b} \tag{8-12}$$

Numerical values of the total emissivity for many engineering materials are tabulated in Appendix Table A-11. Examination of these data reveals that a range of values is reported for each corresponding to a range in temperature, e.g., for mild steel

Temperature range: 450–1950°F

Emissivity range: 0.20–0.32

Intuitively one would suspect the higher emissivity to correspond to the higher temperature, and this is generally true for electric conductors but not for nonconductors. The *trends* indicated in Fig. 8-5 are rather typical.

A very inclusive compilation of emissivity data is given by Gubareff, Janssen, and Torborg [4]. For instructional purposes, linear interpolation of the data in Appendix A is satisfactory; if highly accurate results are required, data should be obtained from Ref. 4, or, better yet, the emissivity of the actual material surface involved should be measured.

Example Determine the radiant heat loss per square foot from a mild-steel plate being formed at 1600°F.

Solution The emittance of a blackbody at 1600°F is given by the Stefan-Boltzmann equation

$$E_b = \sigma T^4 = (0.1714 \times 10^{-8} \text{ Btu/hr-ft}^2\text{-°R}^4)(1600 + 460°\text{R})^4$$
$$= 0.1714 \times 10^{-8}(\tfrac{2060}{100})^4 \times 10^8 \text{ Btu/hr-ft}^2$$
$$= 0.1714 \times 20.6^4 = 30{,}900 \text{ Btu/hr-ft}^2$$

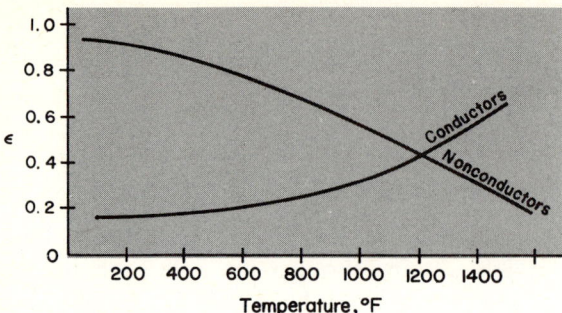

Fig. 8–5 General trend of total emissivity variance with temperature.

Applying a linear interpolation to the data of Appendix A yields

$\epsilon_{1600} \simeq 0.29$

and thus,

$E \simeq 0.29(30,900) \simeq 9000$ Btu/hr-ft^2

A very important relationship between ϵ and α can be determined by considering a black enclosure containing a small body 1 as shown in Fig. 8-6. First, assume that body 1 is black and the amount of energy leaving it per unit time is $E_b A$, where A is its total area. Assuming further that the entire system is in thermal equilibrium, the energy impinging upon body 1 per unit time is also $E_b A$; otherwise a violation of the second law of thermodynamics would occur.

Now, replace body 1 with another body 2, identical in size and shape but having α less than unity. Body 2 must absorb $\alpha_2 E_b A$ since the energy incident is the same as for the previous case. But for thermal equilibrium of the system, the energy emitted by body 2 must be the same as the energy absorbed, and thus

$E_2 A = \alpha_2 E_b A$

Or, since body 2 has completely arbitrary emissive power and absorptivity,

$E = \alpha E_b$ \hfill (8-13)

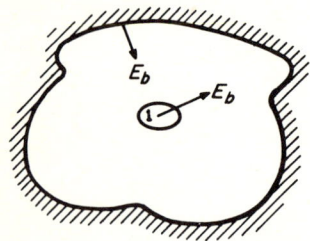

Fig. 8–6 Black enclosure and body exchanging radiant energy.

RADIATIVE HEAT TRANSFER

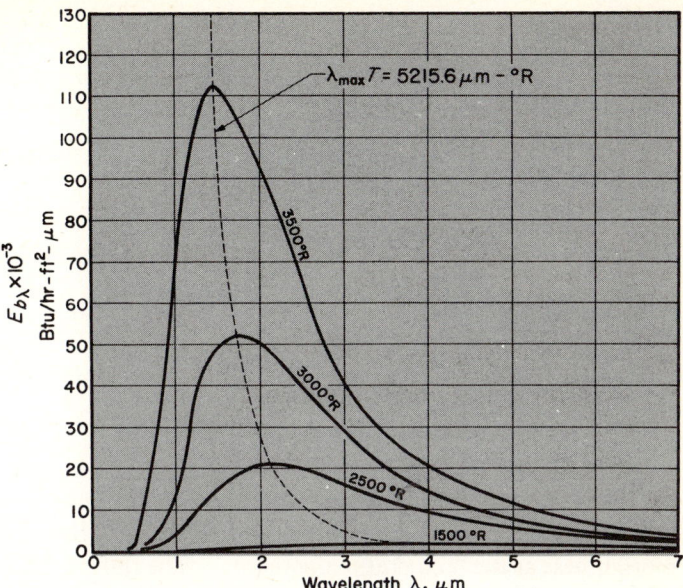

Fig. 8-7 Blackbody spectroradiometric curves.

Rearranging and applying the definition of emissivity yields

$$\alpha = \frac{E}{E_b} = \epsilon \tag{8-14}$$

which is known as *Kirchhoff's law*.

To this point our discussions have concerned *total* radiation properties, for example, E is the total energy emitted over all wavelengths of radiation per unit area and time. Frequently, it is advantageous to use monochromatic values of properties such as emissive power E_λ. The monochromatic emissive power is related to the total emissive power by

$$E = \int_0^\infty E_\lambda \, d\lambda \tag{8-15}$$

and hence E_λ is the derivative of E with respect to wavelength. The engineering units of E_λ are Btu/hr-ft-unit wavelength. For a blackbody we have

$$E_b = \int_0^\infty E_{b\lambda} \, d\lambda = \sigma T^4 \tag{8-16}$$

from Eqs. (8-11) and (8-15).

Figure 8-7 shows *spectroradiometric curves* for a blackbody at several different temperatures plotted directly from Planck's equation based on quantum theory

$$E_{b\lambda} = \frac{C_1 \lambda^{-5}}{\exp(C_2/\lambda T) - 1} \tag{8-17}$$

where

$C_1 = 1.187 \times 10^8$ Btu-μm^4/hr-ft^2

$C_2 = 2.5896 \times 10^4$ μm-°R

with λ in micrometers and temperature in degrees Rankine. There are several significant points to be observed in this figure.

1. The total area under the curve is a function of temperature and is the total emissive power E_b.
2. The peak in monochromatic emissive power shifts to lower wavelength with increasing temperature.
3. Most thermal radiation occurs within the infrared range for the temperatures shown in this figure.

Notice that the wavelength shift of emissive power intensity toward the shorter wavelengths with increasing temperature is in keeping with the fact that a heated object (such as a piece of metal) begins to glow dark red at about 1300°F and increases in brightness as the temperature increases. The shift results in an increasing percentage of the radiation occurring within the narrow visible band.

The shift in wavelength is described mathematically by Wien's displacement law (obtained from classical mechanics)

$$\lambda_{max} T = 5215.6 \quad \mu\text{m-°R} \tag{8-18}$$

where λ_{max} is the wavelength at which the maximum monochromatic radiant emission occurs. Equation (8-18) represents the dotted curve through the maximum values of $E_{b\lambda}$ in Fig. 8-7.

In many engineering problems we require the amount of energy radiated from a blackbody over a specified bandwidth. The energy radiated from zero to λT (or from 0 to λ at a given temperature) is

$$E_{b(0-\lambda T)} \equiv \int_0^{\lambda T} \frac{1}{T} E_{b\lambda} \, d(\lambda T) \tag{8-19}$$

and hence the fraction of the total energy radiated by a blackbody becomes

$$\frac{E_{b(0-\lambda T)}}{\sigma T^4} = \int_0^{\lambda T} \frac{E_{b\lambda}}{\sigma T^5} \, d(\lambda T) \tag{8-20}$$

Dunkle [2] originally integrated this over a wide range of λT and presented a table similar to Table 8-1; the numerical results of this table, however, are slightly more accurate.

Example An instrument box housing a control system is subjected to thermal radiation from molten iron in a smelting operation. The housing emissivity is 0.2 from 0.1 to 4.9 μm and approximately zero for all other wavelengths. Assuming incident blackbody radiation from a source at 2600°F, how much energy is absorbed per square foot of the instrument housing?

Table 8-1 Planck radiation functions†

λT, μm-°R	$\dfrac{E_{b\lambda} \times 10^5}{\sigma T^5}$	$\dfrac{E_{b(0-\lambda T)}}{\sigma T^4}$	λT, μm-°R	$\dfrac{E_{b\lambda} \times 10^5}{\sigma T^5}$	$\dfrac{E_{b(0-\lambda T)}}{\sigma T^4}$
1,000.0	0.000039	0.0000	10,400.0	5.142725	0.7183
1,200.0	0.001191	0.0000	10,600.0	4.921745	0.7284
1,400.0	0.012008	0.0000	10,800.0	4.710716	0.7380
1,600.0	0.062118	0.0000	11,000.0	4.509291	0.7472
1,800.0	0.208018	0.0003	11,200.0	4.317109	0.7561
2,000.0	0.517405	0.0010	11,400.0	4.133804	0.7645
2,200.0	1.041926	0.0025	11,600.0	3.959010	0.7726
2,400.0	1.797651	0.0053	11,800.0	3.792363	0.7803
2,600.0	2.761875	0.0098	12,000.0	3.633505	0.7878
2,800.0	3.882650	0.0164	12,200.0	3.482084	0.7949
3,000.0	5.093279	0.0254	12,400.0	3.337758	0.8017
3,200.0	6.325614	0.0368	12,600.0	3.200195	0.8082
3,400.0	7.519353	0.0507	12,800.0	3.069073	0.8145
3,600.0	8.626936	0.0668	13,000.0	2.944084	0.8205
3,800.0	9.614973	0.0851	13,200.0	2.824930	0.8263
4,000.0	10.463377	0.1052	13,400.0	2.711325	0.8318
4,200.0	11.163315	0.1269	13,600.0	2.602997	0.8371
4,400.0	11.714711	0.1498	13,800.0	2.499685	0.8422
4,600.0	12.123821	0.1736	14,000.0	2.401139	0.8471
4,800.0	12.401105	0.1982	14,200.0	2.307123	0.8518
5,000.0	12.559492	0.2232	14,400.0	2.217411	0.8564
5,200.0	12.613057	0.2483	14,600.0	2.131788	0.8607
5,400.0	12.576066	0.2735	14,800.0	2.050049	0.8649
5,600.0	12.462308	0.2986	15,000.0	1.972000	0.8689
5,800.0	12.284687	0.3234	16,000.0	1.630989	0.8869
6,000.0	12.054971	0.3477	17,000.0	1.358304	0.9018
6,200.0	11.783688	0.3715	18,000.0	1.138794	0.9142
6,400.0	11.480102	0.3948	19,000.0	0.960883	0.9247
6,600.0	11.152254	0.4174	20,000.0	0.815714	0.9335
6,800.0	10.807041	0.4394	21,000.0	0.696480	0.9411
7,000.0	10.450309	0.4607	22,000.0	0.597925	0.9475
7,200.0	10.086964	0.4812	23,000.0	0.515964	0.9531
7,400.0	9.721078	0.5010	24,000.0	0.447405	0.9579
7,600.0	9.355994	0.5201	25,000.0	0.389739	0.9621
7,800.0	8.994419	0.5384	26,000.0	0.340978	0.9657
8,000.0	8.638524	0.5561	27,000.0	0.299540	0.9689
8,200.0	8.290014	0.5730	28,000.0	0.264157	0.9717
8,400.0	7.950202	0.5892	29,000.0	0.233807	0.9742
8,600.0	7.620072	0.6048	30,000.0	0.207663	0.9764
8,800.0	7.300336	0.6197	40,000.0	0.074178	0.9891
9,000.0	6.991475	0.6340	50,000.0	0.032617	0.9941
9,200.0	6.693786	0.6477	60,000.0	0.016479	0.9965
9,400.0	6.407408	0.6608	70,000.0	0.009192	0.9977
9,600.0	6.132361	0.6733	80,000.0	0.005521	0.9984
9,800.0	5.868560	0.6853	90,000.0	0.003512	0.9989
10,000.0	5.615844	0.6968	100,000.0	0.002339	0.9991
10,200.0	5.373989	0.7078			

† From J. A. Wiebelt, "Engineering Radiation Heat Transfer," Holt, Rinehart and Winston, Inc., New York, 1966; used by permission.

Solution

$T = 2600 + 460 = 3060°R$

$\lambda_1 T = 0.1(3,060) = 306$

$\lambda_2 T = 4.9(3,060) = 15,000$

Incident radiation $(0.1\ \mu m < \lambda < 4.9\ \mu m) = E_{b(0-\lambda_2 T)} - E_{b(0-\lambda_1 T)}$

From Table 8-1,

$E_{b(0-\lambda_1 T)} \simeq 0$

$E_{b(0-\lambda_2 T)} = \sigma T^4 (0.8689) = 0.1714(\frac{3060}{100})^4 \times 0.8689 = 131,000\ Btu/hr\text{-}ft^2$

Incident radiation $(0.1\ \mu m < \lambda < 4.9\ \mu m) = 131,000\ Btu/hr\text{-}ft^2$

Total energy absorbed = α(incident radiation)

By Kirchhoff's law,

Total energy absorbed = ϵ(incident radiation) = $0.2(131,000) = 26,000\ Btu/hr\text{-}ft^2$

Another useful term in radiation heat transfer is *radiation intensity*, which is defined as the *radiant energy per unit solid angle per unit area of the emitter projected normal to the line of view of the receiver from the radiating element*. These terms are illustrated with Fig. 8-8 as follows. The total energy radiated from surface element dA_1 is intercepted by the imaginary hemisphere centered upon the emitting element. For a blackbody the energy radiated from element dA_1 to element dA_2 is

$$dq_{1-2} = I_b \cos \phi\ dA_1\ d\omega \qquad (8\text{-}21)$$

where the unit solid angle ω is defined by

$$d\omega \equiv \frac{dA_2}{r^2} \qquad (8\text{-}22)$$

Notice from Fig. 8-8 that $\cos \phi\ dA_1$ is the area of the emitter projected normal to the line of view of the receiver, and this permits the use of the intensity in Eq. (8-21).

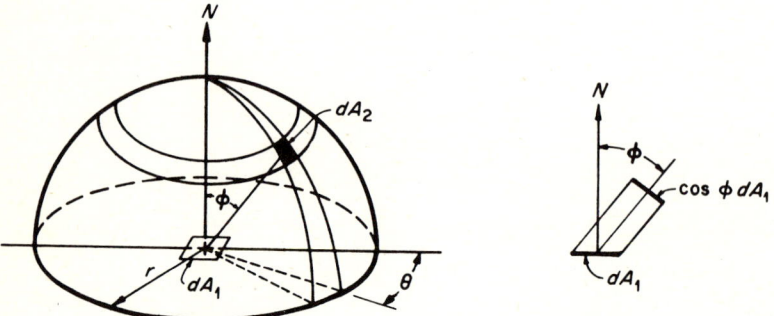

Fig. 8-8 Geometry for hemispherical emittance and solid angle.

RADIATIVE HEAT TRANSFER

Also, the geometry permits use of the expression

$$dA_2 = r\, d\phi(r \sin \phi\, d\theta) \tag{8-23}$$

and hence

$$d\omega = \frac{r\, d\phi(r \sin \phi\, d\theta)}{r^2} = \sin \phi\, d\phi\, d\theta \tag{8-24}$$

Substituting into Eq. (8-21) and integrating over the hemisphere yields

$$q_{1-2} = dA_1 \int_0^{2\pi}\!\!\int_0^{\pi/2} I_b \cos \phi \sin \phi\, d\phi\, d\theta \tag{8-25}$$

which for the blackbody ($I_b = $ const) integrates to

$$\blacksquare \quad \frac{q_{1-2}}{dA_1} = E_b = \pi I_b \tag{8-26}$$

This last relationship is important: it states that the emissive power of a blackbody is equal to π times the intensity of radiation.

8-4 RADIANT EXCHANGE BETWEEN BLACK SURFACES

The simplest case of radiant energy exchange occurs when two infinite parallel planes are maintained at different temperatures, T_1 and T_2 (Fig. 8-9). In this case it is immediately clear that all radiant energy leaving the surface of plane 1 exposed to plane 2 impinges upon surface 2, and a similar statement applies to the energy leaving plane 2. Since both surfaces are black, all incident energy is absorbed. The net heat transfer by radiation from 1 to 2 is

$$q_{1-2} = E_{b_1} A_1 - E_{b_2} A_2 \tag{8-27}$$

or, per unit surface area (since $A_1 = A_2 = A$)

$$\frac{q_{1-2}}{A} = E_{b_1} - E_{b_2} = \sigma(T_1^4 - T_2^4) \tag{8-28}$$

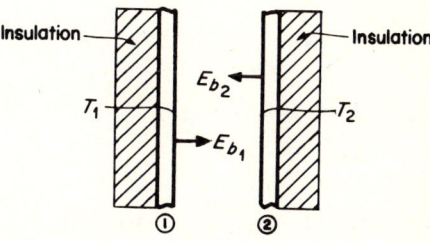

Fig. 8-9 Energy exchange between two infinite parallel black planes.

Configuration factors All problems of practical interest involve at least one finite area, and the heat exchange is heavily influenced by the geometry involved. Thus we must establish the effect of *configuration* upon the net heat transfer. This leads to the definition of the *configuration factor*, which is the *fraction of radiant energy incident upon one surface from another surface, both assumed to be emitting energy diffusely.* Other names frequently used for this term include *view factor*, *angle factor*, and *radiation shape factor*. We shall now develop mathematical expressions for this term.

Consider two surface elements as shown in Fig. 8-10. The energy radiated from dA_1 and incident upon dA_2 is

$$dq_{1-2} = I_b \cos \phi_1 \, dA_1 \, d\omega_{1-2} \tag{8-29}$$

where $d\omega_{1-2}$ is the area of dA_2 seen by dA_1 divided by the square of the distance, i.e.

$$d\omega_{1-2} = \frac{\cos \phi_2 \, dA_2}{r^2} \tag{8-30}$$

Now the total energy radiated from dA_1 is

$$dq = I_b \pi \, dA_1 \tag{8-31}$$

and hence the configuration factor between two infinitesimal areas is by definition the ratio of Eq. (8-29) to (8-31):

$$F_{dA_1-dA_2} = \frac{\cos \phi_1 \cos \phi_2 \, dA_2}{\pi r^2} \tag{8-32}$$

It is significant that the configuration factor involves ϕ_1, ϕ_2, dA_2, and r^2; that is, it is a function of *geometry* only!

Suppose a very small (infinitesimal) emitter transmits energy to a finite surface. Then the ratio of Eq. (8-29) to (8-31) is

$$F_{dA_1-A_2} = \frac{\int_{A_2} I_{b_1} \cos \phi_1 \, dA_1 \cos \phi_2 \, dA_2/r^2}{\pi I_{b_1} \, dA_1} \tag{8-33}$$

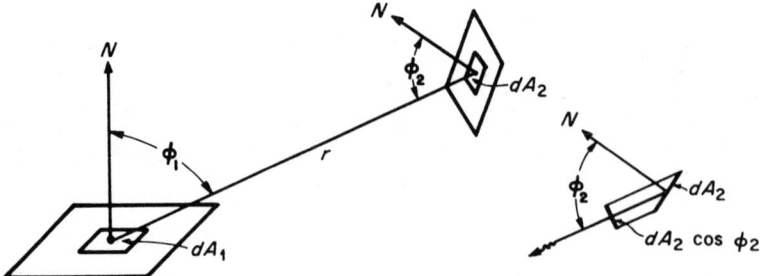

Fig. 8-10 Nomenclature for infinitesimal-area configuration factor.

RADIATIVE HEAT TRANSFER

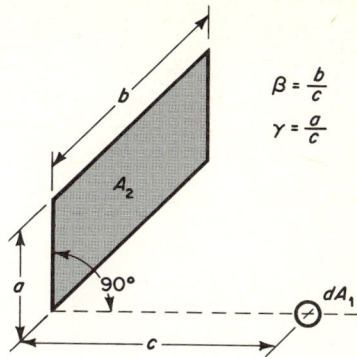

Fig. 8-11 Configuration of a spherical point source and a plane rectangle.

But since both I_{b_1} and dA_1 are independent of dA_2, this simplifies to

$$F_{dA_1-A_2} = \int_{A2} \frac{\cos \phi_1 \cos \phi_2 \, dA_2}{\pi r^2} \tag{8-34}$$

and this expression has been integrated for numerous configurations and reported in the literature, a large number being given by Wiebelt [11], who adapted his results from the work of Hamilton and Morgan [5].

One configuration of practical significance involves the heat exchange from a spherical point source to a plane rectangle, the point source being at one corner of a rectangle that has one common side with plane A_2, as shown in Fig. 8-11. This configuration approximates the physical problem of a small object exposed to the

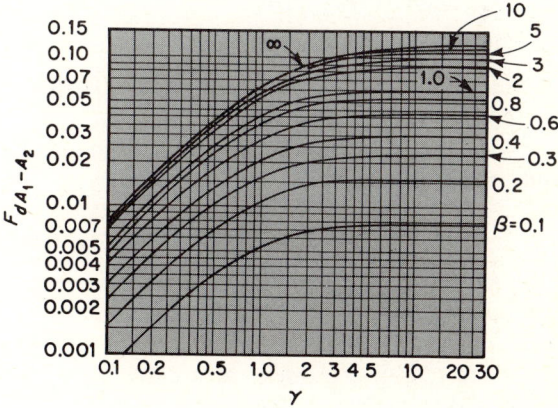

Fig. 8-12 Configuration factors: spherical point source to a plane rectangle. (*Adapted from D. C. Hamilton and W. R. Morgan, NACA Tech. Note TN-2836, 1952.*)

wall of an enclosure. Defining $\beta = b/c$ and $\gamma = a/c$, configuration factors can be presented in terms of the geometrical dimensions and are plotted in Fig. 8-12.

Example Approximate the heat transfer from a 1-in-diam sphere at 2000°R to a 4- by 6-ft wall 4 ft away (Fig. 8-13). The sphere is located on a normal to the wall passing through the wall center as illustrated. All surfaces are assumed to be blackbodies, and the temperature of the

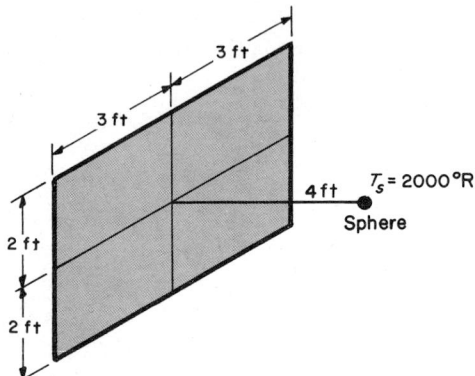

Fig. 8-13 Geometry for sphere radiating to 4- by 6-ft wall.

wall is assumed to be sufficiently low for heat transfer from the wall to the sphere to be negligible.

Solution The radiant energy from the sphere incident upon the wall is

$$q_{1-2} = (E_b A)_{\text{sphere}} (F_{\text{sphere-wall}})$$

The configuration factor for the sphere to one-fourth of the wall can be approximated from Fig. 8-12 with $\beta = 0.75$ and $\gamma = 0.5$. For the entire wall,

$$F_{dA_1-A_2} = 0.021(4)$$

(Note that this is an approximation due to the finite size of the sphere.) Thus,

$$q_{1-2} = 0.1714(\tfrac{2000}{100})^4 (4\pi)(\tfrac{1}{24})^2 (4 \times 0.021) = 50.3 \text{ Btu/hr}$$

Turning to the case of radiant exchange between finite surfaces, we obtain by integration over both surfaces

$$F_{A_1-A_2} = \frac{\int_{A_2} \int_{A_1} I_1 \cos \phi_1 \, dA_1 \cos \phi_2 \, dA_2 / r^2}{\int_{A_1} \pi I_1 \, dA_1} \tag{8-35}$$

and since I_1 is independent of area, it may be taken outside the integral and thus eliminated to yield

$$F_{A_1-A_2} = \frac{1}{\pi A_1} \int_{A_2} \int_{A_1} \frac{\cos \phi_1 \cos \phi_2}{r^2} dA_1 \, dA_2 \tag{8-36}$$

RADIATIVE HEAT TRANSFER

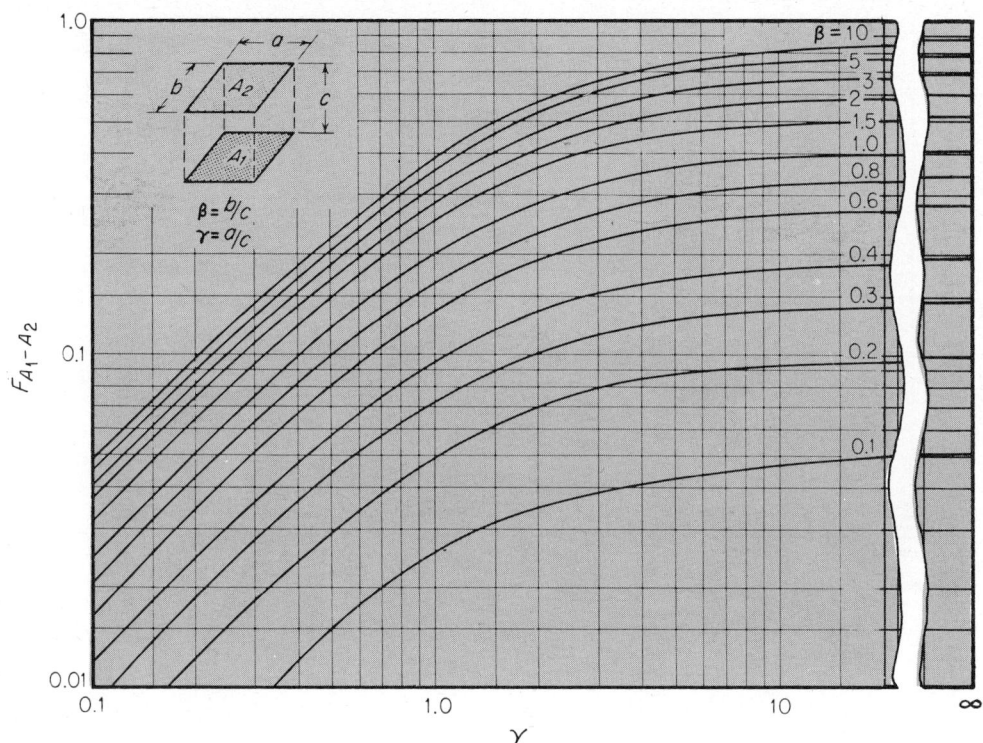

Fig. 8-14 Configuration factors for two identical parallel, directly opposed flat plates. (*Adapted from D. C. Hamilton and W. R. Morgan, NACA Tech. Note TN-2836, 1952.*)

Equation (8-36) has also been evaluated for numerous configurations, many of which are reported in Refs. 5 and 11. Two of particular importance involve (1) two identical parallel directly opposed flat surfaces, as shown in Fig. 8-14, and (2) two rectangles oriented 90° to each other and having one common edge, as shown in Fig. 8-15.

Example Determine the net heat transfer between two blackbody rectangles parallel and directly opposed. The plates are 4 by 6 ft and spaced 6 ft apart. Plate 1 is at 1200°R, and plate 2 is at 1000°R.

Solution Abbreviating[1] $F_{A_1-A_2}$ by F_{1-2}, the radiant heat transfer from body 1 to body 2 is

$$q'_{1-2} = A_1 F_{1-2}(E_{b_1}) = A_1 F_{1-2} \sigma(T_1^4)$$

Likewise, the radiant heat transfer from body 2 to body 1 is

$$q'_{2-1} = A_2 F_{2-1} \sigma(T_2^4)$$

But $A_1 = A_2$ and $F_{1-2} = F_{2-1}$; consequently the net heat transfer from body 1 to body 2 is

---

[1] From this point on in both text and examples, this abbreviation will be employed for finite-area configuration factors.

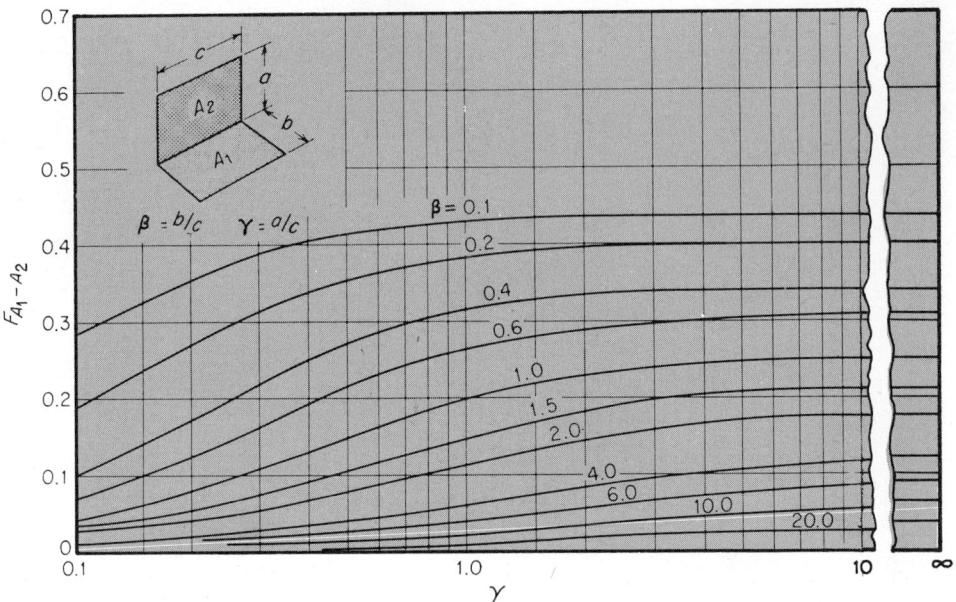

Fig. 8-15 Configuration factors for two perpendicular flat plates with a common edge. (*Adapted from D. C. Hamilton and W. R. Morgan, NACA Tech. Note TN-2836, 1952.*)

$q_{1-2} = q'_{1\ 2} - q'_{2-1} = A_1 F_{1-2} \sigma(T_1^4 - T_2^4)$

F_{1-2} is found from Fig. 8-14

$\beta = \tfrac{6}{6} = 1.0 \qquad \gamma = \tfrac{4}{6} = 0.666 \qquad F_{1-2} = 0.15$

Thus,

$q_{1-2} = 24(0.15)(0.1714)[(\tfrac{1200}{100})^4 - (\tfrac{1000}{100})^4] = 6625$ Btu/hr

There are four useful relationships between configuration factors that should be carefully examined and thoroughly understood. The first concerns the *additive* nature when the receiving or irradiated surface is subdivided. Examination of Eqs. (8-34) and (8-32) reveals for radiation from an infinitesimal area to a finite area

$$F_{dA_1-A_2} = \int_{A_2} \frac{\cos \phi_1 \cos \phi_2 \, dA_2}{\pi r^2} = \int_{A_2} F_{dA_1-dA_2} \tag{8-37}$$

which means that the fractions of the total energy leaving dA_1 and impinging upon small areas dA_2 are simply summed over A_2 to obtain the total. As an illustration of this principle for finite areas, consider radiation from area 1 to areas 2 and 3 of Fig. 8-16. The configuration factor for this is

$$F_{1-(2,3)} = F_{1-2} + F_{1-3} \tag{8-38}$$

The second relationship involves the subdivision of the emitting surface. From the expression for emission from a finite size area to a finite size area [Eqs. (8-36)

RADIATIVE HEAT TRANSFER

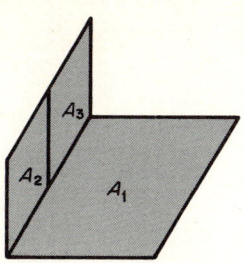

Fig. 8-16 Sketch of areas showing additive nature of configuration factors for subdivided irradiated surface.

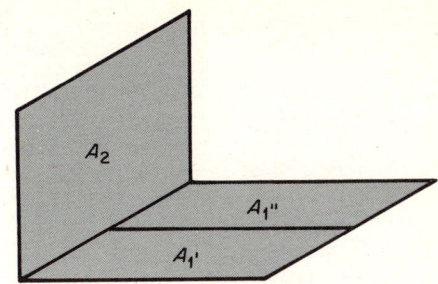

Fig. 8-17 Sketch of areas showing area-weighted average configuration factors for subdivided irradiating surface.

and (8-34)] we have

$$F_{1-2} = \frac{1}{A_1} \int_{A_1} \int_{A_2} \frac{\cos \phi_1 \cos \phi_2 \, dA_2}{\pi r^2} \, dA_1$$

$$= \frac{1}{A_1} \int_{A_1} (F_{dA_1-A_2}) \, dA_1 \tag{8-39}$$

Figure 8-17 illustrates this principle with finite-sized elements. In this case the irradiating surface A_1 is subdivided into two parts. Equation (8-39) can be approximated by

$$F_{1-2} = \frac{1}{A_1} \sum_i \Delta A_i F_{\Delta A_i - A_2} \tag{8-40}$$

which for Fig. 8-17 becomes

$$F_{1-2} \simeq \frac{1}{A_{1'} + A_{1''}} (A_{1'} F_{A_{1'}-A_2} + A_{1''} F_{A_{1''}-A_2}) \tag{8-41}$$

The third useful relationship is that the summation of configuration factors from any surface is unity. Since the definition of the configuration factor is the fraction of energy leaving the emitter i and incident upon a surface j, we have for n surfaces completely enclosing the emitter

$$\sum_{j=1}^{n} F_{A_i - A_j} = 1.0 \tag{8-42}$$

The fourth relationship, and perhaps the most important of all, is known as the *reciprocity theorem*. Multiplying both sides of Eq. (8-36) by A_1, we have

$$A_1 F_{1-2} = \int_{A_2} \int_{A_1} \frac{\cos \phi_1 \cos \phi_2}{\pi r^2} \, dA_1 \, dA_2 \tag{8-43}$$

But since the subscript notation for A_1 and A_2 was completely arbitrary in the derivation, clearly there can be no difference between Eq. (8-43) and a like expression that could be developed beginning again with two arbitrary surface elements as in Fig. 8-10 with subscripts interchanged, and thus

■ $$A_1 F_{1-2} = A_2 F_{2-1} \tag{8-44}$$

This, the reciprocity theorem, is one of the most powerful tools at the disposal of the engineer in making radiant-heat-exchange calculations. This usefulness is illustrated in the following example.

Example A small aircraft component is subjected to accelerated aging tests to evaluate the performance of electric potting compounds. Part of the test requires thermal cycling to be accomplished by placing the unit, originally at 70°F, in a large oven at 540°F. Approximate the initial net instantaneous heat transfer by radiation to the component if all surfaces are blackbodies, the component is approximately spherical in shape with a diameter of 4 in, and the oven is cubical with 4-ft sides.

Solution The net radiative flux from the oven to the component is

$$q_{1-2} = A_1 F_{1-2} \sigma (T_1^4 - T_2^4)$$

where subscript 1 denotes oven and 2 denotes component. F_{1-2} would be quite difficult to evaluate geometrically; however, the configuration factor from component to oven, F_{2-1}, is unity since all radiation leaving the component impinges upon the oven walls. Thus, applying the reciprocity theorem yields

$$F_{1-2} = \frac{A_2}{A_1} F_{2-1} = \frac{4\pi(\frac{2}{12})^2}{6[4(4)]} (1) = 0.545 \times 10^{-2}$$

and

$$q_{1-2} = 6[4(4)] \frac{4\pi(\frac{2}{12})^2}{6[4(4)]} (0.1714)[(\tfrac{1000}{100})^2 - (\tfrac{530}{100})^2] = 4.3 \text{ Btu/hr}$$

Geometric flux algebra Determination of view factors for many configurations is facilitated by using flux algebra. The method is especially advantageous when the problem can be subdivided into parts with known or tabulated configuration factors. Introducing the geometric flux

$$H_{ij} \equiv A_i F_{ij} \tag{8-45}$$

into Eqs. (8-42) and (8-44) yields, respectively,

$$\sum_{j=1}^{n} H_{ij} = A_i \tag{8-46}$$

and

$$H_{ij} = H_{ji} \tag{8-47}$$

The geometric flux obeys certain algebraic rules which will be very useful in application. To obtain the first of them, consider radiation from one finite area A_1 to a receiving area subdivided into two parts, A_2 and A_3. By Eq. (8-38) we can express the product of the view factor and the emitting area as

$$A_1 F_{1-(2,3)} = A_1 F_{1-2} + A_1 F_{1-3} \qquad (8\text{-}48)$$

But in terms of the geometric fluxes, Eq. (8-48) is

Rule 1: $\qquad H_{1-(2,3)} = H_{1-2} + H_{1-3} \qquad (8\text{-}49)$

A second rule involves irradiation of two areas, 3 and 4, from two areas, 1 and 2. The extension of Eq. (8-41) to apply to this problem yields

$$(A_1 + A_2) F_{(1+2)-(3+4)} = \frac{A_1 + A_2}{A_1 + A_2} (A_1 F_{1-(3,4)} + A_2 F_{2-(3,4)})$$

$$= A_1 F_{1-(3,4)} + A_2 F_{2-(3,4)} \qquad (8\text{-}50)$$

which in terms of the geometric fluxes becomes

Rule 2: $\qquad H_{(1,2)-(3,4)} = H_{1-(3,4)} + H_{2-(3,4)} \qquad (8\text{-}51)$

The final rule involves a further decomposition of Eq. (8-51), and is expressed as

Rule 3: $\qquad H_{(1,2)-(3,4)} = H_{1-3} + H_{2-3} + H_{1-4} + H_{2-4} \qquad (8\text{-}52)$

This rule is readily proved by applying rule 1 to obtain

$$H_{(1,2)-(3,4)} = H_{(1,2)-3} + H_{(1,2)-4}$$

and further application of rule 2 to yield rule 3.

Example Determine the configuration factor F_{1-2} for the problem illustrated in the figure where A_2 and A_3 are equal halves of a flat plate directly opposed to an identical flat plate A_1.

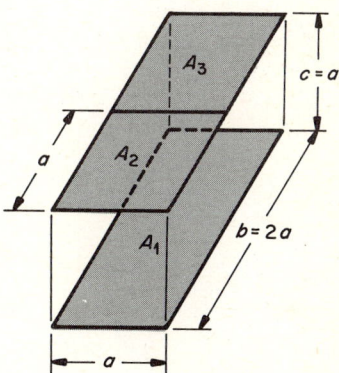

Fig. 8-18 Two opposed flat plates.

Solution The entire configuration factor can be obtained from Fig. 8-14, and is $F_{1-(2,3)}$:

$$\beta = \frac{b}{c} = \frac{2a}{a} = 2 \quad \text{and} \quad \gamma = \frac{a}{c} = 1$$

From Fig. 8-14, $F_{1-(2,3)} = 0.1285$. By rule 1,

$$H_{1-(2,3)} = H_{1-2} + H_{1-3}$$

Since A_2 and A_3 are identical in size and are geometric opposites with respect to orientation from A_1,

$$H_{1-2} = H_{1-3} = \frac{0.1285 A_1}{2} \quad \text{and} \quad F_{1-2} = 0.0642$$

The reader is cautioned to exercise care in applying flux algebra to problems of the preceding type. Consider the additional complication of letting A_2 and A_3 be unequal in area, as shown in Fig. 8-19. In this event, we divide the opposite plate into like areas A_4 and A_5, and we can immediately obtain values for

$$F_{(2,3)-(4,5)} = F_{(4,5)-(2,3)}$$
$$F_{2-4} = F_{4-2}$$
$$F_{3-5} = F_{5-3}$$

from Fig. 8-14. The unknown view factors are then F_{2-5} and F_{3-4}. By rule 3 we have

$$H_{(2,3)-(4,5)} = H_{2-4} + H_{2-5} + H_{3-4} + H_{3-5}$$
$$H_{2-5} + H_{3-4} = H_{(2,3)-(4,5)} - H_{2-4} - H_{3-5}$$

But H_{2-5} is identical by symmetry with H_{4-3}, which by Eq. (8-47) is also $H_{3\ 4}$, and thus

$$2H_{2-5} = H_{(2,3)-(4,5)} - H_{2-4} - H_{3-5}$$

where all terms on the right side are known. This result, together with Eq. (8-47) and rule 1 would permit calculation of the configuration factor between the entire lower plate and either part of the upper one.

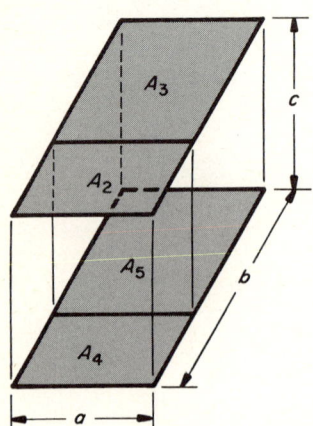

Fig. 8-19 Application of flux algebra.

RADIATIVE HEAT TRANSFER

A similar approach can be applied to determine configuration factors between sections of two planes intersecting at 90°, the complete plane-to-plane factors having been given in Fig. 8-15.

8-5 REAL SURFACES AND GRAY SURFACES

Our discussion to this point has considered radiant exchange between blackbodies only. Real surfaces invariably emit less radiation than blackbodies, as indicated by the emissivity of the surface. Usually the emissivity varies both with wavelength and with temperature. Another idealization involves the concept of a *gray body*, which is defined by

$$(\epsilon_\lambda)_{\text{gray}} \equiv \text{const} \tag{8-53}$$

The advantage afforded by assumption of the gray-body model is apparent upon inspection of the general expression for the emissive power of a body:

$$E = \int_0^\infty \epsilon_\lambda E_{b\lambda}\, d\lambda \tag{8-54}$$

For $\epsilon = \text{const}$, this becomes

$$E = \epsilon \int_0^\infty E_{b\lambda}\, d\lambda = \epsilon \sigma T^4 \tag{8-55}$$

In general, the monochromatic hemispherical emissive power of a real surface is as shown in Fig. 8-20, and the total emissive power can be obtained by planimetering the area under the spectroradiometric curve or by numerical integration of Eq. (8-54).

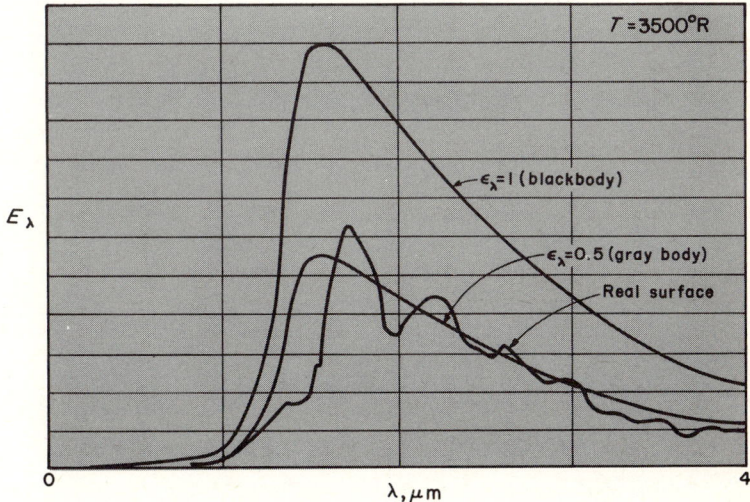

Fig. 8-20 Typical monochromatic emissive power of a real surface.

The gray-body assumption is not only desirable from an analytical viewpoint but is frequently justifiable upon physical grounds. For example, the absorptivity of black paint or aluminum paint is essentially constant for all wavelengths of radiation; the reflectivities of many pure metals, e.g., gold, silver, copper, zinc, and rhodium, approach a constant value over a large part of the infrared spectrum. As a result, most elementary calculations of radiant heat transfer assume gray-body characteristics.

In addition to the variation of properties with wavelength, we should note that radiation properties α and ρ are dependent upon the angle of incidence of the impinging energy. For our purposes we shall consider only total hemispherical properties, and a term such as ρ will imply an average value of reflectivity in all directions for incident radiation of all wavelengths. For a detailed treatment of radiation properties and their dependence upon wavelength, temperature, and angle of incidence, the reader should consult advanced texts devoted solely to radiative phenomena, e.g., Sparrow and Cess [10] or Wiebelt [11].

8-6 RADIANT HEAT EXCHANGE BETWEEN GRAY SURFACES

Net heat-transfer calculations for gray surfaces are considerably more difficult than for blackbodies. To illustrate the nature of the added complexity, consider radiation from a gray surface with emissivity ϵ and receiving irradiation G from other surfaces (Fig. 8-21). Of the energy incident upon the surface, the body absorbs αG, which by Kirchhoff's law is also ϵG. Considering the body to have zero transmissivity, the reflected radiation is the incident minus the absorbed irradiation

$$\rho G = (1 - \alpha)G = (1 - \epsilon)G \tag{8-56}$$

In addition, the body emits ϵE_b, and the radiant heat flux leaving the surface is then the sum of the emitted plus reflected radiation. This energy can undergo multiple reflections which must be considered in the analysis.

Considering the surface[1] to have uniform temperature and properties, the reflection to be diffuse, and the irradiation to be uniform, we obtain the net energy flux per unit area from the surface by an energy balance

$$\frac{q}{A} = (\epsilon E_b + \rho G) - G \tag{8-57}$$

[1] These assumptions will apply to all surfaces throughout the analysis for heat transfer between gray bodies.

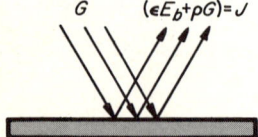

Fig. 8-21 Total radiation leaving a gray surface.

RADIATIVE HEAT TRANSFER

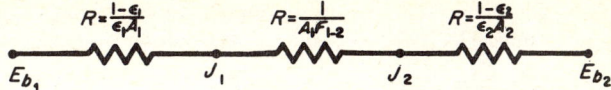

Fig. 8-22 Electrical analog for two-gray-surface system.

In terms of the *radiosity J*, defined as the total radiant energy leaving the surface per unit area per unit time, this is

$$\frac{q}{A} = J - G = \epsilon E_b + \rho G - G \tag{8-58}$$

Eliminating G from this expression yields

$$q = A\left(J - \frac{J - \epsilon E_b}{\rho}\right) = \frac{\epsilon A}{\rho}(E_b - J)$$

or

$$q = \frac{E_b - J}{(1 - \epsilon)/\epsilon A} \tag{8-59}$$

Equation (8-59) is the basis for an electrical analog: the numerator on the right side can be considered as a potential difference and the denominator as a surface resistance to radiative heat transfer. This permits the unknown potential J to be replaced by a known potential E_b connected to node J through the surface resistance.

A second type of resistance is encountered as the radiant flux travels from the emitting to the receiving surface; i.e., only the fraction F_{1-2} of the emitted energy from surface 1 reaches surface 2. For two general gray surfaces i and j, the net heat flux is

$$q_{i-j} = A_i F_{i-j}(J_i - J_j) \tag{8-60}$$

and again this can be considered as analogous to an electric current flow, the potential difference being $J_i - J_j$ and the spatial resistance being $1/A_i F_{i-j}$.

Applying these analog equations to a two-body system in which each surface exchanges radiant energy *only with itself and the other* (sometimes referred to as a two-body enclosure), we have the analog depicted in Fig. 8-22, where the heat flux is from the higher to the lower value of E_b. This net exchange can be calculated from

$$q_{1-2} = \frac{E_{b_1} - E_{b_2}}{\Sigma R} = \frac{E_{b_1} - E_{b_2}}{\dfrac{1 - \epsilon_1}{\epsilon_1 A_1} + \dfrac{1}{A_1 F_{1-2}} + \dfrac{1 - \epsilon_2}{\epsilon_2 A_2}} \tag{8-61}$$

Example Determine the net heat exchange between two infinite gray planes having different temperatures and emissivities.

Solution Since this is a two-body system exchanging heat only between the two planes, Eq. (8-61) is applicable. Therefore the net heat transfer from plane 1 to plane 2 is

$$q_{1-2} = \frac{E_{b_1} - E_{b_2}}{(1 - \epsilon_1)/\epsilon_1 A_1 + 1/A_1 F_{1-2} + (1 - \epsilon_2)/\epsilon_2 A_2}$$

But since the planes are identical, $A_1 = A_2$; and since they are infinite in extent, $F_{1-2} = F_{2-1} = 1.0$. Then

$$\frac{q_{1-2}}{A} = \frac{\sigma(T_1^4 - T_2^4)}{(1 - \epsilon_1)/\epsilon_1 + 1 + (1 - \epsilon_2)/\epsilon_2}$$

Rearranging the denominator, we get

$$\frac{q_{1-2}}{A} = \frac{\sigma(T_1^4 - T_2^4)}{[\epsilon_2(1 - \epsilon_1) + \epsilon_1 \epsilon_2 + \epsilon_1(1 - \epsilon_2)]/\epsilon_1 \epsilon_2} = \frac{\sigma(T_1^4 - T_2^4)}{1/\epsilon_1 + 1/\epsilon_2 - 1}$$

Clearly, the net heat transfer per unit area from plane 2 to plane 1 is

$$\frac{q_{2-1}}{A} = \frac{\sigma(T_2^4 - T_1^4)}{1/\epsilon_1 + 1/\epsilon_2 - 1}$$

Perhaps additional physical insight can be gained by applying the so-called *ray-tracing technique* to this problem. Consider two infinite parallel gray planes with multiple reflections of the radiant energy per unit area leaving surface 1 (this is shown for simplicity as a single ray; the

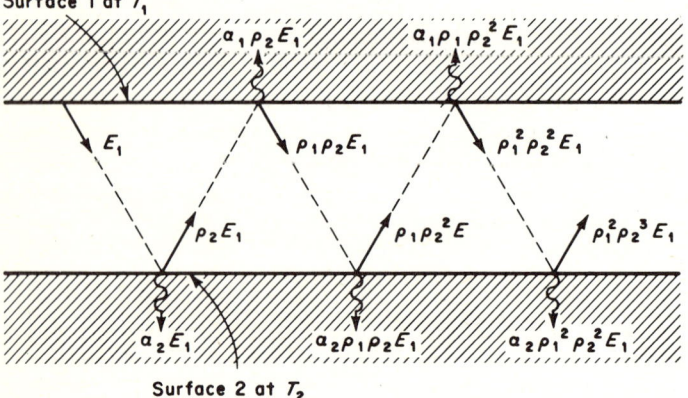

Fig. 8–23 Two infinite parallel gray-body planes.

ray represents diffuse radiant energy in all possible directions), a fraction is absorbed by surface 2, and the remainder is reflected back toward surface 1. At surface 1 a second partial absorption and re-reflection occurs, etc. The radiant transfer per unit area from plane 1 to plane 2 is

$$\frac{q'_{1-2}}{A} = \alpha_2 E_1 + \alpha_2 \rho_1 \rho_2 E_1 + \alpha_2 \rho_1^2 \rho_2^2 E_1 + \cdots + \alpha_2 \rho_1^n \rho_2^n E_1$$

or

$$\frac{q'_{1-2}}{A} = \alpha_2 E_1 (1 + \rho_1 \rho_2 + \rho_1^2 \rho_2^2 + \cdots + \rho_1^n \rho_2^n)$$

RADIATIVE HEAT TRANSFER

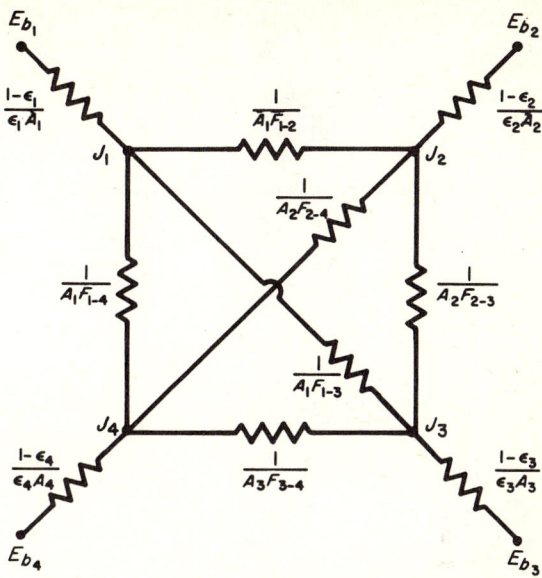

Fig. 8-24 Electrical analog for a four-gray-surface system.

A similar treatment for the radiant transfer per unit area from plane 2 to plane 1 is

$$\frac{q'_{2-1}}{A} = \alpha_1 E_2 (1 + \rho_1 \rho_2 + \rho_1^2 \rho_2^2 + \cdots + \rho_1^n \rho_2^n)$$

Now the infinite series in either of these expressions converges to (binomial series)

$$\frac{1}{1 - \rho_1 \rho_2}$$

Thus, the net heat transfer from plane 1 to plane 2 is

$$\frac{q_{1-2}}{A} = \frac{q'_{1-2}}{A} - \frac{q'_{2-1}}{A} = \frac{\alpha_2 E_1 - \alpha_1 E_2}{1 - \rho_1 \rho_2}$$

But for gray bodies, $E = \epsilon \sigma T^4$, $\epsilon = \alpha$, and $\rho = 1 - \epsilon$, and this becomes

$$\frac{q_{1-2}}{A} = \frac{\epsilon_2 \epsilon_1 \sigma T_1^4 - \epsilon_1 \epsilon_2 \sigma T_2^4}{1 - (1 - \epsilon_1)(1 - \epsilon_2)}$$

or

$$\frac{q_{1-2}}{A} = \frac{\sigma(T_1^4 - T_2^4)}{1/\epsilon_1 + 1/\epsilon_2 - 1}$$

Analog techniques are conveniently applied to multibody enclosure systems. A four-body problem is illustrated in Fig. 8-24. For a system consisting of more than two surfaces, the radiosities at each node must be evaluated in order to determine

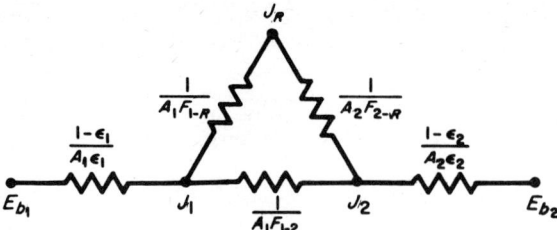

Fig. 8-25 Electrical analog for a two-gray-surface system enclosed by reradiating walls.

the heat flux since, for example, in Fig. 8-24, q_{1-2} is found from

$$q_{1-2} = \frac{J_1 - J_2}{1/A_1 F_{1-2}}$$

A convenient method of determining the J_i's is to apply Kirchhoff's current law, which states that the algebraic sum of the currents entering a node is zero.

Reradiating surfaces When two radiating bodies are exposed to each other in the presence of a reradiating surface, a special type of analog is applicable. A reradiating surface is one which diffusely radiates (reflects and emits) energy at the same rate it receives incident radiation and thereby experiences no net heat transfer. Examples of such surfaces are (1) the inside refractory walls of large industrial furnaces, which experience low conductive-heat-transfer losses to the outside with respect to the incident radiation and in effect act as reradiators, and (2) the walls of well-insulated enclosures containing radiating surfaces. The analog for a two-body system in the presence of a reradiating surface is given in Fig. 8-25. The net heat transfer from body 1 to body 2 is found by dividing the potential difference by the sum of the series resistances. The center series resistance is for the two parallel circuits from J_1 to J_2, and thus

$$q_{1-2} = \frac{A_1 \sigma (T_1^4 - T_2^4)}{\dfrac{1-\epsilon_1}{\epsilon_1} + \dfrac{1}{F_{1-2} + 1/(1/F_{1-R} + A_1/A_2 F_{2-R})} + \dfrac{1-\epsilon_2}{\epsilon_2}\dfrac{A_1}{A_2}} \tag{8-62}$$

Example In an experimental kitchen a beef roast wrapped in aluminum foil is placed in the center of a gas oven to cook. The oven is preheated to 400°F, and heating is accomplished by the gas flame underneath the bottom surface of the oven. The oven side walls and top are well insulated (reradiating), and the physical dimensions are 2 by 2 by 2 ft. The roast can be approximated as a 8-in-diam sphere. Assuming oven-wall emissivity of 0.8, aluminum-foil emissivity of 0.1, and an initial roast temperature of 70°F, calculate the total initial radiant heat flux to the roast.

RADIATIVE HEAT TRANSFER

Solution For this problem Fig. 8-25 and Eq. (8-62) are applicable. Denote the heater surface (bottom of oven) by subscript 1, the roast by subscript 2, and the other five oven walls by R:

$T_1 = 400 + 460 = 860°R \quad\quad \epsilon_1 = 0.8$
$T_2 = 70 + 460 = 530°R \quad\quad \epsilon_2 = 0.1$
$T_R = 860°R \quad\quad\quad\quad\quad\quad \epsilon_R = 0.8$

The configuration factors are best determined from the reciprocity theorem. Since all radiant energy from the roast impinges upon the oven walls and each side receives an equal amount, we have

$$F_{2-1} = \frac{1.0}{6}$$

By reciprocity

$$A_2 F_{2-1} = A_1 F_{1-2}$$

or

$$F_{1-2} = \frac{A_2}{A_1} F_{2-1} = \frac{4(\frac{8}{12})^2 \pi}{2(2)} \frac{1.0}{6} = \frac{64\pi}{4(144)(6)} = 0.058$$

The view factor F_{1-R} is by $\sum_{i=1}^{n} F_{1-i} = 1.0$

$$F_{1-R} = 1.0 - F_{1-2} = 1.0 - 0.058 = 0.942$$

and F_{2-R} is

$$F_{2-R} = \tfrac{5}{6}(1.0) = \tfrac{5}{6}$$

Calculating the terms for the denominator of Eq. (8-62)

$$\frac{1-\epsilon_1}{\epsilon_1} = \frac{1.0 - 0.8}{0.8} = 0.25$$

$$\frac{1}{F_{1-2} + 1/(1/F_{1-R} + A_1/A_2 F_{2-R})} = \frac{1}{0.058 + 1/[1/0.942 + 4/(64\pi/144)(\tfrac{5}{6})]} = 3.57$$

$$\frac{1-\epsilon_2}{\epsilon_2}\frac{A_1}{A_2} = \frac{1-0.1}{0.1}\frac{4}{64\pi/144} = 25.8$$

It should be noted that the resistance to the radiant heat flux is heavily influenced by the low emissivity of the aluminum foil. Substituting into Eq. (8-62), the radiant portion of the total heat transfer is

$$q = \frac{0.1714(4)[(\tfrac{860}{100})^4 - (\tfrac{530}{100})^4]}{0.25 + 3.57 + 25.8} = 109 \text{ Btu/hr}$$

Special configurations For a convex body 1 enclosed by a larger concave surface 2, e.g., a cylinder within a larger cylinder, we note that all radiation leaving body 1 impinges upon 2, and thus $F_{1-2} = 1.0$. In this case, Eq. (8-61) reduces to

$$q_{1-2} = \frac{A_1 \sigma(T_1^4 - T_2^4)}{\dfrac{1}{\epsilon_1} + \dfrac{A_1}{A_2}\left(\dfrac{1}{\epsilon_2} - 1\right)} \quad\quad\quad (8\text{-}63)$$

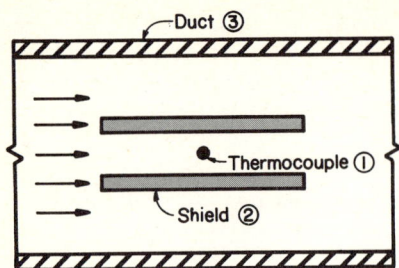

Fig. 8-26 Thermocouple shield.

Further, if body 1 is very small in comparison with the enclosing surface, this becomes

$$q_{1-2} = \sigma A_1 \epsilon_1 (T_1^4 - T_2^4) \tag{8-64}$$

which is useful in calculating the heat flux from a small heated surface in a large room.

For two infinite parallel planes, A_1 and A_2 are equal, and since all radiant energy from A_1 is incident upon A_2, Eq. (8-61) simplifies to

$$q_{1-2} = \frac{A\sigma(T_1^4 - T_2^4)}{1/\epsilon_1 + 1/\epsilon_2 - 1} \tag{8-65}$$

where A is the area of one plate.

Radiation shielding Many thermal designs employ the concept of shielding to reduce radiant heat exchange. Two important specific examples are aluminum-foil-backed fiber-glass insulation in building walls (the foil serves as the shield) and radiation shields for thermocouples employed in high-temperature measurements. Let us consider the latter, of which a typical one is illustrated in Fig. 8-26, a cutaway view of a thermocouple and shield installed in a high-temperature air duct. The thermocouple at temperature T_1 loses heat by radiation to the cooler duct wall, this energy being obtained by convection from the fluid. Consider first the situation with no shield. The appropriate analog is shown in Fig. 8-27, and the heat loss is given by

$$q_{1-3} \simeq \epsilon_1 A_1 \sigma (T_1^4 - T_3^4) \tag{8-63a}$$

since $F_{1-3} = 1.0$ and $A_1/A_3 \simeq 0$. Now examine the problem with a single shield

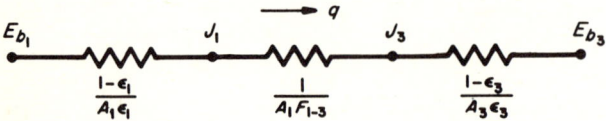

Fig. 8-27 Analog for unshielded thermocouple.

RADIATIVE HEAT TRANSFER

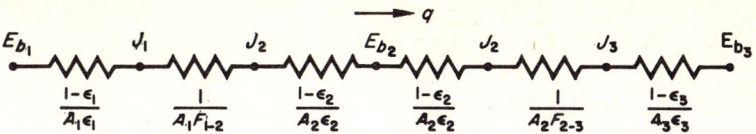

Fig. 8-28 Analog for shielded thermocouple.

added. The appropriate analog is shown in Fig. 8-28, and the heat loss is

$$q_{1-3} = q_{1-2} = q_{2-3} = \frac{A_1 \sigma(T_1^4 - T_2^4)}{\frac{1}{\epsilon_1} + \frac{A_1}{A_2}\left(\frac{1}{\epsilon_2} - 1\right)} = \frac{A_2 \sigma(T_2^4 - T_3^4)}{\frac{1}{\epsilon_2} + \frac{A_2}{A_3}\left(\frac{1}{\epsilon_3} - 1\right)} \quad (8\text{-}63b)$$

which indicates that q can be reduced by appropriate selection of shielding material and sizing.

A particularly informative result is obtained in the case of large parallel walls. In that case, with all areas equal, the single heat shield results in

$$q_{1-3} = q_{1-2} = q_{2-3} = \frac{A\sigma(T_1^4 - T_2^4)}{1/\epsilon_1 + 1/\epsilon_2 - 1} = \frac{A\sigma(T_2^4 - T_3^4)}{1/\epsilon_2 + 1/\epsilon_3 - 1} \quad (8\text{-}63c)$$

where the subscript notation and analog are similar to those for the thermocouple shield. If in addition we consider equal emissivities, we obtain

$$T_2^4 = \tfrac{1}{2}(T_1^4 + T_3^4) \quad (8\text{-}66)$$

and the heat flux is

$$q = \frac{\tfrac{1}{2}A\sigma(T_1^4 - T_3^4)}{1/\epsilon_1 + 1/\epsilon_2 - 1} \quad (8\text{-}63d)$$

which, since $\epsilon_2 = \epsilon_3$, is exactly one-half the value with no heat shield.

Example The wall of a house consists of two large parallel planes. The outer and inner walls have emissivities of 0.9 and 0.4, respectively. If the respective outer and inner temperatures on a summer day are 120°F and 70°F, what will be the radiant heat flux per unit area with and without an aluminum radiation shield having $\epsilon = 0.06$?

Solution Let subscript 1 denote the outer wall, 2 denote the shield, and 3 denote the inner wall. With no shield, Eq. (8-65) yields

$$\frac{q}{A} = \frac{\sigma(T_1^4 - T_3^4)}{1/\epsilon_1 + 1/\epsilon_3 - 1} = \frac{0.1714[(\tfrac{580}{100})^4 - (\tfrac{530}{100})^4]}{1/0.9 + 1/0.4 - 1} = 22.4 \text{ Btu/hr ft}^2$$

With the shield, by Eq. (8-63c),

$$\frac{q}{A} = \frac{\sigma(T_1^4 - T_2^4)}{1/\epsilon_1 + 1/\epsilon_2 - 1} = \frac{\sigma(T_2^4 - T_3^4)}{1/\epsilon_2 + 1/\epsilon_3 - 1}$$

Solving for T_2 gives

$$\frac{(\frac{580}{100})^4 - (T_2/100)^4}{1/0.9 + 1/0.06 - 1} = \frac{(T_2/100)^4 - (\frac{530}{100})^4}{1/0.06 + 1/0.4 - 1}$$

$$\left(\frac{T_2}{100}\right)^4 = 970$$

and thus

$$\frac{q}{A} = \frac{0.1714[(\frac{580}{100})^4 - 970]}{1/0.9 + 1/0.06 - 1} = 1.65 \text{ Btu/hr-ft}^2$$

8-7 RADIANT EXCHANGE WITH GASES AND VAPORS

The treatment of the present section is intended to serve as a brief introduction to this topic and to equip the student to handle only problems commonly encountered in industry. For a comprehensive treatment the reader should consult Hottel and Sarofim [8], Chandrasekhar [1], and Sparrow and Cess [10].

Many common gases and mixtures have a nonpolar symmetrical molecular structure and do not emit or absorb radiant energy within temperature limits of interest. Included among these are oxygen, nitrogen, hydrogen, and mixtures of them, such as dry air. Gases and vapors with nonsymmetrical molecules, however, generally have absorption bands for radiation at certain wavelengths; these gases emit and absorb radiation at wavelengths within these specific bands or limits. Included among these are carbon dioxide, water, sulfur dioxide, and most hydrocarbons. Absorption bands for a typical radiating gas are shown in Fig. 8-29.

Besides this added complication, radiation properties of gases are dependent upon thickness, shape, and pressure, whereas for a solid these are purely surface phenomena. Consider a beam of monochromatic radiation of intensity $I_{\lambda 0}$ passing through a layer of gas as depicted in Fig. 8-30. A decrease in monochromatic radiant

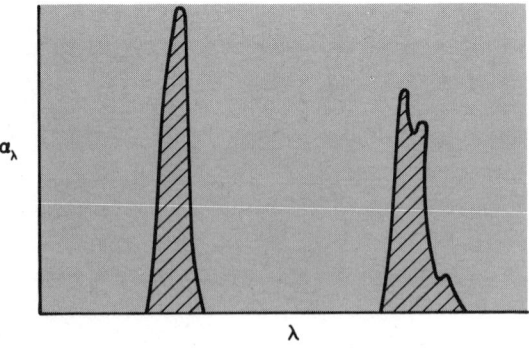

Fig. 8-29 Typical absorption bands for gas.

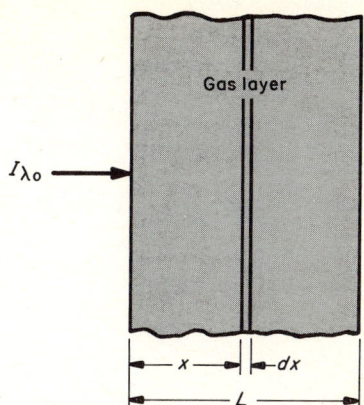

Fig. 8-30 Absorption of monochromatic radiation in a gas layer.

intensity is caused by the absorption as the beam passes through the gas. When the monochromatic absorption coefficient κ_λ is defined as the negative of the product of the reciprocal of the monochromatic intensity and the rate of change of this intensity with distance, the decrease in intensity is given by

$$dI_{\lambda x} = -\kappa_\lambda I_{\lambda x}\, dx \tag{8-67}$$

The monochromatic absorption coefficient depends upon the gas pressure and temperature as well as wavelength. Separating variables and integrating Eq. (8-67) to an arbitrary depth x in the gas yields

$$I_{\lambda x} = I_{\lambda 0} e^{-\kappa_\lambda x} \tag{8-68}$$

which illustrates that the intensity diminishes exponentially with distance in the absorbing gas.

An analytical treatment of radiation in gases and vapors is beyond the scope of this book (for this, Refs. 1, 8, and 10 are suggested). For engineering calculations an approach due to Hottel [6] and Hottel and Egbert [7] is quite practical. They determined the effective emissivity of a hemispherical gas mass of radius L at a partial pressure P_c radiating to a black surface element located at the center of the hemisphere base, and they presented the results in the form of graphs as shown in Figs. 8-31 to 8-34.

Consider first the emissivity of carbon dioxide. Figure 8-31 may be used to obtain the emissivity of a hemispherical gas system of radius L at 1 atm pressure, with partial pressure of CO_2 equal to P_c, and at a given uniform gas temperature. To apply this type of chart to a gas system of other configurations, we obtain appropriate equivalent lengths from Table 8-2.

If the CO_2 gas system is at a total pressure of other than 1 atm, a correction factor C_{pc} (obtained from Fig. 8-32) compensates for the broadening of the absorption bands with increasing pressure.

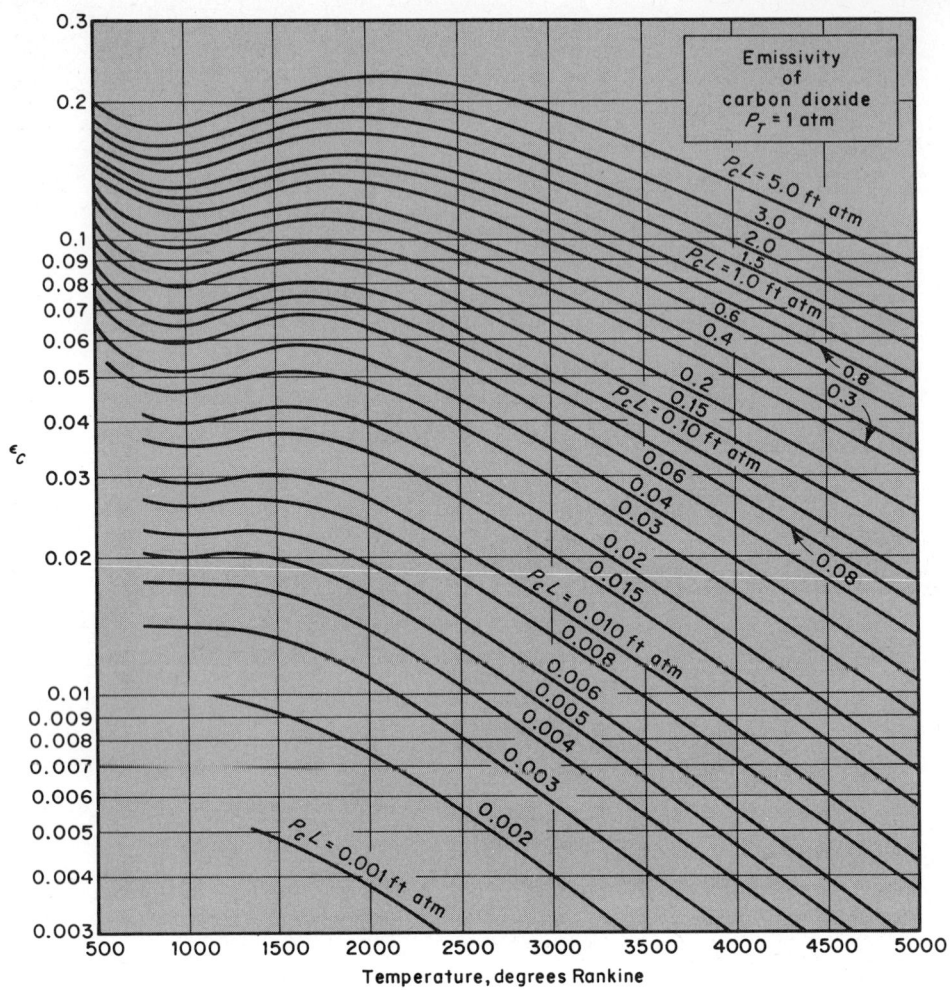

Fig. 8-31 Emissivity of carbon dioxide at 1 atm total pressure. (*Adapted from H. C. Hottel, chap. 4 in W. C. McAdams, "Heat Transmission," 3d ed. Copyright 1954. McGraw-Hill Book Company. Used by permission.*)

Table 8-2†

Shape	L
Space between infinite parallel planes	1.8 × distance between planes
Sphere	$\frac{2}{3}$ × diameter
Infinitely long cylinder	1 × diameter
Cube	$\frac{2}{3}$ × side

† Adapted from [6].

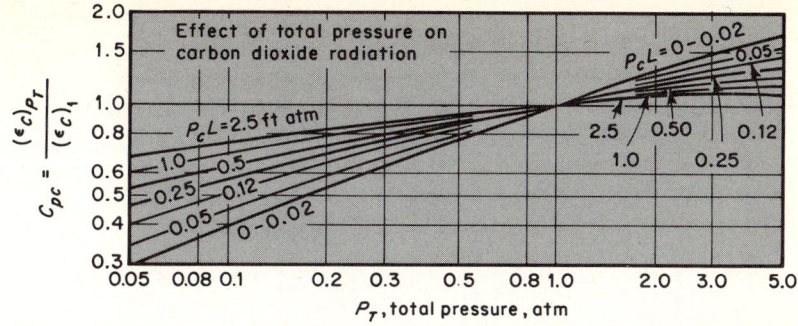

Fig. 8-32 Effect of total pressure on carbon dioxide emissivity. (*Adapted from H. C. Hottel, chap. 4 in W. C. McAdams, "Heat Transmission," 3d ed. Copyright 1954. McGraw-Hill Book Company. Used by permission.*)

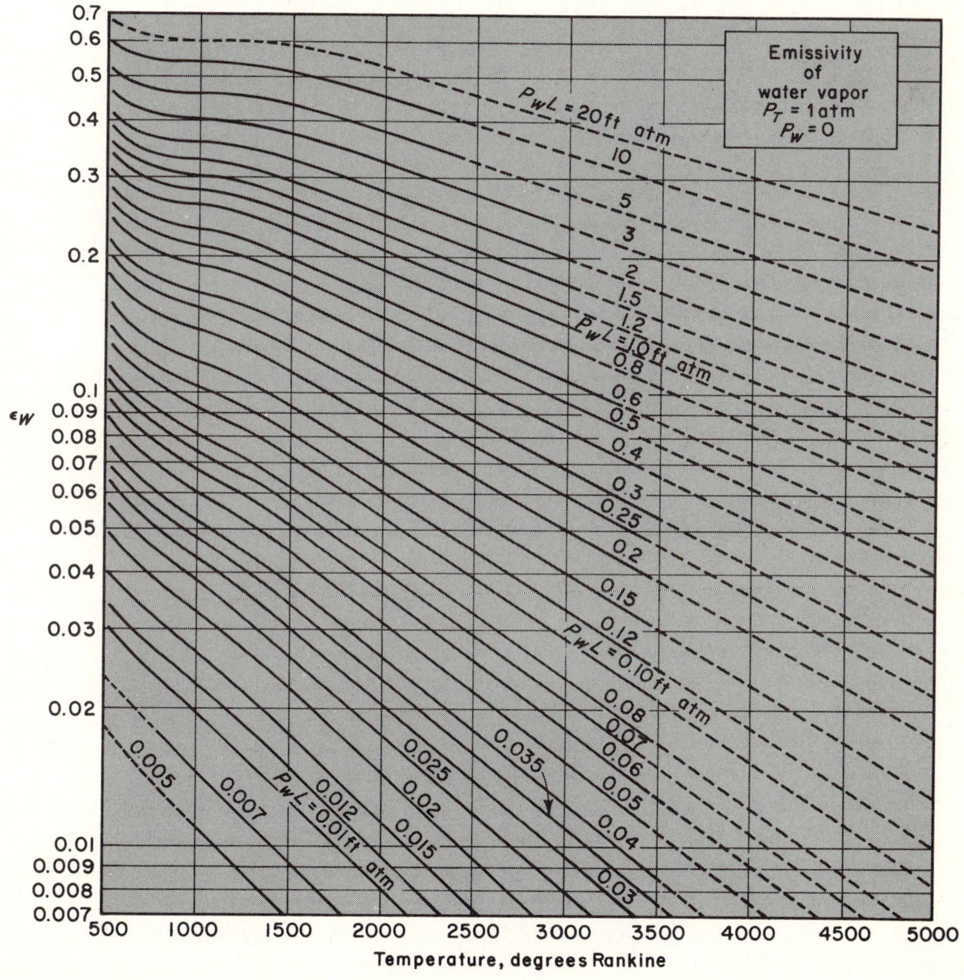

Fig. 8-33 Emissivity of water vapor, hypothetical system at 1 atm total pressure. (*Adapted from H. C. Hottel, chap. 4 in W. C. McAdams, "Heat Transmission," 3d ed. Copyright 1954. McGraw-Hill Book Company. Used by permission.*)

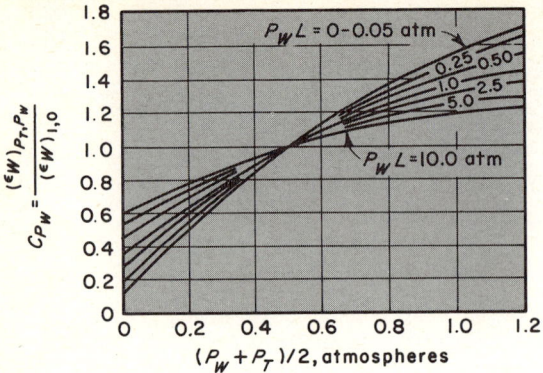

Fig. 8-34 Effect of partial and total pressure on water-vapor emissivity. (*Adapted from H. C. Hottel, chap. 4 in W. C. McAdams, "Heat Transmission," 3d ed. Copyright 1954. McGraw-Hill Book Company. Used by permission.*)

For water vapor, Fig. 8-33 presents the emissivity of a hypothetical system whose partial pressure is zero (later to be corrected) as a function of the actual partial pressure of water vapor, the hemispherical system radius L, and the uniform system temperature. Again, the appropriate dimension L for a system under consideration is taken from Table 8-2, and the emissivity from Fig. 8-33 is modified by an appropriate factor C_{pw} (from Fig. 8-34) which accounts both for the fact that the partial pressure of the water vapor is not zero and that the total system pressure is other than 1 atm.

Example Determine the emissivity of water vapor at 3000°F in a spherical container 1.5 ft in diameter. The partial pressure of the vapor is 0.3 atm, and the total pressure is 1.7 atm.

Solution The equivalent hemispherical beam length is

$L = \frac{2}{3}(1.5) = 1.0 \text{ ft}$

$P_{wL} = 0.3(1.0) = 0.3$

$\dfrac{P_w + P_T}{2} = \dfrac{0.3 + 1.7}{2} = 1.0$

From Fig. 8-34, $C_{pw} = 1.5$. Thus,

$(\epsilon_w)_{\text{actual}} = 1.5(0.055) = 0.0825$

For a mixture containing both carbon dioxide and water vapor an approximate emissivity can be obtained by adding the individually determined emissivities. For a more correct treatment of mixtures and for analyses involving nonblack gas assumptions, Refs. 6, 8, 10, and 11 should be consulted.

RADIATIVE HEAT TRANSFER

PROBLEMS

8-1. A light quantum has an energy of 2.3×10^{-6} erg. Calculate its frequency and wavelength.

8-2. The wavelength of a radiating system is 3.7×10^{-12} cm. Assuming that the velocity of propagation is the same as in a vacuum, that is, $c = 3 \times 10^{10}$ cm/sec, what is the energy of a quantum of this radiation?

8-3. Incident radiation ($G = 500$ Btu/hr-ft^2) strikes an object. The amount of energy absorbed is 150 Btu/hr-ft^2, and the amount of energy transmitted is 25 Btu/hr-ft^2. What is the reflectivity ρ?

8-4. Radiation of 370 Btu/hr-ft^2 is incident upon a shallow layer of liquid which has an absorptivity of 0.3 and a transmissivity of 0.12. Find the rate of energy reflected.

8-5. Incident radiation ($G = 200$ Btu/hr-ft^2) strikes an object with transmissivity of 0.03 and reflectivity of 0.50. Find the amount of energy absorbed.

8-6. Curve a shows the approximate spectral characteristics of the reflectivity of an opaque surface on the Apollo XII command module. Irradiation from the sun can be approximated as shown in the graph of monochromatic irradiation G_λ.

(a) Plot α_λ vs. λ for this surface.
(b) Calculate the total irradiation G.
(c) Find the energy absorption rate in Btu/hr-ft^2.

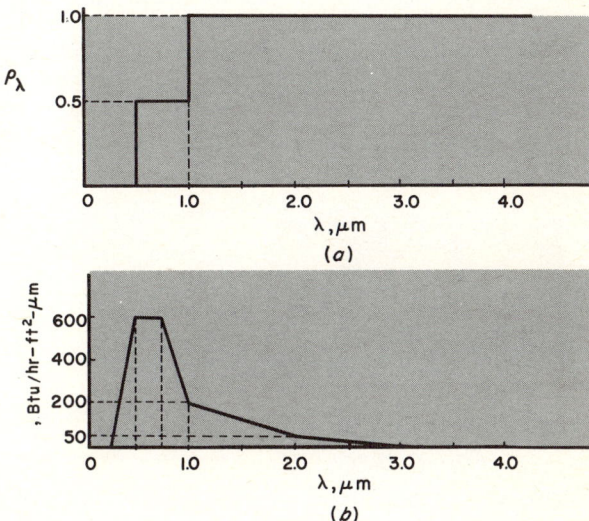

Fig. P8-6

8-7. A blackbody has an emissive power of 2540 Btu/hr-ft^2. Determine its temperature.

8-8. Determine the radiant heat emission per square foot from fireclay brick at 1832°F.

8-9. The average radiant-energy flux incident upon the earth's atmosphere is 444.7 Btu/hr-ft^2 and is known as the *solar constant*. Assuming the sun to radiate as a blackbody, calculate its temperature. *Hint:* The energy originates on the surface of the sun (radius = 433,000 miles) and must be equal to that intercepted by an imaginary sphere surrounding the sun at the earth's orbit distance of approximately 93 million miles. Assume that the heat flux across the sphere is uniformly distributed.

8-10. The heat transferred from a plate of 2 cm² area at 1520°F to an enclosure of a very low temperature is 0.030 watt/cm. Assuming this heat-transfer rate to be constant over the spectrum, what is the emissivity of the plate when radiating at a wavelength of 3 μm?

8-11. Consider two large opaque parallel planes at the same temperature. The first is a blackbody absorbing all incident radiation and emitting E_b. The second is a grey body having an absorptivity α less than unity and emitting $E_b\epsilon$. The grey body absorbs $E_b\alpha$. Assuming that the transmissivity is zero, derive a relation between the absorptivity and the emissivity of a body at thermal equilibrium.

8-12. A large plane, perfectly insulated on one face and maintained at a fixed temperature T_1 on the bare face, which has an emissivity of 0.90, loses 200 Btu/hr-ft² when exposed to surroundings at absolute zero. A second plane the same size as the first is also perfectly insulated on one face, but its bare face has an emissivity of 0.45. When the bare face of the second plane is maintained at a fixed temperature T_2 and exposed to surroundings at absolute zero, it loses 100 Btu/hr-ft². These two planes are brought together so that the parallel bare faces are only 1 in apart, and the heat supply to each is so adjusted that their respective temperatures remain unchanged. What will be the net heat flux between the planes, expressed in Btu/hr-ft²?

8-13. Two large parallel black planes are maintained at temperatures $T_1 = 2000°F$ and $T_2 = 500°F$, respectively. Later a third black plane, which reaches a temperature T_3 after steady-state conditions have been reached, is placed between these two. What is the ratio of heat transferred with this third plane installed to that transferred initially?

8-14. A window in an experimental aircraft has an area of 1 ft² and transmits 3 percent of the incident thermal radiation between 1.5 and 5.0 μm. It is opaque to thermal radiation at all other wavelengths. Determine the heat flux transmitted through the window from a blackbody radiation source at (a) 2340°F and (b) 740°F.

8-15. Which curve of Fig. P8-15 represents the configuration factor F for parallel black planes of finite size?

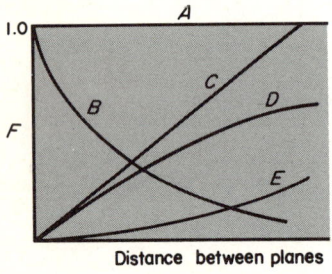

Fig. P8-15

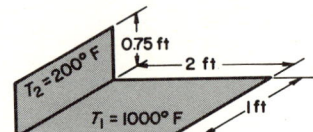

Fig. P8-16

8-16. Assuming the two plane surfaces at right angles to be black, determine the radiant heat transfer from surface 1 to surface 2.

8-17. Two identical 4- by 5-ft rectangular flat plates are separated by a distance of 3 ft. One plate is maintained at 800°F, and the other is maintained at 400°F. What is the net radiant heat transfer between the plates?

8-18. A hemispherical shell and a plane form an enclosure. What is the configuration factor F_{1-2}?

8-19. Two parallel rectangular planes, 1 and 2, of dimensions a by $2a$ are joined on their long edge by a third plane 3 perpendicular to them and of height $1.5a$. Determine the configuration factors F_{1-2} and F_{1-3}.

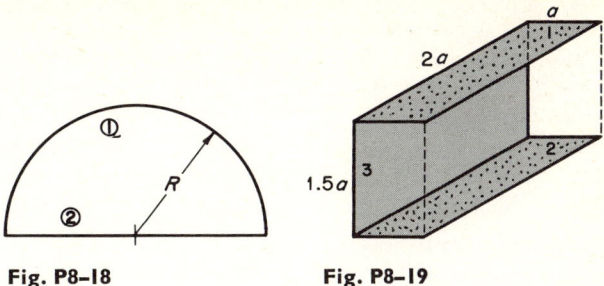

Fig. P8-18   Fig. P8-19

8-20. Assuming the surfaces to be black, determine the radiant heat transfer from surface 1 to surface 2 of the configuration shown.

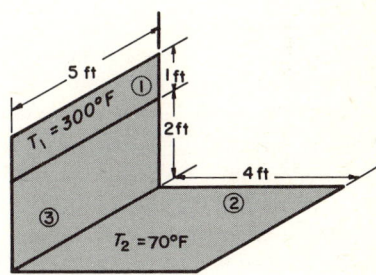

Fig. P8-20

8-21. Find the configuration factors F_{1-2} for the plane wall surfaces shown.

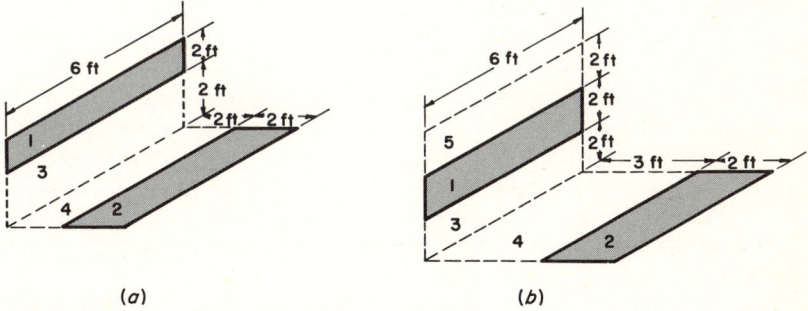

Fig. P8-21

8-22. A polished platinum plate is maintained at 720°F. Assuming that the surroundings act as a blackbody at very low temperature and that the plate is insulated from its support, determine the heat loss from the plate in Btu/hr-ft².

8-23. A plate with an emissivity of 0.4 and an area of 1 ft² is attached to the side of a spaceship so that it is perfectly insulated from the inside of the ship. Assuming that outer space is a blackbody at absolute zero, determine the equilibrium temperature of the plate if the radiant heat flux from the sun is 1000 Btu/hr-ft².

8-24. In a manufacturing operation, a 3- by 5-ft plate of rough carbon at 440°F is placed in a large room where the average temperature is 120°F. Estimate the heat loss from the plate due to radiation.

8-25. A 2- by 4-ft cast-iron oxidized furnace door at 700°F is located in a large room where the average temperature is 100°F. Determine the heat loss by radiation from the door.

8-26. A small heat-treating furnace (2 by 2 by 2 ft) is installed in a large room. The temperature of the outer surface of the furnace is 110°F, and that of the room walls is 65°F. If the emissivity of the furnace surface is the same as that of newly turned cast iron, and if the radiation reflected by the walls is negligible, calculate the heat loss from the furnace by radiation.

8-27. A 2- by 3-ft shallow pan is filled with pure water. It is placed on a well-insulated stand on top of a high building. It is known that the heat loss by radiation to the sky (assumed to be a blackbody at absolute zero) can cause the water to freeze on a calm night. Estimate the rate of heat loss to the sky assuming that the water is at 33°F and that it is a grey body with $\epsilon = 0.6$. Assume further that the water sees only the sky.

8-28. Two perpendicular walls have a common edge. Each wall is 4 by 8 ft, the 4-ft edge being common. Wall 1 is vertical and has an emissivity of 0.7 and a temperature of 1200°F. Wall 2 is horizontal and has an emissivity of 0.4 and a temperature of 800°F. Calculate the net heat transfer from wall 1 to wall 2. Assume no radiation or reflection from other surfaces or surroundings.

8-29. Equal, parallel, and opposite rectangles 8 by 16 ft are located 8 ft apart and are connected by refractory walls. The respective emissivity and temperature of one are 0.90 and 1200°F. The corresponding values for the other plate are 0.80 and 800°F. Determine the net radiant heat transfer between the plates.

8-30. Parallel 3- by 6-ft plates, having temperatures of 1000 and 300°F, respectively, are located 10 ft apart. If the plates have emissivities of 0.8 and 0.3, respectively, what is the net direct radiant-energy interchange if (a) the surfaces exchange radiant energy only with themselves; (b) they are enclosed by reradiating walls?

8-31. In an industrial operation, a wall of fireclay brick is maintained at 1940°F. The wall is 10 ft high and 15 ft wide. The floor of concrete tiles is 15 by 20 ft and is at a temperature of 100°F. Calculate the radiant heat flux from the wall to the floor (a) assuming no radiation or reflection from other surfaces or surroundings; (b) if the surroundings are reradiating surfaces.

8-32. The first and third layers of a three-layer foil insulation are at 80 and 0°F, respectively. The emissivity of the foil is 0.05. Neglecting conduction and convection, determine the radiant heat transmission in Btu/hr-ft².

8-33. Two flat parallel square plates have equal areas of 25 ft² and are separated by a distance of 1 ft. The net radiant heat exchange is 3000 Btu/hr, and the temperature and emissivity of the cooler plate are 140°F and 0.4, respectively. If the emissivity of the other plate is 0.7, determine its temperature.

8-34. The wall of a building consists of two large parallel planes. The emissivity and temperature of the inner wall are 0.5 and 80°F, respectively. A radiation shield with an emissivity of 0.09 is placed between the walls. If the radiant heat loss is 2 Btu/hr-ft², determine the equilibrium temperature of the shield.

8-35. A long, rusty 2-in-OD pipe is in a concrete tunnel 2 ft square. The tunnel wall is 80°F and has an emissivity of 0.9; the corresponding values for the pipe surface are 300°F and 0.73. Determine the radiant heat transfer per linear foot of pipe.

RADIATIVE HEAT TRANSFER

8-36. The radiant heat exchange between two parallel walls is 25 Btu/hr-ft^2. The emissivities of the walls are 0.3 and 0.2, and the temperature of the cooler wall is 70°F. Assuming that the walls are two infinite parallel planes of equal area, determine the temperature of the other wall.

8-37. A cryogen at $-221°F$ is to be stored in the inner of two concentric polished brass spheres with diameters of 9 and 12 in, respectively. If the emissivity of the polished brass is 0.03, calculate the total radiant heat exchange between the spheres. Neglect the temperature drop through the metal and assume that the outer-sphere temperature is 70°F.

8-38. Three thin sheets of polished copper are placed parallel and very close to each other. If the temperature of one of the outside sheets is 720°F and the temperature of the other outside sheet is 100°F, calculate (a) the net radiant heat flow in Btu/hr-ft^2 and (b) the temperature of the middle sheet.

8-39. Two 4-ft-square and parallel flat plates are 2 ft apart. Plate A_1 is maintained at a temperature of 1740°F and A_2 at 720°F. The emissivities are 0.4 and 0.8, respectively. Considering the surroundings black at 0°R and including multiple interreflections, determine (a) the net radiant exchange and (b) the heat input required by surface A_1 to maintain its temperature.

8-40. Determine the emissivity of water vapor at 3540°F between two large parallel walls which are 8 in apart. The partial pressure of the water vapor is 0.2 atm, and the total pressure is 1.4 atm.

8-41. Determine the emissivity of carbon dioxide at 2040°F in a 3-ft cubical container. The partial pressure of the carbon dioxide is 0.06 atm, and the total pressure is 0.2 atm.

8-42. Determine the emissivity of carbon dioxide at 3040°F in a very long 1.5-ft-diam cylindrical passage. The partial pressure of the carbon dioxide is 0.04 atm, and the total pressure is 0.3 atm.

8-43. Calculate the emissive power of water vapor at 1040°F in a spherical container of 3 ft diameter. The partial pressure of the water vapor is 0.8 atm, and the total pressure is 1.0 atm.

8-44. Gaseous products of combustion at 2000°F leave a furnace through a very long cylindrical flue, which is 2.5 ft in diameter. The partial pressure of CO_2 in the mixture is 0.09 atm, and the total pressure is 1.1 atm. Determine the radiative heat transfer per linear foot from the carbon dioxide to the flue wall if the wall emissivity is unity and its temperature is 400°F.

8-45. If the products of combustion in Prob. 8-44 resulted from burning natural gas with an excess of air and the partial pressures of the other major constituents are water vapor = 0.15 atm; oxygen, O_2, = 0.05 atm; nitrogen, N_2, = 0.81 atm, what is the total rate of heat transfer from the hot gases to the flue per lineal foot?

REFERENCES

1. Chandrasekhar, S.: "Radiative Transfer," Oxford University Press, London, 1950.
2. Dunkle, P. V.: Thermal Radiation Tables and Applications, *Trans. ASME*, **76**: 549 (1954).
3. Eckert, E. R. G., and R. M. Drake, Jr.: "Heat and Mass Transfer," McGraw-Hill, New York, 1959.
4. Gubareff, G. G., J. E. Janssen, and R. H. Torborg: "Thermal Radiation Properties Survey," 2d ed., Honeywell Research Center, Minneapolis, Minn., 1960.
5. Hamilton, D. C., and W. R. Morgan: Radiant Interchange Configuration Factors, *NACA Tech. Note* TN-2836, 1952.
6. Hottel, H. C.: chap. 4 in W. C. McAdams, "Heat Transmission," 3d ed., McGraw-Hill, New York, 1954.
7. Hottel, H. C., and R. B. Egbert: Radiant Heat Transmission from Water Vapor, *Trans. AIChE*, **38**: 531 (1942).
8. Hottel, H. C., and A. F. Sarofim: "Radiative Transfer," McGraw-Hill, New York, 1967.
9. Planck, Max: "The Theory of Heat Radiation," English trans. of "Wärmestrahlung," 2d ed. (1913), Dover, New York, 1959.
10. Sparrow, E. M., and R. D. Cess: "Radiation Heat Transfer," Brooks/Cole, Belmont, Calif., 1966.
11. Wiebelt, J. A.: "Engineering Radiation Heat Transfer," Holt, New York, 1966.

PART III

MOVING MEDIA

CHAPTER 9

FLUID FIELDS

The concept of a field was introduced in Chap. 2. The behavior of a field depends upon the interrelation of its fundamental quantities—mass, velocity, and time. How these quantities interact depends upon our viewpoint as well as upon their magnitude (and direction in the case of velocity). This chapter considers these quantities and their relationship when motion occurs.

9-1 DESCRIPTION OF A FLOW FIELD

A *streamline* is an imaginary line in a flow field at an instant of time taken such that the fluid velocity at any point is tangent to it (Fig. 9-1). Since the velocity vector is tangent to the streamline, no matter can cross it. A streamline is analogous to a heat-flow line in the case of heat transfer.

A *stream filament* is a family of streamlines forming a cylindrical passage of infinitesimal cross section. A *stream tube* is bounded by an infinite number of streamlines forming a finite surface across which there is no flow. If there is no creation, storage, or destruction of mass within the stream tube, all fluid which enters must leave.

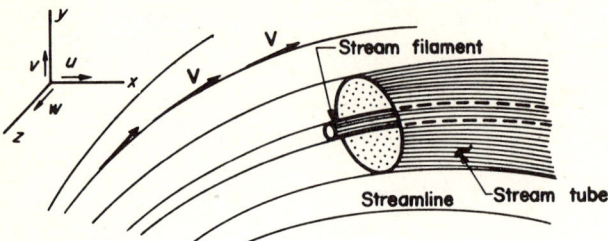

Fig. 9-1 Flow field.

The differential equation for a streamline in two dimensions can be obtained by observing that

$$u = \frac{dx}{dt} \quad \text{and} \quad v = \frac{dy}{dt}$$

Eliminating dt,

$$\frac{dy}{dx} = \frac{v}{u} \tag{9-1}$$

This concept can be extended for a three-dimensional field in cartesian coordinates to give a set of equations

$$\begin{aligned} v\,dx &= u\,dy \\ w\,dx &= u\,dz \\ w\,dy &= v\,dz \end{aligned} \tag{9-2}$$

which are useful later in integrating along a streamline. Two independent equations come from Eqs. (9-2). Any continuous curve which satisfies them simultaneously is a streamline.

Example For a flow field in which the velocity is described by $\mathbf{V}(x,y) = \tfrac{1}{2}y\mathbf{i} + xy^2\mathbf{j}$
(a) determine the streamline equation. (b) What is the equation of the streamline which passes through the point (1,2)?

Solution (a) Equation (9-1) is the two-dimensional differential equation for a streamline; therefore,

$$\frac{dy}{dx} = \frac{v}{u} = 2xy$$

Separating variables and integrating,

$$\int \frac{dy}{y} = 2\int x\,dx$$

$$\ln y = x^2 + \ln C$$

$$y = Ce^{x^2}$$

(b): $\quad C = ye^{-x^2}$

FLUID FIELDS

Therefore $C(1,2) = 2/e$, and

$$y = 2e^{x^2-1}$$

9-2 FLUID MOTION

In the dynamics of solids we are accustomed to describing the motion of particles or rigid bodies by their velocities and accelerations or, more exactly, by the velocities and accelerations of their centers of mass. For a finite number of particles, the velocity of the ith particle can be given by the scalar equations

$$\begin{align} u_i &= f_i(t) \\ v_i &= g_i(t) \\ w_i &= h_i(t) \end{align} \quad (9\text{-}3)$$

where the subscript i identifies the particle. In a fluid, however, there are an infinite number of particles whose character may change continuously, making this approach unfeasible. This technique of describing motion of discrete particles with respect to a fixed set of axes, the lagrangian approach, is not normally used for fluids.

In the lagrangian method the specification of velocity applies only at a given time, locating the particle at some point (a,b,c). Location of the same particle at a subsequent time requires a set of equations:

$$\begin{align} x_i &= F_i(t) \\ y_i &= G_i(t) \\ z_i &= H_i(t) \end{align} \quad (9\text{-}4)$$

The more common approach, the eulerian method, permits us to focus attention on a fixed region in space without regard to the identity of the particles which occupy it at a given time. An observation is an instantaneous picture of the velocities and accelerations of every particle. To accomplish this it is necessary only to take the space coordinates as independent variables, rather than dependent as in the lagrangian method. The eulerian velocity field is given by

$$\mathbf{V} = \mathbf{i}u + \mathbf{j}v + \mathbf{k}w \quad (9\text{-}5)$$

where the respective velocities, in cartesian coordinates, are

$$\begin{align} u &= f(x,y,z,t) \\ v &= g(x,y,z,t) \\ w &= h(x,y,z,t) \end{align} \quad (9\text{-}6)$$

Similarly, in the cylindrical and spherical coordinate systems, respectively, the velocity is

$$\mathbf{V} = \mathbf{V}(r,\phi,z,t) \quad (9\text{-}7)$$
$$\mathbf{V} = \mathbf{V}(r,\phi,\psi,t) \quad (9\text{-}8)$$

With the eulerian approach differential changes in velocities must be expressed in terms of partial derivatives, since each component is affected by both space and time. (Some useful vector identities and operations are given in Appendix B.) From the definition of the total differential, the change in velocity in the x direction, from Eq. (9-6), is

$$du = \frac{\partial u}{\partial x} dx + \frac{\partial u}{\partial y} dy + \frac{\partial u}{\partial z} dz + \frac{\partial u}{\partial t} dt \tag{9-9}$$

Or, using the chain rule for partial differentiation, in three dimensions for an increment of time

$$\frac{d\mathbf{V}}{dt} = \frac{\partial \mathbf{V}}{\partial x} \frac{dx}{dt} + \frac{\partial \mathbf{V}}{\partial y} \frac{dy}{dt} + \frac{\partial \mathbf{V}}{\partial z} \frac{dz}{dt} + \frac{\partial \mathbf{V}}{\partial t} \tag{9-10}$$

If the space-rate components dx/dt, dy/dt, and dz/dt are viewed as the scalar velocity components of the fluid, they can be replaced by their respective velocity components giving

$$\mathbf{a} \equiv \frac{D\mathbf{V}}{Dt} = \left(u \frac{\partial \mathbf{V}}{\partial x} + v \frac{\partial \mathbf{V}}{\partial y} + w \frac{\partial \mathbf{V}}{\partial z} \right) + \frac{\partial \mathbf{V}}{\partial t} \tag{9-11}$$

This is known as the *total*, *substantial*, or *fluid derivative*, designated D/Dt to emphasize that the time derivative is taken as one follows the particle which occupies a particular region in space at a particular time [2]. It is a *system* derivative in the sense that it represents the rate of change at a point which moves with the fluid. The terms in parenthesis give the *convective* acceleration, dependent upon the particle's motion in space. The *local* acceleration, $\partial \mathbf{V}/\partial t$, gives the influence of time on the particle's behavior.

Any fluid property, for example, $\rho = \rho(x,y,z,t)$, can be treated from the eulerian point of view by using the chain rule. To illustrate physically the three kinds of time derivatives cited above, consider the concentration ω of snowflakes in a snowstorm

$$\omega = \omega(x,y,z,t) \tag{9-12}$$

$$\frac{d\omega}{dt} = \frac{\partial \omega}{\partial x} \frac{dx}{dt} + \frac{\partial \omega}{\partial y} \frac{dy}{dt} + \frac{\partial \omega}{\partial z} \frac{dz}{dt} + \frac{\partial \omega}{\partial t} \tag{9-13}$$

$$\frac{D\omega}{Dt} = u \frac{\partial \omega}{\partial x} + v \frac{\partial \omega}{\partial y} + w \frac{\partial \omega}{\partial z} + \frac{\partial \omega}{\partial t} \tag{9-14}$$

1. The *partial time derivative*, $\partial \omega/\partial t$, gives the change in concentration of snowflakes on our head as we stand in the snowstorm; x, y, and z are fixed.
2. Suppose we now race about through the storm in a snowmobile. The change in concentration is given by Eq. (9-13), where dx/dt, dy/dt, and dz/dt are the velocity

FLUID FIELDS

components of the snowmobile. The observer is moving with the reference axes, his head; x, y, and z are variable.
3. Now let us imagine being in the same snowstorm in a balloon, carried about by the same currents which move the snowflakes. The change in concentration is then given by the substantial derivative, Eq. (9-14), the concentration depending upon the local velocity components u, v, w of the air relative to the observer's head [1].

Example A flow field is given by

$$\mathbf{V}(x,y,t) = 2x^2\mathbf{i} + 4xyt\mathbf{j}$$

(a) What is the velocity and acceleration of a particle at (1,3,2)? (b) Sketch the streamline which passes through the point (1,3) when $t = 1$ and when $t = \frac{1}{2}$.

Solution (a) The velocity is readily determined by substituting the given values into the equation, i.e.,

$$\mathbf{V}(1,3,2) = 2\mathbf{i} + 24\mathbf{j}$$

Since the flow depends upon time as well as space, the acceleration of a particle is given by Eq. (9-11):

$$\mathbf{a}(x,y,t) = u\frac{\partial \mathbf{V}}{\partial x} + v\frac{\partial \mathbf{V}}{\partial y} + \frac{\partial \mathbf{V}}{\partial t}$$

$$\mathbf{a}(x,y,t) = 2x^2(4x\mathbf{i} + 4yt\mathbf{j}) + 4xyt(4xt\mathbf{j}) + 4xy\mathbf{j}$$

or

$$\mathbf{a}(x,y,t) = 8x^3\mathbf{i} + (8x^2yt + 16x^2yt^2 + 4xy)\mathbf{j}$$

$$a(1,3,2) = 8\mathbf{i} + 252\mathbf{j}$$

(b) The streamline is given by Eq. (9-1):

$$\frac{dy}{dx} = \frac{v}{u} = \frac{4xyt}{2x^2} = 2\frac{yt}{x}$$

Since t is a constant with respect to x and y, the variables can be separated to give

$$\int \frac{dy}{y} = 2t \int \frac{dx}{x}$$

$$\ln y = 2t \ln x + \ln C$$

$$y = Cx^{2t}$$

For $t = 1$,

$$C = \frac{y}{x^2}$$

At the point (1,3)

$$C(1,3,1) = 3 \quad \text{and} \quad y|_{t=1} = 3x^2$$

and the plot of the streamline is shown as a solid line in Fig. 9-2.

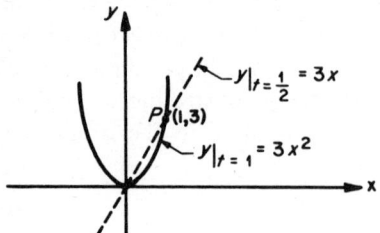

Fig. 9-2 Instantaneous streamlines.

Following the same procedure for $t = \frac{1}{2}$,

$C(1,3,\frac{1}{2}) = 3$ and $y|_{t=\frac{1}{2}} = 3x$

which is shown by the dashed line in Fig. 9-2.

9-3 TYPES OF MOTION

Steady and unsteady flow *If the local acceleration is zero, $\partial \mathbf{V}/\partial t = 0$, the motion is steady.* The velocity does not change with time, although it may change from point to point in space. On the other hand, a flow which is time-dependent is *unsteady*.

Often an unsteady flow can be transformed to steady flow by changing the reference axes. Consider, for example, an airplane moving through the atmosphere at a constant speed of V_0 (Fig. 9-3a). The fluid velocity at a point (x_0,y_0) is unsteady, being zero before the plane reaches the point, varying widely as it passes due to the wake and waves produced by disturbing the air, and finally becoming zero again as the plane disappears.

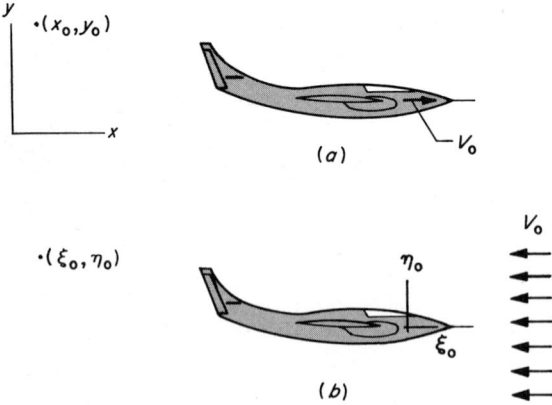

Fig. 9-3 Transformation of unsteady to steady flow.

Fixing the axes with respect to the plane can be accomplished by superimposing a velocity of $-\mathbf{V}_0$ relative to the $\xi\eta$ axes. The effect at the point (ξ_0, η_0), as shown in Fig. 9-3b, is the same at all times, hence steady. This principal is utilized in wind-tunnel testing of aircraft and rocket models; conditioned air is circulated past fixed models. This transformation is valid any time a body is moving with a constant velocity through an initially undisturbed field.

Streamlines remain fixed in steady flow, and they are coincident with *path lines*, lines which describe the path or trajectory of a fluid element. Streamlines and path lines do not coincide, however, in unsteady flow [4].

Uniform and nonuniform flow *If motion is uniform, the convective acceleration is zero.* In uniform flow the velocity vector is identical, in magnitude and direction, at every point in the flow field, that is, $\partial \mathbf{V}/\partial r = 0$, where r is a displacement in any direction. This definition does not require that the velocity itself be constant with respect to time; it requires that any change occur at every point simultaneously; the streamlines must be straight.

A frictionless liquid flowing through a long straight pipe is an example of uniform flow. *Nonuniform flow* is typified by the flow of a frictionless liquid through a pipe of changing cross section or through a pipe which is curved ($\partial \mathbf{V}/\partial r \neq 0$).

Laminar and turbulent flow In 1883, while injecting dyes into flows fed by constant-head tanks similar to that shown in Fig. 9-4, Osborne Reynolds [3] observed two distinct types of flow. At relatively low velocities fluid particles move smoothly, everywhere parallel. Because the fluid moves in a laminated form, it is termed *laminar*. For laminar flow the dye moves in a thin, straight line.

At relatively high velocities, Reynolds noted that the dye would abruptly break up, diffusing throughout the tube. At higher velocities the breaking point moves upstream until it is finally *turbulent* throughout. Turbulent flow is always unsteady flow by our prior strict definition. But to understand the mechanism we

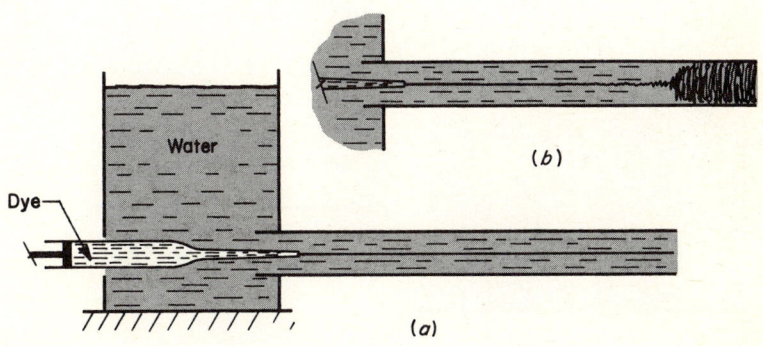

Fig. 9-4 Reynolds experiment: (*a*) laminar; (*b*) turbulent.

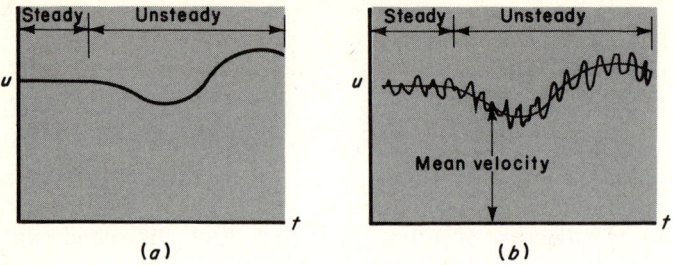

Fig. 9-5 (a) Laminar and (b) turbulent motion.

shall modify that definition and speak in terms of steady and unsteady turbulent flow as depicted in Fig. 9-5.

Exaggerating the turbulent fluctuations for clarity, Fig. 9-6 shows the local velocity u and the fluctuating component u', but it should be noted that random fluctuations occur in all directions as shown in Fig. 9-6b. Therefore, the velocity $\mathbf{V}$ is made up of an average value $\bar{\mathbf{V}}$ and fluctuating components u', v', and w'; that is,

$$\mathbf{V} = \bar{\mathbf{V}} + \mathbf{i}u' + \mathbf{j}v' + \mathbf{k}w' \tag{9-15}$$

or, in the x direction,

$$u = \bar{u} + u' \tag{9-16}$$

The fluctuating components are random; the area under the curve above the mean velocity is equal to that below the curve in Fig. 9-6a.

Taking the time average of Eq. (9-15) over a large period, $\Delta t = t_1 - t_0$, as Δt approaches infinity we get

$$\lim_{\Delta t \to \infty} \frac{1}{\Delta t} \int_{t_0}^{t_1} \mathbf{V}\, dt = \bar{\mathbf{V}} \tag{9-17}$$

Any fluid property can be averaged by this technique for turbulent flow. For example, density

$$\rho = \bar{\rho} + \rho' \tag{9-18}$$

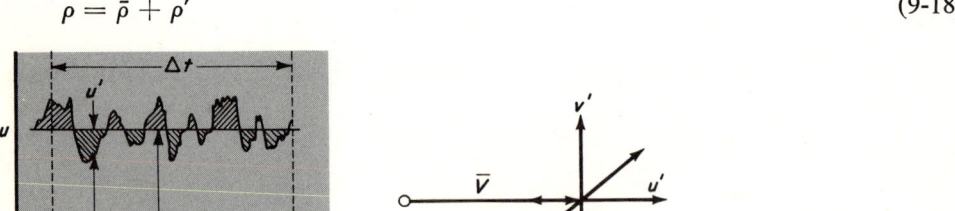

Fig. 9-6 Turbulent motion.

FLUID FIELDS

where

$$\bar{\rho} = \lim_{\Delta t \to \infty} \frac{1}{\Delta t} \int_{t_0}^{t_1} \rho \, dt \qquad (9\text{-}19)$$

Physically, this means that a mean density taken over a period Δt has the same effect on the fluid as the fluctuating density ρ does when integrated (graphically) throughout the period of time.

As one might suspect, the shear stress is complicated by turbulent flow. The simple laminar-flow equation

$$\tau = -\mu \frac{\partial u}{\partial y} \qquad (9\text{-}20)$$

no longer applies in the case of turbulent motion. Eddy viscosity ϵ, resulting from the fluctuations and requiring experimental determination, must be incorporated in the stress equation, namely,

$$\tau = -(\mu + \epsilon) \frac{\partial u}{\partial y} \qquad (9\text{-}21)$$

The value of eddy viscosity depends upon the motion and the density of the fluid; hence, it is not a fluid property.

In his classic experiments, Reynolds discovered that the existence of the two flow types depends not only upon velocity but upon the parameter VD/ν, where V is the average velocity in the pipe, D the pipe diameter, and ν the kinematic viscosity of the fluid. This dimensionless parameter, named in honor of Reynolds, is a basic parameter in the study of fluid motion. In more general terms the *Reynolds number* **Re** is

$$\mathbf{Re} \equiv \frac{Vl}{\nu} \qquad (9\text{-}22)$$

where l is a characteristic length. This formulation permits the Reynolds number to serve as a characteristic parameter in flow situations other than pipe flow. In this form it is a convenient measure of whether a flow is laminar or turbulent. In pipe flow, with the pipe diameter as the characteristic length, the flow may become turbulent when $\mathbf{Re} \geq 2{,}000$. In flow over a flat plate, where the plate length is the characteristic length, the flow may cease to be laminar at $300{,}000 < \mathbf{Re} < 600{,}000$.

9-4 SIMILITUDE

Many important problems in fluid mechanics must be solved experimentally. To correlate the data and apply them to other flow situations often requires modeling. For example, it is obviously not feasible to build a full-scale hydropower plant to test a turbine blade with a new twist. By making use of the concept of similitude, conclusions can be reached regarding the performance of the new type of turbine blade from model studies. But how can a model be built which will yield meaningful results on its prototype counterpart? Certain conditions of similitude must exist. They are discussed here, and the foundation for their existence is detailed in Appendix C.

Two flow fields are *geometrically similar* when both fields and boundaries are in the same geometric proportions and have the same orientation; a dimension of the model is a constant scale of the prototype or vice versa. In equation form, for geometric similarity

$$L_r \equiv \frac{l_M}{l_P} = \text{const} \tag{9-23}$$

The subscript r designates the ratio, in this case the relative size of the model M and prototype P system.

Kinematic similarity requires that the velocity ratios at corresponding points in two flow fields be constant in magnitude and direction, i.e.,

$$\mathbf{V}_r \equiv \frac{\mathbf{V}_M}{\mathbf{V}_P} = \text{const} \tag{9-24}$$

Equations (9-23) and (9-24) may be combined to get a time ratio between the model and prototype, namely,

$$t_r \equiv \frac{t_M}{t_P} = \frac{L_r}{\mathbf{V}_r} \tag{9-25}$$

since $\mathbf{V} \propto L/t$. Kinematic similarity assures that streamline patterns are the same for model and prototype.

Geometric and kinematic similarity are adequate for flows involving fluids which are nearly perfect ($\mu = 0$); however, most fluids are *real* and require the consideration of dynamic similarity; the ratio of corresponding forces in two fields must be constant, i.e.,

$$\mathbf{F}_r \equiv \frac{\mathbf{F}_M}{\mathbf{F}_P} = \text{const} \tag{9-26}$$

Since $a \propto V/t$,

$$a_r \equiv \frac{V_r}{t_r} = \frac{L_r}{t_r^2} \tag{9-27}$$

and Newton's second law gives

$$F_r = M_r \frac{L_r}{t_r^2} \tag{9-28}$$

FLUID FIELDS

But mass is the density-volume product; hence

$$M_r = \rho_r \mathscr{V}_r = \rho_r L_r^3 \tag{9-29}$$

$$F_r = \rho_r L_r^2 V_r^2 \tag{9-30}$$

which must hold for all dynamically similar systems.

Dimensionless groups Common forces which influence the motion of real fluids are inertial, pressure, viscous, gravitational, and elastic. All these forces which are present in flow cases must be in a constant ratio for complete similarity. In most flow fields, however, two of these types are so much larger in comparison to the others that the forces of lesser significance may be neglected. The most common dimensionless groups are listed below with the flow conditions in which they would be expected to be important. The square brackets denote dimensions.

1. Reynolds number **Re**

$$\mathbf{Re} \equiv \frac{Vl}{\nu} = \frac{[\rho V^2 A]}{[lV\mu]} = \frac{\text{inertial force}}{\text{viscous force}} \tag{9-31}$$

Distinguishes between flow regimes, such as laminar or turbulent flow in pipes, in the boundary layer, or around submerged objects. $A = l^2$.

2. Mach number **M**

$$\mathbf{M} \equiv \frac{V}{c} = \frac{[\rho V^2 A]}{[EVA]} = \frac{\text{inertial force}}{\text{elastic force}} \tag{9-32}$$

where c is the speed of sound. Distinguishes between flow regimes, such as subsonic, transonic, supersonic, or hypersonic, in internal or external compressible flow.

3. Froude number **Fr**

$$\mathbf{Fr} \equiv \frac{V}{\sqrt{gl}} = \frac{[\rho V^2 A]}{[\rho(lA)g]} = \frac{\text{inertial force}}{\text{weight}} \tag{9-33}$$

Distinguishes between rapid and tranquil flow in systems having free surfaces; useful in ship design and hydraulic structures.

4. Euler number **Eu**

$$\mathbf{Eu} \equiv \frac{p}{\rho V^2} = \frac{[pA]}{[\rho V^2 A]} = \frac{\text{pressure force}}{\text{inertial force}} \tag{9-34}$$

Equal to twice the *pressure coefficient* $\mathbf{C}_p$,

$$\mathbf{C}_p \equiv \frac{p}{\frac{1}{2}\rho V^2} \tag{9-35}$$

the ratio of static to dynamic pressures; useful in compressible flow.

5. Weber number **We**

$$\mathbf{We} \equiv \frac{V^2 l \rho}{\sigma} = \frac{[\rho V^2 A]}{[\sigma l]} = \frac{\text{inertial force}}{\text{surface tension}} \tag{9-36}$$

Influences liquid-liquid or gas-liquid interfaces.

For dynamic similarity all these dimensionless numbers must be the same for model and prototype. In most cases no more than two of the forces are of the same order of magnitude. For example, similarity in wind-tunnel tests is generally assured when the Mach number and Reynolds number are the same in model and prototype; the Reynolds number is sufficient in incompressible pipe flow when geometric similitude with respect to orientation is observed; the Froude number is sufficient in harbor modeling, although geometric similitude with respect to depth is usually modified because viscous effects (**Re**) become more pronounced in shallow beds.

Example Oil having a kinematic viscosity of 1.9×10^{-5} ft²/sec at 75°F flows through a 10-in-diam pipe at 40 ft/min. At what velocity must water at the same temperature flow for dynamic similitude?

Solution For dynamic similitude

$$\mathbf{Re}_w = \mathbf{Re}_o$$

$$\frac{V_w l_w}{\nu_w} = \frac{V_o l_o}{\nu_o}$$

From Fig. A-2, $\nu_w = 1 \times 10^{-5}$ ft²/sec and $l_w = l_o$; therefore,

$$V_w = \frac{\nu_w}{\nu_o} V_o = \frac{1 \times 10^{-5}}{1.9 \times 10^{-5}} (40) = 21.1 \text{ ft/min}$$

9-5 RELATIVE EQUILIBRIUM

In Chap. 4 we noted an absence of shear stresses in static fluids, making the pressure variation simple to compute. Two interesting cases of motion in which no shear stresses occur will be cited in this section: uniform linear acceleration and uniform angular velocity about a vertical axis. In neither case does a fluid layer move adjacent to another; the fluid acts as a solid body. Pressure variation can be determined by writing the equation of motion, $\sum \mathbf{F} = m\mathbf{a}$, for an appropriate system.

Fluids which behave like solid bodies when in motion are said to be in *relative equilibrium*.

Uniform linear acceleration In Sec. 9-3 we considered the resulting motions when a portion of the fluid's acceleration [Eq. (9-11)] is zero: steady motion when the local acceleration is zero and uniform motion resulting in zero convective acceleration. In this section we shall consider another special case, constant acceleration, that is, $\mathbf{a} = $ const.

FLUID FIELDS

In Fig. 9-7 an open container of liquid is on the back of a truck traveling with a uniform acceleration a. The liquid adjusts with a linear free surface and moves as a solid. The scalar components of the acceleration, a_x and a_y, may be used in Newton's second law in the x and y directions, respectively. For equilibrium of a fluid element of depth Δz

$$F_x = p|_x \Delta y \Delta z - p|_{x+\Delta x} \Delta y \Delta z = \rho \Delta x \Delta y \Delta z \, a_x$$
$$F_y = p|_y \Delta x \Delta z - p|_{y+\Delta y} \Delta x \Delta z - \gamma \Delta x \Delta y \Delta z = \rho \Delta x \Delta y \Delta z \, a_y$$

Dividing by $\Delta x \Delta y \Delta z$ and taking the limit as Δx and Δy approach zero,

$$\frac{\partial p}{\partial x} = -\frac{\gamma}{g} a_x \tag{9-37}$$

$$\frac{\partial p}{\partial y} = -\gamma \frac{a_y + g}{g} \tag{9-38}$$

Since the pressure is a function of x and y, its total differential is

$$dp = \frac{\partial p}{\partial x} dx + \frac{\partial p}{\partial y} dy$$

Using Eqs. (9-37) and (9-38), this becomes

$$dp = -\frac{\gamma}{g} [a_x \, dx + (a_y + g) \, dy] \tag{9-39}$$

Integrating between any two points in the fluid, we get

$$p_2 - p_1 = -\frac{\gamma}{g} [a_x(x_2 - x_1) + (a_y + g)(y_2 - y_1)]$$

or, for a constant pressure,

$$\left. \frac{dy}{dx} \right|_{p=c} = \tan \theta = -\frac{a_x}{a_y + g} \tag{9-40}$$

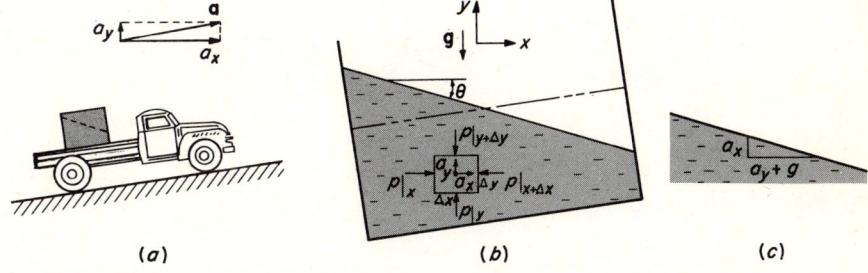

Fig. 9-7 Uniform acceleration.

which is the slope of the free surface, as shown in Fig. 9-7c, since $p_2 - p_1 = 0$; constant-pressure planes are parallel to the free surface.

For no acceleration in the y direction, $a_y = 0$, Eq. (9-38) gives the hydrostatic equation

$$p = \gamma h \tag{4-14}$$

the same as that for a static fluid.

Example How much water spills from an open rectangular tank 4 by 12 by 6 ft wide, initially full, when accelerated by 8.05 ft/sec²?

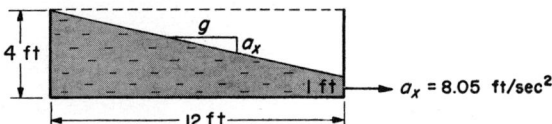

Fig. 9-8 Accelerating tank.

Solution The free surface adjusts as shown in Fig. 9-8, since

$$\tan \theta = -\frac{8.05}{0 + 32.2} = \frac{1}{4}$$

The volume spilled is $\frac{1}{2}[3(12)](6) = 108$ ft³.

Example A U-tube which is closed at one end is filled with water as shown in Fig. 9-9a. Upon being accelerated at 32.2 ft/sec, what are the pressures at A and B?

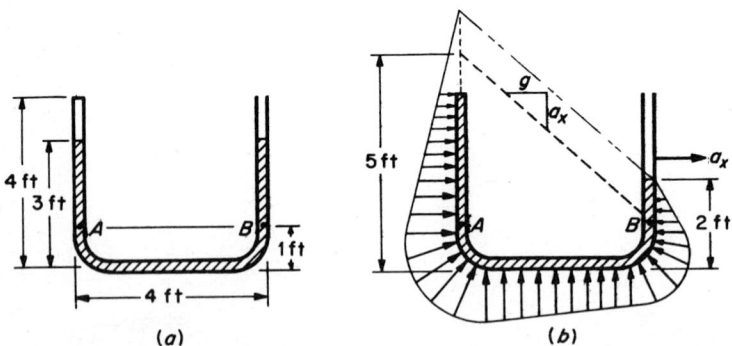

Fig. 9-9 Accelerating U-tube.

Solution The dashed line in Fig. 9-9b represents the free-surface position if the closed end of the tube is extended to accommodate the shift; this is a 45° line since $a_x/(a_y + g) = 1$. It is also a line

FLUID FIELDS

of constant pressure whose value can be determined at point B, that is,

$$p_B = \gamma h_B = (62.4 \text{ lb}_f/\text{ft}^3)(1 \text{ ft}) = 62.4 \text{ lb}_f/\text{ft}^2$$

The pressure at A is then equal to 5 ft of water, or

$$p_A = \gamma h_A = 62.4(5) = 312.0 \text{ lb}_f/\text{ft}^2$$

The pressure distribution throughout the tube is shown in Fig. 9-9b. Note that the plane of zero gauge pressure lies 1 ft above the dashed line.

By using a tube which is open at both ends and by placing a calibrated scale on its legs, a hydrostatic accelerometer results. Since no fluid property appears in Eq. (9-40), the accelerometer would be independent of the fluid used.

Uniform rotation about a vertical axis When a fluid rotates at a constant angular velocity ω about a vertical axis z, as shown in Fig. 9-10, it rotates as a rigid body with circumferential symmetry. Since no motion occurs in the vertical direction, the hydrostatic equation gives the vertical pressure variation when the head is taken from the free surface.

Writing Newton's second law for the fluid element in the radial direction, $\Sigma F_r = ma_r$, the forces are the pressure-area products acting on the arcs plus the components of the side forces. The pressure acting on the element's side is the average of that acting on the two arcs, i.e.,

$$\lim_{\Delta r \to 0} \frac{p + (p + \Delta p)}{2} = p$$

Since the sine of a small angle is equal to the angle, the radial component of this force is

$$p \, \Delta r \, \Delta z \, \frac{\Delta \phi}{2}$$

but the components on opposite faces both act outward. Therefore, the radial equation of motion becomes

$$(pA)_r - (pA)_{r+\Delta r} + p \, \Delta r \, \Delta z \, \Delta \phi = \frac{\gamma \, \Delta \mathscr{V}}{g} a_r \tag{9-41}$$

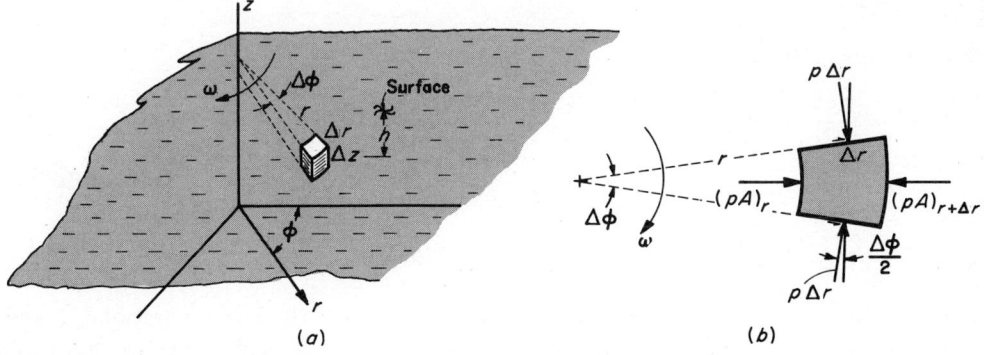

Fig. 9-10 Solid-body rotation.

The radial acceleration $a_r = -r\omega^2$ and

$$\Delta\mathscr{V} = \frac{r\,\Delta\phi\,\Delta z + (r+\Delta r)\,\Delta\phi\,\Delta z}{2}\Delta r = r\,\Delta\phi\,\Delta z\,\Delta r$$

neglecting higher-order terms. Therefore, Eq. (9-41) becomes

$$p|_r\, r\,\Delta\phi\,\Delta z - p|_{r+\Delta r}(r+\Delta r)\,\Delta\phi\,\Delta z + p\,\Delta r\,\Delta z\,\Delta\phi = \frac{\gamma}{g}r\,\Delta\phi\,\Delta z\,\Delta r(-r\omega^2) \tag{9-42}$$

Dividing by $r\,\Delta\phi\,\Delta r\,\Delta z$ and taking the limit as $\Delta r \to 0$,

$$\lim_{\Delta r \to 0}\left(\frac{p_r - p_{r+\Delta r}}{\Delta r} + \frac{p - p|_{r+\Delta r}}{r}\right) = -\frac{\gamma}{g}r\omega^2 \tag{9-43}$$

$$\frac{\partial p}{\partial r} = \frac{\gamma}{g}r\omega^2 \tag{9-44}$$

since the second term in the parentheses reduces to zero in the limit. Integrating Eq. (9-44),

$$p = \frac{\gamma}{g}\frac{r^2}{2}\omega^2 + p(z) + C_1 \tag{9-45}$$

To evaluate the influence of z on pressure, i.e., to find $p(z)$, set the hydrostatic equation

$$\frac{\partial p}{\partial z} = -\rho g \tag{9-46}$$

equal to the partial derivative of Eq. (9-45).

$$\frac{\partial p}{\partial z} = p'(z) = -\rho g$$

$$p(z) = -\rho g z + C_2 \tag{9-47}$$

Substituting into Eq. (9-45),

$$p = \frac{\gamma}{g}\frac{r^2}{2}\omega^2 - \rho g z + C_2 \tag{9-48}$$

For a line of constant pressure

$$z = \frac{r^2\omega^2}{2g} + C \tag{9-49}$$

which is the equation of a parabola. All isobars are therefore parabolic. And since the rotating fluid is circumferentially symmetric, constant-pressure lines are paraboloids of revolution, as shown in Fig. 9-11.

FLUID FIELDS

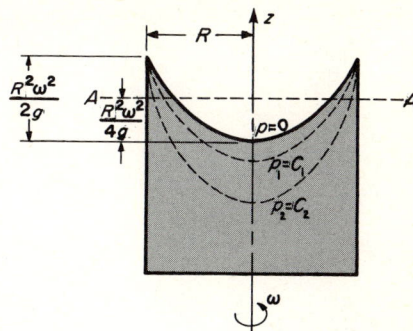

Fig. 9-11 Rotation of circular cylinder.

At the axis $r = 0$, $z = C$. At radius R and $z = 0$, $C = R^2\omega^2/2g$; therefore,

$$z\big|_{r=0} = \frac{R^2\omega^2}{2g} \tag{9-50}$$

Since the volume of a paraboloid of revolution is equal to one-half the volume of its circumscribing cylinder, the volume of fluid above the horizontal plane which passes through the free surface at the vertex is

$$\frac{\pi R^2}{2} \frac{R^2\omega^2}{2g}$$

For zero angular velocity, the free surface of the static liquid would lie at the plane AA of Fig. 9-11.

Example A right-circular cylinder 2 ft in diameter and 4 ft tall, originally full of water, is rotated until the pressure in the center on its bottom is zero. (a) How much water is spilled? (b) What is its angular velocity?

Solution (a) From the preceding analysis and Fig. 9-12, one-half of the water is spilled; i.e.,

$$\mathscr{V}_{\text{spilled}} = \tfrac{1}{2}R^2\pi(4) = 2\pi \text{ ft}^3$$

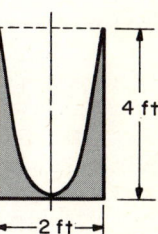

Fig. 9-12 Circular cylinder rotating about its axis.

(b) Knowing the volume above the horizontal plane through the vertex permits determination of angular velocity:

$$\frac{\pi R^4 \omega^2}{4g} = 2\pi \text{ ft}^3$$

$$\omega^2 = 8g$$

$$\omega = 16 \text{ rad/sec}$$

This principle is used to empty the tubs of washing machines. The inside of the tub must be made parabolic to empty all the water at a finite angular velocity.

PROBLEMS

9-1. A two-dimensional velocity field is described in terms of its cartesian components $u = 3xy^2$ and $v = 3x^2y$. Write the equation of the streamline which passes through the point (2,1).

9-2. A two-dimensional velocity field is described by $\mathbf{V}(x,y) = 2x\mathbf{i} + xy\mathbf{j}$. Write the equation of the streamline that passes through the point (1,2).

9-3. A two-dimensional velocity field is described in terms of its cartesian components $u = xy^2$ and $v = xe^x$. Write the equation of the streamline which passes through the point (1,3).

9-4. A two-dimensional velocity field is described by $\mathbf{V} = xy^2\mathbf{i} - x^2y\mathbf{j}$. Determine the equation of the streamline passing through the point (2,3).

9-5. The velocity field of a steady, two-dimensional flow is given by $\mathbf{V}(x,y) = 5x\mathbf{i} - 5y\mathbf{j}$.
  (a) Determine the equation of the streamlines for this flow.
  (b) Determine the acceleration of this flow.

9-6. A temperature field is described by the equation $T = 4x^4y$. At the point (1,2):
  (a) What is the gradient of the temperature?
  (b) Determine the slope of the gradient line.
  (c) Write the equation for an isotherm which passes through the point.

9-7. A flow field is given by $\mathbf{V}(x,y,z,t) = 2x\mathbf{i} + 6xyz\mathbf{j} + x^2zt\mathbf{k}$. What is the velocity and acceleration of a particle at the point (1,2,1,1)?

9-8. If a flow field is given by $\mathbf{V}(x,y,z,t) = xyz\mathbf{i} + y\mathbf{j} - 3zt\mathbf{k}$, what is the total acceleration at the point (2,1,2,1)?

9-9. If $u = x^2yz + yz - 3z$, what is the total differential of u and du/dt?

9-10. Crude oil at 70°F flows through a 6-in-diam pipe at a rate of 200 gal/min. Is the flow laminar or turbulent?

9-11. Calculate the maximum discharge of gasoline at 50°F and standard atmospheric pressure that is sure to have laminar flow in a $\frac{1}{4}$-in-diam tube.

9-12. Air at 100°F flows over a 10-ft-long flat plate at a velocity of 20 ft/sec. What percentage of the flow is laminar?

9-13. A supersonic jet travels at $\mathbf{Ma} = 2$ in air at −40°F. Tests are to be conducted on a one-tenth-scale model in a wind tunnel at −40°F.
  (a) What should be the wind tunnel speed to simulate Mach number effects?
  (b) Does this satisfy all requirements for dynamic similarity?

9-14. The drag of an underwater vehicle is being studied. The pressure at the nose of a one-tenth-scale model, tested in water at 50 ft/sec, is 50 psi. What would be the pressure on the nose of a prototype vehicle? Consider the model to be tested in fresh water and the prototype to run in salt water ($S = 1.025$), both at the same dynamic viscosity.

FLUID FIELDS

9-15. The drag on a one-twenty-fifth-scale model ship is determined to be 0.1 lb$_f$ when tested for a model speed of 5 ft/sec.
  (a) What is the prototype speed corresponding to this model speed?
  (b) What is the drag on the prototype?
The model is tested in fresh water; the prototype is tested in seawater ($S = 1.025$).

9-16. One-fifth-scale water-tunnel tests of lift and drag on hydrofoils for a boat are to be made. The boat is designed for a maximum speed of 40 mph in fresh water at 70°F. What velocity is required for modeling this condition if the water-tunnel temperature is 100°F?

9-17. A model of a river is built to a one-hundredth scale. The surface velocity of the model is 1.5 ft/sec. What is the surface velocity of the prototype river?

9-18. A model dam is 1 ft high and 2 ft long and has a flow 2 in. deep. If 3,000 lb$_m$ of water flows over it in 48.1 sec, what would be the expected flow rate in the prototype in cubic feet per second? The height of the prototype is 20 ft, and the length is 40 ft.

9-19. An open cubical tank 10 ft on each edge and initially full of water is accelerated horizontally parallel to an edge at such a rate that four-tenths of the liquid spills. What is the acceleration?

9-20. A U-tube manometer with both legs open is attached to a vehicle having motion in the direction shown. The manometer fluid is red oil with a specific gravity of 2.97. If the liquid levels are as shown, calculate the acceleration of the vehicle in the x direction.

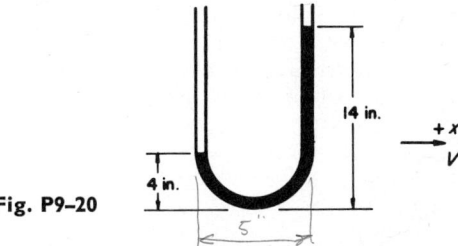

Fig. P9-20

9-21. Show that a paraboloid of revolution has a volume equal to one-half of its circumscribing cylinder.

9-22. An open cylindrical tank 2 ft in diameter and 4 ft deep is filled with water. When the tank is rotated about its axis at 60 rpm, how much water is spilled?

9-23. An open rectangular container of liquid is on a vehicle traveling with a uniform acceleration of 30 ft/sec² up a 20° grade. The container is 4 by 6 by 2 ft. What volume of liquid is spilled from the open container if it is initially filled on a horizontal plane?

9-24. A cylindrical vessel contains oil of specific gravity $S = 0.80$. When it is rotated about its vertical axis, the pressure at a point on the axis is the same as that at a point 3 ft higher and 2 ft from the axis. What is its angular velocity in radians per second?

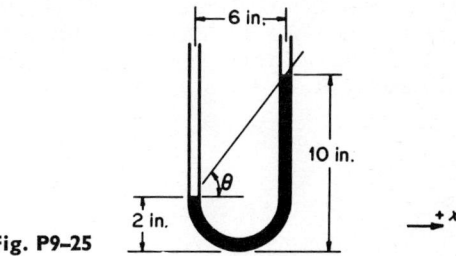

Fig. P9-25

9-25. A U-tube manometer with both legs open is attached to a vehicle having motion in the direction shown. The manometer fluid is olive oil, with a specific gravity of 0.918. If the liquid levels are as shown, what is the acceleration of the vehicle in the x direction?

9-26. A vertical cylindrical tank 3 ft in diameter and 6 ft in height is half full of water. What is the maximum angular velocity the tank can have without spilling any water?

9-27. A U-tube containing water rotates about the vertical axis AA. The angular velocity is 60 rpm, and the atmospheric pressure is assumed to be 14.7 psia. What is the pressure at point 2?

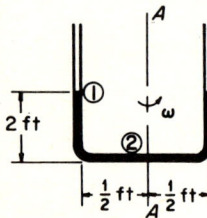

Fig. P9-27

9-28. The impeller of a centrifugal pump has a radius of 4 in. It rotates at 1,800 rpm, and the water in the impeller passages is moving with the same angular velocity as the impeller. Determine the pressure rise from the pump inlet, where $r = 0$, to the point corresponding to the impeller tip, where $r = 4$ in.

9-29. A rectangular container is made to accelerate at a uniform rate **a** along an incline as shown in the figure. Determine the angle θ when the water inside the container assumes a fixed orientation.

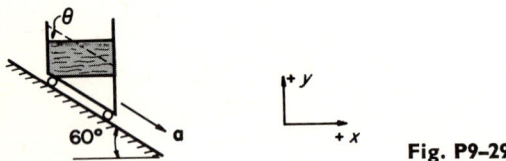

Fig. P9-29

9-30. A vertical cylindrical tank 2 ft in diameter and 5 ft in height contains an oil layer 3 ft deep. The specific gravity of the oil is 0.82. What is the angular velocity that will allow the oil to reach the rim of the tank without spillage?

REFERENCES

1. Bird, R. B., W. E. Stewart, and E. N. Lightfoot: "Transport Phenomena," Wiley, New York, 1960.
2. Daily, J. W., and D. R. F. Harleman: "Fluid Dynamics," Addison-Wesley, Reading, Mass., 1966.
3. Reynolds, O.: An Experimental Investigation of the Circumstances Which Determine Whether the Motion of Water Shall Be Direct or Sinuous, and of the Laws of Resistance in Parallel Channels, *Phil. Trans. Roy. Soc. London*, **174**: 935–982 (1883).
4. Sabersky, R. H., and A. J. Acosta: "Fluid Flow," Macmillan, New York, 1964.

CHAPTER 10

BASIC EQUATIONS

The concept of the *control volume* has been introduced in the preceding chapters, beginning with the elementary form of the conservation laws in Sec. 2-6. Chapter 9 considered the conversion of kinematic and other time-varying quantities to a form applicable to a fixed region in space. Except for these introductory concepts, the preceding analyses have had their foundation in the thermodynamic *system* approach. This is easy to understand, however, when we consider that the system and control volume are coincident in stationary media.

Since we now wish to consider moving media, this chapter presents the fundamental equations applicable to the control volume. The eulerian approach is generally advantageous in the study of transport phenomena in determining forces, pressure, temperature, concentration, viscosity, etc., at a particular location in space without regard for the previous or subsequent history of the fluid. For example, the designer of gas-turbine blades needs to know the temperature of the gas at the blade, but he may not be at all concerned with the temperature of the air entering the combustion chamber or with the temperature of the combustion products leaving the tailpipe.

As a further motivation for the use of the control-volume technique, one need only note that measurement sensors (thermometers, pressure transducers, etc.) are normally fixed with respect to a particular location rather than moving with the fluid. Hence, measurement techniques are based on the control-volume concept. Therefore, the measurements give the eulerian fluid properties, for example, $p(x,y,z,t)$.

We shall consider both finite and infinitesimal (differential) control volumes. The finite control volume requires being able to describe the property variation over the control surface and throughout the control volume (for unsteady processes). The infinitesimal control volume permits fixed-point observations, more amenable to an understanding of the internal transport mechanism. The same approach is applicable to both cases. In the infinitesimal case the control volume approaches a point in the limit.

CONSERVATION OF MASS

10-1 INTEGRAL FORM OF THE CONTINUITY EQUATION

Consider a *nondeformable control volume* at rest with respect to reference axes xyz as shown in Fig. 10-1. The control volume is taken such that it is at all times a part of the system.[1] An arbitrary velocity field $\mathbf{V}(x,y,z,t)$ carries the mass from region I into the control volume while mass leaves the control volume into region III Region II defines the control volume. Region I is defined such that its mass just enters the control volume in the time interval Δt. Similarly, region III is defined such that its mass just leaves the control volume in time Δt. Thus, as Δt approaches zero, region I and region III decrease until in the limit the system and control volume occupy the same space.

Since the mass m, fluid and/or solid bodies, contained in a system remains constant,

$$m_{\text{I},t} + m_{\text{II},t} = m_{\text{II},t+\Delta t} + m_{\text{III},t+\Delta t} \tag{10-1}$$

[1] Suggested by our colleague, Dr. Kenneth R. Purdy, Professor of Mechanical Engineering.

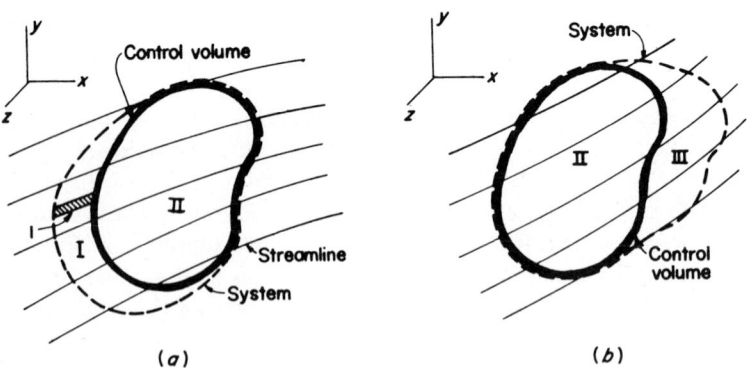

Fig. 10-1 Control volume and system: (*a*) at time t; (*b*) at time $t + \Delta t$.

BASIC EQUATIONS

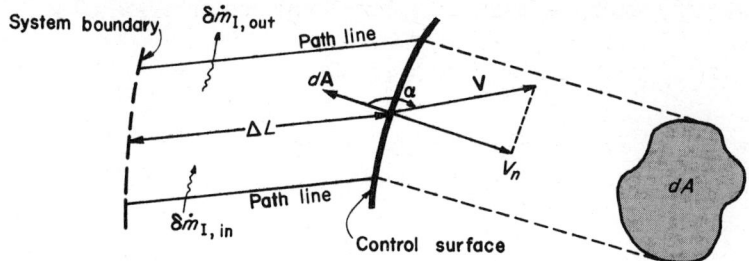

Fig. 10-2 Subregion 1, at time t.

where the left side of the equation is that shown in Fig. 10-1a and the right side is that shown in Fig. 10-1b. Rearranging Eq. (10-1), dividing by Δt, and taking the limit as $\Delta t \to 0$, we get

$$\lim_{\Delta t \to 0} \frac{m_{\text{II},t+\Delta t} - m_{\text{II},t}}{\Delta t} = \lim_{\Delta t \to 0} \frac{m_{\text{I},t} - m_{\text{III},t+\Delta t}}{\Delta t} \tag{10-2}$$

The left side becomes

$$\frac{dm_{\text{cv}}}{dt} = \frac{d}{dt} \int_{\text{cv}} \rho \, d\mathcal{V} \tag{10-3}$$

where m_{cv} is the instantaneous mass within the control volume.

To evaluate the right side of Eq. (10-2), consider the subregion I enlarged as shown in Fig. 10-2. The area vector $d\mathbf{A}$ is taken as the product of the outward-drawn unit normal vector to the control surface and the elemental area dA. Since the fluid is entering the control volume, it should be noted that the angle α will be greater than $\pi/2$ over the control surface common to region I.

Letting the mean length of the subregion be ΔL, its volume is

$$d\mathcal{V}_1 = -(\Delta L \cos \alpha \, dA)_1 \tag{10-4}$$

where $\cos \alpha$ will be negative since $\alpha > \pi/2$. The mass contained in subregion 1 is

$$dm_1 = -(\rho \, \Delta L \cos \alpha \, dA)_1 \tag{10-5}$$

where ρ is the average value of density which, in the limit, is the value of ρ at the control surface, as can be shown by the mean-value theorem[1] [6]. Integrating over the control surface common to region I,

$$m_{\text{I},t} = -\int_{\text{cs I}} (\rho \, \Delta L \cos \alpha \, dA)_t \tag{10-6}$$

where $m_{\text{I},t}$ is the total mass contained in region I at time t.

---

[1] If an entity has an average value, at some time between t and $t + \Delta t$ it must pass through that average value.

By similar reasoning the mass contained in region III is

$$m_{\text{III},t+\Delta t} = \int_{\text{cs III}} (\rho \, \Delta L \cos \alpha \, dA)_{t+\Delta t} \tag{10-7}$$

where $\cos \alpha$ is positive for mass leaving.

Substituting Eqs. (10-6) and (10-7) into Eq. (10-2),

$$\frac{d}{dt}\int_{\text{cv}} \rho \, d\mathscr{V} = \lim_{\Delta t \to 0} \frac{-\int_{\text{cs I}} (\rho \, \Delta L \cos \alpha \, dA)_t - \int_{\text{cs III}} (\rho \, \Delta L \cos \alpha \, dA)_{t+\Delta t}}{\Delta t} \tag{10-8}$$

we now become concerned with the time rate of change of mass within regions I and III. Since the side surface of the subregion depicted in Fig. 10-2 is composed of path lines, ΔL represents the distance traveled by a fluid particle on the system surface in the time interval Δt. But path lines may have fluid crossing them in unsteady flow, represented by $\delta \dot{m}_1$. In the limit, as $\Delta t \to 0$, $(\delta \dot{m}_1)_{\text{in}} \to (\delta \dot{m}_1)_{\text{out}}$, and the only flow out of the subregion passes through the control surface, since the path lines become coincident with streamlines in the limit.

Because the area of the control surface is independent of the time interval, Δt may be taken under the integral of Eq. (10-8) to give

$$\frac{d}{dt}\int_{\text{cv}} \rho \, d\mathscr{V} = \lim_{\Delta t \to 0}\left[-\int_{\text{cs I}}\left(\rho \frac{\Delta L}{\Delta t} \cos \alpha \, dA\right)_t - \int_{\text{cs III}}\left(\rho \frac{\Delta L}{\Delta t} \cos \alpha \, dA\right)_{t+\Delta t}\right] \tag{10-9}$$

Since $A \neq A(t)$ and the functions of the area integrands are continuous, the limit of the integral is equal to the integral of the limit by Leibniz' rule [5], and the limiting process gives

$$\lim_{\Delta t \to 0} \frac{\Delta L}{\Delta t} = V \tag{10-10}$$

where V is the speed of the fluid element. Noting also that the speed normal to the surface V_n is

$$V_n = V \cos \alpha \tag{10-11}$$

Eq. (10-9) becomes

$$\frac{d}{dt}\int_{\text{cv}} \rho \, d\mathscr{V} = -\int_{\text{cs I}} \rho V_n \, dA - \int_{\text{cs III}} \rho V_n \, dA \tag{10-12}$$

or, since the total control surface is made up of regions I and III, i.e.,

$$\int_{\text{cs}} \rho V_n \, dA = \int_{\text{cs I}} \rho V_n \, dA + \int_{\text{cs III}} \rho V_n \, dA \tag{10-13}$$

BASIC EQUATIONS

Eq. (10-12) may be more simply written in terms of the total control surface. Thus

$$\frac{d}{dt}\int_{cv} \rho\, d\mathscr{V} = -\int_{cs} \rho \mathbf{V} \cdot d\mathbf{A} \tag{10-14}$$

Because a nondeformable control volume is being considered, that is, $\mathscr{V} \neq \mathscr{V}(t)$, the differentiation can be performed on the integrand instead of the integral, giving

$$\blacksquare \quad \int_{cv} \frac{\partial \rho}{\partial t} d\mathscr{V} = -\int_{cs} \rho \mathbf{V} \cdot d\mathbf{A} \tag{10-15}$$

This is the *integral form of the continuity equation*. It says that the rate of decrease of mass inside the control volume is equal to the net efflux rate of mass through the control surface. Since no size limitation was imposed on the control volume selected, this equation is valid for any size region, finite or infinitesimal. It is also valid for any fluid—viscous or frictionless, compressible or incompressible, single or multicomponent, single or multiphase, flow with or without heat transfer, etc.

It should be observed that the velocity **V** is measured relative to the control volume, since the control volume is fixed in *xyz*. In many problems, such as a jet engine, a rocket, or a rotating sprinkler, it is advantageous to consider a moving control volume. Equation (10-15) is then valid if the relative velocity is used.

Steady flow If the total mass within the control volume is constant in time, Eq. (10-14) becomes

$$\int_{cs} \rho \mathbf{V} \cdot d\mathbf{A} = 0 \tag{10-16}$$

or

$$\sum (\rho V_n A)_{\text{in}} = \sum (\rho V_n A)_{\text{out}} \tag{10-17}$$

Incompressible flow For either steady or unsteady flow, the general continuity equation reduces to that of Eqs. (10-16) and (10-17) for an incompressible fluid, since $\partial \rho / \partial t = 0$.

One-dimensional steady incompressible flow Flow is one-dimensional when all fluid properties and flow characteristics are expressible as functions of one space coordinate and time; properties are uniform normal to the flow direction. Although such flows are rarely found, a large percentage of problems can be adequately treated one-dimensionally. Figure 10-3 illustrates this for a velocity field in a stream tube or conduit.

Therefore, for a fluid in steady one-dimensional flow, Eq. (10-17) becomes

$$\rho_1 V_1 A_1 = \rho_2 V_2 A_2 \tag{10-18}$$

where the subscripts represent those stations shown in Fig. 10-3a. For an incompressible fluid Eq. (10-18) further simplifies to give

$$V_1 A_1 = V_2 A_2 \tag{10-19}$$

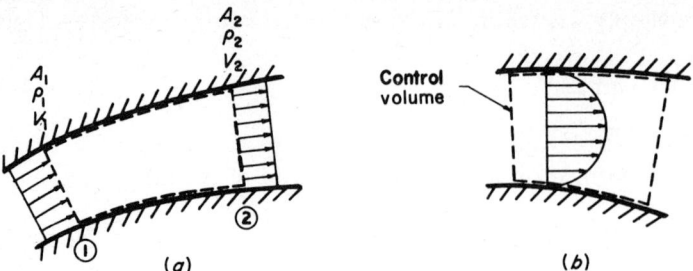

Fig. 10-3 Flow through a stream tube: (*a*) one-dimensional profile; (*b*) actual profile.

Example An incompressible fluid flows steadily through a duct which has two outlets, as shown in Fig. 10-4. The flow is one-dimensional at stations 1 and 2, but the velocity profile is parabolic at station 3. What is the velocity V_1?

Solution Let the surface of the duct be designated as surface 4, and take the control volume such that the control surfaces 1, 2, and 3 are normal to the flow directions of the respective sections.

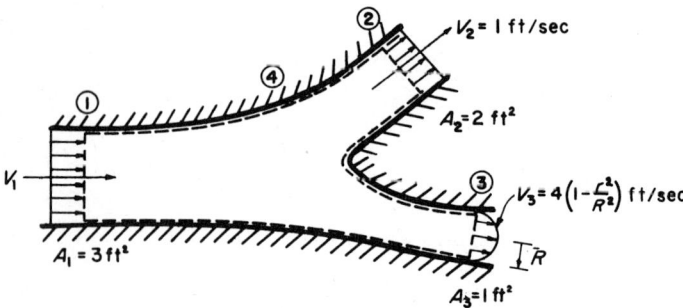

Fig. 10-4 Flow through branching duct.

Since the flow is steady, Eq. (10-15) reduces to

$$\int_{cs} \rho \mathbf{V} \cdot d\mathbf{A} = 0$$

where the integral represents the entire control surface, i.e.,

$$\int_{cs} \rho \mathbf{V} \cdot d\mathbf{A} = \int_{1} \rho \mathbf{V} \cdot d\mathbf{A} + \int_{2} \rho \mathbf{V} \cdot d\mathbf{A} + \int_{3} \rho \mathbf{V} \cdot d\mathbf{A} + \int_{4} \rho \mathbf{V} \cdot d\mathbf{A} = 0$$

and the integral over surface 4 is zero since $\alpha = \pi/2$, giving $\cos \alpha = 0$.

The density ρ divides out since the fluid is incompressible, giving

$$\int_{1} \mathbf{V} \cdot d\mathbf{A} + \int_{2} \mathbf{V} \cdot d\mathbf{A} + \int_{3} \mathbf{V} \cdot d\mathbf{A} = 0$$

At station 1, $\alpha = \pi$, but $\alpha = 0$ at stations 2 and 3, since α is the angle which the outward-

BASIC EQUATIONS

drawn normal makes with the velocity vector. And since the flow is one-dimensional at 1 and 2,

$$\int_1 \mathbf{V} \cdot d\mathbf{A} = -V_1 A_1$$

$$\int_2 \mathbf{V} \cdot d\mathbf{A} = V_2 A_2$$

$$\int_3 \mathbf{V} \cdot d\mathbf{A} = \int_3 V_3 (2\pi r\, dr)$$

Using these simplified relations,

$$-V_1 A_1 + V_2 A_2 + 8\pi \int_0^R \left(r - \frac{r^3}{R^2} \right) dr = 0$$

$$-3V_1 + 2 + 8\pi \left[\frac{r^2}{2} - \frac{r^4}{4R^2} \right] = 0$$

$$3V_1 = 2(1 + \pi R^2)$$

$$V_1 = \tfrac{2}{3}(1 + A_3) = \tfrac{4}{3} \text{ ft/sec}$$

10-2 DIFFERENTIAL FORM OF THE CONTINUITY EQUATION

We now seek a *differential form of the continuity equation* to complement the integral form of Eq. (10-15). To that end we consider an infinitesimal control volume fixed in coordinate space xyz as shown in Fig. 10-5. Since Eq. (10-15) has no size limitation, it may be applied to this differential control volume. Using an average value of density over the volume element, the volume integral becomes

$$\int_{cv} \frac{\partial \rho}{\partial t} d\mathscr{V} = \frac{\partial \rho}{\partial t} \bigg|_{av} \Delta x\, \Delta y\, \Delta z \qquad (10\text{-}20)$$

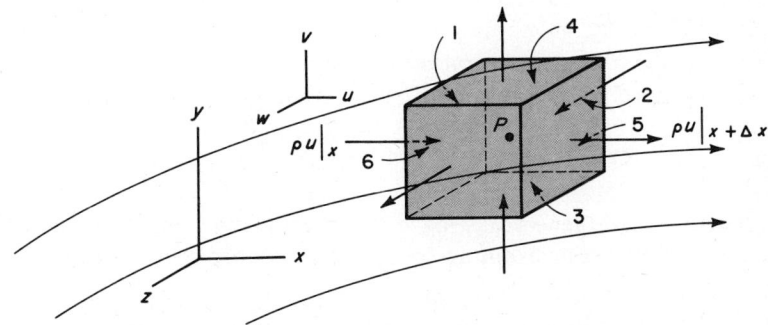

Fig. 10-5 Differential control volume.

Considering the mass flux at each face to be uniform,

$$\int_{\text{cs } 1} \rho \mathbf{V} \cdot d\mathbf{A} = -(\rho u)_x \Delta y \, \Delta z \tag{10-21a}$$

$$\int_{\text{cs } 2} \rho \mathbf{V} \cdot d\mathbf{A} = (\rho u)_{x+\Delta x} \Delta y \, \Delta z \tag{10-21b}$$

$$\int_{\text{cs } 3} \rho \mathbf{V} \cdot d\mathbf{A} = -(\rho v)_y \Delta x \, \Delta z \tag{10-21c}$$

$$\int_{\text{cs } 4} \rho \mathbf{V} \cdot d\mathbf{A} = (\rho v)_{y+\Delta y} \Delta x \, \Delta z \tag{10-21d}$$

$$\int_{\text{cs } 5} \rho \mathbf{V} \cdot d\mathbf{A} = -(\rho w)_z \Delta x \, \Delta y \tag{10-21e}$$

$$\int_{\text{cs } 6} \rho \mathbf{V} \cdot d\mathbf{A} = (\rho w)_{z+\Delta z} \Delta x \, \Delta y \tag{10-21f}$$

Summing over all faces,

$$\int_{\text{cs}} \rho \mathbf{V} \cdot d\mathbf{A} = [(\rho u)_{x+\Delta x} - (\rho u)_x] \Delta y \, \Delta z + [(\rho v)_{y+\Delta y} - (\rho v)_y] \Delta x \, \Delta z$$
$$+ [(\rho w)_{z+\Delta z} - (\rho w)_z] \Delta x \, \Delta y \tag{10-22}$$

Substituting Eqs. (10-20) and (10-22) into Eq. (10-15) and dividing by $\Delta x \, \Delta y \, \Delta z$, we get

$$\left.\frac{\partial \rho}{\partial t}\right|_{\text{av}} + \frac{(\rho u)_{x+\Delta x} - (\rho u)_x}{\Delta x} + \frac{(\rho v)_{y+\Delta y} - (\rho v)_y}{\Delta y} + \frac{(\rho w)_{z+\Delta z} - (\rho w)_z}{\Delta z} = 0 \tag{10-23}$$

Upon letting Δx, Δy, and Δz approach zero, the volume element approaches point P and the average value of density ρ approaches its point value. From the definition of the derivative, the flux terms become gradients in the respective directions in the limit. Taking the limit, Eq. (10-23) reduces to

$$\frac{\partial \rho}{\partial t} + \frac{\partial (\rho u)}{\partial x} + \frac{\partial (\rho v)}{\partial y} + \frac{\partial (\rho w)}{\partial z} = 0 \tag{10-24}$$

which is the *differential form of the continuity equation*.

A more general formulation of the differential continuity equation is

$$\blacksquare \quad \frac{\partial \rho}{\partial t} + \nabla \cdot \rho \mathbf{V} = 0 \tag{10-25}$$

where the divergence may be expressed in any convenient coordinate system (see Appendix B). For example, in cylindrical coordinates (r, ϕ, z)

$$\nabla \cdot \rho \mathbf{V} \equiv \text{div } \rho \mathbf{V} = \frac{1}{r} \frac{\partial}{\partial r} r \rho V_r + \frac{1}{r} \frac{\partial (\rho V_\phi)}{\partial \phi} + \frac{\partial (\rho V_z)}{\partial z} \tag{10-26}$$

BASIC EQUATIONS

Although the divergence in the cartesian and cylindrical coordinate systems appears quite different, the same results are obtained since the properties are invariant, i.e., the same in all coordinate systems.

For steady flow the continuity equation becomes

$$\nabla \cdot \rho \mathbf{V} = 0 \tag{10-27}$$

and for an incompressible fluid

$$\nabla \cdot \mathbf{V} = 0 \tag{10-28}$$

even though the flow may be unsteady, since $\partial \rho / \partial t = 0$.

Example Is the flow described by

$$\mathbf{V}(x,y,z) = x^2\mathbf{i} + (x+z)\mathbf{j} - 2xz\mathbf{k}$$

incompressible?

Solution If incompressible, the velocity components must satisfy the incompressible continuity equation

$$\frac{\partial u}{\partial x} + \frac{\partial v}{\partial y} + \frac{\partial w}{\partial z} = 0$$

Differentiating the respective components,

$$\frac{\partial u}{\partial x} = 2x \qquad \frac{\partial v}{\partial y} = 0 \qquad \frac{\partial w}{\partial z} = -2x$$

which satisfy the continuity equation. Therefore, the flow is incompressible.

Example A fluid flows in a round duct having a velocity of

$$V_z = z\left(1 - \frac{r^2}{R^2}\right) \cos \omega t$$

where R is the duct radius. Strategically placed heating and cooling elements in the duct cause the density to vary with radius r and time t only. At time $t = \pi/\omega$, $\rho = \rho_0$. Determine an expression for density ρ.

Solution For compressible unsteady flow in cylindrical coordinates the continuity equation is

$$\frac{\partial \rho}{\partial t} + \frac{1}{r}\frac{\partial}{\partial r}\rho V_r + \frac{1}{r}\frac{\partial(\rho V_\phi)}{\partial \phi} + \frac{\partial(\rho V_z)}{\partial z} = 0$$

where the middle terms cancel because neither the velocity nor the density varies with ϕ, and $V_r = 0$. Differentiating the product of the last term and rewriting gives

$$\frac{\partial \rho}{\partial t} + \rho \frac{\partial V_z}{\partial z} + V_z \frac{\partial \rho}{\partial z} = 0$$

since $\rho \neq \rho(z)$.

Substituting the velocity into the resulting equation gives

$$\frac{1}{\rho}\frac{\partial \rho}{\partial t} = \left(\frac{r^2}{R^2} - 1\right) \cos \omega t$$

which, upon integration with respect to t, yields

$$\ln \rho = \frac{1}{\omega}\left(\frac{r^2}{R^2} - 1\right)\sin \omega t + f(r)$$

But when $t = \pi/\omega$, $\rho = \rho_0$; therefore,

$$\ln \rho_0 = f(r)$$

Hence

$$\ln \frac{\rho}{\rho_0} = \left(\frac{r^2}{R^2} - 1\right)\frac{\sin \omega t}{\omega}$$

or

$$\frac{\rho}{\rho_0} = \exp\left[\left(\frac{r^2}{R^2} - 1\right)\frac{\sin \omega t}{\omega}\right]$$

CONSERVATION OF MOMENTUM

10-3 LINEAR MOMENTUM OF A SYSTEM

Newton's second law of motion

$$\sum \mathbf{F} = \frac{d\mathbf{M}}{dt} \tag{10-29}$$

vectorially relates the sum of the external forces $\mathbf{F}$ on a body to the rate of change of its momentum $\mathbf{M} = m\mathbf{V}$. Differentiating the $m\mathbf{V}$ product,

$$\sum \mathbf{F} = \frac{d(m\mathbf{V})}{dt} = m\frac{d\mathbf{V}}{dt} + \mathbf{V}\frac{dm}{dt} \tag{10-30}$$

and noting that the absolute acceleration $\mathbf{a}$ is

$$\mathbf{a} \equiv \frac{d\mathbf{V}}{dt} \tag{10-31}$$

we observe that the motion must be measured relative to an inertial (nonaccelerating) frame in order for the equation to be valid. If the acceleration is measured relative to a noninertial reference, e.g., motion referred to a satellite accelerating and rolling in outer space, it must be converted to an inertial reference before Eq. (10-30) is valid.

To review the elementary mechanics, let us apply Eq. (10-30) to a *system of n particles* (or rigid bodies whose motions are determined by the movement of their centers of gravity). Considering all particles simultaneously,

$$\sum_{i=1}^{n} \mathbf{F}_i = \sum_{i=1}^{n} \frac{d}{dt} m_i \mathbf{V}_i \tag{10-32}$$

BASIC EQUATIONS

Since the summation is made on the system of particles independent of time,

$$\sum \mathbf{F}_i = \frac{d}{dt} \sum m_i \mathbf{V}_i \tag{10-33}$$

Expressing the particles' velocities in terms of a generalized space coordinate s_i,

$$\sum \mathbf{F}_i = \frac{d}{dt} \left(\sum m_i \frac{d\mathbf{s}_i}{dt} \right) \tag{10-34}$$

but the particles' masses are independent of time. Therefore

$$\sum \mathbf{F}_i = \frac{d^2}{dt^2} \sum m_i \mathbf{s}_i \tag{10-35}$$

Now the sum of the mass-displacement products is equal to the product of the total mass and the displacement of the center of mass of the system, i.e.,

$$\sum \mathbf{F}_i = \frac{d^2}{dt^2} m\mathbf{s}_c \tag{10-36}$$

For a fixed mass

$$\sum \mathbf{F}_i = \frac{d}{dt} m\mathbf{V}_c \tag{10-37}$$

where $\mathbf{V}_c$ is the velocity of the center of mass of the system and m is the total mass.

Example Equation (10-37) demonstrates that the center of mass of shrapnel from an exploding hand grenade moves on the trajectory which the grenade would travel if no explosion had occurred, since there are no external forces on the particles, i.e.,

$$\frac{d}{dt} m\mathbf{V}_c = 0$$

$$m\mathbf{V}_c = \text{const}$$

Therefore,

$$\mathbf{V}_c = \text{const}$$

Example A tank with no external forces initially at rest on frictionless rollers on a horizontal plane contains 100 lb_m of gas in the left compartment. When the membrane is ruptured, the gas fills the tank. If the tank weighs 1,000 lb_f, does it move? If so, in what direction and how much?

Solution Movement can occur only horizontally, assuming it rolls on a nondeformable plane, and no external forces act horizontally. Therefore,

$$\frac{d}{dt} mu_c = 0$$

$$mu_c = \text{const}$$

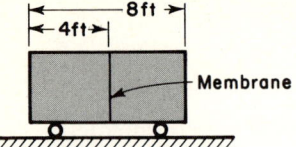

Fig. 10-6 Gas tank separated by membrane.

And since the mass does not change,

$u_c = \text{const}$

But its velocity was initially zero, hence, finally zero; and the center of mass does not move. The cart then must move in such a way that the center of mass of the system remains fixed. Taking moments about the center of the cart at the end of the process,

$100(2) = 1{,}100x$

$x = \tfrac{2}{11}$ ft

The tank moves to the left $\tfrac{2}{11}$ ft.

Equation (10-37) is equally valid for a fluid moving in a flow field relative to an inertial reference. The fluid mass is

$$m = \int_{\mathscr{V}} \rho \, d\mathscr{V} \tag{10-38}$$

Therefore, the momentum equation for a fluid *system* is

$$\sum \mathbf{F} = \frac{d}{dt} \int_{\mathscr{V}} \mathbf{V} \rho \, d\mathscr{V} \tag{10-39}$$

where $\sum \mathbf{F}$ is the resultant of all body and surface forces exerted by the surroundings on the system. Much more useful, however, is the control-volume formulation presented in the following section.

10-4 INTEGRAL FORM OF THE LINEAR-MOMENTUM EQUATION

Inertial control-volume analysis Following the same procedure as that developed in Sec. 10-1 on the conservation of mass, we select a *nondeformable control volume* such that it is at all times a part of the system. It is fixed with respect to its coordinate frame *xyz*, which means that velocities and forces must be in the inertial (nonaccelerating) reference for the formulation to be valid. Region I contains the mass which just passes into the control volume in time Δt, and region III is defined such that its mass just leaves the control volume in the time interval Δt. As Δt becomes smaller and smaller, the system approaches the control volume, becoming coincident with it in the limit.

BASIC EQUATIONS

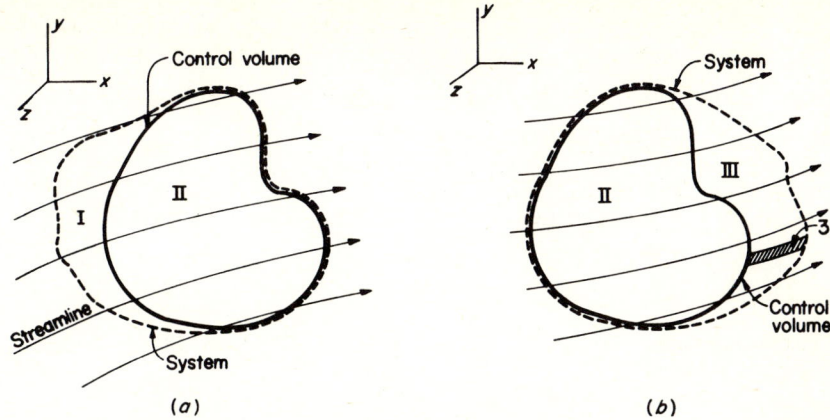

Fig. 10–7 Control volume and system: (*a*) at time t; (*b*) at time $t + \Delta t$.

An arbitrary velocity field $\mathbf{V}(x,y,z,t)$, acting on the fluid elements as they pass into and out of the control volume, gives rise to momentum transport, since momentum $\mathbf{M}$ is the product of mass and velocity ($\mathbf{M} = m\mathbf{V}$). Since the momentum flux is fundamental to Newton's second law, we seek a formulation of this law of motion which is applicable to a control volume.

Referring to Fig. 10-7, the momentum of regions I and II at time t differs from the momentum of regions II and III at time $t + \Delta t$ by the force $\mathbf{F}$ which acts during the time interval Δt. The change in momentum during the time interval Δt is

$$\Delta \mathbf{M} = [(m\mathbf{V})_{\text{II},t+\Delta t} + (m\mathbf{V})_{\text{III},t+\Delta t}] - [(m\mathbf{V})_{\text{I},t} + (m\mathbf{V})_{\text{II},t}] \tag{10-40}$$

Rearranging, dividing by Δt, and taking the limit as Δt approaches zero, we get

$$\lim_{\Delta t \to 0} \frac{\Delta \mathbf{M}}{\Delta t} = \lim_{\Delta t \to 0} \frac{(m\mathbf{V})_{\text{II},t+\Delta t} - (m\mathbf{V})_{\text{II},t}}{\Delta t} + \lim_{\Delta t \to 0} \frac{(m\mathbf{V})_{\text{III},t+\Delta t} - (m\mathbf{V})_{\text{I},t}}{\Delta t} \tag{10-41}$$

Or, from the definition of the derivative,

$$\sum \mathbf{F} = \frac{d\mathbf{M}}{dt} = \frac{d}{dt}(m\mathbf{V})_{\text{cv}} + \lim_{\Delta t \to 0} \frac{(m\mathbf{V})_{\text{III},t+\Delta t} - (m\mathbf{V})_{\text{I},t}}{\Delta t} \tag{10-42}$$

where $(m\mathbf{V})_{\text{cv}}$ is the instantaneous momentum within the control volume (region II). But the mass contained in the control volume is the density-volume product, giving

$$\frac{d}{dt}(m\mathbf{V})_{\text{cv}} = \frac{d}{dt} \int_{\text{cv}} \rho \mathbf{V} \, d\mathscr{V} \tag{10-43}$$

To evaluate the last term of Eq. (10-42), consider an enlargement of subregion 3 as shown in Fig. 10-8. The vector $d\mathbf{A}$ has the magnitude of the element of area dA, and the direction is that of the outward-drawn normal to dA. The angle α is the angle between the velocity vector and the outward-drawn normal, and $\alpha < \pi/2$ over the surface common to region III and the control surface, i.e., in the region where mass leaves the control volume. Accordingly, $\cos \alpha$ is positive. If ΔL is the mean length of the subregion, the momentum within the subregion is

$$d(m\mathbf{V})_3 = (\rho \mathbf{V} \, \Delta L \cos \alpha \, dA)_3 \tag{10-44}$$

where ρ is the average value of density, which, in the limit, is the value of ρ at the control surface. Integrating over the control surface common to region III,

$$(m\mathbf{V})_{\text{III},t+\Delta t} = \int_{\text{cs III}} (\rho \mathbf{V} \, \Delta L \cos \alpha \, dA)_{t+\Delta t} \tag{10-45}$$

where $(m\mathbf{V})_{\text{III}}$ is the total momentum contained in region III at time $t + \Delta t$. Similarly,

$$(m\mathbf{V})_{\text{I},t} = -\int_{\text{cs I}} (\rho \mathbf{V} \, \Delta L \cos \alpha \, dA)_t \tag{10-46}$$

where the negative sign characterizes momentum being transferred into the control volume ($\alpha > \pi/2$).

Substituting Eqs. (10-43), (10-45), and (10-46) into Eq. (10-42) gives

$$\sum \mathbf{F} = \frac{d}{dt} \int_{\text{cv}} \rho \mathbf{V} \, d\mathscr{V} + \lim_{\Delta t \to 0} \left[\int_{\text{cs III}} \left(\rho \mathbf{V} \frac{\Delta L}{\Delta t} \cos \alpha \, dA \right)_{t+\Delta t} + \int_{\text{cs I}} \left(\rho \mathbf{V} \frac{\Delta L}{\Delta t} \cos \alpha \, dA \right)_t \right] \tag{10-47}$$

where Δt has been taken under the integrals, since the area of the control surface is independent of the interval of time.

Observing that the speed V of a fluid particle is

$$\lim_{\Delta t \to 0} \frac{\Delta L}{\Delta t} = V \tag{10-48}$$

and that

$$\int_{\text{cs}} \mathbf{V}(\rho V_n \, dA) = \int_{\text{cs III}} \mathbf{V}(\rho V_n \, dA) + \int_{\text{cs I}} \mathbf{V}(\rho V_n \, dA) \tag{10-49}$$

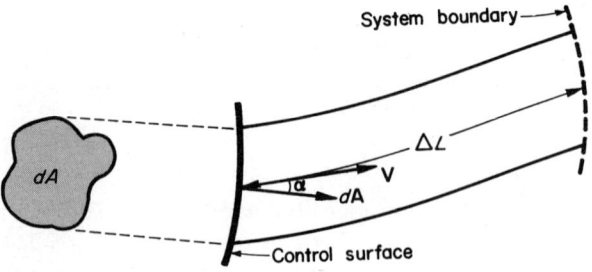

Fig. 10-8 Subregion 3 at time $t + \Delta t$.

BASIC EQUATIONS

where $V_n = V \cos \alpha$, Eq. (10-47) becomes

■ $$\sum \mathbf{F} = \frac{d}{dt} \int_{cv} \rho \mathbf{V} \, d\mathscr{V} + \int_{cs} \mathbf{V}(\rho \mathbf{V} \cdot d\mathbf{A}) \tag{10-50}$$

The force $\sum \mathbf{F}$ is the sum of all external forces acting on the fluid—surface forces acting on the control surface and body forces, such as weight, acting on the distributed mass within the control volume. This equation states that the sum of these forces is equal to the rate of change of momentum inside the control volume plus the rate of efflux of momentum across the control surface. For steady flow it reduces to

$$\sum \mathbf{F} = \int_{cs} \mathbf{V}(\rho \mathbf{V} \cdot d\mathbf{A}) \tag{10-51}$$

Use of the control-volume momentum equation Both $\mathbf{F}$ and $\mathbf{V}$ in Eq. (10-50) are measured with respect to an inertial reference. For engineering calculations the vector equation is rewritten into its three orthogonal scalar momentum equations. In cartesian coordinates

$$\sum F_x = \frac{d}{dt} \int_{cv} \rho u \, d\mathscr{V} + \int_{cs} u(\rho \mathbf{V} \cdot d\mathbf{A}) \tag{10-52a}$$

$$\sum F_y = \frac{d}{dt} \int_{cv} \rho v \, d\mathscr{V} + \int_{cs} v(\rho \mathbf{V} \cdot d\mathbf{A}) \tag{10-52b}$$

$$\sum F_z = \frac{d}{dt} \int_{cv} \rho w \, d\mathscr{V} + \int_{cs} w(\rho \mathbf{V} \cdot d\mathbf{A}) \tag{10-52c}$$

Of paramount importance in using these equations is the sign convention. The signs of F_x, F_y, F_z and u, v, w depend upon the positive directions chosen for the xyz coordinates, respectively. But the sign for the dot product $\mathbf{V} \cdot d\mathbf{A}$ depends upon the orientation of the control surface relative to the velocity at the surface. To illustrate, consider a centrifugal fan, exhausting leftward as shown in Fig. 10-9. The control volume is shown by a dotted line. Air is taken into the fan parallel to the z axis; therefore, it has no inlet component of velocity in the x direction. Since x was chosen positive to the right, the exhaust velocity is negative, $-u_e$. This sign convention is arbitrary, depending upon one's choice of coordinates. On the contrary,

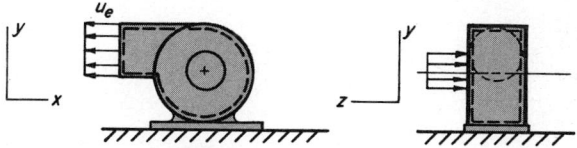

Fig. 10-9 Centrifugal fan, illustrating the sign convention for the momentum equation.

the sign of $\mathbf{V} \cdot d\mathbf{A}$ has nothing to do with the coordinates chosen but depends upon the control volume. For example, the mass flow at the outlet of the fan is $+\rho_e u_e A_e$, assuming one-dimensional flow. This occurs because both the exhaust velocity $\mathbf{V}$ and area vector $\mathbf{A}$ are in the same direction (since $\alpha = 0$, $\cos \alpha = 1$). But since the exhaust-velocity component is negative, the surface integration, the last term of Eq. (10-52a), yields $(-u_e)(+\rho_e u_e A_e)$. To avoid any problem with signs, the mass-flow term in the parentheses of Eq. (10-52) should be considered as an entity itself with respect to the control volume.

The control volume should be chosen to involve the best relation between forces and velocities for getting a solution to a given problem. The choice of the most advantageous control volume is analogous to choosing the best free-body diagram in a dynamics problem to solve for desired forces or stresses. A variety of choices is available just as a variety of free-body diagrams may be chosen in a dynamics problem.

Example What force is exerted in the x direction by the steady exhausting of 300 lb_m/min of air from the centrifugal fan in Fig. 10-9 if the exhaust velocity u_e is 40 ft/sec?

Solution The x-direction momentum equation is

$$\sum F_x = \frac{d}{dt} \int_{cv} \rho u \, d\mathscr{V} + \int_{cs} u(\rho \mathbf{V} \cdot d\mathbf{A})$$

where the change in momentum inside the control volume is zero because of steady flow. Hence,

$$F_x = (-u_e)(+\rho_e u_e A_e) = (-u_e)(m_e)$$

$$F_x = \left(-40 \frac{\text{ft}}{\text{sec}}\right)\left(300 \frac{lb_m}{\text{min}}\right) \frac{lb_f\text{-sec}^2}{32.2 \, lb_m\text{-ft}} \frac{\text{min}}{60 \, \text{sec}} = -6.22 \, lb_f$$

This is the force on the control volume. It may be thought of as the force required to hold the fan in place.

Example An incompressible fluid flows steadily through a reducing elbow as illustrated in Fig. 10-10. Assuming one-dimensional flow, determine the force which must be exerted by the elbow on the fluid.

Solution Let the interior of the elbow be the control volume, cutting the cross sections normal to the duct at stations 1 and 2. The pressure distribution is shown in Fig. 10-11a, where the pressure of the duct wall on the fluid is p_w and τ_w is the shear stress resulting from viscous shear. The force distribution Fig. 10-11b replaces the fluid pressure forces and shear forces, resulting from p_w and τ_w, by equivalent components R_x and R_y, which are assumed positive.

The momentum equation

$$\sum \mathbf{F} = \frac{d}{dt} \int_{cv} \rho \mathbf{V} \, d\mathscr{V} + \int_{cs} \mathbf{V}(\rho \mathbf{V} \cdot d\mathbf{A})$$

reduces to the following component equations for steady flow:

$$\sum F_x = \int_{cs} u(\rho \mathbf{V} \cdot d\mathbf{A})$$

$$\sum F_y = \int_{cs} v(\rho \mathbf{V} \cdot d\mathbf{A})$$

BASIC EQUATIONS

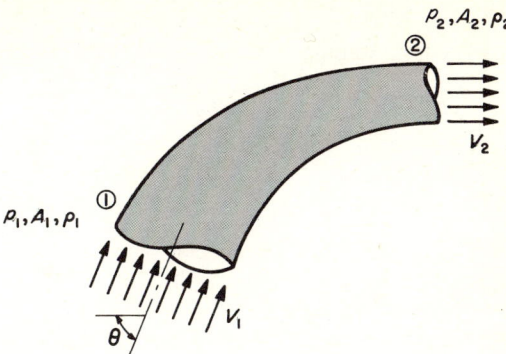

Fig. 10-10 Flow through reducing elbow.

Note that the vector formulation has been retained for the mass-flow rate, since the algebraic sign depends upon the control surfaces in question. Referring to Fig. 10-11b,

$$\sum F_x = p_1 A_1 \cos\theta - p_2 A_2 + R_x = V_1 \cos\theta(-\rho_1 V_1 A_1) + V_2(\rho_2 V_2 A_2)$$
$$\sum F_y = p_1 A_1 \sin\theta - W + R_y = V_1 \sin\theta(-\rho_1 V_1 A_1)$$

Since the flow is steady and incompressible the continuity equation is

$$\rho_1 V_1 A_1 = \rho_2 V_2 A_2$$

Using this in the x-momentum equation and solving for R_x and R_y gives

$$R_x = p_2 A_2 - p_1 A_1 \cos\theta + \rho_1 V_1 A_1(V_2 - V_1 \cos\theta)$$
$$R_y = W - p_1 A_1 \sin\theta - \rho_1 V_1 A_1(V_1 \sin\theta)$$

These are the force components of the elbow *on the fluid*. The vector sum of these components is the desired result. That is,

$$\mathbf{R} = R_x \mathbf{i} + R_y \mathbf{j}$$

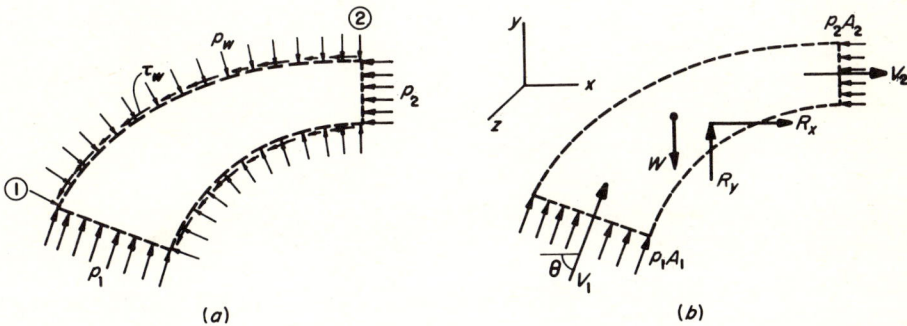

Fig. 10-11 Control volume for reducing elbow: (a) pressure distribution; (b) force distribution.

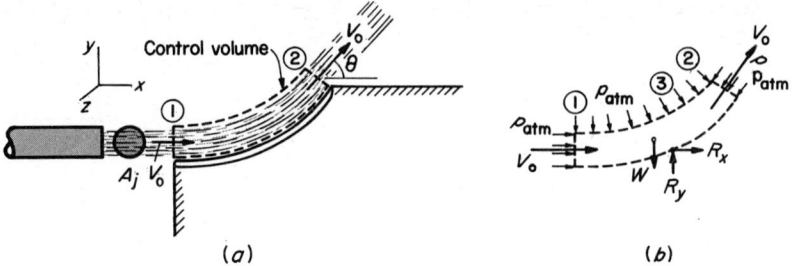

Fig. 10-12 Free jet striking a fixed vane.

Example A jet of water impinges on a smooth fixed vane as illustrated. Find the force **F** exerted on the vane. The flow is steady.

Solution The applicable momentum equations are

$$\sum F_x = \int_{cs} u(\rho \mathbf{V} \cdot d\mathbf{A}) \qquad \sum F_y = \int_{cs} v(\rho \mathbf{V} \cdot d\mathbf{A})$$

The fluid pressure forces and the shear forces of the vane on the jet are replaced by R_x and R_y. Letting A_v be the projected wetted area of the vane,

$$\sum F_x = p_{atm}(A_v)_x + R_x = V_0(-\rho_1 V_0 A_1) + V_0 \cos\theta(\rho_2 V_0 A_2)$$
$$\sum F_y = -p_{atm}(A_v)_y - W + R_y = V_0 \sin\theta(\rho_2 V_0 A_2)$$

since the velocity vector is parallel to control surface 3 ($\cos\alpha = 0$). From continuity

$$\rho V_0 A_j = \rho_1 V_0 A_1 = \rho_2 V_0 A_2$$

where A_j is the area of the jet. Therefore,

$$R_x = -p_{atm}(A_v)_x + \rho V_0^2 A_j (\cos\theta - 1)$$
$$R_y = p_{atm}(A_v)_y + \rho V_0^2 A_j \sin\theta + W$$

The force components *on the vane* are equal and opposite, i.e.,

$$\mathbf{F}_v = -(R_x \mathbf{i} + R_y \mathbf{j})$$

If the vane is exposed to the atmosphere on all sides, as shown in Fig. 10-13, the atmospheric pressure does not enter into the equations and the resulting equations are

$$R_x = \rho V_0^2 A_j (\cos\theta - 1)$$
$$R_y = \rho V_0^2 A_j \sin\theta + W$$

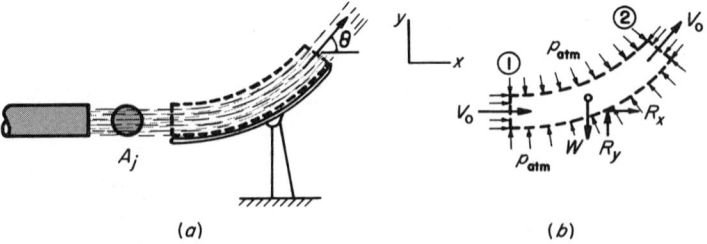

Fig. 10-13 Fixed vane exposed to atmosphere on all sides.

BASIC EQUATIONS

Example Consider a vane and jet of the same configuration as that of the preceding example but let the vane move to the right with a velocity of V_v. For steady flow, what are the force components R_x and R_y of the vane on the fluid?

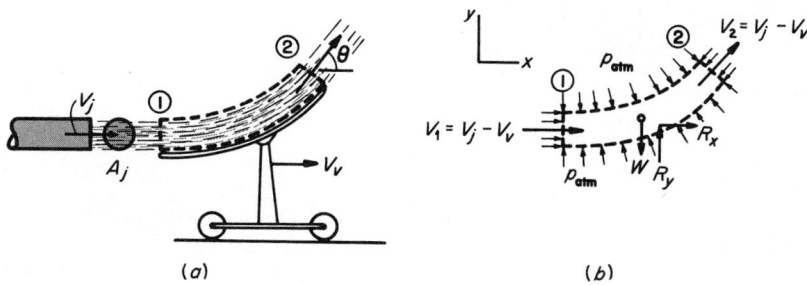

Fig. 10-14 Free jet impinging on moving vane.

Solution Since the vane is moving with a constant (nonaccelerating) velocity, the momentum equation (10-50) applies when the momentum fluxes include the relative velocities as shown in Fig. 10-14b. Proceeding as before,

$$\sum F_x = \int_{cs} u(\rho \mathbf{V} \cdot d\mathbf{A})$$

or

$$R_x = (V_j - V_v)[-\rho_1(V_j - V_v)A_1] + (V_j - V_v)[\rho_2(V_j - V_v)\cos\theta \, A_2]$$

since the atmospheric pressure cancels out. The continuity equation for the control volume is

$$\rho(V_j - V_v)A_j = \rho_1(V_j - V_v)A_1 = \rho_2(V_j - V_v)A_2$$

Therefore,

$$R_x = (V_j - V_v)^2 \rho A_j (\cos\theta - 1)$$

Similarly, for the y direction

$$\sum F_y = \int_{cs} v(\rho \mathbf{V} \cdot d\mathbf{A}) = -W + R_y = (V_j - V_v)[\rho_2(V_j - V_v)\sin\theta \, A_2]$$

$$R_y = (V_j - V_v)^2 \rho A_j \sin\theta + W$$

Noninertial control-volume analysis There are many problems which cannot be reduced to the nonaccelerating control-volume form for which Eq. (10-50) is applicable. Notable examples are found in rocket mechanics. Therefore, we seek a formulation of the momentum principle which will hold for accelerating fluids. Figure 10-15 shows a *control nondeformable volume* moving arbitrarily, translating and rotating, relative to an inertial reference *XYZ*. (Recall that any motion may be described by translation and pure rotation.) The control volume is either fixed in the noninertial reference *xyz* or moving at a constant velocity relative to it.

To understand the kinematics of this relative motion, let us consider the relationship between the derivative of a vector in the two reference systems, since its

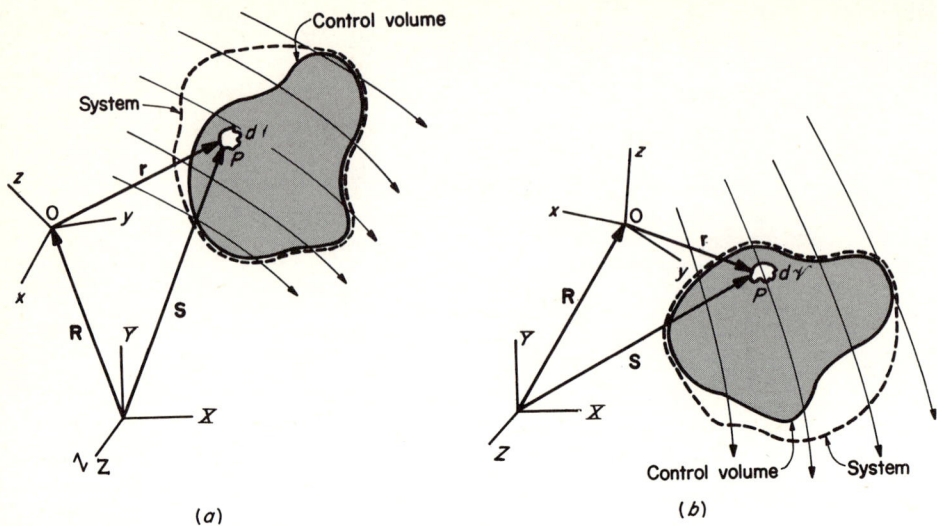

Fig. 10-15 Noninertial control volume and system: (*a*) at time t; (*b*) at time $t + \Delta t$.

value is dependent upon the point of observation. The position vector **r** in the xyz reference may be expressed as

$$\mathbf{r} = \mathbf{i}x + \mathbf{j}y + \mathbf{k}z \tag{10-53}$$

Differentiating with respect to time in the xyz frame

$$\left.\frac{d\mathbf{r}}{dt}\right|_{xyz} = \mathbf{i}\dot{x} + \mathbf{j}\dot{y} + \mathbf{k}\dot{z} \tag{10-54}$$

If we differentiate with respect to time in the XYZ frame, **i**, **j**, and **k** are also functions of time, giving

$$\left.\frac{d\mathbf{r}}{dt}\right|_{XYZ} = (\mathbf{i}\dot{x} + \mathbf{j}\dot{y} + \mathbf{k}\dot{z}) + (\dot{\mathbf{i}}x + \dot{\mathbf{j}}y + \dot{\mathbf{k}}z) \tag{10-55}$$

In order to understand the last term, consider planar rotation as shown in Fig. 10-16, the angular velocity $\boldsymbol{\omega}$ being a vector out of the paper in the positive z direction

$$d\mathbf{i} = \mathbf{j}\, d\theta \tag{10-56}$$

$$\frac{d\mathbf{i}}{dt} = \dot{\mathbf{i}} = \mathbf{j}\frac{d\theta}{dt} = \boldsymbol{\omega} \times \mathbf{i} \tag{10-57a}$$

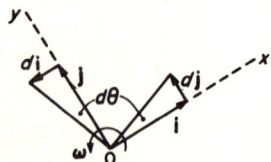

Fig. 10-16 Rotation in a plane.

BASIC EQUATIONS

Similarly,

$$\dot{\mathbf{j}} = \boldsymbol{\omega} \times \mathbf{j} \tag{10-57b}$$

$$\dot{\mathbf{k}} = \boldsymbol{\omega} \times \mathbf{k} \tag{10-57c}$$

Therefore, Eq. (10-55) becomes

$$\left.\frac{d\mathbf{r}}{dt}\right|_{XYZ} = \left.\frac{d\mathbf{r}}{dt}\right|_{xyz} + \boldsymbol{\omega} \times (\mathbf{i}x + \mathbf{j}y + \mathbf{k}z) \tag{10-58}$$

$$\left.\frac{d\mathbf{r}}{dt}\right|_{XYZ} = \left.\frac{d\mathbf{r}}{dt}\right|_{xyz} + \boldsymbol{\omega} \times \mathbf{r} \tag{10-59}$$

Defining the particle velocity as the time derivative of the position vector of the particle in the same reference, the velocities in the two reference systems are

$$\mathbf{V}_{XYZ} = \left.\frac{d\mathbf{s}}{dt}\right|_{XYZ} \tag{10-60a}$$

$$\mathbf{V}_{xyz} = \left.\frac{d\mathbf{r}}{dt}\right|_{xyz} \tag{10-60b}$$

We can relate these velocities by noting that

$$\mathbf{s} = \mathbf{R} + \mathbf{r} \tag{10-61}$$

which upon differentiation with respect to time in the XYZ frame gives

$$\left.\frac{d\mathbf{s}}{dt}\right|_{XYZ} = \left.\frac{d\mathbf{R}}{dt}\right|_{XYZ} + \left.\frac{d\mathbf{r}}{dt}\right|_{XYZ} \tag{10-62}$$

But $(d\mathbf{R}/dt)_{XYZ}$ is the velocity $\dot{\mathbf{R}}$ of some convenient point O in the xyz frame relative to the XYZ reference axes, and $(d\mathbf{r}/dt)_{XYZ}$ is given by Eq. (10-59); therefore

$$\blacksquare \quad \mathbf{V}_{XYZ} = \mathbf{V}_{xyz} + \dot{\mathbf{R}} + \boldsymbol{\omega} \times \mathbf{r} \tag{10-63a}$$

Or, as sometimes written,

$$\mathbf{V}_{\text{abs}} = \mathbf{V}_{\text{rel}} + \dot{\mathbf{R}} + \boldsymbol{\omega} \times \mathbf{r} \tag{10-63b}$$

which is the fundamental equation relating the velocities in the two reference systems.

Defining acceleration in a manner analogous to that of velocity, i.e.,

$$\mathbf{a}_{XYZ} = \frac{d}{dt}(\mathbf{V}_{XYZ})_{XYZ} = \left.\frac{d^2\mathbf{s}}{dt^2}\right|_{XYZ} \tag{10-64a}$$

$$\mathbf{a}_{xyz} = \frac{d}{dt}(\mathbf{V}_{xyz})_{xyz} = \left.\frac{d^2\mathbf{r}}{dt^2}\right|_{xyz} \tag{10-64b}$$

and proceeding as before, the acceleration is [1, 3]

■ $\quad \mathbf{a}_{XYZ} = \mathbf{a}_{xyz} + \ddot{\mathbf{R}} + [\boldsymbol{\omega} \times (\boldsymbol{\omega} \times \mathbf{r})] + (\dot{\boldsymbol{\omega}} \times \mathbf{r}) + (2\boldsymbol{\omega} \times \mathbf{V}_{xyz})$ (10-65)

where $\ddot{\mathbf{R}}$ = acceleration of a convenient point O (origin) in xyz relative to XYZ
$\boldsymbol{\omega}$ = angular velocity of the control volume about an axis through point O
$\dot{\boldsymbol{\omega}}$ = angular acceleration of the control volume about an axis through O

Now we shall apply this formidable equation to a fluid system. Newton's law of motion for a fluid system, Eq. (10-39), can be rewritten

$$\sum \mathbf{F} = \int_{\mathscr{V}} \mathbf{a}_{XYZ} \rho \, d\mathscr{V} \qquad (10\text{-}66)$$

But for convenience, we seek a formulation involving motion with respect to the arbitrarily moving control volume rather than a reference system XYZ (often fixed). Since the right side of Eq. (10-65) involves only quantities which are measured relative to xyz, it can be used in Eq. (10-66) to give the desired formulation. Upon making the substitution and rearranging, we get for the fluid *system*

$$\sum \mathbf{F} - \int_{\mathscr{V}} [\ddot{\mathbf{R}} + \boldsymbol{\omega} \times (\boldsymbol{\omega} \times \mathbf{r}) + (\dot{\boldsymbol{\omega}} \times \mathbf{r}) + (2\dot{\boldsymbol{\omega}} \times \mathbf{V}_{xyz})] \rho \, d\mathscr{V} = \int_{\mathscr{V}} \mathbf{a}_{xyz} \rho \, d\mathscr{V}$$
(10-67)

Noting that

$$\int_{\mathscr{V}} \mathbf{a}_{xyz} \rho \, d\mathscr{V} = \frac{d}{dt} \int_{\mathscr{V}} \mathbf{V}_{xyz} \, dm = \left.\frac{d\mathbf{M}_{xyz}}{dt}\right|_{\text{sys}} \qquad (10\text{-}68)$$

the right-hand side of Eq. (10-68) can be cast in the form applicable to a control volume by returning to the formulation of Eq. (10-42). Then by analogy with Eq. (10-50), we can write the *integral form of the linear-momentum equation for a noninertial control volume:*

■ $\quad \sum \mathbf{F} - \int_{\text{cv}} [\ddot{\mathbf{R}} + \boldsymbol{\omega} \times (\boldsymbol{\omega} \times \mathbf{r}) + (\dot{\boldsymbol{\omega}} \times \mathbf{r}) + (2\boldsymbol{\omega} \times \mathbf{V}_{xyz})] \rho \, d\mathscr{V}$

$$= \frac{d}{dt} \int_{\text{cv}} \rho \mathbf{V}_{xyz} \, d\mathscr{V} + \int_{\text{cs}} \mathbf{V}_{xyz} (\rho \mathbf{V}_{xyz} \cdot d\mathbf{A}) \qquad (10\text{-}69)$$

The limits of integration have been taken in terms of the control volume, since the *system* of Eq. (10-67) occupies the same volume as the *control volume* of Eq. (10-69) as Δt approaches zero.

The use of this equation is analogous to that of Eq. (10-50) for an inertial control volume. $\sum \mathbf{F}$ includes the body and surface forces independent of those arising from particle action and reaction and independent of coordinate motion. The terms under the volume integral on the left-hand side of Eq. (10-69) produce

BASIC EQUATIONS

imaginary forces, since

$m\ddot{\mathbf{R}}$ = inertial force due to mass of fluid inside control volume at any time relative to XYZ

$m[\boldsymbol{\omega} \times (\boldsymbol{\omega} \times \mathbf{r})]$ = centrifugal (normal) force relative to O

$m(\dot{\boldsymbol{\omega}} \times \mathbf{r})$ = tangential force relative to O

$m(2\boldsymbol{\omega} \times \mathbf{V}_{xyz})$ = Coriolis force relative to xyz

Example A rocket having an initial mass m_0 (including fuel) is fired from rest at the earth's surface. The velocity of the exhaust gases relative to the rocket V_e and the mass efflux rate $\dot{m}$ are constant until burnout. The discharge pressure p_e of the rocket exhausts the gases, through a discharge area A_e, into ambient surroundings of pressure p_a. (a) Assuming that the aerodynamic resistance (pressure and shear drag) is proportional to the time after firing, that is, $F_D = Kt$, where K is a constant and t is in seconds, derive an expression for the velocity of the rocket relative to the earth V_R. (b) Neglecting air resistance, what is the rocket's acceleration with respect to the earth, dV_R/dt, after 10 sec for the following conditions?

$\dot{m} = 20 \text{ lb}_m/\text{sec}$ $\quad p_e = p_a = 12 \text{ psia}$

$V_e = 3{,}000 \text{ ft/sec}$ $\quad m_0 = 300 \text{ lb}_m$

Solution (a) Selecting a control volume which includes and moves with the rocket, Eq. (10-69) reduces to

$$\sum F_y - m\ddot{R} = \frac{d}{dt}\int_{cv} \rho V_{xyz}\, d\mathscr{V} + \int_{cs} V_e(\rho \mathbf{V}_e \cdot d\mathbf{A})$$

since the rocket's acceleration $\ddot{R}$ is not a function of the control volume (coordinates xyz) and there is no rotation, that is, $\boldsymbol{\omega} = 0$ and $\dot{\boldsymbol{\omega}} = 0$. Since $\mathbf{V}_{xyz}$ remains constant relative to the

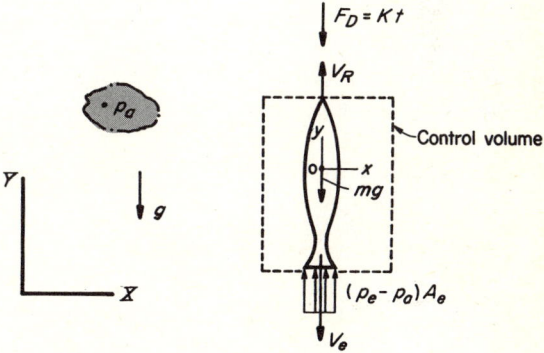

Fig. 10–17 Rocket fired from rest.

noninertial reference frame, the derivative of the volume integral is zero. The vertical forces are

$$\sum F_y = (p_e - p_a)A_e - mg - Kt$$

Therefore, the equation simplifies to

$$(p_e - p_a)A_e - mg - Kt - m\ddot{R} = -\dot{m}V_e$$

since $\dot{m} = \rho_e V_e A_e$. Noting that

$$\ddot{R} \equiv \frac{dV_{XYZ}}{dt} = \frac{dV_R}{dt}$$

and from continuity

$$m = m_0 - \dot{m}t$$

we get

$$\frac{dV_R}{dt} = \frac{(p_e - p_a)A_e}{m_0 - \dot{m}t} - g - \frac{Kt}{m_0 - \dot{m}t} + \frac{\dot{m}V_e}{m_0 - \dot{m}t}$$

Assuming that the acceleration of gravity is constant during the time under consideration, we can integrate the equation with the initial condition $V = 0$ at $t = 0$, getting

$$V_R = \left[V_e + \frac{(p_e - p_a)A_e}{\dot{m}} - \frac{Km_0}{\dot{m}^2}\right]\ln\frac{m_0}{m_0 - \dot{m}t} + \left(\frac{K}{\dot{m}} - g\right)t$$

(b) For the given conditions the expression for the acceleration simplifies to

$$\frac{dV_R}{dt} = \frac{\dot{m}V_e}{m_0 - \dot{m}t} - g$$

Assuming that the rocket is sufficiently close to the earth for g to be 32 ft/sec²,

$$\frac{dV_R}{dt} = \frac{(20\ \text{lb}_m/\text{sec})(3{,}000\ \text{ft/sec})}{(300 - 200)\text{lb}_m} - 32.2\ \text{ft/sec}^2 = 567.8\ \text{ft/sec}^2$$

10-5 DIFFERENTIAL FORM OF THE LINEAR-MOMENTUM EQUATION

Following the procedure established in the derivation of the differential form of the continuity equation, let us apply Eq. (10-52a) to an infinitesimal control volume fixed with respect to frame xyz. Using an average value of the density-velocity product, the control-volume term applied to the differential element becomes

$$\frac{d}{dt}\int_{cv}\rho u\,d\mathcal{V} = \frac{\partial}{\partial t}(\rho u)\bigg|_{av}\Delta x\,\Delta y\,\Delta z \tag{10-70}$$

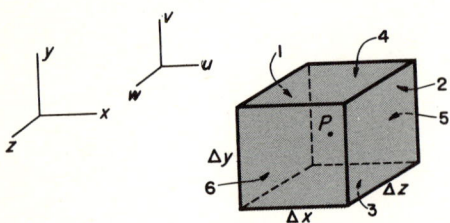

Fig. 10-18 Differential control volume.

BASIC EQUATIONS

It should be noted that as the volume element approaches point P in the limit, the average value of ρu approaches the exact value at P.

Assuming the momentum flux to be uniform on each face, the x-direction flux terms become

$$\int_{\text{cs }1} u(\rho \mathbf{V} \cdot d\mathbf{A}) = -u_x(\rho u)_x \, \Delta y \, \Delta z \tag{10-71a}$$

$$\int_{\text{cs }2} u(\rho \mathbf{V} \cdot d\mathbf{A}) = u_{x+\Delta x}(\rho u)_{x+\Delta x} \, \Delta y \, \Delta z \tag{10-71b}$$

$$\int_{\text{cs }3} u(\rho \mathbf{V} \cdot d\mathbf{A}) = -u_y(\rho v)_y \, \Delta x \, \Delta z \tag{10-71c}$$

$$\int_{\text{cs }4} u(\rho \mathbf{V} \cdot d\mathbf{A}) = u_{y+\Delta y}(\rho v)_{y+\Delta y} \, \Delta x \, \Delta z \tag{10-71d}$$

$$\int_{\text{cs }5} u(\rho \mathbf{V} \cdot d\mathbf{A}) = -u_z(\rho w)_z \, \Delta x \, \Delta y \tag{10-71e}$$

$$\int_{\text{cs }6} u(\rho \mathbf{V} \cdot d\mathbf{A}) = u_{z+\Delta z}(\rho w)_{z+\Delta z} \, \Delta x \, \Delta y \tag{10-71f}$$

Adding the momentum-flux terms over the entire control surface gives

$$\int_{\text{cs}} u(\rho \mathbf{V} \cdot d\mathbf{A}) = [(\rho u u)_{x+\Delta x} - (\rho u u)_x] \, \Delta y \, \Delta z + [(\rho v u)_{y+\Delta y} - (\rho v u)_y] \, \Delta x \, \Delta z$$
$$+ [(\rho w u)_{z+\Delta z} - (\rho w u)_z] \, \Delta x \, \Delta y \tag{10-72}$$

Letting $\mathbf{f}$ be the average force per unit volume,

$$\sum F_x = (\sum f_x)_{\text{av}} \, \Delta x \, \Delta y \, \Delta z \tag{10-73}$$

Upon substituting Eqs. (10-70), (10-72), and (10-73) into Eq. (10-52a) and dividing by $\Delta x \, \Delta y \, \Delta z$, we get

$$(\sum f_x)_{\text{av}} = \frac{\partial}{\partial t} \rho u \bigg|_{\text{av}} + \frac{(\rho u u)_{x+\Delta x} - (\rho u u)_x}{\Delta x}$$
$$+ \frac{(\rho v u)_{y+\Delta y} - (\rho v u)_y}{\Delta y} + \frac{(\rho w u)_{z+\Delta z} - (\rho w u)_z}{\Delta z} \tag{10-74}$$

Taking the limit as Δx, Δy, and Δz approach zero,

$$\sum f_x = \frac{\partial}{\partial t} \rho u + \frac{\partial}{\partial x} \rho u u + \frac{\partial}{\partial y} \rho v u + \frac{\partial}{\partial z} \rho w u \tag{10-75}$$

By differentiating the products and rearranging, the equation becomes

$$\sum f_x = \rho \left(\frac{\partial u}{\partial t} + u \frac{\partial u}{\partial x} + v \frac{\partial u}{\partial y} + w \frac{\partial u}{\partial z} \right) + u \left(\frac{\partial \rho}{\partial t} + \frac{\partial}{\partial x} \rho u + \frac{\partial}{\partial y} \rho v + \frac{\partial}{\partial z} \rho w \right)$$
$$\tag{10-76}$$

But the first term in parentheses is the substantial derivative, and the second term in parentheses is zero from the continuity equation (10-24). Hence

$$\sum f_x = \rho \frac{Du}{Dt} \qquad (10\text{-}77a)$$

Similarly,

$$\sum f_y = \rho \frac{Dv}{Dt} \qquad (10\text{-}77b)$$

$$\sum f_z = \rho \frac{Dw}{Dt} \qquad (10\text{-}77c)$$

Or, in vector form,

$$\sum \mathbf{f} = \rho \frac{D\mathbf{V}}{Dt} \qquad (10\text{-}78)$$

Note that this result could have been deduced directly from Newton's second law of motion [Eq. (10-30)], since the system and control volume are coincident for infinitesimal considerations [4]. The substantial derivative indicates that the time derivative is taken as one follows the element. Another method for getting the same result is to convert the surface integral of Eq. (10-50) into the volume integral by Gauss' theorem, letting **f** be the average force per unit volume and equating integrands rather than integrals [2]. This method, however, requires tensor operations, which are considered unessential to the understanding of the material presented in this text.

Euler's equation of motion   To complete the formulation of the differential equation of motion, we note that the element of Fig. 10-18 is subjected to the same forces as that of Fig. 4-2—body force due to gravity and surface forces due to pressure only (no shear) i.e.,

$$\sum \mathbf{f} = -\nabla p + \rho \mathbf{g} \qquad (10\text{-}79)$$

Substituting Eq. (10-79) into Eq. (10-78), we get

$$-\nabla p + \rho \mathbf{g} = \rho \frac{D\mathbf{V}}{Dt} \qquad (10\text{-}80)$$

Or, in cartesian coordinates,

$$-\frac{\partial p}{\partial x} + \rho g_x = \rho \left(\frac{\partial u}{\partial t} + u \frac{\partial u}{\partial x} + v \frac{\partial u}{\partial y} + w \frac{\partial u}{\partial z} \right) \qquad (10\text{-}81a)$$

$$-\frac{\partial p}{\partial y} + \rho g_y = \rho \left(\frac{\partial v}{\partial t} + u \frac{\partial v}{\partial x} + v \frac{\partial v}{\partial y} + w \frac{\partial v}{\partial z} \right) \qquad (10\text{-}81b)$$

$$-\frac{\partial p}{\partial z} + \rho g_z = \rho \left(\frac{\partial w}{\partial t} + u \frac{\partial w}{\partial x} + v \frac{\partial w}{\partial y} + w \frac{\partial w}{\partial z} \right) \qquad (10\text{-}81c)$$

BASIC EQUATIONS 315

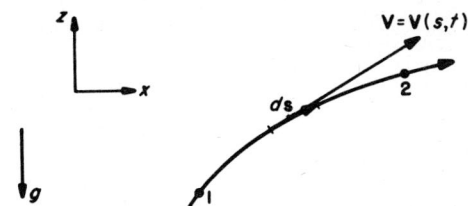

Fig. 10-19 Streamline coordinates.

For application in other coordinate systems, Eq. (10-80) can be rewritten

$$-\frac{1}{\rho}\nabla p + \mathbf{g} = (\mathbf{V}\cdot\nabla)\mathbf{V} + \frac{\partial \mathbf{V}}{\partial t} \qquad (10\text{-}82)$$

where the indicated vector operations are equally valid in other orthogonal systems (see Appendix B). This is the classical *Euler equation of motion*.

Bernoulli's equation along a streamline Many problems in steady or unsteady flow can be solved by considering the instantaneous motion along a streamline. Letting the positive z axis be vertically upward, opposite to g, Eq. (10-80) becomes

$$-\frac{1}{\rho}\nabla p - g\,\nabla z = \frac{D\mathbf{V}}{Dt} \qquad (10\text{-}83)$$

where the gravity force is expressed as $-\rho g\,\nabla z$. Since the velocity is a function of the streamline direction and time, that is, $\mathbf{V} = \mathbf{V}(s,t)$

$$\frac{D\mathbf{V}}{Dt} = V\frac{\partial \mathbf{V}}{\partial s} + \frac{\partial \mathbf{V}}{\partial t} \qquad (10\text{-}84)$$

which may be used in Eq. (10-83), modifying it in terms of the streamline coordinates, i.e.,

$$\frac{1}{\rho}\nabla p + g\,\nabla z + V\frac{\partial \mathbf{V}}{\partial s} + \frac{\partial \mathbf{V}}{\partial t} = 0 \qquad (10\text{-}85)$$

Recalling from vector calculus that the projection of the gradient in any direction is the derivative in that direction, that is, $\nabla\phi\cdot d\mathbf{s} = d\phi$, Eq. (10-85) can be transformed into an equation applicable along a streamline by taking the dot product of each term with the displacement vector $d\mathbf{s}$. Accordingly,

$$\frac{1}{\rho}\nabla p\cdot d\mathbf{s} + g\,\nabla z\cdot d\mathbf{s} + V\frac{\partial \mathbf{V}}{\partial s}\cdot d\mathbf{s} + \frac{\partial \mathbf{V}}{\partial t}\cdot d\mathbf{s} = 0 \qquad (10\text{-}86)$$

$$\frac{dp}{\rho} + g\,dz + d\frac{V^2}{2} + \frac{\partial V}{\partial t}ds = 0 \qquad (10\text{-}87)$$

The last term becomes a scalar product since **V** and d**s** are collinear. With g as a constant we can integrate between a reference point O and any other point along a streamline to get

$$\int_0^p \frac{dp}{\rho} + gz + \frac{V^2}{2} + \int_0^s \frac{\partial V}{\partial t} ds = B(t) \tag{10-88}$$

where $B(t)$, the *Bernoulli function*, is an arbitrary function of time. At any instant the Bernoulli function is equal at all points on the same streamline, although it will vary from streamline to streamline. Carrying out the integration for any two points along the same streamline, we get

$$\blacksquare \quad \int_{p_1}^{p_2} \frac{dp}{\rho} + g(z_2 - z_1) + \frac{V_2^2 - V_1^2}{2} + \int_{s_1}^{s_2} \frac{\partial V}{\partial t} ds = 0 \tag{10-89}$$

which is *Bernoulli's equation for the steady or unsteady flow of a frictionless compressible fluid along a streamline*. When the functional relationship of density-pressure and velocity-time is known, the integration can be carried out.

For the common case of *steady incompressible* flow, Eq. (10-89) reduces to

$$\frac{p_1}{\gamma} + z_1 + \frac{V_1^2}{2g} = \frac{p_2}{\gamma} + z_2 + \frac{V_2^2}{2g} = \text{const} \tag{10-90}$$

where $\gamma = \rho g$. The constant is commonly known as the *total head H*, that is,

$$\blacksquare \quad H = \frac{p}{\gamma} + z + \frac{V^2}{2g} \tag{10-91}$$

having dimensions of length (feet of the fluid). The individual terms p/γ,

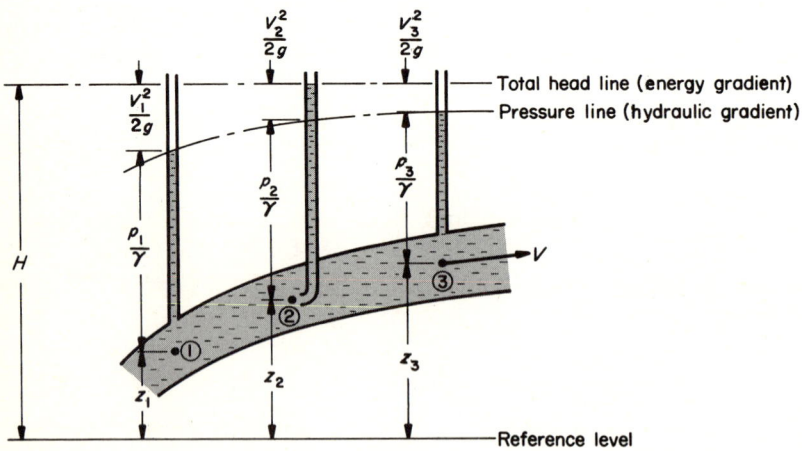

Fig. 10-20 Hydraulic and energy grade lines.

BASIC EQUATIONS

z, and $V^2/2g$ are respectively called *pressure head*, *elevation head*, and *velocity head*. The head is more precisely given in energy per unit mass (ft-lb$_f$/lb$_m$) upon using the gravitational constant g_c. Figure 10-20 graphically depicts the physical significance of the head terms. The sum of pressure head and elevation head, $p/\gamma + z$, is called *piezometric head*.

Example An incompressible frictionless fluid flows unsteadily with velocity $V = V_0 \sin \omega t$ through a constant-area expansion joint. What is the difference in pressure $(p_1 - p_2)$ when time $t = \pi/2\omega$?

Solution Written for a streamline coincident with the centerline of the pipe, Eq. (10-89) reduces to

$$\frac{1}{\rho} \int_{p_1}^{p_2} dp + \frac{\partial V}{\partial t} \int_{s_1}^{s_2} ds = 0$$

since $V \neq V(s)$. Letting $s_2 - s_1 = L$, integration of the equation gives

$$(p_2 - p_1) + \rho L \frac{\partial V}{\partial t} = 0$$

or

$$p_1 - p_2 = \rho L V_0 \omega \cos \omega t$$

When $t = \pi/2\omega$,

$$p_1 - p_2 = 0$$

 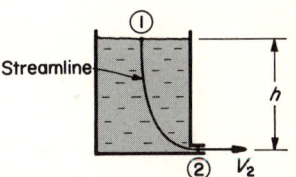

Fig. 10-21 Unsteady flow through expansion joint.

Fig. 10-22 Constant-head tank.

Example What is the efflux velocity of an incompressible fluid from a small orifice in a large constant-head tank?

Solution If we select a streamline which extends from the free surface to the exit orifice, Eq. (10-90) applies, i.e.,

$$\frac{p_1}{\gamma} + z_1 + \frac{V_1^2}{2g} = \frac{p_2}{\gamma} + z_2 + \frac{V_2^2}{2g}$$

The pressure at both 1 and 2 is that of the atmosphere, that is, $p_1 = p_2$, and V_1 is negligible compared to V_2, that is, $V_1 \ll V_2 \rightarrow V_1 \simeq 0$. Hence

$$z_1 - z_2 = \frac{V_2^2}{2g} = h$$

$$V_2 = \sqrt{2gh}$$

Example Water flows steadily through an inclined reducing pipe as shown in Fig. 10-23. What pressure is required at station 1 to deliver 20 ft³/sec to station 2 at a pressure of 1,500 psfa?

Solution Bernoulli's equation for this case of steady incompressible flow is

$$\frac{p_1}{\gamma} + z_1 + \frac{V_1^2}{2g} = \frac{p_2}{\gamma} + z_2 + \frac{V_2^2}{2g}$$

$$p_1 = p_2 + \gamma\left[(z_2 - z_1) + \frac{V_2^2 - V_1^2}{2g}\right]$$

The velocities can be determined directly from the continuity equation, assuming one-dimensional flow, i.e.,

$$Q = VA = 20 \text{ ft}^3/\text{sec} \qquad V_1 = 10 \text{ ft/sec} \qquad V_2 = 40 \text{ ft/sec}$$

Solving for p_1,

$$p_1 = 1{,}500 \text{ psfa} + (62.4 \text{ lb}_f/\text{ft}^3)\left[100 \text{ ft} + \frac{(1{,}600 - 100)\text{ft}^2/\text{sec}^2}{64.4 \text{ ft/sec}^2}\right]$$

$$= 1{,}500 + 6{,}240 + 1{,}455 = 9{,}195 \text{ psfa}$$

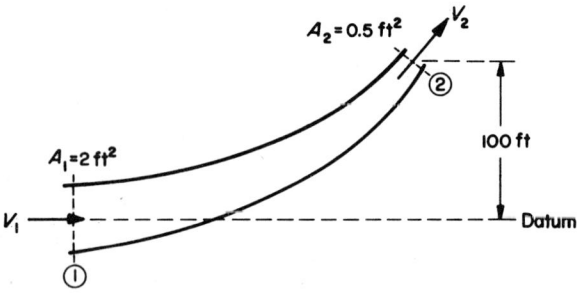

Fig. 10-23 Incompressible flow through reducing pipe.

10-6 MOMENT OF MOMENTUM

Letting **r** be the position vector which locates a particle with respect to a fixed point O in an inertial reference, the moment of the resultant force on the body is given by **r** × **F**. But the resultant force arises because of the time rate of change of the momentum of the body, i.e.,

$$\sum \mathbf{F} = \frac{D}{Dt}\sum m\mathbf{V} \tag{10-92}$$

Hence,

$$\sum \mathbf{M}_0 \equiv \mathbf{r} \times \sum \mathbf{F} = \mathbf{r} \times \frac{D}{Dt}\sum m\mathbf{V} \tag{10-93}$$

which becomes

$$\sum \mathbf{M}_0 = \frac{D}{Dt}(\mathbf{r} \times \sum m\mathbf{V}) \tag{10-94}$$

BASIC EQUATIONS

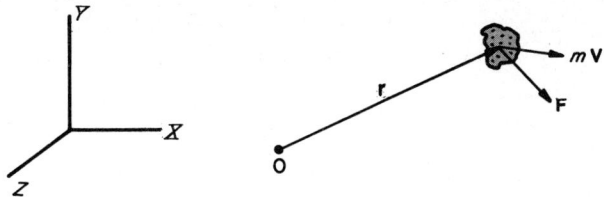

Fig. 10–24 Nomenclature for the moment of momentum of a system.

since

$$\frac{D}{Dt}(\mathbf{r} \times m\mathbf{V}) = \frac{D\mathbf{r}}{Dt} \times m\mathbf{V} + \mathbf{r} \times \frac{D}{Dt} m\mathbf{V} \qquad (10\text{-}95a)$$

$$\frac{D}{Dt}(\mathbf{r} \times m\mathbf{V}) = (\mathbf{V} \times m\mathbf{V}) + \mathbf{r} \times \frac{D}{Dt} m\mathbf{V} \qquad (10\text{-}95b)$$

and $\mathbf{V} \times m\mathbf{V} = 0$.

Equation (10-94) gives the moment of momentum of a particle, a system of particles, or a fluid system if the mass can be considered as concentrated at a point. For an extended fluid *system*

$$\sum \mathbf{M}_0 = \frac{D}{Dt} \int_{\mathcal{V}} (\mathbf{r} \times \mathbf{V}) \rho \, d\mathcal{V} \qquad (10\text{-}96)$$

where $\sum \mathbf{M}_0$ is the sum of all external moments, external meaning not produced by forces within the system resulting from actions and reactions of the system's particles.

By following the procedures which were developed in detail in Sec. 10-4 for converting from a system to a control volume, we get the desired *moment-of-momentum equation for an inertial control volume*:

$$\sum \mathbf{M}_0 = \frac{d}{dt} \int_{cv} (\mathbf{r} \times \mathbf{V}) \rho \, d\mathcal{V} + \int_{cs} (\mathbf{r} \times \mathbf{V})(\rho \mathbf{V} \cdot d\mathbf{A}) \qquad (10\text{-}97)$$

This equation is particularly valuable in the study of turbomachines, where it is more feasible to consider external torques than forces. It is generally more useful, however, to take moments about an axis rather than a point. By taking moments about, say, the Z axis, we can express the position vector and velocities by their counterparts in cylindrical coordinates, giving the *scalar form of the moment-of-momentum equation for an inertial control volume*

$$\blacksquare \quad \sum M_Z = \frac{d}{dt} \int_{cv} (rV_\phi) \rho \, d\mathcal{V} + \int_{cs} (rV_\phi)(\rho \mathbf{V} \cdot d\mathbf{A}) \qquad (10\text{-}98)$$

Figure 10-25 graphically depicts the nomenclature of Eq. (10-98). V_ϕ and r form a plane normal to the Z axis.

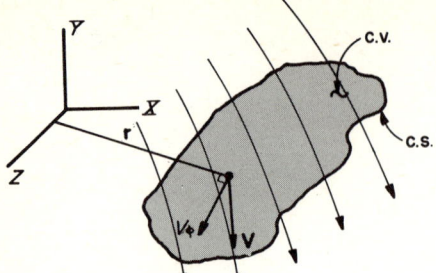

Fig. 10-25 Nomenclature for Eq. (10-98).

For completeness, we give the *moment-of-momentum equation for a noninertial control volume*:

∎ $$\sum \mathbf{M}_0 - \int_{\text{cv}} \{\mathbf{r} \times [\ddot{\mathbf{R}} + \boldsymbol{\omega} \times (\boldsymbol{\omega} \times \mathbf{r}) + (\dot{\boldsymbol{\omega}} \times \mathbf{r}) + (2\boldsymbol{\omega} \times \mathbf{V}_{xyz})]\} \rho \, d\mathscr{V}$$
$$= \frac{d}{dt} \int_{\text{cv}} (\mathbf{r} \times \mathbf{V}_{xyz}) \rho \, d\mathscr{V} + \int_{\text{cs}} (\mathbf{r} \times \mathbf{V}_{xyz})(\rho \mathbf{V}_{xyz} \cdot d\mathbf{A}) \quad (10\text{-}99)$$

Although no derivation or plausibility argument is offered, its development follows the same procedures as previous sections. Application of Eq. (10-99) is identical to that of the other control-volume equations when its terms are evaluated with respect to the moving axes.

Example A constant-head tank feeds a lawn sprinkler which is free to rotate about the Z axis. At time $t = 0$, $\boldsymbol{\omega} = 0$. Neglecting all losses and friction, find an expression for the velocity of the end of the sprinkler arm as a function of time.

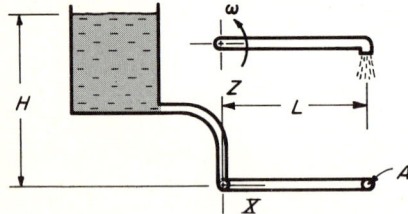

Fig. 10-26 Constant-head lawn sprinkler.

Solution Taking the control volume on the inside of the rotating arm, Eq. (10-98) applies:

$$M_Z = \frac{d}{dt} \int_{\text{cv}} (rV_\phi) \rho \, d\mathscr{V} + \int_{\text{cs}} (rV_\phi)(\rho \mathbf{V} \cdot d\mathbf{A})$$

In the absence of friction at the axis, there are no external moments, that is, $M_Z = 0$. At every point in the control volume at a distance r from the axis an element of fluid has a velocity $V_\phi = r\omega$, where r varies from 0 to L. The fluid velocity at the sprinkler axis, resulting from

BASIC EQUATIONS

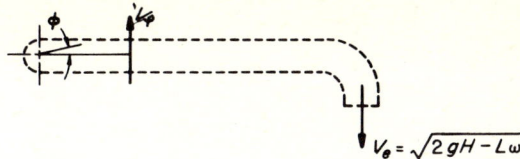

Fig. 10-27 Sprinkler control volume.

the steady-flow incompressible Bernoulli equation, is $\sqrt{2gH}$, as shown in a preceding example. But the fluid does not exit with this velocity, since its exit velocity varies with the motion of the sprinkler arm; it is decreased by the arm velocity $L\omega$. Hence, $V_e = \sqrt{2gH} - L\omega$.

Choosing clockwise as the positive direction and noting that $d\mathscr{V} = A\,dr$, we get for the volume integral

$$\frac{d}{dt}\int_{cv}(rV_\phi)\rho\,d\mathscr{V} = -\rho A\frac{d\omega}{dt}\int_0^L r^2\,dr$$

since the space integration is independent of time. The integral over the control surface at the axis is 0 since the tangential velocity is zero at the axis, i.e.,

$$\int_{cs\ inlet}(rV_\phi)(\rho\mathbf{V}\cdot d\mathbf{A}) = 0$$

The mass flow rate $\dot{m} = \rho\mathbf{V}\cdot d\mathbf{A}$ is the same at the exit as at the entrance, since all fluid which enters must leave, i.e.,

$$\rho\mathbf{V}\cdot d\mathbf{A} = \rho A\sqrt{2gH}$$

and we shall assume that the exit area is small compared to the length of the arm, so that

$$\int_{cs\ exit}(rV_\phi)(\rho\mathbf{V}\cdot d\mathbf{A}) = L(\sqrt{2gH} - L\omega)\rho A\sqrt{2gH}$$

The exit momentum is positive since the exit-velocity vector and the outward-drawn normal are in the same direction, that is, $\cos\alpha = 1$.

Substitution of these values into the moment-of-momentum equation gives

$$0 = -\rho A\frac{d\omega}{dt}\int_0^L r^2\,dr + \rho AL(\sqrt{2gH} - L\omega)\sqrt{2gH}$$

$$\frac{L^3}{3}\frac{d\omega}{dt} = L(\sqrt{2gH} - L\omega)\sqrt{2gH}$$

Separating variables and integrating, we get

$$-\frac{1}{L}\int_0^\omega \frac{-L\,d\omega}{\sqrt{2gH} - L\omega} = \frac{3\sqrt{2gH}}{L^2}\int_0^t dt$$

$$-\ln[\sqrt{2gH} - L\omega]_0^\omega = \frac{3\sqrt{2gH}}{L}t$$

$$\frac{\sqrt{2gH} - L\omega}{\sqrt{2gH}} = \exp\left(\frac{-3\sqrt{2gH}}{L}t\right)$$

But $\omega = V_e/L$; therefore,

$$V_e = \sqrt{2gH}\left[1 - \exp\left(\frac{-3\sqrt{2gH}}{L}t\right)\right]$$

Note that as $t \to \infty$, that is, as a steady state is reached, the velocity reduces to that from a constant-head tank $\sqrt{2gH}$.

CONSERVATION OF ENERGY

10-7 THE FIRST LAW OF THERMODYNAMICS: CONTROL-VOLUME ANALYSIS

In Chap. 3 we learned that the amount of energy a *system* receives from, or yields to, its surroundings is equal to the variation in energy of the system during any transformation, i.e.,

$$\frac{\delta Q}{\delta t} - \frac{\delta W}{\delta t} = \frac{dE}{dt}\bigg|_{\text{sys}} = \frac{DE}{DT} \qquad (10\text{-}100)$$

where the substantial derivative is used to indicate that we must follow the system. We recall that the heat and work terms on the left-hand side of the equation are energy in transition. The energy term on the right-hand side of the equation is proportional to the mass of the system, that is, $E = me$, where e is the "stored" specific (unit) energy. This equation is valid for any substance or combination of substances, solid, liquid, or vapor, "sawdust, rusty nails, or old rubber boots."[1]

Returning to our familiar technique of converting from a system to a nondeformable control volume, a bladed shaft is included in the control volume of Fig. 10-28 to account for the torque transmitted by the part of the shaft inside the control

[1] Often quoted by our colleague Dr. Charles O. Glisson, Professor Emeritus of Mechanical Engineering.

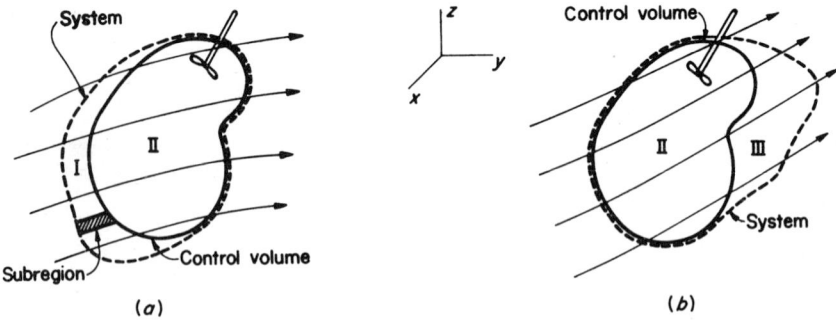

Fig. 10–28 Control volume and system: (*a*) at time t; (*b*) at time $t + \Delta t$.

BASIC EQUATIONS

volume to the part which is outside. This torque results from the fluid forces acting on the rotating shaft.

Observing the analogy between the energy equation for a system [Eq. (10-100)] and Newton's second law for a system

$$\sum \mathbf{F} = \frac{d\mathbf{M}}{dt} \tag{10-29}$$

we note that

$$\mathbf{M} = m\mathbf{V} \qquad \text{is analogous to} \qquad E = me$$

or

$$\mathbf{M} = \mathbf{V}(\rho \mathbf{V} \cdot d\mathbf{A}) \qquad \text{is analogous to} \qquad E = e(\rho \mathbf{V} \cdot d\mathbf{A})$$

Therefore, we can write down the energy equation directly by completing the analogy with the integral form of the momentum equation (10-50), i.e.,

$$\frac{\delta Q}{\delta t} - \frac{\delta W}{\delta t} = \frac{d}{dt} \int_{cv} \rho e \, d\mathscr{V} + \int_{cs} e(\rho \mathbf{V} \cdot d\mathbf{A}) \tag{10-101}$$

where $\delta Q/\delta t$ is the heat-transfer rate for the control volume, since

$$\lim_{t_\Delta \to 0} \frac{\delta Q}{\delta t}\bigg|_{\text{sys}} = \frac{\delta Q}{\delta t}\bigg|_{\text{cv}} \tag{10-102}$$

and

$$\frac{\delta W}{\delta t} = \frac{\delta W_{\text{shaft}}}{\delta t} + \frac{\delta W_{\text{surf}}}{\delta t} \tag{10-103}$$

This work-rate term, consisting of the rate of shaft work and the rate of surface work, requires elaboration. (1) The shaft work is that work done by the system on the surroundings (outside of the control volume) as a result of work being transferred through the control surface. (2) The surface work, sometimes called flow work since it arises as a result of flow, is the work required in deforming the system boundary to make it coincident with the control surface.

To evaluate the surface work rate, consider a subregion of region I as shown in Fig. 10-29. The normal to the surface is designated by n; s designates the tangential direction. The stress vector $\boldsymbol{\tau}$ is

$$\boldsymbol{\tau} = \tau_{nn}\mathbf{n} + \tau_{ns}\mathbf{s} \tag{10-104}$$

where τ_{nn} is the normal stress component (positive for tension) and τ_{ns} is the shear (tangential) stress component; $\mathbf{n}$ and $\mathbf{s}$ are unit vectors. The external force on the element of area dA of the system is

$$d\mathbf{F} = \boldsymbol{\tau} \, dA \tag{10-105}$$

Letting ΔL be the mean length of the subregion, the *work done by the system* in displacing the system's boundary element to coincide with the control-surface element in time Δt is

$$_t(\Delta W_{\text{surf}})_{t+\Delta t} = -d\mathbf{F} \cdot \Delta \mathbf{L} \tag{10-106}$$

The negative sign results from the work being done on the system. Upon substituting Eq. (10-105) into Eq. (10-106), dividing by Δt, and taking the limit as $\Delta t \to 0$, we get

$$\lim_{\Delta t \to 0} \frac{_t(\Delta W_{\text{surf}})_{t+\Delta t}}{\Delta t} = -\boldsymbol{\tau}\, dA \cdot \frac{\Delta \mathbf{L}}{\Delta t} \tag{10-107}$$

$$\delta \frac{\delta W_{\text{surf}}}{\delta t} = -\boldsymbol{\tau} \cdot \mathbf{V}\, dA \tag{10-108}$$

since $\mathbf{V} = \lim_{\Delta t \to 0}(\Delta \mathbf{L}/\Delta t)$. Summing over the entire system boundary, which coincides with the control surface as $\Delta t \to 0$,

$$\frac{\delta W_{\text{surf}}}{\delta t} = -\int_{\text{cs}} \boldsymbol{\tau} \cdot \mathbf{V}\, dA \tag{10-109}$$

Combining Eqs. (10-101), (10-103), (10-104), and (10-109), we get the general form of the energy equation:

$$\frac{\delta Q}{\delta t} - \frac{\delta W_{\text{shaft}}}{\delta t} + \int_{\text{cs}} (\tau_{nn}\mathbf{n} + \tau_{ns}\mathbf{s}) \cdot \mathbf{V}\, dA = \frac{d}{dt}\int_{\text{cv}} \rho e\, d\mathscr{V} + \int_{\text{cs}} e(\rho \mathbf{V} \cdot d\mathbf{A}) \tag{10-110}$$

Because of the difficulty in evaluating the shear work, this form of the energy equation is rarely used. If we choose the control volume such that there is no tangential component of velocity at the control surface, $\int_{\text{cs}} \tau_{ns}\mathbf{s} \cdot \mathbf{V}\, dA = 0$. This requires that the control surface at any section through which mass passes be normal to the velocity vector. If we further require that the streamlines be straight and parallel at the control surface, $\tau_{nn} = -p$, where p is the thermodynamic (static) pressure. It should also be noted that *shear forces perform no work on solid*

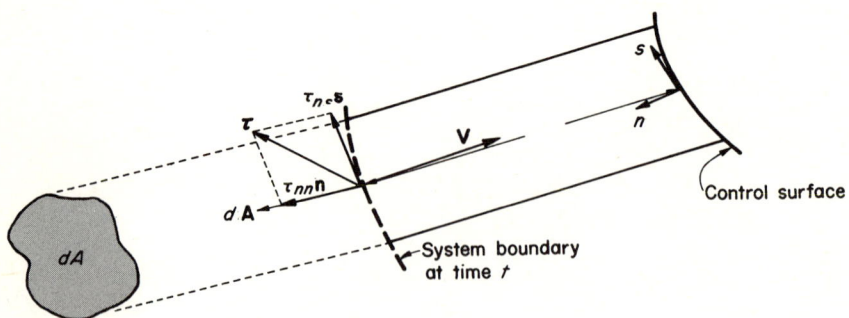

Fig. 10-29 Subregion at time t.

BASIC EQUATIONS

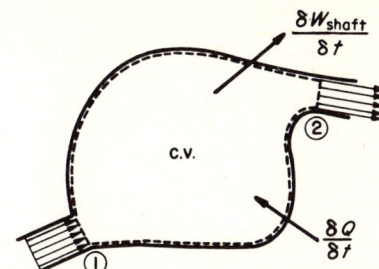

Fig. 10-30 Steady one-dimensional flow.

boundaries, since the velocity of the flow relative to the solid boundary is zero. Under these conditions the surface work rate is

$$\int_{cs} (\tau_{nn}\mathbf{n} + \tau_{ns}\mathbf{s}) \cdot \mathbf{V}\, dA = -\int_{cs} \frac{p}{\rho} (\rho \mathbf{V} \cdot d\mathbf{A}) \tag{10-111}$$

which is valid *for frictionless flows and in cases where the flow enters and leaves the control volume normal to the control surface.* Substituting into Eq. (10-110) and rearranging, we get

$$\frac{\delta Q}{\delta t} - \frac{\delta W_{\text{shaft}}}{\delta t} = \frac{d}{dt} \int_{cv} \rho e\, d\mathscr{V} + \int_{cs} \left(e + \frac{p}{\rho}\right)(\rho \mathbf{V} \cdot d\mathbf{A}) \tag{10-112}$$

Recalling that

$$e \equiv u + \frac{V^2}{2} + gz$$

$$h \equiv u + \frac{p}{\rho}$$

the energy equation can be expressed in its more familiar form as

$$\blacksquare \quad \frac{\delta Q}{\delta t} - \frac{\delta W_{\text{shaft}}}{\delta t} = \frac{d}{dt} \int_{cv} \rho e\, d\mathscr{V} + \int_{cs} \left(h + \frac{V^2}{2} + gz\right)(\rho \mathbf{V} \cdot d\mathbf{A}) \tag{10-113}$$

A large class of problems involves steady one-dimensional flow through a control volume with only one inlet and one outlet, as depicted in Fig. 10-30. For this case, the unsteady term is zero and

$$\int_{cs_1} \left(h + \frac{V^2}{2} + gz\right)(\rho \mathbf{V} \cdot d\mathbf{A}) = -\left(h_1 + \frac{V_1^2}{2} + gz_1\right)\rho_1 V_1 A_1 \tag{10-114}$$

where z_1 is the z coordinate of the centroid of the inlet area. Similarly,

$$\int_{cs_2} \left(h + \frac{V^2}{2} + gz\right)(\rho \mathbf{V} \cdot d\mathbf{A}) = \left(h_2 + \frac{V_2^2}{2} + gz_2\right)\rho_2 V_2 A_2 \tag{10-115}$$

and from continuity

$$\dot{m} = \rho_1 V_1 A_1 = \rho_2 V_2 A_2 \tag{10-116}$$

Substituting Eqs. (10-114) to (10-116) into Eq. (10-113), we get the simplified energy equation

$$\frac{\delta Q}{\delta t} - \frac{\delta W_{\text{shaft}}}{\delta t} = \dot{m}\left[(h_2 - h_1) + \frac{V_2^2 - V_1^2}{2} + g(z_2 - z_1)\right] \tag{10-117}$$

or, per unit mass basis,

$$\delta q - \delta w_{\text{shaft}} = (h_2 - h_1) + \frac{V_2^2 - V_1^2}{2} + g(z_2 - z_1) \tag{10-118}$$

where

$$\delta q = \frac{\delta Q/\delta t}{\dot{m}}$$

$$\delta w_{\text{shaft}} = \frac{\delta W_{\text{shaft}}/\delta t}{\dot{m}}$$

Example Water flows through a turbomachine at the rate of 10 ft³/sec. For the given data, what is the work rate of the turbomachine? Is the turbomachine a pump or a turbine?

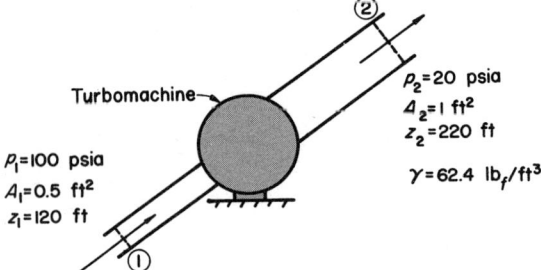

Fig. 10-31 Incompressible flow through a turbomachine.

Solution Assuming one-dimensional flow, we can apply Eq. (10-117) between stations 1 and 2:

$$\underset{\text{assume adiabatic}}{\cancel{\frac{\delta Q}{\delta t}}} - \frac{\delta W_{\text{shaft}}}{\delta t} = \dot{m}\left[(h_2 - h_1) + \frac{V_2^2 - V_1^2}{2} + g(z_2 - z_1)\right]$$

or

$$\frac{\delta W_{\text{shaft}}}{\delta t} = \gamma A V \left[\frac{p_1 - p_2}{\gamma} + \frac{V_1^2 - V_2^2}{2g} + (z_1 - z_2)\right]$$

since there is no change in the water's internal energy.

$$VA = 10 \text{ ft}^3/\text{sec} \qquad V_1 = \frac{10}{A_1} = 20 \text{ ft/sec} \qquad V_2 = \frac{10}{A_2} = 10 \text{ ft/sec}$$

BASIC EQUATIONS

Therefore,

$$\frac{\delta W_{shaft}}{\delta t} = \left(10 \frac{ft^3}{sec}\right) \frac{62.4 \; lb_f}{ft^3} \left[\frac{80(144)}{62.4} ft + \frac{300}{64.4} ft - 100 \; ft\right]$$

$$= \left(55{,}600 \frac{ft\text{-}lb_f}{sec}\right) \frac{sec\text{-}hp}{550 \; ft\text{-}lb_f} = 101 \; hp$$

Since the work rate is positive, the turbomachine is a turbine; i.e., the fluid produces work.

Example A small tank is filled from a very large reservoir which maintains the pressure, temperature, and velocity at the inlet at a constant value. (Liquid petroleum products, such as propane, are sometimes transferred in this manner.) The tank initially contains mass m_i at p_i, ρ_i, and T_i and fills to a final mass m_f at p_f, ρ_f, and T_f. Associated with the initial mass m_i and final mass m_f is internal energy u_i and u_f, respectively. Accounting for the heat which is transferred during the process, derive an equation relating the initial and final quantities.

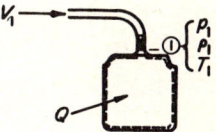

Fig. 10–32 Control-volume analysis for tank filling.

Solution Choosing the interior of the tank as the control volume, it is necessary to use Eq. (10-113) since the process is unsteady with respect to the control volume. Therefore,

$$\frac{\delta Q}{\delta t} - \frac{\delta W_{shaft}}{\delta t} = \frac{dE}{dt} + \int_{cs} \left(h + \frac{V^2}{2} + gz\right)(\rho \mathbf{V} \cdot d\mathbf{A})$$

where the volume integral, $\int_{cv} \rho e \, d\mathscr{V}$, has been replaced by its equivalent $E = me$. The shaft work is zero (although it may appear trivial to say so, no shaft crosses the control surface). Mass crosses the control surface at only one point, station 1, and the properties there are constant with time. Also, for the unsteady flow process,

$$\int_{cs} (\rho \mathbf{V} \cdot d\mathbf{A}) = - \frac{dm}{dt}\bigg|_{cv}$$

Therefore, the resulting equation is

$$\frac{\delta Q}{\delta t} = \frac{dE}{dt} - \left(h_1 + \frac{V_1^2}{2}\right)\frac{dm}{dt}$$

where the potential energy has been neglected. Multiplying by dt and integrating, we get

$$_iQ_f = (E_f - E_i) - \left(h_1 + \frac{V_1^2}{2}\right)(m_f - m_i)$$

But the only energy the fluid in the tank has is internal energy; therefore, $E = mu$. Hence for the *tank-filling process*

$$_iQ_f = (m_f u_f - m_i u_i) - \left(h_1 + \frac{V_1^2}{2}\right)(m_f - m_i)$$

If the tank is initially empty and insulated, the equation reduces to

$$u_f = h_1 + \frac{V_1^2}{2}$$

If the process occurs very slowly, or if the fill line is large enough such that the kinetic energy is negligible, the equation further reduces to

$$u_f = h_1$$

i.e., the final internal energy is equal to the enthalpy in the fill line.

Following the same procedures the energy equation for a *tank-discharge process* is

$$_iQ_f = m_f u_f - m_i u_i + \left(h_2 + \frac{V_2^2}{2}\right)(m_i - m_f)$$

where the subscript 2 refers to the tank exit.

10-8 COMPARISON OF THE FIRST LAW AND BERNOULLI'S EQUATION

For the case of steady incompressible flow in which the shaft work is zero, the Bernoulli equation is

$$\frac{dp}{\rho} + g\,dz + d\frac{V^2}{2} = 0 \qquad (10\text{-}119)$$

from Eq. (10-87). But from Eq. (10-118), the energy equation can be expressed in its differential form as

$$\delta q = dh + g\,dz + d\frac{V^2}{2} \qquad (10\text{-}120)$$

Recalling that

$$h \equiv u + pv \qquad (3\text{-}31)$$

and writing Eq. (3-70) per unit mass

$$dh = du + p\,dv + v\,dp$$

the energy equation becomes

$$\delta q = du + p\,dv + v\,dp + g\,dz + d\frac{V^2}{2} \qquad (10\text{-}121)$$

But upon combining Eqs. (3-66) and (3-69),

$$\delta q = du + p\,dv \qquad (10\text{-}122)$$

and Eq. (10-121) becomes

$$v\,dp + g\,dz + d\frac{V^2}{2} = 0 \qquad (10\text{-}123)$$

BASIC EQUATIONS

which is identical to Eq. (10-119), since $v \equiv 1/\rho$. What is the difference in these two equations?

Bernoulli's equation, which was derived from Newton's second law of motion, is valid only for frictionless flow—in fluids which do not deform under shear. The energy equation is valid for any substance or combination of substances. Therefore, *the Bernoulli and energy equations are equivalent only for the case of steady incompressible flow of a frictionless fluid.* It should be noted that no further comments are necessary regarding shaft work, since a rotating shaft would have no effect on a frictionless fluid; i.e., no torque would be transmitted across a control surface. If any of these conditions are not met, the energy equation (first law of thermodynamics) and the Bernoulli equation (from the momentum equation) are independent equations and must be satisfied separately.

PROBLEMS

10-1. A 2-in-diam nozzle is attached to the end of a 4-in-diam pipe. The velocity in the pipe is 10 ft/sec. What is the discharge velocity?

10-2. Castor oil flows upward through a vertical 1-in-diam pipe at 2 ft/sec. At a section of the same line 40 ft higher the cross section of the pipe is rectangular, $\frac{1}{4}$ by π in. What is the velocity at this point in feet per second?

10-3. A 4- by 4-in conduit discharges gasoline into a 3-in-diam pipe, and both flow full. What is the relationship between the velocities in the two pipes?

10-4. A pair of 12-in-diam cymbals are brought together coaxially at a relative speed of 20 ft/sec. At what radial velocity does the air pass the perimeter of the cymbals when they are $\frac{1}{4}$ in apart?

10-5. An incompressible fluid flows steadily through a duct. It strikes a smooth flat plate inclined 45° from the inlet flow direction. Assume one-dimensional flow. What is the inlet velocity V_1?

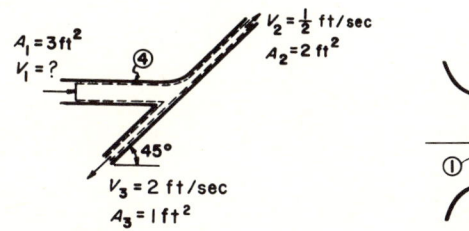

Fig. P10-5

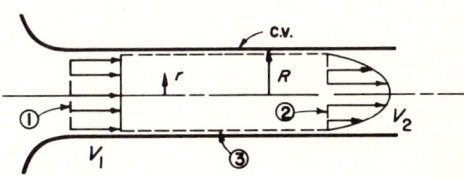

Fig. P10-6

10-6. An inlet-region laminar flow in a round duct usually begins with slug flow at the inlet, and the velocity profile develops to a parabolic form. The velocity at station 2 is given by $V_2 = 6(1 - r^2/R^2)$ ft/sec. The duct radius is 4 in, and the fluid is incompressible. Find V_1.

10-7. Consider water flow over a wide plate. The inlet flow has a slug velocity profile V_1. The velocity V_2 has a linear profile of the form $V_2 = 4y$ ft/sec. The fluid depth is 1 ft, and plate width is 1 ft. What is the inlet velocity V_1?

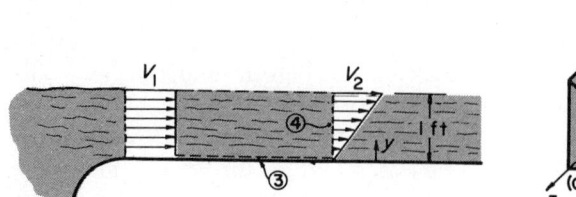

Fig. P10-7   Fig. P10-8

10-8. A two-dimensional velocity field is represented by
$$V = (-2x - 4t^2)i + (2x + 2y)j \quad \text{ft/sec}$$
The control volume to be analyzed is shown; x, y, and z are given in units of feet, and t is in seconds. Is the incompressible continuity equation satisfied?

10-9. The velocity field for a two-dimensional incompressible steady flow is given by
$$V = (-2x - 4t^2)i + (2x + 2y)j$$
The control volume of interest is shown.
  (a) What is the mass flow rate out of the control volume through surface A_1?
  (b) What is the x-momentum flux out of the control volume through surface A_2?

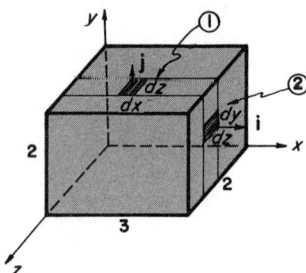

Fig. P10-9

10-10. An incompressible fluid flows radially outward from a point source. Assume that the velocity depends only upon r, the distance from the source. Derive an expression for the velocity as a function of the radius and the flow rate.

10-11. Water is flowing through a large pipe having a diameter of 5 ft. The velocity of the water relative to the pipe is given by $V = 6.25 - r^2$ ft/sec. What is the average velocity of the water leaving by the smaller pipe having an ID of 1 ft?

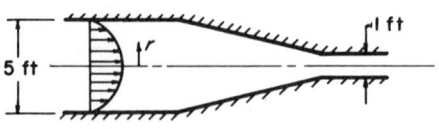 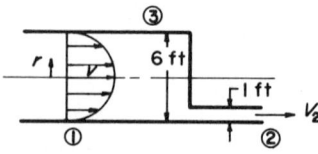

Fig. P10-11   Fig. P10-12

BASIC EQUATIONS

10-12. Water is flowing through a large tank having a diameter of 6 ft. The velocity of the water relative to the tank is $V = 9 - r^2$ ft/sec. What is the average velocity of the water leaving the tank through the small pipe having a diameter of 1 ft?

10-13. Water enters a 1-ft square pipe at a rate of 100 ft³/sec. Two of the faces of the duct are porous. On the upper face the water that is added varies parabolically as shown. The leak rate on the bottom surface also varies parabolically as shown. What is the average velocity of the water leaving the pipe?

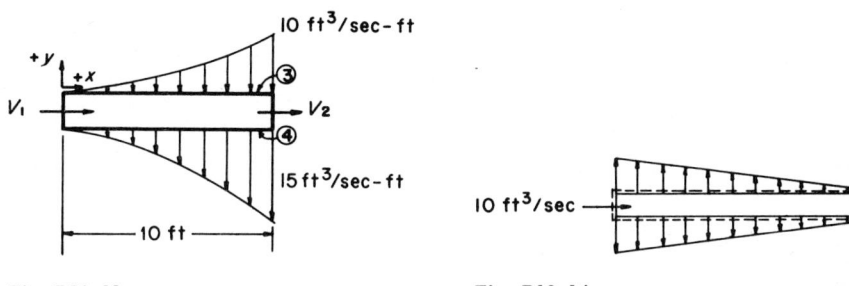

Fig. P10-13   Fig. P10-14

10-14. As shown in Fig. P10-14, 10 ft³/sec of water enters a porous pipe. Water leaves the pipe radially with a velocity that decreases linearly with distance from the pipe inlet. The pipe has a cross section of 0.1 ft² and is 20 ft long. What is the radial velocity 10 ft from the inlet?

10-15. Water enters a porous cylindrical pipe of 1 ft ID at a rate of 500 ft³/sec. The leak rate of the pipe varies parabolically as shown. What is the average velocity of the water leaving the pipe?

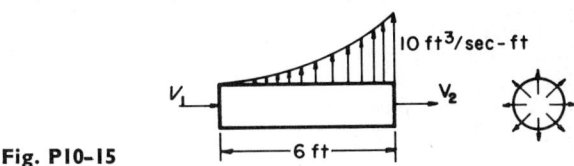

Fig. P10-15

10-16. Does the flow described by $\mathbf{V}(x,y,z) = 2xy\mathbf{i} + (x - z)\mathbf{j} + (y - 2xy)\mathbf{k}$ satisfy the continuity equation for steady incompressible flow?

10-17. Which of the following satisfy continuity for an incompressible fluid flow?

(a) $u = x \cos^2 y \qquad v = -2x \sin y$

(b) $u = x^2 yt \qquad v = x^3 - \dfrac{y^2 t}{2}$

(c) $u = \ln(x + y) \qquad v = \dfrac{y}{x + y}$

(d) $u = x + y \qquad v = x - y$

10-18. A two-dimensional incompressible flow is described by the velocity $\mathbf{V} = 4x\mathbf{i} - (3x^2 + 4y)\mathbf{j}$. Does this equation satisfy continuity?

10-19. A compressible fluid flows steadily through the stream tube shown. What is the differential continuity equation for this flow?

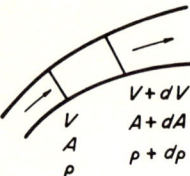

Fig. P10-19

10-20. Derive the differential form of the continuity equation for two-dimensional steady incompressible flow in cylindrical coordinates.

10-21. The velocity of the blade shown is 7.07 ft/sec. The velocity of the water relative to the blade is 10 ft/sec. What is the magnitude and direction of the absolute exit velocity?

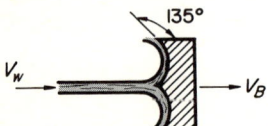

Fig. P10-21

10-22. The gauge pressure in a 6-in-diam fire hose (on a fireboat) just upstream from a 2-in-diam nozzle is 5 psig. The exit velocity is 100 ft/sec.
  (a) What force is exerted on the hose?
  (b) Is the hose in tension or compression?
  (c) Using the jet from the nozzle to aid in maneuvering the fireboat, can more force be obtained by directing the jet against a solid such as a wharf than by allowing it to discharge into the surroundings?

10-23. Gas leaves a jet engine at 4,500 ft/sec relative to the jet pipe, and 1 lb_m of fuel is burned for each 10 lb_m of air. What is the speed of the jet engine for zero thrust?

10-24. A jet engine in flight at 600 mph inducts air at 120 lb_m/sec and burns fuel at 4,500 lb_m/hr. The exhaust jet velocity is 2,000 ft/sec relative to the engine. Assuming $p_1 A_1 = p_2 A_2$, determine the thrust of the engine.

10-25. A circular duct has a mixing section where fluid 2 is injected into fluid 1 by means of two injectors; the densities of both fluids are the same. Fluid 1 enters with velocity V_1 and is thoroughly mixed with fluid 2 before reaching section 3, where the velocity V_3 is assumed to be uniform. If the difference in pressure between fluids 1 and 2 at station 2 is negligible and the wall shear stresses in the mixing section are negligible, determine (a) the velocity V_3 in feet per second and (b) the pressure difference $p_3 - p_2$ in pounds per square inch.

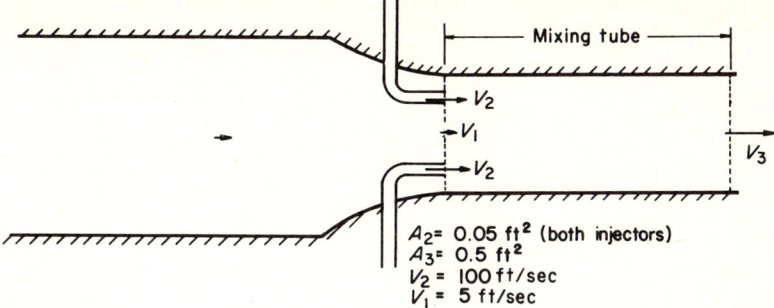

Fig. P10–25

10–26. A jet engine is in flight at 600 mph. It inducts air at 120 lb$_m$/sec ($\rho = 0.076$ lb$_m$/ft^3) and uses fuel at 4,500 lb$_m$/hr ($S = 0.85$). The exhaust velocity is 1,800 ft/sec. Disregarding the pressure-area correction, that is, $p_1A_1 \simeq p_2A_2$, compute the gross thrust of the engine in pounds force.

10–27. Consider the flow of water over a wide plate. The inlet has a slug velocity profile with $V_1 = 2$ ft/sec. Assuming a linear profile of the form $V_2 = 0 + cy$, application of the continuity equation yields $V_2 = 4y$ ft/sec. The fluid depth is 1 ft, and the fluid is water at 70°F. Calculate the force per foot of plate width the water exerts on the plate in the developing region.

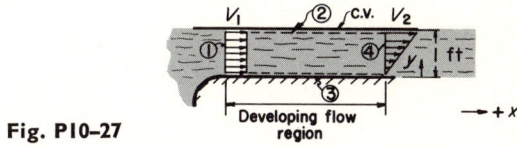

Fig. P10–27

10–28. A high-speed experimental rocket sled utilizes two large water scoops, one on each side, to decelerate. After deployment, each scoop is 4 ft wide and extends 6 in below the water surface of the water troughs. If the sled has a speed of 120 mph when the scoops are deployed, determine (a) the volumetric water flow rate through both scoops in cubic feet per second and (b) the initial braking force due to both scoops.

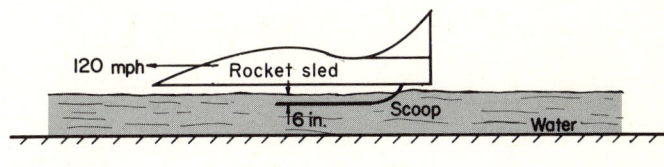

Fig. P10–28

10–29. An incompressible fluid enters a pipe of radius R with a uniform velocity V_1. At section 2 downstream, where the flow is laminar and fully developed, the velocity varies with the radius, that is, $V = V_{\max}(1 - r^2/R^2)$. Show that $V_{\max} = 2V_1$.

10-30. Can a rocket travel faster than the velocity of ejecting gas? Discuss.

10-31. A rocket exhausts gas at 100 lb_m/sec at an exit velocity of 4,800 ft/sec relative to the rocket. The exit pressure is 14.7 psia, and the exit diameter is 1 ft. What is the thrust produced:
  (a) When the rocket is held at rest on the test stand?
  (b) When the rocket travels in a 14.7-psia atmosphere at 800 ft/sec?
  (c) If the rocket flies at an altitude where the ambient pressure is 4 psia?

10-32. A commercial aircraft has three engines in the tail of the craft. To slow the plane when landing, thrust reversers are extended behind each engine to divert the thrust from the engines at an angle of 30°, as shown. On touching down at the edge of a runway the plane is traveling at a velocity of 186 ft/sec, and the brakes fail. The total mass of the plane is 131,000 lb_m. We know that $|\dot{m}V|$ out of each engine is 14,400 lb_f and $|\dot{m}V|$ at the engine inlets is negligible. Neglecting the drag force on the plane, find (a) the time required to bring the plane to a stop in seconds and (b) the distance the plane will travel in that time in feet.

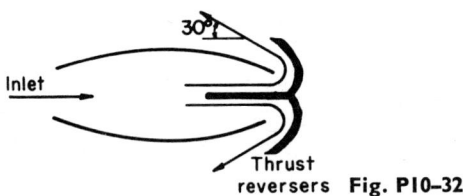

Fig. P10–32

10-33. Wind-tunnel velocity measurements made at a large distance from a cylinder of diameter d, where the pressure was uniform, are as shown. For low velocities (assume incompressible), determine the drag force per unit length on the cylinder in terms of ρ, V_∞, and d.

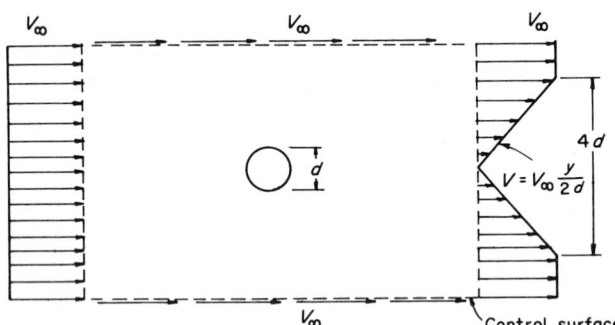

Fig. P10–33

10-34. A helicopter having weight W hovers in midair by imparting momentum downward to a column of air defined by the slipstream boundary shown. Derive an expression for the downward velocity v given to the air by the rotor at a section in the stream below the motor where the pressure is atmospheric and the stream radius is r.

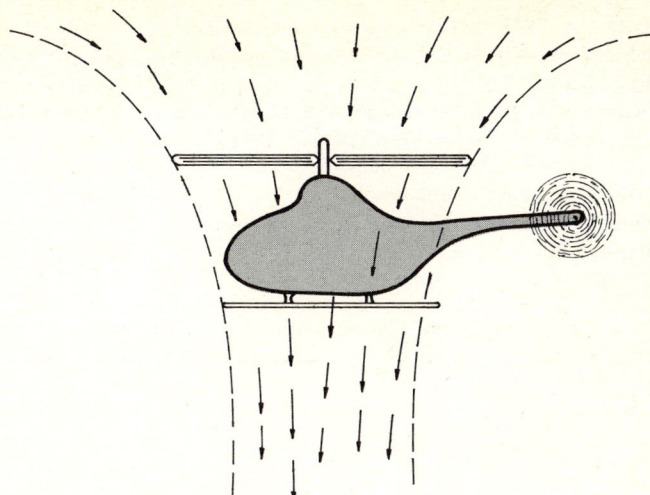

Fig. P10-34

10-35. An incompressible fluid of density ρ flows in an infinitely wide channel of constant depth $2h$. At the inlet section of the channel the velocity is uniform while it is parabolic downstream at distance L. If the pressures at sections 1 and 2 are p_1 and p_2, respectively, determine the total frictional force of the fluid on the channel in terms of p_1, p_2, L, h, and V.

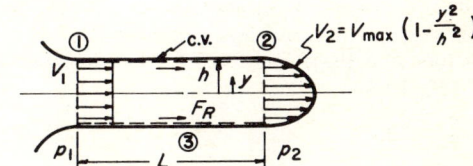

Fig. P10-35

10-36. Water leaves a horizontal nozzle of 0.07-in² area with a velocity of 200 ft/sec and strikes a vertical flat plate. The stream is perpendicular to the plate. Assume all water leaves parallel with the plate. What is the force of the water on the plate?

10-37. The cross section of a sluice gate, used to control the flow of water in channels, is shown. Determine the force per unit width on the gate.

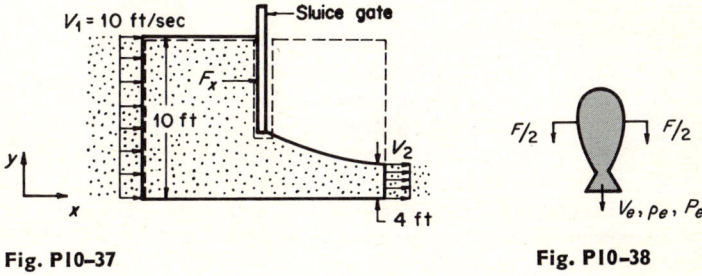

Fig. P10-37

Fig. P10-38

10-38. A rocket is held in position as shown. The atmospheric pressure is p_a and the pressure, density, and velocity at the exit from the nozzle are p_e, ρ_e, and V_e, respectively. What is the thrust of the rocket?

10-39. A gas flows through an expanding section in a pipe. The area increases from 2 to 6 in^2, the velocity decreases from 30 to 10 ft/sec, the pressure increases from 20 to 40 psi, and the gas density at the smaller section is 2 slugs/ft^3; $p_{atm} = 14.7$ psi. What is the net force exerted by the wall in the expanding section on the fluid?

10-40. What force F is required to prevent the sprinkler from rotating about a vertical axis. The density of the fluid is ρ.

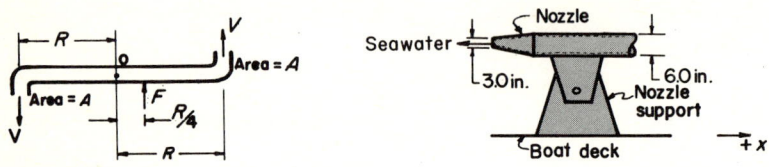

Fig. P10-40   Fig. P10-41

10-41. A fireboat in a harbor is executing a fire drill. It is directing a stream of seawater, $\rho = 64$ lb$_m$/ft^3, from a 3-in-ID nozzle in a horizontal direction $-x$. The exit velocity is 250 ft/sec. Assuming negligible pressure forces, determine the magnitude and direction of the force which the nozzle-support system exerts on the boat deck.

10-42. A jet of water issues from a nozzle at a speed of 10 ft/sec and strikes a stationary flat plate oriented normal to the jet. The exit area of the nozzle is 0.10 ft^2. Determine the horizontal force that the fluid exerts on the plate.

10-43. A liquid jet strikes a curved blade and is deflected through an angle of 75°. The jet velocity is 100 ft/sec, the jet area is 0.2 ft^2, and the liquid density is 62.4 lb$_m$/ft^3. Determine the force exerted by the jet on the vane.

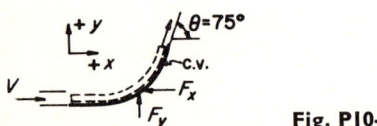

Fig. P10-43

10-44. Air enters a constant right-angle elbow in the vertical direction at a mass flow rate of 5 lb$_m$/sec. It discharges to the atmosphere horizontally through a nozzle having an exit area of 5 ft^2. The density of the air is 0.075 lb$_m$/ft^3. Determine the horizontal force F_x required to keep the elbow stationary.

10-45. Water issues from a pipe embedded in a concrete wall as shown. The velocity leaving the pipe is 30 ft/sec, and the cross-sectional area is 2 in^2. What is the resultant force exerted on the pipe by the wall? Assume that the pressure drop from the wall to the end of the pipe is negligible.

10-46. Water flows steadily through a reducing elbow as shown. The inlet velocity is 5 ft/sec, and the inlet diameter is 6 in. The exit velocity is 10 ft/sec. The pressure p_1 is 20 psia and p_2 is atmospheric pressure. The elbow weighs 5 lb. Neglecting the weight of the water, what force must be applied to the elbow to hold it stationary?

BASIC EQUATIONS

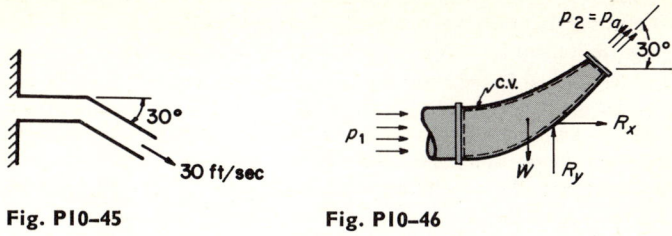

Fig. P10-45   Fig. P10-46

10-47. An aircraft has two engines, each producing 20,000 lb of thrust. The aircraft is flying at 300 ft/sec, and the engine gas discharge is 1,000 ft/sec relative to the aircraft. What is the fuel consumption rate in pounds mass per second if the combustion ratio is 100 parts air to 1 part fuel by mass?

10-48. A jet of water issues from a 1-in-ID nozzle at a velocity of 20 ft/sec and strikes a flat plate on a cart moving away from the jet at 5 ft/sec as shown. What is the force of the water jet on the cart?

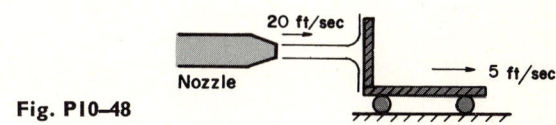

Fig. P10-48

10-49. A 6-in-diam 180° bend carries water ($\rho = 62.4$ lb$_m$/ft^3) at 10 ft/sec and at a pressure of zero gauge. What is the force required to hold the bend in place?

10-50. A cart with frictionless wheels is moving on a horizontal plane with an initial velocity V_i. At time $t = 0$ the cart passes under a pipe, directed vertically downward, through which a fluid of density ρ is flowing at a constant-volume flow rate Q. The flow continues for a total time t_1. If m_c is the mass of the cart and wheels, derive an expression for the velocity V of the cart in terms of V_i, Q, ρ, and t for $0 \leq t \leq t_1$. Neglect air resistance.

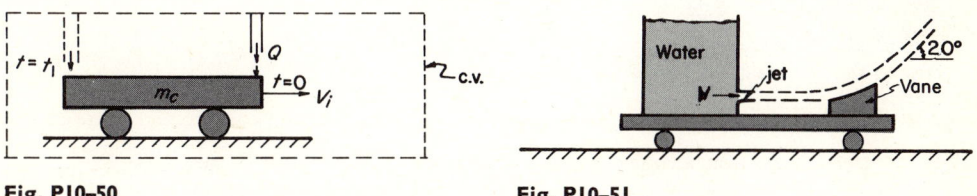

Fig. P10-50   Fig. P10-51

10-51. Water issues from a large tank through a 2.0-in^2 nozzle at a speed of 20 ft/sec relative to the cart to which the tank is attached. The jet then strikes a vane, which turns the direction of flow by an angle of 20° while maintaining the jet speed and cross-sectional area.

(a) Assuming steady flow, determine the force exerted on the cart.

(b) If the cart-tank system has a mass of 1,000 lb$_m$, determine the velocity increase during a 10-sec period. Rolling friction is negligible.

10-52. A rocket which weighs 100,000 lb$_f$, including 60,000 lb$_f$ of fuel, is fired vertically upward. It burns fuel at the rate of 1,000 lb$_m$/sec while exhausting gases at 18,000 ft/sec relative to the rocket. Neglect drag and consider g to be constant at 32.2 ft/sec^2. Determine (a) the speed of the rocket at burnout in feet per second and (b) the maximum height reached by the rocket.

10-53. A rocket is fired from rest along a straight line in outer space where we can neglect air friction and gravitational effects. The fuel has an initial mass m_0, and the mass efflux rate $\dot{m}$ is constant. The mass at any time t after firing is $m = m_0 - \dot{m}t$. The velocity of the exhaust gases relative to the rocket is V_e. The discharge pressure p_e of the rocket exhausts the gases through a discharge area A_e, and the density is ρ_e. The velocity of the rocket relative to the inertial reference is V_R. Determine the motion of the rocket relative to the inertial reference.

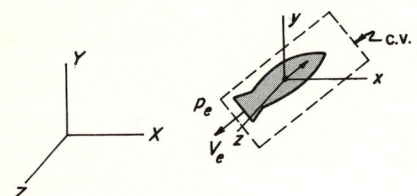

Fig. P10–53

10-54. Water flows steadily through a 6-in-diam fire hose. The nozzle diameter is 3 in. The mass flow rate is 5 ft^3/sec. What is the pressure upstream of the nozzle if the discharge pressure is 100 psfa?

10-55. Water flows in a rectangular channel 5 ft wide at a depth of 2 ft. The channel bottom gradually rises 1 ft. The water surface rises 1.5 ft as the water passes over the raised portion of the channel. Determine the flow in the channel.

10-56. The width of a rectangular channel is reduced from 10 to 6 ft. The depth upstream is 5 ft, and the surface drops 6 in at the contracted section. Determine the flow rate.

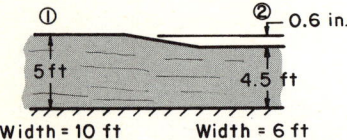

Fig. P10–56

10-57. Gasoline flows from a large open tank through a 1-in-diam hole in its side. The free surface of the gasoline is 25 ft above the centerline of the hole. What is the velocity issuing from the hole? What is the flow rate?

10-58. Water issues from the tank shown as a free jet. A water spray maintains a constant volume in the tank. What is the discharge rate?

10-59. The pressure on the nose of a football moving through air is 25 psia. The air is at $p = 14.7$ psia. What is the velocity of the football?

BASIC EQUATIONS

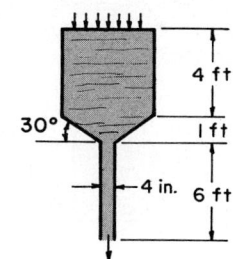

Fig. P10-58

10-60. For the flow of water shown, what is the pressure at point B?

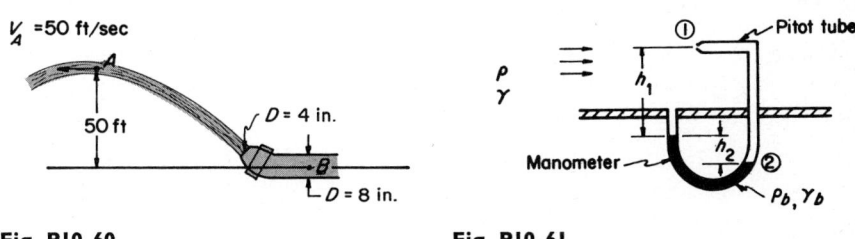

Fig. P10-60

Fig. P10-61

10-61. Derive an expression for the velocity measured by a pitot tube inserted into an airflow as shown.

10-62. One end of a U-tube is oriented directly into the flow of water as shown. The velocity at this point is zero. The pressure at this point is the stagnation pressure. The other end of the U-tube measures the undisturbed pressure at some section in the flow. What is the volumetric flow rate in the pipe?

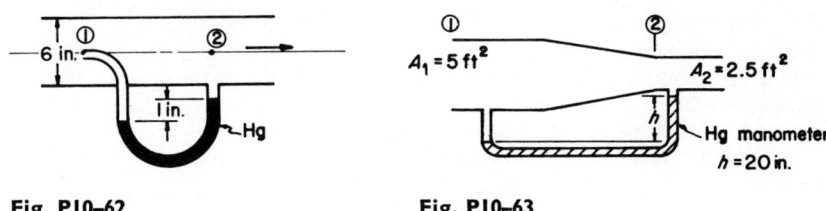

Fig. P10-62

Fig. P10-63

10-63. A duct with an area of 5 ft² gradually contracts to an area of 2.5 ft². The pressure drop between the two sections is measured with a mercury manometer which has a deflection of 20 in. Determine the flow rate through the duct.

10-64. Water flows from a large open reservoir and discharges horizontally into the atmosphere. Find (a) velocity V_3 and (b) velocity V_2.

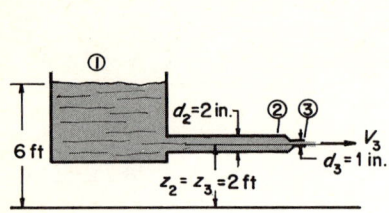

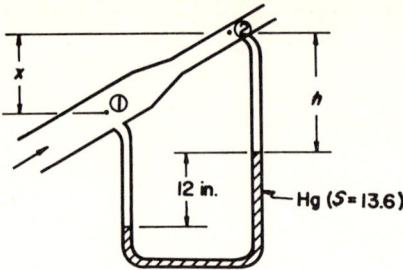

Fig. P10-64   Fig. P10-65

10-65. A 6-in-diam pipeline inclined as shown is connected to a 4-in-diam pipe by a reducer. Water is flowing through the pipe. What is the average velocity V_2?

10-66. Neglecting frictional effects, determine an expression for the discharge Q of an incompressible fluid from a flow nozzle of exit area A_2 in a pipe of area A_1 with a pressure differential of $p_1 - p_2$. Express in terms of the height h of the mercury column.

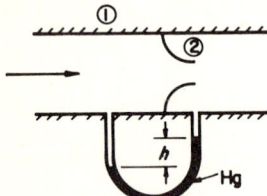

Fig. P10-66

10-67. A 2-in-diam flow nozzle is installed on the end of a 6-in-diam pipe. If the pressure in the pipe is 20 psi, what is the discharge of water in cubic feet per second?

10-68. A 1-in-diam siphon is used to drain gasoline ($S = 0.75$) from a large tank. The highest point in the siphon is 4 ft above the surface of the gasoline, and the siphon discharges at a point 9 ft below the surface.
  (a) What is the discharge rate in cubic feet per second?
  (b) What is the pressure at the highest point in the siphon?

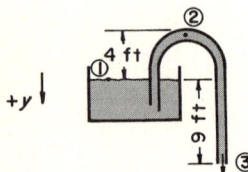

Fig. P10-68

BASIC EQUATIONS 341

10-69. A 2-in-diam jet of water leaves a nozzle at an angle of 45° with the horizontal. If the discharge from the jet is 4 ft³/sec, (a) how high does the jet go and (b) what is the diameter of the jet at its maximum elevation?

10-70. A suction pump operates by evacuating a chamber into which fluid flows and is used to pump water from a well (sea-level conditions exist). What is the maximum depth from which water can be pumped?

10-71. Thieves commonly use a siphon for stealing gasoline ($S = 0.8$). They reason that the siphon time to get a given amount of gasoline will be cut in half by doubling the height h. Is this true? Show by calculation.

10-72. Water at a rate of 5 ft³/sec flows with no frictional loss through the expansion shown. The pressure at 1 is 12 psig. Assume one-dimensional flow. Find the pressure at 2.

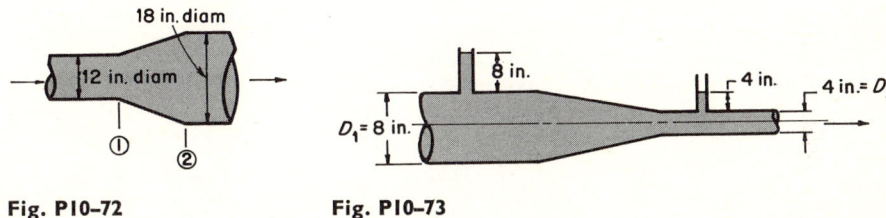

Fig. P10-72   Fig. P10-73

10-73. A frictionless incompressible fluid flows through the device shown. The fluid density is 50 lb$_m$/ft³. Assume one-dimensional flow. What is the mass flow rate?

10-74. Point B is 7,000 ft from A and 40 ft higher. A pipe which has a head loss of 10 ft per 1,000 ft is used to pipe oil ($S = 0.70$) from A to B.

(a) For no negative pressure (gauge) in the line, i.e., the pressure gradient cannot fall below the pipeline, what is the smallest horsepower required for pumping 10 ft³/sec if the pump is 85 percent efficient? Assume that the velocity head is negligible.

(b) What pressure (psig) exists at B?

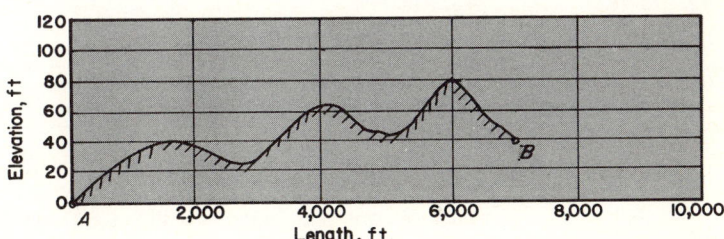

Fig. P10-74

10-75. Prove that $\mathbf{g} = g\nabla z$ when the positive z direction is taken opposite to the direction of $\mathbf{g}$.

10-76. Show that Eq. (10-82) reduces to Eq. (10-80) in the cartesian coordinate system.

10-77. A constant-area U-tube manometer connects two reservoirs containing gas. The valves are initially closed so that the fluid starts at rest at the same height in each leg. The total length of the

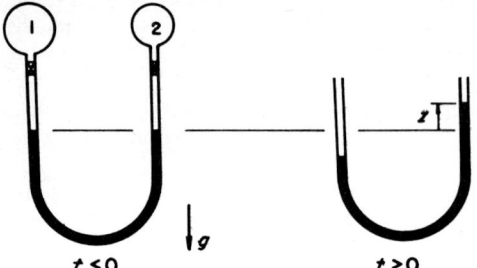

Fig. P10-77

incompressible frictionless-fluid column is L. At time $t = 0$ both valves are suddenly opened, and a pressure differential $p_1 - p_2$ exists from that time on.

(a) For $p_1 > p_2$ derive the differential equation that z must satisfy.
(b) What are the necessary boundary conditions?
(c) Letting $h \equiv (p_1 - p_2)/\rho g$ and transforming variables by letting $y = 2 - h/z$, determine an expression for z.

10-78. An explosion occurs under the ocean causing a spherical wavefront of radius R to move outward as a function of time. No mass crosses this spherical front. Determine an expression for the difference between the pressure at any point r and the initial pressure, $p_r - p_\infty$. *Hint:* Write Bernoulli's equation along a streamline.

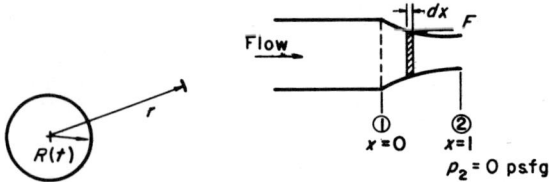

Fig. P10-78   **Fig. P10-79**

10-79. An incompressible frictionless fluid of density $\rho = 2$ slugs/ft^3 flows through a converging section whose area varies as $A = e^{-x}$, where x is in feet and A has units of square feet. For a certain time, at $x = 0$, $V_1 = 10$ ft/sec and $dV_1/dt = 25$ ft/sec. The fluid discharges into the atmosphere at $x = 1.0$ ft. Determine the force of the fluid on the walls of the converging section at that particular time.

10-80. A lawn sprinkler has two arms of 10-in radius with tangentially opposed $\frac{1}{16}$-in-diam nozzles which discharge at 45° with the horizontal. Water enters at the axis of rotation with no tangential component of velocity.

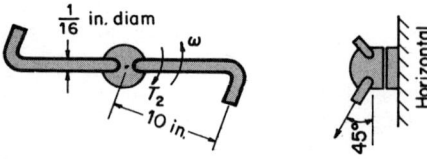

Fig. P10-80

BASIC EQUATIONS

(a) For a flow rate of 1.0 gal/min, what is the net torque on the rotor when held stationary?

(b) If the pivot friction causes a resisting torque of 0.25 in-lb$_f$, what will be the angular velocity of the rotor when the water leaves at a relative velocity of 20 ft/sec?

10–81. Water flows through the impeller of a centrifugal pump as shown. Find the torque exerted on the impeller and the power required to drive it. Assume that the absolute velocity of the water at the inlet has no tangential component.

$Q = 2 \text{ ft}^3/\text{sec}$ $r_2 = 6$ in
$\omega = 100$ rad/sec $b = \frac{1}{2}$ in
$r_1 = 2$ in $\beta = 120°$

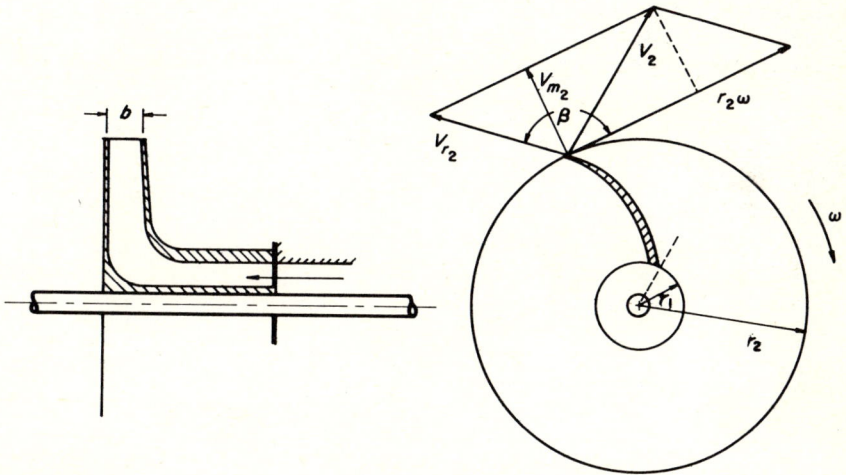

Fig. P10–81

10–82. A circular platform can rotate about the axis O. A jet of water is directed out from the center of the platform while it is stationary and strikes a vane at the outside of the platform. The vane turns the jet 90°. The velocity of the jet is 5 ft/sec, and the radius of the platform is 1 ft. What is the torque developed about the axis O?

10–83. A simple impulse type of turbomachine is shown. A single jet of water issues out of a nozzle and impinges on the system of buckets attached to a wheel. The runner, which is the assembly

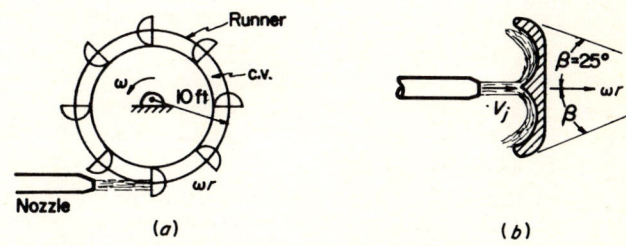

Fig. P10–83

of buckets and wheel, has a radius of 10 ft to the center of the buckets. Note the shape of the bucket. The jet is split into two parts by the bucket and is rotated 155° relative to the bucket. The speed of the water relative to the bucket is a constant 20 ft/sec. The flow rate from the nozzle is 100 ft³/sec, and the runner is loaded by a generator so as to rotate at a constant angular speed of 5 rad/sec. What torque does the water exert on the wheel?

10-84. A Pelton waterwheel has a mean radius of 5 ft, and each bucket has an angle β of 10°. The bucket velocity is 50 percent of the jet velocity. The jet has a diameter of 2 in and a speed of 50 ft/sec. Considering the bucket to be of the type shown in Fig. P10-83, what is the torque developed by the wheel?

10-85. In a reaction turbine (or Francis turbine) (Fig. P10-85) a fluid enters at A and is directed by stationary blades toward the blades on the runner. Here the flow is essentially radial relative to the waterwheel. For a known Q_i and ρ_i at the inlet and a constant mass flow, what is the torque developed on the runner? The inlet velocity is V_1, and the final velocity relative to the runner is V_{r_2}.

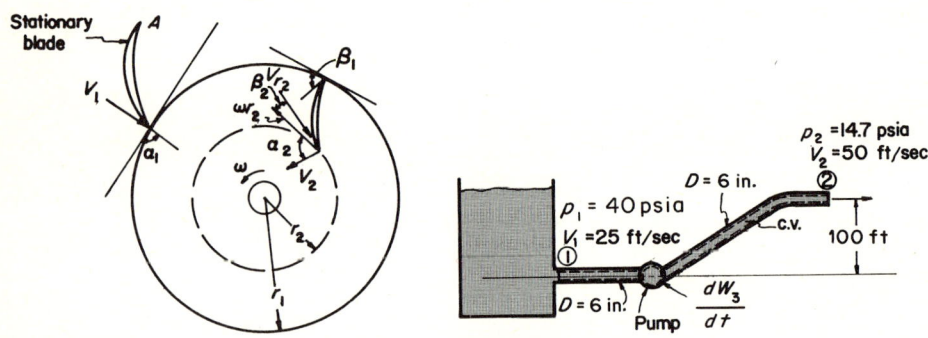

Fig. P10-85   **Fig. P10-86**

10-86. Water is pumped through a pipe to a higher elevation as shown. The inlet and outlet conditions are specified in the figure. Determine the power requirement of the pump which causes the flow.

10-87. The flow rate through a turbine is 10,000 lb$_m$/hr, and the heat loss through the casing is 50,000 Btu/hr. The inlet and exit enthalpies are 1000 and 500 Btu/lb$_m$, respectively. The inlet and exit velocities are 500 and 1,000 ft/sec, respectively. Calculate the shaft horsepower of the turbine.

10-88. Water flows through a pipe with a velocity profile given by $V = 2(1 - r^2/R^2)$ ft/sec. Find the heat transfer in 1 hr to the walls of a 4-in-diam pipe over a section where the pressure drop is 50 psfa.

10-89. Air enters a gas turbine with an average velocity of 500 ft/sec and a temperature of 2000°F; it leaves with a velocity of 800 ft/sec and a temperature of 1500°F. The mass flow rate through the turbine is 50 lb$_m$/sec. The heat loss from the turbine is 250 Btu/hr. Determine the theoretical horsepower of the turbine.

10-90. Fluid is flowing through a pipe which is connected to an initially evacuated tank. When the valve is opened, fluid enters the tank, but this does not appreciably affect the flow in the pipe. If the flow conditions in the pipe are known and the initial and final conditions in the tank are also known, what is the heat transfer during the filling process?

BASIC EQUATIONS

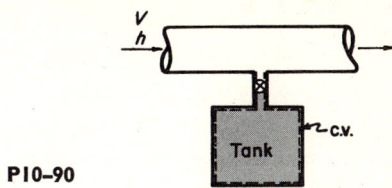

Fig. P10-90

10-91. A steam turbine uses 10,000 lb$_m$/hr of steam while delivering 500 hp to the turbine shaft. The inlet and outlet velocities are 100 and 500 ft/sec, respectively. The inlet and outlet enthalpies have been measured and are 100 and 750 Btu/lb$_m$, respectively. Determine the rate at which heat is lost from the turbine casing.

10-92. A tank of volume $\mathscr{V}$ contains a perfect gas with constant specific heats at a temperature T_i. Heat is to be added to the gas. A pressure relief valve allows gas to escape, maintaining the pressure inside the tank at a constant value. Determine an expression relating the heat addition to the initial temperature T_i and the final temperature T_f.

10-93. Water flows through a system. The inlet has a 1-in diameter, and the 2-in-diam discharge is 5 ft above the inlet. The pressure and velocity of the flow at the inlet are $p_1 = 15$ psia and $V_1 = 20$ ft/sec. The pressure at the outlet is 50 psia. Heat is transferred to the system at the rate of 500 Btu/hr. Assuming that the flow is uniform at the inlet and outlet, find the external shaft power to (or from) the system.

10-94. A compressible-flow machine, using air, has a shaft input power of 750 hp and a heat transfer into the machine of 1000 Btu/sec. The other conditions are shown on the figure. Assuming a steady rate of heat transfer and work input, find the time rate of change of energy stored in the machine.

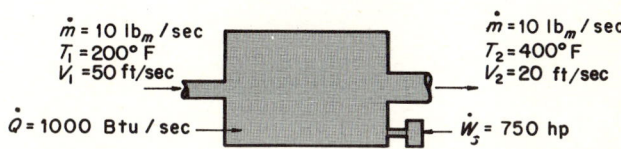

Fig. P10-94

10-95. A gas at an initial temperature T_i flows into an initially evacuated vessel where it reaches a final temperature T_f. The vessel is insulated so that the process is adiabatic. Derive a general expression relating the initial temperature and the final temperature in terms of the thermodynamic properties of the gas.

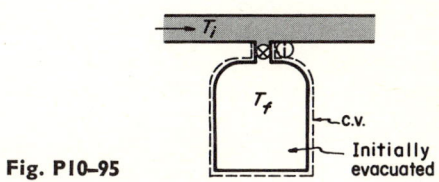

Fig. P10-95

REFERENCES

1. Downey, G. L., and G. M. Smith: "Advanced Dynamics," International Textbook, Scranton, Pa., 1960.
2. Pao, R. H. F.: "Fluid Dynamics," Merrill, Columbus, Ohio, 1967.
3. Shames, I. H.: "Engineering Mechanics," Prentice-Hall, Englewood Cliffs, N.J., 1960.
4. Shames, I. H.: "Mechanics of Fluids," McGraw-Hill, New York, 1962.
5. Sokolnikoff, I. S., and R. M. Redheffer: "Mathematics of Physics and Modern Engineering," 2d ed., McGraw-Hill, New York, 1966.
6. Thomas, G. B.: "Calculus and Analytic Geometry," Addison-Wesley, Reading, Mass., 1953.

CHAPTER 11
PERFECT FLUIDS

Before they can be solved, most engineering problems require simplifying assumptions because of the inherent complexities in the mathematical models which describe the physical situation. In Chap. 10 the control-volume technique was introduced to facilitate the prediction of certain overall characteristics of flow, often adequate for engineering purposes. For detailed studies, however, the differential equations which accurately describe motion at every point in a flow field are frequently insolvable, because of such mathematical complexities as that found in nonlinear partial differential equations. For example, the classical Navier-Stokes equations, developed in the nineteenth century and derived in Chap. 12, have not yet been solved for the general case although they adequately describe the fluid motion in all cases where solutions have been obtained.

In this chapter we shall consider techniques which will permit us to solve a large class of problems involving perfect fluids. A *perfect fluid* is one which *has no viscosity* or one which behaves as if the effects due to viscosity were negligible. Obviously, no fluid fits the inviscid portion of this definition, but in many practical cases the flow of a real fluid can be accurately analyzed in terms of the perfect-fluid theory

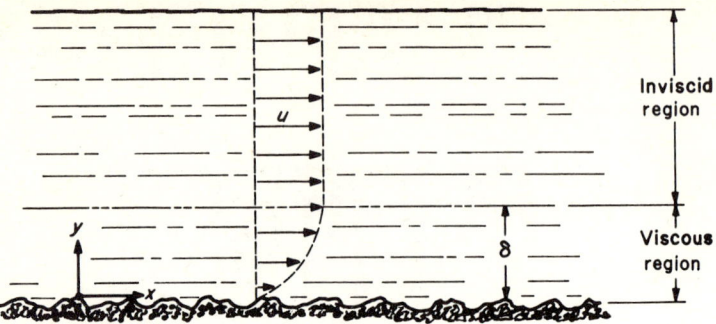

Fig. 11-1 River flowing over rough bed.

introduced in this chapter. To motivate our study, let us consider an illustration in which real and perfect fluid flows occur in the same problem.

Figure 11-1 depicts a deep river flowing over a rough gravel bed. Although the viscosity of water is relatively low (when compared to molasses in January), the stream is retarded by the rough gravel bed. Its velocity profile may be as shown in the figure, changing rapidly in a region near the bed and becoming rather uniform at a distance away from the bed. In the viscous region, known as the *boundary layer*, the shear stresses are high since the velocity gradient is significant, $\tau = \mu \, du/dy$. The same fluid is flowing (same viscosity) in the inviscid region, but the velocity gradient is negligible, resulting in negligible shear stresses. The flow in the inviscid region can be analyzed by perfect-fluid theory, but the flow in the viscous region cannot be so treated. Technically, the viscous effects are not the predominant criteria in this case; however, the flow behaves as if they were and can be treated accordingly. Note that any solution in the inviscid region, resulting in a pressure or velocity distribution, for example, must match with that of the viscous region at the edge of the boundary layer because of the continuous nature of the physical problem.

Before proceeding with the development of the perfect-fluid theory, we must consider a kinematical characteristic of the deformation of fluids in motion, rotation, and the mathematical concept of circulation.

11-1 ROTATION AND VORTICITY

Rotation ω is the average angular velocity of any two mutually perpendicular line elements in the plane of the flow. A typical incompressible fluid element of dimensions Δx, Δy, and Δz may in time Δt deform in the xy plane as shown in Fig. 11-2 by translation and rotation. Letting the velocity components at a general point P be u and v, the displacements are shown in the figure by the velocity-time products. By the definition of rotation the angular velocity about the z axis is

$$\omega_z \equiv \lim_{\Delta t \to 0} \frac{1}{2} \frac{\Delta \alpha - \Delta \beta}{\Delta t} \tag{11-1}$$

PERFECT FLUIDS

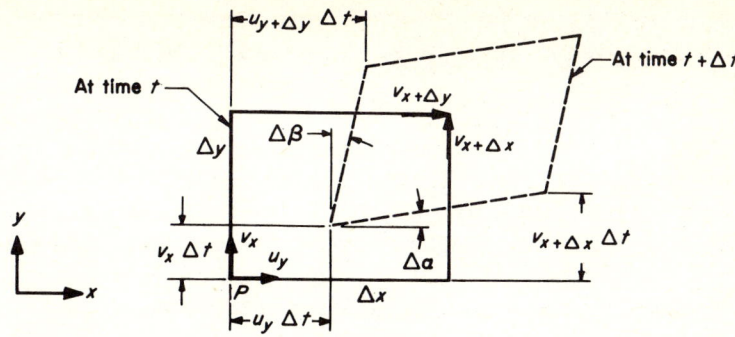

Fig. 11-2 Deformation of a fluid element.

where we have assumed counterclockwise rotation as positive. For small angular deformation, which becomes exact in the limit,

$$\Delta\alpha \simeq \tan \Delta\alpha \quad \text{and} \quad \Delta\beta \simeq \tan \Delta\beta$$

Therefore,

$$\omega_z \simeq \tfrac{1}{2} \lim_{\substack{\Delta x \to 0 \\ \Delta y \to 0}} \left[\frac{(v_{x+\Delta x} - v_x)\Delta t}{\Delta x \, \Delta t} - \frac{(u_{y+\Delta y} - u_y)\Delta t}{\Delta y \, \Delta t} \right] \tag{11-2}$$

$$\omega_z = \frac{1}{2}\left(\frac{\partial v}{\partial x} - \frac{\partial u}{\partial y}\right) \tag{11-3a}$$

Similarly,

$$\omega_y = \frac{1}{2}\left(\frac{\partial u}{\partial z} - \frac{\partial w}{\partial x}\right) \tag{11-3b}$$

$$\omega_x = \frac{1}{2}\left(\frac{\partial w}{\partial y} - \frac{\partial v}{\partial z}\right) \tag{11-3c}$$

In three dimensions

$$\boldsymbol{\omega} = \frac{1}{2}\left[\left(\frac{\partial w}{\partial y} - \frac{\partial v}{\partial z}\right)\mathbf{i} + \left(\frac{\partial u}{\partial z} - \frac{\partial w}{\partial x}\right)\mathbf{j} + \left(\frac{\partial v}{\partial x} - \frac{\partial u}{\partial y}\right)\mathbf{k}\right] = \tfrac{1}{2}(\nabla \times \mathbf{V}) \tag{11-4}$$

where $\nabla \times \mathbf{V}$ is the curl operator, which can be expressed in any orthogonal coordinate system. In cartesian coordinates

$$\nabla \times \mathbf{V} = \operatorname{curl} \mathbf{V} = \begin{vmatrix} \mathbf{i} & \mathbf{j} & \mathbf{k} \\ \dfrac{\partial}{\partial x} & \dfrac{\partial}{\partial y} & \dfrac{\partial}{\partial z} \\ u & v & w \end{vmatrix} \tag{11-5}$$

which readily gives Eq. (11-4). The curl operator for other common coordinate systems is given in Appendix B. *Vorticity* Ω is defined as twice the rotation, i.e.,

$$\Omega \equiv 2\omega \tag{11-6}$$

A flow is irrotational when curl $\mathbf{V} \equiv 0$. The flow of a perfect irrotational fluid is called *potential flow*. Its mathematical formulation is identical to that in other potential fields, such as thermal, electric, and magnetic fields. The differential equations for incompressible potential flow are linear and may be superimposed. Only incompressible fluids will be treated in this introductory chapter. From the streamline patterns, which are given by perfect-fluid theory, velocity and pressure variations can be obtained throughout a flow field. Lift and drag on a body can then be determined from the pressure distribution. It is this result which we seek in our study of perfect fluids.

Example A flow field is described by $\mathbf{V} = (x^2 - y^2 + x)\mathbf{i} - (2xy + y)\mathbf{j}$. Is it irrotational?
Solution We must subject this velocity to the "curl test." Equation (11-5) reduces to

$$\nabla \times \mathbf{V} = \mathbf{k} \begin{vmatrix} \dfrac{\partial}{\partial x} & \dfrac{\partial}{\partial y} \\ u & v \end{vmatrix} = \mathbf{k}\left(\dfrac{\partial v}{\partial x} - \dfrac{\partial u}{\partial y}\right)$$

for this two-dimensional case, giving

$$\nabla \times \mathbf{V} = \mathbf{k}(-2y + 2y) = 0$$

Therefore, the flow is irrotational. Note that we could have used Eq. (11-3*a*).

What does being irrotational mean in a physical problem? To answer this, recall that our analysis pertains to an element of fluid and not to the motion of a body of fluid as a whole. To illustrate, compare the rotation of a solid body with that of a seat on a Ferris wheel. As shown in Fig. 11-3, the orientation of a line element in a rotating solid body changes with time, whereas a line element on the seat of a Ferris wheel retains its initial orientation. In irrotational flow, a fluid element, like the Ferris wheel seat, retains its initial orientation.

Returning to our illustration of a deep river flowing over a rough gravel bed, let us imagine placing a toothpick on the surface of the water. If the flow is tranquil, the toothpick can be expected to have the same orientation downstream even though the river may wind in an irregular fashion in its flow. *Irrotationality is then a measure*

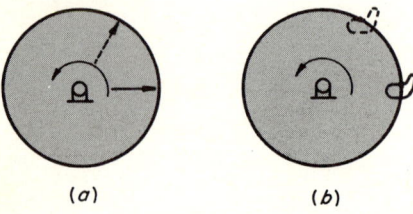

Fig. 11-3 Rotational and irrotational motion: (*a*) solid body: rotational; (*b*) Ferris wheel: irrotational (seat).

PERFECT FLUIDS

of orientation and not path. The toothpick may be thought of as a curl meter or vorticity meter.

11-2 CIRCULATION

Circulation Γ is the summation, taken over the entire length of a closed path, of the dot product of the tangential velocity and the infinitesimal length of the path, a line integral. If the path is traversed with the enclosed region on the left, as shown in Fig. 11-4, the circulation is positive. In equation form, the circulation is

$$\Gamma \equiv \oint \mathbf{V} \cdot d\mathbf{s} = \oint V \cos \alpha \, ds \tag{11-7}$$

The symbol $\oint$ is used to emphasize that the line integration must be made over a closed path. The integration is performed at time t. In cartesian coordinates the circulation is

$$\Gamma = \oint (u \, dx + v \, dy + w \, dz) \tag{11-8}$$

By considering its analogy with other line integrals, we may feel more comfortable with the concept of circulation. For example, work and electric potential are similarly determined, i.e.,

$$_1W_2 = \int_1^2 \mathbf{F} \cdot d\mathbf{s} \tag{11-9}$$

$$\text{Potential} = \int_1^2 \text{intensity} \cdot d\mathbf{s} \tag{11-10}$$

As an example of its usefulness, circulation accounts for the lift force on an airfoil, as will be noted later in this chapter.

Relationship between circulation and vorticity The theorem of Stokes [1, 8] is a convenient means of relating circulation and vorticity. In terms of our nomenclature, it is

$$\oint_C \mathbf{V} \cdot d\mathbf{s} = \int_A (\nabla \times \mathbf{V}) \cdot d\mathbf{A} \tag{11-11}$$

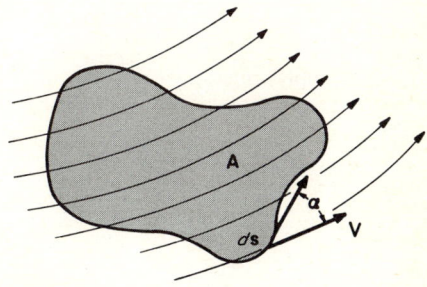

Fig. 11-4 Nomenclature for circulation.

where **A** is the area of the surface bounded by curve C. Therefore, the circulation can be expressed in terms of the curl:

$$\Gamma = \int_A (\nabla \times \mathbf{V}) \cdot d\mathbf{A} \tag{11-12}$$

It is the integral of the normal component of the curl of the velocity over an area for which C is the path enclosing the surface.

Example Compute the circulation around an infinitesimal rectangular path in the xy plane. Velocity components are given in Fig. 11-5.

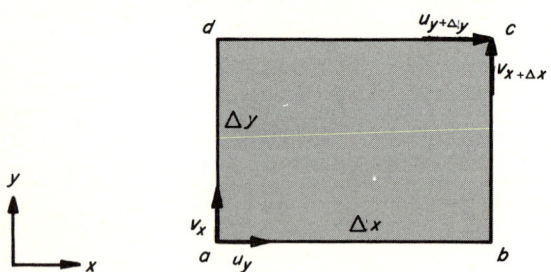

Fig. 11-5 Velocities around elemental path.

Solution Starting at point a and proceeding counterclockwise around the path, Eq. (11-7)

$$\Gamma = \oint \mathbf{V} \cdot d\mathbf{s}$$

can be evaluated by considering each segment of the path individually, giving

$$\Delta\Gamma|_{abcda} = u_y \, \Delta x|_{ab} + v_{x+\Delta x} \, \Delta y|_{bc} - u_{y+\Delta y} \, \Delta x|_{cd} - v_x \, \Delta y|_{da}$$

Or since each velocity component makes no contribution to the contour integral except on its own segment,

$$\Delta\Gamma|_{abcda} = (v_{x+\Delta x} - v_x)\,\Delta y - (u_{y+\Delta y} - u_y)\,\Delta x$$

Multiplying and dividing the right-hand side by $\Delta x \, \Delta y$ and taking the limit as $\Delta x \to 0$ and $\Delta y \to 0$, we get the differential expression (by employing the definition of the derivative), i.e.,

$$d\Gamma = \lim_{\substack{\Delta x \to 0 \\ \Delta y \to 0}} \frac{(v_{x+\Delta x} - v_x)\,\Delta y - (u_{y+\Delta y} - u_y)\,\Delta x}{\Delta x \, \Delta y} \Delta x \, \Delta y$$

$$d\Gamma = \left(\frac{\partial v}{\partial x} - \frac{\partial u}{\partial y}\right) dA$$

PERFECT FLUIDS

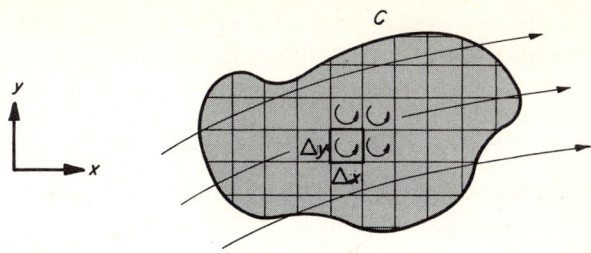

Fig. 11-6 General curve in a plane.

But this is the two-dimensional equivalent of Stokes' theorem [Eq. (11-11)]

$$\oint \mathbf{V} \cdot d\mathbf{s} = \int_A \left(\frac{\partial v}{\partial x} - \frac{\partial u}{\partial y} \right) dA = \int_A (\nabla \times \mathbf{V})_z \, dA = \int_A \Omega_z \, dA$$

and the result could have been determined by direct application of Stokes' theorem.

The preceding example can be readily generalized to any closed path in a plane. Consider curve C in Fig. 11-6, which has been divided into elemental areas $\Delta x \, \Delta y$. Applying the final equation of the preceding example to each of the area elements and summing,

$$\sum_{n=1}^{\infty} \oint \mathbf{V} \cdot d\mathbf{s} = \sum_{n=1}^{\infty} \int_A (\nabla \times \mathbf{V})_z \, dA \qquad (11\text{-}13)$$

In integrating around all elements in the same direction, we note that integration along a segment common to each element cancels. This occurs in all interior curve segments, leaving only the outer boundary C, since the elements become infinitesimal in the limit.

Very little argument is necessary to generalize further to the three-dimensional case as given by Eq. (11-11).

In an irrotational flow field $\nabla \times \mathbf{V} = 0$, and the circulation is zero from Eq. (11-11). There is *an exception*, however, when the boundary curve C surrounds *a singular point* in the flow field. For example, consider the case of vortex flow (discussed in detail in Sec. 11-7) described by the velocity

$$V = \frac{k}{r} \qquad (11\text{-}14)$$

where k is a constant and r is the radius from the origin. Mathematically, the origin is a singular point since it gives an infinite velocity. Applying Eq. (11-11) to this case, the circulation is zero everywhere except at the origin, the singular point. Because of the mathematical singularity at the origin, the circulation must be evaluated by some other technique, as illustrated in Sec. 11-7 for this case. The circulation at the origin in vortex flow is not zero, a fact which accounts for the lift on airfoils as we shall see later.

At the present, we may summarize by stating that the *circulation is zero in an irrotational flow field for every path which does not include a singular point.*

11-3 THE STREAM FUNCTION

Figure 11-7 represents a family of streamlines, introduced in Sec. 9-1, in a steady incompressible flow field. Since no mass crosses a streamline, we can arbitrarily assign a value ψ' to a streamline which is a measure of the flow rate. Choosing the mass flow rate between adjacent streamlines to be $d\psi'$, the continuity equation applied to the triangular element gives

$$\rho u \, dy - \rho v \, dx = d\psi' \tag{11-15}$$

where the efflux face of the element is perpendicular to the streamlines and unit depth is assumed. Dividing by ρ and defining the stream function $\psi \equiv \psi'/\rho$, we get

$$u \, dy - v \, dx = d\psi \tag{11-16}$$

Recalling from Chap. 9 that the equation for a streamline in the xy plane is

$$u \, dy - v \, dx = 0 \tag{9-2}$$

we observe a striking similarity between these equations. They would be equivalent if we could determine a stream function ψ such that $d\psi = 0$. But this is the condition for an exact differential when ψ is uniquely defined in terms of the field coordinates, x and y, and is equal to a constant [4], i.e.,

$$\psi = \psi(x,y) = \text{const} \tag{11-17}$$

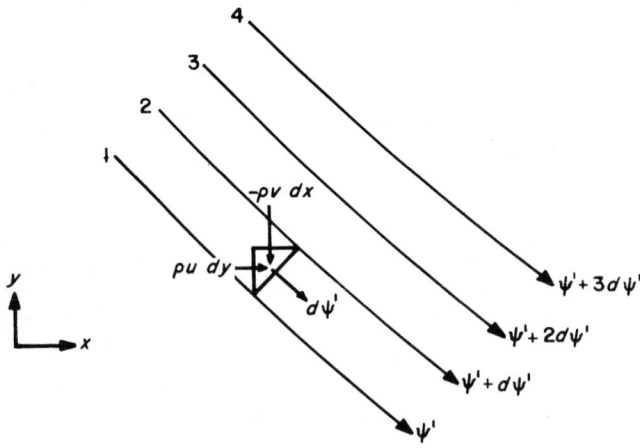

Fig. 11-7 Family of streamlines.

PERFECT FLUIDS

Differentiation by the chain rule gives

$$d\psi = \frac{\partial \psi}{\partial x} dx + \frac{\partial \psi}{\partial y} dy = 0 \tag{11-18}$$

and the condition of exactness is satisfied when

■ $$u = \frac{\partial \psi}{\partial y} \tag{11-19a}$$

■ $$v = -\frac{\partial \psi}{\partial x} \tag{11-19b}$$

which is easily observed upon comparing Eqs. (11-16) and (11-18).

We must now ask whether this function satisfies the differential continuity equation

$$\frac{\partial u}{\partial x} + \frac{\partial v}{\partial y} = 0 \tag{11-20}$$

Substituting Eqs. (11-19a) and (11-19b) into Eq. (11-20), we get

$$\frac{\partial}{\partial x}\frac{\partial \psi}{\partial y} + \frac{\partial}{\partial y}\left(-\frac{\partial \psi}{\partial x}\right) \stackrel{?}{=} 0 \tag{11-21}$$

which yields the identity

$$\frac{\partial^2 \psi}{\partial x\, \partial y} - \frac{\partial^2 \psi}{\partial x\, \partial y} = 0 \tag{11-22}$$

upon interchanging the order of differentiation, which is always valid with continuous functions.

Up to this point in our analysis the flow may be either rotational or irrotational. *Satisfaction of the continuity equation guarantees the existence of a stream function,* but for irrotationality

$$\frac{\partial v}{\partial x} - \frac{\partial u}{\partial y} = 0 \tag{11-23}$$

Upon substituting Eqs. (11-19a) and (11-19b) into this equation, we get

$$\frac{\partial}{\partial x}\left(-\frac{\partial \psi}{\partial x}\right) - \frac{\partial}{\partial y}\frac{\partial \psi}{\partial y} = 0$$

$$\frac{\partial^2 \psi}{\partial x^2} + \frac{\partial^2 \psi}{\partial y^2} = 0 \tag{11-24a}$$

or, in general,

$$\nabla^2 \psi = 0 \tag{11-24b}$$

which is Laplace's equation. This was discussed in detail in Sec. 5-6, where linearity was noted as its most important characteristic. We shall make use of this property in the superposition of flows in Sec. 11-6, as was done for the thermal field in Sec. 5-6. Mathematically, the stream function reduces the number of variables in a flow problem by 1, but it increases the order of the differential equation, requiring an additional boundary condition.

In polar coordinates, if increasing values of θ and r are measured counterclockwise and outward, respectively, as shown in Fig. 11-8, the velocity components are

■ $$V_\theta = -\frac{\partial \psi}{\partial r} \tag{11-25a}$$

■ $$V_r = \frac{1}{r}\frac{\partial \psi}{\partial \theta} \tag{11-25b}$$

The details of the derivation are left as an exercise at the end of the chapter. Transformations between the cartesian and the polar coordinate systems are facilitated by the equations

$$y = r \sin \theta \tag{11-26a}$$

$$x = r \cos \theta \tag{11-26b}$$

$$\theta = \tan^{-1} \frac{y}{x} \tag{11-26c}$$

$$r = \sqrt{x^2 + y^2} \tag{11-26d}$$

Stream functions in three dimensions for flows with axial symmetry, e.g., flow about a body of revolution, for unsteady flow and for compressible flow are presented

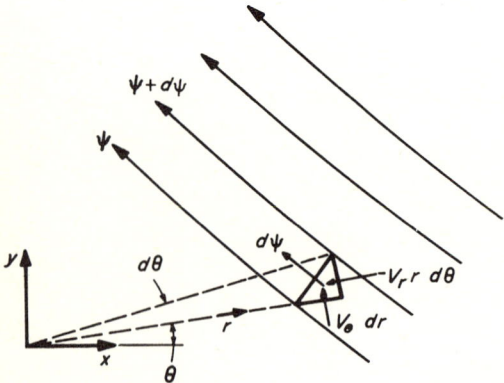

Fig. 11–8 Stream function in polar coordinates.

PERFECT FLUIDS

in more extensive treatments of perfect fluids. See, for example, Milne-Thomson [3], Kuethe and Schetzer [2], and Pao [5].

11-4 THE VELOCITY POTENTIAL

No physical analog to the stream function exists for the velocity potential, but the mathematical concept is straightforward. For irrotationality,

$$\frac{\partial v}{\partial x} = \frac{\partial u}{\partial y} \tag{11-23}$$

is the necessary and sufficient condition for

$$u(x,y)\,dx + v(x,y)\,dy$$

to be an exact differential. A function ϕ, defined as velocity potential, exists then such that

$$u\,dx + v\,dy = d\phi \tag{11-27}$$

where

$$\phi = \phi(x,y) = \text{const}$$

$$d\phi = \frac{\partial \phi}{\partial x}\,dx + \frac{\partial \phi}{\partial y}\,dy = 0 \tag{11-28}$$

Comparison of Eqs. (11-27) and (11-28) gives

■ $\quad u = \dfrac{\partial \phi}{\partial x}$ \hfill (11-29a)

■ $\quad v = \dfrac{\partial \phi}{\partial y}$ \hfill (11-29b)

which, upon substitution into the continuity equation, gives

$$\frac{\partial^2 \phi}{\partial x^2} + \frac{\partial^2 \phi}{\partial y^2} = 0 \tag{11-30a}$$

or

$$\nabla^2 \phi = 0 \tag{11-30b}$$

The comments made about the Laplace equation of the stream function apply also to the Laplace equation of the velocity potential.

In polar coordinates the velocity components in terms of the velocity potential are

■ $\quad V_r = \dfrac{\partial \phi}{\partial r}$ \hfill (11-31a)

■ $\quad V_\theta = \dfrac{1}{r}\dfrac{\partial \phi}{\partial \theta}$ \hfill (11-31b)

which can be written down by analogy with the stream-function equations since the functions are orthogonal.

11-5 RELATIONSHIP BETWEEN STREAM FUNCTION AND VELOCITY POTENTIAL

1. Because of the analogous nature of Laplace's equations for the stream function and velocity potential, we suspect a geometrical relationship. From Eq. (11-18) the slope of the line $\psi = C$ is

$$\left.\frac{dy}{dx}\right|_{\psi=C} = -\frac{\partial \psi/\partial x}{\partial \psi/\partial y} \tag{11-32}$$

and from Eq. (11-28) the slope of the line $\phi = K$ is

$$\left.\frac{dy}{dx}\right|_{\phi=K} = -\frac{\partial \phi/\partial x}{\partial \phi/\partial y} \tag{11-33}$$

But

$$\frac{\partial \phi}{\partial x} = \frac{\partial \psi}{\partial y} \tag{11-34a}$$

$$\frac{\partial \phi}{\partial y} = -\frac{\partial \psi}{\partial x} \tag{11-34b}$$

which, upon substituting into Eq. (11-33), gives

$$\left.\frac{dy}{dx}\right|_{\phi=K} = \frac{\partial \psi/\partial y}{\partial \psi/\partial x} = -\frac{1}{(dy/dx)_{\psi=C}} \tag{11-35}$$

Therefore, *the slope of the streamline is the negative reciprocal of the slope of the equipotential line*—orthogonal as illustrated in Fig. 11-9.

2. Equations (11-34a) and (11-34b) are known as the *Cauchy-Riemann equations*. The relationship between ψ and ϕ permits determination of the velocity field if

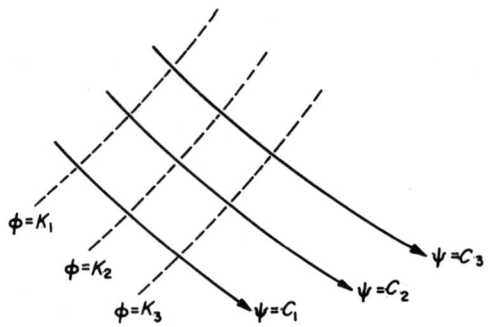

Fig. 11-9 Equipotential lines and streamlines in irrotational flow.

PERFECT FLUIDS

either is known, since

$$|V| = \sqrt{u^2 + v^2} = \sqrt{\left(\frac{\partial \phi}{\partial x}\right)^2 + \left(\frac{\partial \phi}{\partial y}\right)^2} \tag{11-36}$$

and

$$|V| = \sqrt{\left(\frac{\partial \psi}{\partial x}\right)^2 + \left(\frac{\partial \psi}{\partial y}\right)^2} \tag{11-37}$$

3. In reality ψ and ϕ represent isoflow and equipotential *surfaces*, respectively, since we have assumed unit depth in our analysis. No generality is lost, however, by thinking in terms of lines rather than surfaces since no gradients exist in the third dimension.
4. If ψ exists, continuity is satisfied. The flow may be either rotational or irrotational.
5. If ϕ exists, continuity is satisfied and the flow is irrotational.
6. If $\nabla^2 \psi = 0$ or $\nabla^2 \phi = 0$ is satisfied, the flow is irrotational and continuity is guaranteed.
7. If $\nabla \times V = 0$, a velocity potential ϕ exists, $V = \nabla \phi$, and $\nabla^2 \phi = 0$.
8. The existence of a velocity potential ϕ guarantees continuity, but the existence of a stream function ψ does not assure the existence of a velocity potential ϕ.

Example A steady incompressible flow field is defined by $\psi = 2x^2 - 2y^2$. (a) Is the flow irrotational? If so, what is the velocity potential? (b) What is the flow rate per unit depth between streamlines which pass through the points (4,6) and (2,4)?

Solution (a) For irrotationality

$$\frac{\partial v}{\partial x} = \frac{\partial u}{\partial y}$$

which requires knowing the velocity components. But the velocity components can be determined from Eqs. (11-19a) and (11-19b) as follows:

$$u = \frac{\partial \psi}{\partial y} = -4y \qquad \frac{\partial u}{\partial y} = -4$$

$$v = -\frac{\partial \psi}{\partial x} = -4x \qquad \frac{\partial v}{\partial x} = -4$$

Therefore, the flow is irrotational.

Using Eqs. (11-29a) and (11-29b), we can get the velocity potential, i.e.,

$$u = \frac{\partial \phi}{\partial x} = -4y$$

$$v = \frac{\partial \phi}{\partial y} = -4x$$

Integrating the expression for u,

$$\phi = -4xy + C(y)$$

which, upon differentiation with respect to y, gives

$$\frac{\partial \phi}{\partial y} = -4x + C'(y)$$

But this is equal to v, that is,

$$-4x + C'(y) = -4x$$
$$C'(y) = 0$$
$$C(y) = \text{const}$$

Hence,

$$\phi = -4xy + C$$

It might be instructive to note that alternatively we could have integrated the expression for v, getting

$$\phi = -4xy + C(x)$$

Following the same procedure as before,

$$\frac{\partial \phi}{\partial x} = -4y + C'(x) = -4y$$

$$C'(x) = 0$$
$$C(x) = \text{const}$$
$$\phi = -4xy + C$$

which is the same result. This provides a check on our first solution.

(b) Since the streamlines are a measure of the flow rate ψ' as given by $\psi' = \rho\psi$, the flow is determined as follows:

$$d\psi = \frac{\partial \psi}{\partial x} dx + \frac{\partial \psi}{\partial y} dy$$

$$\psi = 4 \int_4^2 x \, dx - 4 \int_6^4 y \, dy = 2x^2 \big|_4^2 - 2y^2 \big|_6^4 = 2(4 - 16) - 2(16 - 36) = 16$$

Therefore the flow rate is

$$\psi' = 16\rho$$

11-6 BOUNDARY CONDITIONS

Any scalar function which satisfies $\nabla^2 \psi = 0$ or $\nabla^2 \phi = 0$ can represent the stream function ψ or velocity potential ϕ, respectively, of a steady incompressible irrotational flow. But a particular solution which represents the flow about a given body must satisfy the boundary conditions, conditions at the body and at infinity.

Consider the flow of a perfect fluid over the arbitrary body of Fig. 11-10. Due to the absence of friction, the ψ_0 streamline conforms to the shape of the body, and tangential velocity is not affected by the body. Since no flow can cross a streamline (stream surface), the normal component of velocity at the body is zero. (The body may then be either real or imaginary.) Stated in terms of the normal n and tangential

PERFECT FLUIDS

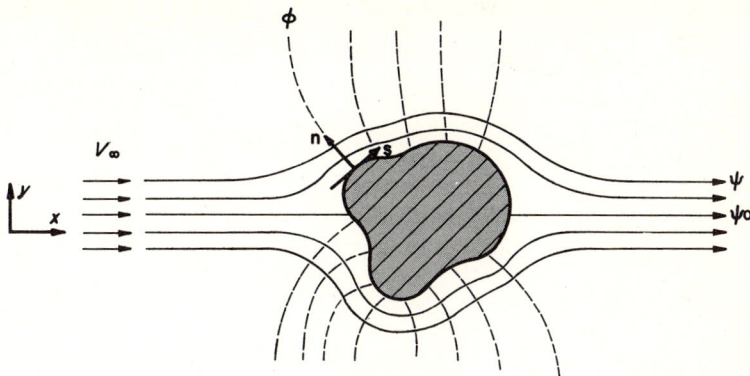

Fig. 11-10 Illustration for boundary conditions.

s directions, the *boundary conditions at the body* are

$$V_n = \left.\frac{\partial \phi}{\partial n}\right|_0 = \left.\frac{\partial \psi}{\partial s}\right|_0 = 0 \tag{11-38a}$$

where the zero subscript denotes the ψ_0 streamline at the body. Conditions at infinity refer to those at a sufficient distance from the body for the flow to be undisturbed. *At infinity* the boundary conditions are

$$\left.\frac{\partial \phi}{\partial y}\right|_\infty = \left.\frac{\partial \psi}{\partial x}\right|_\infty = 0 \tag{11-38b}$$

$$V_\infty = \left.\frac{\partial \phi}{\partial x}\right|_\infty = \left.\frac{\partial \psi}{\partial y}\right|_\infty \tag{11-38c}$$

since the only flow which occurs will be in the direction of the mainstream. Equipotential lines will be parallel to the y axis at infinity.

In irrotational flow without circulation, a function which satisfies Laplace's equation and the boundary conditions at the body and at infinity is a unique solution. In flow fields with circulation, uniqueness also requires specification of the circulation [3]. Under these conditions, the kinematical problem of finding a flow pattern in a perfect-fluid flow field reduces to a mathematical problem, the elegance of which is not seen in other areas of fluid dynamics.

11-7 TWO-DIMENSIONAL SIMPLE FLOWS

Any physically possible irrotational flow has a stream function and velocity potential which satisfy the Laplace equation. Conversely, any solution of Laplace's equation represents the stream function or velocity potential of a physically possible irrotational flow. Since Laplace's equation is linear, the sum of any number of solutions is also a solution. To illustrate, let ψ_1 and ψ_2 represent the stream functions of flow fields which have velocities $\mathbf{V}_1$ and $\mathbf{V}_2$, respectively. The two velocity fields can be added, point by point, to get

$$\mathbf{V} = \mathbf{V}_1 + \mathbf{V}_2$$
$$u = u_1 + u_2 \tag{11-39}$$
$$v = v_1 + v_2$$

Writing the velocity components in terms of their stream functions gives

$$\frac{\partial \psi}{\partial y} = \frac{\partial \psi_1}{\partial y} + \frac{\partial \psi_2}{\partial y} = \frac{\partial}{\partial y}(\psi_1 + \psi_2) \tag{11-40a}$$

$$-\frac{\partial \psi}{\partial x} = -\frac{\partial \psi_1}{\partial x} - \frac{\partial \psi_2}{\partial x} = -\frac{\partial}{\partial x}(\psi_1 + \psi_2) \tag{11-40b}$$

The directional derivatives have the same value in an arbitrary direction η, that is,

$$\frac{\partial \psi}{\partial \eta} = \frac{\partial}{\partial \eta}(\psi_1 + \psi_2) \tag{11-41}$$

and

$$\psi = \psi_1 + \psi_2 + \text{const} \tag{11-42}$$

The constant is made zero by choosing the zero streamline of the resultant flow along the path for which the sum of the stream functions of the component flows vanishes.

Under the same conditions the velocity potential for a resultant flow is equal to the sum of the velocity potentials of the component flows

$$\phi = \phi_1 + \phi_2 \tag{11-43}$$

This principle of superposition will be used in this section to synthesize complicated flow patterns by adding elementary flow patterns.

Uniform flow The most elementary flow pattern is that produced by rectilinear flow, as illustrated in Fig. 11-11. Referring to Eq. (11-19a), the stream function is

$$\psi_{\text{uniform}} = V_\infty y + \text{const} \tag{11-44}$$

The velocity potential can be determined by the method illustrated in the example of Sec. 11-5, i.e.,

$$u = \frac{\partial \psi}{\partial y} = \frac{\partial \phi}{\partial x}$$

Therefore, since $u = V_\infty$,

$$\phi_{\text{uniform}} = V_\infty x + \text{const} \tag{11-45}$$

There are no singular points in uniform flow; hence, the flow is everywhere irrotational.

PERFECT FLUIDS

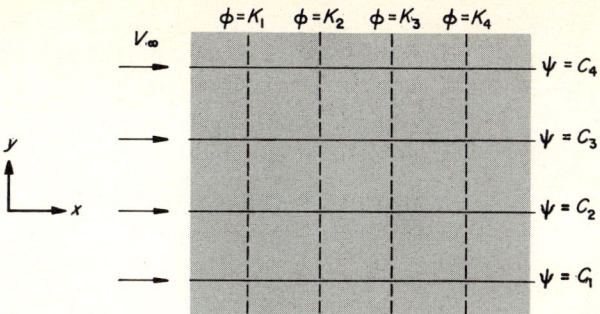

Fig. 11-11 Rectilinear flow.

In polar coordinates

$$\psi_{\text{uniform}} = V_\infty r \sin \theta \tag{11-46a}$$

$$\phi_{\text{uniform}} = V_\infty r \cos \theta \tag{11-46b}$$

where the constant of integration has been tacitly set equal to zero. In the remainder of this section the constant will be taken as zero.

Source flow A source is a point (actually a line coincident with the z axis) from which a fluid issues at a constant flow rate, $\Lambda = AV$, along radial paths as depicted in Fig. 11-12. From continuity, the flow rate crossing any circle (cylinder) whose center is at the source is the product of the area and velocity, i.e.,

$$\Lambda = 2\pi r V_r = \text{const} \tag{11-47}$$

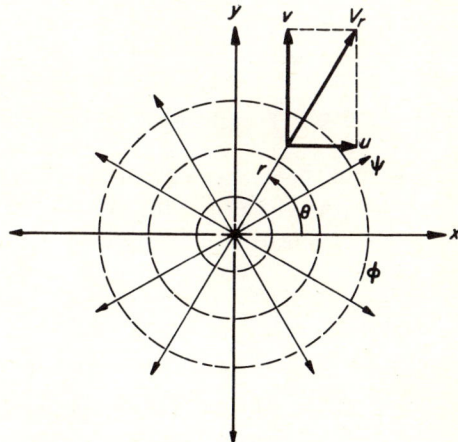

Fig. 11-12 Two-dimensional source.

where Λ is defined as the source strength, flow rate per unit depth. It is obvious that as $r \to 0$, $V_r \to \infty$ (a physically impossible case), making the origin a point of discontinuity.[1]

By direct integration of the velocity components, we get the stream function and velocity potential. In polar coordinates

$$V_r = \frac{\Lambda}{2\pi r} = \frac{1}{r}\frac{\partial \psi}{\partial \theta} = \frac{\partial \phi}{\partial r} \tag{11-48a}$$

$$V_\theta = 0 = -\frac{\partial \psi}{\partial r} = \frac{1}{r}\frac{\partial \phi}{\partial \theta} \tag{11-48b}$$

Omitting the constants of integration, the resulting functions are

$$\psi_{\text{source}} = \frac{\Lambda}{2\pi}\theta \tag{11-49a}$$

$$\phi_{\text{source}} = \frac{\Lambda}{2\pi}\ln r \tag{11-49b}$$

Therefore, the family of streamlines are radial lines, and the equipotential lines are concentric circles.

A *sink* is a point located at the origin into which a fluid issues at a constant flow rate $-\Lambda$, where the negative sign represents inward flow. Hence,

$$\psi_{\text{sink}} = -\frac{\Lambda}{2\pi}\theta \tag{11-50a}$$

$$\phi_{\text{sink}} = -\frac{\Lambda}{2\pi}\ln r \tag{11-50b}$$

Because of the singular point at the origin of a source (or sink), we must determine the circulation in the flow field if the flow is to be uniquely defined. Choosing a circle of radius r about the origin as the path of integration, the circulation is

$$\Gamma = \oint \mathbf{V} \cdot d\mathbf{s} = \int_0^{2\pi} V_\theta r\, d\theta = 0 \tag{11-51}$$

since $V_\theta = 0$. Therefore, the flow is irrotational except at the origin.

Source-sink flow Now let us superimpose a source and sink located at equal distances a from the origin on the x axis as shown in Fig. 11-13. Making use of the property of linearity of Laplace's equations, we have

$$\psi_{ss} = \psi_{\text{source}} + \psi_{\text{sink}} \tag{11-52a}$$

$$\psi_{ss} = -\frac{\Lambda}{2\pi}(\theta_2 - \theta_1) \tag{11-52b}$$

---

[1] A point of discontinuity affords no complication in synthesizing complex flow patterns as long as the discontinuity is kept outside the flow field.

PERFECT FLUIDS

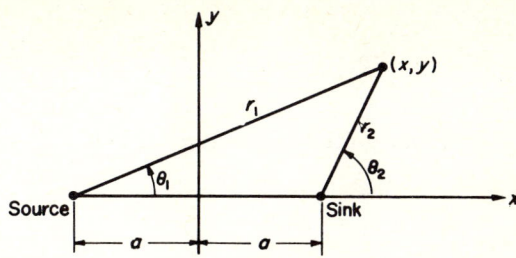

Fig. 11-13 Two-dimensional source-sink flow.

Noting that

$$\tan \theta_2 = \frac{y}{x - a} \tag{11-53a}$$

$$\tan \theta_1 = \frac{y}{x + a} \tag{11-53b}$$

we can simplify Eq. (11-52b) by using the trigonometric identity

$$\theta_2 - \theta_1 = \tan^{-1} \frac{\tan \theta_2 - \tan \theta_1}{1 + \tan \theta_2 \tan \theta_1} \tag{11-54}$$

Therefore,

$$\psi_{ss} = -\frac{\Lambda}{2\pi} \tan^{-1} \frac{\dfrac{y}{x - a} - \dfrac{y}{x + a}}{1 + \dfrac{y}{x - a}\dfrac{y}{x + a}} \tag{11-55}$$

Upon simplification, we get

$$\psi_{ss} = -\frac{\Lambda}{2\pi} \tan^{-1} \frac{2ay}{x^2 + y^2 - a^2} \tag{11-56a}$$

or, in terms of the cotangent,

$$x^2 + y^2 + 2ay \cot \frac{2\pi \psi_{ss}}{\Lambda} = a^2 \tag{11-56b}$$

Completing the square and rearranging, we have

$$x^2 + y^2 + 2ay \cot \frac{2\pi \psi_{ss}}{\Lambda} + a^2 \cot^2 \frac{2\pi \psi_{ss}}{\Lambda} = a^2 \left(1 + \cot^2 \frac{2\pi \psi_{ss}}{\Lambda}\right) \tag{11-56c}$$

$$x^2 + \left(y + a \cot \frac{2\pi \psi_{ss}}{\Lambda}\right)^2 = \left(a \csc \frac{2\pi \psi_{ss}}{\Lambda}\right)^2 \tag{11-56d}$$

which is the equation of a circle of radius

$$a \csc \frac{2\pi\psi_{ss}}{\Lambda}$$

with centers on the y axis at a distance of

$$a \cot \frac{2\pi\psi_{ss}}{\Lambda}$$

If $y = 0$, from Eq. (11-56a), $x = \pm a$ for all streamlines; i.e., all circles pass through the source-sink pair as shown in Fig. 11-14.

By solving for the velocity potential

$$\phi_{ss} = \phi_{\text{source}} + \phi_{\text{sink}} \tag{11-57a}$$

$$\phi_{ss} = \frac{\Lambda}{2\pi} \ln \frac{r_1}{r_2} \tag{11-57b}$$

it can readily be shown that the equipotential lines are circles with centers on the x axis and orthogonal to the streamlines [5] as indicated in Fig. 11-14.

Doublet A doublet results when a source and sink of equal strength are allowed to approach each other while the product of their strength and distance apart are held constant, i.e.,

$$\mu \equiv 2a\Lambda = \text{const} \tag{11-58}$$

where μ is the doublet strength. The stream function for the doublet ψ_D is

$$\psi_D = -\lim_{a \to 0} \left[\frac{(2a)\Lambda}{(2a)2\pi} \tan^{-1} \frac{2ay}{x^2 + y^2 - a^2} \right] \tag{11-59}$$

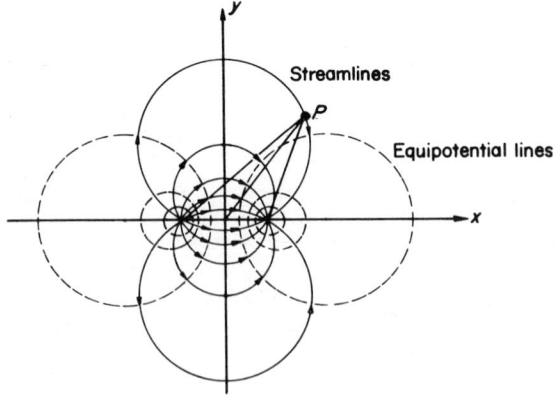

Fig. 11-14 Source-sink flow.

PERFECT FLUIDS

As $a \to 0$, $\tan(\theta_1 - \theta_2) \to (\theta_1 - \theta_2)$. Therefore,

$$\psi_D = -\lim_{a \to 0} \frac{\mu}{4\pi a} \frac{2ay}{x^2 + y^2 - a^2} \tag{11-60}$$

$$\psi_D = -\frac{\mu}{2\pi} \frac{y}{x^2 + y^2} \tag{11-61a}$$

Rearranging, we get

$$x^2 + y^2 = -\frac{\mu y}{2\pi \psi_D} \tag{11-61b}$$

or

$$x^2 + \left(y + \frac{\mu}{4\pi \psi_D}\right)^2 = \left(\frac{\mu}{4\pi \psi_D}\right)^2 \tag{11-61c}$$

which is the equation of a circle of radius $\mu/4\pi\psi_D$ and located at $(0, \mu/4\pi\psi_D)$. Expressing Eq. (11-61a) in polar form,

$$\psi_D = -\frac{\mu}{2\pi r} \sin \theta \tag{11-62}$$

it becomes more convenient to determine the velocity potential as

$$rV_r = \frac{\partial \psi_D}{\partial \theta} = -\frac{\mu}{2\pi r} \cos \theta \tag{11-63}$$

But $V_r = \partial \phi / \partial r$; hence,

$$\frac{\partial \phi_D}{\partial r} = -\frac{\mu}{2\pi r^2} \cos \theta \tag{11-64a}$$

$$\phi_D = \frac{\mu}{2\pi r} \cos \theta \tag{11-64b}$$

or

$$\phi_D = \frac{\mu}{2\pi} \frac{x}{x^2 + y^2} \tag{11-64c}$$

which are clearly orthogonal to the streamlines as illustrated in Fig. 11-15.

Flow about a stationary cylinder So far the flow patterns have been physically unreal but mathematically elegant. By combining a uniform flow and a doublet, however, as shown in Fig. 11-16a, a flow pattern results which is of great practical significance: flow over a circular cylinder, as shown in Fig. 11-16b. The student should be cautioned at this point to note that the cylinder is not physically real but results from the combination of the flow patterns. It is equally important to note,

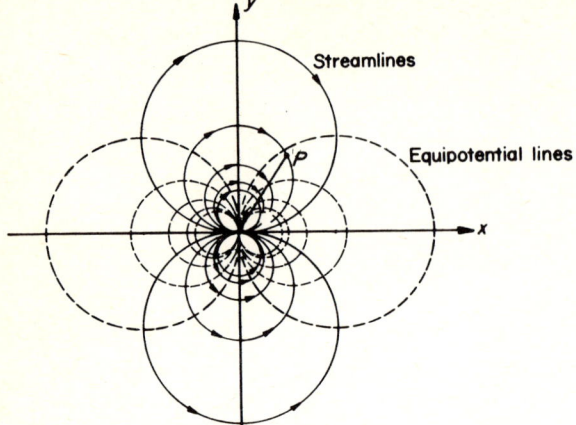

Fig. 11-15 Doublet.

however, that the velocity and pressure at the radius r_0 are just as real as if a real cylinder were immersed in a flowing perfect fluid. Therefore, this mathematical model becomes a "real" situation.

Upon adding the stream functions of uniform flow and a doublet centered on the y axis, we get in polar coordinates

$$\psi_{\text{cyl}} = V_\infty r \sin \theta - \frac{\mu}{2\pi r} \sin \theta \qquad (11\text{-}65)$$

The streamline pattern shown in Fig. 11-16b results from assigning different values to $\psi_{\text{cyl}} = \text{const}$. By setting $\psi_{\text{cyl}} = 0$ we get

$$r_0 = \sqrt{\frac{\mu}{2\pi V_\infty}} \qquad (11\text{-}66)$$

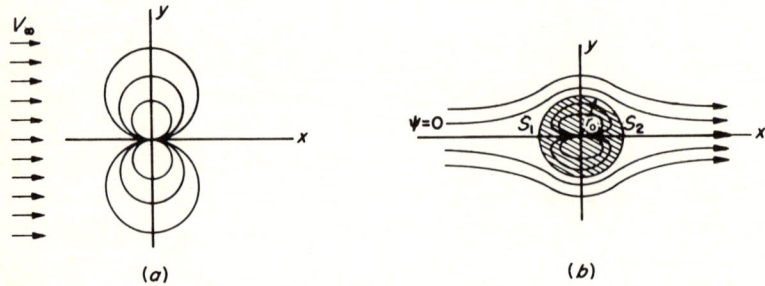

Fig. 11-16 Flow about a cylinder without circulation.

PERFECT FLUIDS

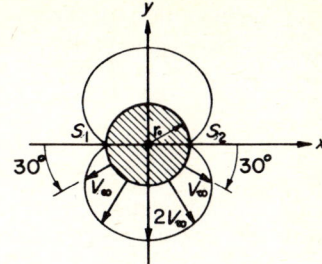

Fig. 11-17 Velocity distribution at surface of cylinder.

where the zero subscript designates the radius of the cylinder whose boundary is the streamline $\psi_{\text{cyl}} = 0$. For $\theta = 0$ and π, $\psi_{\text{cyl}} = 0$ is a solution to Eq. (11-65) for all radii. The $\psi_{\text{cyl}} = 0$ streamline is shown in the figure; it is a line of symmetry. The careful student will note that for $r_0 > 0$ the singularity of the source which was used to synthesize the doublet is outside of the flow field, being at the origin, and the entire flow field is irrotational.

It is often more convenient to express the stream function in terms of the cylinder radius r_0 (a constant) rather than the strength of the doublet μ. Therefore, by combining Eqs. (11-65) and (11-66), we have

$$\psi_{\text{cyl}} = V_\infty \left(r - \frac{r_0^2}{r} \right) \sin \theta \tag{11-67}$$

from which the velocity components are found to be

$$V_r = \frac{1}{r}\frac{\partial \psi}{\partial \theta} = V_\infty \left(1 - \frac{r_0^2}{r^2} \right) \cos \theta \tag{11-68a}$$

$$V_\theta = -\frac{\partial \psi}{\partial r} = -V_\infty \left(1 + \frac{r_0^2}{r^2} \right) \sin \theta \tag{11-68b}$$

At the surface of the cylinder, $r = r_0$; hence,

$$V_r \big|_{r=r_0} = 0 \tag{11-69a}$$

$$V_0 = V_\theta \big|_{r=r_0} = -2V_\infty \sin \theta \tag{11-69b}$$

Figure 11-17 is a polar plot of the velocity distribution at the surface of the cylinder. The points where the velocity is zero, S_1 and S_2, are *stagnation points*.

Force on a stationary cylinder By taking the dot product of each term in the Euler equation of motion [Eq. (10-82)] with an incremental distance vector $d\mathbf{r}$ taken in any direction, it can readily be shown that the Bernoulli equation is valid between any two points (excluding singular points) in a steady irrotational flow field [5, 9]. This removes the previous restriction of applying Bernoulli's equation along a streamline.

Writing Bernoulli's equation between a point located at a large distance from the cylinder, where $p = p_\infty$, $V = V_\infty$, etc., and a point on the surface of the cylinder, we have

$$p_\infty + \frac{\rho V_\infty^2}{2} = p_0 + \frac{\rho V_0^2}{2} \tag{11-70}$$

where the zero subscript designates quantities at the surface of the cylinder. Using V_0 from Eq. (11-69b), the pressure distribution on the surface is

$$p_0 = p_\infty + \frac{\rho V_\infty^2}{2} - \frac{\rho(-2V_\infty \sin \theta)^2}{2} \tag{11-71a}$$

or

$$p_0 = p_\infty + \frac{\rho V_\infty^2}{2}(1 - 4\sin^2 \theta) \tag{11-71b}$$

In aerodynamics a dimensionless pressure coefficient $\mathbf{C}_p$, defined as

$$\mathbf{C}_p \equiv \frac{p_0 - p_\infty}{\rho V_\infty^2/2} \tag{11-72}$$

is sometimes employed, simplifying the expression of Eq. (11-71b), i.e., for a cylinder

$$\mathbf{C}_p = 1 - 4\sin^2 \theta \tag{11-73}$$

The drag force F_0, defined as the force per unit length on the cylinder in the direction parallel to the uniform flow, is the product of pressure and area integrated around the cylinder, i.e.,

$$F_0 = -\int_0^{2\pi} p_0 r_0 \cos \theta \, d\theta \tag{11-74}$$

where the negative sign indicates that the pressure force is always directed toward the surface and p_0 is given by Eq. (11-71b). Carrying out the integration, we find that *the resultant drag force is zero*—perhaps a surprise, but perfect fluids behave contrary to our experience with real fluids. This result, which conflicts with our experience, is known as *d'Alembert's paradox* and came about because of our complete disregard of viscous effects throughout the flow field. This conflict will be resolved in our study of the boundary layer in subsequent chapters.

The lift force F_L, defined as the force per unit length on the cylinder in the direction normal to the uniform flow, is also zero for the cylinder without circulation. This might have been suspected because of the symmetry of the streamline pattern.

Free vortex A two-dimensional vortex has tangential velocity only as given by

$$V = \frac{k}{r} \tag{11-14}$$

PERFECT FLUIDS

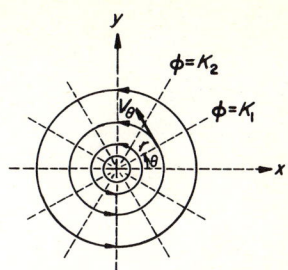

Fig. 11-18 Free vortex.

having streamlines which are concentric circles about the origin (a singular point) as shown in Fig. 11-18. Since the origin is a singularity, we must specify the circulation for uniqueness. Integrating around a closed path of radius r, the circulation is

$$\Gamma = \oint \mathbf{V} \cdot d\mathbf{s} = \int_0^{2\pi} V_\theta r \, d\theta = \int_0^{2\pi} \frac{k}{r} r \, d\theta = 2\pi k \tag{11-75}$$

Therefore,

$$\mathbf{V} = V_\theta = \frac{\Gamma}{2\pi r} = \frac{1}{r}\frac{\partial \phi}{\partial \theta} = -\frac{\partial \psi}{\partial r} \tag{11-76a}$$

$$V_r = 0 = \frac{\partial \phi}{\partial r} = \frac{1}{r}\frac{\partial \psi}{\partial \theta} \tag{11-76b}$$

from which the velocity potential and stream function are

$$\phi_V = \frac{\Gamma}{2\pi}\theta \tag{11-77a}$$

$$\psi_V = -\frac{\Gamma}{2\pi}\ln r \tag{11-77b}$$

and the orthogonal flow pattern is opposite that of the source, with the equipotential lines being radially outward. Counterclockwise is positive.

Flow about a rotating cylinder Consider setting in motion a lightweight cardboard cylinder by pulling a string which has been wrapped around it as indicated in Fig. 11-19. If it is done carefully, the cylinder will rise as a result of the lift produced by rotation. An analogous flow field can be produced in a perfect fluid by adding the effect of a vortex to a stationary cylinder. In polar coordinates

$$\psi_{\text{rot cyl}} = V_\infty \left(r - \frac{r_0^2}{r} \right) \sin\theta + \frac{\Gamma}{2\pi}\ln r \tag{11-78}$$

where the vortex has been taken as clockwise. The streamline patterns which are added are shown in Fig. 11-20.

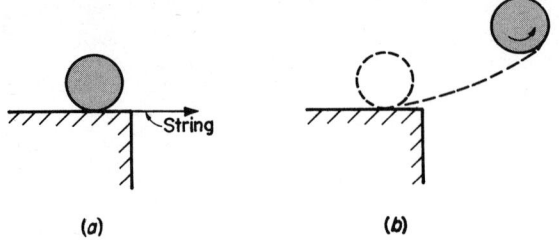

Fig. 11-19 Lift produced by rotation.

Proceeding in the usual manner, the velocity components are

$$V_\theta = -\frac{\partial \psi}{\partial r} = -V_\infty \left(1 + \frac{r_0^2}{r^2}\right) \sin \theta - \frac{\Gamma}{2\pi r} \qquad (11\text{-}79a)$$

$$V_r = \frac{1}{r}\frac{\partial \psi}{\partial \theta} = V_\infty \left(1 - \frac{r_0^2}{r^2}\right) \cos \theta \qquad (11\text{-}79b)$$

On the streamline $r = r_0$, that is, at the cylinder surface, $V_r = 0$ and

$$V_0 = V_\theta|_{r=r_0} = -2V_\infty \sin \theta - \frac{\Gamma}{2\pi r_0} \qquad (11\text{-}80)$$

At the stagnation point $V_\theta = V_r = 0$; hence,

$$\sin \theta_s = -\frac{\Gamma}{4\pi r_0 V_\infty} \qquad (11\text{-}81)$$

where the subscript s designates the stagnation point. The stagnation points are located symmetrically about the y axis in the third and fourth quadrants. If a counter-clockwise vortex had been taken, the stagnations points would lie in the first and second quadrants. Figure 11-21 shows the location of the stagnation points for different values of circulation Γ.

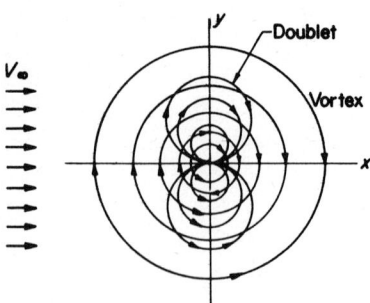

Fig. 11-20 Streamline patterns of a uniform flow, a doublet, and a clockwise vortex.

PERFECT FLUIDS

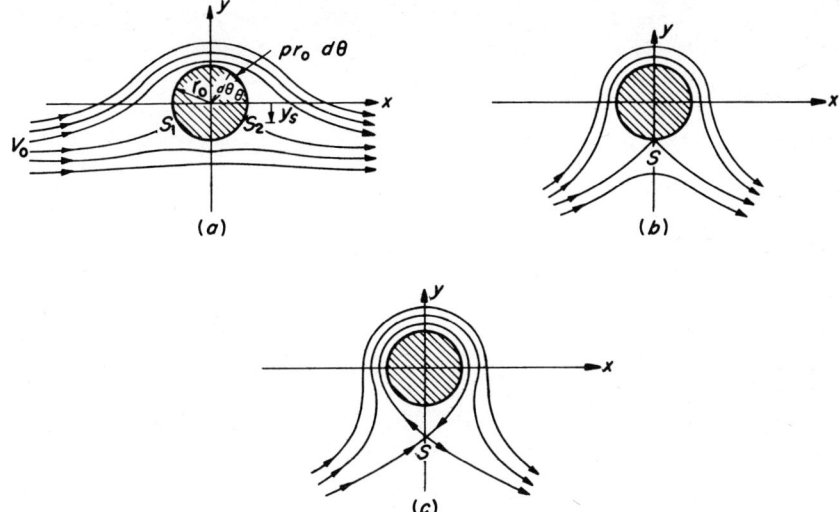

Fig. 11-21 Flow about a cylinder with circulation: (a) $\Gamma < 4\pi r_0 V_\infty$; (b) $\Gamma = 4\pi r_0 V_\infty$; (c) $\Gamma > 4\pi r_0 V_\infty$.

Force on a rotating cylinder At the cylinder surface the pressure can be determined by substituting the expression for velocity, Eq. (11-80), into the Bernoulli equation (11-70), getting

$$p_0 = p_\infty + \frac{\rho V_\infty^2}{2} - \frac{\rho}{2}\left(4V_\infty^2 \sin^2\theta + \frac{2V_\infty \Gamma \sin\theta}{\pi r_0} + \frac{\Gamma^2}{4\pi^2 r_0^2}\right) \tag{11-82}$$

We can integrate the product of this pressure and the area in the respective directions to get lift and drag as indicated in Fig. 11-22. By symmetry, the drag force is zero.

Carrying out the integration for lift gives

$$F_L = -2r_0 \int_{-\pi/2}^{\pi/2} \left(p_\infty + \frac{\rho V_\infty^2}{2} - \frac{\rho \Gamma^2}{8\pi^2 r_0^2}\right) \sin\theta\, d\theta - \frac{2\rho V_\infty \Gamma}{\pi} \int_{-\pi/2}^{\pi/2} \sin^2\theta\, d\theta$$

$$- 4\rho V_\infty^2 \int_{-\pi/2}^{\pi/2} \sin^3\theta\, d\theta \tag{11-83}$$

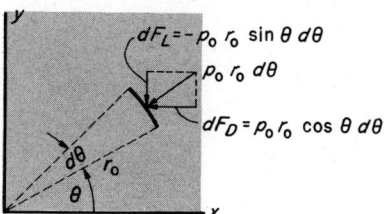

Fig. 11-22 Lift and drag forces on element of cylinder.

The limits were judiciously chosen to shorten the work of integration. The first and third integrals are zero; hence, from the second integral

$$F_L = \rho V_\infty \Gamma \quad \text{per unit length} \tag{11-84}$$

a simple result from such a formidable looking equation. *This result, known as the Kutta-Joukowski theorem, is valid for a boundary of any shape* although we have proved it only for a circular cylinder [6, 7]. This force accounts for the curve of a spinning baseball.

11-8 CLOSURE

In more extensive treatments of perfect-fluid theory a wide variety of flow patterns, limited only by ingenuity, might be synthesized by the combination of the elementary flow patterns presented in this chapter. Conformal-mapping techniques from complex-variable theory facilitate the determination of forces on airfoils by transformation to the circular cylinder for which the results are presented herein.

We shall see in subsequent chapters how potential-flow theory can be coupled with boundary-layer theory in solving "real" problems.

PROBLEMS

11-1. The velocity $\mathbf{V} = (y^2 - x^2)t\mathbf{i} + 2xyt\mathbf{k}$ describes the flow of a two-dimensional incompressible fluid. Is this flow irrotational? Find the vorticity.

11-2. If an incompressible flow is described by $\mathbf{V} = 2xyz\mathbf{i} + 16x^2yz^2\mathbf{j} + 4z\mathbf{k}$, find the three components of rotation and the three components of vorticity.

11-3. A three-dimensional incompressible flow field is given by $\mathbf{V} = (2x + 3y + 3)\mathbf{i} + (2xyz)\mathbf{j} + (xy/3)\mathbf{k}$. Find the vorticity.

11-4. For an incompressible fluid, which of the following flows satisfy irrotationality?

(a) $u = x^3 \sin y \quad v = 3x^2 \cos y$

(b) $V_r = 0 \quad V_\theta = \dfrac{k}{r}$

(c) $u = \ln x \quad v = xy(1 - \ln x)$

(d) $u = \ln xy + \sin yt \quad v = \cos xt - \dfrac{y}{x}$

11-5. Show that the circulation about a circular path with a vortex at the center is a constant.

11-6. If a vortex has a strength $\Gamma = 500$ ft²/sec, what is the velocity at a point 16 ft from the center of the vortex? Find the constant for this vortex flow.

11-7. A two-dimensional flow field is described by $\mathbf{V} = 2x^2y\mathbf{i} + \tfrac{2}{3}x^3\mathbf{j}$. For a circular path of 2-ft radius, find the circulation.

11-8. For an incompressible two-dimensional flow described by $\mathbf{V} = 2xy\mathbf{i} + 4xy^2\mathbf{j}$, what is the circulation around a 4-ft-diam circular path?

11-9. A two-dimensional velocity field is described by $\mathbf{V} = (x^2 + y^2)\mathbf{i} + (2xy^2)\mathbf{j}$. Find the integral along the path C between the points $(0,0)$ and $(1,3)$ of the component of $\mathbf{V}$ in the direction of C for the following cases:

(a) C is a straight line.

PERFECT FLUIDS

   (b) C is a parabola with vertex at the origin and opening to the right.
   (c) C is a portion of the y axis and a straight line perpendicular to it.

11-10. For an incompressible fluid, which of the following flows satisfy continuity?
   (a) $u = x^3 \sin y \qquad v = 3x^2 \cos y$
   (b) $V_r = 0 \qquad V_\theta = \dfrac{k}{r}$
   (c) $u = \ln x \qquad v = xy(1 - \ln x)$
   (d) $u = \ln xy + \sin yt \qquad v = \cos xt - \dfrac{y}{x}$

11-11. For $\psi = 3x - \frac{3}{2}xy^2$, (a) is this stream function possible and (b) does it represent potential flow?

11-12. In a two-dimensional planar flow of an incompressible fluid, the x component of velocity is $u = \frac{1}{2}x^2 + x - 2y$. Derive an expression for v.

11-13. For a certain frictionless flow, the only body force is gravity (the z axis is directed vertically upward), and the velocity (in feet per second) is given by $\mathbf{V} = x\mathbf{i} - y\mathbf{j}$ and $\mathbf{g} = -g\mathbf{k}$.
   (a) Is the flow steady or unsteady? Why?
   (b) Is the flow one-, two-, or three-dimensional? Explain your answer.
   (c) What is the acceleration of a particle at $(x,y,z) = (2,3,2)$?
   (d) Is this a possible incompressible flow field? Justify your answer.

11-14. A two-dimensional flow is described by the velocity $\mathbf{V} = 4x\mathbf{i} - (3x^2 + 4y)\mathbf{j}$.
   (a) Does this equation satisfy continuity?
   (b) Is the flow irrotational?

11-15. For a two-dimensional incompressible flow the velocity is given by $\mathbf{V} = 20xy\mathbf{i}$. Does a stream function exist? If so, what is it?

11-16. A two-dimensional incompressible flow is described by $\mathbf{V} = (xy + x)\mathbf{i} - \frac{1}{2}(y^2 + 2y)\mathbf{j}$.
   (a) Does this flow satisfy continuity?
   (b) Find the average angular velocity.

11-17. For the velocity field given by $\mathbf{V} = (\frac{3}{2}x^2 + y^2)\mathbf{i} - (x + 3xy)\mathbf{j}$, find the stream function.

11-18. A two-dimensional incompressible flow is described by the velocity $\mathbf{V} = 2x\mathbf{i} - (6x + 2y)\mathbf{j}$.
   (a) Does this flow satisfy continuity?
   (b) Is the flow irrotational?

11-19. For a steady two-dimensional flow of an incompressible fluid around a circular cylinder the stream function is $\psi = V_\infty \sin \theta\, r_0^2/r - V_\infty y$. Taking $V_\infty = 1$ ft/sec and the cylinder radius $r_0 = 4$ ft, obtain the streamlines $\psi = 0$ and $\psi = -1$ for this flow.

11-20. Find the velocity field described by the velocity potential $\phi = x + y$.

11-21. The velocity field for a two-dimensional incompressible flow is given by $\mathbf{V} = (1/r) \cos \theta\, \mathbf{e}_r + y\, \mathbf{e}_\theta$. If it exists, find the velocity potential for this flow.

11-22. Find the velocity potential for the flow given by the stream function $\psi = 5x^2 + y$.

11-23. Determine the potential functions for the flows of Prob. 11-4 which are irrotational.

11-24. For the cartesian velocity components $u = 3y$ and $v = 6x$:
   (a) Does a stream function exist? If so, what is it?
   (b) Are the velocity components derivable from a potential function? If so, what is the potential function?

11-25. A two-dimensional flow field is described by the velocity $\mathbf{V} = -2x\mathbf{i} + (2y + 2)\mathbf{j}$. Does a velocity potential exist? If so, what is it?

11-26. For steady frictionless incompressible flow show that Bernoulli's equation for flow of a fluid along a streamline is equally valid between points not on the same streamline. *Hint:* Write

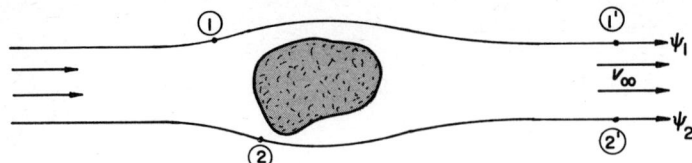

Fig. P11-26

Bernoulli's equation between the respective points on each streamline and solve simultaneously with the pressure-variation equation, $\partial p/\partial z = -\rho g$, at infinity.

11-27. Consider an incompressible flow field $\psi = 3x^2 + 3y^2$.
  (a) Determine the velocity $\mathbf{V}(x,y)$.
  (b) Determine the rotation ω_z.
  (c) If possible evaluate the pressure difference between the two points $\mathbf{r} = \mathbf{i} + \mathbf{j}$ and $\mathbf{r} = 2\mathbf{i} + 2\mathbf{j}$ if the fluid is water.
  (d) Where are the stagnation points in the flow located?

11-28. A steady incompressible flow field is defined by $\phi = x - y$.
  (a) Is the flow field irrotational? If so, find the stream function.
  (b) What is the mass flow rate per unit depth between (0,0) and (3,3)?

11-29. Show that the two-dimensional velocity potential $\phi = r \sin \theta$ satisfies continuity and irrotationality. Find the stream function.

11-30. If $u = 3x$ and $v = -3y$, find the stream function. Does a velocity potential exist? If so, what is it?

11-31. Explain why the existence of a stream function does not assure the existence of a velocity potential. Is the converse true?

11-32. For the stream function $\psi = 2x^2 - 2y^2$, find the slope of the equipotential lines through the point (2,2). What is the slope of the streamline at this point?

11-33. The stream function for a two-dimensional, incompressible flow is given by $\psi = x^2 + y^3$.
  (a) What is the velocity at the point (2,4)?
  (b) Is the flow irrotational?

11-34. For the velocity potential $\phi = x^3/3 - x^2 - xy^2 + y^2$:
  (a) Determine the velocity which describes the flow.
  (b) Show that the vorticity is everywhere zero.
  (c) What is the stream function?

11-35. Assuming that an incompressible steady two-dimensional flow fluid is irrotational and satisfies continuity, use the Cauchy-Riemann equations to find the stream function if $\phi = \ln x + e^y$. Comment on whether the assumption made concerning irrotationality and continuity was valid. Back up your comments with mathematical proof.

11-36. The stream function for a cylinder is given by $\psi = V_\infty (r - r_0^2/r) \sin \theta$, where V_∞ and r_0 are constants. Find the velocity potential.

11-37. Explain why high winds sometimes lift the roofs from buildings.

11-38. Indicate by sketch and explanation why a baseball will curve.

11-39. A source of strength 4π is located at (4,0), and a source of strength π is located at (0,0).
  (a) Locate the stagnation point.
  (b) What is the velocity at the point (0,4)?

11-40. (a) For a source strength of 2 ft³/sec-ft, plot the streamlines for two equal sources located 2 ft apart on the x axis.
  (b) Explain the significance of this flow.

PERFECT FLUIDS

11-41. Describe the velocity field produced by a source and a uniform flow field.

11-42. Show that the potential function for a three-dimensional source of strength Λ is $\phi = \Lambda/4\pi r$.

11-43. What is the stream function for potential flow past a circular cylinder of radius 1 ft normal to a free-stream velocity of $V_\infty = 15$ ft/sec?

11-44. For a circular cylinder with a uniform flow field ($V_\infty = 15$ ft/sec) perpendicular to it, find the velocity components at $(r,\theta) = (2 \text{ ft}, \pi/6)$. The radius of the cylinder is 1 ft.

11-45. A 2-ft-diam circular cylinder is normal to a free stream of velocity $V_\infty = 200$ ft/sec. For a circulation of 600π ft²/sec, find the location of the stagnation points and the largest negative pressure.

11-46. A circular cylinder has water flowing perpendicular to it at 10 ft/sec. What is the change in pressure from the free-stream flow field to the point $(r,\theta) = (2,0)$? The cylinder radius is 2 ft.

11-47. At what point on the surface of a circular cylinder in a uniform stream is the tangential velocity one-half of the free-stream velocity?

11-48. A 4-ft-diam cylinder 10 ft long is rotated about its axis at 240 rpm in a stream of standard air flowing at 60 ft/sec. The rotation produces a circulation of 200π ft²/sec at the surface of the cylinder. What is the lift force on the cylinder?

REFERENCES

1. Karamcheti, K.: "Principles of Ideal-fluid Aerodynamics," Wiley, New York, 1966.
2. Kuethe, A. M., and J. D. Schetzer: "Foundations of Aerodynamics," Wiley, New York, 1959.
3. Milne-Thomson, L. M.: "Theoretical Hydrodynamics," Macmillan, New York, 1960.
4. Nelson, A. L., K. W. Folley, and M. Coral: "Differential Equations," Heath, Boston, 1952.
5. Pao, R. H. F.: "Fluid Dynamics," Merrill, Columbus, Ohio, 1967.
6. Pope, A.: "Basic Wing and Airfoil Theory," McGraw-Hill, New York, 1951.
7. Rauscher, M.: "Introduction to Aeronautical Dynamics," Wiley, New York, 1953.
8. Stein, F. M.: "An Introduction to Vector Analysis," Harper & Row, New York, 1963.
9. Yuan, S. W.: "Foundations of Fluid Mechanics," Prentice-Hall, Englewood Cliffs, N.J., 1967.

CHAPTER 12

LAMINAR FLOW OF INCOMPRESSIBLE VISCOUS FLUIDS

The previous chapters have introduced all the basic concepts of momentum and energy transport necessary for a study of convective processes in a real fluid. In the present chapter we shall consider laminar flow of an incompressible viscous fluid, first treating momentum transport in an isothermal flow field and then extending to the more general case involving both momentum and heat transfer in a forced flow field. The study of natural convective processes is deferred until Chap. 16.

12-1 ISOTHERMAL FLOW

Chapter 11 consisted of a study of inviscid or perfect-fluid flow, a situation of considerable practical importance, as we shall see in the study of the flow of real fluids. Every real fluid, however, has a finite viscosity which gives rise to shear forces not considered in perfect-fluid flow theory. In many flow fields, this viscosity is quite small; e.g., the kinematic viscosity of water at room temperature is of the order of 10^{-5} ft²/sec, and it would appear that viscous effects in such fields would be negligible compared with other forces in the momentum equation. This condition is generally true in fields far from solid bodies, as we shall see later.

LAMINAR FLOW OF INCOMPRESSIBLE VISCOUS FLUIDS

The small viscosities of such important fluids as air and water presented a formidable barrier to the early study of fluid mechanics. The early Greek mathematicians, familiar with the diminishing velocity of a spear in flight, erroneously concluded that it was necessary to apply a force continually to sustain the velocity of a body in motion subjected to no opposing forces! The low viscosity of air prevented them from recognizing the existence of an opposing force, drag, and this hampered their progress in the study of mechanics.

During the last half of the nineteenth century the study of fluid dynamics was sharply divided between theoretical and experimental efforts. A complete[1] formulation of the equations of motion of a viscous fluid has been available since 1845. Known as the *Navier-Stokes equations*, they are largely attributable to the contributions of Navier [13], Poisson [19], St. Venant [21], and Stokes [24] during the period from 1827 to 1845. These equations form a set of nonlinear partial differential equations the solution of which is a formidable task. This fact, coupled with the very small viscosity of air and water, led many theoreticians to conclude that the inviscid-fluid assumption was justifiable, and the mathematical theory of perfect-fluid flow was highly developed before the turn of the twentieth century.

Practical engineers, on the other hand, were not enthusiastic supporters of mathematical efforts which yielded such absurd results as zero pressure loss for flow of water through a pipe or zero drag for a cylinder subjected to a cross flow of air. It is certainly not surprising that engineering efforts were heavily concentrated toward experimental programs and correlation efforts to obtain maximum applicability of the measured data.

At this time the field of fluid mechanics was divided into *theoretical hydrodynamics* and *hydraulics*, the former being a mathematical science, the latter an empirical one. The reunification of these two branches was largely due to the contribution of Prandtl [20], who in 1904 presented a paper "On Fluid Motion with Very Small Friction" before the Third International Mathematical Congress in Heidelburg. In this work, Prandtl showed both experimentally and analytically that the flow over a solid body is divided into two regions, a *boundary layer* adjacent to the body, in which viscous effects are important, and an outer flow field, in which perfect-fluid flow theory is applicable. A typical boundary layer is shown in Fig. 12-1. The importance of his work cannot be overstated. The boundary-layer theory permitted Prandtl to mathematically analyze several simple flow problems with meaningful results. Nonetheless, boundary-layer theory was applied and developed only in Prandtl's own institute in Göttingen for the next 20 years. Following this period of development and demonstrated success, the theory was accepted, applied, and developed throughout Germany, Great Britain, and the United States. Today it is recognized as one of the most important concepts in fluid mechanics.

Before we can attempt to treat any real isothermal-flow problem, we must have at our disposal the momentum equations containing the viscous forces. Of

---
[1] Complete is here intended to include viscous, gravitational, and pressure forces in the momentum equation; certainly more complete equations could be formulated.

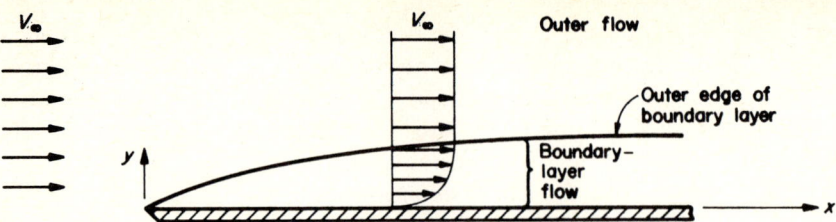

Fig. 12-1 Typical isothermal velocity boundary layer for laminar flow over a flat plate.

particular importance to the study of isothermal momentum transport is a working knowledge of the significance of each term in the viscous momentum (Navier-Stokes) equations, which are derived in the following section.

12-2 NAVIER-STOKES EQUATIONS

In Chap. 10, the linear momentum equation for a finite control volume was shown to be

$$\sum \mathbf{F} = \frac{d}{dt} \int_{cv} \rho \mathbf{V} \, d\mathscr{V} + \int_{cs} \mathbf{V}(\rho \mathbf{V} \cdot d\mathbf{A}) \tag{10-50}$$

and this was applied in Sec. 10-5 to a differential control volume. The results in cartesian coordinates were

$$\sum f_x = \rho \frac{Du}{Dt} \tag{10-77a}$$

$$\sum f_y = \rho \frac{Dv}{Dt} \tag{10-77b}$$

$$\sum f_z = \rho \frac{Dw}{Dt} \tag{10-77c}$$

It is important to realize that these equations place no restrictions on the forces $f_x, f_y,$ and f_z. Thus, Eqs. (10-77a), (10-77b), and (10-77c) apply to an inviscid- or a viscous-fluid flow.

Consider a differential control volume as shown in Fig. 12-2 and imagine it to be a small element of fluid removed from a boundary layer such as the viscous flow of Fig. 12-1. The surfaces normal to the x and z directions have been shown removed from the control volume for clarity in depicting all stresses on the element. The forces and stresses acting on the element are these:

1. Normal stresses, for example, τ_{xx} is the normal stress in the x direction, and this includes the static-pressure force.

LAMINAR FLOW OF INCOMPRESSIBLE VISCOUS FLUIDS

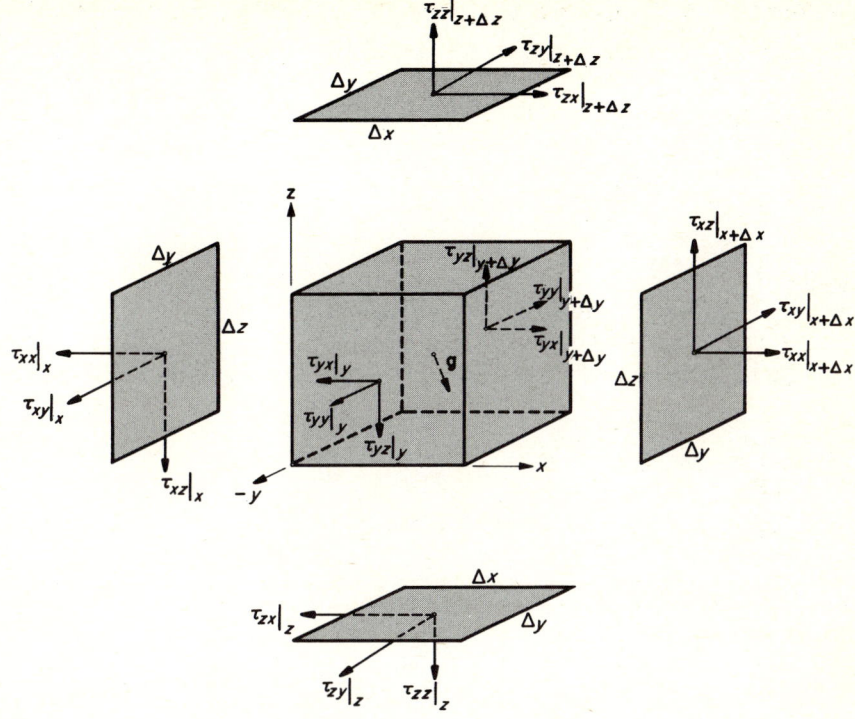

Fig. 12-2 Differential control volume for a viscous fluid showing stress acting on the six faces.

2. Shear stresses, for example, τ_{xy} is the shear stress on the surface normal to the x axis and acting in the y direction. Note that the second subscript denotes the axis of the direction of the stress. The sign convention for shear stress is positive if the stress acts in a positive coordinate direction on a surface whose outward-drawn normal is in a positive coordinate direction and vice versa. For example, the outward-drawn normal to the surface $\Delta x\, \Delta y$ at z is in the negative z-coordinate direction. $\tau_{zy}|_z$ is shown in the direction of the negative y-coordinate axis and thus is a positive shear stress. (All shear stresses in Fig. 12-2 are positive.)
3. Body forces, e.g., gravitational, electric, or magnetic forces. We shall consider only gravitational body forces. In Fig. 12-2, the gravitational-force vector is **g**, and we shall use g_x, g_y, and g_z for the gravitational force per unit mass in the x-, y-, and z-coordinate directions, respectively.

For simplicity we shall evaluate the forces acting on the infinitesimal control volume in the x-coordinate direction only, and apply them to Eq. (10-77a) to obtain one of the Navier-Stokes equations in cartesian coordinates. The remaining two can then be written by analogy. The stresses in the x direction give rise to forces

$$(\tau_{xx}|_{x+\Delta x} - \tau_{xx}|_x)\, \Delta y\, \Delta z \qquad (12\text{-}1)$$

$$(\tau_{yx}|_{y+\Delta y} - \tau_{yx}|_y)\, \Delta x\, \Delta z \qquad (12\text{-}2)$$

$$(\tau_{zx}|_{z+\Delta z} - \tau_{zx}|_z)\, \Delta x\, \Delta y \qquad (12\text{-}3)$$

The x-direction gravitational (body) force is

$$\rho g_x\, \Delta x\, \Delta y\, \Delta z \qquad (12\text{-}4)$$

Adding Eqs. (12-1) through (12-4), we have

$$\sum (f_x)_{\text{av}}\, \Delta x\, \Delta y\, \Delta z = \rho g_x\, \Delta x\, \Delta y\, \Delta z + (\tau_{xx}|_{x+\Delta x} - \tau_{xx}|_x)\, \Delta y\, \Delta z$$
$$+ (\tau_{yx}|_{y+\Delta y} - \tau_{yx}|_y)\, \Delta x\, \Delta z + (\tau_{zx}|_{z+\Delta z} - \tau_{zx}|_z)\, \Delta x\, \Delta y$$

Dividing by $\Delta x\, \Delta y\, \Delta z$ yields

$$\sum (f_x)_{\text{av}} = \rho g_x + \frac{\tau_{xx}|_{x+\Delta x} - \tau_{xx}|_x}{\Delta x} + \frac{\tau_{yx}|_{y+\Delta y} - \tau_{yx}|_y}{\Delta y} + \frac{\tau_{zx}|_{z+\Delta z} - \tau_{zx}|_z}{\Delta z}$$

which in the limit as Δx, Δy, and Δz approach zero becomes

$$\sum f_x = \rho g_x + \frac{\partial \tau_{xx}}{\partial x} + \frac{\partial \tau_{yx}}{\partial y} + \frac{\partial \tau_{zx}}{\partial z} \qquad (12\text{-}5)$$

Substituting Eq. (12-5) into (10-77a) gives

$$\rho \frac{Du}{Dt} = \rho g_x + \frac{\partial \tau_{xx}}{\partial x} + \frac{\partial \tau_{yx}}{\partial y} + \frac{\partial \tau_{zx}}{\partial z} \qquad (12\text{-}6)$$

In a similar fashion the y- and z-coordinate direction equations of motion are

$$\rho \frac{Dv}{Dt} = \rho g_y + \frac{\partial \tau_{xy}}{\partial x} + \frac{\partial \tau_{yy}}{\partial y} + \frac{\partial \tau_{zy}}{\partial z} \qquad (12\text{-}7)$$

$$\rho \frac{Dw}{Dt} = \rho g_z + \frac{\partial \tau_{xz}}{\partial x} + \frac{\partial \tau_{yz}}{\partial y} + \frac{\partial \tau_{zz}}{\partial z} \qquad (12\text{-}8)$$

and the problem is reduced to obtaining useful expressions for the normal and shear stresses.

By Stokes' hypothesis, the general relationships between normal stresses and *rate of strain* are

$$\begin{aligned}\tau_{xx} &= -p + \lambda \nabla \cdot \mathbf{V} + 2\mu \xi_x \\ \tau_{yy} &= -p + \lambda \nabla \cdot \mathbf{V} + 2\mu \xi_y \\ \tau_{zz} &= -p + \lambda \nabla \cdot \mathbf{V} + 2\mu \xi_z\end{aligned} \qquad (12\text{-}9)$$

LAMINAR FLOW OF INCOMPRESSIBLE VISCOUS FLUIDS

and between shearing stresses and rate of strain are

$$\tau_{xy} = \mu \gamma_{xy} = \tau_{yx}$$
$$\tau_{yz} = \mu \gamma_{yz} = \tau_{zy} \tag{12-10}$$
$$\tau_{zx} = \mu \gamma_{zx} = \tau_{xz}$$

as shown by Pai [16] and Schlichting [23]. In these expressions the rates of *normal strain* are

$$\xi_x = \text{rate of normal strain in } x \text{ direction} = \frac{\partial u}{\partial x}$$

$$\xi_y = \text{rate of normal strain in } y \text{ direction} = \frac{\partial v}{\partial y} \tag{12-11}$$

$$\xi_z = \text{rate of normal strain in } z \text{ direction} = \frac{\partial w}{\partial z}$$

and the rates of *shearing strain* are

$$\gamma_{xy} = \text{shearing strain on } xy \text{ plane} = \frac{\partial u}{\partial y} + \frac{\partial v}{\partial x}$$

$$\gamma_{yz} = \text{shearing strain on } yz \text{ plane} = \frac{\partial v}{\partial z} + \frac{\partial w}{\partial y} \tag{12-12}$$

$$\gamma_{zx} = \text{shearing strain on } xz \text{ plane} = \frac{\partial w}{\partial x} + \frac{\partial u}{\partial z}$$

Now the expressions for the normal stresses contain the parameter λ, sometimes referred to as the second coefficient of viscosity. This is not a directly measurable fluid property, and the numerical value assigned to this term in the Navier-Stokes equations is due to the hypothesis made by Stokes in 1845 which requires that

$$3\lambda + 2\mu = 0 \tag{12-13}$$

or

$$\lambda = -\tfrac{2}{3}\mu \tag{12-13a}$$

This expression has been shown to be valid for a monatomic gas using kinetic theory; it has not been shown for other fluids. It should be noted that for an incompressible fluid, the divergence of **V** vanishes by the continuity equation and this term is not a part of the equations. Further, the validity of this hypothesis for common fluids can hardly be doubted considering the number of verifications afforded by excellent agreement between the solutions of the resulting equations with experimental data. For additional discussions of Stokes' hypothesis see Schlichting [23] and Pai [16].

Substituting Eqs. (12-11) and (12-13) into (12-9) yields

$$\tau_{xx} = -p - \tfrac{2}{3}\mu \nabla \cdot \mathbf{V} + 2\mu \frac{\partial u}{\partial x}$$

$$\tau_{yy} = -p - \tfrac{2}{3}\mu \nabla \cdot \mathbf{V} + 2\mu \frac{\partial v}{\partial y} \qquad (12\text{-}14)$$

$$\tau_{zz} = -p - \tfrac{2}{3}\mu \nabla \cdot \mathbf{V} + 2\mu \frac{\partial w}{\partial z}$$

and substituting Eqs. (12-12) into Eqs. (12-10) gives

$$\tau_{xy} = \mu\left(\frac{\partial u}{\partial y} + \frac{\partial v}{\partial x}\right)$$

$$\tau_{yz} = \mu\left(\frac{\partial v}{\partial z} + \frac{\partial w}{\partial y}\right) \qquad (12\text{-}15)$$

$$\tau_{zx} = \mu\left(\frac{\partial w}{\partial x} + \frac{\partial u}{\partial z}\right)$$

With these expressions for normal and shearing stresses, Eqs. (12-6) to (12-8) become

$$\rho \frac{Du}{Dt} = \rho g_x - \frac{\partial p}{\partial x} + \frac{\partial}{\partial x}\left[\mu\left(2\frac{\partial u}{\partial x} - \frac{2}{3}\nabla\cdot\mathbf{V}\right)\right] + \frac{\partial}{\partial y}\left[\mu\left(\frac{\partial u}{\partial y} + \frac{\partial v}{\partial x}\right)\right]$$
$$+ \frac{\partial}{\partial z}\left[\mu\left(\frac{\partial w}{\partial x} + \frac{\partial u}{\partial z}\right)\right] \qquad (12\text{-}16)$$

$$\rho \frac{Dv}{Dt} = \rho g_y - \frac{\partial p}{\partial y} + \frac{\partial}{\partial y}\left[\mu\left(2\frac{\partial v}{\partial y} - \frac{2}{3}\nabla\cdot\mathbf{V}\right)\right] + \frac{\partial}{\partial z}\left[\mu\left(\frac{\partial v}{\partial z} + \frac{\partial w}{\partial y}\right)\right]$$
$$+ \frac{\partial}{\partial x}\left[\mu\left(\frac{\partial u}{\partial y} + \frac{\partial v}{\partial x}\right)\right] \qquad (12\text{-}17)$$

$$\rho \frac{Dw}{Dt} = \rho g_z - \frac{\partial p}{\partial z} + \frac{\partial}{\partial z}\left[\mu\left(2\frac{\partial w}{\partial z} - \frac{2}{3}\nabla\cdot\mathbf{V}\right)\right] + \frac{\partial}{\partial x}\left[\mu\left(\frac{\partial w}{\partial x} + \frac{\partial u}{\partial z}\right)\right]$$
$$+ \frac{\partial}{\partial y}\left[\mu\left(\frac{\partial v}{\partial z} + \frac{\partial w}{\partial y}\right)\right] \qquad (12\text{-}18)$$

These are the fundamental equations of motion for a viscous compressible fluid and are usually referred to as the Navier-Stokes equations. For a compressible fluid, we are generally concerned with temperature changes also, but in the present section we restrict our consideration to the case of isothermal flow. To complete the formulation of the appropriate equations then, we need only add the continuity equation from Sec. 10-2

$$\frac{\partial \rho}{\partial t} + \frac{\partial(\rho u)}{\partial x} + \frac{\partial(\rho v)}{\partial y} + \frac{\partial(\rho w)}{\partial z} = 0 \qquad (10\text{-}24)$$

and an equation of state relating pressure, temperature, and density. As an example,

LAMINAR FLOW OF INCOMPRESSIBLE VISCOUS FLUIDS

for a perfect gas this can be taken from Sec. 1-5 and is

$$p = \rho RT \tag{1-7}$$

It is informative to note that isothermal flow in the cartesian coordinate system involves five parameters, u, v, w, ρ, and p and that there are five equations, three Navier-Stokes, the continuity, and the equation of state. The Navier-Stokes equations are second-order nonlinear partial differential equations, and the solution for all but the simplest flows presents a formidable mathematical problem.

If the density is constant, which is true for most isothermal flows, the flow is termed *incompressible*. For this case, Eq. (10-24) reduces to

$$\frac{\partial u}{\partial x} + \frac{\partial v}{\partial y} + \frac{\partial w}{\partial z} = \nabla \cdot \mathbf{V} = 0 \tag{12-19}$$

and the terms involving the divergence of $\mathbf{V}$ in the momentum equations vanish, giving for constant viscosity

■ $$\rho \frac{Du}{Dt} = \rho g_x - \frac{\partial p}{\partial x} + \mu \left(\frac{\partial^2 u}{\partial x^2} + \frac{\partial^2 u}{\partial y^2} + \frac{\partial^2 u}{\partial z^2} \right) \tag{12-16a}$$

■ $$\rho \frac{Dv}{Dt} = \rho g_y - \frac{\partial p}{\partial y} + \mu \left(\frac{\partial^2 v}{\partial x^2} + \frac{\partial^2 v}{\partial y^2} + \frac{\partial^2 v}{\partial z^2} \right) \tag{12-17a}$$

■ $$\rho \frac{Dw}{Dt} = \rho g_z - \frac{\partial p}{\partial z} + \mu \left(\frac{\partial^2 w}{\partial x^2} + \frac{\partial^2 w}{\partial y^2} + \frac{\partial^2 w}{\partial z^2} \right) \tag{12-18a}$$

It should be noted that each of these equations involves the substantial or fluid derivative and ∇^2 of the velocity component for the cartesian coordinate direction in which the body and pressure forces are taken.

Cylindrical coordinates: constant fluid properties It is frequently desirable to formulate a flow problem in the cylindrical coordinate system, both for internal tube flow and for external flows over bodies of revolution. For the r, θ, and z radial, aximuthal, and axial coordinates the equations are [7]

$$\rho \left(\frac{\partial V_r}{\partial t} + V_r \frac{\partial V_r}{\partial r} + \frac{V_\theta}{r} \frac{\partial V_r}{\partial \theta} - \frac{V_\theta^2}{r} + V_z \frac{\partial V_r}{\partial z} \right) = \rho g_r - \frac{\partial p}{\partial r}$$
$$+ \mu \left(\frac{\partial^2 V_r}{\partial r^2} + \frac{1}{r} \frac{\partial V_r}{\partial r} - \frac{V_r}{r^2} + \frac{1}{r^2} \frac{\partial^2 V_r}{\partial \theta^2} - \frac{2}{r^2} \frac{\partial V_\theta}{\partial \theta} + \frac{\partial^2 V_r}{\partial z^2} \right) \tag{12-20}$$

$$\rho \left(\frac{\partial V_\theta}{\partial t} + V_r \frac{\partial V_\theta}{\partial r} + \frac{V_\theta}{r} \frac{\partial V_\theta}{\partial \theta} + \frac{V_r V_\theta}{r} + V_z \frac{\partial V_\theta}{\partial z} \right) = \rho g_\theta - \frac{1}{r} \frac{\partial p}{\partial \theta}$$
$$+ \mu \left(\frac{\partial^2 V_\theta}{\partial r^2} + \frac{1}{r} \frac{\partial V_\theta}{\partial r} - \frac{V_\theta}{r^2} + \frac{1}{r^2} \frac{\partial^2 V_\theta}{\partial \theta^2} + \frac{2}{r^2} \frac{\partial V_r}{\partial \theta} + \frac{\partial^2 V_\theta}{\partial z^2} \right) \tag{12-21}$$

$$\rho \left(\frac{\partial V_z}{\partial t} + V_r \frac{\partial V_z}{\partial r} + \frac{V_\theta}{r} \frac{\partial V_z}{\partial \theta} + V_z \frac{\partial V_z}{\partial z} \right) = \rho g_z - \frac{\partial p}{\partial z}$$
$$+ \mu \left(\frac{\partial^2 V_z}{\partial r^2} + \frac{1}{r} \frac{\partial V_z}{\partial r} + \frac{1}{r^2} \frac{\partial^2 V_z}{\partial \theta^2} + \frac{\partial^2 V_z}{\partial z^2} \right) \tag{12-22}$$

The continuity equation is obtained from div **V** in cylindrical coordinates and is

$$\frac{\partial V_r}{\partial r} + \frac{V_r}{r} + \frac{1}{r}\frac{\partial V_\theta}{\partial \theta} + \frac{\partial V_z}{\partial z} = 0 \tag{12-23}$$

12-3 FLOW IN TUBES

One of the simplest (yet one of the most important) cases of viscous incompressible flow is flow in a circular tube, pipe, or duct. In accordance with our previous discussion, a boundary-layer flow begins at the entrance of the duct, and this boundary layer develops in thickness until it completely fills the tube, as shown in Fig. 12-3.

At station 1, most of the flow area is filled with potential (inviscid) flow, the portion near the wall being boundary-layer flow with a typical velocity profile due to viscous retardation. At station 2, the boundary layer has developed or grown considerably, and a large part of the flow is within the boundary layer. Notice that the velocity in the core or potential flow has increased, as required by the continuity equation, which requires the average velocity to remain constant for incompressible flow in a duct of constant cross-sectional area. At station 3, the flow is *fully developed* with respect to axial position; i.e., the boundary layer completely fills the tube. From this point on, the velocity profile is unchanged, and the distance from entrance to station 3 is usually called the *entry length*.

For fully developed laminar incompressible flow in a circular tube, the axial symmetry and absence of rotation mean that the tangential and radial velocity components are nonexistent, i.e.,

$$V_\theta = V_r = 0 \tag{12-24}$$

and thus the Navier-Stokes equations in cylindrical coordinates are reduced to one equation. Specifically, Eqs. (12-20) and (12-21) show that $p \neq p(r,\theta)$ and thus Eq. (12-22) simplifies to

$$\mu\left(\frac{d^2 V_z}{dr^2} + \frac{1}{r}\frac{dV_z}{dr}\right) = \frac{dp}{dz} \tag{12-25}$$

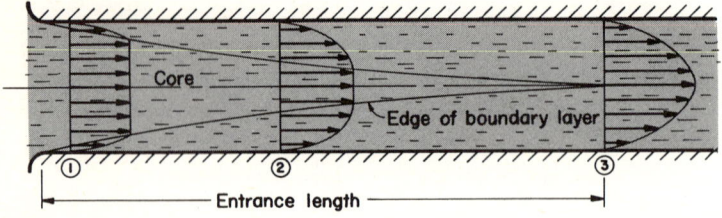

Fig. 12-3 Entry-length laminar incompressible flow in a circular tube.

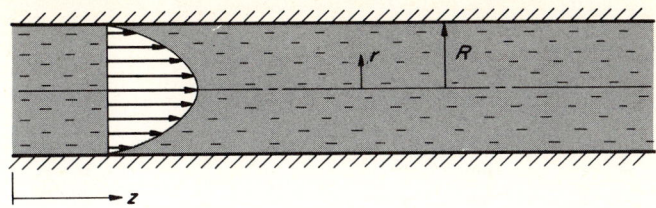

Fig. 12-4 Fully developed laminar flow in a constant-area tube.

where the body force ρg_z is assumed to be negligible. The flow is depicted in Fig. 12-4, from which an obvious boundary condition is

$$V_z = 0 \quad \text{at } r = R \tag{12-26a}$$

Less obvious, but equally important, is the requirement that

$$V_z \text{ be finite} \quad \text{at } r = 0 \tag{12-26b}$$

This will become evident as the constants for this second-order linear ordinary differential equation are evaluated.

To effect a solution, let $dV_z/dr = \xi$, and Eq. (12-25) is reduced to

$$\frac{d\xi}{dr} + \frac{1}{r}\xi = \frac{1}{\mu}\frac{dp}{dz} \tag{12-27}$$

where $\frac{1}{\mu}\frac{dp}{dz}$ is independent of r and may be regarded as a known constant. The associated homogeneous equation is

$$\frac{d\xi_c}{dr} + \frac{1}{r}\xi_c = 0 \tag{12-28}$$

which can be solved by separating the variables and integrating. This yields

$$\ln \xi_c = -\ln r + \ln C_1$$

or

$$\xi_c = \frac{C_1}{r} \tag{12-29}$$

By inspection we see that

$$\xi_p = \frac{r}{2}\frac{1}{\mu}\frac{dp}{dz} \tag{12-30}$$

is a particular solution of Eq. (12-27), and the general solution for ξ is

$$\xi = \frac{1}{2\mu}\frac{dp}{dz}r + \frac{C_1}{r} \tag{12-31}$$

Since $\xi = dV_z/dr$, we have

$$\frac{dV_z}{dr} = \frac{1}{2\mu}\frac{dp}{dz}r + \frac{C_1}{r} \qquad (12\text{-}31a)$$

and integration yields

$$V_z = \frac{1}{4\mu}\frac{dp}{dz}r^2 + C_1 \ln r + C_2 \qquad (12\text{-}32)$$

The problem has thus been reduced to evaluation of the two constants, C_1 and C_2. First, examine C_1 for $r = 0$. The term $\ln r$ becomes infinite, a physically impossible situation, and we conclude that C_1 must be zero. The other constant is found by application of the boundary condition of Eq. (12-26) and is

$$C_2 = -\frac{1}{4\mu}\frac{dp}{dz}R^2 \qquad (12\text{-}33)$$

The final expression for the velocity profile is

$$V_z = -\frac{1}{4\mu}\frac{dp}{dz}(R^2 - r^2) \qquad (12\text{-}34)$$

This is the equation of a parabola, and since the flow is axially symmetric, the velocity profile generates a paraboloidal surface. The maximum value of the velocity is readily seen to occur at $r = 0$, and thus

$$(V_z)_{\max} = -\frac{1}{4\mu}\frac{dp}{dz}R^2 \qquad (12\text{-}35)$$

The minus sign in this equation is physically correct, since the pressure in a viscous flow decreases with distance; that is, dp/dz is negative. The mean velocity can be obtained by integrating to obtain the volume under the paraboloidal surface and dividing this by the cross-sectional area of the tube. Doing so gives

$$V = \frac{1}{\pi R^2}\int_0^R V_z 2\pi r\, dr$$

$$V = (V_z)_{\mathrm{av}} = \tfrac{1}{2}(V_z)_{\max} = -\frac{1}{8\mu}\frac{dp}{dz}R^2 \qquad (12\text{-}36)$$

In engineering practice it is customary to express the pressure gradient in terms of the friction factor f defined by

$$-\frac{dp}{dz} = \frac{f}{D}\frac{\rho V^2}{2} \qquad (12\text{-}37)$$

where $\rho V^2/2$ is the dynamic pressure of the mean flow. If Δp is the pressure drop $p_1 - p_2$ in a length $L = l_2 - l_1$, this can be expressed as

■ $$\frac{\Delta p}{L} = \frac{f}{D}\frac{\rho V^2}{2} \qquad (12\text{-}37a)$$

LAMINAR FLOW OF INCOMPRESSIBLE VISCOUS FLUIDS

frequently referred to as the *Darcy-Weisbach equation*. Substituting the expression for dp/dz from Eq. (12-37) into (12-36) yields

$$V = \frac{1}{8\mu} \frac{f}{D} \frac{\rho V^2}{2} \left(\frac{D}{2}\right)^2$$

which upon solving for f gives

$$f = \frac{64}{\mathbf{Re}_D} \tag{12-38}$$

This result, which was obtained by solving the Navier-Stokes equations, has been experimentally verified in both smooth and rough tubes and pipes for $\mathbf{Re}_D$ up to 2,000.

It is frequently useful to have working equations in terms of the volumetric flow rate Q. Since

$$Q = \frac{\pi D^2}{4} V \tag{12-39}$$

we can obtain

$$\frac{dp}{dz} = -\frac{1280\mu}{\pi D^4} \tag{12-40}$$

from Eq. (12-36). Integrating over a finite length L yields the pressure loss due to the viscous-fluid motion

$$\Delta p = p_1 - p_2 = \frac{1280 L \mu}{\pi D^4} \tag{12-41}$$

or dividing by ρ, we have the head loss

$$\frac{\Delta p}{\rho} \equiv h_l = 128 \frac{QL\mu}{\pi D^4 \rho} \tag{12-42}$$

It is important to note that this result is valid for any pipe orientation, vertical, horizontal, or at an angle.

Combining the expression for the average velocity for a finite length L from Eq. (12-36) with Eq. (12-39), we have

$$Q = \frac{\pi}{8\mu} \frac{R^4}{L} (p_1 - p_2) \tag{12-43}$$

which is known as the *Hagen-Poiseuille equation* for laminar flow through a pipe. It is interesting to note that Hagen [4] first deduced this experimentally in 1839, and Poiseuille [18] arrived at the same result independently in 1840.

We may note that the Navier-Stokes equations are not required for the analytical solution of this problem. Consider the control volume illustrated in Fig. 12-5, which represents a section of fluid in fully developed steady laminar flow. The fluid

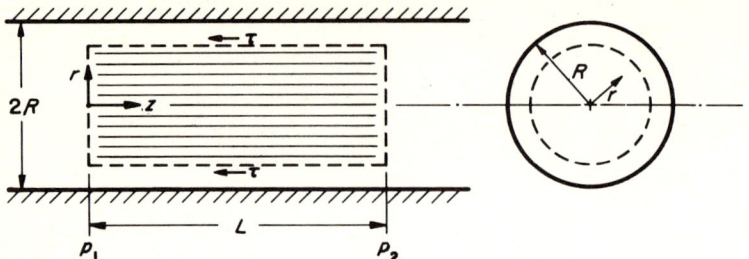

Fig. 12-5 Control volume for fully developed steady laminar flow in a tube.

is acted upon by two types of force, a pressure gradient with $p_1 > p_2$ tending to cause acceleration to the right and a viscous shear tending to retard the motion. These are in balance for this problem. Recalling that the shearing stress is simply the product of the viscosity and the velocity gradient in the r direction, we may speak of shear stress acting at any radius r within the fluid control volume. Since the flow is axisymmetric, the velocity at a given radius is constant; hence, there is no change of momentum in the axial direction and consequently no inertia force. For steady flow, the equilibrium condition requires that the pressure force be equal and opposite to the shear force. The pressure force is given by

$$F_p = (p_1 - p_2)\pi r^2 \tag{12-44}$$

and the shear force is

$$F_s = 2\pi r L \tau \tag{12-45}$$

For a newtonian fluid, τ is given by Eq. (1-2), which for the present case is $-\mu \, dV_z/dr$. A summation of external forces is $F_p + F_s = 0$, which yields

$$(p_1 - p_2)\pi r^2 = -2\pi r L \mu \frac{dV_z}{dr}$$

or

$$\frac{dV_z}{dr} = -\frac{p_1 - p_2}{\mu L} \frac{r}{2} \tag{12-46}$$

Integrating, we have

$$V_z = \frac{p_1 - p_2}{\mu L}\left(C - \frac{r^2}{4}\right) \tag{12-47}$$

The constant of integration C is evaluated with the no-slip boundary condition at the wall, i.e., Eq. (12-26a). This requires

$$C = \frac{R^2}{4}$$

LAMINAR FLOW OF INCOMPRESSIBLE VISCOUS FLUIDS

and thus

$$V_z = \frac{1}{4\mu} \frac{p_1 - p_2}{L} (R^2 - r^2) \tag{12-48}$$

which is identical with Eq. (12-34) with dp/dz evaluated over a finite length L.

Example Calculate the pressure drop in psi per 100 linear feet for water flowing at an average velocity 0.5 ft/sec and 70°F in a ½-in-ID tube. Assume fully developed flow.

Solution The Reynolds number for this flow is

$$\mathbf{Re}_D = \frac{DV}{\nu} = \frac{(\tfrac{1}{24} \text{ ft})(0.5 \text{ ft/sec})}{1.06 \times 10^{-5} \text{ ft}^2/\text{sec}} = 1{,}962$$

where the kinematic viscosity is from Table A-3. Since $\mathbf{Re}_D < 2{,}000$, the flow is laminar and the friction factor is, by Eq. (12-38),

$$f = \frac{64}{\mathbf{Re}_D} = \frac{64}{1{,}962} = 3.26 \times 10^{-2}$$

The pressure drop per unit length is given by Eq. (12-37a) and is

$$\frac{\Delta p}{L} = \frac{f}{D} \rho \frac{V^2}{2} = \frac{3.26 \times 10^{-2}}{\tfrac{1}{24} \text{ ft}} \left(\frac{62.3 \text{ lb}_f\text{-sec}^2}{32.2 \text{ ft}^4} \right) \frac{(0.5 \text{ ft/sec})^2}{2}$$
$$= 18.9 \times 10^{-2} \text{ lb}_f/\text{ft}^2\text{-ft}$$

Thus the pressure drop for 100 ft of pipe is

$$\Delta p = 18.9 \times 10^{-2} \frac{\text{lb}_f/\text{ft}^2}{\text{ft}} (100 \text{ ft}) \frac{\text{ft}^2}{144 \text{ in}^2} = 0.131 \text{ psi}$$

Example Repeat the previous example except the fluid is hydraulic fluid at 100°F.

Solution The kinematic viscosity from Fig. A-1 is 1.75×10^{-4} ft²/sec, and the specific gravity is 0.848. The Reynolds number is

$$\mathbf{Re}_D = \frac{DV}{\nu} = \frac{(\tfrac{1}{24} \text{ ft})(0.5 \text{ ft/sec})}{1.75 \times 10^{-4} \text{ ft}^2/\text{sec}} = 119$$

The friction factor is

$$f = \frac{64}{\mathbf{Re}_D} = \frac{64}{119} = 0.538$$

and the pressure drop per unit length is

$$\frac{\Delta p}{L} = \frac{f}{D} \rho \frac{V^2}{2} = \frac{0.538}{\tfrac{1}{24} \text{ ft}} 0.838 \left(\frac{62.4 \text{ lb}_f\text{-sec}^2}{32.2 \text{ ft}^4} \right) \frac{(0.5 \text{ ft/sec})^2}{2}$$
$$= 2.62 \text{ lb}_f/\text{ft}^2\text{-ft}$$

For 100 linear feet the pressure drop is

$$\Delta p = 2.62 \frac{\text{lb}_f/\text{ft}^2}{\text{ft}} (100 \text{ ft}) \frac{\text{ft}^2}{144 \text{ in}^2} = 1.81 \text{ psi}$$

Note that this last result could have been obtained from the answer for the water flow since the Δp is directly proportional to the kinematic viscosity and to the density. The kinematic viscosity of the hydraulic fluid is 16.5 times that of the water, while its density is 0.838 that of water. Thus,

$$(\Delta p)_{\text{hyd}} = (\Delta p)_{\text{H}_2\text{O}}(16.5)(0.838) = 1.81 \text{ psi}$$

Noncircular ducts Several investigations have been directed toward determination of the friction factor for flow in noncircular ducts. Before discussing them, however, it is necessary to define the Reynolds number for such flow situations. Consider, for example, flow in a rectangular duct having cross-sectional sides a and b, where a is one-half the length of b. Which then is the appropriate dimension to use in formulating the Reynolds number? The answer is neither a nor b but an effective or *hydraulic diameter* D_h defined by

$$D_h = \frac{4A}{P} \tag{12-49}$$

where A is the cross-sectional flow area and P is the wetted duct perimeter. This definition has been found to best correlate such data as pressure loss, heat tranfers, etc., when used to evaluate the Reynolds number. Note that for a circular duct flowing full, Eq. (12-49) results in D_h equal to the geometric diameter.

Table 12-1 Friction factors for concentric annuli†

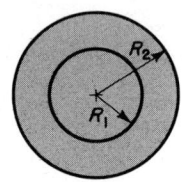

$\dfrac{R_1}{R_2}$	$f\,\text{Re}$
0.001	74.68
0.01	80.11
0.05	86.27
0.10	89.37
0.20	92.35
0.40	94.71
0.60	95.59
0.80	95.92
1.00	96.00

† Data from Lundgren [12].

Table 12-2 Friction factors for rectangular ducts†

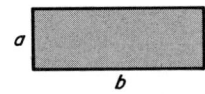

$\dfrac{a}{b}$	$f\,\text{Re}$
0.05	89.91
0.10	84.68
0.125	82.34
0.166	78.81
0.25	72.93
0.40	65.47
0.50	62.19
0.75	57.89
1.00	56.91

† Data from Lundgren [12].

LAMINAR FLOW OF INCOMPRESSIBLE VISCOUS FLUIDS

The two most important industrial configurations involving noncircular ducts are the concentric annulus and the rectangular duct, and the friction factors for these cases are dependent upon the ratios of radii and height to width respectively. The products of friction factor and Reynolds number are tabulated as functions of these geometric ratios in Tables 12-1 and 12-2.

Entry length All the discussions and results of this section have dealt with fully developed laminar flow, a condition existing only after the velocity profile becomes invariant with respect to axial position as shown in Fig. 12-3, station 3. This entry length has been investigated by Langhaar [11], Nikuradse [14], and Schiller [22]. According to Langhaar, the entry length for fully developed laminar flow in a round duct is

$$Z_E = 0.05 \text{Re}_D D \qquad (12\text{-}50)$$

While there is no simple representation for the pressure loss in the entry length, this total length is often short when compared with the pipe length having fully developed flow, and engineering calculations are usually made assuming the entire length to have fully developed flow.

12-4 CHANNEL FLOW

Consider steady incompressible fully developed isothermal flow in a channel formed by two parallel walls (assumed to extend to infinity in both the positive and negative z directions) as shown in Fig. 12-6. In this case $v = w = 0$, and the Navier-Stokes equations reduce to

$$\frac{dp}{dx} = \mu \frac{d^2 u}{dy^2} \qquad (12\text{-}51)$$

where the steady character ensures the absence of inertia terms. It should be noted that this and the preceding case of flow in a tube are members of a group known as *parallel flows*, this term being applicable to any flow where only one velocity component is different from zero, i.e., all fluid particles are moving in the same direction.

The boundary conditions on Eq. (12-51) are the no-slip condition

$$(12\text{-}52)$$

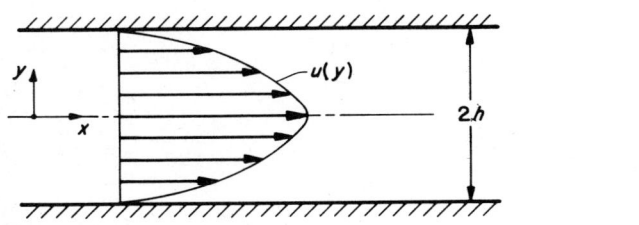

Fig. 12-6 Laminar fully developed incompressible flow in a channel.

Integrating Eq. (12-51) twice, we have

$$u = \frac{1}{\mu}\frac{dp}{dx}\left(\frac{y^2}{2} + C_1 y + C_2\right) \tag{12-53}$$

Applying the two boundary conditions yields

$$0 = \frac{1}{\mu}\frac{dp}{dx}\left(\frac{h^2}{2} + C_1 h + C_2\right)$$

$$0 = \frac{1}{\mu}\frac{dp}{dx}\left(\frac{h^2}{2} - C_1 h + C_2\right)$$

which are satisfied if $C_1 = 0$ and $C_2 = -h^2/2$. Substituting these values into Eq. (12-53) gives the velocity profile

$$u = -\frac{1}{2\mu}\frac{dp}{dx}(h^2 - y^2) \tag{12-54}$$

where the viscosity and pressure gradient are considered to be known. Again, the velocity profile is parabolic, as was the case for circular tube flow.

The maximum velocity occurs at the centerline; thus, at $y = 0$

$$u_{\max} = -\frac{1}{2\mu}\frac{dp}{dx}h^2 \tag{12-55}$$

The average velocity is determined by integrating the volume under the parabolic profile for unit depth in the z direction and dividing by the flow cross-sectional area for this same z-direction depth. This yields

$$u_{\mathrm{av}} = \tfrac{2}{3} u_{\max} \tag{12-56}$$

or

$$u_{\mathrm{av}} = -\frac{1}{3\mu}\frac{dp}{dx}h^2 \tag{12-57}$$

In terms of the volumetric flow rate per unit depth Q the average velocity is $Q/2h$, and this permits determination of the pressure gradient due to viscous motion from the expression

$$\frac{dp}{dx} = -\frac{3}{2}\frac{Q\mu}{h^3} \tag{12-58}$$

Integrating this expression over a finite length L from station 1 to station 2 yields

$$p_1 - p_2 = \frac{3}{2}\frac{Q\mu L}{h^3} \tag{12-59}$$

LAMINAR FLOW OF INCOMPRESSIBLE VISCOUS FLUIDS

which when divided by the fluid density yields for the head loss

$$\frac{\Delta p}{\rho} \equiv h_l = \frac{3}{2}\frac{Q\mu L}{\rho h^3} \qquad (12\text{-}60)$$

12-5 FLOW OVER A FLAT PLATE

To this point in our study of isothermal flows, we have considered only *internal* incompressible laminar flow. We will now direct our attention to *external* flow. Referring again to the laminar boundary layer on a flat plate as shown in Fig. 12-2, we recognize immediately that the problem is two-dimensional; i.e., the velocity vector at any location depends upon both the x and y coordinates. The problem is assumed to be invariant in the z direction, and w is identically zero. Thus, neglecting body forces, the Navier-Stokes equations for incompressible flow reduce to

x direction:
$$\frac{\partial u}{\partial t} + u\frac{\partial u}{\partial x} + v\frac{\partial u}{\partial y} = -\frac{1}{\rho}\frac{\partial p}{\partial x} + \nu\left(\frac{\partial^2 u}{\partial x^2} + \frac{\partial^2 u}{\partial y^2}\right) \qquad (12\text{-}61)$$
$$\quad 1 \qquad 1\ 1 \qquad \delta\ 1/\delta \qquad\qquad \delta^2\ 1 \qquad 1/\delta^2$$

y direction:
$$\frac{\partial v}{\partial t} + u\frac{\partial v}{\partial x} + v\frac{\partial v}{\partial y} = -\frac{1}{\rho}\frac{\partial p}{\partial y} + \nu\left(\frac{\partial^2 v}{\partial x^2} + \frac{\partial^2 v}{\partial y^2}\right) \qquad (12\text{-}62)$$
$$\quad \delta \qquad 1\ \delta \qquad \delta\ 1 \qquad\qquad \delta^2\ \delta \qquad 1/\delta$$

Continuity:
$$\frac{\partial u}{\partial x} + \frac{\partial v}{\partial y} = 0 \qquad (12\text{-}63)$$
$$\quad 1 \qquad 1$$

where the quantities 1, δ, etc., result from an order-of-magnitude analysis as discussed in the following paragraphs.

We now attempt a rough *order-of-magnitude estimation* for each of the terms of these equations with the purpose of eliminating terms which appear to be negligible. Admittedly the results will be approximations, but careful approximations are the essence of sound engineering.[1] To begin this analysis, we assign a comparative value of 1 to the maximum length and maximum velocity involved in this problem and attempt to evaluate the extreme value of all terms compared with these. Thus, the maximum velocity is $u_{\max}$, and we say u is of the order of 1, written

$$u = O(1) \qquad (12\text{-}64)$$

Now x can vary from zero to 1, and thus the maximum change in x is also of the order of unity. The change in u can be from zero to unity, and writing the derivative in

---

[1] In an analogous fashion one applies order-of-magnitude approximations to numerous engineering problems without actually calling it that. As a specific example, the wind drag on cables for suspension bridges is usually negligible compared with the bridge load and is omitted in the calculations.

finite-difference form

$$\frac{\partial u}{\partial x} \simeq \frac{\Delta u}{\Delta x} = \frac{O(1)}{O(1)} = O(1) \tag{12-65}$$

Likewise we can readily show that the order of magnitude of the second derivative is

$$\frac{\partial^2 u}{\partial x^2} = O(1) \tag{12-66}$$

Examining the y direction, the extreme value of y is the very thin boundary-layer thickness δ. From the continuity equation with $\partial u/\partial x = O(1)$, we conclude that $\partial v/\partial y = O(1)$. Since the extreme value of Δy is $O(\delta)$, we require that

$$v = O(\delta) \tag{12-67}$$

Then the derivatives of u with respect to y are

$$\frac{\partial u}{\partial y} = O\left(\frac{1}{\delta}\right) \tag{12-68}$$

$$\frac{\partial^2 u}{\partial y^2} = O\left(\frac{1}{\delta^2}\right) \tag{12-69}$$

Proceeding with similar arguments, we obtain

$$\frac{\partial^2 v}{\partial y^2} = O\left(\frac{1}{\delta}\right) \tag{12-70}$$

$$\frac{\partial v}{\partial x} = O(\delta) \tag{12-71}$$

$$\frac{\partial^2 v}{\partial x^2} = O(\delta) \tag{12-72}$$

These orders of magnitude are given below the respective terms in Eqs. (12-61) to (12-63).

Assuming the extreme change of time to be of the order of unity, we have

$$\frac{\partial u}{\partial t} = O(1) \tag{12-73}$$

$$\frac{\partial v}{\partial t} = O(\delta) \tag{12-74}$$

and this also results in individual inertia terms being of the same order of magnitude in each Navier-Stokes equation. The order of ν must be δ^2 for the viscous terms in these equations to be of the same order as the inertia terms.

Examination of Eqs. (12-61) and (12-62) with the orders of the terms indicated beneath now reveals a very important fact. The order of magnitude of the density

LAMINAR FLOW OF INCOMPRESSIBLE VISCOUS FLUIDS

p does not exceed unity, and assuming it to be $O(1)$, we see that

$$\frac{\partial p}{\partial x} = O(1) \tag{12-75}$$

for the pressure term to be significant. By the same argument

$$\frac{\partial p}{\partial y} = O(\delta) \tag{12-76}$$

and since $\delta \ll 1$, we conclude that *the vertical pressure gradient is negligible*! This allows us to calculate the pressure inside the boundary layer using perfect-fluid flow theory by considering the pressure gradient in the x direction as being imposed from the outer flow. Also, since every term in the y-direction momentum equation is of the order of δ and each term in the x-direction equation is of the order of unity, the entire y-direction equation is insignificant; and taking note that

$$\frac{\partial^2 u}{\partial x^2} \ll \frac{\partial^2 u}{\partial y^2}$$

we have

$$\blacksquare \quad \frac{\partial u}{\partial t} + u\frac{\partial u}{\partial x} + v\frac{\partial u}{\partial y} = -\frac{1}{\rho}\frac{dp}{dx} + \nu\frac{\partial^2 u}{\partial y^2} \tag{12-61a}$$

$$\blacksquare \quad \frac{\partial p}{\partial y} = 0 \tag{12-62a}$$

$$\blacksquare \quad \frac{\partial u}{\partial x} + \frac{\partial v}{\partial y} = 0 \tag{12-63a}$$

These are known as *Prandtl's boundary-layer equations*.

In summary, the order analysis has reduced the mathematical problem from three simultaneous equations to two and has determined the pressure gradient, since it can be readily found by application of potential-flow theory. Also, one of the second-order partial derivatives in the viscous term has been eliminated.

Blasius solution In 1908, Blasius [1] presented an outstanding early successful application of boundary-layer theory in which he solved the problem of steady-state isothermal gas flow with zero pressure gradient over a flat plate using Prandtl's boundary-layer equations. A brief outline of his solution follows. Consider the developing boundary-layer flow of Fig. 12-7. It is reasonable to expect the velocity profiles at different stations (x locations) along the plate to be geometrically similar, i.e., to differ only by a *stretching factor* in the y direction. The thickness δ of the boundary layer has a significant effect upon the velocity profile, and the *similarity principle* can be expressed as

$$\frac{u}{V_\infty} = \phi_1\left(\frac{y}{\delta}\right) \tag{12-77}$$

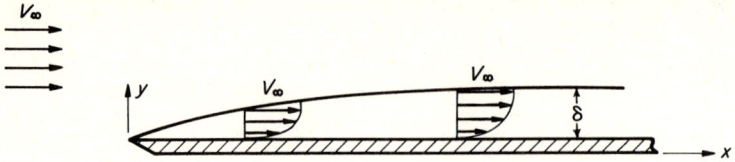

Fig. 12-7 Similarity profiles in boundary-layer flow over a flat plate.

It is known from experimental measurements with laminar boundary layers that the boundary-layer thickness (1) is directly proportional to the square root of the product of distance along the plate and the fluid kinematic viscosity and (2) is inversely proportional to the square root of the free-stream velocity. Thus the similarity principle becomes

$$\frac{u}{V_\infty} = \phi_2\left(y\sqrt{\frac{V_\infty}{\nu x}}\right) \qquad (12\text{-}77a)$$

where the term in parentheses is a nondimensional coordinate. For simplicity this is given the symbol η

$$\eta \equiv y\sqrt{\frac{V_\infty}{\nu x}} \qquad (12\text{-}78)$$

and is sometimes called the *similarity parameter*. Notice that it contains both x and y, and the possibility exists of using this variable to transform the partial differential equation (12-61a) into an ordinary differential equation in one independent variable. The mathematical consequences of this are significant (it is generally far simpler to solve an ordinary differential equation than a partial differential equation), and this has resulted in a classification of flows as *similar* or *nonsimilar*.[1]

It is convenient to express the velocity components and their derivatives in terms of the stream function ψ. A dimensionless stream function in the η coordinate system which satisfies the continuity equation is $f(\eta) = \psi/\sqrt{\nu x V_\infty}$, and consequently

$$\psi = \sqrt{\nu x V_\infty}\, f(\eta) \qquad (12\text{-}79)$$

The velocity components are related to the stream function by

$$u = \frac{\partial \psi}{\partial y} \qquad (11\text{-}19a)$$

and

$$v = -\frac{\partial \psi}{\partial x} \qquad (11\text{-}19b)$$

[1] For a treatment of the general problem of similarity of velocity profiles the reader should consult chap. 8 of Schlichting [23].

and consequently, the velocities and their derivatives can be expressed in terms of $f(\eta)$. Thus

$$u = \frac{\partial}{\partial y}[\sqrt{\nu x V_\infty}\, f(\eta)] = \sqrt{\nu x V_\infty}\, \frac{\partial f}{\partial \eta}\frac{d\eta}{dy}$$

or

$$u = \sqrt{\nu x V_\infty}\, \frac{V_\infty}{\nu x}\frac{\partial f}{\partial \eta} = V_\infty f' \tag{12-80}$$

In a similar fashion we obtain

$$\frac{\partial u}{\partial x} = -\frac{\eta V_\infty}{2x} f'' \tag{12-81}$$

$$\frac{\partial u}{\partial y} = V_\infty \sqrt{\frac{V_\infty}{\nu x}}\, f'' \tag{12-82}$$

$$\frac{\partial^2 u}{\partial y^2} = \frac{V_\infty^2}{\nu x}\, f''' \tag{12-83}$$

$$v = \frac{1}{2}\sqrt{\frac{\nu V_\infty}{x}}\,(\eta f' - f) \tag{12-84}$$

where primes denote differentiation with respect to η. Substituting these expressions into Eq. (12-61a) and simplifying yields for steady flow

$$2f''' + ff'' = 0 \tag{12-85}$$

which is a third-order *ordinary* differential equation. The boundary conditions in the xy coordinate system are

(1) at $y = 0$: $\quad u = v = 0$
(2) at $y \to \infty$: $\quad u = V_\infty \quad v = 0$

which become

$$f = f' = 0 \quad \text{at } \eta = 0 \tag{12-86}$$

$$f' = 1 \quad \text{at } \eta \to \infty \tag{12-87}$$

Blasius obtained a series solution for Eq. (12-85) subject to these boundary conditions; the details of his solution are quite tedious. Recalling that this was several decades before the advent of high-speed digital computers, we readily recognize this to have been a formidable task. His numerical values were later improved by several independent workers, and quite accurate results were presented in 1938 by Howarth [6].

Table 12-3 The function $f(\eta)$ and its derivatives†

$\eta = y\sqrt{\dfrac{V_\infty}{\nu x}}$	f	$f' = \dfrac{u}{V_\infty}$	f''
0	0	0	0.33206
0.2	0.00664	0.06641	0.33199
0.4	0.02656	0.13277	0.33147
0.6	0.05974	0.19894	0.33008
0.8	0.10611	0.26471	0.32739
1.0	0.16557	0.32979	0.32301
1.2	0.23795	0.39378	0.31659
1.4	0.32298	0.45627	0.30787
1.6	0.42032	0.51676	0.29667
1.8	0.52952	0.57477	0.28293
2.0	0.65003	0.62977	0.26675
2.2	0.78120	0.68132	0.24835
2.4	0.92230	0.72899	0.22809
2.6	1.07252	0.77246	0.20646
2.8	1.23099	0.81152	0.18401
3.0	1.39682	0.84605	0.16136
3.2	1.56911	0.87609	0.13913
3.4	1.74696	0.90177	0.11788
3.6	1.92954	0.92333	0.09809
3.8	2.11605	0.94112	0.08013
4.0	2.30576	0.95552	0.06424
4.2	2.49806	0.96696	0.05052
4.4	2.69238	0.97587	0.03897
4.6	2.88826	0.98269	0.02948
4.8	3.08534	0.98779	0.02187
5.0	3.28329	0.99155	0.01591
5.2	3.48189	0.99425	0.01134
5.4	3.68094	0.99616	0.00793
5.6	3.88031	0.99748	0.00543
5.8	4.07990	0.99838	0.00365
6.0	4.27964	0.99898	0.00240
6.2	4.47948	0.99937	0.00155
6.4	4.67938	0.99961	0.00098
6.6	4.87931	0.99977	0.00061
6.8	5.07928	0.99987	0.00037
7.0	5.27926	0.99992	0.00022
7.2	5.47925	0.99996	0.00013
7.4	5.67924	0.99998	0.00007
7.6	5.87924	0.99999	0.00004
7.8	6.07923	1.00000	0.00002
8.0	6.27923	1.00000	0.00001
8.2	6.47923	1.00000	0.00001
8.4	6.67923	1.00000	0.00000
8.6	6.87923	1.00000	0.00000
8.8	7.07923	1.00000	0.00000

† Data from Howarth [6].

LAMINAR FLOW OF INCOMPRESSIBLE VISCOUS FLUIDS

The boundary-layer thickness δ is arbitrarily defined as the location where the velocity is 99 percent of the free-stream value, i.e.,

$$\frac{u}{V_\infty} = 0.99 \tag{12-88}$$

Using this, we find from Table 12-3 that $\eta \simeq 5.0$. Thus,

$$\eta = \delta\sqrt{\frac{V_\infty}{\nu x}} \simeq 5.0 \tag{12-89}$$

or

$$\frac{\delta}{x} \simeq \frac{5.0}{\sqrt{\mathbf{Re}_x}} \tag{12-90}$$

which expresses the boundary-layer thickness along a flat plate in steady incompressible laminar flow of a viscous fluid.

Example A pitot tube is to be located 3 in aft of the leading edge of the bottom of the undercarriage of a blimp. This bottom surface is very nearly flat, and the pressure gradient is negligible. The speed of the blimp is from 25 to 100 mph, and the air temperature may be assumed to be 32°F. The pitot is to be used to measure the airspeed and thus must be outside the boundary layer. What is the maximum boundary-layer thickness at this location?

Solution The maximum Reynolds number is obtained at 100 mph. At 32°F, the kinematic viscosity is 0.145×10^{-3} ft²/sec. The maximum Reynolds number is

$$\mathbf{Re}_{max} = \frac{(\tfrac{1}{4}\text{ ft})(146.8\text{ ft/sec})}{0.145 \times 10^{-3}\text{ ft}^2/\text{sec}} = 253{,}000$$

and thus the boundary-layer flow is laminar. The maximum boundary-layer thickness will occur at a minimum value of $\mathbf{Re}_x$, according to Eq. (12-90), which occurs at 25 mph:

$$(\mathbf{Re}_x)_{min} = \frac{(\tfrac{1}{4}\text{ ft})(36.7\text{ ft/sec})}{0.145 \times 10^{-3}\text{ ft}^2/\text{sec}} = 63{,}300$$

Then by Eq. (12-90)

$$\delta_{max} = \frac{(\tfrac{1}{4}\text{ ft})(5.0)}{\sqrt{63{,}300}} = 0.00496\text{ ft} = 0.06\text{ in}$$

The Blasius solution also affords an "exact answer" for the drag on a flat plate. The local shear stress $\tau_s(x)$ is given by

$$\tau_s(x) = \mu\left.\frac{\partial u}{\partial y}\right|_{y=0} = \mu V_\infty \sqrt{\frac{V_\infty}{\nu x}} f''(0) \tag{12-91}$$

where the gradient of u with respect to y is obtained from Eq. (12-82). From Table 12-3 we obtain $f''(0) = 0.332$ and consequently

$$\tau_s(x) = 0.332\sqrt{\frac{V_\infty^3 \rho \mu}{x}} \tag{12-92}$$

The local skin-friction coefficient is defined as

$$c_f \equiv \frac{\tau_s}{\rho_\infty V_\infty^2/2} \tag{12-93}$$

thus,

$$c_f = \frac{2(0.332)\sqrt{V_\infty^3 \rho \mu/x}}{\rho_\infty V_\infty^2}$$

or

$$c_f = \frac{0.664}{\sqrt{\mathbf{Re}_x}} \tag{12-94}$$

Applying the usual technique of averaging, we can integrate $f(x)$ for a finite length L and divide by the integrated length to obtain

$$C_f = \frac{0.664 \int_0^L \left(\frac{\mu}{V_\infty \rho x}\right)^{\frac{1}{2}} dx}{\int_0^L dx} = \frac{1.328}{\sqrt{\frac{V_\infty \rho L}{\mu}}}$$

■ $$C_f = \frac{1.328}{\sqrt{\mathbf{Re}_L}} \tag{12-95}$$

Example A model airfoil is tested in a wind tunnel using helium gas. The airfoil is slender and symmetrical and to a first approximation may be treated as a flat plate. (This simplified treatment neglects the pressure gradient due to curvature.) At a tunnel velocity of 400 ft/sec and a fluid temperature of 100°F what is the drag per foot of length on the model?

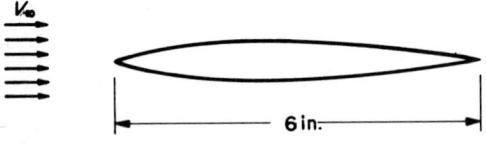

Fig. 12-8 Model airfoil in a wind tunnel.

Solution The Reynolds number at the trailing edge is

$$\mathbf{Re}_L = \frac{\frac{1}{2}(400)}{1.36 \times 10^{-3}} = 147{,}000$$

and thus we expect the entire flow to be laminar since $\mathbf{Re}_L < 300{,}000$. The drag coefficient is

$$C_f = \frac{1.328}{\sqrt{147{,}000}} = 0.347 \times 10^{-2}$$

LAMINAR FLOW OF INCOMPRESSIBLE VISCOUS FLUIDS

The average shear stress then is

$$(\tau_s)_{\text{av}} = C_f \frac{\rho_\infty V_\infty^2}{2}$$

$$= \frac{0.347 \times 10^{-2}}{2} \left(\frac{0.010}{32.2} \frac{\text{lb}_f\text{-sec}^2}{\text{ft}^4} \right) \frac{160{,}000 \text{ ft}^2}{\text{sec}^2}$$

$$= 0.0862 \text{ lb}_f/\text{ft}^2$$

Since the model is slender, the area per foot of length is approximately

$$A \simeq 2(\tfrac{6}{12} \text{ ft})(1 \text{ ft}) = 1 \text{ ft}^2$$

and the skin-friction drag force is

$$F_f \simeq (1 \text{ ft}^2)(0.0862 \text{ lb}_f/\text{ft}^2) = 0.0862 \text{ lb}_f$$

Integral momentum equation The preceding solutions to the Navier-Stokes equations were obtained for very simple flows. The flow in a tube and the channel flow involved reduced forms of the momentum equations; the inertia terms, which are nonlinear and hence most troublesome, were eliminated. The simplest case involving retention of the nonlinear inertia terms is flow over a flat plate, and the solution of this problem is formidable, as we have seen in the preceding discussion.

Fortunately, an approximate method due to von Kármán [8] is available. This method is considerably simpler than the solution of the differential equations and offers the mathematical advantage of already being integrated through the boundary-layer thickness, requiring only satisfaction of certain boundary conditions. The essence of the method involves equating the net momentum outflux from a control volume to the summation of the forces acting on this control volume. Consider a two-dimensional boundary-layer flow as shown in Fig. 12-9 and a control volume of length Δx, height L (where L extends outside the boundary layer), and unit depth in the z direction.

Recall that the x-direction linear-momentum equation is

$$\sum F_x = \frac{d}{dt} \int_{\text{cv}} \rho u \, d\mathscr{V} + \int_{\text{cs}} u(\rho \mathbf{V} \cdot d\mathbf{A}) \tag{10-52a}$$

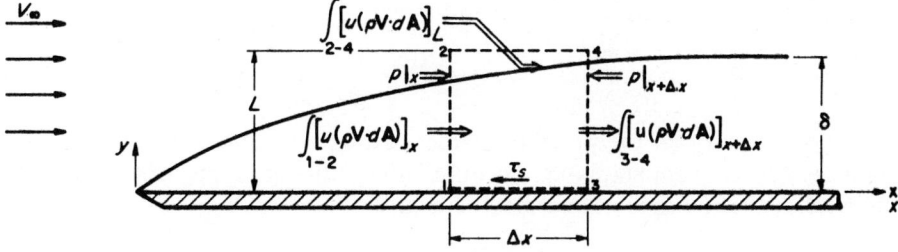

Fig. 12-9 Control volume for integral momentum equation.

which for steady flow reduces to

$$\sum F_x = \int_{cs} u(\rho \mathbf{V} \cdot d\mathbf{A}) \tag{12-96}$$

Determining the momentum fluxes, we have:

$$\text{Across surface 1-2:} \quad \int_{cs_{1\text{-}2}} [u(\rho \mathbf{V} \cdot d\mathbf{A})]_x = -\int_0^L (\rho u^2 \, dy)_x \tag{12-97}$$

$$\text{Across surface 3-4:} \quad \int_{cs_{3\text{-}4}} [u(\rho \mathbf{V} \cdot d\mathbf{A})]_{x+\Delta x} = \int_0^L (\rho u^2 \, dy)_{x+\Delta x}$$

Across surface 2-4 the mass flow rate is given by

$$\rho \mathbf{V} \cdot d\mathbf{A} = \int_0^L [(\rho u \, dy)|_{x+\Delta x} - (\rho u \, dy)|_x]$$

and the x component of velocity is V_∞. Consequently the momentum flux is

$$\int_{cs_{2\text{-}4}} [u(\rho \mathbf{V} \cdot d\mathbf{A})] = -V_\infty \int_0^L [(\rho u \, dy)|_{x+\Delta x} - (\rho u \, dy)|_x] \tag{12-98}$$

Across surface 1-3 the dot product $\mathbf{V} \cdot d\mathbf{A}$ is identically zero; hence there is no x-momentum contribution. The total forces in the x direction are the pressure forces on surfaces 1-2 and 3-4 and the shearing force due to τ_s acting along the lower surface. Substituting these and the momentum fluxes into Eq. (12-96), we have

$$(p|_x)L - \tau_s \Delta x - (p|_{x+\Delta x})L = \int_0^L (\rho u^2 \, dy)_{x+\Delta x} - \int_0^L (\rho u^2 \, dy)_x$$

$$- V_\infty \int_0^L [(\rho u \, dy)|_{x+\Delta x} - (\rho u \, dy)|_x]$$

Dividing by Δx and taking the limit as Δx approaches zero yields for an incompressible fluid

$$\tau_s + L\frac{dp}{dx} = -\rho \frac{d}{dx} \int_0^L u^2 \, dy + \rho \frac{d}{dx}\left(V_\infty \int_0^L u \, dy\right) \tag{12-99}$$

Since V_∞ is independent of y, this can be rearranged as

$$\tau_s + L\frac{dp}{dx} = \rho \frac{d}{dx} \int_0^L (V_\infty - u)u \, dy - \rho \frac{dV_\infty}{dx} \int_0^L u \, dy \tag{12-100}$$

Now recall that the pressure gradient along the plate is impressed through the boundary layer from the potential outer flow. In the undisturbed outer flow, the Bernoulli equation applies and may be expressed as

$$p + \rho \frac{V_\infty^2}{2} = \text{const}$$

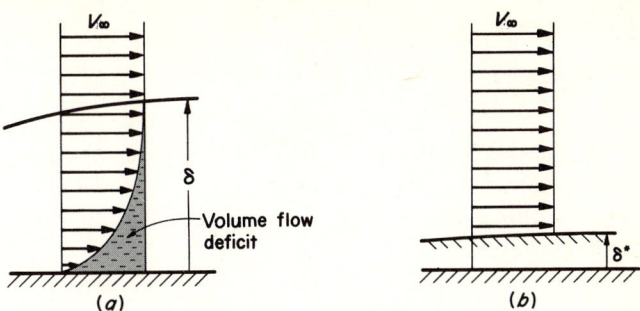

Fig. 12-10 Displacement thickness: (*a*) typical volume-flow deficit due to boundary layer; (*b*) wall displacement for potential flow rate to equal viscous flow rate.

Differentiating with respect to x, we have

$$\frac{dp}{dx} + \rho V_\infty \frac{dV_\infty}{dx} = 0$$

and consequently

$$L\frac{dp}{dx} = -L\rho V_\infty \frac{dV_\infty}{dx} = -\rho \frac{dV_\infty}{dx}\int_0^L V_\infty\, dy \tag{12-101}$$

since V_∞ is independent of y. Substituting into Eq. (12-100), we obtain

$$\blacksquare\quad \tau_s = \rho\frac{d}{dx}\left[\int_0^\delta (V_\infty - u)u\, dy\right] + \rho\frac{dV_\infty}{dx}\int_0^\delta (V_\infty - u)\, dy \tag{12-102}$$

where the upper limit of integration has been changed from L to δ since $V_\infty - u$ vanishes outside the boundary layer.

In addition to the boundary-layer thickness δ defined as the distance from the boundary to the point where $u = 0.99V_\infty$, there are two other useful definitions of boundary-layer thickness. One of these is the *displacement thickness* δ^* defined as the distance the physical boundary would have to be displaced for the flow rate to be the same for a perfect fluid flowing over the displaced body as for the viscous fluid flowing over the real surface. The decrease in volume flow rate due to friction as shown in Fig. 12-10 is

$$\int_0^\delta (V_\infty - u)\, dy$$

and consequently

$$\delta^* V_\infty = \int_0^\delta (V_\infty - u)\, dy$$

Thus,

$$\delta^* = \int_0^\delta \left(1 - \frac{u}{V_\infty}\right) dy \qquad (12\text{-}103)$$

Another meaningful thickness is the *momentum thickness* δ_i defined as the distance from the solid boundary such that the momentum flux in potential flow through this distance is equal to the momentum deficit due to viscous effects in the boundary layer. Thus,

$$\rho V_\infty^2 \delta_i = \rho \int_0^\delta u(V_\infty - u)\, dy$$

since the right side of this expression is the loss of momentum within the boundary layer when compared with potential flow. Rearranging, we obtain

$$\delta_i = \int_0^\delta \frac{u}{V_\infty}\left(1 - \frac{u}{V_\infty}\right) dy \qquad (12\text{-}104)$$

With these expressions we can express the integral momentum equation as

$$\tau_s = \rho \frac{d}{dx} V_\infty^2 \delta_i + \rho V_\infty \delta^* \frac{dV_\infty}{dx} \qquad (12\text{-}105)$$

Solution of the integral momentum equation Consider again the same problem for which Blasius solved the Prandtl boundary-layer equations, specifically laminar incompressible flow along a flat plate with zero pressure gradient. For no pressure gradient, the free-stream velocity V_∞ is constant, and Eq. (12-102) reduces to

$$\blacksquare \quad \tau_s = \rho \frac{d}{dx}\left[\int_0^\delta (V_\infty - u)u\, dy\right] \qquad (12\text{-}106)$$

The general procedure followed in solving this equation requires (1) assumption of a reasonable velocity profile, $u = u(y/\delta)$, within the boundary layer (accomplished by selecting a profile which satisfies the boundary conditions with a reasonable degree of accuracy) and (2) the use of Newton's expression for the wall shear, $\tau_s = \mu(du/dy)_s$.

Before attempting to establish a profile, let us list all known boundary conditions for this flow. At the wall, we have the no-slip condition, $u = 0$. Also, by inspection of Prandtl's boundary-layer equations it is readily apparent that $d^2u/dy^2 = 0$ here. Then at the outer edge of the boundary layer we have $u = V_\infty$, and the gradient du/dy is zero. Thus

$$\begin{aligned} u &= 0 & \frac{d^2u}{dy^2} &= 0 & \text{at } y &= 0 \\ u &= V_\infty & \frac{du}{dy} &= 0 & \text{at } y &= \delta \end{aligned} \qquad (12\text{-}107)$$

LAMINAR FLOW OF INCOMPRESSIBLE VISCOUS FLUIDS

Now assume the velocity profile to be of the form.

$$\frac{u}{V_\infty} = C + C_1 \frac{y}{\delta} + C_2 \left(\frac{y}{\delta}\right)^2 + C_3 \left(\frac{y}{\delta}\right)^3 \qquad (12\text{-}108)$$

Since $u = 0$ for $y = 0$, we have $C = 0$. At $y = \delta$, $u = V_\infty$ yields

$$C_1 + C_2 + C_3 = 1 \qquad (12\text{-}109)$$

The first derivative is

$$\frac{1}{V_\infty} \frac{du}{dy} = \frac{C_1}{\delta} + \frac{2C_2}{\delta} \frac{y}{\delta} + \frac{3C_3}{\delta} \left(\frac{y}{\delta}\right)^2$$

and at $y = \delta$, $du/dy = 0$ yields

$$C_1 + 2C_2 + 3C_3 = 0 \qquad (12\text{-}110)$$

The second derivative is

$$\frac{1}{V_\infty} \frac{d^2u}{dy^2} = \frac{2C_2}{\delta^2} + \frac{6C_3}{\delta^2} \frac{y}{\delta}$$

from which the remaining boundary condition, $d^2u/dy^2 = 0$ at $y = 0$, gives

$$C_2 = 0$$

Then, Eqs. (12-109) and (12-110) yield

$$C_1 = \tfrac{3}{2} \qquad C_3 = -\tfrac{1}{2}$$

and the profile satisfying the boundary conditions is

$$\frac{u}{V_\infty} = \frac{3}{2} \frac{y}{\delta} - \frac{1}{2}\left(\frac{y}{\delta}\right)^3 \qquad (12\text{-}111)$$

Applying this velocity profile and Newton's expression for shear stress to Eq. (12-106), we have

$$\tfrac{3}{2}\mu \frac{V_\infty}{\delta} = \rho V_\infty^2 \frac{d}{dx} \left\{ \int_0^\delta \left[\frac{3}{2}\frac{y}{\delta} - \frac{1}{2}\left(\frac{y}{\delta}\right)^3\right]\left[1 - \frac{3}{2}\frac{y}{\delta} + \frac{1}{2}\left(\frac{y}{\delta}\right)^3\right] dy \right\}$$

which integrates to

$$\tfrac{3}{2}\mu \frac{V_\infty}{\delta} = \tfrac{39}{280} \rho V_\infty^2 \frac{d\delta}{dx}$$

Separating variables, we obtain

$$\frac{140}{13} \frac{\mu}{\rho V_\infty} dx = \delta \, d\delta$$

which can readily be integrated to give

$$\delta = 4.64 \sqrt{\frac{\mu x}{\rho V_\infty}} + \text{const}$$

Since δ is zero for $x = 0$, the constant of integration is zero and the dimensionless boundary-layer thickness is

$$\frac{\delta}{x} = \frac{4.64}{\sqrt{\text{Re}_x}} \quad (12\text{-}112)$$

The agreement between this and Blasius' result is very good indeed, the present constant being approximately 7 percent below that obtained by Blasius. Looking at the local skin-friction coefficient, we have by Eq. (12-93)

$$c_f = \frac{\tau_s}{\rho V_\infty^2/2}$$

but

$$\tau_s = \tfrac{3}{2}\mu \frac{V_\infty}{\delta} = 0.323 \sqrt{\frac{V_\infty^3 \rho \mu}{x}} \quad (12\text{-}113)$$

and

$$c_f = \frac{0.646}{\sqrt{\text{Re}_x}} \quad (12\text{-}114)$$

Table 12-4 Results of flat-plate laminar boundary-layer calculations

Velocity profile	Boundary conditions		$\dfrac{\delta}{x}\sqrt{\text{Re}_x}$	$C_f\sqrt{\text{Re}_l}$
	$y = 0$	$y = \delta$		
$\dfrac{u}{V_\infty} = \dfrac{y}{\delta}$	$u = 0$	$u = V_\infty$	3.46	1.156
$\dfrac{u}{V_\infty} = 2\dfrac{y}{\delta} - \left(\dfrac{y}{\delta}\right)^2$	$u = 0$	$u = V_\infty$	5.47	1.462
		$\dfrac{du}{dy} = 0$		
$\dfrac{u}{V_\infty} = \dfrac{3}{2}\dfrac{y}{\delta} - \dfrac{1}{2}\left(\dfrac{y}{\delta}\right)^3$	$u = 0$	$u = V_\infty$	4.64	1.292
	$\dfrac{d^2u}{dy^2} = 0$	$\dfrac{du}{dy} = 0$		
Blasius solution (exact)			5.0	1.328

LAMINAR FLOW OF INCOMPRESSIBLE VISCOUS FLUIDS

Determining the average skin-friction coefficient as we did for the Blasius solution yields

■ $$C_f = \frac{1.292}{\sqrt{\text{Re}_L}} \qquad (12\text{-}115)$$

The agreement of Eqs. (12-113) to (12-115) with their counterparts from the Blasius solution, Eqs. (12-92), (12-94), and (12-95), is remarkable.

It is very instructive to consider the possibility of using other velocity profiles in the integral momentum equation. Many authors have considered first- and second-order profiles, two of which are

$$\frac{u}{V_\infty} = \frac{y}{\delta} \qquad \frac{u}{V_\infty} = 2\frac{y}{\delta} - \left(\frac{y}{\delta}\right)^2$$

Results using these profiles together with the boundary conditions they satisfy and results using the cubic parabola of Eq. (12-111) are presented for comparison with Blasius' solution in Table 12-4.

The agreement of the solution to the integral momentum equation with Blasius' solution is quite good for any of the assumed profiles; the better the agreement with the boundary conditions, however, the better the solution.

General observations that should be made regarding the results of the present section are: (1) The boundary-layer thickness increases with the square root of distance from the leading edge and decreases with the $\frac{1}{2}$ power of the free-stream velocity. At a given location, it decreases with the $\frac{1}{2}$ power of the local Reynolds number. (2) The skin-friction coefficient decreases with the square root of both distance from the leading edge and free-stream velocity. The total drag force per unit width, however, is the product of C_f and $x\rho V_\infty^2/2$ and consequently increases with the $\frac{3}{2}$ power of the free-stream velocity and the $\frac{1}{2}$ power of the distance from the leading edge.

12-6 NONISOTHERMAL FLOW

In numerous physical problems, fluid motion occurs over surfaces which differ significantly in temperature from the fluid. For forced incompressible flow, the resulting heat transfer has little effect upon the flow field other than that caused by the temperature dependence of the fluid viscosity. Consequently, we can frequently solve the momentum equation independently of the energy equation; the converse, however, is generally not possible. This fact should be apparent from our earlier treatment of the basic energy equation for a control volume [Eq. (10-111)]. The term

$$\int_{cs} e(\rho \mathbf{V} \cdot d\mathbf{A})$$

in that equation is clearly *convective* in nature and is strongly dependent upon the *flow field*.

For high-speed gas flow, situations are frequently encountered wherein the viscous shear within the boundary layer gives rise to considerable heating of the gas and a resulting heat transfer. This is not an incompressible-flow problem, however, and will be discussed in a later chapter. It is quite possible to have incompressible flow with the energy transport having a significant effect upon the flow field. Such is the case in free-convective problems, as discussed in Chap. 16.

To illustrate the primary differences between forced and free convection, consider Fig. 12-11. In Fig. 12-11a the imposed flow field V_∞ is of the type existing in a wind tunnel or over an aircraft surface in flight at constant speed. For reasonably low velocity, the boundary-layer flow can be calculated by the isothermal-flow methods using an average temperature (which will be discussed later) to evaluate fluid properties.

In Fig. 12-11b we have a flow pattern typical of that resulting from a heated vertical plate in an otherwise quiescent body of fluid. The flow field results from buoyant forces due to density changes caused by heating of the fluid, and the momentum and energy equations are said to be coupled. We can summarize the important features of these as follows:

1. Forced convection
    a. Flow patterns are determined by externally applied forces.
    b. For constant properties, i.e., relatively low-speed flow, velocity profiles are determined and then used to calculate temperature profiles. (The heat transfer does not affect the velocity profile.)
    c. For variable properties, such as encountered in high-speed compressible flow, the velocity and temperature profiles are interdependent.
2. Free (natural) convection
    a. Flow patterns are determined by density changes in the fluid.

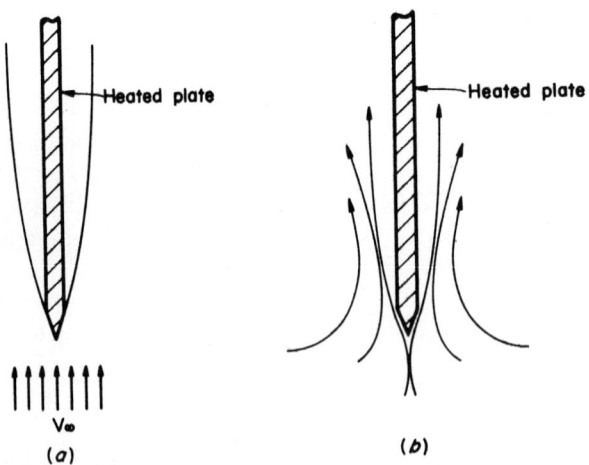

Fig. 12-11 Heated plate with convective flow field: (a) forced convection; (b) free convection.

LAMINAR FLOW OF INCOMPRESSIBLE VISCOUS FLUIDS

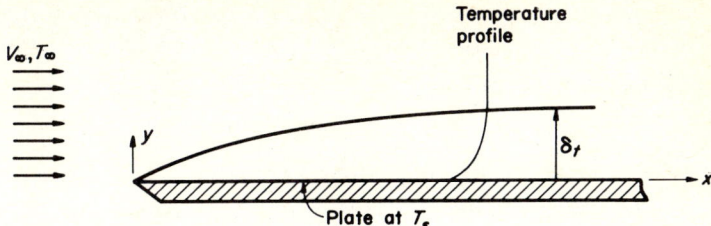

Fig. 12-12 Thermal boundary layer.

b. The velocity and temperature profiles are always interdependent, and consequently the momentum and energy equations must be solved simultaneously.

From our study of the flow or momentum boundary layer and its development along a solid wall, we would intuitively suspect that a thermal counterpart exists. Such is the case whenever the body and the fluid are at different temperatures. A thermal field develops along the body and in the *wake*, or trailing flow, which is quite analogous to the momentum boundary layer. This is called the *thermal boundary layer* and is shown in Fig. 12-12 for a plate at a temperature lower than that of the free stream. The thermal-boundary-layer thickness is defined arbitrarily as the length required for the temperature to attain 99 percent of the free-stream value T_∞. This definition is convenient due to the asymptotic development of the thermal profile. Recalling that the fluid velocity is zero at the wall, we would expect the energy transport there to be primarily conductive in nature, with convective transport becoming important as we move out from the body.

This chapter considers only forced laminar incompressible flow, and we shall be concerned for the present with the case where the momentum equation can be solved independently of the energy equation. The remainder of the chapter will be devoted to the energy equation and its solution. The reader should note that the rate of energy transport is frequently the term of primary interest in a particular engineering problem, the solution to the momentum equation being required merely as a prerequisite to solving the energy equation. Such is the case, for example, when the problem is to design a heat exchanger to maintain the proper temperature for the lubricating oil of an aircraft jet engine. Conversely, the drag of this same heat exchanger, if it extends into the aircraft external boundary layer, is fundamentally an isothermal-flow problem.

12-7 ENERGY EQUATION

In Chap. 10 the first law of thermodynamics applied to a control volume yielded an energy equation in the form

$$\frac{\delta Q}{\delta t} - \frac{\delta W_{\text{shaft}}}{\delta t} + \int_{\text{cs}} (\tau_{nn}\mathbf{n} + \tau_{ns}\mathbf{s}) \cdot \mathbf{V}\, dA = \frac{d}{dt}\int_{\text{cv}} \rho e\, d\mathscr{V} + \int_{\text{cs}} e(\rho \mathbf{V} \cdot d\mathbf{A})$$

(10-110)

We briefly review the significance of these terms.

1. $\delta Q/\delta t$ is the heat-transfer rate, other than convective, to the control volume.
2. $\delta W_{\text{shaft}}/\delta t$ is the rate of mechanical work done by the system on its surroundings; this work is transferred across the control surface.
3. $\int_{\text{cs}} (\tau_{nn}\mathbf{n} + \tau_{ns}\mathbf{s}) \cdot \mathbf{V}\, dA$ represents the rate of work done by the system of fluid within the control volume along its surface. This consists of a friction work due to viscous shear and a work of expansion because of volume changes opposed by the normal stresses. It is frequently termed *flow work* since it exists only during fluid motion.
4. The time rate of change of the product ρe integrated throughout the control volume

$$\frac{d}{dt} \int_{\text{cv}} \rho e\, d\mathscr{V}$$

represents the rate of energy storage. (This term vanishes identically for steady-state problems.)
5. The surface integral $\int_{\text{cs}} e(\rho \mathbf{V} \cdot d\mathbf{A})$ represents the rate of efflux of energy from the control volume; i.e., it is the net rate of energy *convected* out of the control volume.

Now we apply Eq. (10-110) to a differential control volume for which no mechanical or shaft work crosses the control surface. The control volume is shown in Fig. 12-13 and is identical, except for orientation, to that formerly used in the derivation of the conduction equation (Sec. 5-1) and the Navier-Stokes equations (Sec. 12-2). Consider the terms of Eq. (10-110) individually as they apply to this control volume.

1. The heat-transfer rate to the control volume, $\delta Q/\delta t$, is assumed[1] to be purely conductive and due to the temperature gradient in the fluid. Then, from Sec. 5-1

[1] This is generally sufficient; the next most commonly encountered heat flux (other than convective) is the radiative transport, which we shall neglect.

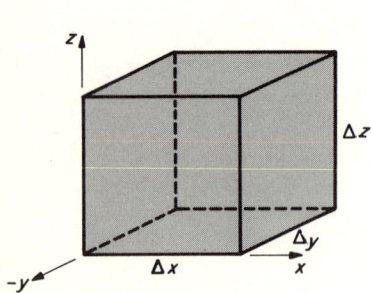

Fig. 12-13 Differential control volume for energy equation.

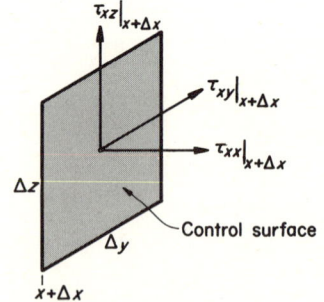

Fig. 12-14 Normal and shear stresses acting on one face of the differential control volume.

LAMINAR FLOW OF INCOMPRESSIBLE VISCOUS FLUIDS 413

we have

$$\lim_{\substack{\Delta x \to 0 \\ \Delta y \to 0 \\ \Delta z \to 0}} \frac{\delta Q/\delta t}{\Delta x\, \Delta y\, \Delta z} = -\left[\frac{\partial}{\partial x}\left(k\frac{\partial T}{\partial x}\right) + \frac{\partial}{\partial y}\left(k\frac{\partial T}{\partial y}\right) + \frac{\partial}{\partial z}\left(k\frac{\partial T}{\partial z}\right)\right]$$

or

$$\text{Heat-transfer rate per unit volume} = -\nabla \cdot \mathbf{q}'' \tag{12-116}$$

2. The rate of shaft work per unit volume is zero by our stated assumptions.
3. The flow-work term is evaluated for each of the six faces of the control volume. Let us consider the face $\Delta y\, \Delta z$ at $x + \Delta x$ as shown in Fig. 12-14 and evaluate

$$\int_{cs} (\tau_{nn}\mathbf{n} + \tau_{ns}\mathbf{s}) \cdot \mathbf{V}\, dA$$

for it. Thus,

$$\int_{cs_1} (\tau_{nn}\mathbf{n} + \tau_{ns}\mathbf{s}) \cdot \mathbf{V}\, dA = \int_{\Delta y \Delta z} (\tau_{xx}\mathbf{i} + \tau_{xy}\mathbf{j} + \tau_{xz}\mathbf{k})|_{x+\Delta x} \cdot (u\mathbf{i} + v\mathbf{j} + w\mathbf{k})\, dy\, dz$$

which yields

$$\int_{cs_1} (\tau_{nn}\mathbf{n} + \tau_{ns}\mathbf{s}) \cdot \mathbf{V}\, dA = (u\tau_{xx} + v\tau_{xy} + w\tau_{xz})_{x+\Delta x}\, \Delta y\, \Delta z \tag{12-117}$$

In a similar manner the flow-work term evaluated for the surface $\Delta y\, \Delta z$ at x would yield

$$\int_{cs_2} (\tau_{nn}\mathbf{n} + \tau_{n}\mathbf{s}) \cdot \mathbf{V}\, dA = (u\tau_{xx} + v\tau_{xy} + w\tau_{xz})_{x}\, \Delta y\, \Delta z \tag{12-118}$$

where we have arbitrarily designated this as control surface 2.

Forming similar terms for the other four surfaces (two each normal to the y and z directions) and writing the flow-work term per unit volume, we obtain

$$\frac{\int_{cs} (\tau_{nn}\mathbf{n} + \tau_{ns}\mathbf{s}) \cdot \mathbf{V}\, dA}{\Delta x\, \Delta y\, \Delta z} = \frac{(u\tau_{xx} + v\tau_{xy} + w\tau_{xz})_{x+\Delta x} - (u\tau_{xx} + v\tau_{xy} + w\tau_{xz})_{x}}{\Delta x}$$

$$+ \frac{(u\tau_{yx} + v\tau_{yy} + w\tau_{yz})_{y+\Delta y} - (u\tau_{yx} + v\tau_{yy} + w\tau_{yz})_{y}}{\Delta y}$$

$$+ \frac{(u\tau_{zx} + v\tau_{zy} + w\tau_{zz})_{z+\Delta z} - (u\tau_{zx} + v\tau_{zy} + w\tau_{zz})_{z}}{\Delta z}$$

Taking the limit as Δx, Δy, and Δz approach zero, this becomes

$$\frac{\text{Flow work}}{\text{Unit volume}} = \frac{\partial}{\partial x}(u\tau_{xx} + v\tau_{xy} + w\tau_{xz}) + \frac{\partial}{\partial y}(u\tau_{yx} + v\tau_{yy} + w\tau_{yz})$$

$$+ \frac{\partial}{\partial z}(u\tau_{zx} + v\tau_{zy} + w\tau_{zz}) \tag{12-119}$$

4. The time rate of energy storage within the control volume is

$$\frac{d}{dt}\int_{cv} \rho e \, d\mathcal{V} = \left(\frac{\partial}{\partial t} \rho e\right) \Delta x \, \Delta y \, \Delta z$$

Consequently we have

Rate of change of stored energy per unit volume $= \dfrac{\partial}{\partial t} \rho e$ \hfill (12-120)

5. The rate of energy efflux from the control volume can be obtained by evaluation of the surface integral for each of the six surfaces of the control volume. For example, for the surface of Fig. 12-14, the energy convected across is

$$\int_{cs_1} e(\rho \mathbf{V} \cdot d\mathbf{A}) = (e\rho u)_{x+\Delta x} \Delta y \, \Delta z$$

Forming similar terms for the remaining five surfaces, adding them, and dividing by the volume yields

$$\int_{cs} e(\rho \mathbf{V} \cdot d\mathbf{A}) = \frac{(e\rho u)_{x+\Delta x} - (e\rho u)_x}{\Delta x} + \frac{(e\rho v)_{y+\Delta y} - (e\rho v)_y}{\Delta y} + \frac{(e\rho w)_{z+\Delta z} - (e\rho w)_z}{\Delta z}$$

In the limit as the control volume dimensions approach zero we have

Net energy efflux per unit volume $= \dfrac{\partial}{\partial x} e\rho u + \dfrac{\partial}{\partial y} e\rho v + \dfrac{\partial}{\partial z} e\rho w$ \hfill (12-121)

Substituting Eqs. (12-116) to (12-121) into Eq. (10-110) yields the following form of the first law of thermodynamics for a differential control volume with no shaft work:

$$\frac{\partial}{\partial t}\rho e + \frac{\partial}{\partial x} u\rho e + \frac{\partial}{\partial y} v\rho e + \frac{\partial}{\partial z} w\rho e = -\nabla \cdot \mathbf{q}'' + \frac{\partial}{\partial x}(u\tau_{xx} + v\tau_{xy} + w\tau_{xz})$$
$$+ \frac{\partial}{\partial y}(u\tau_{yx} + v\tau_{yy} + w\tau_{yz}) + \frac{\partial}{\partial z}(u\tau_{zx} + v\tau_{zy} + w\tau_{zz}) \quad (12\text{-}122)$$

This however, is not a usable form since it still contains shear and normal stresses which are not easily measured. We can eliminate these unknowns by the following method, which is given in outline form.

First, perform the indicated differentiations of the terms on the right side of Eq. (12-119). Then group the resulting terms to permit substitutions of $u\rho\, Du/Dt$, $v\rho\, Dv/Dt$, and $w\rho\, Dw/Dt$ from Eqs. (12-6) to (12-8) with zero body (gravity) forces. After slight rearrangement

$$\frac{\text{Flow work}}{\text{Unit volume}} = \rho \frac{D}{Dt}\frac{u^2 + v^2 + w^2}{2} + \left(\tau'_{xx}\frac{\partial u}{\partial x} + \tau_{xy}\frac{\partial v}{\partial x} + \tau_{xz}\frac{\partial w}{\partial x}\right)$$
$$+ \left(\tau_{yx}\frac{\partial u}{\partial y} + \tau'_{yy}\frac{\partial v}{\partial y} + \tau_{yz}\frac{\partial w}{\partial y}\right) + \left(\tau_{zx}\frac{\partial u}{\partial z} + \tau_{zy}\frac{\partial v}{\partial z} + \tau'_{zz}\frac{\partial w}{\partial z}\right)$$
$$- p\left(\frac{\partial u}{\partial x} + \frac{\partial v}{\partial y} + \frac{\partial w}{\partial z}\right) \quad (12\text{-}123)$$

where $\tau_{xx} = -p + \tau'_{xx}$, etc. Now we have included in the energy term e the internal energy per unit mass here designated by U/ρ, the kinetic energy, and the potential energy. Rewriting the left side of Eq. (12-122), we have for an incompressible fluid

$$\rho \frac{De}{Dt} = \rho \left[\frac{D(U/\rho)}{Dt} + \frac{D(V^2/2)}{Dt} + \frac{DP}{Dt} \right] \tag{12-124}$$

where P is the specific potential energy. Thus

$$\frac{DP}{Dt} = \frac{\partial P}{\partial t} + u \frac{\partial P}{\partial x} + v \frac{\partial P}{\partial y} + w \frac{\partial P}{\partial z}$$

and assuming potential energy to be independent of time, we have

$$\frac{DP}{Dt} = -\mathbf{V} \cdot \mathbf{g} \tag{12-125}$$

Next substituting Eq. (12-125) into (12-124) and then substituting (12-123) and (12-124) into (12-122) yields

$$\rho \frac{D(U/\rho)}{Dt} = -\nabla \cdot \mathbf{q}'' + \rho \mathbf{V} \cdot \mathbf{g} - p(\nabla \cdot \mathbf{V}) + \mu \Phi \tag{12-126}$$

where Φ is the viscous-dissipation function defined for any fluid by

$$\Phi \equiv \left(\tau'_{xx} \frac{\partial u}{\partial x} + \tau'_{yy} \frac{\partial v}{\partial y} + \tau'_{zz} \frac{\partial w}{\partial z} \right) + \tau_{xy} \left(\frac{\partial v}{\partial x} + \frac{\partial u}{\partial y} \right)$$

$$+ \tau_{xz} \left(\frac{\partial w}{\partial x} + \frac{\partial u}{\partial z} \right) + \tau_{yz} \left(\frac{\partial w}{\partial y} + \frac{\partial v}{\partial z} \right) \tag{12-127}$$

With the aid of the continuity equation the energy equation can be further simplified. Differentiating Eq. (10-24), grouping terms, and multiplying by p/ρ yields

$$\frac{p}{\rho} \frac{D\rho}{Dt} = -p \nabla \cdot \mathbf{V} \tag{12-128}$$

This can be expressed as

$$\frac{Dp}{Dt} - \rho \frac{D}{Dt} \left(\frac{p}{\rho} \right) = -p \nabla \cdot \mathbf{V} \tag{12-128a}$$

which, when substituted into Eq. (12-126), yields

$$\rho \frac{Dh}{Dt} = -\nabla \cdot \mathbf{q}'' + \rho \mathbf{V} \cdot \mathbf{g} + \frac{Dp}{Dt} + \mu \Phi \tag{12-129}$$

where h is the enthalpy, $U/\rho + p/\rho$. By expressing h as a function of temperature and pressure and applying a well-known thermodynamic relationship, the substantial derivative of h can be expressed as

$$\frac{Dh}{Dt} = c_p \frac{DT}{Dt} + \left[v - T\left(\frac{\partial v}{\partial T}\right)_p\right]\frac{Dp}{Dt} \qquad (12\text{-}130)$$

With this, the energy equation can be written

$$\rho c_p \frac{DT}{Dt} = -\nabla \cdot \mathbf{q}'' + \rho \mathbf{V} \cdot \mathbf{g} + \frac{T}{v}\left(\frac{\partial v}{\partial T}\right)_p \frac{Dp}{Dt} + \mu \Phi \qquad (12\text{-}131)$$

which for constant thermal conductivity is

$$\blacksquare \quad \rho c_p \frac{DT}{Dt} = k \nabla^2 T + \rho \mathbf{V} \cdot \mathbf{g} + \frac{T}{v}\left(\frac{\partial v}{\partial T}\right)_p \frac{Dp}{Dt} + \mu \Phi \qquad (12\text{-}132)$$

This is one of the most useful general forms of the energy equation. To this point we have made no assumptions concerning the viscous-dissipation term Φ. The general expression for this is given by Eq. (12-127), and if we apply Stokes' hypothesis, we can express Φ in the cartesian coordinate system by substituting Eqs. (12-14) and (12-15) into (12-127), again noting that $\tau_{xx} = \tau'_{xx} - p$, etc. This yields

$$\Phi = 2\left[\left(\frac{\partial u}{\partial x}\right)^2 + \left(\frac{\partial v}{\partial y}\right)^2 + \left(\frac{\partial w}{\partial z}\right)^2\right] + \left(\frac{\partial u}{\partial y} + \frac{\partial v}{\partial x}\right)^2 + \left(\frac{\partial u}{\partial z} + \frac{\partial w}{\partial x}\right)^2$$
$$+ \left(\frac{\partial v}{\partial z} + \frac{\partial w}{\partial y}\right)^2 - \frac{2}{3}\left(\frac{\partial u}{\partial x} + \frac{\partial v}{\partial y} + \frac{\partial w}{\partial z}\right)^2 \qquad (12\text{-}133)$$

For Φ in other coordinate systems, the reader should consult Hughes and Gaylord [7].

From our previous discussions of boundary-layer flow, we realize that for many problems the only significant velocity gradient is $\partial u/\partial y$, and then the dissipation function reduces to

$$\Phi \simeq \left(\frac{\partial u}{\partial y}\right)^2 \qquad (12\text{-}134)$$

This expression is quite valid for flow over flat plates (and inside tubes if y is replaced with r). Further, the entire viscous-dissipation term is frequently negligible for low-speed incompressible flow.

The energy equation can readily be transformed to the cylindrical or spherical coordinate systems by use of the vector identities and operations of Appendix B. These forms can also be obtained from Ref. 7.

12-8 CONVECTIVE HEAT TRANSPORT ON A FLAT PLATE

In this section we consider the very important case of heat transfer resulting from laminar flow over external surfaces; the simplest configuration, the flat plate, will be treated in detail. It should be noted at the outset that there are many important

LAMINAR FLOW OF INCOMPRESSIBLE VISCOUS FLUIDS

applications of external laminar-boundary-layer flow—flow over airfoils, turbine blades, inside very large ducts such as wind tunnels, and over plate-type heat exchangers, to mention a few. Also, many of these situations are reasonably well approximated by a flat plate.

Consider the problem of a constant free-stream velocity flow along a constant-temperature plate, as shown in Fig. 12-12. This is the companion thermal problem to the momentum-transport problem of Blasius in Sec. 12-5.

Under the assumptions of (1) negligible body forces, (2) constant fluid properties, (3) low velocity, i.e., viscous dissipation negligible, (4) plate infinite in $\pm z$ directions, and (5) steady flow, the energy Eq. (12-132) reduces to

$$u \frac{\partial T}{\partial x} + v \frac{\partial T}{\partial y} = \alpha \frac{\partial^2 T}{\partial y^2} \tag{12-135}$$

where α is the thermal diffusivity of the fluid, $k/\rho c_p$. The boundary conditions are

$$T = \begin{cases} T_s & \text{at } y = 0 \\ T_\infty & \text{at } y = \infty \end{cases} \tag{12-136}$$

At this point it is informative to nondimensionalize the temperature by letting

$$\theta = \frac{T - T_s}{T_\infty - T_s} \tag{12-137}$$

Then Eq. (12-135) becomes

$$u \frac{\partial \theta}{\partial x} + v \frac{\partial \theta}{\partial y} = \alpha \frac{\partial^2 \theta}{\partial y^2} \tag{12-138}$$

with boundary conditions

$$\theta = \begin{cases} 0 & \text{at } y = 0 \\ 1 & \text{at } y = \infty \end{cases} \tag{12-139}$$

Now return to Prandtl's boundary-layer equations, and for the flat plate in steady flow the momentum equation reduces to

$$u \frac{\partial u}{\partial x} + v \frac{\partial u}{\partial y} = \nu \frac{\partial^2 u}{\partial y^2} \tag{12-140}$$

Nondimensionalizing the velocity u by letting

$$u^* = \frac{u}{V_\infty} \tag{12-141}$$

this becomes

$$u \frac{\partial u^*}{\partial x} + v \frac{\partial u^*}{\partial y} = \nu \frac{\partial^2 u^*}{\partial y^2} \tag{12-142}$$

with boundary conditions

$$u^* = \begin{cases} 0 & \text{at } y = 0 \\ 1 & \text{at } y = \infty \end{cases} \qquad (12\text{-}143)$$

A comparison of Eq. (12-138) and its boundary conditions with Eq. (12-142) and its boundary conditions reveals a very important fact. *The thermal and momentum equations are identical if $\nu = \alpha$.* The Prandtl number is defined by

$$\mathbf{Pr} \equiv \frac{\mu c_p}{k} = \frac{\nu}{\alpha} \qquad (12\text{-}144)$$

and consequently if the Prandtl number is unity, the thermal and velocity profiles for the flat plate in low-speed steady incompressible flow are identical. The Prandtl number is a fluid property and ranges from 0.6 to 1.0 for most gases. Thus, the assumption of identical thermal and velocity profiles and boundary layers is frequently justifiable for gas flow. Most liquids, however, have Prandtl numbers considerably different from unity, the number being very large for viscous fluids such as oil and very small for liquid metals with their attendant high thermal conductivities.

Pohlhausen [17] obtained a solution to Eq. (12-138) subject to the boundary conditions of Eqs. (12-139) for fluids with arbitrary Prandtl number. As a result of the similarity of the thermal and velocity boundary layers, he assumed the existence of *similar* temperature profiles along the plate, where again this simply means that the profiles differ by a stretching factor, as discussed in the Blasius problem. The details of his solution are quite lengthy, and only the results are of primary interest to us. Assuming the similarity parameter η and the stream function ψ defined by Eqs. (12-77) and (12-79),

$$\left(\frac{d\theta}{d\eta}\right)_{\eta=0} = \frac{1}{\int_0^\infty \exp\left(-\frac{\mathbf{Pr}}{2}\int_0^\eta f\,d\eta\right)d\eta} \qquad (12\text{-}145)$$

Pohlhausen numerically integrated this last expression, and his results are well represented by

$$\left(\frac{d\theta}{d\eta}\right)_{\eta=0} = 0.332\mathbf{Pr}^{0.343} \qquad (12\text{-}146)$$

for fluids with $0.5 < \mathbf{Pr} < 10$. For simplicity, the exponent of $\mathbf{Pr}$ is taken as $\tfrac{1}{3}$, and the last expression is written as

$$\left(\frac{d\theta}{d\eta}\right)_{\eta=0} = 0.332\mathbf{Pr}^{\frac{1}{3}} \qquad (12\text{-}147)$$

Now the local heat-transfer coefficient h_x is defined by

$$h_x \equiv \frac{q_s''}{T_s - T_\infty} = \frac{-k(T_\infty - T_s)(d\theta/dy)_{y=0}}{T_s - T_\infty}$$

LAMINAR FLOW OF INCOMPRESSIBLE VISCOUS FLUIDS

where q'' is the heat flux per unit area, and thus

$$h_x = k\sqrt{\frac{V_\infty}{\nu x}}\left(\frac{d\theta}{d\eta}\right)_{\eta=0}$$

Combining this with Eq. (12-147) yields

$$h_x = 0.332k\sqrt{\frac{V_\infty}{\nu x}}\,\mathbf{Pr}^{\frac{1}{3}} \tag{12-148}$$

which can be expressed in terms of the nondimensional Nusselt modulus

$$\blacksquare \quad \mathbf{Nu}_x \equiv \frac{h_x x}{k} = 0.332\mathbf{Pr}^{\frac{1}{3}}\mathbf{Re}_x^{\frac{1}{2}} \tag{12-149}$$

Recall that by Eq. (12-80) the x-direction velocity within the boundary layer is

$$\frac{u}{V_\infty} = u^* = f'$$

and the gradient is

$$\left(\frac{du^*}{d\eta}\right)_{\eta=0} = f''(0) = 0.332 \tag{12-150}$$

where the numerical value is from Table 12-3. Now for a fluid with $\mathbf{Pr} = 1.0$, the nondimensional velocity gradient of Eq. (12-150) is identical with the nondimensional temperature gradient, and substituting this into Eq. (12-48) yields

$$\mathbf{Nu}_x = 0.332\mathbf{Re}_x^{\frac{1}{2}} \tag{12-151}$$

which agrees with the result directly obtainable from Eq. (12-149). Now consider the approximations to the nondimensional thermal and velocity gradients given by

$$\left(\frac{d\theta}{d\eta}\right)_{\eta=0} \simeq \frac{\theta_\infty - \theta_s}{\delta_t} = \frac{1}{\delta_t} \tag{12-152a}$$

$$\left(\frac{du^*}{d\eta}\right)_{\eta=0} \simeq \frac{u_\infty^* - u_s^*}{\delta} = \frac{1}{\delta} \tag{12-152b}$$

The first of these must be equivalent to Eq. (12-147), consequently,

$$\frac{1}{\delta_t} \simeq 0.332\mathbf{Pr}^{\frac{1}{3}}$$

and the second, by Eq. (12-150), is

$$\frac{1}{\delta} \simeq 0.332$$

Therefore, Pohlhausen's results indicate that

■ $\quad \delta_t = \dfrac{\delta}{\Pr^{\frac{1}{3}}}$ (12-153)

From Eq. (12-149) we observe that for a given fluid, the Nusselt modulus is proportional to the square root of the Reynolds number based on length. This may be expressed as

$$\dfrac{h_x x}{k} \simeq \left(\dfrac{xV_\infty}{\nu}\right)^{\frac{1}{2}}$$

or, since ν is also a fluid property and V_∞ is constant,

$$h_x = Cx^{-\frac{1}{2}} \qquad (12\text{-}154)$$

where C is kV_∞/ν, a constant. Determining the average value of h for a plate of length L and unit width, we have

$$\bar{h} \equiv h_{\mathrm{av}} = \dfrac{C\displaystyle\int_0^L x^{-\frac{1}{2}}\,dx}{L} = 2CL^{-\frac{1}{2}}$$

or

$$\bar{h} = 2h_x\big|_{x=L} \qquad (12\text{-}155)$$

This shows the average value of the convective-heat-transfer coefficient for a plate at constant temperature over its entire length to be exactly twice the local value at the end of the plate.

Using this, we can express the average Nusselt modulus as

■ $\quad \overline{\mathrm{Nu}}_L = 0.664\,\mathrm{Re}_L^{\frac{1}{2}}\,\mathrm{Pr}^{\frac{1}{3}}$ (12-156)

and this expression is convenient in the solution of practical problems.

Example In a continuous cooking operation, light oil flows at 0.1 ft/sec over a 30-ft-long heated surface that is very wide. The oil temperature T_∞ is 200°F, and the heated surface is 300°F. Evaluate all fluid properties at a film temperature (to be discussed later) of 250°F and calculate the following at $x = 30$ ft: (a) the velocity-boundary-layer thickness δ, (b) the total drag per unit width of the surface, (c) the thermal-boundary-layer thickness δ_t, (d) the local heat-transfer coefficient at the end of the surface, and (e) the total heat flux from the plate per unit width.

Solution The Reynolds number at the end of the plate is

$$\mathrm{Re}_L = \dfrac{30(0.1)}{2.6 \times 10^{-5}} = 115{,}300$$

and the flow is completely laminar.

(a) The boundary-layer thickness is, by Eq. (12-90),

$$\delta = L\dfrac{5.0}{\sqrt{\mathrm{Re}_L}} = \dfrac{30(5.0)}{\sqrt{115{,}300}} = 0.442 \text{ ft}$$

LAMINAR FLOW OF INCOMPRESSIBLE VISCOUS FLUIDS

(b) From Eq. (12-95), the drag coefficient is

$$C_f = \frac{1.328}{\sqrt{\text{Re}_L}} = \frac{1.328}{\sqrt{115{,}300}} = 0.392 \times 10^{-2}$$

The total frictional drag for unit width is then

$$\text{Drag} = C_f(\text{area})\frac{\rho_\infty V_\infty^2}{2}$$

$$= 30(1.0)(0.392 \times 10^{-2})\frac{53.0}{32.2}\frac{0.1^2}{2} = 9.66 \times 10^{-4}\ \text{lb}_f$$

(c) The thermal-boundary-layer thickness can be obtained from Eq. (12-153) and is

$$\delta_t = \frac{\delta}{\text{Pr}^{\frac{1}{3}}} = \frac{0.442}{35^{\frac{1}{3}}} = 0.135\ \text{ft}$$

at the end of the plate. Notice that the thermal boundary layer is considerably thinner than the velocity boundary layer.

(d) From Eq. (12-148) the local heat-transfer coefficient at the end of the plate is

$$h_x = 0.332(0.074)\sqrt{\frac{0.1 \times 10^5}{2.6 \times 30}}\ 35^{\frac{1}{3}} = 0.91\ \text{Btu/hr-ft}^2\text{-}°\text{F}$$

(e) To determine the total heat flux from the plate per unit width, we note by Eq. (12-155) that

$$\bar{h} = 2h_x|_{30}$$

or

$$\bar{h} = 2(0.91) = 1.82\ \text{Btu/hr-ft}^2\text{-}°\text{F}$$

Since the wall temperature is 300°F while the free-stream temperature is 200°F,

$$q_s = \bar{h}A(T_s - T_\infty) = 1.82(30)(100) = 5460\ \text{Btu/hr}$$

We note in passing that this value of heat flux is appropriate only if the fluid is not experiencing phase change with resultant boiling and increased fluid motion. In actuality, most commercial cooking operations involve water liberation from the substance being processed and a resulting boiling action within the fluid bed.

Integral boundary-layer energy equation In a manner quite analogous to that employed in developing the integral momentum equation, we can derive an integral energy equation. The advantages of such an equation over its differential counterpart are the same as those of the integral momentum equation: primarily it is integrated through the boundary layer and contains derivatives only with respect to x. It must be admitted at the outset that the integral energy equation is considerably less exact than its differential counterpart, but the success it has enjoyed in many applications and the simplifications it affords justify the necessary approximations.

The first law of thermodynamics applied to a control volume for an incompressible steady-state flow with no shaft work and negligible viscous shear and normal stresses can be written from Eq. (10-110) as

$$\frac{\delta Q}{\delta t} = \int_{cs} e(\rho \mathbf{V} \cdot d\mathbf{A}) - \int_{cs} (\tau_{nn}\mathbf{n} + \tau_{ns}\mathbf{s}) \cdot \mathbf{V}\, dA \qquad (12\text{-}157)$$

This equation expresses an energy balance between the rate of energy conducted into the control volume and the net rate of energy convected out. A suitable control volume illustrating the energy-flux terms is shown in Fig. 12-15. Recalling that e includes internal energy, kinetic energy, and potential energy and that the term τ_{nn} includes the static pressure p, we can effect a combination of the two surface integrals of Eq. (12-157) to obtain

$$\frac{\delta Q}{\delta t} = \int_{cs} h(\rho \mathbf{V} \cdot d\mathbf{A}) \qquad (12\text{-}158)$$

which is valid for negligible kinetic-energy changes, potential-energy difference, pressure gradients, and viscous shear and normal stresses. The interested reader can readily verify this with the aid of the material of Sec. 12-7. Proceeding to evaluate the energy fluxes, we have:

Across surface 1-2:
$$\int_{cs_{1\text{-}2}} [h(\rho \mathbf{V} \cdot d\mathbf{A})]_x = -c_p \left(\int_0^L \rho T u\, dy \right)_x \qquad (12\text{-}159)$$

Across surface 3-4:
$$\int_{cs_{3\text{-}4}} [h(\rho \mathbf{V} \cdot d\mathbf{A})]_{x+\Delta x} = c_p \left(\int_0^L \rho T u\, dy \right)_{x+\Delta x} \qquad (12\text{-}160)$$

Across surface 2-4 the mass flow rate is

$$\rho \mathbf{V} \cdot d\mathbf{A} = \int_0^L [(\rho u\, dy)_{x+\Delta x} - (\rho u\, dy)_x]$$

and the temperature is T_∞. Consequently the energy flux is

$$\int_{cs_{2\text{-}4}} h(\rho \mathbf{V} \cdot d\mathbf{A}) = -c_p T_\infty \int_0^L [(\rho u\, dy)_{x+\Delta x} - (\rho u\, dy)_x] \qquad (12\text{-}161)$$

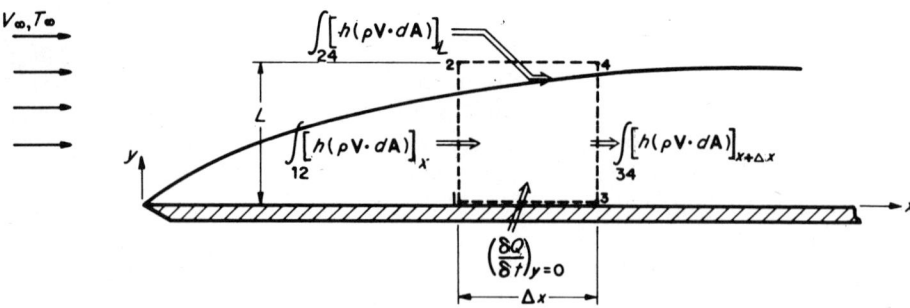

Fig. 12-15 Control volume for integral energy equation.

LAMINAR FLOW OF INCOMPRESSIBLE VISCOUS FLUIDS

Across surface 1-3 the dot product $\mathbf{V} \cdot d\mathbf{A}$ vanishes, resulting in no enthalpy flux. There is a conductive flux, however, given by

$$\left(\frac{\partial Q}{\partial t}\right)_{y=0} = -k\,\Delta x \left(\frac{\partial T}{\partial y}\right)_{y=0} \tag{12-162}$$

The conductive fluxes along surfaces 1-2 and 3-4 are negligible because of the very small temperature gradient in the x direction, and the conductive flux along surface 2-4 is identically zero because the temperature is constant.

Substituting Eqs. (12-159) to (12-162) into Eq. (12-158) yields

$$-k\,\Delta x\left(\frac{\partial T}{\partial y}\right)_{y=0} = c_p\left[\left(\int_0^L \rho T u\, dy\right)_{x+\Delta x} - \left(\int_0^L \rho T u\, dy\right)_x\right]$$
$$- c_p T_\infty \int_0^L [(\rho u\, dy)_{x+\Delta x} - (\rho u\, dy)_x]$$

Dividing by Δx and taking the limit as Δx approaches zero gives for an incompressible fluid

$$\rho c_p \frac{d}{dx}\left(T_\infty \int_0^L u\, dy - \int_0^L Tu\, dy\right) = k\left(\frac{\partial T}{\partial y}\right)_{y=0}$$

which, since T_∞ is independent of y, can be written as

$$\blacksquare \quad \frac{d}{dx}\int_0^{\delta_t} u(T_\infty - T)\, dy = \alpha\left(\frac{\partial T}{\partial y}\right)_{y=0} \tag{12-163}$$

The upper limit of integration is δ_t since $T_\infty - T$ vanishes for $y > \delta_t$. This is the integral energy equation of the boundary layer for an incompressible flow with negligible viscous effects.

The general procedure for solving this equation involves the assumption of reasonable profiles for both u and T; "reasonable" again implies satisfaction of the boundary conditions with a reasonable degree of accuracy. Consider heat transfer from a heated plate, as shown in Fig. 12-16. From our study of the integral momentum equation we recall that the boundary conditions on the velocity are best satisfied by

$$\frac{u}{V_\infty} = \frac{3}{2}\frac{y}{\delta} - \frac{1}{2}\left(\frac{y}{\delta}\right)^3 \tag{12-111}$$

The usual boundary conditions on temperature are

$$\begin{aligned} T &= T_s & \frac{\partial^2 T}{\partial y^2} &= 0 & \text{at } y &= 0 \\ T &= T_\infty & \frac{\partial T}{\partial y} &= 0 & \text{at } y &= \delta \end{aligned} \tag{12-164}$$

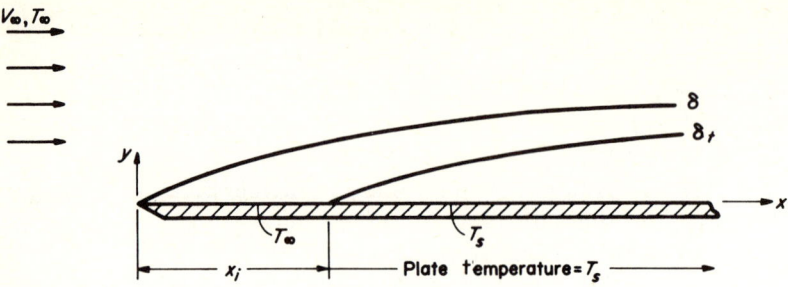

Fig. 12-16 Thermal and velocity boundary layers on a flat plate with an unheated initial length.

where the condition $\partial^2 T/\partial y^2 = 0$ at the wall is obtained by inspection of the differential form of the energy equation (12-135).

Since there are four known boundary conditions, it is logical to assume a temperature profile of the form

$$\theta = \frac{T - T_s}{T_\infty - T_s} = a + a_1 \frac{y}{\delta_t} + a_2 \left(\frac{y}{\delta_t}\right)^2 + a_3 \left(\frac{y}{\delta_t}\right)^3 \tag{12-165}$$

Then by direct analogy with the solution for the constants in Eq. (12-108) (note the similarity of the boundary conditions) we have

$$\theta = \frac{3}{2} \frac{y}{\delta_t} - \frac{1}{2} \left(\frac{y}{\delta_t}\right)^3 \tag{12-166}$$

With the profiles of Eqs. (12-111) and (12-166), the integral in Eq. (12-163) becomes

$$(T_\infty - T_s) V_\infty \int_0^{\delta_t} \left[1 - \frac{3}{2}\frac{y}{\delta_t} + \frac{1}{2}\left(\frac{y}{\delta_t}\right)^3\right]\left[\frac{3}{2}\frac{y}{\delta} - \frac{1}{2}\left(\frac{y}{\delta}\right)^3\right] dy$$

Carrying out the indicated multiplication and grouping according to the exponent of y, we have

$$(T_\infty - T_s) V_\infty \int_0^{\delta_t}\left[\frac{3}{2\delta} y - \frac{9}{4\delta\delta_t} y^2 - \frac{1}{2\delta^3} y^3 + \left(\frac{3}{4\delta\delta_t^3} + \frac{3}{4\delta^3\delta_t}\right) y^4 - \frac{1}{4\delta^3\delta_t^3} y^6\right] dy$$

which can be integrated to obtain

$$(T_\infty - T_s) V_\infty \left[\frac{3}{4}\frac{\delta_t^2}{\delta} - \frac{3}{4}\frac{\delta_t^2}{\delta} - \frac{1}{8}\frac{\delta_t^4}{\delta^3} + \frac{3}{20}\left(\frac{\delta_t^2}{\delta} + \frac{\delta_t^4}{\delta^3}\right) - \frac{1}{28}\frac{\delta_t^4}{\delta^3}\right]$$

Letting $\zeta = \delta_t/\delta$, the ratio of thermal- to velocity-boundary-layer thickness, the integral of Eq. (12-163) can be written as

$$\int_0^{\delta_t} u(T_\infty - T)\, dy = V_\infty (T_\infty - T_s)\delta\left(\tfrac{3}{20}\zeta^2 - \tfrac{3}{280}\zeta^4\right) \tag{12-167}$$

LAMINAR FLOW OF INCOMPRESSIBLE VISCOUS FLUIDS

Recall that one result of Pohlhausen's solution of the differential form of the energy equation was $\delta_t = \delta/\mathbf{Pr}^{\frac{1}{3}}$; and since $\mathbf{Pr}$ is very close to unity for most gases and quite large for most viscous fluids, it is reasonable to assume ζ equal to or less than unity.[1] Then the term involving ζ^4 in Eq. (12-167) can be neglected, and Eq. (12-163) becomes

$$\tfrac{3}{20} V_\infty (T_s - T_\infty) \frac{d}{dx} \zeta^2 \delta = \tfrac{3}{2}\alpha \frac{T_s - T_\infty}{\delta}$$

Performing the differentiation where ζ can depend upon x and noting from our earlier solution of the integral momentum equation that

$$\delta\, d\delta = \frac{140}{13} \frac{\mu}{\rho V_\infty} dx$$

we obtain after rearrangement

$$\frac{14}{13}\frac{\nu}{\alpha}\left(\zeta^3 + 4x\zeta^2 \frac{d\zeta}{dx}\right) = 1$$

(The variation of ζ with distance along the plate is significant only for the case of an unheated starting length or leading edge.) This can be written as

$$\tfrac{4}{3} x \frac{d(\zeta^3)}{dx} + \zeta^3 = \tfrac{13}{14}\mathbf{Pr} \tag{12-168}$$

which is a first-order ordinary differential equation in ζ^3. The boundary condition is

$$\delta_t = 0 \quad \zeta = 0 \quad \text{at } x = x_i \tag{12-169}$$

The solution to Eq. (12-168) is

$$\zeta^3 = C x^{-\frac{3}{4}} + \tfrac{13}{14}\mathbf{Pr}^{-1}$$

which when C is evaluated by the boundary condition of Eq. (12-169) yields

$$\zeta = \frac{\delta_t}{\delta} = \frac{0.975}{\mathbf{Pr}^{\frac{1}{3}}}\left[1 - \left(\frac{x_i}{x}\right)^{\frac{3}{4}}\right]^{\frac{1}{3}} \tag{12-170}$$

The local heat-transfer coefficient can be obtained from

$$h_x = \frac{q_s''}{T_s - T_\infty} = \frac{-k(T_\infty - T_s)(d\theta/dy)_{y=0}}{T_s - T_\infty} = \frac{3}{2}\frac{k}{\zeta\delta}$$

Substituting the local boundary-layer thickness δ from Eq. (12-112) and the ratio ζ from Eq. (12-170), we obtain

$$h_x = \frac{3}{2}\frac{k}{x}\frac{\mathbf{Pr}^{\frac{1}{3}}\mathbf{Re}_x^{\frac{1}{2}}}{0.975(4.64)}\left[1 - \left(\frac{x_i}{x}\right)^{\frac{3}{4}}\right]^{-\frac{1}{3}} \tag{12-171}$$

---

[1] Of course, this assumption is invalid for liquid metals, which have small Prandtl numbers.

Nondimensionalizing by multiplying both sides by x/k yields

$$\mathbf{Nu}_x = 0.332 \mathbf{Pr}^{\frac{1}{3}} \mathbf{Re}_x^{\frac{1}{2}} \left[1 - \left(\frac{x_i}{x}\right)^{\frac{3}{4}} \right]^{-\frac{1}{3}}$$

which for a plate with no unheated starting length yields a result identical with that obtained by Pohlhausen.

The agreement between the technique using the integral method and the solution of the differential equation depends upon the profiles chosen. As with the momentum equation, the better the satisfaction of known boundary conditions, the better the final solution, in general. The method is very important, since it has been applied with a high degree of success to many problems for which closed-form or exact solutions to the energy equation are not possible.

Film temperature In the solutions of the differential and integral forms of the energy equation we have assumed constant fluid properties. In effect, we have obtained solutions in which the temperature has been assumed invariant. Yet we were first concerned in the solution to the differential energy equation with a determination of temperature as a function of x and y along a flat plate, and later we employed an assumed function of y to represent the temperature profile in the integral energy equation.

The question naturally arises: Is it valid to assume constant fluid properties in order to effect a solution for temperature dependence upon position?

To answer this question, the variation of the various fluid properties with temperature must be examined. To put the energy equation in the usable form of Eq. (12-132), we assumed constant thermal conductivity. An examination of the property data of Appendix A reveals that k is a weak function of temperature for liquids and a moderate function of temperature for gases. For either, the change over a temperature range of 100°F is quite small, and for most problems the assumption of constant thermal conductivity is reasonable.

For the solution of the momentum equation, the primary assumption was that of constant viscosity. The data of Appendix A reveal viscosity to be a weak function of temperature for gases and a moderate function of temperature for liquids. In either event, the assumption of constant viscosity is acceptable for moderate temperature changes.

The alternative to using constant-property assumptions is to apply analytic expressions for the variance of the properties with temperature or to use purely numerical solutions of the momentum and energy equations. The simplifications resulting from the constant-property assumptions are significant and permit a more understandable study of these problems. Further, the results obtained assuming constant properties have been experimentally confirmed in many cases. Invariably a suitable choice of temperature for evaluation of fluid properties is necessary, however. Recall in the example concerning the flow of cooking oil that the properties were evaluated at a film temperature different from the wall or stream temperature,

LAMINAR FLOW OF INCOMPRESSIBLE VISCOUS FLUIDS

T_s or T_∞. It has been found that the use of a film temperature defined by

$$T_f \equiv \frac{T_s + T_\infty}{2} \qquad (12\text{-}172)$$

gives good results with the constant-property analyses for flow over external surfaces with moderate temperature differences between the free stream and the wall.

12-9 CONVECTIVE HEAT TRANSPORT IN A TUBE

Energy transport due to laminar flow inside tubes, pipes, and ducts has been studied extensively. The applications are quite important and include compact gas-flow and liquid-metal heat exchangers, the latter having been applied in numerous nuclear-energy systems. The energy transport in fully developed laminar flow is purely by conduction, and the resulting heat-transfer coefficients are lower than for turbulent flow. They are high for liquid metals with their very high thermal conductivities, however. For other fluids, laminar flow is sometimes necessary for small passage diameters because of the high attendant pumping losses for turbulent flow.

Exact solutions of the differential energy equation in cylindrical coordinates for fully developed laminar flow have been obtained by Graetz [3], Callandar [2], and Nusselt [15]. The mathematical details of their solutions are not of primary importance to our study, and their results can be summarized as follows:

Constant heat transfer along tube: $\qquad \mathbf{Nu}_D \equiv \dfrac{hD}{k} = 4.36 \qquad (12\text{-}173)$

Constant wall temperature along tube: $\qquad \mathbf{Nu}_D \equiv \dfrac{hD}{k} = 3.66 \qquad (12\text{-}174)$

where the Nusselt modulus is constant for any station along the tube. These results, while mathematically correct, are not suitable for most engineering calculations. There are several reasons why the mathematical model assumed in obtaining Eqs. (12-173) and (12-174) is inappropriate; these include:

1. Fully developed velocity and temperature profiles are seldom encountered in laminar tube flow.
2. Many cases of laminar tube flow involve liquids with high viscosities (oils), where the viscosity is rather strongly dependent upon temperature. This results in serious departure from the parabolic velocity profile.
3. These fluids (oils) also have relatively large Prandtl numbers, and the thermal profile develops much slower than the velocity profile. The previous results assumed both profiles to be fully developed.
4. The relatively small momentum terms in forced laminar flow are frequently of the same order of magnitude as the free convective terms, and hence the flow can be a combined forced- and free-convection problem.

Recall that the velocity entry length for fully developed laminar flow is

$$z_E \simeq 0.05 \text{Re}_D D$$

The thermal entry length can be shown (see, for example, Kays [9], p. 125) to be

$$(z_{E_t}) \simeq 0.05 \text{Re}_D \text{Pr} D \tag{12-175}$$

where this is defined as the length required for establishment of the fully developed temperature profile under the conditions of constant wall temperature and developing velocity profile. The differences between these two entry lengths are illustrated in the following example.

Example Light oil at a bulk temperature (to be discussed later) of 200°F flows at 1 ft/sec velocity in a 1-in-ID tube. Calculate the velocity and thermal entry lengths.

Solution The Reynolds number is

$$\text{Re}_D = \frac{VD}{\nu} = \frac{1(\frac{1}{12})}{4.6 \times 10^{-5}} = 1{,}812$$

and the flow is laminar. The velocity entry length is

$$z_E = 0.05(\tfrac{1}{12})(1{,}812) = 7.55 \text{ ft}$$

The thermal entry length, however, is

$$z_{E_t} = 0.05(\tfrac{1}{12})(1{,}812)(62) = 468 \text{ ft}$$

The developing thermal profile is qualitatively similar to the developing velocity profile and is shown, together with a typical variation of local Nusselt modulus, in Fig. 12-17. This illustrates the important fact that the Nusselt modulus is strongly

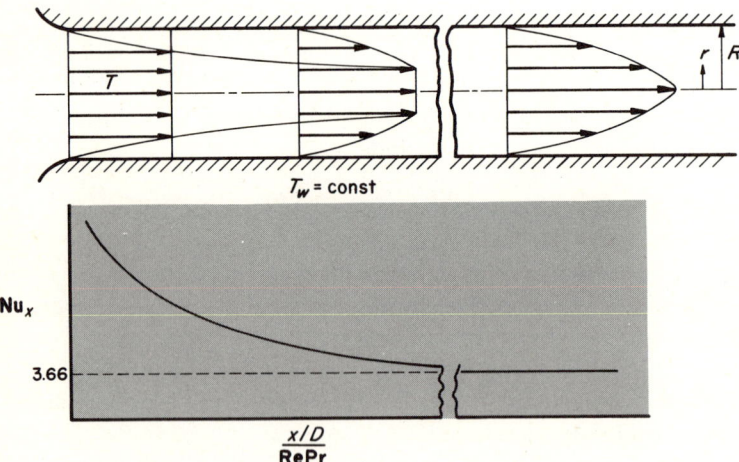

Fig. 12–17 Temperature and Nusselt modulus variation in the inlet region of a tube with fluid being cooled.

Table 12-5

Wall condition	Velocity	Pr	Nu	Nu$_\infty$	K_1	K_2	n
Constant T	Parabolic	Any	Av.	3.66	0.0668	0.04	$\frac{2}{3}$
Constant T	Developing	0.7	Av.	3.66	0.104	0.016	0.8
Uniform q/A	Parabolic	Any	Local	4.36	0.023	0.0012	1.0
Uniform q/A	Developing	0.7	Local	4.36	0.036	0.0011	1.0

dependent upon the length of a short tube and can be much larger than the value obtained by assuming fully developed flow. The appropriate dimensionless groups for correlating the Nusselt modulus are x/D and the product **RePr**. This last group is also known as the *Peclet number*, **Pe = RePr**. The grouping (D/x)**RePr** is sometimes referred to as the *Graetz number* for flow in a circular tube; for other than circular tube flow this is $\dot{m}c_p/kx$.

For engineering calculations, Kays [10] presented the following equation due to Hausen [5] for the cases listed in Table 12-1:

■ $$\mathbf{Nu} = \mathbf{Nu}_\infty + \frac{K_1[(D/x)\mathbf{Re}_D\mathbf{Pr}]}{1 + K_2[(D/x)\mathbf{Re}_D\mathbf{Pr}]^n} \qquad (12\text{-}176)$$

The fluid properties for using this equation (or any equation for flow in a tube or duct) should be evaluated at the bulk or mixing-cup temperature. This is a volume-weighted average temperature which for an incompressible fluid is identical with a mass-weighted average. To clearly understand the meaning of this term, it is advantageous to consider a laminar-flow temperature profile like that depicted for the fully developed section of Fig. 12-17. The bulk temperature is that which would be obtained by directing such a flow into an adiabatic container of negligible thermal capacity, mixing the fluid thoroughly, and then measuring the resulting fluid temperature. Mathematically, this can be accomplished by integrating the product of velocity and temperature over the surface of a control volume consisting of a section of the tube and then dividing by the volumetric flow rate through this control volume. Thus,

$$T_b = \frac{\int_{cs} uT\,dA}{\int_{cs} u\,dA} = \frac{\int_0^R uTr\,dr}{\int_0^R ur\,dr} \qquad (12\text{-}177)$$

In general, the bulk temperature is difficult to obtain by this equation. In practice, the customary objective is to obtain a prescribed temperature drop (or increase) in a given problem. This temperature change, coupled with the initial temperature, permits a first approximation to the bulk temperature for use in determining average heat-transfer coefficients from Eq. (12-176).

Example Liquid Freon-12 at 0°F enters a 2-ft-long $\frac{1}{2}$-in-ID tube having a constant wall temperature of 50°F. The preceding flow is in an adiabatic section of the same tube; hence the velocity profile is developed upstream from this section. The average liquid velocity is 0.1 ft/sec. Assume that the fluid remains liquid and use saturated-liquid properties to determine the exit temperature of the Freon.

Solution We must assume an exit temperature for initial determination of fluid properties. Try $T_e = 20°F$. Then the average bulk temperature is

$$T_b = \frac{T_i + T_e}{2} = 10°F$$

From Appendix A, we have

$\nu = 0.246 \times 10^{-5}$ ft²/sec $\text{Pr} = 4.2$

$c_p = 0.2195$ Btu/lb$_m$-°F $k = 0.0415$ Btu/hr-ft-°F

$\rho = 90.2$ lb$_m$/ft³

The Reynolds number is

$$\text{Re}_D = \frac{(\frac{1}{24} \text{ ft})(0.1 \text{ ft/sec})}{0.246 \times 10^{-5} \text{ft}^2/\text{sec}} = 1{,}693$$

and the flow is laminar. Obviously this is an entry-region problem with a developing thermal profile. (The velocity profile is fully developed by the problem statement.) The average Nusselt modulus is, by Eq. (12-176),

$$\overline{\text{Nu}} = \frac{\bar{h}D}{k} = 3.66 + \frac{0.0668\left[\left(\dfrac{1}{2 \times 24}\right)(1{,}693)(4.2)\right]}{1 + 0.04\left[\left(\dfrac{1}{2 \times 24}\right)(1{,}693)(4.2)\right]^{\frac{2}{3}}}$$

$$= 3.66 + 4.67 = 8.33$$

and

$$\bar{h} = \frac{8.33 \times 0.0415}{\frac{1}{24}} = 8.3 \text{ Btu/hr-ft}^2\text{-°F}$$

Then the total heat-transfer rate to the Freon is

$q = \bar{h}A(T_s - T_b) = 8.3[\pi(\frac{1}{24})(2)](50 - 10) = 86.9$ Btu/hr

The exit Freon temperature is obtained by an energy balance on the fluid, which is

$q = \dot{m}c_p(T_e - T_i) = \rho A V c_p(T_e - T_i)$

$$T_e - 0°F = \frac{86.9/3{,}600}{90.2(\pi/4)(\frac{1}{24})^2(0.1)(0.2195)}$$

$T_e = 9°F$

and our original assumption of $T_e = 20°F$ is considerably high. To obtain a more accurate solution, the exit temperature should now be assumed as 9°F, and the properties reevaluated at $T_b = 4.5°F$. This iteration is left as an exercise and will yield $T_e = 10.1°F$.

LAMINAR FLOW OF INCOMPRESSIBLE VISCOUS FLUIDS

It should be noted that not all combinations of inlet conditions and Prandtl numbers are represented by Eq. (12-176). For conditions not listed, the interested reader should consult Kays [9].

PROBLEMS

12-1. For steady incompressible laminar flow over a flat plate, reduce Eq. (12-16a) as much as possible.

12-2. An incompressible fluid flows steadily through a horizontal square duct. Determine the axial pressure gradient for fully developed flow in terms of the fluid dynamic viscosity and the second derivative of velocity with respect to one coordinate direction normal to the flow direction.

12-3. Consider fully developed laminar incompressible steady flow in a tube. Using the Navier-Stokes equations, derive an expression for the axial pressure gradient.

12-4. A filter in a large industrial plant operates by letting a solution flow across a large sheet of filter paper. Assume the flow to be steady, incompressible, laminar, and fully developed across the paper; neglect z-directional terms and body forces. Determine the velocity gradients.

12-5. For the steady laminar fully developed flow of an incompressible fluid of constant viscosity in a porous annulus with constant injection and withdrawal of fluid:
  (a) Obtain the simplified continuity and momentum equations.
  (b) What are the boundary conditions?
  (c) Find V_r.

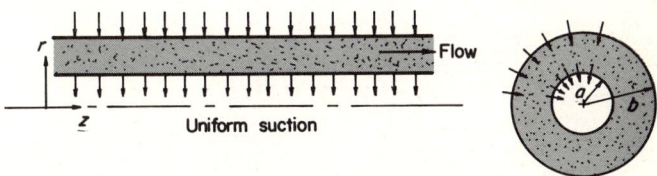

Fig. P12–5

12-6. An incompressible fluid of constant viscosity fills the space h between a horizontal oscillating flat plate and a fixed plane. There is no net flow in the x direction, and $w = 0$.
  (a) Beginning with the most general form of the governing equations, derive the simplified continuity and momentum equations for laminar flow.
  (b) State the boundary conditions.

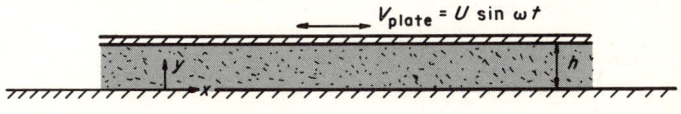

Fig. P12–6

12-7. A fluid coupling consists of two flat smooth circular disks a distance h apart. With respect to the other, one disk (the driver) rotates at ω rad/sec and accelerates at α rad/sec². The system is confined such that there is no radial velocity of an incompressible viscous fluid between the disks. If $h = 0.05$ ft and $\nu = 1.7 \times 10^{-4}$ ft²/sec, find the angular velocity of the driven disk after 5 sec.

12-8. A concentric-cylinder viscometer as illustrated is used to measure the viscosity of fluids. Assuming that (*a*) the inner cylinder rotates at a constant angular velocity ω sufficiently slow for laminar flow to occur between the cylinders, (*b*) fluid properties are constant, (*c*) the flow is axially symmetric and does not vary in the z direction, (*d*) the radial and vertical velocity components are zero, and (*e*) end effects are negligible, determine an expression for the fluid viscosity in terms of the torque T required to rotate the cylinder, the angular velocity, and the cylinder geometry.

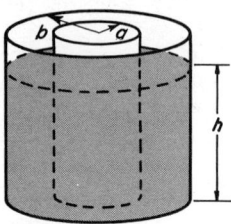

h = fluid height   **Fig. P12–8**

12-9. A viscous incompressible fluid flows steadily between infinite porous parallel planes separated by a distance $2h$. Fluid is uniformly injected at the lower plane at a velocity of v_{-h} while uniform suction occurs at the upper surface with a velocity of v_h. There is no flow in the z direction and the

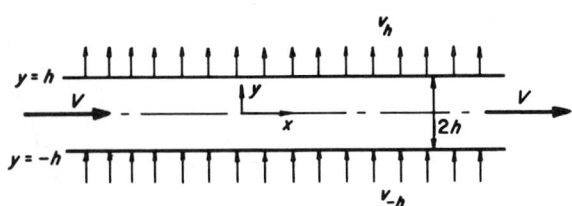

Fig. P12–9

no-slip boundary condition holds for the x direction at the porous surfaces. For fully developed laminar flow of an incompressible fluid:

(*a*) Find the reduced continuity equation.
(*b*) What are the simplified momentum equations?
(*c*) State the boundary conditions.
(*d*) Determine the y component of velocity v.
(*e*) Find the x component of velocity u.
(*f*) Sketch the streamlines.

12-10. Crude oil at 90°F flows steadily in a smooth 1-in-ID tube.

(*a*) What is the maximum flow rate in gallons per minute for which the flow may be considered laminar?
(*b*) For the maximum flow rate, what is the pressure drop in 100 ft of pipe?

12-11. Butyl alcohol at 80°F flows through a pipe which is 73 ft long and 2 in ID. The pipe joins a large tank above a level floor and empties into another tank 16 ft below the floor. Calculate the head loss if the velocity is 0.13 ft/sec.

LAMINAR FLOW OF INCOMPRESSIBLE VISCOUS FLUIDS

12-12. Freon-12 in a household refrigerator flows through condenser tubes with a diameter of $\frac{3}{8}$ in. The kinematic viscosity of Freon-12 is 3×10^{-6} ft^2/sec.

(a) What is the maximum flow rate in cubic inches per second for which the flow may be considered laminar?

(b) What is the friction factor for this flow rate?

12-13. At what radial distance from the centerline of a pipe axis will the velocity be equal to the mean velocity for fully developed laminar flow?

12-14. For a flow of nitrogen gas at 75°F and 1 atm pressure, calculate the pressure drop in 49 ft of 1.0-in-diam circular duct. The dynamic head at the duct axial centerline is held constant at 0.0486 lb$_f$/ft^2.

12-15. Fuel oil at 50°F flows from a large open tank, under the influence of gravity, through a 60-ft length of smooth $\frac{1}{4}$-in-diam pipe. What is the flow rate (gallons per minute) when the lower end of the pipe is 40 ft below the surface of the oil? Neglect entrance effects.

12-16. Kerosene at an inlet temperature of 50°F flows in a 0.5-in-ID pipe. The average velocity is 0.5 ft/sec, and the wall temperature is at 50°F. Find (a) the pressure drop per 10 ft of pipe and (b) the heat transfer per 100 ft of pipe.

12-17. Glycerin at a bulk temperature of 70°F flows through a 100-ft-long 1-in-ID smooth horizontal tube at an average velocity of 12 ft/sec. What is the pressure drop per linear foot for fully developed flow, assuming: (a) frictionless flow; tube at 70°F; (b) flow with friction; tube at 70°F; (c) flow with friction; tube at 130°F?

12-18. For an ammonia flow at 85°F in a horizontal tube, the velocity profile is measured and determined to be $w = 0.172 - 160r^2$ ft/sec, where $r = 0$ is the centerline. The flow is steady and may be assumed axially symmetric. Calculate the pressure drop by direct use of the Navier-Stokes equations and compare with the results of the Hagen-Poiseuille equation.

12-19. A circular tube carries 5.43 in^3/min of water at 200°F. The pressure loss in the 35-ft length is 6.17 lb$_f$/ft^2. What is the inside radius of the tube?

12-20. During installation of additional equipment in an industrial plant, it was necessary to run two pipes through a thick concrete floor. To avoid cutting two holes, however, it was decided that the pipes would be run concentrically. The inside pipe has a 3.5-in OD. The outer pipe required a flow rate of 0.16 ft/sec and a pressure loss of not more than 0.06 lb$_f$/ft^2 in 75 ft. For laminar flow of aniline at 60°F, is an outer ID of 3.7 in satisfactory?

12-21. Kerosene at 50°F flows in a 1- by 3-in rectangular duct at an average velocity of 0.10 ft/sec. What is the pressure drop per 100 ft of duct length?

12-22. Helium gas at 100°F and 30 psia flows through a rectangular duct at 0.4 ft/sec. The duct is 3 by 7 in. Calculate the pressure loss per foot of duct.

12-23. An electric-resistance heating element consists of a smooth cylindrical shell placed in a pipe such that the two are concentric. The thickness of the heating element may be neglected. When the element is off, air flows through the system at 140°F and has an average velocity of 2.80 ft/sec with a pressure loss in the annular section of 0.00452 lb$_f$/ft^2-ft. The element has a radius equal to one-half that of the pipe. What is the total flow rate (through both the annular and the circular cross sections)?

12-24. Air at 50°F flows through an air-conditioning duct with a flow rate of 13.8 ft^3/min. The 6-in by 2-ft duct goes through a smooth transition to 1 by 1 ft. Assume incompressible flow and neglect the transition effects. If there is 100 ft of 6-in by 2-ft duct and 30 ft of 1-ft duct, calculate the head loss.

12-25. Fuel oil at 50°F flows through an octagonal duct at a velocity of 3.5 ft/sec. The duct is $\frac{1}{2}$ in on a side. Approximate the flow rate. Check your answer by $Q = VA$.

12-26. Consider steady incompressible fully developed viscous isothermal flow in a channel formed by two parallel walls. For no flow in the y and z directions, reduce the Navier-Stokes equations to their simplest forms.

12-27. Water is kept for a small town in a natural reservoir having sides of slate. The slate is layered. There is a horizontal crack between layers at an average depth of 75 ft; it is 0.25 in wide and 400 ft long. The water seeping through the crack discharges to a natural river at an average

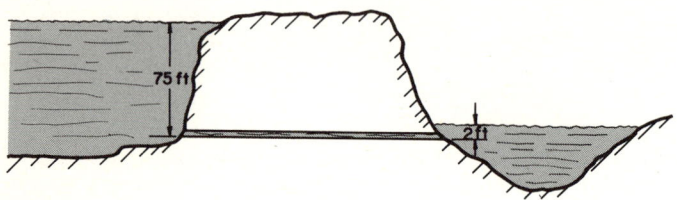

Fig. P12–27

depth of 2 ft. If the water is 55°F and the distance from the reservoir to the river is 400 ft, what is the flow rate per foot of crack width, assuming the layered rock to be level?

12-28. Air at 80°F and 14.7 psia flows along a smooth flat plate with a velocity of 50 ft/sec. Assuming that the flow remains laminar, at what length from the plate leading edge does the boundary-layer thickness reach $\frac{1}{8}$ in?

12-29. A sheet of $8\frac{1}{2}$- by 11-in paper lies flat on a smooth table in an office. As the door opens, a draft blows across the table at 3.1 ft/sec. The temperature in the room is 65°F, and the draft blows along the 11-in direction of the paper. Air pressure holds the paper down so that the total shear bond between the paper and table is 0.0001 lb_f/ft^2. Will the paper slide?

12-30. A probe is positioned near a flat plate. Water at 90°F flows over the plate, $V_\infty = 0.5$ ft/sec. The flow is laminar, and the velocity profile is determined to be $u = 36.65y - 337y^2$, $0 < y < \delta$. How far from the leading edge is the probe?

12-31. Air at 300°F flows along the outside of a large cylinder. The velocity is 0.3 ft/sec. Assuming the effects of curvature to be negligible, what is the local skin-friction coefficient where the boundary layer is 1 in thick?

12-32. Approximate the skin-friction drag on a blimp 500 ft long and 50 ft in average diameter. The airspeed is 0.04 mph (hovering). The air is at 12 psia and 75°F.

12-33. In terms of the boundary-layer thickness δ and the length along the plate x, determine an expression for the drag coefficient C_f per unit depth for a flat plate parallel to a flow with a linear distribution of velocity across the laminar boundary layer.

12-34. Assuming the velocity across the laminar boundary layer of a flat plate to be linear, that is, $u/V_\infty = y/\delta$, derive an expression for the boundary-layer thickness δ making use of the integral technique for steady incompressible flow.

12-35. An approximate relation for boundary-layer flow of air over a flat plate is $u/V_\infty = \frac{3}{2}(y/\delta) - \frac{1}{2}(y/\delta)^3$. The boundary-layer thickness may be approximated by $\delta(x) = 0.0623 x^{\frac{4}{5}}$
    (a) What is the wall shear stress at $x = 8$ ft in terms of V_∞ and μ?
    (b) What is the local skin-friction coefficient at $x = 8$ ft in terms of V_∞ and ν?

12-36. Compare the results of the Blasius solution with the results of the integral momentum equation with respect to skin-friction coefficient and boundary-layer thickness at the end of a flat plate subjected to a flow of light oil at 60°F having a velocity of 2 ft/sec. The plate is 30 ft long.

LAMINAR FLOW OF INCOMPRESSIBLE VISCOUS FLUIDS

12–37. Glycerin flows over a smooth flat plate at 2.5 ft/sec. The kinematic viscosity is 1.27×10^{-2} ft²/sec. Graphically determine the average skin friction for $L = 30$ ft and compare your answer with the result of Eq. (12-95).

12–38. Air flows through the air-conditioning duct in a large building with a velocity of 0.5 ft/sec. The duct is 9 ft square and 200 ft long. At 100 ft, what velocity is found 40 percent of the way through the boundary layer? The air is at 45°F.

12–39. From the boundary-layer solution of Blasius determine an approximate relation for the displacement thickness δ^*.

12–40. A large rectangular duct is 6 by 8 ft, and 15 ft from an inlet the free-stream velocity is found to be 4.5 ft/sec. The fluid in the duct is carbon monoxide at 150°F and 1 atm pressure. What is the volumetric flow rate? *Hint:* Use the result of δ^* obtained in Prob. 12-39.

12–41. A triangular fin projects from the under surface of a rowboat. If the boat moves at 1.2 ft/sec in water at 60°F, calculate the friction drag on the fin. The fin is a right triangle with legs of 4 and 3 ft. The 3-ft leg is perpendicular to the bottom of the boat. (Assume a critical Reynolds number of 500,000.)

12–42. A model rocket ascends vertically at 200 ft/sec. The rocket body is small and has thin fins 3 in long by ½-in chord mounted at an angle of 45°F. The air is at 80°F, and the fins are made of balsa. What is the drag on four fins? Assume two-dimensional chordwise flow over the thin fins with no spanwise flow.

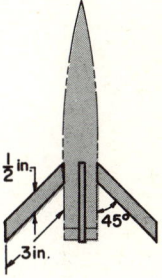

Fig. P12-42

12–43. Crude oil at 90°F flows over one side of a plate that is 12 ft long with a free-stream velocity of 0.5 ft/sec. What is the total drag force on the plate if it is 40 ft wide?

12–44. An incompressible viscous fluid flows steadily down an inclined plane. At a position where the flow is fully developed, the depth is h, and the velocity of the fluid parallel to the plane is not a function of the distance down the plane. There is no flow in the z direction. For unit width, what is the mass flow rate for laminar flow? (Answer in terms of fluid density, fluid viscosity, h and θ.)

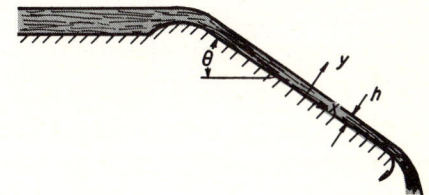

Fig. P12–44

12-45. A slowly moving viscous fluid flows steadily with negligible body forces and constant fluid properties over a flat plate with uniform surface temperature. Reduce the general energy equation for this situation.

12-46. Derive Eq. (12-123) beginning with Eq. (12-119).

12-47. Molten lead at an average temperature of 950°F flows slowly along a horizontal trough. The volume expansivity is

$$\left(\frac{1}{v}\frac{\partial v}{\partial T}\right)_p = 1.7 \times 10^{-2} \text{ °R}^{-1}$$

The pressure in the lead is found to be given by p (lb$_f$/ft) $= 1,800 - 5.9y - 2t$, where y is in feet and t is in seconds. Assuming constant fluid properties and neglecting body forces, estimate the heat loss per pound mass per hour.

12-48. Beginning with Eq. (12-129), show that Eq. (12-131) is valid.

12-49. Kerosene at 70°F and 40 psia flows through a rectangular duct of 2-ft^2 cross section with an average velcoity of 3 ft/sec. The duct necks down in a one-dimensional contraction from 1 by 2 ft to $\frac{1}{2}$ by 2 ft in accordance with the relations $y = 0.5 - \sqrt{x}/16$ and $v = \sqrt{x}/8$, where y is measured from the centerline. Assuming constant fluid properties, what is the viscous-dissipation function at $x = 4$ ft?

12-50. Light oil at 60°F flows over a flat plate at 1.5 ft/sec. The plate is at 100°F. At what point does the thermal-boundary-layer thickness reach 1 in?

12-51. Nitrogen gas at 160°F and 1 atm pressure flows along a smooth flat plate at 80°F. The average velocity is 3 ft/sec. Plot the velocity and thermal-boundary-layer thicknesses for 20 ft of the plate. Use 5-ft increments of plate length.

12-52. Air at 80°F approaches a 3-ft-long flat plate of 2-ft width at 15 ft/sec.
  (a) Estimate the boundary-layer thickness at the end of the plate.
  (b) What is the local skin-friction coefficient c_f 1.5 ft from the leading edge?
  (c) Calculate the shear stress on the surface of the plate at a distance of 1.5 ft from the leading edge.
  (d) Calculate the total drag (both sides) on the plate.
  (e) Calculate the local heat-transfer coefficient h at a distance of 1.5 ft from the leading edge.
  (f) For a plate at 250°F, what is the total heat transfer to the air?
  (g) Assuming a Prandtl number of unity for the air, what is the total heat transfer? Compare with the answer to part (f).

12-53. Water flows over a smooth flat plate at 0.6 ft/sec. The water is at 65°F and the plate is 4 ft long. The temperature of the plate is 115°F. Find the average heat transfer to the water per square foot.

12-54. Water at 250°F and a pressure of 4 atm flows through a large rectangular duct, one side of which is at 470°F. The average velocity is 0.1 ft/sec. What is the local heat-transfer coefficient at $x = 5$ ft based on flat-plate theory?

12-55. Glycerin at 85°F flows along a flat plate with a velocity of 12 ft/sec. If the plate is 6 ft long, compare the average convective heat-transfer coefficient $\bar{h}$ with the local value h_x halfway along the plate. The plate is at 115°F.

12-56. Air moving at 1.0 ft/sec blows over the top of a chest-type freezer. The top of the freezer is 3 by 5 ft and is poorly insulated, so that the surface remains at 50°F. If the average ambient temperature is 80°F, what is the maximum heat transfer by forced convection from the top of the freezer?

12-57. The oil pan of an internal-combustion engine protrudes below the framework of an automobile. It approximates a flat plate 1 ft wide by 1.5 ft long. The ambient temperature of the air is 100°F; the engine oil is at 200°F. Assume that the resistance to conduction of the pan is negligible.

LAMINAR FLOW OF INCOMPRESSIBLE VISCOUS FLUIDS 437

The car travels at 20 mph. How much heat is convected from the oil-pan surface? (Assume two-dimensional flow and flat-plate theory.)

12-58. A heat-treated steel plate is air-cooled after oil immersion. Air at 50°F blows along both sides of the plate at 6 mph. The plate is initially at 80°F. The plate is 10 by 3.3 ft with the air flowing parallel to the 3.3-ft side. Calculate the initial heat-transfer rate.

12-59. A gentle breeze blows along the roof of a house at 3 ft/sec. The air is at 40°F, and the roof is at 80°F. Calculate the heat transfer from the roof per square foot if the air moves in a direction such that the roof is 15 ft long.

12-60. A kitchen in a restaurant has a large flat burner plate for frying. Since a great deal of heat rises from the plate, the cook decides to let a small fan blow over the burner, which is 4 ft long and positioned some 5 ft down a level smooth counter from the fan. (The total length is 9 ft.) If 80°F air blows at 7 ft/sec and the plate is at 250°F, what is the heat transfer per square foot?

12-61. Heating of benzene is accomplished with a heated section 6 ft long in a wide shallow trough through which the fluid must pass in processing. When the fluid reaches the heating section, it has been flowing along the trough for 3 ft and flows at 0.2 ft/sec. (The total length, unheated plus heated, is 9 ft.) The average temperature of the benzene is 80°F, and the heated trough is at 120°F. The trough surface upstream of the heated section is at 80°F. Approximate the heat transfer per square foot.

12-62. A thin 1-ft² plate at 100°F is immersed parallel with a flowing stream of glycerin at 70°F. The measured drag force on the plate is 2.0 lb$_f$ (both sides). What is the total heat transfer?

12-63. A thin sheet of mild steel weighing 1.53 lb$_f$ is dipped into a vat of aniline at 60°F. The sheet is 2 by 3 ft and moves with the 3-ft edge downward at a constant speed of 1 ft/sec (restrained by cables). The steel is initially at 200°F. Calculate the initial heat-transfer rate in Btu per hour. If this rate were constant, how far would the leading edge travel before the plate in the fluid reaches 100°F?

12-64. Glycerin flows through a 1-in-ID horizontal pipe which is 100 ft long. The average velocity is 12 ft/sec. The bulk temperature of the glycerin is 70°F. Calculate the heat transfer for (a) frictionless flow; tube wall at 70°F; (b) flow with friction; tube wall at 70°F; and (c) flow with friction; tube wall at 130°F.

12-65. Fuel oil at an average temperature of 60°F flows through a smooth 1-in-ID tube at 2 ft/sec. If the inner surface of the tube is held at 140°F and the tube is 8 ft long, determine for a fully developed velocity profile the (a) average heat-transfer coefficient $\bar{h}$, (b) heat transfer per foot of tube, (c) increase in bulk temperature per foot, and (d) pressure drop per foot (psi). This oil may be assumed to have properties k, ρ, and c_p similar to those of light oil; however, viscosity and Prandtl numbers are very different.

12-66. The wall of a 0.5-in-ID pipe is heated to 200°F. Kerosene with an inlet temperature of 50°F enters the pipe with a velocity of 0.5 ft/sec. Assume c_p, k, and ρ to be the same as for light oil but not μ, ν, or **Pr**. Assume an exit temperature of 150°F and calculate (a) the pressure drop in 10 ft of pipe and (b) the heat transfer to the kerosene for 10 ft of pipe. Check the validity of the assumed exit temperature at 10 ft.

12-67. Alcohol at a bulk temperature of 60°F flows with a velocity of 2 ft/sec in a long 0.5-in-ID tube with a wall temperature of 100°F. Assuming a well-rounded inlet, what is the length of the developing velocity profile? Thermal profile?

12-68. Glycerin with an average bulk temperature of 50°F flows through a long 8-in square duct. A 2-ft-long section of the duct is heated to a temperature of 200°F. The fluid velocity is 10 ft/sec. Determine the heat transferred.

12-69. Water at 60°F enters a 4-ft-long section of 0.5-in-ID pipe. The pipe wall is at 100°F, and the fluid velocity is 0.4 ft/sec. What is the exit temperature (± 1°F)?

12-70. Glycerin at an average inlet temperature of 100°F enters an 8-ft-long section of 1-in-ID pipe at 0.3 ft/sec. The pipe wall is at 60°F. What is the exit temperature ($\pm 0.3°F$)?

12-71. A 1- by 2-in rectangular duct has nitrogen flowing through it at 5 ft/sec and a bulk temperature of 400°F. The walls are at a constant temperature of 100°F, and the duct is 20 ft long. What is the heat transfer from the nitrogen?

12-72. Water at 70°F enters a 12-ft length of heated 0.5-in-ID pipe with a velocity of 0.4 ft/sec. What constant wall temperature is required for an exit water temperature of 90°F?

REFERENCES

1. Blasius, H.: Grenzschichten in Flüssigkeiten mit kleiner Reibung, *Z. Math. Phys.*, **56**: 1 (1908).
2. Callandar, H. L.: *Phil. Trans. Roy. Soc. London, Ser. A*, **199**: 55 (1902).
3. Graetz, L.: Über die Wärmeleitungsfähigkeit von Flüssigkeiten, *Ann. Physik. Chem.*, **25**: 337–357 (1885).
4. Hagen, G.: Über die Bewegung des Wassers in engen zylindrischen Röhren, *Pogg. Ann.*, **46**: 439 (1839).
5. Hausen, H.: Darstellung des Wärmeüberganges in Rohren durch verallgemeinerte Potenzbeziehungen, *Z. VDI. Beih. Verfarenstech.* 4, 1943, pp. 91–98.
6. Howarth, L.: On the Solution of the Laminar Boundary Layer Equations, *Proc. Roy. Soc. London, Ser. A*, **164**: 547 (1938).
7. Hughes, W. F., and E. W. Gaylord: "Basic Equations of Engineering Science," Schaum's Outline Series, McGraw-Hill, New York, 1964.
8. Kármán, T. von: Über laminare und turbulente Reibung, *Z. Angew. Math. Mech.*, **1**: 233–252 (1921); reprinted in *NACA Tech. Mem.* 1092, 1946.
9. Kays, W. M.: "Convective Heat and Mass Transfer," McGraw-Hill, New York, 1966.
10. Kays, W. M.: Numerical Solutions for Laminar-flow Heat Transfer in Circular Tubes, *Trans. ASME*, **77**: 1265 (1955).
11. Langhaar, H. L.: Steady Flow in the Transitional Length of a Straight Tube, *J. Appl. Mech.*, **64**: A-55 (1942).
12. Lundgren, T. S., E. M. Sparrow, and J. B. Starr: Pressure Drop Due to the Entrance Region in Ducts of Arbitrary Cross Section, *Trans. ASME, Ser. D*, **86**(3): 620–626 (1964).
13. Navier, M.: Mémoire sur les lois du mouvements des fluides, *Mém. Acad. Sci.*, **6**: 389 (1827).
14. Nikuradse, J.: reported in L. Prandtl and O. Tietjens in Hydro- und Aeromechanik, p. 28, Springer-Verlag, Berlin, 1931.
15. Nusselt, W.: Die Abhängigkeit der Wärmeübergangszahl von der Rohrlänge, *Z. VDI.* **54**: 1154 (1910).
16. Pai, Shih-I: "Viscous Flow Theory," vol. I, "*Laminar Flow*," Van Nostrand Company, Princeton, N.J., 1956.
17. Pohlhausen, E.: Der Wärmeaustauch zwischen festen Körpen und Flüssigkeiten mit kleiner Reibung und kleiner Wärmeleitung, *Z. Angew. Math. Mech.*, **1**: 115 (1921).
18. Poiseuille, J.: Récherches expérimentelles sur le mouvement des liquides dans les tubes de très petits diamètres, *Mém. Savants Étrangers*, **9** (1846).
19. Poisson, S. D.: Mémoire sur les équations générales de l'équilibre et du mouvement des corps solides élastique et des fluides, *J. École Polytech.*, **13**: 1 (1831).
20. Prandtl, L.: Über Flüssigkeitsbewegung bei sehr kleiner Reibung, *Proc. 3d Intern. Math. Congr., Heidelberg,* 1904; reprinted in *NACA Tech. Mem.* 452, 1928.
21. St. Venant, B. de: Note à joindre un mémoire sur la dynamique des fluides, *Compt. Rend.*, **17**: 1240–1244 (1843).

22. Schiller, L.: Untersuchungen über laminare und turbulente Strömung, *VDI. Forschungsh.* **248**: 29–33 (1922).
23. Schlichting, H.: "Boundary Layer Theory," 6th ed., J. Kestin, trans., McGraw-Hill, New York, 1968.
24. Stokes, G. G.: On the Theories of the Internal Friction of Fluids in Motion, *Trans. Cambridge Phil. Soc.*, **8** (1845).

CHAPTER 13

TURBULENT FLOW OF INCOMPRESSIBLE ISOTHERMAL FLUIDS

Most flows encountered in practical applications are *turbulent*, and the subject matter of this chapter is momentum and heat transfer in turbulent forced convective flow. Following the pattern established in Chap. 12, we first consider isothermal flow and then discuss energy transfer in nonisothermal flow. As a further parallelism with the treatment of laminar flow, these discussions will be subdivided into internal and external flow.

13-1 ISOTHERMAL FLOW

The basic differences between *laminar* and *turbulent* flow were discussed in Chap. 9. The fundamental difference between these two types of flow is the existence of completely *random* fluctuations in the velocity components for the turbulent case. In addition to purely laminar and purely turbulent flow, we find that *transition* flow usually exists whenever we have the turbulent case. In the development of any boundary layer, internal or external, there normally exists a laminar leading section which becomes turbulent as the fluid moves downstream. This results in a flow

TURBULENT FLOW OF INCOMPRESSIBLE VISCOUS FLUIDS

regime between the completely laminar and completely turbulent areas in which the fluid motion is highly unstable, fluctuating between laminar and turbulent characteristics. This will be discussed in more detail for particular flow models.

For the fully turbulent regime, the velocity vector is given by

$$\mathbf{V} = \bar{\mathbf{V}} + \mathbf{i}u' + \mathbf{j}v' + \mathbf{k}w' \qquad (9\text{-}15)$$

where u', v', and w' are the velocity *fluctuations* in the x-, y-, and z-coordinate directions, respectively, and $\bar{\mathbf{V}}$ is the mean velocity. Or for each coordinate direction we may express the velocity component by Eq. (9-16) as

$$u = \bar{u} + u' \qquad v = \bar{v} + v' \qquad w = \bar{w} + w'$$

We recall that fluid viscosity was defined for a newtonian fluid in purely laminar flow by Eq. (2-5)

$$\tau = \mu \frac{du}{dy}\bigg|_{y=0}$$

Since the streamlines in laminar flow are parallel and the fluid particles slide over each other in layers, or lamina, without mixing, the newtonian viscosity represents the fluid-surface interaction on a *microscopic scale*.

Now consider the turbulent flow in the x direction as shown in Fig. 13-1. As a result of the vertical velocity fluctuation v', a lump of fluid particles designated as 1 moves upward. This is replaced by lump 2 moving downward. This type of interchange is occurring frequently throughout the flow field, resulting in interaction between the fluid layers on a *macroscopic scale*. In fact, the layers may tend to be completely obliterated by these random fluctuations. Since these velocity fluctuations give rise to additional stress terms in the equations of motion, we must reexamine the momentum equations.

First, however, we shall examine the concept of mean time averages, introduced in Chap. 9, and extend it for use in developing the turbulent-flow equations of motion. From Eq. (9-17), we can form the average of any variable over a suitably long interval

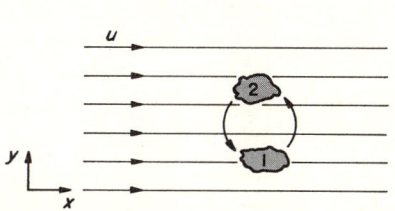

Fig. 13-1 Interchange of fluid particles resulting from turbulence.

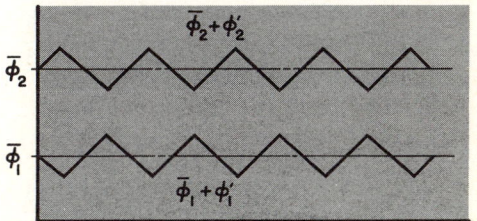

Fig. 13-2 Functions with nonzero time-averaged perturbation product.

of time as

$$\lim_{\Delta t \to \infty} \frac{1}{\Delta t} \int_0^t \phi \, dt = \bar{\phi} \qquad (13\text{-}1)$$

In this expression infinity merely represents a suitably long time for the average to be meaningful, i.e., repeatable and representative. The choice of infinity is analogous to choosing $y = \infty$ as representing locations outside the boundary layer for flow over an external surface.

Application of Eq. (13-1) yields the following set of rules for time-averaging:

$$\overline{\phi_1 + \phi_2} = \bar{\phi}_1 + \bar{\phi}_2 \qquad (13\text{-}2)$$

$$\overline{C\phi_1} = C\bar{\phi}_1 \qquad C = \text{const} \qquad (13\text{-}3)$$

$$\overline{\bar{\phi}_1} = \bar{\phi}_1 \qquad (13\text{-}4)$$

$$\overline{(\phi_1)(\phi_2)} = (\overline{\phi_1})(\overline{\phi_2}) \qquad (13\text{-}5)$$

$$\overline{\frac{\partial \phi_1}{\partial s}} = \frac{\partial \bar{\phi}_1}{\partial s} \qquad (13\text{-}6)$$

where s is a directional parameter. Now, by definition, the time average of a perturbation term is zero. Thus

$$\overline{\phi'} \equiv 0 \qquad (13\text{-}7)$$

but

$$\overline{\phi'_1 \phi'_2} \neq 0 \qquad (13\text{-}8)$$

As an illustration of Eq. (13-8), consider the time average of the perturbation components of ϕ_1 and ϕ_2 shown in Fig. 13-2. For these two out-of-phase but otherwise identical sawtooth waves, the product $\phi'_1 \phi'_2$ is zero only at a finite number of points over any fixed time span. This product is finite and negative for all other time, and $\overline{\phi'_1 \phi'_2}$ is nonzero and negative for this case.

13-2 EQUATIONS OF MOTION FOR TURBULENT FLOW

In turbulent flow, the Navier-Stokes equations are assumed to be valid if the instantaneous velocities are used rather than time averages. In particular, this is based upon the assumption that Stokes' hypothesis remains valid for the microscopic-scale stresses and that the momentum terms account for the fluid interaction on a macroscopic scale. Returning to the Navier-Stokes equations for an incompressible fluid of constant viscosity, we have from Eqs. (12-16a), (12-17a), and (12-18a)

$$\rho\left(\frac{\partial u}{\partial t} + u\frac{\partial u}{\partial x} + v\frac{\partial u}{\partial y} + w\frac{\partial u}{\partial z}\right) = -\frac{\partial p}{\partial x} + \mu\left(\frac{\partial^2 u}{\partial x^2} + \frac{\partial^2 u}{\partial y^2} + \frac{\partial^2 u}{\partial z^2}\right) \qquad (13\text{-}9)$$

$$\rho\left(\frac{\partial v}{\partial t} + u\frac{\partial v}{\partial x} + v\frac{\partial v}{\partial y} + w\frac{\partial v}{\partial z}\right) = -\frac{\partial p}{\partial y} + \mu\left(\frac{\partial^2 v}{\partial x^2} + \frac{\partial^2 v}{\partial y^2} + \frac{\partial^2 v}{\partial z^2}\right) \qquad (13\text{-}10)$$

$$\rho\left(\frac{\partial w}{\partial t} + u\frac{\partial w}{\partial x} + v\frac{\partial w}{\partial y} + w\frac{\partial w}{\partial z}\right) = -\frac{\partial p}{\partial z} + \mu\left(\frac{\partial^2 w}{\partial x^2} + \frac{\partial^2 w}{\partial y^2} + \frac{\partial^2 w}{\partial z^2}\right) \qquad (13\text{-}11)$$

TURBULENT FLOW OF INCOMPRESSIBLE VISCOUS FLUIDS

where body forces are considered negligible. Now assuming that velocity or pressure can be represented as a time-averaged value plus a fluctuation term and substituting into the *x*-direction equation, we obtain:

Inertia terms:

$$\rho\left[\frac{\partial(\bar{u}+u')}{\partial t} + (\bar{u}+u')\frac{\partial(\bar{u}+u')}{\partial x} + (\bar{v}+v')\frac{\partial(\bar{u}+u')}{\partial y} + (\bar{w}+w')\frac{\partial(\bar{u}+u')}{\partial z}\right] \quad (13\text{-}12)$$

Pressure terms:
$$-\frac{\partial}{\partial x}(\bar{p}+p') \quad (13\text{-}13)$$

Viscous terms:
$$\mu\left[\frac{\partial^2(\bar{u}+u')}{\partial x^2} + \frac{\partial^2(\bar{u}+u')}{\partial y^2} + \frac{\partial^2(\bar{u}+u')}{\partial z^2}\right] \quad (13\text{-}14)$$

The only significant results obtainable with the momentum equations involve time-averaged quantities. Time-averaging the preceding terms and applying the rules of Eqs. (13-2) to (13-8) yields:

Inertia terms:

$$\rho\left[\left(\bar{u}\frac{\partial \bar{u}}{\partial x} + \bar{v}\frac{\partial \bar{u}}{\partial y} + \bar{w}\frac{\partial \bar{u}}{\partial z}\right) + \left(\overline{u'\frac{\partial u'}{\partial x}} + \overline{v'\frac{\partial u'}{\partial y}} + \overline{w'\frac{\partial u'}{\partial z}}\right)\right] \quad (13\text{-}12a)$$

Pressure term:
$$-\frac{\partial \bar{p}}{\partial x} \quad (13\text{-}13a)$$

Viscous terms:
$$\mu\left(\frac{\partial^2 \bar{u}}{\partial x^2} + \frac{\partial^2 \bar{u}}{\partial y^2} + \frac{\partial^2 \bar{u}}{\partial z^2}\right) \quad (13\text{-}14a)$$

Before regrouping the expressions to form the turbulent momentum equation, let us first rearrange the inertia terms by means of the continuity equation. Assuming the velocity vector of Eq. (9-15) to satisfy the continuity equation for an incompressible fluid, $\nabla \cdot \mathbf{V} = 0$, we readily obtain

$$\left(\frac{\partial \bar{u}}{\partial x} + \frac{\partial \bar{v}}{\partial y} + \frac{\partial \bar{w}}{\partial z}\right) + \left(\frac{\partial u'}{\partial x} + \frac{\partial v'}{\partial y} + \frac{\partial w'}{\partial z}\right) = 0 \quad (13\text{-}15)$$

Now taking the time average of Eq. (13-15) by applying rules (13-2), (13-4), (13-6), and (13-7) results in

$$\frac{\partial \bar{u}}{\partial x} + \frac{\partial \bar{v}}{\partial y} + \frac{\partial \bar{w}}{\partial z} = 0 \quad (13\text{-}16)$$

which shows that the average values of the velocity components satisfy the continuity equation, an anticipated result. But this result together with Eq. (13-15) yields

$$\frac{\partial u'}{\partial x} + \frac{\partial v'}{\partial y} + \frac{\partial w'}{\partial z} = 0 \quad (13\text{-}17)$$

which shows that the instantaneous fluctuation terms also satisfy the continuity equation. Using this last expression we note that

$$\overline{u'\frac{\partial u'}{\partial x}} + \overline{v'\frac{\partial u'}{\partial y}} + \overline{w'\frac{\partial u'}{\partial z}} = \frac{\partial \overline{(u')^2}}{\partial x} + \frac{\partial \overline{(v'u')}}{\partial y} + \frac{\partial \overline{(w'u')}}{\partial z} \tag{13-18}$$

and the inertia terms of Eq. (13-12a) become

$$\rho\left[\left(\bar{u}\frac{\partial \bar{u}}{\partial x} + \bar{v}\frac{\partial \bar{u}}{\partial y} + \bar{w}\frac{\partial \bar{u}}{\partial z}\right) + \left(\frac{\partial \overline{(u')^2}}{\partial x} + \frac{\partial \overline{(v'u')}}{\partial y} + \frac{\partial \overline{(w'u')}}{\partial z}\right)\right] \tag{13-12b}$$

Regrouping the inertia, pressure, and viscous terms as required by the original Navier-Stokes equations yields the *momentum equations for turbulent flow of a viscous incompressible fluid* with negligible body forces:

$$\rho\left(\bar{u}\frac{\partial \bar{u}}{\partial x} + \bar{v}\frac{\partial \bar{u}}{\partial y} + \bar{w}\frac{\partial \bar{u}}{\partial z}\right) = -\frac{\partial \bar{p}}{\partial x} + \mu \nabla^2 \bar{u} - \rho\left(\frac{\partial \overline{u'^2}}{\partial x} + \frac{\partial \overline{v'u'}}{\partial y} + \frac{\partial \overline{w'u'}}{\partial z}\right) \tag{13-19}$$

$$\rho\left(\bar{u}\frac{\partial \bar{v}}{\partial x} + \bar{v}\frac{\partial \bar{v}}{\partial y} + \bar{w}\frac{\partial \bar{v}}{\partial z}\right) = -\frac{\partial \bar{p}}{\partial y} + \mu \nabla^2 \bar{v} - \rho\left(\frac{\partial \overline{u'v'}}{\partial x} + \frac{\partial \overline{v'^2}}{\partial y} + \frac{\partial \overline{w'v'}}{\partial z}\right) \tag{13-20}$$

$$\rho\left(\bar{u}\frac{\partial \bar{w}}{\partial x} + \bar{v}\frac{\partial \bar{w}}{\partial y} + \bar{w}\frac{\partial \bar{w}}{\partial z}\right) = -\frac{\partial \bar{p}}{\partial z} + \mu \nabla^2 \bar{w} - \rho\left(\frac{\partial \overline{u'w'}}{\partial x} + \frac{\partial \overline{v'w'}}{\partial y} + \frac{\partial \overline{w'^2}}{\partial z}\right)$$

$$\tag{13-21}$$

where the *y*- and *z*-direction equations are written by analogy with Eq. (13-19).

Reynolds stresses From a comparison of these momentum equations with their laminar-flow counterparts, it is clear that the turbulent fluctuations give rise to additional forces represented by the terms containing derivatives of products of velocity-component fluctuations such as $\rho\, \partial \overline{u'v'}/\partial x$. All other terms in these equations are identical with those of the Navier-Stokes equations which were derived for laminar flow. Now by analogy with Eq. (12-5), we can express the forces acting upon a control volume such as that of Fig. 12-2 by

$$\sum (f_x)_{\text{app}} = -\rho\left(\frac{\partial \overline{u'^2}}{\partial x} + \frac{\partial \overline{v'u'}}{\partial y} + \frac{\partial \overline{w'u'}}{\partial z}\right)$$

$$\sum (f_y)_{\text{app}} = -\rho\left(\frac{\partial \overline{u'v'}}{\partial x} + \frac{\partial \overline{v'^2}}{\partial y} + \frac{\partial \overline{w'v'}}{\partial z}\right) \tag{13-22}$$

$$\sum (f_z)_{\text{app}} = -\rho\left(\frac{\partial \overline{u'w'}}{\partial x} + \frac{\partial \overline{v'w'}}{\partial y} + \frac{\partial \overline{w'^2}}{\partial z}\right)$$

TURBULENT FLOW OF INCOMPRESSIBLE VISCOUS FLUIDS

which are frequently referred to as *Reynolds*, *apparent*, or *virtual forces*. Since they result from turbulent fluctuations and are additive to the laminar forces, they are sometimes said to be caused by *eddy viscosity*. Because these forces for a differential area are equivalent to apparent stresses, they may be considered as

$$-\rho \left(\frac{\partial \overline{u'^2}}{\partial x} + \frac{\partial \overline{v'u'}}{\partial y} + \frac{\partial \overline{w'u'}}{\partial z} \right) = \frac{\partial}{\partial x}(\tau_{xx})_{\text{app}} + \frac{\partial}{\partial y}(\tau_{xy})_{\text{app}} + \frac{\partial}{\partial z}(\tau_{xz})_{\text{app}}$$

$$-\rho \left(\frac{\partial \overline{u'v'}}{\partial x} + \frac{\partial \overline{v'^2}}{\partial y} + \frac{\partial \overline{w'v'}}{\partial z} \right) = \frac{\partial}{\partial x}(\tau_{xy})_{\text{app}} + \frac{\partial}{\partial y}(\tau_{yy})_{\text{app}} + \frac{\partial}{\partial z}(\tau_{zy})_{\text{app}} \quad (13\text{-}23)$$

$$-\rho \left(\frac{\partial \overline{u'w'}}{\partial x} + \frac{\partial \overline{v'w'}}{\partial y} + \frac{\partial \overline{w'^2}}{\partial z} \right) = \frac{\partial}{\partial x}(\tau_{xz})_{\text{app}} + \frac{\partial}{\partial y}(\tau_{yz})_{\text{app}} + \frac{\partial}{\partial z}(\tau_{zz})_{\text{app}}$$

In most turbulent flows, the apparent stresses are considerably larger than the viscous stresses, and the terms involving viscosity can usually be omitted in the application of Eqs. (13-19) to (13-21). It is these apparent stresses which cause the pressure gradients in turbulent flow to be many times those experienced in laminar flow at essentially similar flow conditions.

Recall that we have discussed the physics of laminar viscous flow and how the molecular interactions (microscopic level) give rise to macroscopic stresses. Correspondingly to this, the macroscopic interactions between fluid particles resulting from turbulent fluctuations (Fig. 13-1) result in large momentum exchanges. It is the time average of these momentum exchanges which we term Reynolds or apparent stresses, and they are generally an order of magnitude larger than laminar viscous stresses.

Mixing-length theories In order to apply the turbulent-flow momentum equations (sometimes called Reynolds equations), some assumptions are necessary with respect to the Reynolds stresses. This is purely a mathematical requirement; for laminar incompressible constant-viscosity flow there are five variables and five equations, as discussed in Sec. 12-1. The corresponding turbulent flow involves three added unknowns, u', v', and w', but the number of equations is unchanged. Clearly then the solution for eight variables (unknowns) with only five equations is mathematically impossible, and further simplifying assumptions are required.

Apparently the first such assumption was put forward by Boussinesq [2], who introduced a turbulent exchange coefficient ϵ defined by

$$-\rho \overline{u'v'} = \rho \epsilon \frac{d\bar{u}}{dy} = (\tau_{xy})_{\text{app}} \quad (13\text{-}24)$$

in analogy with the newtonian viscosity of laminar flow. While this approach has enjoyed some success, it is difficult to assign a value to ϵ. Note in particular that this term is dependent upon flow conditions, whereas the viscosity μ is a property

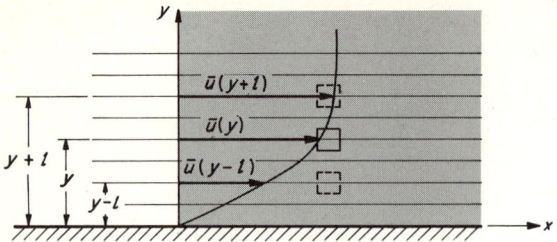

Fig. 13-3 Turbulent boundary-layer flow illustrating the mixing length.

of the fluid. For example, if a suitable value of ϵ is determined for a square-duct flow, there is no assurance that it will be appropriate for flow in a round duct.

A more successful approach has been the application of various *mixing-length theories*, one of the earliest and best known being Prandtl's [24] mixing length. In this semiempirical theory, Prandtl introduced the concept of a group of fluid particles moving as a body and retaining a property such as velocity over a distance ℓ which he termed the *mixing length*. To clarify, consider the turbulent boundary-layer flow of Fig. 13-3. A body of fluid at layer $y + \ell$ is assumed to have a y-direction perturbation $-v'$ which causes it to move downward to layer y. This distance of movement ℓ is that required for the momentum exchange to result in velocity fluctuations u' of the same magnitude as those occurring in the real turbulent flow. In other words, the fluid body at $y + \ell$ is assumed to move to y with no change in x-direction velocity until the vertical motion ceases. Then at the new position its original velocity $\bar{u}(y + \ell)$ results in a fluctuation at y of magnitude

$$|u'| = |\bar{u}(y + \ell) - \bar{u}(y)| \tag{13-25}$$

and this is the value of $|u'|$ in the real flow field. Thus ℓ is a physically meaningful dimension directly related to the real turbulent flow.

Now replacing ℓ with Δy we can express Eq. (13-25) as

$$\frac{|u'|}{\ell} = \frac{|\bar{u}(y + \Delta y) - \bar{u}(y)|}{\Delta y}$$

which in the limit as Δy approaches zero becomes

$$|u'| = \ell \left| \frac{d\bar{u}}{dy} \right| \tag{13-26}$$

This same result could have been obtained for a body of fluid particles moving from $y - \ell$ to y. The time average of $|u'|$ evaluated at y is found by averaging the fluctuation due to both upward and downward movement, i.e.,

$$\overline{|u'|} = \tfrac{1}{2}\left(\ell \left|\frac{d\bar{u}}{dy}\right| + \ell \left|\frac{d\bar{u}}{dy}\right| \right) = \ell \left|\frac{d\bar{u}}{dy}\right| \tag{13-27}$$

Prandtl also assumed the time average of the fluctuations in the vertical velocity component to be of the same order of magnitude as those in the x direction; i.e., he assumed

$$\overline{|v'|} = C_1 \overline{|u'|} = C_1 \ell \left| \frac{d\bar{u}}{dy} \right| \tag{13-28}$$

where C_1 is a constant. From the relationships in (13-23) the apparent shear stress due to velocity fluctuations u' and v' is

$$(\tau_{xy})_{\text{app}} = -\rho \overline{u'v'}$$

and we must examine the time average $\overline{u'v'}$. From Fig. 13-3, a positive value of v' will move a body of fluid to a region of higher $\bar{u}$ and will consequently result in a negative product of u' and v'. Likewise a negative v' results in a positive u', and the average $\overline{u'v'}$ is seen to be usually negative and nonzero. As a consequence Prandtl assumed

$$\overline{u'v'} = -C_2 \overline{|u'|}\,\overline{|v'|} \tag{13-29}$$

where C_2 is a correlation factor with values greater than zero and less than or equal to 1.

Then applying Eqs. (13-27) and (13-28), Eq. (13-29) can be written as

$$\overline{u'v'} = -C_3 \ell^2 \left(\frac{d\bar{u}}{dy}\right)^2 \tag{13-30}$$

where the absolute-value sign on the derivative is now superfluous due to the squaring operation. Since we do not have measurements for ℓ, we may include the constant C_3 and express the apparent shear stress as

$$(\tau_{xy})_{\text{app}} = \rho \ell^2 \left(\frac{d\bar{u}}{dy}\right)^2 \tag{13-31}$$

Since the stress must change sign with the sign of $d\bar{u}/dy$, it is better to express this as

$$\blacksquare \quad (\tau_{xy})_{\text{app}} = \rho \ell^2 \left|\frac{d\bar{u}}{dy}\right| \frac{d\bar{u}}{dy} \tag{13-31a}$$

This is Prandtl's mixing-length hypothesis. While the assumption of small ℓ made in the derivation is not physically justifiable, the result has been used with considerable success in the prediction of mean velocity distributions for numerous problems in turbulent flow.

Other mixing-length theories include Taylor's vorticity transport theory and von Kármán's similarity hypothesis, as discussed by Pai [23] and Schlichting [29].

13-3 FUNDAMENTALS OF TURBULENT FLOW IN PIPES AND CHANNELS

We next apply the preceding concepts to a study of important cases of internal turbulent flow, and in particular we shall find that Prandtl's mixing-length hypothesis yields satisfactory results for the limiting case of very large Reynolds numbers. At small to moderate Reynolds numbers for turbulent flow, a partially empirical theory is necessary.

Velocity profiles for large Reynolds numbers Consider fully developed turbulent flow in a two-dimensional channel (or in a circular pipe). Assuming the mixing length near the wall to be linearly proportional to the distance from the wall, we have

$$\underline{l} = ky \tag{13-32}$$

In the vicinity of the wall it is reasonable to assume the shearing stress to be constant and equal to τ_s, the stress at the wall. Prandtl assumed this value to hold throughout the channel flow field. Introducing this assumption and Eq. (13-32) into Prandtl's mixing-length hypothesis, we can solve for the gradient of $\bar{u}$, which is

$$\frac{d\bar{u}}{dy} = \frac{\sqrt{\tau_s/\rho}}{ky} \tag{13-33}$$

Notice that the term $\sqrt{\tau_s/\rho}$ has the units of velocity; this is usually called the *shear velocity* v_*. Using this notation and separating variables, we have

$$d\bar{u} = \frac{1}{k} v_* \frac{dy}{y} \tag{13-34}$$

which yields upon integration

$$\bar{u} = \frac{1}{k} v_* \ln y + \ln C \tag{13-35}$$

For the case of flow in a channel of width $2h$, the maximum velocity occurs at the center, $y = h$. Then forming a difference equation we can eliminate the constant of integration of Eq. (13-35). Thus

$$\bar{u}_{\max} - \bar{u} = \frac{1}{k} v_* \ln \frac{h}{y} \tag{13-36}$$

Next it is necessary to compare this expression with experimental results to determine a value of k. The data of Nikuradse [21] for flow in smooth and rough pipes are well correlated with $k = 0.4$, resulting in

$$\bar{u}_{\max} - \bar{u} = 2.5 v_* \ln \frac{R}{y} \tag{13-37}$$

TURBULENT FLOW OF INCOMPRESSIBLE VISCOUS FLUIDS

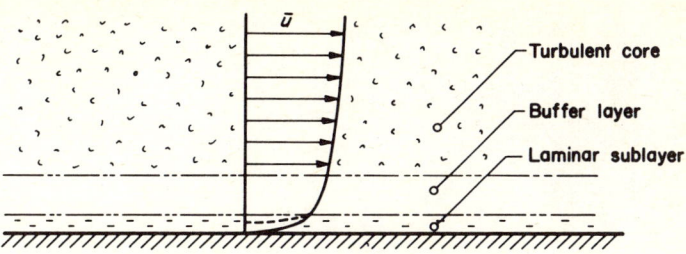

Fig. 13-4 Turbulent flow near a solid boundary.

where R is the pipe radius. Equation (13-37) is known as the *universal velocity-difference distribution law for pipes*. The term universal arises from its applicability for all values of surface roughness.

Returning to the velocity distribution given by Eq. (13-35), let us determine the constant of integration so that we obtain the *velocity distribution* rather than the *velocity-difference distribution*. It has been experimentally verified that the velocity distribution is dependent upon the surface roughness. It is convenient to consider a turbulent-flow model as shown in Fig. 13-4 and to classify three flow roughness regimes: (1) hydraulically smooth pipe-flow regime, (2) completely rough pipe-flow regime, and (3) transition roughness between smooth and rough regime. These regimes will be discussed in detail later. Notice that the flow model of Fig. 13-4 is applicable to all three of these roughness regimes, and the buffer layer of that figure is not to be misconstrued as the transition roughness regime. Before we consider flow in rough pipes in detail, it will be advantageous to direct our attention to a study of flow in smooth pipes. The results, as we shall see later, will be applicable to the hydraulically smooth regime of rough pipe flow.

Flow in smooth pipes: velocity distribution Numerous velocity-profile measurements have been conducted in smooth pipes. Among the earliest accurate measurements are those due to Nikuradse [22], which have been supplemented with data very close to the wall by Reichardt [27] (see Fig. 13-5).

In the laminar sublayer, Newton's viscosity law

$$\tau_s = \mu \frac{du}{dy}$$

can be integrated to yield the velocity

$$u = \frac{\tau_s y}{\mu} + \text{const} \tag{13-38}$$

The no-slip boundary condition at the wall requires the constant of integration to be zero, and if both sides of the last equation are divided by v_*, we obtain

$$\frac{u}{v_*} = \frac{y v_*}{v} \tag{13-39}$$

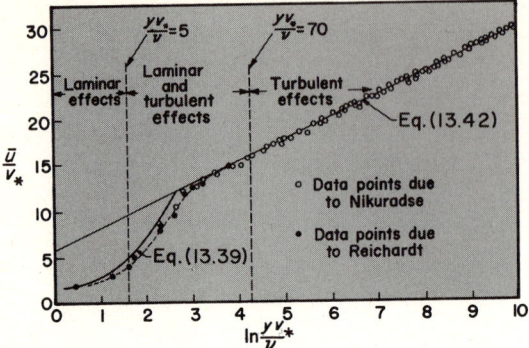

Fig. 13-5 Universal velocity profile for turbulent flow in smooth pipes. (*Adapted from H. Schlichting, "Boundary Layer Theory," 6th ed. Copyright 1968. McGraw-Hill Book Company. Used by permission.*)

This expression is plotted in Fig. 13-5 and represents the experimental data reasonably well for

$$0 \leq \frac{yv_*}{\nu} \leq 5 \tag{13-40}$$

Clearly, the viscous effects in this region are purely laminar.

In the completely rough pipe-flow regime, we expect Eq. (13-35) obtained from Prandtl's turbulent mixing-length hypothesis to apply. This equation can be rewritten

$$\frac{\bar{u}}{v_*} = \frac{1}{k}\left(\ln \frac{yv_*}{\nu} - \ln \beta\right) \tag{13-41}$$

since v_* and ν are constants. Recall that in obtaining the universal velocity-difference distribution law for pipes, it was stated that the data of Nikuradse yielded $k = 0.4$. With this value, we find that $\ln \beta = -2.2$, or

$$\frac{\bar{u}}{v_*} = 2.5 \ln \frac{yv_*}{\nu} + 5.5 \tag{13-42}$$

which represents the data in Fig. 13-5 exceptionally well for

$$70 \leq \frac{yv^*}{\nu} \tag{13-43}$$

It should be noted that the numerical value of 70 is quite arbitrary; numbers reported in the literature vary from 20 to 70. In summary, for flow in smooth pipes at high

TURBULENT FLOW OF INCOMPRESSIBLE VISCOUS FLUIDS

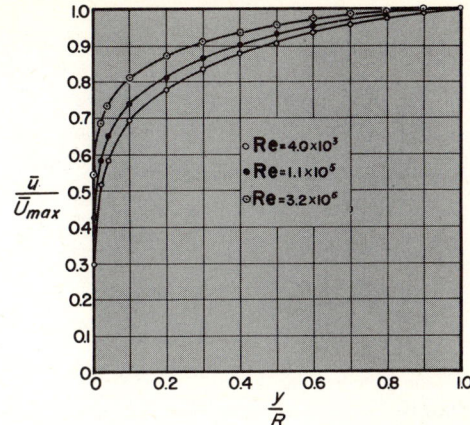

Fig. 13-6 Velocity distribution in smooth pipes for these Reynolds numbers; $\bar{U}_{max}$ = centerline velocity in pipe, $\bar{u}$ = local time-mean average velocity; R = pipe radius, y = distance from pipe wall. (*Adapted from H. Schlichting, "Boundary Layer Theory," 6th ed. Copyright 1968. McGraw-Hill Book Company. Used by permission.*)

Reynolds numbers the viscous effects in the three layers of Fig. 13-4 are characterized by

$$0 \leq \frac{yv_*}{\nu} < 5 \qquad \text{laminar friction}$$

$$5 < \frac{yv_*}{\nu} < 70 \qquad \text{laminar-turbulent friction} \tag{13-44}$$

$$70 < \frac{yv_*}{\nu} \qquad \text{turbulent friction}$$

where the velocity distribution in the laminar sublayer is given by Eq. (13-39) and in the turbulent core by Eq. (13-42).

Nikuradse [22] has also carried out extensive measurements of turbulent flow in smooth pipes at low to moderate Reynolds numbers. His results for three Reynolds numbers are presented in Fig. 13-6.

From these experiments it is apparent that the velocity profile is a function of the Reynolds number and it becomes flatter as **Re** increases. Further, if these data for each **Re** are plotted in the form of $(\bar{u}/\bar{U}_{max})^n$ vs. y/R, it is found that by a suitable choice of n the plot is a straight line for most of the flow area. To illustrate this the three profiles of Fig. 13-6 are plotted in Fig. 13-7. This irrefutable evidence has led to a general acceptance of a *power-law profile* of the form

$$\frac{\bar{u}}{\bar{U}_{max}} = \left(\frac{y}{R}\right)^{1/n} \tag{13-45}$$

for turbulent flow in smooth pipes at moderate Reynolds numbers. This is sometimes referred to as the one-seventh-power law, but generalization to include a specific value for n is not justifiable. The value of n is itself dependent upon the Reynolds

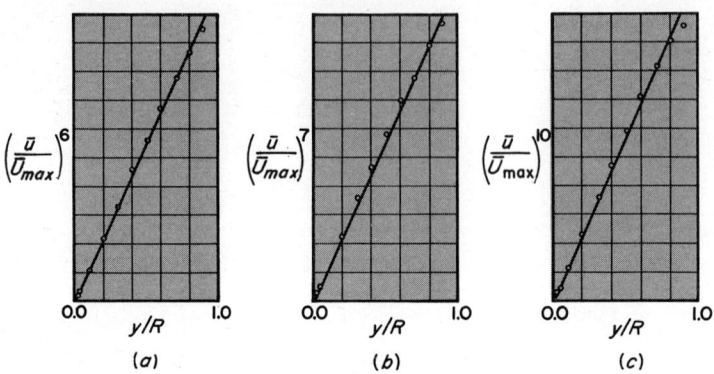

Fig. 13-7 Velocity distribution in smooth pipes: (a) $\text{Re} = 4 \times 10^3$; (b) $\text{Re} = 1.1 \times 10^5$; (c) $\text{Re} = 3.2 \times 10^6$. (*Adapted from H. Schlichting, "Boundary Layer Theory," 6th ed. Copyright 1968. McGraw-Hill Book Company. Used by permission.*)

number and must be chosen accordingly. Experimental values of n for several Reynolds numbers are given in Table 13-1.

Flow in smooth pipes: resistance From the definition of the friction factor given by Eq. (12-37a) we obtain

$$-\frac{\Delta p}{L} = \frac{f}{D} \frac{\rho V^2}{2} = \frac{f}{D} \frac{\rho (\bar{u}_{\text{av}})^2}{2} \tag{13-46}$$

for turbulent flow simply by substituting the time-mean average velocity $\bar{u}_{\text{av}}$ for V.

Prandtl [25] obtained an expression for the friction factor for turbulent flow in smooth pipes. Beginning with Eq. (13-37), he integrated this over the pipe cross-sectional area and divided by the area to obtain

$$(\bar{u})_{\text{av}} = \bar{U}_{\max} - 3.75 v_* \tag{13-47}$$

From a force balance on a cylindrical element of fluid in a pipe the wall shear stress is

$$\tau_s = -\frac{\Delta p}{L} \frac{R}{2} \tag{13-48}$$

Table 13-1 Values of n for various Reynolds numbers†

Re	4.0×10^3	2.3×10^4	1.1×10^5	1.1×10^6	2.0×10^6	3.2×10^6
n	6.0	6.6	7.0	8.8	10	10

† Data of Nikuradse [22].

TURBULENT FLOW OF INCOMPRESSIBLE VISCOUS FLUIDS

which can be combined with Eq. (13-46) to yield

$$f = 8\frac{\tau_s/\rho}{\bar{u}_{av}^2} = 8\left(\frac{v_*}{\bar{u}_{av}}\right)^2 \tag{13-49}$$

From the universal velocity distribution of Eq. (13-42) the maximum velocity occurs at $y = R$ and is

$$\bar{U}_{max} = v_*\left(2.5 \ln \frac{Rv_*}{\nu} + 5.5\right)$$

Combining this with Eq. (13-47) yields

$$\bar{u}_{av} = v_*\left(2.5 \ln \frac{Rv_*}{\nu} + 1.75\right) \tag{13-50}$$

Eliminating v_* with the aid of Eq. (13-49) and rearranging, Prandtl obtained

$$\frac{1}{\sqrt{f}} = 2.035 \log\left(\frac{\bar{u}_{av}D}{\nu}\sqrt{f}\right) - 0.91 \tag{13-51}$$

where D is the pipe diameter. This has been compared with experimental results of numerous investigators. Schlichting [29] gives a graphical comparison with experimental results over a wide range of Reynolds numbers. A straight line representing the data in a plot of $1/\sqrt{f}$ vs. log $(\mathbf{Re}/\sqrt{f})$ differs slightly in the constants from Eq. (13-51) and can be written

$$\frac{1}{\sqrt{f}} = 2.0 \log\left(\frac{\bar{u}_{av}D}{\nu}\sqrt{f}\right) - 0.8 \tag{13-52}$$

This is known as *Prandtl's universal law of friction for smooth pipes*.

To summarize our results for turbulent flow in smooth pipes, we found that Prandtl's mixing-length hypothesis led directly to the velocity-distribution law [Eq. (13-43)] and to Prandtl's universal law of friction [Eq. (13-52)]. The first of these is valid only at very high Reynolds numbers. The second is valid for all Reynolds numbers. We have also seen that experimental velocity profiles in turbulent flow in smooth pipes at low to moderate Reynolds numbers are well represented by the power law of Eq. (13-45), where n is dependent upon **Re**.

Flow in rough pipes: definitions In engineering practice few pipes are used which are hydraulically smooth, and we must consider the effect of roughness elements on the velocity profile and resistance to flow. We have already mentioned the three types of flow roughness regimes; they are illustrated in Fig. 13-8. In Fig. 13-8a, the roughness-element height e does not extend beyond the laminar sublayer. It is said to be buried within the sublayer, and the roughness has little or no effect upon the turbulent-flow profiles or the resistance to flow. This is the *hydraulically smooth*

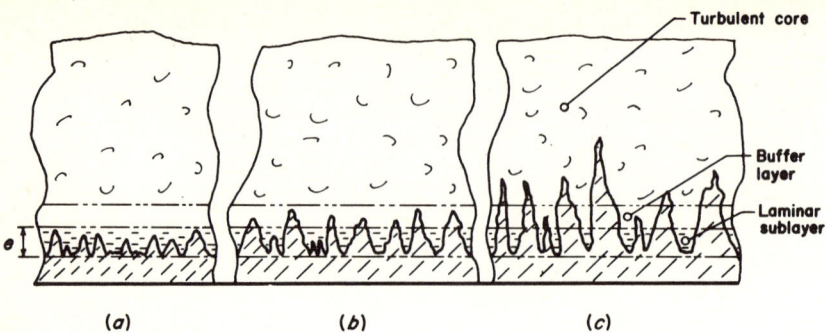

Fig. 13-8 The three regimes of roughness: (*a*) hydraulically smooth; (*b*) transition; (*c*) completely rough.

regime, and it exists for

$$0 < \frac{ev_*}{\nu} < 5$$

The previous results for f obtained for turbulent flow in a smooth pipe are applicable in this regime. Thus, f is dependent only upon **Re**.

In Fig. 13-8*b*, the roughness extends partly into the buffer layer, and there is an added resistance to flow due to the form drag acting on these protrusions into the flow. This is the *transition roughness regime* and exists for

$$5 < \frac{ev_*}{\nu} < 70$$

The friction factor is dependent upon both e and **Re** in this regime.

The *completely rough regime* is characterized by all wall roughness protrusions extending outside the laminar sublayer, with some extending into the turbulent boundary layer. Here

$$70 < \frac{ev_*}{\nu}$$

and f is a function of e and **Re**.

Flow in rough pipes: velocity distribution Measurements by Nikuradse for flow in roughened pipes indicate that the velocity profile near a rough wall is less steep than for a smooth wall. His data, obtained at a Reynolds number of 10^6 with the roughness parameter e/R varying from 0.00197 to 0.0327, indicate that the power-law exponent for Eq. (13-45) would lie between $\frac{1}{5}$ and $\frac{1}{4}$.

Assuming Prandtl's mixing-length hypothesis to hold for this case and trying to represent Nikuradse's data by Eq. (13-41), we find that

$$\ln \beta = \ln \frac{ev_*}{\nu} - 3.4 \tag{13-53}$$

TURBULENT FLOW OF INCOMPRESSIBLE VISCOUS FLUIDS

which results in

$$\frac{\bar{u}}{v_*} = 2.5 \ln \frac{y}{e} + 8.5 \tag{13-54}$$

This equation is quite accurate in the *completely rough* pipe flow regime. Note that the expression of Eq. (13-53) is analogous to assuming the constant of integration of Eq. (13-35) to be proportional to the roughness height e. Then with $C = \gamma e$ Eq. (13-35) becomes

$$\frac{\bar{u}}{v_*} = 2.5(\ln y - \ln \gamma e) = 2.5\left(\ln \frac{y}{e} - \ln \gamma\right) \tag{13-55}$$

which yields Eq. (13-54) if $\ln \gamma = -3.4$.

Flow in rough pipes: resistance In a manner similar to that used in obtaining Eqs. (13-51) and (13-52), we can use Eqs. (13-47), (13-49), and (13-54) to obtain

$$f = \frac{1}{[2 \log (R/e) + 1.68]^2} \tag{13-56}$$

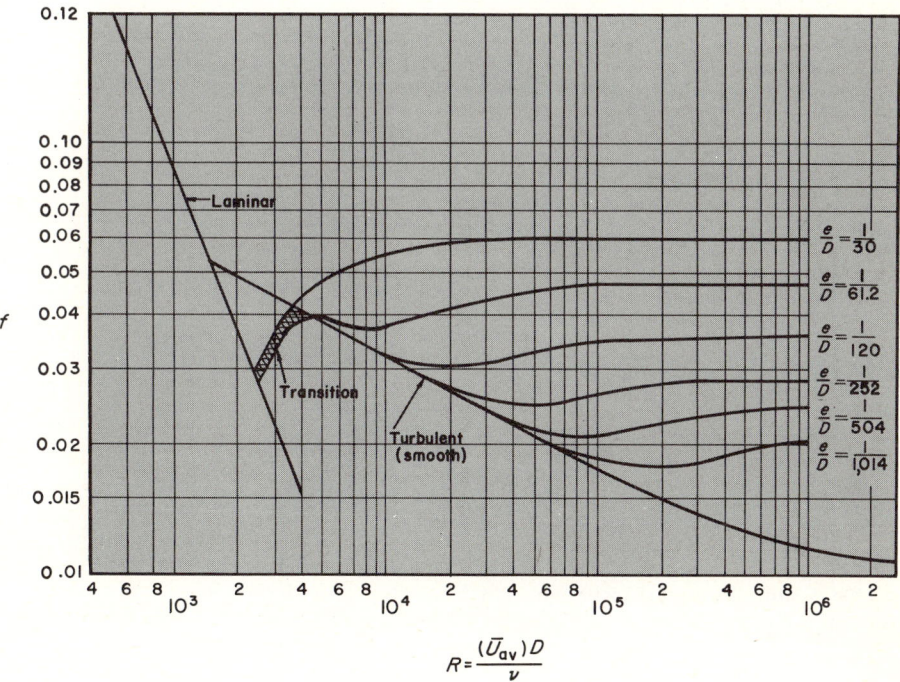

Fig. 13-9 Friction factors for pipe flow.

for the *completely rough* regime. Comparison with the data of Nikuradse indicates the constant should be changed to 1.74, according to Schlichting [29], and thus

$$f = \frac{1}{[2 \log (R/e) + 1.74]^2} \tag{13-57}$$

The friction-factor experiments of Nikuradse consisted of an extensive set of very careful measurements obtained with sand grains of uniform size glued to the inside of circular pipes. The inner surface was completely covered with the sand grains. By using different size pipes and changing the grain size, he was able to vary the ratio of e/D from 0.000986 to 0.0333, where D is the pipe diameter. His results are given in Fig. 13-9. While these data are unquestionably accurate, they are not directly related to flow in commercial piping, and Fig. 13-9 is not usable for engineering calculations.

13-4 EMPIRICAL RELATIONS FOR TURBULENT FLOW IN PIPES

It has been determined that the pressure loss for turbulent flow in a straight pipe depends upon

1. The length of the pipe L
2. The pipe diameter D
3. The cross-sectional average of the time-mean velocity $\bar{u}_{av}$
4. The pipe-wall roughness height e
5. The fluid density ρ
6. The fluid viscosity μ

Thus we can express the pressure loss as a function of these variables by

$$\Delta p = F_1(L, D, \bar{u}_{av}, e, \rho, \mu)$$

Performing a dimensional analysis by the procedures of Appendix C, we obtain four dimensionless groups

$$\frac{\Delta p}{\rho V^2} = F_2\left(\frac{L}{D}, \frac{e}{D}, \frac{DV\rho}{\mu}\right) \tag{13-58}$$

where V is used to represent $\bar{u}_{av}$. This is as far as we can proceed by dimensional analysis. At this point it is well to introduce a very evident bit of empiricism, namely, that the pressure loss is *directly proportional* to the length of the pipe. Then we can write

$$\frac{\Delta p}{\rho V^2} = \frac{L}{D} F_3\left(\frac{DV\rho}{\mu}, \frac{e}{D}\right) \tag{13-59}$$

TURBULENT FLOW OF INCOMPRESSIBLE VISCOUS FLUIDS

or inserting the constant 2, since $\rho V^2/2$ is the fluid kinetic energy and calling the unknown function $H(DV\rho/\mu, e/D)$ the friction factor f, we have

$$\Delta p = f \frac{L}{D} \frac{\rho V^2}{2} \tag{13-60}$$

Since the pressure loss divided by the fluid density is the head loss, we can write an alternative form of this equation as

$$h_f = \frac{\Delta p}{\rho} = f \frac{L}{D} \frac{V^2}{2} \tag{13-60a}$$

This is the same as the Darcy-Weisbach equation discussed in Chap. 12 for laminar flow. For turbulent flows, however, the friction factor f must be determined experimentally, whereas this was derived analytically for the laminar case.

Turbulent flow in smooth tubes Blasius formulated an expression for the friction factor for smooth tubes in 1913. He conducted a critical review of the data existing at that time and recommended the following empirical equation:

$$f = 0.3164 \mathbf{Re}_D^{-\frac{1}{4}} \tag{13-61}$$

Note that this is in agreement with the results of the dimensional analysis which requires f to be dependent upon **Re** and e/D. For smooth tubes e is zero, and f is a function solely of the Reynolds number.

The range of experiments Blasius considered extended to a Reynolds number of 100,000, and Eq. (13-61) should not be used beyond that value.

Later studies of turbulent flow inside smooth ducts indicate that a better representation of the friction factor over the range of Reynolds numbers from 10,000 to 100,000 is given by

$$f = 0.184 \mathbf{Re}_D^{-0.2} \tag{13-62}$$

While development of the turbulent velocity profile may require 40 to 50 diameters (or more) of pipe length, the distance required for the friction factor to become constant is usually much less. This is easily understood since the friction factor is determined by the shear stress at the wall; this in turn is controlled by the boundary-layer flow, and the velocity profile adjacent to the wall develops much quicker than the entire velocity profile. For turbulent flow in a smooth circular tube, the dimensionless distance required for the friction factor to become constant, according to Latzko [16], is

$$\left(\frac{L}{D}\right)_c = 0.623 \mathbf{Re}_D^{\frac{1}{4}} \tag{13-63}$$

Beyond this distance, Eq. (13-62) may be used to calculate the friction factor within the range of its validity.

Example Calculate the pressure loss in psi for a flow of light oil at 80°F and 100 ft³/min in a 1-in-ID smooth drawn tube 20 ft long.

Solution The average velocity is

$$\bar{u}_{av} = \frac{Q}{A} = \frac{(100 \text{ ft}^3/\text{min})(\text{min}/60 \text{ sec})}{(\pi/4)(\frac{1}{12})^2 \text{ ft}^2} = 306 \text{ ft/sec}$$

The Reynolds number is

$$\text{Re}_D = \frac{\bar{u}_{av} D}{\nu} = \frac{(306 \text{ ft/sec})\frac{1}{12} \text{ ft}}{49 \times 10^{-5} \text{ ft}^2/\text{sec}} = 52{,}000$$

This is clearly turbulent and within the range of Eq. (13-62). Checking the distance for the friction factor to become constant by Eq. (13-63),

$$\left(\frac{L}{D}\right)_c = 0.623(52{,}000)^{\frac{1}{4}}$$

$$L_c = 9.97(\tfrac{1}{12} \text{ ft}) = 0.83 \text{ ft}$$

This should be insignificant for a piping system 20 ft long, and in the calculation we shall assume constant f over the entire length. By Eq. (13-62),

$$f = 0.184(52{,}000)^{-0.2} = 0.021$$

Then, by Eq. (13-60),

$$\Delta p = f \frac{L}{D} \rho \frac{V^2}{2g_c}$$

$$= 0.021 \frac{20 \text{ ft}}{\frac{1}{12} \text{ ft}} \frac{(56.8 \text{ lb}_m/\text{ft}^3)(306 \text{ ft/sec})^2}{2(32.2 \text{ lb}_m\text{-ft}/\text{lb}_f\text{-sec}^2)}$$

$$= 4.17 \times 10^5 \frac{\text{lb}_f}{\text{ft}^2} \frac{\text{ft}^2}{144 \text{ in}^2} = 2{,}895 \text{ lb}_f/\text{in}^2$$

Turbulent flow in rough pipes Moody [20] has made an extensive study of pipe roughness which permitted him to apply the work of Nikuradse (with artificially roughened pipes) to commercial piping. His results are presented in Figs. 13-10 and 13-11. It is apparent from the plot of relative roughness that the value of e/D is approximate. This depends upon manufacturing conditions, time in service, etc. As a consequence, in performing calculations of pressure drop in piping using these friction factors and Eq. (13-60), allowances for errors of approximately 10 percent should be considered.

Example Calculate the pressure loss in psi for flow of 80°F light oil at 100 ft³/min in a 1-in-ID commercial steel pipe 20 ft long.

Solution From the previous example the Reynolds number is

Re = 52,000

Then from Fig. 13-11

$$\frac{e}{D} = 0.00175$$

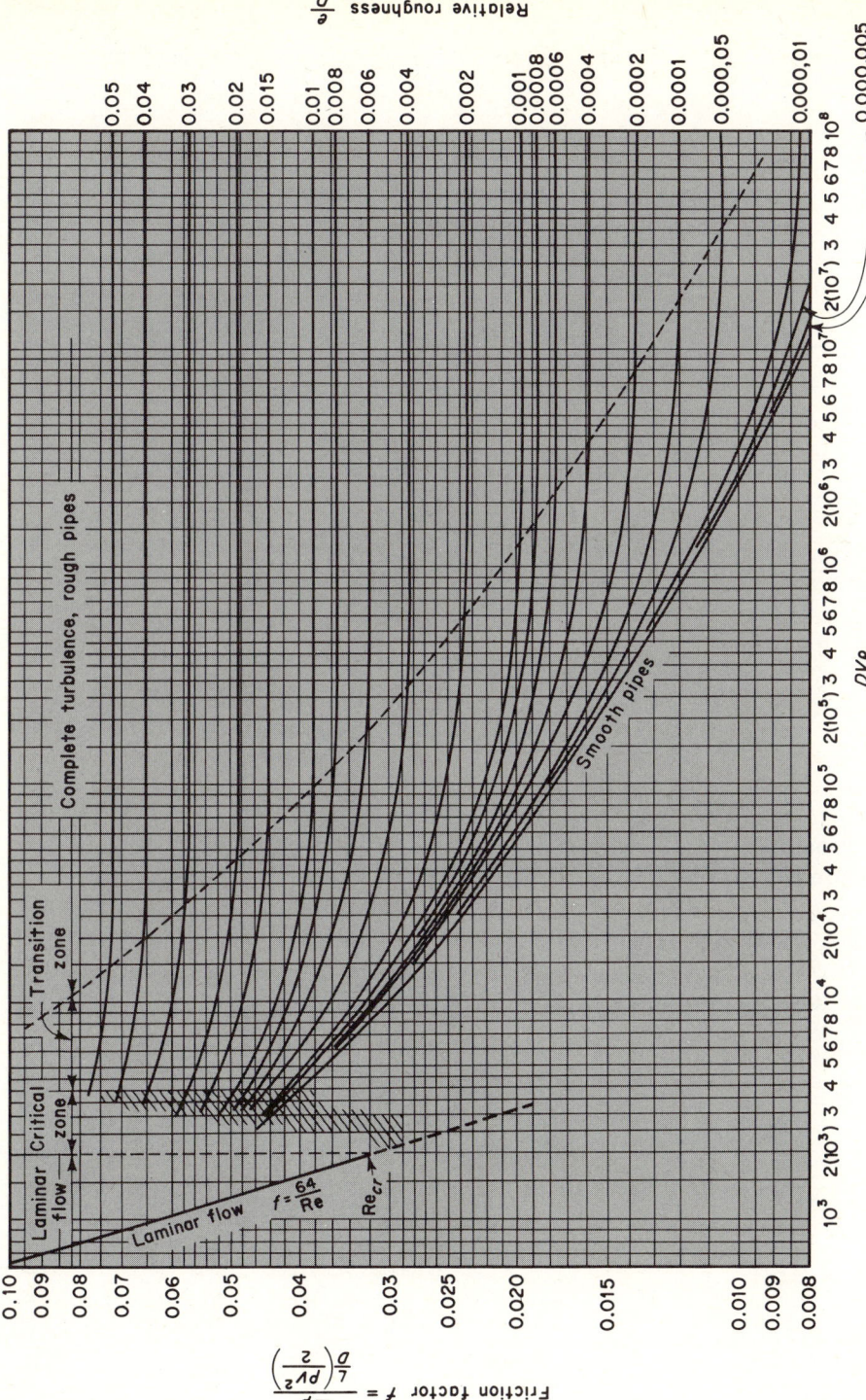

Fig. 13-10 Friction factors for commercial piping. [*Adapted from L. M. Moody, Trans. ASME,* **66**:672 (1944).]

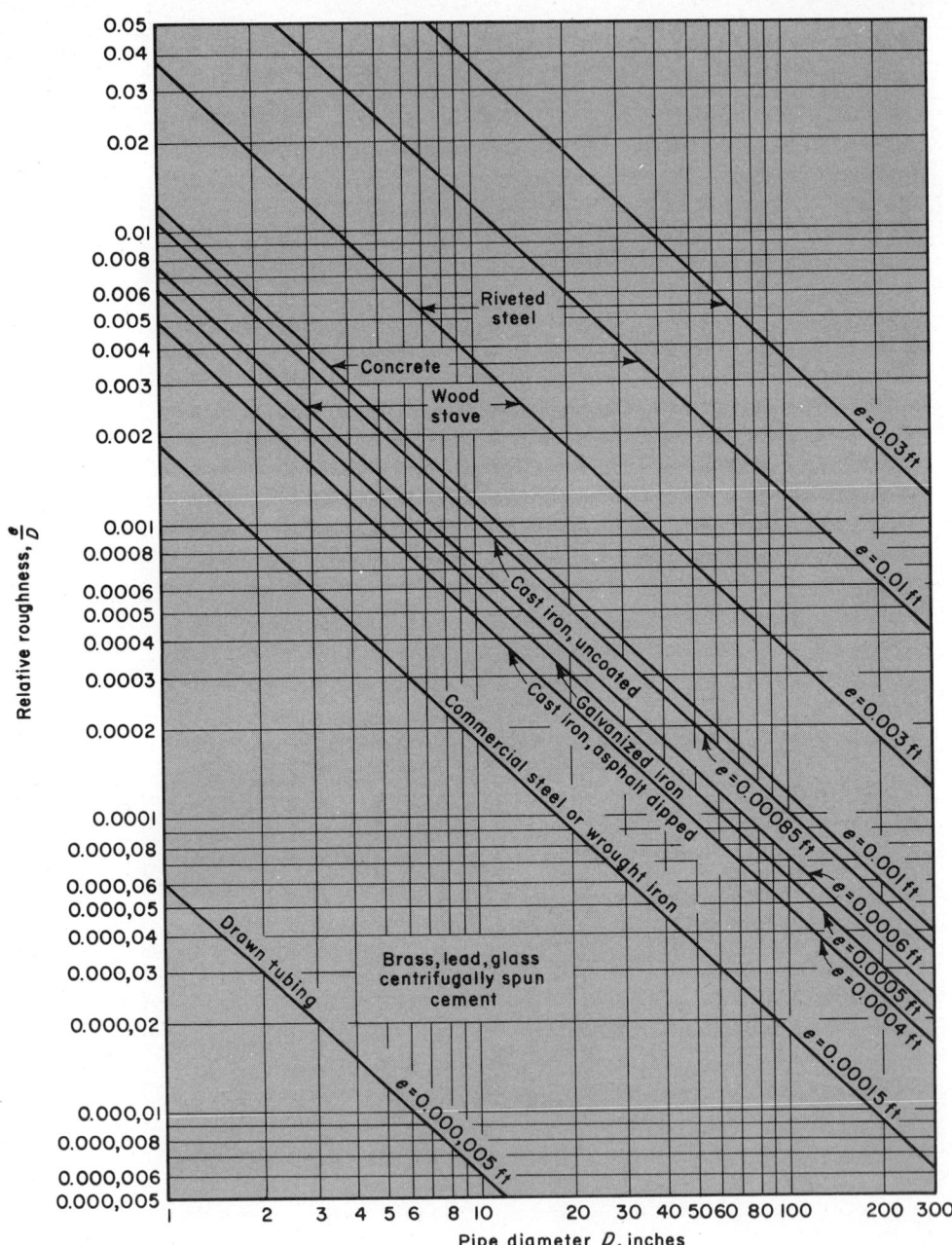

Fig. 13-11 Relative roughness for new commercial piping. [*Adapted from L. M. Moody, Trans. ASME,* **66**:*673 (1944).*]

TURBULENT FLOW OF INCOMPRESSIBLE VISCOUS FLUIDS

and from Fig. 13-10

$f = 0.0258$

By Eq. (13-61)

$$\Delta p = 0.0258 \frac{20 \text{ ft}}{\frac{1}{12} \text{ ft}} \frac{(56.8 \text{ lb}_m/\text{ft}^3)(306 \text{ ft/sec})^2}{64.4 \text{ lb}_m\text{-ft}/\text{lb}_f\text{-sec}^2}$$

$$= 5.12 \times 10^5 \frac{\text{lb}_f}{\text{ft}^2} \frac{\text{ft}^2}{144 \text{ in}^2} = 3{,}560 \text{ psi}$$

which is 22.9 percent greater than the pressure loss in a smooth tube.

There are other types of losses in piping. The common cases include those due to fittings and those caused by expansion or contraction of the piping configuration. These are sometimes referred to as *minor losses*, though they are by no means always small. For short piping systems, they are frequently larger than the loss due to the straight piping. These losses are usually due to *separation* of the fluid from the wall as it passes through the device and thus constitute a form drag

A convenient method of handling these losses is to use experimental data in the form of loss coefficients k_L, where

$$h_m = k_L \frac{V^2}{2g_c} \qquad (13\text{-}64)$$

A second method is to use the concept of an equivalent length defined by

$$L_{eq} = \frac{k_L D}{f} \qquad (13\text{-}65)$$

This latter approach is particularly well suited to calculations where a piping network is being evaluated as a unit, since it permits treatment as though the system consisted only of an equivalent length of straight piping

Pipe fittings Loss coefficients for various valves and fittings are published in piping handbooks and catalogs. A few selected values are reproduced in Table 13-2. It is important to note that the loss coefficients in Table 13-2 apply to a single item in a flow loop with sufficient straight pipe both upstream and downstream to permit establishment of fully developed flow. This is well illustrated by the fact that k_L for a single-flanged regular-radius elbow is 0.3, whereas that for two of them connected to form a flanged return bend is 0.38. Since many piping systems involve closely spaced fittings and valves, using loss coefficients such as these will result in approximate answers at best. Generally speaking, a length of 40 to 50 pipe diameters is required downstream from any flow obstruction to permit a return to fully developed turbulent flow. Spacing of fittings and valves closer than this will generally result in actual loss coefficients smaller than the published values, and consequently calculated losses are usually higher (more conservative) than those actually encountered.

Table 13-2 Loss coefficients for valves and fittings

Item	k_L
Angle valve, fully open	3.1–5.0
Ball check valve, fully open	4.5–7.0
Gate valve, fully open	0.19
Globe valve, fully open	10
Swing check valve, fully open	2.3–3.5
Regular-radius elbow, screwed	0.9
Flanged	0.3
Long-radius elbow, screwed	0.6
Flanged	0.23
Close return bend, screwed	2.2
Flanged return bend, two elbows, regular radius	0.38
Long radius	0.25
Standard tee, screwed, flow through run	0.6
Flow through side	1.8

Contractions Abrupt contractions are frequently encountered in piping systems. Consider the general contraction situation shown in Fig. 13-12 and note in particular that there is a necking down of the flow area downstream from the abrupt pipe change. This is due to the fluid momentum, and the minimum-flow area is known as the *vena contracta*.

This problem can be analyzed by applying the momentum, energy, and continuity equations to the control volume surrounding the fluid between sections 2 and 3, as shown in Fig. 13-12b. For steady flow with shear stresses along the curved boundaries neglected and taking the average pressure along the curved boundaries to be p_2, the momentum equation is

$$p_2 A_3 + \frac{\rho V_2^2 A_2}{g_c} = p_3 A_3 + \frac{\rho V_3^2 A_3}{g_c}$$

By the continuity equation V_2 is equal to $(A_3/A_2)V_3$ for an incompressible fluid, and thus the momentum equation yields

$$\frac{p_2 - p_3}{\rho} = \frac{V_3^2}{g_c}\left(1 - \frac{A_3}{A_2}\right) \tag{13-66}$$

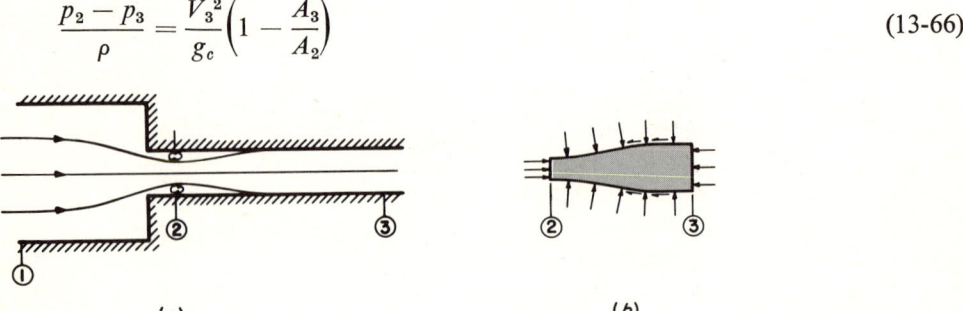

Fig. 13–12 Abrupt pipe contraction.

TURBULENT FLOW OF INCOMPRESSIBLE VISCOUS FLUIDS

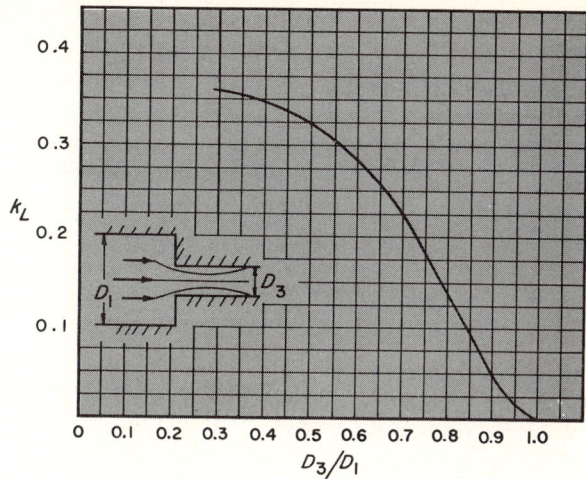

Fig. 13-13 Loss coefficients for water flow through a sharp-edged contraction.

The energy equation written for this control volume is

$$\frac{V_2^2}{2g_c} + \frac{p_2}{\rho} = \frac{V_3^2}{2g_c} + \frac{p_3}{\rho} + h_L \qquad (13\text{-}67)$$

Rearranging this and again applying the continuity equation yields

$$h_L = \frac{p_2 - p_3}{\rho} + \frac{V_3^2}{2g_c}\left[\left(\frac{A_3}{A_2}\right)^2 - 1\right]$$

which, combined with Eq. (13-66), results in

$$h_L = \frac{V_3^2}{2g_c}\left(1 - \frac{A_3}{A_2}\right)^2 \qquad (13\text{-}68)$$

At this point we note that calculation of the head loss through the contraction requires a knowledge of the vena contracta flow area A_2. This is difficult to determine analytically, and resort is usually made to experimental data. Upon comparing Eqs. (13-64) and (13-68) we observe that for the contraction

$$k_L = \left(1 - \frac{A_3}{A_2}\right)^2 = \left(1 - \frac{1}{C_c}\right)^2 \qquad (13\text{-}69)$$

where C_c is the contraction ratio at the vena contracta, A_2/A_3.

Using the experimental data of Weisbach [33] for water, loss coefficients are plotted as a function of the ratio of downstream to upstream pipe diameters, D_3/D_1, in Fig. 13-13. If the curve of Fig. 13-13 were extrapolated to a value of zero for D_3/D_1

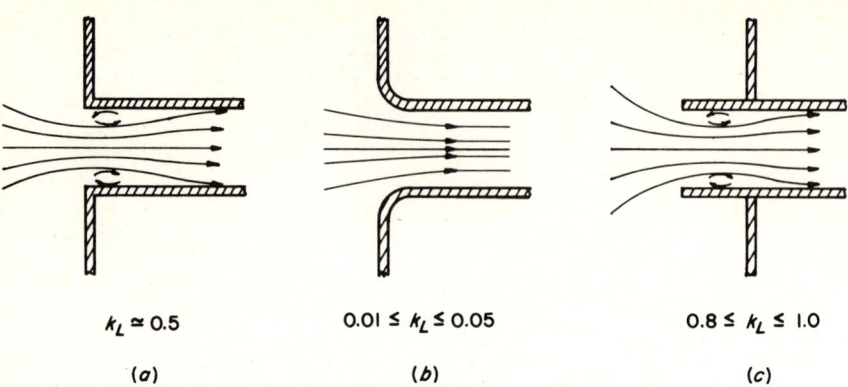

Fig. 13-14 Loss coefficients for pipe-entrance configurations: (*a*) square-edged; (*b*) well-rounded; (*c*) reentrant.

(this case approximates a pipe entrance from a large reservoir), the value of k_L would be about 0.4. Customary practice is to take this to be 0.5 for a square-edged opening, which is slightly conservative. If the entrance is well rounded with the radius of curvature approximately one-seventh of the pipe diameter, the loss coefficient is from 0.01 to 0.05. For reentrant pipe openings from a reservoir, such as a thin-walled pipe extending into the reservoir, the loss coefficient is between 0.8 and 1.0. This last case is sometimes called a *Borda mouthpiece*. The loss coefficients for these configurations are summarized in Fig. 13-14.

Abrupt expansions Flow through a sudden expansion can be analyzed in exactly the same manner as for the contraction. Referring again to Fig. 13-12*b*, it is clear that the flow from the vena contracta to the discharge in a contraction is identical to the flow from the inlet to the discharge of an expansion, as shown in Fig. 13-15.

Then we can rewrite Eq. (13-68) for this configuration as

$$h_L = \frac{V_2^2}{2g_c}\left(1 - \frac{A_2}{A_1}\right)^2 \tag{13-70}$$

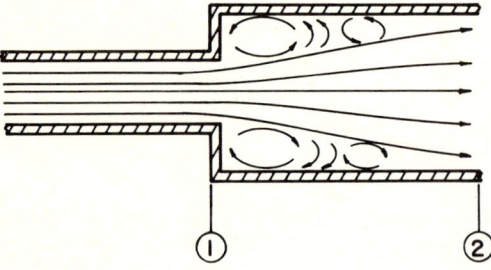

Fig. 13-15 Sudden expansion in a pipeline.

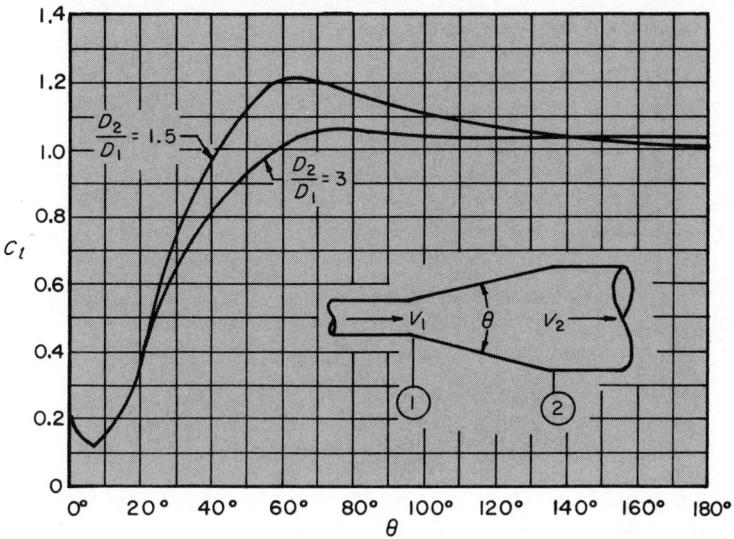

Fig. 13-16 Loss coefficients for flow through a diffuser.

or in terms of diameters as

$$h_L = \frac{V_2^2}{2g_c}\left[1 - \left(\frac{D_2}{D_1}\right)^2\right]^2 = \frac{V_1^2}{2g_c}\left[1 - \left(\frac{D_1}{D_2}\right)^2\right]^2 \qquad (13\text{-}71)$$

In this case, the areas or diameters are known, and no empiricism is necessary to determine the loss. Recalling that A_2/A_1 is also V_1/V_2 by the continuity equation, Eq. (13-70) can readily be expressed as

$$h_L = \frac{(V_1 - V_2)^2}{2g_c} \qquad (13\text{-}72)$$

From this form, the loss incurred when a pipe discharges into a very large reservoir ($V_2 = 0$) is readily seen to be one velocity head, $V_1^2/2g_c$.

The reader will recall that in the analysis of the contractive flow which we have applied directly to the expansion, the energy dissipated in the corner (or irregular-flow area of Fig. 13-15) has been neglected. For this reason the results are in error by as much as 20 percent.

Diffusers Diffusers, or gradual expansion devices, are frequently used to minimize the losses incurred in expanding an internal flow. Such a device is depicted in Fig. 13-16, along with a loss coefficient C_l defined by

$$h_L = C_l \frac{V_1^2}{2g_c}\left[1 - \left(\frac{D_1}{D_2}\right)^2\right]^2 \qquad (13\text{-}73)$$

These experimental data are by Gibson [6, 7] and were obtained with water at high Reynolds numbers. By inspection of Fig. 13-16, it is apparent that the optimum diffuser included angle θ is approximately 7°.

At least two disturbing facts are evident in these data. (1) One would intuitively expect the maximum value of C_l to be unity; then the loss would be identical with that for an abrupt expansion. But the measured values exceed unity. (2) The fact that C_l increases with decreasing values of D_2/D_1 is rather puzzling. Nonetheless, these data are considered adequate for preliminary design estimates.

Piping systems In this section we shall discuss the solution to frequently encountered problems of *single-path* pipe flow. There are the three basic types:

1. The entire pipe-system configuration, relative roughness, fluid properties, and flow rate are known. The head loss or pressure drop is to be determined.
2. The entire pipe-system configuration, relative roughness, fluid properties, and head loss are specified. The flow rate is to be determined.
3. The head loss, flow rate, fluid properties, pipe material, and centerline layout of the piping system are specified. The pipe diameter is to be determined.

In the following examples the first law of thermodynamics, the Darcy-Weisbach equation, the Moody diagram, and the loss coefficients for fittings, expansions, and contractions will be used to obtain solutions for the three basic types of pipe-flow problems. Throughout this chapter, nominal pipe sizes will be used as the inside pipe diameter. Actually, inside diameters may be either larger or smaller than nominal and depend upon the pipe schedule or wall thickness. For precise engineering calculations, manufacturer's catalogs should be consulted.

Example Type 1. A manufacturing company has an emergency fire-protection system consisting of a well, a gasoline-driven pump, and a piping system to their building. If the flow rate of 70°F water is 0.3 ft³/sec and the pump discharge pressure is 120 psig, what is the delivery pressure to the sprinkler system in the roof of the plant?

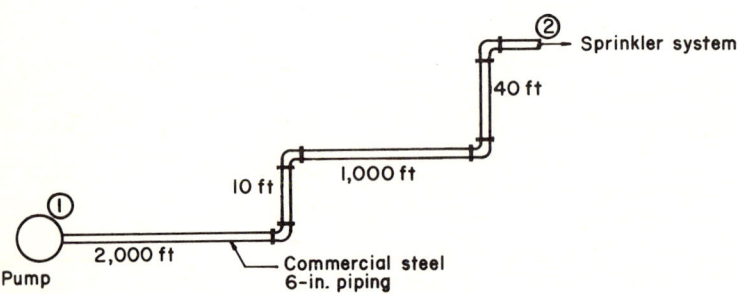

Fig. 13–17 Fire-protection system.

TURBULENT FLOW OF INCOMPRESSIBLE VISCOUS FLUIDS

Solution The governing equation is the first law of thermodynamics

$$\frac{V_1^2}{2g_c} + \frac{p_1}{\rho} + \frac{g}{g_c} y_1 = \frac{V_2^2}{2g_c} + \frac{p_2}{\rho} + \frac{g}{g_c} y_2 + h_L$$

where h_L is the head loss due to pipe friction and fittings. Since the fluid is incompressible and the pipe size is constant, $V_1 = V_2$. Thus

$$\frac{p_1}{\rho} + \frac{g}{g_c} y_1 = \frac{p_2}{\rho} + \frac{g}{g_c} y_2 + h_L$$

or

$$p_2 = p_1 + \rho \frac{g}{g_c} (y_1 - y_2) - \rho h_L$$

Now $p_1 = 120$ psig, $\rho = 62.4$ lb$_m$/ft^3, and $y_1 - y_2 = -50$ ft. Assuming standard gravitational acceleration, $g = 32.2$ ft/sec^2, the problem is reduced to finding ρh_L, which is the Δp due to pipe friction and fitting losses.

The flow Reynolds number is

$$\text{Re} = \frac{\frac{Q}{A} D}{\nu} = \frac{\frac{0.3}{\pi/16} \frac{1}{2}}{1.06 \times 10^{-5}} = 72,000$$

and the flow is turbulent. From Fig. 13-11, the relative roughness factor is found to be 0.00028, and then from Fig. 13-10 the friction factor is 0.0199. By Eq. (13-60) the frictional loss due to the straight piping is

$$(\Delta p)_{\text{pipe}} = 0.0199 \frac{3{,}050 \text{ ft}}{0.5 \text{ ft}} \left(62.4 \frac{\text{lb}_m}{\text{ft}^3}\right) \frac{\left(\frac{0.3}{\pi/16} \frac{\text{ft}}{\text{sec}}\right)^2}{64.4 \text{ lb}_m\text{-ft/lb}_f\text{-sec}^2}$$

$$= 274 \text{ lb}_f/\text{ft}^2 = 1.9 \text{ psi}$$

The loss coefficient for each elbow is given as 0.9 (assuming regular-radius screwed fittings). Then using Eq. (13-64), the frictional loss due to the four elbows is

$$(\Delta p)_{\text{elbows}} = \rho(h_L)_{\text{elbows}} = 4(0.9)\left(62.4 \frac{\text{lb}_m}{\text{ft}^3}\right) \frac{\left(\frac{0.3}{\pi/16} \frac{\text{ft}}{\text{sec}}\right)^2}{64.4 \text{ lb}_m\text{-ft/lb}_f\text{-sec}^2}$$

$$= 8.16 \text{ lb}_f/\text{ft}^2 = 0.057 \text{ psi}$$

Then the total frictional losses are

$$\rho h_L = (\Delta p)_{\text{pipe}} + (\Delta p)_{\text{elbows}} \approx 2.0 \text{ psi}$$

and

$$p_2 = 120 \text{ psig} + \frac{(62.4 \text{ lb}_m/\text{ft}^3)(32.2 \text{ ft/sec}^2)}{32.2 \text{ lb}_m\text{-ft/lb}_f\text{-sec}^2} \frac{0 - 50 \text{ ft}}{144 \text{ in}^2/\text{ft}^2} - 2.0 \text{ psi}$$

$$= 120 - 21.6 - 2.0 = 96.4 \text{ psig}$$

Example Type 2. Light oil at 150°F is pumped through 1,000 ft of straight 8-in-diam galvanized iron pipe. The pump output pressure is 60 psig, and the delivery pressure is atmospheric. Find the flow rate.

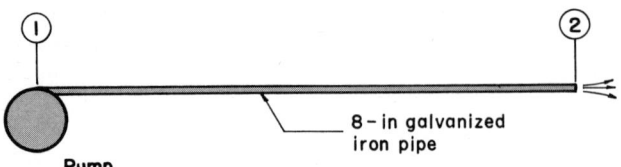

Fig. 13-18 Straight pipe system.

Solution Applying the first law of thermodynamics to the horizontal pipeline of constant diameter, we obtain for an incompressible fluid

$$p_1 - p_2 = \rho h_L$$

From Table A-3, the fluid properties are

$$\rho = 54.3 \text{ lb}_m/\text{ft}^3 \qquad \nu = 9.8 \times 10^{-5} \text{ ft}^2/\text{sec}$$

Since the discharge pressure is atmospheric, the pressure loss due to pipe friction is

$$\Delta p = 60 - 0 = 60 \text{ psi}$$

and the head loss is

$$h_L = \frac{\Delta p}{\rho} = \frac{60(144) \text{ lb}_f/\text{ft}^2}{54.3 \text{ lb}_m/\text{ft}^3} = 159 \text{ ft-lb}_f/\text{lb}_m$$

From Fig. 13-11 the relative roughness is $e/D = 0.00075$. Since the flow rate is unknown, the Reynolds number cannot be calculated and hence the friction factor must be estimated to solve the problem by iteration. From Fig. 13-10 a trial f of 0.023 (corresponding to **Re** $\simeq$ 50,000) is assumed. Then by the Darcy-Weisbach equation

$$V^2 = \frac{2 g_c h_L D}{fL} = \frac{(64.4 \text{ lb}_m\text{-ft/lb}_f\text{-sec}^2)(159 \text{ ft-lb}_f/\text{lb}_m)(0.667 \text{ ft})}{0.023(1,000 \text{ ft})}$$

$$V = 18.05 \text{ ft/sec}$$

Now we must check to see whether the assumed Reynolds number was suitable for determination of f. If not, we must resort to iteration.

$$\mathbf{Re} = \frac{DV}{\nu} = \frac{0.667(18.05)}{9.8 \times 10^{-5}} = 123{,}000$$

Using this, a more accurate value for f, 0.021, is obtained from Fig. 13-10. Then

$$V^2 = \frac{(64.4 \text{ lb}_m\text{-ft/lb}_f\text{-sec}^2)(159 \text{ ft-lb}_f/\text{lb}_m)(0.667 \text{ ft})}{0.021(1,000 \text{ ft})}$$

$$V = 17.2 \text{ ft/sec}$$

and clearly further iteration is unnecessary. Thus,

$$Q = AV = \frac{\pi(0.667 \text{ ft})^2}{4} (17.2 \text{ ft/sec}) = 6.0 \text{ ft}^3/\text{sec}$$

Problems of type 3 are frequently encountered in system design. Notice that since the pipe diameter is to be determined, the velocity, Reynolds number, and friction factor are all unknown. An iterative solution to this type problem is always necessary. Consider first the simplest problem of this type—specifically one for which the minor losses are negligible. Here the procedure is to apply the first law of thermodynamics to an appropriate control volume to determine h_f numerically.

Then, rearrangement of the Darcy-Weisbach equation with the velocity replaced by means of the continuity equation yields

$$h_f = \frac{f}{2g_c} \frac{L}{D} \left(\frac{Q}{\pi D^2/4} \right)^2$$

$$D^5 = \frac{8fLQ^2}{\pi^2 g_c h_f} \qquad (13\text{-}74)$$

This last expression for D contains only one unknown, f, which is an implicit function of D. The suggested iterative procedure is as follows:

1. Assume a reasonable value for f. (A value of 0.025 is as good as any other initial guess.)
2. Calculate D from Eq. (13-74).
3. With this value of D, calculate the Reynolds number.
4. Using this value of D, obtain a relative roughness e/D from Fig. 13-11. Then using this and the calculated Reynolds number, obtain a corrected value of f from Fig. 13-10.
5. Repeat the preceding steps beginning with the corrected value of f. The solution is completed when f does not change appreciably as the result of an iteration.

For problems in which the minor or fitting losses are not negligible, the suggested approach is to neglect these minor losses *initially* and proceed as though the system consisted only of the straight piping. Upon finding the pipe diameter for this hypothetical case, one then selects the nearest larger nominal pipe diameter and checks to determine whether it is adequate by the complete first-law equation (containing head loss due both to pipe friction and fittings). This, of course, may require a trial-and-error solution, especially in cases where the so-called minor losses are of the same order of magnitude as pipe-friction losses. In this regard, a plot of the results of the trial solutions is helpful.

Example Type 3. Determine the size of commercial steel pipe required for a minimum flow of 1 ft³/sec of light oil at 80°F from the pump to the receiver tank.

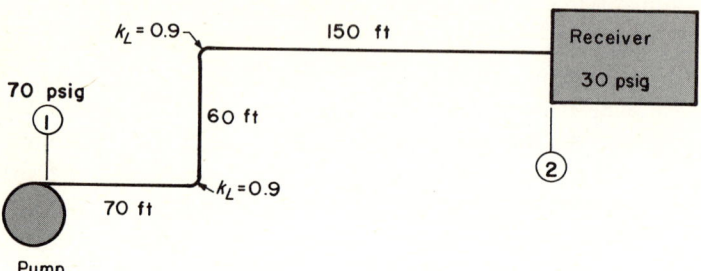

Fig. 13-19 Flow schematic for determination of pipe diameter.

Solution The first law of thermodynamics applied to a control volume consisting of the piping system is

$$\frac{V_1^2}{2g_c} + \frac{p_1}{\rho} + \frac{g}{g_c} y_1 = \frac{V_2^2}{2g_c} + \frac{p_2}{\rho} + \frac{g}{g_c} y_2 + h_L$$

For uniform pipe size throughout and an incompressible fluid, $V_1 = V_2$. The term h_L includes both h_f and h_m. Initially neglecting h_m, we find

$$h_f \simeq \frac{p_1 - p_2}{\rho} + \frac{g}{g_c}(y_1 - y_2)$$

From Table A-3 the fluid properties are

$\rho = 56.8 \text{ lb}_m/\text{ft}^3 \qquad \nu = 49 \times 10^{-5} \text{ ft}^2/\text{sec}$

Thus

$$h_f \simeq \frac{70-30}{56.8 \text{ lb}_m/\text{ft}^3} \frac{\text{lb}_f}{\text{in}^2}\left(144 \frac{\text{in}^2}{\text{ft}^2}\right) + \frac{32.2 \text{ ft/sec}^2}{32.2 \text{ ft-lb}_m/\text{lb}_f\text{-sec}^2}(0-60)\text{ft}$$

$$\simeq 101.5 - 60 = 41.5 \text{ ft-lb}_f/\text{lb}_m$$

Now assuming $f = 0.025$, we have by Eq. (13-74)

$$D^5 = \frac{8(0.025)(280 \text{ ft})(1 \text{ ft}^6/\text{sec}^2)}{\pi^2(32.2 \text{ ft-lb}_m/\text{lb}_f\text{-sec}^2)(41.5 \text{ ft-lb}_f/\text{lb}_m)} = 4.22 \times 10^{-3}$$

$D = 0.335 \text{ ft} = 4.02 \text{ in}$

Assuming a 4-in-ID pipe, the Reynolds number will be

$$\text{Re} = \frac{D(4Q/\pi D^2)}{\nu}$$

$$= \frac{4(1 \text{ ft}^3/\text{sec})}{\pi(\frac{4}{12} \text{ ft})(49 \times 10^{-5} \text{ ft}^2/\text{sec})} = 7{,}800$$

TURBULENT FLOW OF INCOMPRESSIBLE VISCOUS FLUIDS

From Fig. 13-11, the relative roughness for a 4-in commercial steel pipe is 0.00042, and then from Fig. 13-10 at the calculated Reynolds number we find

$$f \simeq 0.033$$

Repeating the above procedure, we find

$$D^5 = 5.57 \times 10^{-3}$$
$$D = 0.354 \text{ ft} = 4.25 \text{ in}$$

The next larger pipe size is 4.5 in. Using this, the corrected Reynolds number is

$$\mathbf{Re} = \frac{4(1)}{\pi(4.5/12)(4.9 \times 10^{-5})} = 6{,}930$$

From Fig. 13-10

$$f = 0.034$$

Further iteration using Eq. (13-74) yields a value of D less than 4.5 in and is unnecessary. We have omitted the two elbows in the system. To account for them we have

$$h_m = 2k_L \frac{V^2}{2g_c}$$

$$= \frac{0.9(4Q/\pi D^2)^2}{g_c} = \frac{0.9}{32.2 \text{ ft-lb}_m/\text{lb}_f\text{-sec}^2} \left[\frac{4(1) \text{ ft}^3/\text{sec}}{\pi(4.5/12 \text{ ft})^2} \right]^2$$

$$= 0.323 \text{ ft-lb}_f/\text{lb}_m$$

Then the corrected h_f is

$$h_f = 41.5 - 0.323 \simeq 41.18$$

Using this in Eq. (13-74) together with the value of f obtained by the last iteration yields

$$D^5 = \frac{8(0.034)(280)(1)}{\pi^2(32.2)(41.18)} = 5.83 \times 10^{-3}$$

or

$$D = 0.357 \text{ ft} = 4.29 \text{ in}$$

Therefore, the 4.5-in-ID pipe is suitable.

In this last example, the fitting or minor losses in the system were, in fact, negligible. The student is warned that such is not always the case. In many industrial applications piping lengths are quite short, and sometimes the minor losses exceed the pipe-friction loss.

For very short systems with numerous valves and fittings, the number of iterations can become quite large. It is sometime advantageous to solve such problems with a digital computer, and then it is necessary to use an analytical expression for the friction factor rather than the Moody diagram. An equation by Colebrook and White

$$\frac{1}{\sqrt{f}} = 1.74 - 2 \log \left(\frac{2e}{D} + \frac{18.7}{\mathbf{Re}_D \sqrt{f}} \right) \qquad (13\text{-}75)$$

which is valid for all flow regimes, eliminates the need for graphical or tabulated friction-factor data.

Piping networks Multiple-path piping systems are also common. The simplest of such systems is shown in Fig. 13-20. Our discussion will be centered on a two-branch, two-node system, though the solution can be applied to a system with any number of branches.

For cases where the pipe diameter remains constant in each individual branch (not necessarily the same for both branches, however), the first law of thermodynamics combined with the continuity equation yields for either branch

$$(h_L)_{\text{branch}} = \frac{p_a - p_b}{\rho} + \frac{g}{g_c}(y_a - y_b) \tag{13-76}$$

In this expression y_a and y_b are the elevations of the two nodes a and b, respectively. Clearly this expression applies to either branch, and we have

$$(h_L)_1 = (h_L)_2 \tag{13-77}$$

The remaining general relationship, which is applicable to any type of network problem, is an expression of the continuity equation

$$Q_{\text{in}} = Q_{\text{out}} = Q_1 + Q_2 \tag{13-78}$$

There are two general types of network problems which frequently arise: (1) The entire network configuration, relative roughnesses, fluid properties, total flow rate, and pressure at nodal point a are known. The pressure at nodal point b is to be determined. (2) The entire network configuration, relative roughnesses, fluid properties, and both nodal pressures are known. The flow rate in each branch and the total flow rate are to be determined.

For the first type an iterative method of solution is required. A suggested procedure is as follows:

1. Assume a flow rate Q_1' through branch 1. For both branches having similar lengths, a first guess is to divide the total flow proportional to the cross-sectional

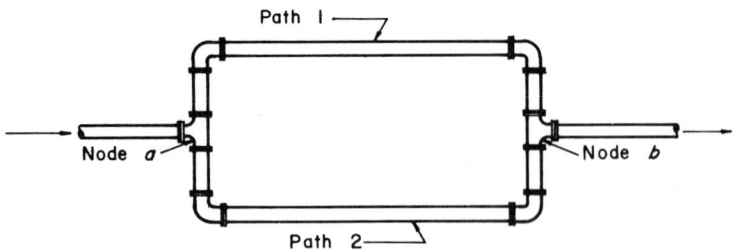

Fig. 13–20 A two-path piping network.

area of the branch pipe size. Solve for the head loss for branch 1 as done under type 1 of single-path systems.
2. Using the head loss thus calculated, solve for the flow Q_2' through branch 2. (See type 2 of single-path systems.)
3. Now assume that the given total flow divides in the same proportions as the assumed Q_1' and calculated Q_2'. [If Q_1' and Q_2' satisfy Eq. (13-78), the problem is finished; this occurs very infrequently, and iteration is usually required.] This division of the total flow will yield corrected flow terms, Q_1'' and Q_2''.
4. Using Q_1'' and Q_2'', recalculate the head loss for each branch. Modify these flow rates as necessary until the pressure losses for both branches are equal and the continuity equation (13-78) is satisfied.

For highly complex multiple-path piping systems, the iteration methods of Hardy Cross [5] or digital-computer techniques are suggested.

13-5 TURBULENT FLOW OVER SURFACES

We now direct our attention to incompressible external boundary-layer flow. Our primary objective is to develop solutions for boundary-layer thickness, drag, and lift for various simple configurations.

Transition to turbulent flow: flat plate The development of the turbulent boundary layer can be explained best for flow along a flat plate. Consider a free-stream uniform-velocity flow approaching a flat plate at zero incidence, as shown in Fig. 13-21. As the fluid approaches the leading edge, large shear forces result in the fluid velocity being altered or slowed near the plate. This always results in the development of an initial section of laminar boundary layer. This boundary layer thickens with distance (solutions for the thickness as a function of the length Reynolds number were presented in Chap. 12), and eventually instabilities cause the boundary layer to become turbulent. The turbulent boundary layer is much thicker, and because of the velocity

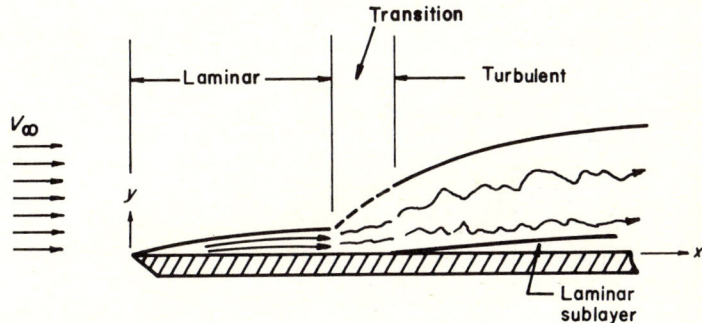

Fig. 13-21 Developing turbulent boundary layer on a flat plate.

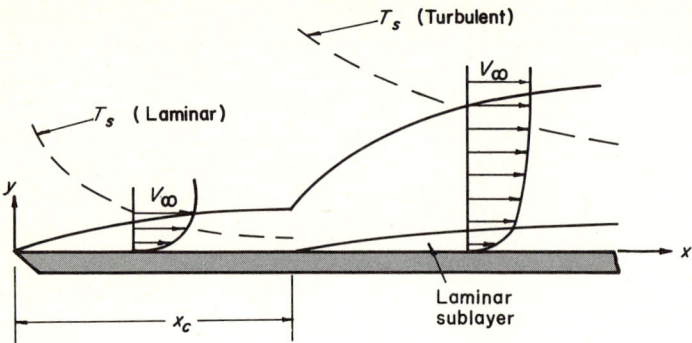

Fig. 13-22 Simplified model of transition from laminar to turbulent boundary-layer flow along a flat plate.

perturbations in the y direction it has a much flatter velocity profile than laminar flow over most of its thickness. In the laminar sublayer, however, there is a very steep gradient. As a consequence, the shear stress at the wall is much greater for the turbulent boundary layer than for the laminar boundary layer.

The total drag on a plate is highly dependent upon the location of the *transition* from laminar to turbulent boundary-layer flow. Transition-region flow is highly oscillatory in nature, appearing at one instant in time to be laminar and slightly later to be turbulent. This transition region is actually a finite length, but since we are unable to analyze transitional flow mathematically, we shall simplify our model to consider transition to occur at a single location, the boundary-layer flow ahead of this being laminar and that downstream being turbulent, as shown in Fig. 13-22, which includes qualitative curves for the shearing stress.

The transition to turbulent boundary-layer flow depends upon many parameters; the more significant ones are (1) the critical Reynolds number $V_\infty x_c/\nu$, (2) the wall roughness, (3) the free-stream turbulence, and (4) the external-flow pressure gradient. For the flat plate at zero incidence to the direction of flow the pressure gradient is zero. For many other practical problems (such as airfoils) this is not the case, and the pressure gradient is not only important with regard to transition but also has a decided influence upon *separation*. This will be discussed later.

A normal airstream has an *intensity of turbulence* $T \simeq 0.5$ percent, where T is defined by

$$T \equiv \frac{\sqrt{\tfrac{1}{3}(\overline{u'^2} + \overline{v'^2} + \overline{w'^2})}}{V_\infty} \qquad (13\text{-}79)$$

For this level of free-stream turbulence, transition on a smooth flat isothermal plate with a sharp leading edge occurs at

$$3.5 \times 10^5 \leq \frac{V_\infty x_c}{\nu} \leq 10^6 \qquad (13\text{-}80)$$

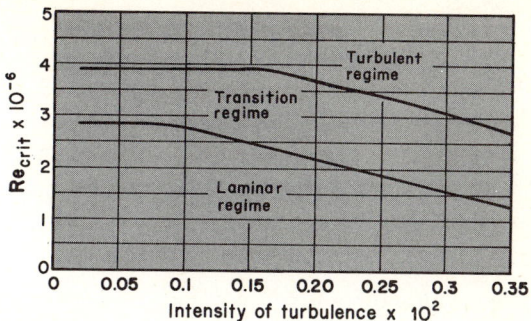

Fig. 13-23 Effect of free-stream turbulence upon transition from laminar to turbulent flow on a flat plate.

according to Schlichting [29]. For engineering calculations it is usually assumed that transition occurs at a critical Reynolds number somewhere between 300,000 and 500,000. A conservative calculation of the drag normally dictates the use of the lower value of critical Reynolds number, which results, of course, in the largest calculated frictional drag.

A better understanding of the effect of free-stream turbulence on the transition to turbulent flow can be obtained by a study of Fig. 13-23, which shows the results of Schubauer and Skramstad [30]. At a given intensity of turbulence, the lower curve indicates the maximum critical Reynolds number based on length for which laminar boundary-layer flow exists on a smooth isothermal flat plate. The upper curve yields the minimum value of the critical Reynolds number for the boundary-layer flow to be turbulent. These results indicate an upper limit of approximately 2.8×10^6 for the critical Reynolds number, and this occurs at $T \simeq 0.0008$. Since their plate was smooth and isothermal, a larger value of critical Reynolds number is unattainable. (Both plate roughness and heating would cause transition to occur in a shorter distance.)

Completely turbulent boundary layer: flat plate We shall now apply the integral momentum equation to obtain expressions for the boundary-layer thicknesses. From the previous section it is evident that the determination of an exact location for the beginning of a turbulent boundary layer would be a formidable task. Further, the turbulent regime begins with a finite thickness, as shown in Figs. 13-21 and 13-22. We shall eliminate this obstacle by imagining, after Prandtl, that the turbulent boundary layer begins at the leading edge of the plate. The results obtained in this way are quite good for distances beyond x_c in Fig. 13-22 but are invalid for the laminar leading section.

The control volume is identical with that shown in Fig. 12-9 for a laminar boundary layer, and the momentum relation given by Eq. (12-106) is applicable.

This is

$$\tau_s = \rho \frac{d}{dx}\left[\int_0^\delta (V_\infty - \bar{u})\bar{u}\, dy\right] \tag{13-81}$$

for the turbulent boundary layer. For laminar flow the reader will recall that we obtained the wall shear stress τ_s from Newton's definition of fluid viscosity. Also we used an assumed velocity profile to obtain the gradient at the wall and for evaluation of the integral term.

In turbulent flow we have seen that

$$\frac{\bar{u}}{U_{\max}} = \left(\frac{y}{\delta}\right)^{1/n}$$

where $U_{\max}$ is the centerline velocity for pipe flow and is analogous to V_∞ for the external flow. The exponent $1/n$ is a function of the Reynolds number. In order to proceed with an integration of Eq. (13-81), we shall choose $n = 7$. (The reader will recall that the choice of the velocity profile for laminar flow was also quite arbitrary.) Thus,

$$\frac{\bar{u}}{V_\infty} = \left(\frac{y}{\delta}\right)^{\frac{1}{7}} \tag{13-82}$$

There is an added complication in turbulent flow, however, since the one-seventh velocity profile is a fair approximation for the bulk of the boundary layer but is totally unacceptable near the wall. Thus we cannot use this to obtain a gradient for determination of the wall shear stress.

The experimental expression of Blasius for the shear stress

$$\tau_s = 0.0225 \rho V_\infty^2 \left(\frac{\nu}{V_\infty \delta}\right)^{\frac{1}{4}} \tag{13-83}$$

will be used. This was obtained with turbulent pipe flow by assuming the boundary-layer thickness to be the pipe radius, and it has been shown by Schultz-Grunow to be valid for Reynolds numbers up to 10^7 for flat smooth plates.

Substituting Eqs. (13-82) and (13-83) into (13-81) yields

$$0.0225 V_\infty^2 \left(\frac{\nu}{V_\infty \delta}\right)^{\frac{1}{4}} = \frac{d}{dx} \int_0^\delta V_\infty^2 \left[\left(\frac{y}{\delta}\right)^{\frac{1}{7}} - \left(\frac{y}{\delta}\right)^{\frac{2}{7}}\right] dy$$

Integrating, we have

$$0.0225 \left(\frac{\nu}{V_\infty \delta}\right)^{\frac{1}{4}} = \frac{7}{72} \frac{d\delta}{dx}$$

Separation of variables and a second integration results in

$$\left(\frac{\nu}{V_\infty}\right)^{\frac{1}{4}} x = 3.45 \delta^{\frac{5}{4}} + C$$

TURBULENT FLOW OF INCOMPRESSIBLE VISCOUS FLUIDS

where C is the constant of integration. This can be evaluated by application of the fictitious boundary condition $\delta = 0$ at $x = 0$, since we have assumed the turbulent boundary layer to begin at the leading edge. Thus, $C = 0$, and

$$\frac{\delta}{x} = \frac{0.376}{(V_\infty x/\nu)^{\frac{1}{5}}} = \frac{0.376}{\mathbf{Re}_x^{\frac{1}{5}}} \qquad (13\text{-}84)$$

By comparison of this with Eq. (12-112), we see that the turbulent boundary-layer thickness grows with the $\frac{4}{5}$ power of x, whereas the laminar boundary-layer thickness grows with the $\frac{1}{2}$ power of x. Thus, a turbulent boundary layer grows faster or is thicker than a laminar boundary layer at a given distance from the leading edge.

The *displacement thickness* δ^* can easily be obtained using the one-seventh-power law and Eq. (12-103). These yield

$$\delta^* = \int_0^\delta \left[1 - \left(\frac{y}{\delta}\right)^{\frac{1}{7}}\right] dy = \frac{\delta}{8} \qquad (13\text{-}85)$$

and thus

$$\frac{\delta^*}{x} = \frac{0.047}{\mathbf{Re}_x^{\frac{1}{5}}} \qquad (13\text{-}86)$$

Likewise, the *momentum thickness* from Eq. (12-104) and the one-seventh-power law is

$$\delta_i = \int_0^\delta \left(\frac{y}{\delta}\right)^{\frac{1}{7}} \left[1 - \left(\frac{y}{\delta}\right)^{\frac{1}{7}}\right] dy = \frac{7\delta}{72}$$

or

$$\frac{\delta_i}{x} = \frac{0.036}{\mathbf{Re}_x^{\frac{1}{5}}} \qquad (13\text{-}87)$$

To obtain the local skin-friction coefficient from Eq. (12-83) we need an expression for the wall shear stress. From the empirical expression of Blasius and Eq. (13-84) we obtain

$$\tau_s = 0.0225 \rho V_\infty^2 \left(\frac{\nu}{V_\infty \delta}\right)^{\frac{1}{4}} = 0.0288 \rho V_\infty^2 \left(\frac{\nu}{V_\infty x}\right)^{\frac{1}{5}} \qquad (13\text{-}88)$$

and then

$$c_f = \frac{\tau_s}{\rho V_\infty^2/2} = \frac{0.0576}{\mathbf{Re}_x^{\frac{1}{5}}} \qquad (13\text{-}89)$$

To obtain the average drag per unit width of a plate of finite length L, we employ the average friction coefficient C_f, defined by

$$C_f = \frac{\text{friction drag per unit width of plate}}{(\rho V_\infty^2/2)L} \qquad (13\text{-}90)$$

This can be evaluated by integrating the shear-stress expression (13-88) over the length of the plate and using it in Eq. (13-90) or by applying the usual technique of averaging to the local skin-friction coefficient. By either method we find

■ $$C_f = \frac{0.072}{\text{Re}_L^{\frac{1}{5}}} \qquad (13\text{-}91)$$

This result is in good agreement with experimental data if the constant is increased to 0.074.

Example A model airfoil is tested in a wind tunnel using air. The airfoil is smooth, slender, and symmetrical and may be approximated by a smooth flat plate. The chord length is 6 in. At a tunnel velocity of 400 ft/sec, an intensity of turbulence of 0.5 percent, and a temperature of 32°F what is the drag per foot of width on the model? (This problem is similar to the laminar-flow example of Sec. 12-5.)

Solution The Reynolds number at the trailing edge is

$$\text{Re}_L = \frac{\frac{1}{2}(400)}{0.145 \times 10^{-3}} = 1{,}370{,}000$$

and by Eq. (13-80) this is turbulent. Assuming the entire boundary layer to be turbulent, we have

$$C_f = \frac{0.074}{\text{Re}_L^{\frac{1}{5}}} = \frac{0.074}{13.7^{\frac{1}{5}}(10)} = 0.439 \times 10^{-2}$$

and thus, the average shear stress is

$$(\tau_s)_{\text{av}} = C_f \frac{\rho V_\infty^2}{2 g_c}$$

$$= 0.439 \times 10^{-2} \frac{0.081 \text{ lb}_m/\text{ft}^3}{64.4 \text{ lb}_m\text{-ft/lb}_f\text{-sec}^2} \frac{160{,}000 \text{ ft}^2}{\text{sec}^2}$$

$$= 0.884 \text{ lb}_f/\text{ft}^2$$

For a slender airfoil, the area per foot of width is approximately

$$A = 2(\tfrac{6}{12} \text{ ft})(1 \text{ ft}) = 1 \text{ ft}^2$$

and the total frictional drag is

$$F_f = (1 \text{ ft}^2)(0.884 \text{ lb}_f/\text{ft}^2) = 0.884 \text{ lb}_f$$

We note in passing that this much higher frictional drag force than obtained in Sec. 12-5 was due primarily to the difference in density of the fluid in the tunnel. For very large plates, however, the frictional-coefficient difference between the laminar and turbulent cases predominates.

Laminar and turbulent boundary layer This last example illustrates the fact that frequently the laminar leading portion of the boundary layer should be accounted for. In that problem a significant length of the model experiences laminar boundary-layer flow. Considering the simplified model of Fig. 13-22, the drag could be

TURBULENT FLOW OF INCOMPRESSIBLE VISCOUS FLUIDS

calculated assuming laminar flow up to x_c and turbulent from this point downstream. According to Prandtl, the turbulent region behaves as though it were turbulent from the leading edge. Substracting the turbulent drag for the critical length and adding the laminar drag for it, we obtain the following expression for the average frictional coefficient:

$$C_f = \frac{0.074}{\mathbf{Re}_L^{\frac{1}{5}}} - \frac{A}{\mathbf{Re}_L} \qquad (13\text{-}92)$$

where values of A are:

$\mathbf{Re}_{crit}$	10^5	3×10^5	5×10^5	10^6
A	360	1,050	1,700	3,300

Example Reconsider the last problem and assume $\mathbf{Re}_{crit} = 5 \times 10^5$. Calculate the total drag on the model assuming a laminar leading edge.

Solution From the previous example

$\mathbf{Re}_L = 1{,}370{,}000$

Thus

$$C_f = \frac{0.074}{\mathbf{Re}_L^{\frac{1}{5}}} - \frac{A}{\mathbf{Re}_L}$$

$$= \frac{0.074}{13.7^{\frac{1}{5}}(10)} - \frac{1{,}700}{13.7 \times 10^5} = (0.439 \times 10^{-2}) - (0.124 \times 10^{-2})$$

$$= 0.315 \times 10^{-2}$$

The average shear stress is

$$(\tau_s)_{av} = C_f \frac{\rho V_\infty^2}{2g_c} = 0.315 \times 10^{-2} \frac{0.081}{64.4} 160{,}000 = 0.635 \text{ lb}_f/\text{ft}^2$$

The total frictional drag force is then

$$F_f = (1 \text{ ft}^2)(0.635 \text{ lb}_f/\text{ft}^2) = 0.635 \text{ lb}_f$$

which is significantly less than that obtained assuming a turbulent boundary layer to exist over the entire model.

Boundary-layer separation In many external-flow problems, the analysis is further complicated by the phenomenon of *separation*. Consider the flight of a projectile, as illustrated in Fig. 13-24. For the blunt-nosed configuration, the boundary layer detaches and leaves the surface near the front edge. By simply rounding the leading portion, the boundary-layer flow remains attached to the wall along all the length, with separation occurring at the back edge. The region following the point of separation is known as the *wake*, and this results in a *form drag*. Frequently the form drag is larger than the skin-friction drag on a body.

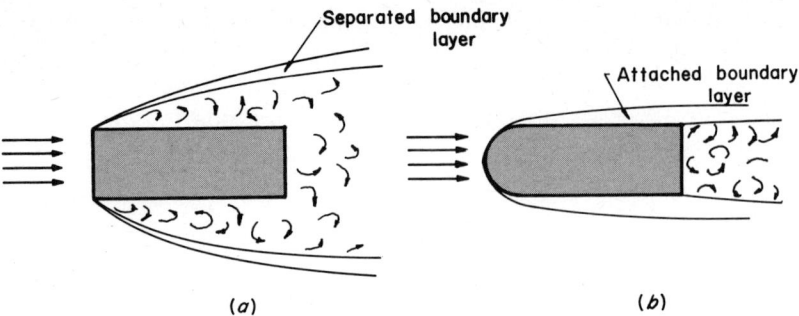

Fig. 13-24 Flow along a projectile: (*a*) blunt-nosed (separation); (*b*) rounded-nosed (no separation).

The nature of the form drag is perhaps clarified by considering the total drag as consisting of skin-friction drag and pressure drag. Since the pressure is greater in the wake due to the decreased kinetic energy, a body having a large (wide) wake flow experiences a larger pressure drag than one having a small wake. Thus the round-nosed projectile of Fig. 13-24 experiences a much smaller form drag than the blunt-nosed specimen does.

Curved bodies have numerous applications; e.g., wing airfoils are curved. For such surfaces, the pressure gradient differs from zero. Consider boundary-layer flow over an airfoil as shown in Fig. 13-25. Focusing attention upon the upper surface, we see that at the outer edge of the boundary layer the flow accelerates from A to B and then decelerates from B to D. The pressure distribution along the body is given by Bernouilli's equation for the external flow *up to the point of separation*. Thus, the boundary-layer flow is subjected to a decreasing pressure from point A to B. This is called a *favorable* pressure gradient.

From point B onward, however, the decreasing velocity of the external flow results in an increasing pressure. This is an *adverse* pressure gradient. The fluid near the wall, having lost energy due to the viscous drag, is slowed until finally at point C reverse flow sets in and the boundary-layer flow detaches: it is forced away from the wall.

Enlarging a portion of the boundary-layer flow along an airfoil in Fig. 13-26,

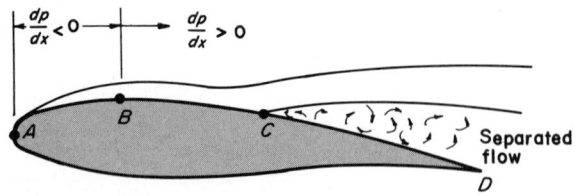

Fig. 13-25 Separated flow along an airfoil.

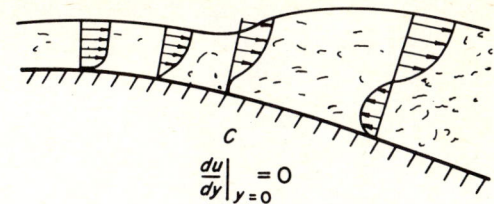

Fig. 13-26 Velocity profiles along an airfoil.

we see that the point of separation corresponds to a zero velocity gradient at the wall. This must exist before the backflow required by the continuity equation for separated flow can exist.

Schlichting [29] shows that separation can occur only for an adverse pressure gradient which requires a decelerating flow. It is very difficult to predict the location of the point of separation for either laminar or turbulent boundary-layer flow. For the laminar case, solution of Prandtl's boundary-layer equation (12-51a) at the wall, where $u = v = 0$, and with the pressure gradient determined by potential-flow theory can be carried out by numerical methods to find the point where

$$\left. \frac{du}{dy} \right|_{y=0} = 0 \tag{13-93}$$

which is the position of the onset of separation. Beyond this position, the flow separation invalidates Prandtl's boundary-layer-equation assumptions. For turbulent flow, the problem is far more complicated and is a subject of current research.

Engineering calculations for separated flows are usually accomplished with empirical equations or graphical data. A few common cases will be considered below.

Flow over bluff bodies Flow over immersed bodies in both the laminar and turbulent regimes has been purposely delayed until after the discussion of separation. The cylindrical and spherical configurations will be considered. For either, the total drag coefficient is defined by

$$C_D = \frac{F_D}{(\rho V^2 / 2g_c) A} \tag{13-94}$$

where A is the frontal area, i.e., the area of the body projected normal to the free-stream velocity.

A dimensional analysis shows that for geometrically similar bodies the drag coefficient is a function of the Reynolds number. The results of several investigators

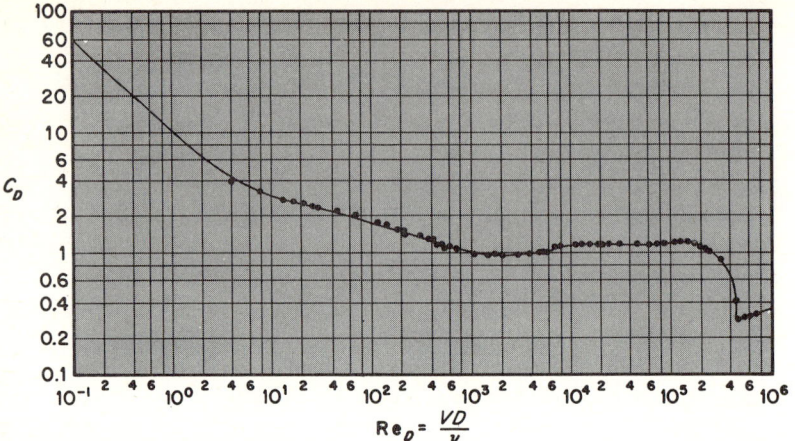

Fig. 13-27 Drag coefficients for cylinders. [*Adapted from H. Schlichting "Boundary Layer Theory," 6th ed. Copyright 1968. McGraw-Hill Book Company. Used by permission.*]

are presented as plots of C_D as a function of the Reynolds number in Figs. 13-27 and 13-28.

The laminar boundary-layer regime for the cylinder extends to a diameter Reynolds number of approximately 5×10^5, whereas for the sphere this is about 3×10^5. At the transition to turbulent boundary-layer flow there is a marked

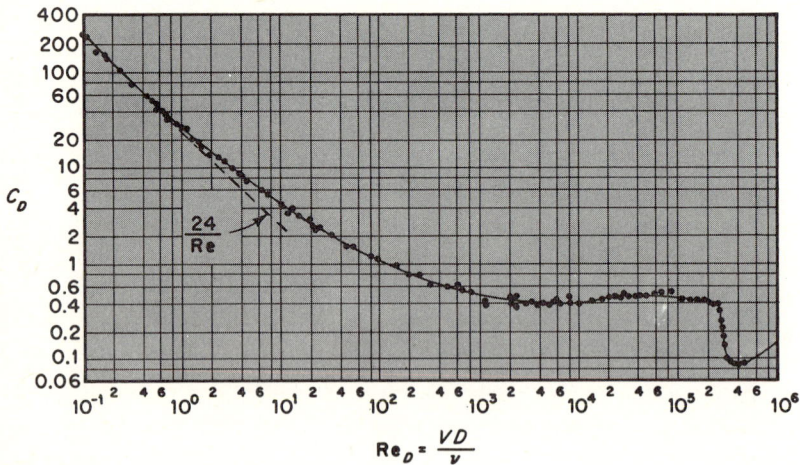

Fig. 13-28 Drag coefficients for spheres. (*Adapted from H. Schlichting, "Boundary Layer Theory," 6th ed. Copyright 1968. McGraw-Hill Book Company. Used by permission.*)

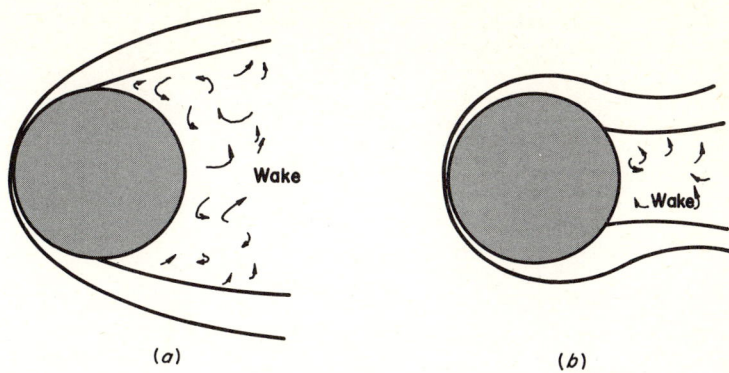

Fig. 13-29 Boundary-layer separation in (*a*) laminar and (*b*) turbulent flow over a sphere.

decrease in the drag coefficient as a result of the point of separation being further back along the body for turbulent than for laminar flow, as shown in Fig. 13-29. This effect was first demonstrated by Prandtl [26], who used a tripping wire mounted on the front half of a sphere to induce transition to turbulent flow with a resulting delayed separation and lowered drag coefficient.

This delay of separation in turbulent boundary-layer flow is readily understood when one considers both the cause of separation and the nature of turbulent flow. Separation occurs when the fluid near the surface does not have sufficient kinetic energy to overcome the pressure gradient. By its very nature the turbulent motion in the boundary layer adds kinetic energy to the fluid adjacent to the wall and thus delays separation. As a final comment, laminar flow over cylinders and spheres is more common than turbulent flow since these large transition Reynolds numbers are calculated with diameter as the significant length.

Drag coefficients for short, elliptical, square, and triangular cylinders as well as hemispherical shells, disks, and rectangular plates are given in *NACA Tech. Rept. 619,* 1938. The total drag for an object such as an elliptical cylinder or an airfoil is influenced to a significant extent by the viscous or skin-friction drag, whereas that for a completely bluff body (such as a disk) is not. Consequently, the drag coefficient for a streamline body is strongly dependent upon the Reynolds number, but that for a bluff body is essentially constant over a wide range of Reynolds numbers.

Two bluff bodies of importance are the circular disk and the rectangular plate. Drag coefficients for these configurations suitable for **Re** $> 10^3$ are presented in Table 13-3, which also contains drag coefficients for circular cylinders of finite length obtained at a Reynolds number of 10^5. The application of these latter data for other Reynolds numbers should be undertaken with caution, however, since the viscous drag on a cylinder may be significant.

Table 13-3 Drag coefficients for bluff bodies and short cylinders†

Configuration		L/D	$\text{Re} = \dfrac{V_\infty D}{\nu}$	C_D
Circular disk	$V_\infty \rightarrow$ ▯	0	$>10^3$	1.12
Rectangular plate, D = height L = length	$V_\infty \rightarrow$ ▯	1	$>10^3$	1.16
		5	$>10^3$	1.20
		20	$>10^3$	1.50
		∞	$>10^3$	1.90
Circular cylinder, D = diameter L = length	$V_\infty \rightarrow$ ◯	1	10^5	0.63
		5	10^5	0.74
		20	10^5	0.90
		∞	10^5	1.20

† Abstracted from *NACA Tech. Rept.* 619, 1938.

13-6 NONISOTHERMAL FLOW

In the remainder of this chapter turbulent viscous incompressible flow with heat transfer will be considered. This is a subject of considerable practical interest, since most convective-heat-transfer problems of industrial importance involve turbulent flow.

In Sec. 13-2 we discussed *apparent*, or *Reynolds*, *stresses* of the form

$$(\tau_{xy})_{\text{app}} = -\rho \overline{v'u'}$$

which arise as a consequence of the fluctuating turbulent velocity components. The reader will recall that these terms caused a marked increase in the mathematical difficulties associated with solution of the Navier-Stokes equations: the number of unknowns became larger than the number of equations. As a direct consequence, we employed the integral momentum equation in Sec. 13-5 for flow over a flat plate, and we have not attempted a solution for instantaneous turbulent velocity information.

To analyze nonisothermal forced-convection problems mathematically, it was shown in Chap. 12 that it is necessary to have velocity information to solve the energy equation. For the laminar flows of that chapter we found that integral techniques could be employed, and a reasonable representation of the profile of the average velocity was satisfactory.

For turbulent flow, however, the fluctuating velocity components also give rise to very large energy transfers, and thus a knowledge of average velocities is not adequate for solving the energy equation. Though many outstanding investigators have contributed to the problem, no one has yet succeeded by analysis alone in determining the heat transfer in turbulent flow. Fortunately it is possible to derive expressions for turbulent forced-convection heat transfer from hydrodynamic data, as first

TURBULENT FLOW OF INCOMPRESSIBLE VISCOUS FLUIDS

suggested by Reynolds. In the following section this approach will be used for flow over the flat plate.

13-7 REYNOLDS ANALOGY FOR TURBULENT FLOW

With the postulation of a laminar sublayer in turbulent flow (see Figs. 13-22 and 13-23), the heat transfer at the wall is by conduction only. Thus by Eq. (2-6) we have

$$q_s = -kA \frac{dT}{dy}\bigg|_{y=0}$$

and this also is valid throughout the sublayer. Also by Eq. (2-5) we have

$$\tau_s = \frac{\mu}{g_c} \frac{du}{dy}\bigg|_{y=0}$$

for this laminar two-dimensional flow. Combining these two expressions yields

$$\frac{q_s}{A} = -g_c \tau_s \frac{k}{\mu} \frac{dT}{du}\bigg|_{y=0} \tag{13-95}$$

While this expression is valid for the laminar sublayer, it is as yet of little use. The physical problem involves boundary conditions at the wall and outside the turbulent boundary layer. Hence we need an expression which can be integrated through the turbulent boundary layer.

Consider again the turbulent-flow model of Fig. 13-1 but with a thermal gradient. Then the fluid at y_1 has velocity u_1 and temperature T_1, and that at y_2 has velocity u_2 and temperature T_2. A vertical velocity perturbation v' results in a local mass transfer per unit area of $\rho v'$, and the resulting momentum exchange is

$$\tau_t = -\frac{\rho v'}{g_c}(u_1 - u_2) \tag{13-96}$$

where τ_t is a turbulent shear stress. This same perturbation causes a net heat transfer of

$$q_t = A \rho v' c_p (T_1 - T_2) \tag{13-97}$$

Combining these two expressions yields

$$\frac{q_t}{A} = -g_c \tau_t c_p \frac{T_1 - T_2}{u_1 - u_2}$$

or, in differential form,

$$\frac{q_t}{A} = -g_c \tau_t c_p \frac{dT}{du} \tag{13-98}$$

By comparing Eqs. (13-95) and (13-98) we find that

$$\left(\frac{q_s}{\tau_s}\right)_{\text{lam}} = \left(\frac{q_t}{\tau_t}\right)_{\text{turb}} \tag{13-99}$$

if $c_p = k/\mu$. This is true for a fluid with a Prandtl number of unity, a condition approximated by many real fluids.

To determine the heat transfer in turbulent flow over a surface we replace q_s/τ_s in Eq. (13-95) by its turbulent counterpart and k/μ by c_p, separate variables, and integrate

$$\frac{q_t}{A\tau_t c_p g_c} \int_0^{V_\infty} du = -\int_{T_s}^{T_\infty} dT \tag{13-100}$$

The limits for the integrations are established by the boundary conditions $u = 0$ and $T = T_s$ at the wall, and $u = V_\infty$ and $T = T_\infty$ outside the boundary layer. Carrying out the integrations yields

$$\frac{q_t V_\infty}{A\tau_t c_p g_c} = T_s - T_\infty \tag{13-101}$$

Rearranging, we obtain

$$\underbrace{\frac{q_t}{A(T_s - T_\infty)}}_{h_x} \frac{1}{c_p \rho V_\infty} = \underbrace{\frac{2 g_c \tau_t}{\rho V_\infty^2}}_{c_f} \frac{1}{2}$$

or

$$\frac{h_x}{c_p \rho V_\infty} = \frac{c_f}{2} \tag{13-102}$$

The left side of this equation is also the Nusselt number divided by the product of the Reynolds and Prandtl number. This is known as the *Stanton number*, and Eq. (13-102) is frequently expressed as

■ $$\frac{\text{Nu}_x}{\text{Re}_x \text{Pr}} = \text{St}_x = \frac{c_f}{2} \tag{13-103}$$

which is the result of the Reynolds analogy for flow over an external surface.

The most important restriction concerning the use of Eq. (13-103) is the Prandtl number limitation. It was obtained for a Prandtl number of unity. Colburn [4] has shown that by modifying this equation to

$$j_H \equiv \text{St}_x \text{Pr}^{\frac{2}{3}} = \frac{c_f}{2} \tag{13-104}$$

it is in accordance with experimental results for fluids with Prandtl numbers from 0.6 to 50; j_H is the *Colburn factor*, or simply the *j* factor, for heat transfer.

TURBULENT FLOW OF INCOMPRESSIBLE VISCOUS FLUIDS

In most practical problems an average heat-transfer coefficient is required to make total heat-transfer calculations possible. From Eq. (13-102) with c_p, ρ, and V_∞ constant, it is readily seen that the usual technique of averaging over a finite length would yield

$$\frac{\bar{h}}{c_p \rho V_\infty} = \frac{\bar{C}_f}{2} \tag{13-105}$$

where $\bar{C}_f$ is the average skin-friction coefficient. Also, Colburn's equation can be written for average values as

$$\blacksquare \quad \bar{j}_H \equiv \overline{\mathrm{St}}\mathrm{Pr}^{\frac{2}{3}} = \frac{\bar{C}_f}{2} \tag{13-106}$$

where $\bar{j}_H$ is the average Colburn factor. At this point it should be noted that Eqs. (13-104) and (13-106) are applicable for external flows in general. We now apply them to the familiar flat-plate geometry.

Flat plate To determine the local heat-transfer coefficient on a flat plate we substitute the local friction coefficient of Eq. (13-89) into (13-104) and obtain

$$\frac{\mathrm{Nu}_x}{\mathrm{Re}_x \mathrm{Pr}} \mathrm{Pr}^{\frac{2}{3}} = \frac{0.0576}{2\mathrm{Re}_x^{\frac{1}{5}}}$$

or

$$\mathrm{Nu}_x = \frac{h_x x}{k} = 0.0288 \mathrm{Pr}^{\frac{1}{3}} \mathrm{Re}_x^{0.8} \tag{13-107}$$

Likewise the average Nusselt number for a plate of length L is obtained by combining Eqs. (13-91) and (13-106). This gives

$$\mathrm{Nu}_L = \frac{\bar{h}L}{k} = 0.036 \mathrm{Pr}^{\frac{1}{3}} \mathrm{Re}_L^{0.8} \tag{13-108}$$

These last two results are based on the assumption of a turbulent boundary layer beginning at the leading edge of the plate. These are applicable to two cases: (1) very long plates with early transition resulting in a very short laminar leading edge and (2) the turbulent region only of short plates considering the critical length x_c to be the zero starting length of the plate. To determine an average heat-transfer coefficient for a plate with an appreciable laminar leading section, the appropriate laminar and turbulent expressions must be used. The average is

$$\bar{h} = \frac{\int_0^{x_c} h_\mathrm{lam}\, dx + \int_{x_c}^{L} h_\mathrm{turb}\, dx}{L} \tag{13-109}$$

where h_{lam} is obtained from Eq. (12-149) and h_{turb} is from Eq. (13-107). Substituting these expressions into (13-109) and integrating yields

■ $$\overline{\text{Nu}} = \frac{\bar{h}L}{k} = 0.036 \text{Pr}^{\frac{1}{3}}(\text{Re}_L^{0.8} - 23{,}200) \qquad (13\text{-}110)$$

for a critical Reynolds number of 5×10^5. This expression is recommended for turbulent flow over a flat plate with heating beginning at the leading edge.

For all heat-transfer calculations over external surfaces, the fluid properties should be evaluated at the film temperature defined by Eq. (12-172).

Example Commercial aniline at a free-stream temperature $T_\infty = 60°F$ flows at a velocity of 10 ft/sec over a flat plate 3 ft long heated to 140°F over its entire length. Calculate the average heat-transfer coefficient assuming that transition to turbulence occurs at $\text{Re}_L = 5 \times 10^5$.

Solution The film temperature is

$$T_f = \frac{T_\infty + T_s}{2} = \frac{60 + 140}{2} = 100°F$$

Properties of aniline from Appendix A are

$k = 0.10$ Btu/hr-ft-°F $\nu = 2.7 \times 10^{-5}$ ft²/sec $\text{Pr} = 30$

The Reynolds number at the end of the plate is

$$\text{Re}_L = \frac{(10 \text{ ft/sec})(3 \text{ ft})}{2.7 \times 10^{-5} \text{ ft}^2/\text{sec}} = 11.1 \times 10^5$$

and we conclude that transition will occur. (Slightly over half the plate is subjected to a turbulent boundary layer.) By Eq. (13-110)

$$\frac{\bar{h}L}{k} = 0.036(30^{\frac{1}{3}})[(11.1 \times 10^5)^{0.8} - 23{,}200] = 5{,}075$$

Thus

$$\bar{h} = \frac{5{,}075 \times 0.1}{3 \text{ ft}} \text{ Btu/hr-ft-°F} = 169 \text{ Btu/hr-ft}^2\text{-°F}$$

It should be noted that Eq. (13-108) would not be suitable for this problem since a significant part of the plate experiences laminar flow.

13-8 CONVECTIVE HEAT TRANSPORT IN TUBES

The analytical prediction of heat transfer for turbulent flow inside tubes is impeded by the same difficulties encountered in flow over external surfaces; specifically, the convective terms are highly dependent upon the fluctuating velocity components. Again we shall resort to the analogy between momentum and energy transport. Historically, such analogies were first forwarded for the case of turbulent flow through tubes and later extended to the case of external flow.

The development of Eqs. (13-95) to (13-99) is independent of whether the flow is external or internal, and these equations are applicable for heat transport inside tubes.

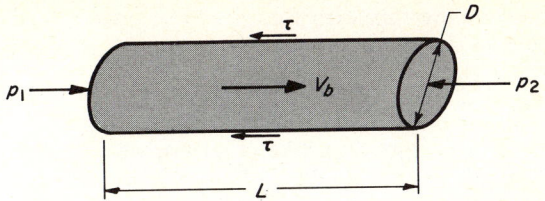

Fig. 13-30 Forces acting on a fluid element in motion in a tube.

Integrating the turbulent counterpart of Eq. (13-95) from the tube wall, where $u = 0$ and $T = T_s$, to the bulk of the fluid, where $u = V_b$ and $T = T_b$, we have

$$\frac{q_t}{A\tau_t c_p g_c} \int_0^{V_b} du = -\int_{T_s}^{T_b} dT \tag{13-111}$$

or

$$\frac{q_t V_b}{A\tau_t c_p g_c} = T_s - T_b \tag{13-112}$$

Rearrangement of Eq. (13-112) yields

$$\frac{q_t}{A(T_s - T_b)} \frac{1}{c_p \rho V_b} = \frac{2 g_c}{\rho V_b^2} \frac{\tau_t}{2}$$

or

$$\text{St} = \frac{g_c}{\rho V_b^2} \tau_t \tag{13-113}$$

Consider a force balance on a fluid element within a tube as shown in Fig. 13-30. The motion of the fluid element gives rise to the shear stress τ along the sides, this being τ_t for a turbulent flow. The force balance is

$$\tau_t \pi D L = (p_1 - p_2) \frac{\pi D^2}{4}$$

and consequently

$$\tau_t = \frac{(p_1 - p_2) D}{4L} \tag{13-114}$$

Obtaining an expression for $p_1 - p_2$ from Eq. (13-60), the turbulent shear stress can be written as

$$\tau_t = f \frac{\rho V^2}{8 g_c} \tag{13-115}$$

Combining this with Eq. (13-113), we obtain

■ $$\overline{St} = \frac{f}{8} \qquad (13\text{-}116)$$

This equation, like (13-103), was obtained for a fluid with a Prandtl number of unity. The Colburn modification is also applicable to internal tube flow, and this yields

■ $$\overline{St}Pr^{\frac{2}{3}} = \frac{f}{8} \qquad (13\text{-}117)$$

Equation (13-117) is valid for fluids with Prandtl numbers from 0.5 to 100 according to Colburn.

The empirical expression for the friction factor for turbulent flow in smooth pipes or tubes as given by Eq. (13-62) is

$$f = 0.184 Re_D^{-0.2}$$

Substituting this value into Eq. (13-117) yields

$$\overline{St} = 0.023 Pr^{-\frac{2}{3}} Re_D^{-\frac{1}{5}} \qquad (13\text{-}118)$$

which is known as the *Colburn equation*. The average Nusselt number can be obtained by multiplying both sides of this expression by the product of the Reynolds and Prandtl number:

■ $$\overline{Nu}_D = \frac{\bar{h}D}{k} = 0.023 Pr^{\frac{1}{3}} Re_D^{0.8} \qquad (13\text{-}119)$$

The ranges of validity are

$$10{,}000 < Re_D < 100{,}000$$

$$0.5 < Pr < 100$$

$$\frac{L}{D} > 60$$

Heating or cooling of fluid

and all fluid properties except the specific heat are to be evaluated at the average *film temperature*, $(T_s + T_b)/2$. The specific heat is evaluated at the fluid *mean bulk temperature*, $(T_i + T_e)/2$.

Entrance-region modifications The restriction on the last equation concerning the ratio L/D is a result of the development of the velocity and temperature profiles in the tube. As discussed in Sec. 13-4, the development of the turbulent velocity profile over the entire tube cross section usually requires a length of 40 to 50 tube diameters. The velocity profile in the sublayer, however, develops much more rapidly. This is

Table 13-4 Values of the coefficient C in Eq. (13-121)

$(L/D > 5; 26{,}000 < \mathbf{Re}_D < 56{,}000)$

Inlet configuration	C
Bell-mouthed with screen	1.4
Calming section, $L/D = 11.2$	1.4
$L/D = 2.8$	3.0
45° bend	5.0
90° bend	7.0

shown by Eq. (13-63), which indicates the distance required for development of a constant value for the friction factor.

It is this development of both the thermal and velocity profiles in the entry region which must be considered in short tubes. As a rule of thumb, the average heat transfer is not markedly dependent upon length if the dimensionless parameter L/D is greater than 60.

In many cases L/D is less than 60. Latzko [16] recommends the following expression for L/D less than that predicted by Eq. (13-63):

$$\frac{\bar{h}_e}{\bar{h}} = 1.11 \left[\frac{\mathbf{Re}_D^{\frac{1}{5}}}{(L/D)^{\frac{4}{5}}} \right]^{0.275} \qquad (13\text{-}120)$$

In this expression $\bar{h}_e$ is the average heat-transfer coefficient in the entry region, $L < L_c$, and $\bar{h}$ is the asymptotic heat-transfer coefficient for the fully developed flow.

For L/D greater than $(L/D)_c$ but less than 60, Latzko recommends

$$\frac{\bar{h}_e}{\bar{h}} = 1 + \frac{C}{L/D} \qquad (13\text{-}121)$$

where C is a weak function of the Reynolds number. Boelter et al. [1] obtained experimental values for C over a Reynolds number range from 26,000 to 56,000. These are summarized in Table 13-4.

Experimental correlation equations One of the most widely used correlation equations for heat transfer inside tubes with constant wall temperature is the Dittus-Boelter equation

$$\overline{\mathbf{Nu}}_D = \frac{\bar{h}D}{k} = 0.023 \mathbf{Pr}^n \mathbf{Re}_D^{0.8} \qquad (13\text{-}122)$$

where

$$n = \begin{cases} 0.4 & \text{heating the fluid} \\ 0.3 & \text{cooling the fluid} \end{cases}$$

The recommended ranges of application are

$$10{,}000 < \mathbf{Re}_D < 100{,}000$$
$$0.7 < \mathbf{Pr} < 120$$
$$\frac{L}{D} > 60$$

and all fluid properties are evaluated at the *mean bulk temperature* of the fluid. The careful student will note the change in temperature between Eqs. (13-119) and (13-122) for evaluation of fluid properties.

Another widely recognized experimental equation is the modification of Eq. (13-119) due to Sieder and Tate. This is for high Prandtl number fluids and accounts for the large changes in viscosity which occur from the tube wall to the bulk of the fluid. Their equation is

$$\overline{\mathbf{Nu}}_D = \frac{\bar{h}D}{k} = 0.023 \mathbf{Pr}^{\frac{1}{3}} \mathbf{Re}_D^{0.8} \left(\frac{\mu_b}{\mu_w}\right)^{0.14} \qquad (13\text{-}123)$$

and the restrictions are

$$10{,}000 < \mathbf{Re}_D < 100{,}000$$
$$0.7 < \mathbf{Pr} < 16{,}700$$
$$\frac{L}{D} > 60$$

Heating or cooling of fluid

All fluid properties with the exception of μ_w are evaluated at the fluid arithmetic bulk temperature; μ_w is evaluated at the wall temperature.

Example Calculate the length of 3-in-ID smooth tubing required to raise the bulk temperature of light oil from 60 to 100°F if the wall temperature is constant at 200°F and the fluid average velocity is 30 ft/sec.

Solution The arithmetic mean bulk fluid temperature is

$$T_{b,\text{av}} = \frac{60 + 100}{2} = 80°F$$

From Appendix A the fluid properties are

$c_{p,b} = 0.44 \text{ Btu/lb}_m\text{-°F}$

$k_b = 0.077 \text{ Btu/hr-ft-°F}$ $\qquad \mu_w = 250 \times 10^{-5} \text{ lb}_m/\text{ft-sec}$

$\mathbf{Pr}_b = 570$ $\qquad \nu_b = 49 \times 10^{-5} \text{ ft}^2/\text{sec}$

$\mu_b = 2{,}780 \times 10^{-5} \text{ lb}_m/\text{ft-sec} \qquad \rho_b = 56.8 \text{ lb}_m/\text{ft}^3$

Since the Prandtl number is large, the Sieder-Tate equation should be used. The Reynolds number is

$$\mathbf{Re}_{D,b} = \frac{(\tfrac{3}{12} \text{ ft})(30 \text{ ft/sec})}{49 \times 10^{-5} \text{ ft}^2/\text{sec}} = 15{,}300$$

TURBULENT FLOW OF INCOMPRESSIBLE VISCOUS FLUIDS

which is also within the range of validity of Eq. (13-123). Thus

$$\bar{h} = \frac{k_b}{D} 0.023 \, \mathrm{Pr}_b^{\frac{1}{3}} \mathrm{Re}_D^{0.8} \left(\frac{\mu_b}{\mu_w}\right)^{0.14}$$

$$= \frac{0.077 \, \mathrm{Btu/hr\text{-}ft\text{-}°F}}{\frac{3}{12} \, \mathrm{ft}} (0.023)(570^{\frac{1}{3}})(15{,}300^{0.8}) \left(\frac{2{,}780 \times 10^{-5}}{250 \times 10^{-5}}\right)^{0.14}$$

$$= 183.2 \, \mathrm{Btu/hr\text{-}ft^2\text{-}°F}$$

Assuming the simple model of a linear increase in the fluid bulk temperature with distance, an energy balance on the fluid passing through a control volume of length L and diameter D is

Heat transfer from tube = energy gained by the fluid

$$\bar{h}(\pi DL)(T_w - T_{b,\mathrm{av}}) = \rho \frac{\pi D^2}{4} V c_{p,b} \Delta T_b$$

Thus

$$L = \frac{\rho D V c_p (T_{b,\mathrm{outlet}} - T_{b,\mathrm{inlet}})}{4\bar{h}(T_w - T_{b,\mathrm{av}})}$$

$$= \frac{(56.8 \, \mathrm{lb}_m/\mathrm{ft}^3)(\frac{3}{12} \, \mathrm{ft})(30 \, \mathrm{ft/sec})(0.44 \, \mathrm{Btu/lb}_m\text{-}°F)(40°F)}{4(183.2 \, \mathrm{Btu/hr\text{-}ft^2\text{-}°F})(\mathrm{hr}/3{,}600 \, \mathrm{sec})(120°F)} = 307 \, \mathrm{ft}$$

and obviously with this very long pipe $L/D \gg 60$.

Transition-region heat transfer The heat transfer in the *transition* flow regime, $2{,}100 < \mathrm{Re}_D < 10{,}000$, has been investigated by Sieder and Tate [31], who recommended the family of correlation curves of Fig. 13-31. The very nature of transition

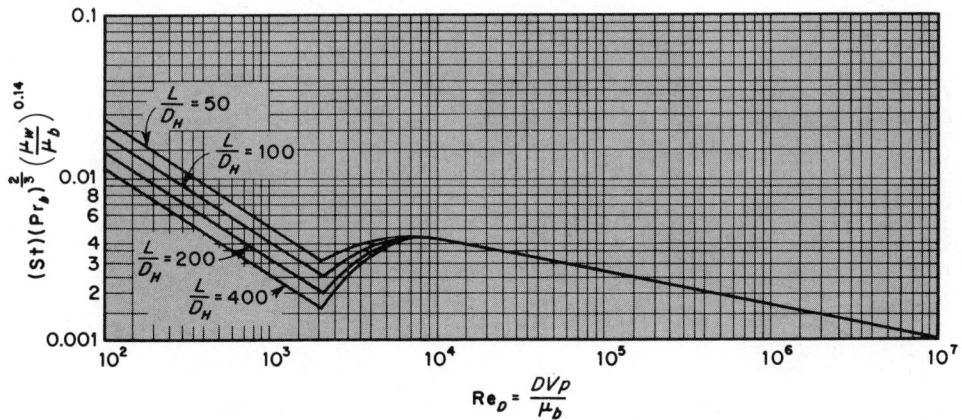

Fig. 13-31 Heat transfer in the transition regime. [*Adapted from E. N. Sieder and C. E. Tate, Ind. Eng. Chem.*, **28**:1429 (1936). Copyright 1936 by the American Chemical Society. Reprinted by permission of the copyright owner.]

flow, however, prevents such experimental data from being universally applicable, and calculations in this flow regime are not very reliable.

Noncircular ducts Heat-transfer calculations for flow inside noncircular ducts can be made with the equations of this section. The duct diameter must be replaced with the hydraulic diameter defined by Eq. (12-49). The results using this approach must be considered as approximate. For more accurate methods, the reader should consult Knudsen and Katz [15].

13-9 ENGINEERING CORRELATIONS FOR EXTERNAL FLOW[1]

Heat transfer to or from bodies subjected to forced convective external flow is encountered in many practical applications. Many heat-exchanger designs depend upon external convective heat transfer from cylinders subjected to cross flow, and heat transfer in packed beds and fluidized systems frequently involves convective heat transport from spheres.

From our discussion of external flow around bodies in Sec. 13-5, it is readily apparent that calculations of heat transfer solely by analysis would be most difficult. The flow field for the simplest configuration usually involves separation; such a flow field defies attempts at mathematical modeling or representation. It is also this problem of separation that invalidates attempts to apply the Reynolds analogy. The Reynolds analogy permits calculation of heat transfer from the *skin-friction factor*. It does not apply to the total measured drag-loss coefficient which involves form drag due to separation. As a consequence, heat-transfer calculations for problems of this type are based on empirical correlation equations, some of the more useful of which will be discussed in the following subsections.

The single cylinder in cross flow Heat transfer from a single heated cylinder subjected to a cross flow of air at a uniform approach temperature and velocity has been investigated by Giedt [8]. His experimental data are summarized in Fig. 13-32. There are several important points to be observed with respect to this figure. An examination of either of the curves for a single Reynolds number reveals that the local heat flux is highly dependent upon position. The front portion (approximately 80° in these experiments) was subjected to a laminar boundary layer with a resulting high initial heat flux that diminishes with distance from the stagnation point. For the lowest Reynolds number experiment, there is a minimum value of the Nusselt modulus at the point of boundary-layer separation. The increasing heat transfer in the wake region is attributable to the high eddy-velocity terms there. At the higher Reynolds numbers, there are two minimum points in each curve. The first of these occurs where the boundary-layer flow becomes turbulent and then there

---

[1] In much of this section the boundary-layer flow is laminar. Since most of these correlations are highly dependent upon separation, this topic has been included in turbulent flow.

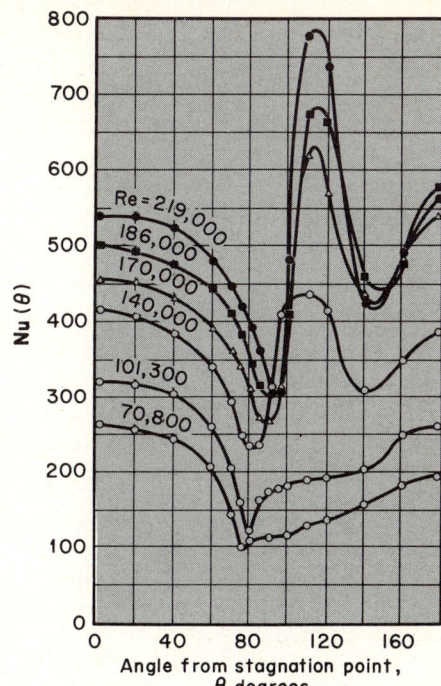

Fig. 13-32 Local Nusselt number for a circular cylinder in a cross flow of air. [*Adapted from W. H. Giedt, Trans. ASME,* **71**:378 (1949).]

results a very large increase in heat transfer. Farther back along the cylinder, separation occurs, causing the second minimum, which is followed by an increase in the wake region.

From these curves it is evident that the average value of the Nusselt number for a complete cylinder is a function of the Reynolds number. The experimental data of numerous investigators are presented in Fig. 13-33 for air flowing normal to a single cylinder. These data are well represented by several short straight-line segments of the form

$$\overline{\mathbf{Nu}}_{Df} = \frac{\bar{h}D}{k_f} = C_1 \left(\frac{V_\infty D}{\nu_f}\right)^n \tag{13-124}$$

This type of equation can be modified for fluids with a Prandtl number other than 0.72 by inclusion of the Prandtl number to the $\frac{1}{3}$ power. This results in

$$\overline{\mathbf{Nu}}_{Df} = \frac{\bar{h}D}{k_f} = C_2 \mathbf{Pr}_f^{\frac{1}{3}} \mathbf{Re}_{Df}{}^n \tag{13-125}$$

The values of C_1, C_2, and n for representation of a Reynolds number flow range of 0.4 to 400,000 with five straight-line segments are given in Table 13-5.

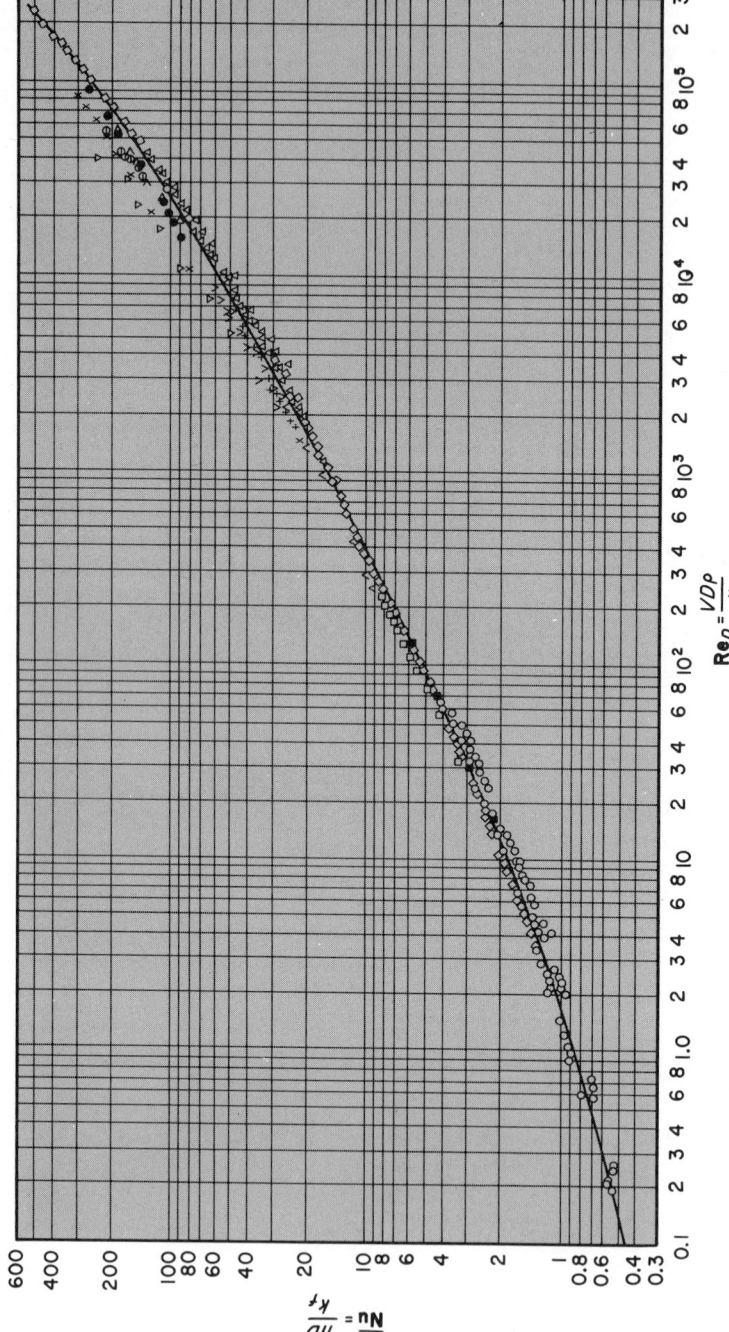

Fig. 13-33 Experimental data for heating and cooling air flowing normal to a single cylinder. (*Adapted from W. H. McAdams, "Heat Transmission," 3d ed., p. 259. Copyright 1954. McGraw-Hill Book Company. Used by permission.*)

TURBULENT FLOW OF INCOMPRESSIBLE VISCOUS FLUIDS

Table 13-5 Constants for Eqs. (13-124) and (13-125)

Re_{Df}	C_1	C_2	n
0.4–4	0.891	0.989	0.330
4–40	0.821	0.911	0.385
40–4,000	0.615	0.683	0.466
4,000–40,000	0.174	0.193	0.618
40,000–400,000	0.0239	0.0266	0.805

The reader should note that all fluid properties for use in these two equations are to be evaluated at the film temperature. Also, the free-stream turbulent intensity should not exceed 2 percent for the constants of Table 13-5 to remain valid.

For heat transfer to single cylinders of other than circular configuration, Jakob [12] has summarized values of C_1 and n for use with Eq. (13-124) for several configurations. These are presented in Table 13-6.

Example Determine the heat transfer per foot of length to a ½-in circular refrigerant line at 0°F subjected to a cross flow of 60°F air at 70 ft/sec.

Solution Equation (13-124) is applicable to either heating or cooling airflow across a singular cylinder. The fluid properties of air are to be evaluated at

$$T_f = \frac{60 + 0}{2} = 30°F$$

Since Appendix A contains air-property data at 32°F, these will be used. Thus

$\nu_f = 0.145 \times 10^{-3}$ ft²/sec $k_f = 0.0140$ Btu/hr-ft-°F $Pr_f = 0.72$

Table 13-6 Constants for noncircular cylinders in cross flow of air†

Configuration	Re_{Df}	C_1	n
$V_\infty \rightarrow$ ◇ D	$5 \times 10^3 - 10^5$	0.222	0.588
$V_\infty \rightarrow$ ▢ D	$5 \times 10^3 - 10^5$	0.092	0.675
$V_\infty \rightarrow$ ⬡ D	$5 \times 10^3 - 1.95 \times 10^4$ $1.95 \times 10^4 - 10^5$	0.144 0.0347	0.638 0.782
$V_\infty \rightarrow$ ⬢ D	$5 \times 10^3 - 10^5$	0.138	0.638
$V_\infty \rightarrow$ ∣ D	$4 \times 10^3 - 1.5 \times 10^4$	0.205	0.731

† From M. Jakob, "Heat Transfer," Vol. I, John Wiley & Sons, Inc., New York, 1959, by permission.

The Reynolds number is

$$\mathbf{Re}_{Df} = \frac{(\frac{1}{24} \text{ ft})(70 \text{ ft/sec})}{0.145 \times 10^{-3} \text{ ft}^2/\text{sec}} = 20{,}100$$

From Table 13-5

$C_1 = 0.174 \qquad n = 0.618$

Substituting into Eq. (13-124),

$$\bar{h} = 0.174 \frac{k_f}{D}(20{,}100^{0.618})$$

$$= 0.174 \frac{0.0140 \text{ Btu/hr-ft-}°\text{F}}{\frac{1}{24} \text{ ft}} 20{,}100^{0.618}$$

$$= 26.9 \text{ Btu/hr-ft}^2\text{-}°\text{F}$$

$$q = \bar{h}A(T_w - T_\infty)$$

$$= \left(26.9 \frac{\text{Btu}}{\text{hr-ft}^2\text{-}°\text{F}}\right)\left[\frac{\pi}{24}(1 \text{ ft}^2)(0 - 60°\text{F})\right] = -211 \text{ Btu/hr}$$

Note that the heat flux is negative; this is in agreement with usual thermodynamic convention since q represents the heat-transfer rate to the air.

Heat transfer to a single sphere The phenomenon of flow over a sphere is similar to that over a cylinder in cross flow in that the boundary layer usually experiences separation. Consequently heat-transfer predictions for this configuration are also based upon empirical equations.

For the flow of air over a single sphere, McAdams [19] recommends

$$\frac{\bar{h}D}{k_f} = 0.37\left(\frac{V_\infty D}{\nu_f}\right)^{0.6} \tag{13-126}$$

where

$$17 < \frac{V_\infty D}{\nu_f} < 70{,}000$$

Note that all properties are evaluated at the film temperature. This same expression can be used for other gases since the Prandtl number does not usually differ greatly from that of air.

For liquid flow over a single sphere, Vliet and Leppert [32] recommend

$$\frac{\bar{h}D}{k_\infty} = \left[1.2 + 0.53\left(\frac{V_\infty D}{\nu_\infty}\right)^{0.54}\right]\mathbf{Pr}^{0.3}\left(\frac{\mu_\infty}{\mu_w}\right)^{0.25} \tag{13-127}$$

where

$$1 < \frac{V_\infty D}{\nu_\infty} < 200{,}000$$

In this expression all properties except μ_w are evaluated at the free-stream temperature.

TURBULENT FLOW OF INCOMPRESSIBLE VISCOUS FLUIDS

For very slow flow over a sphere, Johnston et al. [13] have shown theoretically that

$$\mathbf{Nu}_D \rightarrow 2 \tag{13-128}$$

for Reynolds numbers less than 1 and a Prandtl number of unity unless the sphere diameter is so small that it is of the order of the mean free path of the molecules in the gas.

Cross flow over tube bundles Banks or bundles of closely spaced cylinders subjected to a normal cross flow are frequently employed in heat-exchanger designs. A typical bundle is shown in Fig. 13-34. From our previous discussion of flow over a single cylinder, we would expect the wake flow of an upstream body to have a pronounced effect upon the flow field of a closely located downstream object. This in turn influences the heat transfer to or from the downstream body.

Two general geometric classifications are widely used in heat-exchanger designs. One consists of rows (parallel with direction of V_∞) of in-line tubes; the other has staggered tube rows. They are illustrated, together with geometric nomenclature, in Fig. 13-35.

Grimson [9] has evaluated the results of several investigators and found that the heat-transfer coefficient can be obtained from Eq. (13-124) with C_1 and n as given in Table 13-7. As we would expect, the heat transfer is dependent upon dimensionless geometric ratios of tube spacing to diameter and whether the tubes are in line or staggered.

The Reynolds number in a tube bank for use in Eq. (13-124) is

$$\mathbf{Re}_{\max} = \frac{V_{\max} D}{\nu}$$

where $V_{\max}$ is the maximum velocity in the bank. This occurs at the minimum flow passage, which is either between two adjacent tubes in a row or between two diagonally opposed tubes (staggered tube banks only). Per unit depth this is the smaller of $S_t - D$ or $\sqrt{S_t^2 + S_l^2} - D$, and the maximum velocity is $V_\infty S_t$ divided by this smaller number. The value of V_∞ is obtained by assuming no tubes in the shell container, i.e., free-passage flow.

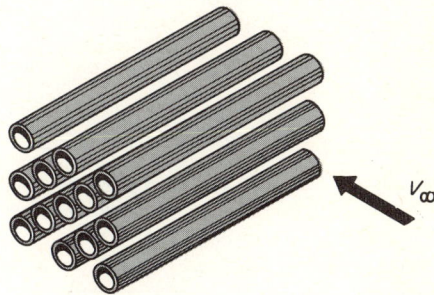

Fig. 13-34 Tube bundle in cross flow.

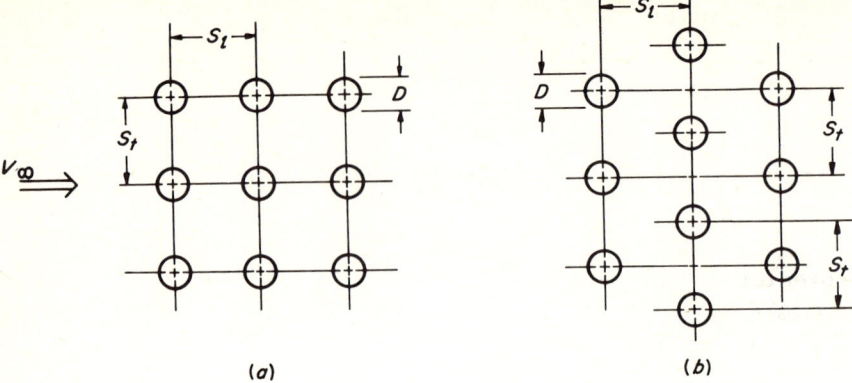

Fig. 13-35 Geometric arrangements for tube bundles: (*a*) in-line tube bundle; (*b*) staggered tube bundle.

The Nusselt modulus obtained with Eq. (13-124) and the appropriate constant and exponent of Table 13-7 are satisfactory for tube banks which are 10 or more tubes deep (in the direction of V_∞). For fewer tubes, the average Nusselt modulus is affected less by the lower turbulence in the flow over the first few tubes. Kays and Lo [14] have obtained correction coefficients for less than 10 tubes, given in Table 13-8. No change in the average heat-transfer coefficient is discernible after the tenth tube in a row.

Table 13-7 Coefficients for calculation of heat transfer from tube bundles 10 or more rows deep†

$\dfrac{S_l}{D}$	S_t/D							
	1.25		1.5		2		3	
	C_1	n	C_1	n	C_1	n	C_1	n
In-line tubes:								
1.25	0.348	0.592	0.275	0.608	0.100	0.704	0.0633	0.752
1.5	0.367	0.586	0.250	0.620	0.101	0.702	0.0678	0.744
2	0.418	0.570	0.299	0.602	0.229	0.632	0.198	0.648
3	0.290	0.601	0.357	0.584	0.374	0.581	0.286	0.608
Staggered tubes:								
0.6							0.213	0.636
0.9					0.446	0.571	0.401	0.581
1			0.497	0.558				
1.125					0.478	0.565	0.518	0.560
1.25	0.518	0.556	0.505	0.554	0.519	0.556	0.522	0.562
1.5	0.451	0.568	0.460	0.562	0.452	0.568	0.488	0.568
2	0.404	0.572	0.416	0.568	0.482	0.556	0.449	0.570
3	0.310	0.592	0.356	0.580	0.440	0.562	0.421	0.574

† Data from Grimson [9].

TURBULENT FLOW OF INCOMPRESSIBLE VISCOUS FLUIDS

Table 13-8 Ratio of $\bar{h}/\bar{h}_{10}$†

	\multicolumn{10}{c}{Number of tubes}
	1	2	3	4	5	6	7	8	9	10
Staggered	0.68	0.75	0.83	0.89	0.92	0.95	0.97	0.98	0.99	1.0
In line	0.64	0.80	0.87	0.90	0.92	0.94	0.96	0.98	0.99	1.0

† Data from Kays and Lo [14].

13-10 HEAT TRANSFER TO LIQUID METALS

Liquid metals are particularly well suited for use as convective fluids when large energy-transfer rates must be achieved in relatively small spaces. This situation is encountered in nuclear reactors, and the advent of nuclear power plants has spurred interest in liquid-metal heat transfer. Liquid metals frequently used include mercury, sodium, and lead-bismuth alloys. In general, liquid metals have the advantages of high thermal capacity (product of specific heat and density), low viscosity, and low vapor pressure at high temperature; i.e. they remain in the liquid state at high temperature. In addition to toxicity, they have other disadvantages: some react violently with air or water; most solidify at room temperature; and all are relatively expensive. These disadvantages must be considered in any application, and specially designed pumps, valves, and tubing in addition to elaborate safety precautions are usually required. Nonetheless, their advantages justify their use in certain applications, and suitable handling techniques and equipment have been developed.

The mechanism of heat transfer to a flowing liquid metal is significantly different from that to other fluids. An examination of the physical properties in Appendix A reveals that the thermal conductivities are very high and that the Prandtl numbers range from 0.004 to 0.027. This results in a very thick thermal boundary layer. Recall that for laminar flow over a flat plate, the relationship between the velocity and thermal boundary layers is

$$\delta_t = \frac{\delta}{\mathbf{Pr}^{\frac{1}{3}}} \qquad (12\text{-}153)$$

and the low Prandtl numbers of liquid metals result in boundary layers typified by Fig. 13-36.

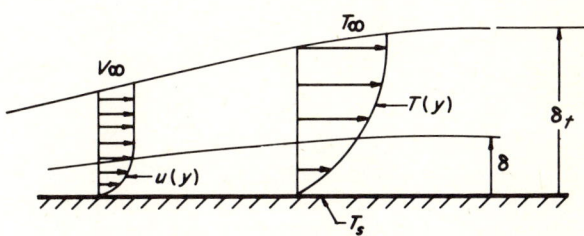

Fig. 13-36 Typical thermal- and velocity-boundary-layer profiles for liquid-metal heat transfer.

Flat-plate analysis Because the thermal conductivity is so high, the primary mode of energy transfer in the boundary layer is by conduction. This is true for both laminar and turbulent flow, and for purposes of analysis we may assume that turbulent eddies add little to the transport mechanism. Consequently a laminar analysis such as the Pohlhausen solution of Sec. 12-8 is valid for laminar or turbulent flow of liquid metals over a flat plate.

Recall that Pohlhausen's numerical solution of Eq. (12-145) can be expressed as $0.332\mathrm{Pr}^{\frac{1}{3}}$ only in the range of $0.5 < \mathrm{Pr} < 10$. This result does not hold for the very low Prandtl numbers encountered in liquid-metal heat transfer. Grosh and Cess [10] numerically integrated Eq. (12-145) for the low Prandtl number regime and compared the results with a pure conduction solution obtained as follows.

Since the velocity profile is flat over a large portion of the thermal boundary layer, slug flow is a reasonable approximation. Then the energy equation (12-135) reduces to

$$V_\infty \frac{\partial T}{\partial x} = \alpha \frac{\partial^2 T}{\partial y^2} \tag{13-129}$$

With V_∞ and α considered constants, this is a partial differential equation treated in the conduction-heat-transfer literature. The solution from Ref. 3 for a constant surface temperature is

$$\frac{q}{A} = k(T_s - T_\infty)\left(\frac{V_\infty}{\pi \alpha x}\right)^{\frac{1}{2}} \tag{13-130}$$

and since this heat flux is convected away by the liquid metal, we have

$$\frac{q}{A} = h_x(T_s - T_\infty) = k(T_s - T_\infty)\left(\frac{V_\infty}{\pi \alpha x}\right)^{\frac{1}{2}} \tag{13-131}$$

Thus

$$\frac{h_x x}{k} = \left(\frac{V_\infty k}{\pi \alpha}\right)^{\frac{1}{2}} = \left(\frac{1}{\pi}\right)^{\frac{1}{2}} \left(\frac{V_\infty x}{\nu}\right)^{\frac{1}{2}} \left(\frac{\nu}{\alpha}\right)^{\frac{1}{2}}$$

or

$$\mathrm{Nu}_x = \frac{\sqrt{\mathrm{Re}_x \mathrm{Pr}}}{\sqrt{\pi}} = 0.564\sqrt{\mathrm{Pe}_x} \tag{13-132}$$

where the local Peclet number Pe_x is the product of the local Reynolds number and the fluid Prandtl number. From a physical viewpoint, the Peclet number can be thought of as the ratio of energy transport by convection to energy transport by conduction.

Grosh and Cess compared their numerical results (Pohlhausen type of solution) with the simplified approach [Eq. (13-132)] and found the latter to be 7 to 12 percent high over the range $0.005 < \mathrm{Pr} < 0.025$. They also considered the effects of longitudinal conduction (since the fluids under consideration have very high thermal conductivity) and presented an example which indicates that this is insignificant for $\mathbf{Pr}_L > 50$.

The same authors treated the case of a flat plate with constant heat flux and a circular cylinder in cross flow with either surface tempe ature or heat flux constant. For details the reader should consult Ref. 10.

Correlation equations From the preceding discussion of the mechanism of heat transfer with these fluids, it is apparent that the usual correlation equations for gases and moderate to high Prandtl number liquids do not apply. Most design equations for liquid-metal heat transfer involve flow inside tubes or cross flow over tube banks. We shall now discuss some of the more frequently used correlations.

Cross flow over tube banks Hoe, Dropkin, and Dwyer [11] reported experimental results with a staggered tube bank 10 rows deep of $\frac{1}{2}$-in-OD tubes arranged in an equilateral-triangular array with a 1.375 pitch-to-diameter ratio. Liquid mercury flowed normal to the tubes. Their correlation for the Nusselt modulus in the interior of the tube bank is

$$\overline{\mathbf{Nu}}_D = 4.03 + 0.228(\mathbf{Re}_{\max}\mathbf{Pr})^{0.67} \tag{13-133}$$

where

$$20{,}000 < \mathbf{Re}_{\max} < 80{,}000 \quad \text{and} \quad \mathbf{Pr} = 0.022 \text{ (mercury)}$$

and fluid properties are evaluated at the film temperature T_f. The maximum Reynolds number is based on the tube OD and the maximum velocity in the passage, i.e., the velocity at the minimum flow area between tubes.

Flow inside tubes Lyon [18] gave a simplified equation which represents Martinelli's analytical results for smooth tubes with constant wall flux. This is

$$\overline{\mathbf{Nu}}_D = 7 + 0.025\mathbf{Pe}_D^{0.8} \tag{13-134}$$

where

$$\frac{L}{D} > 60 \quad \text{and} \quad \mathbf{Pe} > 100$$

The fluid properties are evaluated at the average fluid bulk temperature. Another equation which fits some of the experimental data better than Eq. (13-134) is that of Lubarsky and Kaufman [17]

$$\overline{\mathbf{Nu}}_D = 0.625\mathbf{Pe}_D^{0.4} \tag{13-135}$$

and this is recommended for

$$\frac{L}{D} > 60 \quad 100 < \mathbf{Pe}_D < 10{,}000$$

with all fluid properties evaluated at the average fluid bulk temperature.

For constant wall temperature, Seban and Shimazaki [28] proposed

$$\overline{Nu}_D = 5 + 0.025 Pe_D^{0.8} \qquad (13\text{-}136)$$

where the limitations and temperature for property evaluation are the same as for Lyon's equation. Whether constant heat flux or constant wall temperature is more appropriate depends upon several factors, one of the more important being the ratio of the flow heat capacity to the wall heat capacity. Generally speaking, the heat capacity of liquid metals is high, and the wall temperature may be altered significantly by the fluid.

The reader is cautioned that larger deviations from predicted results occur with liquid metals than with other fluids. No one equation appears to be suitable for general use, and errors as large as 80 percent are not uncommon. Data scatter in experimental results is extremely bad for condensation and boiling phenomena and when nonwetting of the tube surface by the liquid metal occurs. Also, the entry-length modifications of Sec. 13-8 do not apply to liquid-metal heat transfer. The entry regions are somewhat shorter with liquid metals because of the high thermal conductivity and relative unimportance of turbulent exchange.

PROBLEMS

13-1. A fluid particle at $y - l$ moves to y. Show that $|u'| = l|d\bar{u}/dy|$.

13-2. Recall that the Reynolds stresses are generally an order of magnitude larger than the laminar viscous stresses. In terms of ν, determine an expression for the constant k in the mixing length $\ell = ky$ for the velocity distribution $\bar{u} = 28(y/\delta)^{\frac{1}{8}}$ ft/sec. Assume the turbulent stresses to be 10 times greater than the laminar viscous stresses.

13-3. Assuming a linear variation of the mixing length ($\ell = ky$, where y is the distance from the wall) and a velocity given by $\bar{u} = 14(y/\delta)^{\frac{1}{7}}$, approximate the apparent stress on a control volume at $y/\delta = 0.5$ for $k = 0.4$.

13-4. The term v_* is commonly referred to as the shear velocity and is given by $v_* = \sqrt{\tau_{app}/\rho}$. For a linear expression of the mixing length $\ell = ky$, show that $\bar{u} = (1/k)v_* \ln y + \text{const}$.

13-5. Water at 80°F flows with an average velocity of 20 ft/sec in a 4-in-ID pipe. Calculate the friction factor ± 0.0005.

13-6. Given $\bar{u}_{max} - \bar{u} = 2.5v_* \ln (R/y)$ [Eq. (13-37)], show that $\bar{u}_{avg} = \bar{u}_{max} - 3.75v_*$ [Eq. (13-47)].

13-7. Freon at 60°F flows through a 3-in-ID pipe with a maximum velocity of 25 ft/sec and a measured wall shear stress of 0.82 lb$_f$/ft². What is the velocity 1 in from the centerline?

13-8. Air at 70°F flows through a 6-in-diam smooth pipe. The velocity profile at a point along the pipe where the flow is fully developed is given by $\bar{u} = 34(y/R)^{0.145}$ ft/sec, $0 < y < \frac{1}{4}$ ft. What is the local shear stress at the pipe wall?

13-9. Alcohol at 60°F flows at an average velocity of 6 ft/sec in a 1-in-diam tube. Is Eq. (13-62) valid for determining f for this smooth tube if the tube is 18 in long?

13-10. Water at 70°F flows through a 2-in-ID smooth hose with a flow rate of 300 gal/min. What is the pressure loss in 70 ft of hose?

13-11. For the flow of glycerin at 100°F in a 12-in-ID tube ($e = 0.0001$), determine the friction factor f to three decimal places by iteration of Eq. (13-75). It flows with an average velocity of 30 ft/sec.

13-12. Aniline at 60°F flows through a 3-in-ID copper tube at 15 ft/sec. At a point in the piping system the 3-in tube transforms smoothly into three 1-in-ID tubes made of steel. What is the head loss per tube in 30 ft of this 1-in tubing?

13-13. Steam at 1000°F and 100 psia flows through a riveted steel boiler duct with a velocity of 200 ft/sec; the duct has a 20-in ID. What is the pressure loss for an equivalent length of 140 ft?

13-14. The following data were taken in a pipe-flow experiment: $f = 0.025$; $\mathbf{Re} = 50,000$; $D = 14$ in; water flows in the piping system.

    (a) What is the average roughness height e?

    (b) If the measured pressure loss for a given length of pipe is 20.6 psi in this experiment, what would it be if the pipe length were doubled?

    (c) For the longer pipe length of part (b) what would be the Δp if the flow rate were doubled?

13-15. Water at 70°F is pumped through a 6-in-diam wrought-iron pipe to a residential subdivision. If the maximum flow rate is 500 gal/min calculate (per mile of pipe) (a) the head loss and (b) the horsepower (theoretical) required.

13-16. An oil line is contemplated for the newly discovered Alaskan oil fields. Crude oil at 0°F is to be pumped through a 24-in cast-iron pipeline at 8.0 ft³/sec. If pumps with 100-psi pressure rise are used, how far apart can they be located?

13-17. The volume flow rate of water at 70°F from a kitchen faucet is 20 gal/min. Find (a) the Reynolds number in the supply tubing if it has a 1.0-in ID, (b) the pressure drop per foot of tubing if it is smooth, and (c) the pressure drop per foot of tubing if it is galvanized iron.

13-18. A large wooden aqueduct ($e = 0.003$ ft) carries water to a generating station. There is an energy loss through the station of 80 ft-lb$_f$/lb$_m$, and the power output is 50,000 kw. The aqueduct is 10 ft in diameter, and the ambient temperature is 60°F. Assuming the generating station to be 100 percent efficient, what is the pressure loss due to friction in 400 ft of duct leading to the generating station?

13-19. Compare the friction factors of Moody and Nikuradse for the following:

    (a) $\dfrac{e}{D} = 0.01635$ $\mathbf{Re} = 300,000$

    (b) $\dfrac{e}{D} = 0.00397$ $\mathbf{Re} = 20,000$

    (c) $\dfrac{e}{D} = 0.000985$ $\mathbf{Re} = 600,000$

13-20. What is the friction factor for an extremely rough 8-in-diam concrete pipe?

13-21. The Reynolds number in a smooth 6-in-ID pipe is 2,000. For the flow of kerosene at 50°F, what is the pressure drop between sections 1 and 4? Section 3-4 is cast iron, and $p_2 - p_3 = 0.002$ psi.

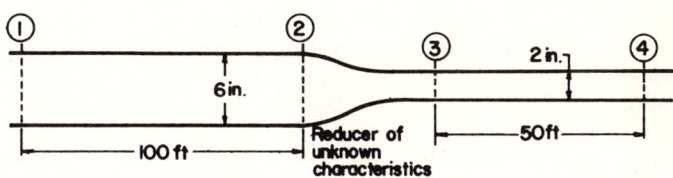

Fig. P13-21

13-22. Determine the size of clean wrought-iron pipe required to carry 3,000 gal/min of benzene at 80°F 6,000 ft with a head loss of 50 ft-lb$_f$/lb$_m$. There are no elevation changes in the pipe.

13-23. A 24-in riveted steel pipe ($e = 0.003$ ft) 30,000 ft long connects to open reservoirs (water at 60°F) whose levels differ by 150 ft.

(a) What is the flow rate from the higher to the lower reservoir in cubic feet per second?

(b) What horsepower would be required to pump 10,000 gal/min from the upper to the lower reservoir?

13-24. Hydraulic fluid is cooled by passing it through a 1-in-diam tube with a wall temperature at 180°F. The average velocity is 24 ft/sec. For an average fluid bulk temperature of 200°F, find the loss in a 20-ft heated length, assuming fully developed flow at the start of the heated tube length. Assume fluid properties the same as for light oil.

13-25. To transport water from a river to a filtration plant 900 ft of piping is required. The filtering station is 75 ft above the river, and there is to be no more than 50 psi pressure loss. There is one gate valve and one 45° bend. What diameter of asphalt-dipped cast-iron pipe should be used for a water flow rate of 675 ft^3/min at 70°F?

13-26. A 6-in-ID pipeline consists of 2,000 ft of asphalt-dipped cast-iron pipe, three long-radius flanged elbows, two gate valves, and a long-radius flanged return bend. For the flow of water at 40°F, what is the head loss at a velocity of 30 ft/sec?

13-27. A 3-in-ID lead pipeline is to be used to carry 3 ft^3/sec of irradiated ammonia from the treating area to the storage vat, a distance of 600 ft. The pipeline is straight but must have one gate valve. Cost analysis shows that lead pipe is available in 20- and 40-ft lengths, but transportational problems make the 40-ft lengths cost 43 cents more per foot. If the maximum pressure loss tolerable is 425 lb$_f$/in^2 and the loss coefficient k_L for each union between pipe lengths is 0.17, which pipe length is more economical? The temperature at the plant is never below 0°F or above 120°F.

13-28. A member of a city planning commission is charged with the task of designing a water-supply pipeline from a lake to a purification plant, a distance of 20 miles. Another member of the commission advises him that a 2-ft-diam concrete pipeline ($e = 0.001$) could transport 8 million gallons of water per day with a total head loss of less than 700 ft, assuming the pipeline is level. Is his advice sound?

13-29. A 30-in-ID asphalt-coated water line is being built to supply a city from a nearby lake. The maximum flow rate is 6 ft^3/sec, and the water temperature is 70°F. There are seven 90° bends and nine 45° bends. The elevation increase is 200 ft, and 17 miles of pipe are required. Calculate the total pressure change.

13-30. A viscous fluid with properties approximately those of water flows through a sharp-edged contraction. The velocity on one side is 17 ft/sec, and that on the other side is 53 ft/sec.

(a) What is the approximate value of the contraction coefficient C_c?

(b) What is the approximate average velocity in the vena contracta?

13-31. A diffuser is to be designed to slow the velocity of water in a piping system from 36 to 4 ft/sec. The pipe is round, and the diffusing section is to be conical. If the diameter of the smaller pipe is 3 in, how far apart should two brackets be placed so that one is at each end of the tapered portion? All material is 0.1 in thick. Optimum results are desired.

13-32. Compare the values of the friction factor as given by Eqs. (13-61) and (13-62) for 80°F water flowing at 5 ft/sec in a 2-in-ID glass tube to that obtained from Figs. 13-10 and 13-11.

13-33. A gasoline delivery truck is to be designed to pump 50 gal/min through the piping system shown schematically. The 2-in ID commercial steel pipe has screwed fittings. What is the required discharge pressure of the pump? Assume a temperature of 60°F.

13-34. Sketch the hydraulic and energy gradients for the flow system shown.

13-35. Determine the water elevation required to result in an average velocity of 4.2 ft/sec in the device shown. The pipe is 2.0-in-ID wrought iron with a total length of 30 ft.

TURBULENT FLOW OF INCOMPRESSIBLE VISCOUS FLUIDS 507

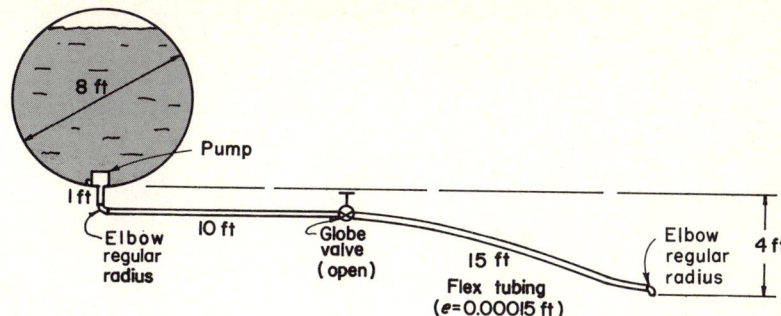

Fig. P13-33

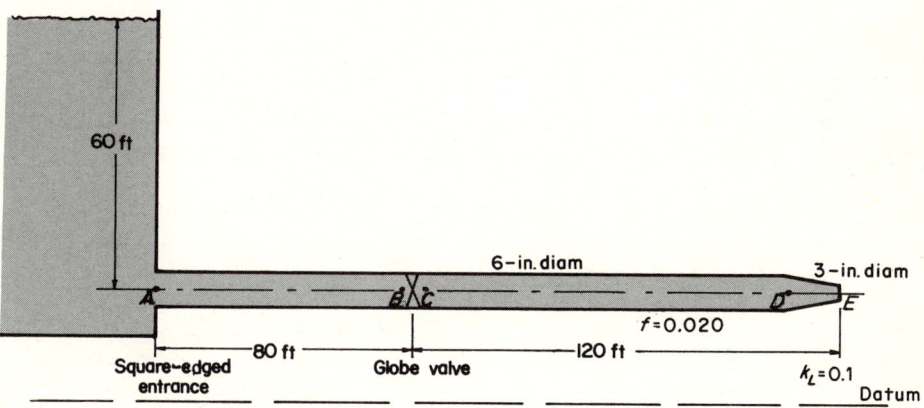

Fig. P13-34

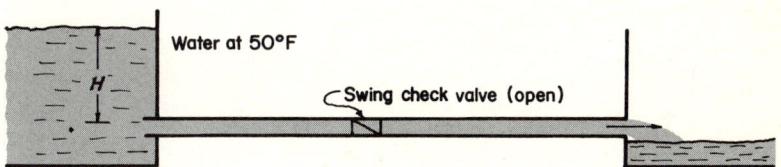

Fig. P13-35

13-36. Water at 70°F is pumped from a 20-ft-deep well (below ground) to a storage tank located 30 ft above ground and 110 ft horizontally distant from the well. The pump is to be positioned at the top of the well as shown.

(a) For a flow rate of 30 gal/min, what nominal size pipe ID should be used if the velocity is not to exceed 10 ft/sec? (Answer to nearest $\frac{1}{2}$ in.)

(b) Find the total pressure rise across the pump in psi for commercial steel pipe of the size selected in part (a).

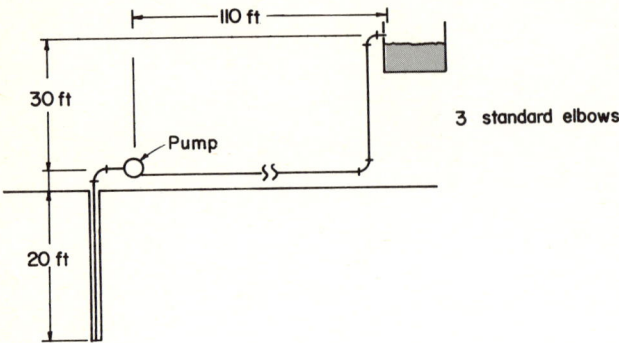

Fig. P13-36

13-37. What pressure loss is caused by the piping arrangement shown if the fluid is light oil at 80°F and flows with a velocity of 2 ft/sec in the larger tube? The pipe is commercial steel and has screwed fittings.

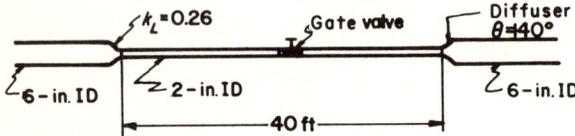

Fig. P13-37

13-38. Determine the maximum flow rate for the piping system shown. There are 240 ft of 18-in-ID commercial steel pipe before the reducer and 190 ft of 9-in-ID pipe after the reducer. All fittings are flanged, and the head loss from the supply to the tank is 3.18 ft for the flow of ammonia at 80°F.

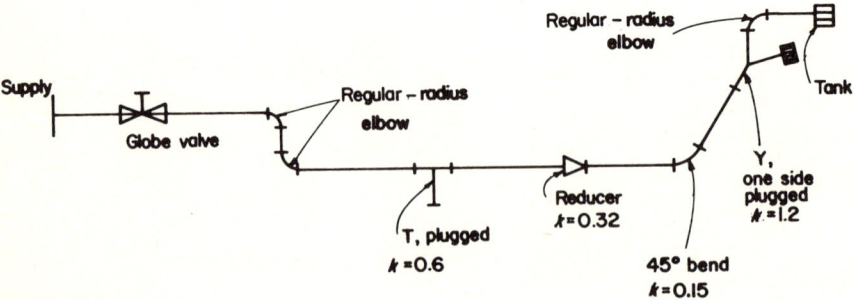

Fig. P13-38

13-39. What is the equivalent length for the system shown? Water at 70°F flows at 10 ft/sec.

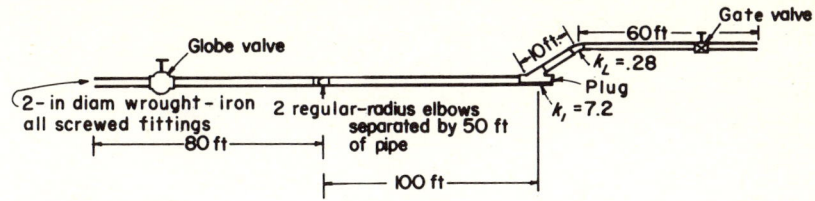

Fig. P13-39

13-40. Water flows through a reentrant from a large reservoir as shown. The average friction factor is 0.03, and the pipe has screwed fittings. What is the head loss through the system?

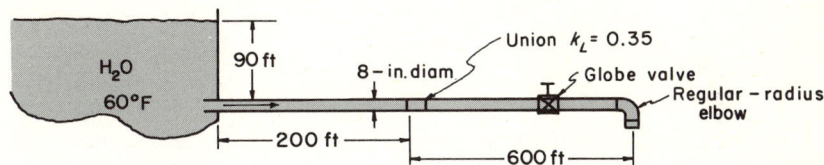

Fig. P13-40

13-41. For the piping network shown, the numbers in parentheses indicate head loss in feet and in the braces indicate velocities in feet per second. Determine the pressure at stations B and C.

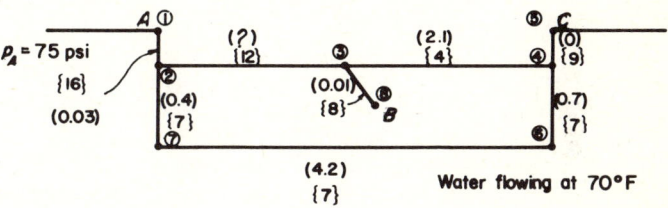

Fig. P13-41

13-42. Alcohol at 100°F flows through the 6-in-ID network shown at the rate of 12 ft³/min. Stations A and B are at the same elevation; fittings are screwed. Find the pressure at station B.

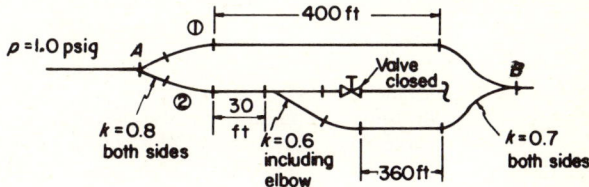

Fig. P13-42

13-43. For the network shown, determine the flow rate in each branch. All piping is 12-in-diam galvanized iron; $v = 0.213 \times 10^{-5}$ ft²/sec, $\rho = 83$ lb$_m$/ft³. A and B are at the same elevation.

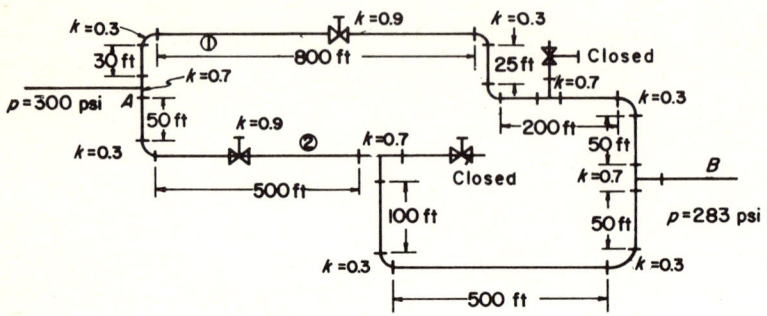

Fig. P13-43

13-44. Compare the boundary-layer thickness of turbulent to laminar flow over a flat plate for:
(a) $\text{Re} = 300,000$ $x = 30$
(b) $\text{Re} = 400,000$ $x = 40$
(c) $\text{Re} = 500,000$ $x = 50$

13-45. Estimate the skin-friction drag on an airship 1,000 ft long and 80 ft average diameter at a velocity of 40 mph through air at 14.7 psia and 60°F.

13-46. At what point on a flat plate is the local skin friction maximum? Assume constant velocity and viscosity.

13-47. A 20-ft smooth wax model of a cargo ship is towed through 60°F fresh water at a speed of 5 knots (8.45 ft/sec). The wetted hull area is 70 ft². What is the skin-friction drag, assuming a completely turbulent boundary layer equivalent to that on a flat plate?

13-48. Water at 60°F flows along a smooth 3-ft-long plate with a free-stream velocity of 1.22 ft/sec. A tripping wire is mounted near the leading edge of the plate to artificially generate turbulent flow over the entire length. What is the ratio of the resulting drag to the drag which would occur without the tripping wire? Assume $\text{Re}_c = 300,000$.

13-49. Helium at 400°F flows over an isothermal flat plate at 20 ft/sec. What is the boundary-layer thickness 70 ft along the plate, assuming the flow is everywhere turbulent?

13-50. A barge 20 by 200 ft moves along the surface of a river with a speed of 6 ft/sec relative to the water. The river is at 60°F. Estimate the skin-friction drag.

13-51. Air at 80°F and 14.7 psia flows parallel to a flat plate at a velocity of 100 ft/sec.
(a) What is the boundary-layer thickness at 2 ft from the leading edge?
(b) What is the drag on the plate if it is 2 ft wide and 4 ft long?

13-52. A wind tunnel has an 8- by 8-ft inlet and an 8:1 linear contraction ratio to a 1- by 1-ft test section, which consists of flat plates. The test position is at 3, where the velocity profile is essentially slug flow except for the boundary layer. This velocity is 300 ft/sec. Assuming that fluid properties are constant and that flat-plate boundary-layer development begins at 2, calculate the (a) boundary-layer thickness at 3, (b) displacement thickness at 3, (c) velocity at 2, and (d) velocity at 1.

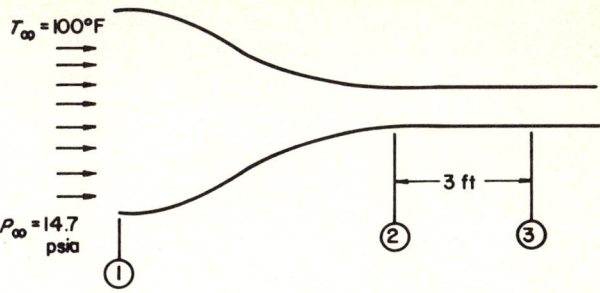

Fig. P13-52

13-53. A submarine moves at 15 ft/sec underwater. The length is 120 ft. Using flat-plate theory, approximate the average skin-friction drag per square foot of surface if seawater has properties similar to those of fresh water at 32°F.

13-54. A new type of annular torpedo has been developed for use as an aimed depth charge. It is 20 ft long, 18 in ID, and 21 in OD, and weighs 1,000 lb_f. If this torpedo breaks the surface of a body of fresh water (70°F) and retains a velocity of 58 ft/sec, what is the deceleration the instant the torpedo is submerged?

13-55. Air flows through a 3-ft square duct which is 9 ft long. Neglecting corner effects and using flat-plate theory, find for 32°F airflow with $V_\infty|_{x=9} = 120$ ft/sec (a) the boundary-layer thickness at the end of the duct, (b) the displacement thickness at the end of the duct, (c) the inlet velocity V_1 at the leading edge of the duct.

13-56. A smooth heated flat plate in a wind tunnel is subjected to approaching air at $V_\infty = 20$ ft/sec. The plate is 2 ft long and is heated to 300°F in a 100°F airstream. Estimate the maximum drag per foot of width from one side of the plate.

13-57. The sides of a large octagonal duct can be approximated by flat plates. Saturated liquid ammonia at 50°F flows through the duct with an inlet free-stream velocity of 17 ft/sec.
(a) What is the boundary-layer thickness at the end of the duct if it is 20 ft long?
(b) How large must it be to carry 230 ft^3/sec?

13-58. Calculate the skin-friction drag on the fin of a model rocket moving through 70°F air at 400 ft/sec. The average length of the fin is 6 in, and the total area is 1 ft^2.

13-59. Some new cars are equipped with spoilers over the trunk. What is the drag on a spoiler that approximates a flat plate 10 in long and 30 in wide when the air is at 80°F and the car travels at 80 mph?

13-60. Develop Eq. (13-91) from Eq. (13-89).

13-61. Explain the difference between boundary-layer transition and boundary-layer separation.

13-62. A 6-ft-diam sign mounted on top of a 6-in-diam pipe 10 ft long is struck perpendicularly by a 45-mph wind which is at 80°F. What is the moment (foot-pounds force) at the base of the pipe?

13-63. A major oil company uses a circular disk sign to advertise its service stations. One of these signs located along an expressway is 10 ft in diameter and is supported on each side by a 10-in-diam steel pipe 70 ft long. Assuming a maximum wind velocity of 80 mph, calculate the wall thickness of the pipe if the allowable stress is 10,000 psi. Assume the load is equally divided between the two supports.

13-64. A rectangular sign is 8 ft high and 16 ft wide. Approximate the maximum drag force on this sign in a 50-mph wind at 100°F.

13–65. Compare the drag coefficient obtained by Eq. (13-92) to that of laminar-flow theory for a critical **Re** of 500,000 over a smooth flat plate.

13–66. (a) A 1-in-diam golf ball is hit straight down the fairway at 180 ft/sec on a hot 100°F day. If the ball is smooth, what is the drag?
    (b) Why are golf balls dimpled?

13–67. Several long bolts pass through a very large duct to brace the sides. Light oil at 100°F flows through the duct at 30 ft/sec so that the bolts are subjected to a small shearing force. If the bolts have a 1-in diameter what is the force on the bolts per unit projected area? Is this force significant for a small factor of safety?

13–68. A large rectangular sign 30 ft above the ground is 12 ft tall and 24 ft wide. It is supported by four posts. What is the maximum moment at the base if the maximum wind expected is 30 mph and the average temperature is 60°F? The posts have a 9-in diameter and pass behind the sign 24 ft from the ground.

13–69. Calculate the force on a weather balloon 3 ft in diameter. The balloon is in air at 40°F which moves at 3 ft/sec with respect to the balloon. The balloon weighs 8 oz and is initially at rest. (Neglect the drag on the instrumentation package.) How fast does it accelerate when it is first released?

13–70. Calculate the drag on a submarine net across a river. The average water temperature is 70°F, and the average velocity is $\frac{1}{2}$ ft/sec. The net is made of 6-in-diam steel cables.

13–71. A radio-controlled model airplane is equipped with a whip antenna which may be treated as a 2-ft $\frac{1}{8}$-in-OD cylinder with a $\frac{1}{2}$-in-diam flat disk attached, normal to the fluid motion, to its end. What is the drag on the antenna on an 80°F day? The plane flies at 6 ft/sec.

13–72. An 8-in-diam ball is dropped from a helicopter at low altitude. The measured terminal velocity is 116 ft/sec. If the air temperature is 32°F and there is no crosswind, estimate the weight of the ball. The air density can be assumed constant at 1 atm pressure and 32°F.

13–73. A nominal 2-in-diam schedule 40 steel pipe is used for a flagpole. The pole is 30 ft high and is embedded at ground level in concrete. With no flag flying, estimate for a 70-mph wind (invariant over the pole height) at 80°F (a) the deflection at the top and (b) the maximum stress in the pipe.

13–74. A football-field scoreboard 18 ft high and 34 ft long is being erected in southern Florida, where hurricane wind velocities reach 110 mph. The air temperature is approximately 100°F during such tropical storms. What total force should be considered in designing the support structure?

13–75. Alcohol is heated by passing it over a 25-ft-long flat plate at 100°F. The alcohol is initially at 40°F and flows at 3 ft/sec. Assume that the free-stream temperature of the alcohol remains at 40°F and that the film temperature remains constant. Determine the average heat-transfer coefficient.

13–76. For benzene at 120°F flowing with a velocity of 1 ft/sec over a 20-ft-long flat plate which is maintained at 90°F, determine the average heat-transfer coefficient.

13–77. How much heat is convected away from the hood of an automobile on a 60°F day if the hood is at 100°F, assuming it approximates a 4- by 4-ft flat plate? The average velocity of the car is 65 mph.

13–78. Freon at 70°F flows over a smooth plate at 12 ft/sec. The plate is 5 ft long and held at 130°F. Compare the skin-friction coefficient to that which would exist for a 70°F plate. What is the heat transfer (for the 130°F plate) per square foot to the Freon?

13–79. Light oil at 60°F flows over a flat surface and parallel to it at a velocity of 10 ft/sec. The surface is at 340°F.
    (a) What is the thickness of the hydrodynamic boundary layer at a distance of 4 ft from the leading edge?
    (b) What is the heat transfer for the 4-ft section?

13–80. Benzene at an average bulk temperature of 100°F flows at 11.0 ft/sec through a 3-in-ID tube. The tube wall is at 60°F. Estimate the heat transfer per foot of tube.

TURBULENT FLOW OF INCOMPRESSIBLE VISCOUS FLUIDS 513

13–81. Water at 150°F flows through a 1-in-diam uninsulated pipe at 4 ft/sec. Assuming a pipe temperature of 60°F and negligible resistance in the pipe, estimate the energy loss from 20 ft of pipe.

13–82. Commercial aniline flows through a 2-in-ID pipe at 2 ft/sec. The pipe wall is held at 60°F, and the aniline is initially at 300°F. If the pipe is 9 ft 4.2 in long, determine the exit temperature of the aniline (to ± 1°F).

13–83. Saturated liquid Freon-12 flows through a 4-ft-long $\frac{1}{2}$-in-ID tube which is at 100°F. The Freon is initially at 0°F and flows with an average velocity of 1 ft/sec. Determine the final temperature of the Freon (to ± 1°F).

13–84. Water at 60°F enters a section of 1-in-ID pipe at 10 ft/sec. The pipe is heated to 120°F, and the pipe is 30 ft long. Estimate the temperature of the water as it leaves the pipe (to ± 1°F).

13–85. Light oil at an average bulk temperature of 100°F flows through a 1-in-diam pipe at 12 ft/sec. The pipe wall is at 80°F, and the pipe is 4 ft long. Estimate the heat transfer from the oil. The pipe has a bell-mouthed inlet.

13–86. In a delicate machine developed to replace the human heart during operations, blood leaves the body, is pumped through the machine to the lungs, back to the machine, and then through the body. During testing it is found that the blood temperature falls below critical so that it is necessary to heat a section of tubing in each stage just before the blood returns to the body. To what temperature must a 2-ft section of tubing be heated to raise the temperature of the blood from 92 to 98°F if the blood flows through the $\frac{1}{4}$-in tube at 4 ft/sec? Assume blood to have the properties of water.

13–87. Using the transition-regime results of Sieder and Tate, determine the heat-transfer coefficient for the flow of water at an average bulk temperature of 90°F through a 1-in-diam pipe at 0.5 ft/sec if the pipe is at 200°F and is 8 ft long.

13–88. Water flows through a 2-in-diam pipe at an average bulk temperature of 70°F and at a velocity of 2 ft/sec. The pipe wall is at 100°F. Compare the Sieder-Tate equation to the results of their analysis in the transition regime.

13–89. Light oil at an average bulk temperature of 150°F passes through a cooling tube before being recycled. The tube wall is held at 80°F, and the oil flows through the 2-in-diam tube at 7.5 ft/sec. The cooled portion of the long cycling tube is 9 ft long. Estimate the heat transfer from the oil.

13–90. Benzene enters a heated section of 1-in-diam tubing through a 45° bend. The benzene flows at 2.5 ft/sec through the 130°F tube with an average bulk temperature of 80°F. What is the heat-transfer coefficient if the tube is 4 ft long?

13–91. How long must a heated section of 3-in-diam tubing be at 130°F to raise the temperature of liquid Freon from 40 to 80°F if it flows at an average velocity of 0.5 ft/sec?

13–92. Water at 70°F flows at 2 ft/sec through a 1-in-diam tube. A 20-ft-long section of the tube is to be heated to raise the temperature of the water to 90°F. How hot must the tube be?

13–93. Determine the heat-transfer coefficient for the flow of alcohol at 60°F in a 1-in-diam tube which is at 100°F. The velocity is 20 ft/sec. Is the change in viscosity significant?

13–94. Benzene enters a 3-in-diam tube at 2.5 ft/sec from a large vat. The vat is slightly less than 40°F, and the tube wall is at 80°F. The tube is heated for 12 ft. The first 9 in of the tube consists of a calming section. Estimate the heat transfer to the benzene if the average bulk temperature is 40°F.

13–95. (*a*) Determine the heat transfer per square foot to a 16-in pipe at 800°F from 1200°F steam flowing at 10 ft³/sec, making use of an analytically derived result.

(*b*) Compare the result of part (*a*) with that obtained by using an experimental-correlation equation.

13–96. Water at an average Reynolds number of 30,000 flows in a 1-in-ID tube. The wall temperature is 100°F, and the average fluid bulk temperature is 60°F. If the tube is very long, what is the average heat-transfer coefficient? Use an experimental-correlation equation.

13–97. Water at an average bulk temperature of 60°F flows in a 1-in-ID tube with an average

Reynolds number of 30,000. The tube wall is at 100°F. What is the average heat-transfer coefficient if (a) the tube is 6 in long; (b) the tube is 18 in long and immediately downstream from a 90° elbow?

13-98. Air at a bulk temperature of 200°F flows through a 1-ft-diam duct which is at 150°F. The air velocity is 180 ft/sec. Estimate the heat loss from the air per linear foot of duct.

13-99. Consider the flow of water through a $\frac{1}{2}$-in-ID stainless-steel tube, 2 ft long. The tube wall is at 200°F, and the inlet water is at 70°F. The Reynolds number is 40,000.

(a) What is the mass flow rate?

(b) Calculate the Nusselt modulus based on fully developed turbulent flow at the entrance, using 70°F as the average bulk temperature.

(c) Calculate the heat transfer to the water for part (b) and determine the exit temperature.

(d) Iterate part (c) using the exit and inlet temperatures to obtain the average fluid bulk temperature until the exit temperature is correct to within 1°F.

(e) Repeat part (c) for a bell-mouthed inlet with a screen at the tube. (No iteration required.)

13-100. Air at an average bulk temperature of 200°F enters a 1-ft-ID duct at 180 ft/sec. The first 5 ft of the duct is heated to 150°F. Estimate the heat loss to the duct. *Hint:* The diameter is large compared to the length of the duct. Flat-plate theory *might* give a more accurate result.

13-101. Carbon monoxide at 200°F flows at 3 ft/sec through a cooling chamber which is a duct 1 ft square with walls at 32°F. Estimate the heat-transfer coefficient.

13-102. Nitrogen at an average bulk temperature of 200°F flows through a square duct having 6-in sides at a flow rate of 300 ft^3/min. The duct is at 70°F. Estimate the heat transfer from the nitrogen per foot of length.

13-103. At what temperature must a 3-in square duct 25 ft long be kept to raise the temperature of oxygen flowing at 11 ft/sec from 90 to 110°F?

13-104. Air moving at 20 ft/sec flows normal to a $\frac{1}{2}$-in-diam cylinder. The air is at 100°F, and the cylinder wall is at 300°F.

(a) Calculate the average heat transfer per square foot of cylinder surface.

(b) Is the cylinder surface based on frontal profile or complete periphery? Why?

13-105. Air at 60°F blows normal to a 180°F 1-in-OD water pipe. The air moves at 3 ft/sec. Determine the heat transfer from the pipe per foot of length.

13-106. A 1-in square shaft rotates slowly at constant velocity. The shaft temperature is 400°F, and 75°F air flows normal to it at 22 ft/sec. Estimate the heat transfer from the shaft.

13-107. A stiff strip of steel held at 20°F moves through 90°F air as shown. The angular velocity is 3 rad/sec. Estimate the initial heat transfer to the strip of steel if it is 3 ft long and 2 in wide.

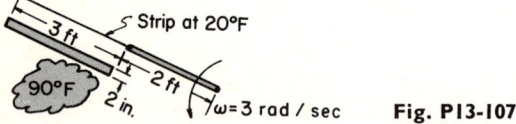

Fig. P13-107

13-108. Estimate the heat transfer from a 25-watt incandescent bulb which is constantly at 230°F to a 70°F airstream moving at 1 ft/sec. Assume the bulb is a 2.4-in-diam sphere. What percent of the power is lost as heat by convection?

13-109. A $\frac{1}{2}$-in-diam marble drops through a liquid of very high viscosity with a velocity of 1 ft/min. If the coefficient of thermal conductivity is 0.71 Btu/hr-ft-°F and the temperature difference between the marble and the liquid is 40°F, determine the theoretical heat transfer.

13-110. An egg at 40°F is conveyed at 2 ft/sec through a large vat of boiling water in a food-processing plant. Determine the initial heat transfer to the egg. (Assume the egg is a 1.25-in-diam sphere with unrestricted flow over it.)

13-111. Air at 14.7 psia and 100°F flows with a maximum velocity of 30 ft/sec through an in-line tube bundle which is 10 rows wide with 12 tubes in each row. The tube surfaces are at 200°F. The diameter of the tubes is 1.00 in, and the spacing is 1.5 and 2.0 in in the direction normal to the flow and parallel with it, respectively. Calculate the total heat transfer from the tube bundle per foot of length.

13-112. Combustion products at an average temperature of 800°F and an approach velocity of 4 ft/sec flow through a heat exchanger consisting of a staggered tube bundle, with eight columns of tubes, each column being normal to the flow direction and containing 14 tubes. The tubes are 2 in OD, and spacing in both directions is 2.5 in. Assuming that the average tube surface temperature is 400°F and that the properties of the products of combustion are the same as for air, calculate the net heat transfer to the tube bundle if it is 9 ft long.

13-113. Water flows over an in-line tube bundle with a free-volume velocity of $\frac{2}{3}$ ft/sec. The water is at an average bulk temperature of 90°F. The 1-in-diam tubes are at 210°F, spaced 1.5 in apart (both directions). There are 12 tubes in each direction. Determine the average heat-transfer coefficient.

13-114. Light oil at an average bulk temperature of 70°F flows over a staggered tube bundle which has a surface temperature of 230°F. The bundle has six tubes in both directions, and they are spaced $2\frac{1}{4}$ in apart parallel to the flow and 3 in apart perpendicular to the flow. The tubes have a 1.5-in diameter. Determine the average heat-transfer coefficient for a velocity of 1 ft/sec based on the minimum flow area.

13-115. Water flows over an in-line tube bundle at a temperature of 120°F. The tubes are at 30°F and spaced 2 in apart in both directions. The water reaches the bundle flowing at 2 ft/sec in a duct 14 in by 1 ft and the 1-in-diam tubes are 1 ft long, the width of the duct. There are 6 tubes in each column perpendicular to flow and 15 tubes in each row parallel to the flow. Determine the temperature of the water leaving the bundle. Is this a practical way to heat the tube surfaces?

REFERENCES

1. Boelter, L. M. K., G. Young, and H. W. Iverson: *NACA Tech. Note* 141, 1948.
2. Boussinesq, J.: Théorie de l'écoulement tourbillant, *Mém. Prés. Acad. Sci.*, **23**: 46 (1877).
3. Carslaw, H. S., and J. C. Jaeger: "Conduction of Heat in Solids," 2d ed., Oxford University Press, London, 1959.
4. Colburn, A. P.: A Method of Correlating Forced Convection Heat Transfer Data and a Comparison with Fluid Friction, *Trans. AIChE*, **29**: 174–210 (1933).
5. Cross, H.: Analysis of Flow in Networks of Conduits or Conductors, *Univ. Ill. Eng. Exp. Sta. Bull.* 286, 1936.
6. Gibson, A. H.: The Conversion of Kinetic to Pressure Energy in the Flow of Water through Passages Having Divergent Boundaries, *Engineering*, **93** (1912).
7. Gibson, A. H.: "Hydraulics and Its Applications," 5th ed., Constable, London, 1952.
8. Giedt, W. H.: Investigation of Variation of Point Unit-heat-transfer Coefficient around a Cylinder Normal to an Air Stream, *Trans. ASME*, **71** (1949).
9. Grimson, E. D.: Correlation and Utilization of New Data on Flow Resistance and Heat Transfer for Cross Flow of Gases over Tube Banks, *Trans. ASME*, **59**: 583 (1937).
10. Grosh, R. J., and R. D. Cess: Heat Transfer to Fluids with Low Prandtl Numbers for Flow across Plates and Cylinders of Various Cross Sections, *Trans. ASME*, **80**: 667 (1958).
11. Hoe, R. J., D. Dropkin, and O. E. Dwyer: Heat Transfer Rates to Crossflowing Mercury in a Staggered Tube Bank, I, *Trans. ASME*, **79**: 899 (1957).
12. Jakob, M.: "Heat Transfer," vol. 1, Wiley, New York, 1959.
13. Johnston, H. F., R. L. Pigford, and J. H. Chapin: Heat Transfer to Clouds of Falling Particles, *Univ. Ill. Bull.*, **38** (43) (1941).
14. Kays, W. M., and R. K. Lo: Basic Heat Transfer and Flow Friction Data for Gas Flow Normal

to Banks of Staggered Tubes: Use of a Transient Technique, *Stanford Univ. Tech. Rept.* 15 (Navy Contract N6-ONR-251 T.L.6), 1952.
15. Knudsen, J. G., and D. L. Katz: "Fluid Dynamics and Heat Transfer," McGraw-Hill, New York, 1958.
16. Latzko, H.: *Z. Angew. Math. Mech.* **1**: 268 (1921); English trans. *NACA Tech. Mem.* 1068, 1944.
17. Lubarsky, B., and S. J. Kaufman: Review of Experimental Investigations of Liquid-metal Heat Transfer, *NACA Tech. Note* 3336, 1955.
18. Lyon, R. N. (ed.): "Liquid Metals Handbook," 3d ed., Atomic Energy Commission and Department of the Navy, Washington, D.C., 1952.
19. McAdams, W. H.: "Heat Transmission," 3d ed., McGraw-Hill, New York, 1954.
20. Moody, L. F.: Friction Factors for Pipe Flow, *Trans. ASME*, **66**: 641 (1944).
21. Nikuradse, J.: Strömungsgesetze in rauhen Rohran, *Forsch. Arb. Ingenieurw.* 361, 1933.
22. Nikuradse, J.: Gesetzmässigkeit der turbulenten Strömung in glatten Rohren, *Forsch. Arb. Ingenieurw.*, No. 356, 1932.
23. Pai, Shih-I: "Viscous Flow Theory," vol. II, "Turbulent Flow," Van Nostrand, Princeton, N.J., 1957.
24. Prandtl, L.: Über die ausgebildete Turbulenz, *Z. Angew. Math. Mech.*, **5**: 136–139 (1925).
25. Prandtl, L.: The Mechanics of Viscous Fluids, in W. F. Durand, (ed.), "Aerodynamic Theory," vol. III., Durand Reprinting Committee, Pasadena, Calif., 1943; reprinted by Dover, New York, 1963.
26. Prandtl, L.: Über den Luftwiderstand von Kageln, *Nachr. Ges. Wiss. Göttingen, Math. Physik Kl.*, **1914**: 177–190, as cited by Schlichting [29].
27. Reichardt, H.: Die Wärmeübertragung in turbulenten Reibungs, *Z. Angew. Math. Mech.*, **20**: 297–328 (1940).
28. Seban, R. A., and T. T. Shimazaki: Heat Transfer to a Fluid Flowing Turbulently in a Smooth Pipe with Walls at Constant Temperature, *Trans. ASME*, **73**: 803 (1951).
29. Schlichting, H.: "Boundary Layer Theory," 6th ed., J. Kestin, trans., McGraw-Hill, New York, 1968.
30. Schubauer, G. B., and H. K. Skramstad: Laminar Boundary Layer Oscillations and Stability of Laminar Flow, *Natl. Bur. Std. (U.S.) Res. Paper* 1772 and *J. Aeron. Sci.*, **14** (1947).
31. Sieder, E. N., and C. E. Tate: Heat Transfer and Pressure Drop of Liquids in Tubes, *Ind. Eng. Chem.*, **28**: 1429 (1936).
32. Vliet, G. C., and G. Leppert: Forced Convection Heat Transfer from an Isothermal Sphere to Water, *Trans. ASME, ser. C*, **83**: 163 (1961).
33. Weisbach, J.: "Die Experimental-Hydraulik," J. S. Englehardt, Freiburg, 1855.

CHAPTER 14
CONVECTIVE MASS TRANSFER

Convective mass transfer occurs when mass is transported between the boundary of a surface and a moving fluid or between two moving fluids which are relatively immiscible. The convective process is either *forced* or *natural*, depending upon the existence of a gradient of pressure or density, respectively, in the medium.

The concept of convective mass transfer was introduced in Chap. 7 as a boundary condition in the case of mass diffusion. At that time it was assumed that the convective-mass-transfer coefficient k_c, defined by

$$N_A = k_c(\rho_A - \rho_{A_\infty}) \tag{14-1}$$

was known. This chapter is concerned with the determination of k_c, which depends upon the fluid properties, the dynamics of the flow, and the geometry of the flow system, analogous to the convective-heat-transfer coefficient h defined by

$$q'' = h(T - T_\infty) \tag{14-2}$$

Four methods of evaluating the convective-mass-transfer coefficient k_c will be considered in this chapter: (1) exact boundary-layer analysis (laminar), (2) approximate boundary-layer analysis, (3) analogy between momentum, heat, and mass transfer, and (4) dimensional analysis. In each case the similarities between the transport processes will be pointed out.

14-1 EXACT ANALYSIS IN A LAMINAR BOUNDARY LAYER

Forced convection with mass transfer over a flat plate Consider an airstream flowing over the surface of a quiescent lake, as shown in Fig. 14-1. Extending the boundary layer to the point $x = 0$, the figure might also represent the flow of an incompressible fluid over a porous flat plate. In either case it is assumed that the boundary does not move in the x direction, that is, $u_s = 0$.

Assuming constant physical properties ν, α, and D_{AB}, the following equations describe the behavior in the steady-state incompressible two-dimensional laminar flow in the boundary layer over a flat plate:

$$\text{Continuity:} \qquad \frac{\partial u}{\partial x} + \frac{\partial v}{\partial y} = 0 \qquad (14\text{-}3)$$

$$\text{Momentum:} \qquad u\frac{\partial u}{\partial x} + v\frac{\partial u}{\partial y} = \nu \frac{\partial^2 u}{\partial y^2} \qquad (14\text{-}4)$$

$$\text{Energy:} \qquad u\frac{\partial T}{\partial x} + v\frac{\partial T}{\partial y} = \alpha \frac{\partial^2 T}{\partial y^2} \qquad (14\text{-}5)$$

$$\text{Conservation of species } A: \qquad u\frac{\partial \rho_A}{\partial x} + v\frac{\partial \rho_A}{\partial y} = D_{AB} \frac{\partial^2 \rho_A}{\partial y^2} \qquad (14\text{-}6)$$

Equations (14-3) to (14-5) were introduced in Chap. 12 for the case of no mass transfer. Assuming that they are valid when mass is transferred implies that any additional momentum and energy fluxes which arise as a result of the mass transfer are negligible, a good assumption, from an engineering point of view. Equation (14-6) comes from

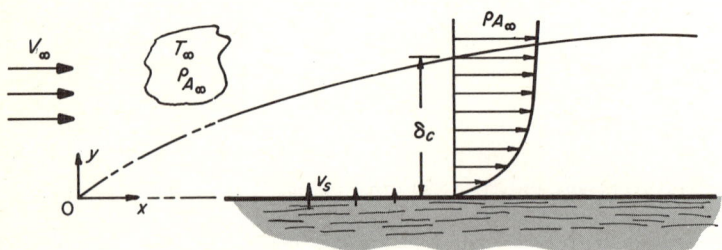

Fig. 14-1 Concentration boundary layer.

CONVECTIVE MASS TRANSFER

Eq. (7-25a) for the case of no chemical reaction, that is, $r_A''' = 0$, and by noting that

$$\frac{\partial^2 \rho_A}{\partial x^2} \ll \frac{\partial^2 \rho_A}{\partial y^2}$$

which follows by direct analogy with the order-of-magnitude analysis of Sec. 12-5. The thickness of the concentration boundary layer δ_c is defined analogous to that of the hydrodynamic and thermal boundary layers.

In nondimensional form, the boundary conditions for the governing equations are

(1) at $y = 0$: $\quad \dfrac{u}{V_\infty} = 0 \quad \dfrac{T - T_s}{T_\infty - T_s} = 0 \quad \dfrac{\rho_A - \rho_{A_s}}{\rho_{A_\infty} - \rho_{A_s}} = 0 \quad v = v_s \quad$ (14-7a)

(2) at $y = \infty$: $\quad \dfrac{u}{V_\infty} = 1 \quad \dfrac{T - T_s}{T_\infty - T_s} = 1 \quad \dfrac{\rho_A - \rho_{A_s}}{\rho_{A_\infty} - \rho_{A_s}} = 1 \quad$ (14-7b)

In Sec. 12-8 the nondimensional forms of the momentum and energy equations were shown to be equivalent when $\nu = \alpha$, that is, when

$$\mathbf{Pr} \equiv \frac{\text{momentum diffusivity}}{\text{thermal diffusivity}} = \frac{\nu}{\alpha} = 1$$

Upon nondimensionalizing Eq. (14-6) we note that complete similarity exists when $\nu = D_{AB}$, or

$$\mathbf{Sc} \equiv \frac{\text{momentum diffusivity}}{\text{mass diffusivity}} = \frac{\nu}{D_{AB}} = 1$$

Because of this similarity, a solution to one equation is a solution to either of the others when $\nu = \alpha = D_{AB}$, that is, when the Prandtl and Schmidt numbers are unity. Therefore, *Blasius' hydrodynamic results may be applied directly to the convective-mass-transfer problem when the Schmidt number is unity;* this is analogous to the application of Blasius' solution to the convective-heat-transfer problem when the Prandtl number is unity.

Using the nomenclature of Sec. 12-5,

$$\eta \equiv y\sqrt{\frac{V_\infty}{\nu x}} \qquad (12\text{-}78)$$

$$f(\eta) \equiv \frac{\psi}{\sqrt{\nu x V_\infty}} \qquad (12\text{-}79)$$

$$f' = \frac{u}{V_\infty} \qquad (12\text{-}80)$$

$$\frac{\partial u}{\partial y} = V_\infty \sqrt{\frac{V_\infty}{\nu x}} f'' \qquad (12\text{-}82)$$

multiplying Eq. (12-82) by x/V_∞, and evaluating at $y = 0$, we get

$$\left.\frac{\partial}{\partial y}\left(\frac{u}{V_\infty}\right)\right|_{y=0} = \frac{\sqrt{\mathbf{Re}_x}}{x} f''(0) = 0.332 \frac{\sqrt{\mathbf{Re}_x}}{x} \qquad (14\text{-}8)$$

where $f''(0) = 0.332$ is taken from Table 12-3. Since the nondimensional forms of the transport equations are identical when $\mathbf{Pr} = \mathbf{Sc} = 1$, Eq. (14-8) can be expressed in terms of the concentration gradient, giving

$$\left.\frac{\partial}{\partial y}\left(\frac{\rho_A - \rho_{A_s}}{\rho_{A_\infty} - \rho_{A_s}}\right)\right|_{y=0} = 0.332 \frac{\sqrt{\mathbf{Re}_x}}{x} \qquad (14\text{-}9)$$

or

$$\left.\frac{\partial \rho_A}{\partial y}\right|_{y=0} = (\rho_{A_\infty} - \rho_{A_s})\left(0.332 \frac{\sqrt{\mathbf{Re}_x}}{x}\right) \qquad (14\text{-}10)$$

If the mass transfer at the surface is so small that it does not alter the velocity profile given by the Blasius solution, we may assume that the bulk motion in the y direction is negligible, giving

$$\left.N_A\right|_{y=0} = -D_{AB} \left.\frac{\partial \rho_A}{\partial y}\right|_{y=0} \qquad (14\text{-}11)$$

from Eq. (7-17). Combining Eqs. (14-10) and (14-11),

$$N_{A_s} = -D_{AB}(\rho_{A_\infty} - \rho_{A_s})\left(0.332 \frac{\sqrt{\mathbf{Re}_x}}{x}\right) \qquad (14\text{-}12)$$

In terms of the mass-transfer coefficient, Eq. (14-1) gives another expression for the mass transfer:

$$N_{A_s} = k_c(\rho_{A_s} - \rho_{A_\infty}) \qquad (14\text{-}13)$$

which combined with Eq. (14-12) yields

■ $\qquad \mathbf{Sh}_x \equiv \dfrac{k_c x}{D_{AB}} = 0.332 \sqrt{\mathbf{Re}_x} \qquad (14\text{-}14)$

This new dimensionless combination is known as the *Sherwood number*. It should be noted that if the fluid which is injected (evaporates or condenses) at the surface is the same as the fluid flowing in the boundary layer, there is no mass transfer insofar as the governing equations can tell; hence, the requirement of a low mass-transfer rate is met. The only other requirement is that the Schmidt number be unity.

Figure 14-2 graphically depicts the solution of Eqs. (14-4) to (14-6) for Prandtl and Schmidt numbers of 0.7 and 1.0. The *blowing (suction) parameter*

$$-\frac{f_s}{2} = \frac{v_s}{V_\infty} \sqrt{\mathbf{Re}_x} \qquad (14\text{-}15)$$

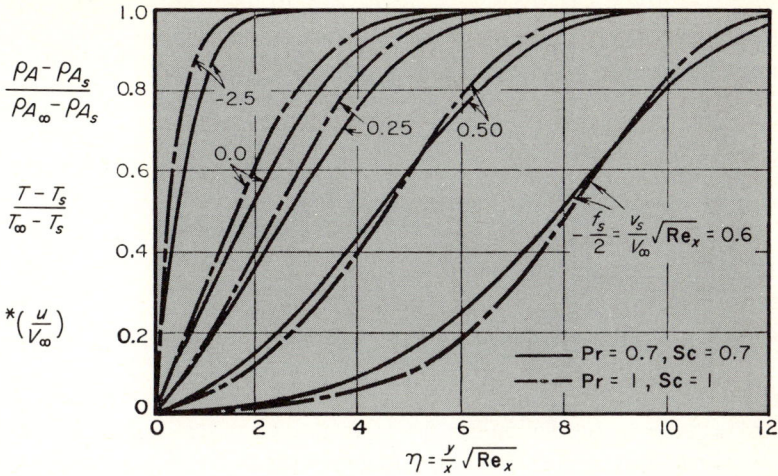

Fig. 14-2 Concentration, temperature, and velocity profiles in a laminar boundary layer on a flat plate with mass transfer at the surface. (*Given by broken-line curves only.) [*From J. P. Hartnett and E. R. G. Eckert, Trans. ASME.* **79**: 247–254 (1957).]

comes from an evaluation of Eq. (12-84) at the surface. Positive values of the parameter give a mass transfer into the stream (blowing); negative values indicate that mass is transferred from the stream to the plate (suction).

Since $f'(\eta)$ must be zero in order to satisfy the no-slip boundary condition, $f(\eta)$ must be constant; hence,

$$v_s = \frac{\text{const}}{\sqrt{x}} \tag{14-16}$$

from Eq. (14-15). This restriction is acceptable since it agrees in most cases with the condition of constant temperature and concentration at the wall [6]. In the hydrodynamic case of Blasius $f(\eta) = 0$ in keeping with the solid-boundary condition, that is, $v_s = 0$.

The solid lines of Fig. 14-2 give the temperature profiles for a fluid of Prandtl number equal to 0.7 and the mass-concentration profiles for a fluid of **Sc** = 0.7. The broken lines present the velocity profiles as well as the temperature and concentration profiles. The temperature and concentration profiles are similar when **Sc** = **Pr** or when

$$\mathbf{Le} \equiv \frac{\mathbf{Sc}}{\mathbf{Pr}} \equiv \frac{\text{thermal diffusivity}}{\text{mass diffusivity}} = \frac{\alpha}{D_{AB}} = 1$$

which is often encountered in gas mixtures.

The Pohlhausen solution in the case of convective heat transfer in laminar flow over a flat plate, outlined in Sec. 12-8, gave the relation between the hydrodynamic and the thermal boundary layer, namely,

$$\frac{\delta}{\delta_t} = \mathbf{Pr}^{\frac{1}{3}} \tag{12-153}$$

Similar reasoning gives the analogous relation

$$\frac{\delta}{\delta_c} = \mathbf{Sc}^{\frac{1}{3}} \tag{14-17}$$

between the hydrodynamic and concentration boundary layers.

By multiplying the dimensionless transformation η by the *stretching factor* $\mathbf{Sc}^{\frac{1}{3}}$ the concentration profiles for any range of Schmidt numbers can be collapsed into a single curve, as typified in Fig. 14-3 for the case of low mass-transfer rates. Following the same procedure as that taken in developing Eqs. (14-10) to (14-14), the convective-mass-transfer coefficient for $\mathbf{Sc} > 0.6$ is given by

$$\mathbf{Sh}_x \equiv \frac{k_c x}{D_{AB}} = 0.332 \sqrt{\mathbf{Re}_x}\, \mathbf{Sc}^{\frac{1}{3}} \tag{14-18}$$

This of course reduces to Eq. (14-14) for a Schmidt number of unity.

While Eq. (14-18) is applicable to a wide range of Schmidt numbers, the restriction to low mass-transfer rates remains. For mass-transfer rates which are not negligible, such as the vaporization of a volatile fluid, a curve similar to that of Fig. 14-3 can be plotted for any desired blowing parameter. Equation (14-18) is then valid when the constant (0.332) is replaced by the respective slope (at $y = 0$); that is, *for mass-transfer rates which are not negligible*

$$\frac{k_c x}{D_{AB}} = (\text{slope})_s \sqrt{\mathbf{Re}_x}\, \mathbf{Sc}^{\frac{1}{3}} \tag{14-19}$$

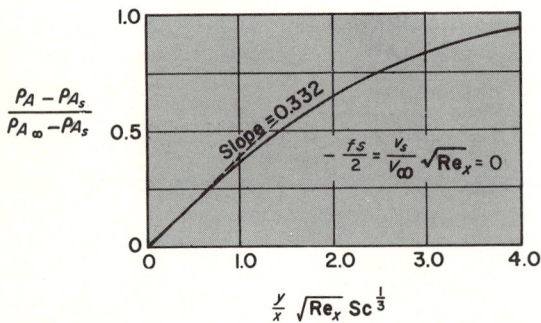

Fig. 14-3 Concentration profile for laminar flow over a flat plate with negligible mass transfer at the surface.

CONVECTIVE MASS TRANSFER

Approximate values of the slopes for the blowing (suction) parameters given in Fig. 14-2 are:

$\dfrac{v_s}{V_\infty}\sqrt{\mathrm{Re}_x}$	0.6	0.50	0.25	0.0	−2.5
(Slope)$_s$	0.01	0.06	0.17	0.332	1.64

By employing the averaging technique introduced in Sec. 12-8, the average mass-transfer coefficient for a plate of finite length L is found to be double the local value, i.e.,

$$\bar{k}_c = 2k_c\big|_{x=L} \tag{14-20}$$

In terms of the average mass-transfer coefficient, Eq. (14-18) becomes

■ $$\overline{\mathrm{Sh}}_L \equiv \frac{\bar{k}_c L}{D_{AB}} = 0.664\sqrt{\mathrm{Re}_L}\,\mathrm{Sc}^{\tfrac{1}{3}} \tag{14-21}$$

Christian and Kezios [5] experimentally studied the validity of Eqs. (14-18) and (14-21) for the case of sublimating naphthalene into air. Their results agreed very well with the equations when the fluid properties were evaluated at a *film* condition defined by

$$T_f \equiv \frac{T_s + T_\infty}{2} \tag{12-172}$$

and

$$\rho_{A_f} \equiv \frac{\rho_{A_s} + \rho_{A_\infty}}{2} \tag{14-22}$$

Forced convection with mass transfer in a tube For no chemical reaction Eq. (7-25a), expressed in cylindrical coordinates, is valid for laminar flow with mass transfer in a tube, as shown in Fig. 14-4. When axial diffusion is neglected, the equation reduces to

$$u\frac{\partial \rho_A}{\partial x} = D_{AB}\left[\frac{1}{r}\frac{\partial}{\partial r}\left(r\frac{\partial \rho_A}{\partial r}\right)\right] \tag{14-23}$$

for fully developed flow (shown to the right of the break in the figure). But this is of the same form, with the same nondimensional boundary conditions, as the Graetz [9] equation for convective heat transfer (Sec. 12-9). The heat-transfer results can then be used for the mass-transfer case upon substituting the Sherwood number for the Nusselt number and the Schmidt number for the Prandtl number in Eqs. (12-173) and

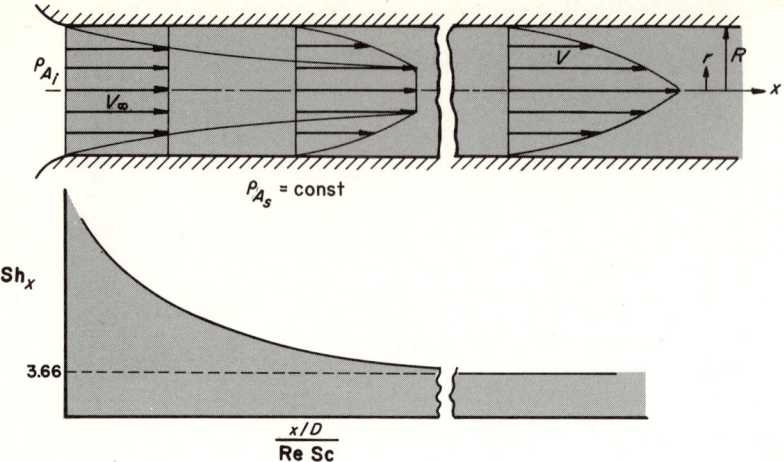

Fig. 14-4 Concentration and Sherwood modulus variation in the inlet region of a tube with mass transfer.

(12-174), giving

$$\text{Constant mass transfer along tube:} \quad \mathbf{Sh}_D \equiv \frac{k_c D}{D_{AB}} = 4.36 \quad (14\text{-}24)$$

$$\text{Constant concentration along tube:} \quad \mathbf{Sh}_D \equiv \frac{k_c D}{D_{AB}} = 3.66 \quad (14\text{-}25)$$

for a fully developed parabolic velocity distribution and low mass-transfer rates.

For engineering calculations, more accurate results can be obtained from the Hausen equation (12-176) by making the appropriate substitution of parameters, resulting in

$$\blacksquare \quad \mathbf{Sh} = \mathbf{Sh}_\infty + \frac{K_1[(D/x)\mathbf{Re}_D\mathbf{Sc}]}{1 + K_2[(D/x)\mathbf{Re}_D\mathbf{Sc}]^n} \quad (14\text{-}26)$$

where the pertinent parameters are shown in Table 14-1 for four common cases. Applicable relations for other cases of engineering interest can be taken analogously

Table 14-1

Wall condition	Velocity	Sc	Sh	$\mathbf{Sh}_\infty$	K_1	K_2	n
Constant ρ_A	Parabolic	Any	Avg.	3.66	0.0668	0.04	$\tfrac{2}{3}$
Constant ρ_A	Developing	0.7	Avg.	3.66	0.104	0.016	0.8
Uniform N_A	Parabolic	Any	Local	4.36	0.023	0.0012	1.0
Uniform N_A	Developing	0.7	Local	4.36	0.036	0.0011	1.0

CONVECTIVE MASS TRANSFER

from heat-transfer equations when mass-transfer rates are low enough for the hydrodynamic and thermal boundary layers to be influenced only negligibly by the mass transfer. See, for example, Bennett and Myers [1] and Kays [13]. In all cases the fluid properties should be evaluated at the *bulk* temperature and mass concentration defined by

$$T_b = \frac{T_i + T_e}{2} \tag{14-27}$$

$$\rho_A|_b = \frac{\rho_{A_i} + \rho_{A_e}}{2} \tag{14-28}$$

where the subscripts i and e represent the inlet and exit stations, respectively.

14-2 APPROXIMATE ANALYSIS OF THE CONCENTRATION BOUNDARY LAYER

The exact analysis of the preceding section was limited to laminar flow and simple configurations, neither of which is commonplace. The integral technique of von Kármán, introduced in Sec. 12-5, is valid for either laminar or turbulent flow and may be used for a variety of configurations. The accuracy of the von Kármán integral analysis depends upon one's ingenuity in assuming accurate velocity and concentration profiles. To capitalize on our previous experience in analysis, we shall apply the technique to the concentration boundary layer over a flat plate, as illustrated in Fig. 14-5. The mass fluxes per unit depth are shown in the figure.

Assuming steady-state incompressible flow, a mass balance yields

$$\int_0^L \rho_A u \, dy\Big|_x + \rho_{A_s} v_s \, \Delta x + \rho_{A_\infty} v\Big|_L \Delta x = \int_0^L \rho_A u \, dy\Big|_{x+\Delta x} \tag{14-29}$$

Rearranging, dividing by Δx, and taking the limit as Δx approaches zero, we get

$$\frac{d}{dx} \int_0^L \rho_A u \, dy = \rho_{A_s} v_s + \rho_{A_\infty} v\Big|_L \tag{14-30}$$

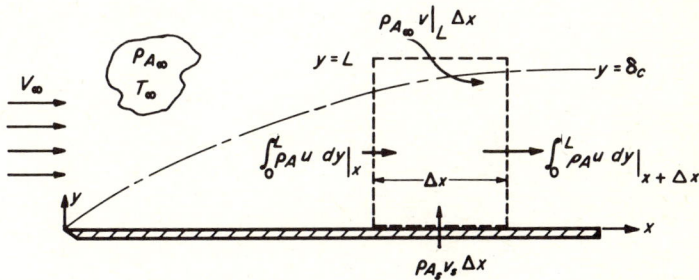

Fig. 14-5 Control volume for integral mass-transfer equation.

But the mass transfer at the surface can be expressed by the convective equation (14-1), giving

$$\rho_{A_s} v_s = k_c(\rho_{A_s} - \rho_{A_\infty}) \tag{14-31}$$

When we neglect the mass flux at the surface compared with the flux of the mainstream, continuity requires

$$\rho_\infty v|_L \simeq \frac{d}{dx} \int_0^L \rho u \, dy \tag{14-32a}$$

If we restrict our analysis to dilute solutions in which the concentration of the diffusing component is so low that we can assume the density of the mixture to be uniform throughout the boundary layer, $\rho_\infty \simeq \rho$; hence,

$$v|_L \simeq \frac{d}{dx} \int_0^L u \, dy \tag{14-32b}$$

Substituting Eqs. (14-31) and (14-32b) into Eq. (14-30), the integral mass-transfer equation results:

$$\frac{d}{dx} \int_0^L \rho_A u \, dy = k_c(\rho_{A_s} - \rho_{A_\infty}) + \frac{d}{dx} \int_0^L \rho_{A_\infty} u \, dy$$

We have so far tacitly taken the upper control surface of the control volume to be outside both the hydrodynamic and concentration boundary layers. If we now assume that the boundary layers coincide, valid for a Schmidt number of unity, the upper limit on the integrals may be replaced by $\delta_c = \delta$ since $\rho_A - \rho_{A_\infty}$ vanishes outside the boundary layer. Hence,

■ $$\frac{d}{dx} \int_0^{\delta_c} (\rho_A - \rho_{A_\infty}) u \, dy = k_c(\rho_{A_s} - \rho_{A_\infty}) \tag{14-33}$$

This resulting equation is analogous to the integral momentum equation (12-106) and the integral energy equation (12-163). Its solution requires knowing (or assuming) the velocity and concentration profiles.

The boundary conditions which the profiles must satisfy are

	Velocity	*Concentration*
(1) at $y = 0$:	$u = 0$	$\rho_A - \rho_{A_s} = 0$
	$\dfrac{\partial^2 u}{\partial y^2} = 0$	$\dfrac{\partial^2}{\partial y^2}(\rho_A - \rho_{A_s}) = 0$
(2) at $y = \delta$:	$u = V_\infty$	
	$\dfrac{\partial u}{\partial y} = 0$	
at $y = \delta_c$:		$\rho_A - \rho_{A_s} = \rho_{A_\infty} - \rho_{A_s}$
		$\dfrac{\partial}{\partial y}(\rho_A - \rho_{A_s}) = 0$

Laminar flow over a flat plate In Sec. 12-5 we assumed the velocity distribution to be described by a polynomial of the form

$$\frac{u}{V_\infty} = C + C_1 \frac{y}{\delta} + C_2 \left(\frac{y}{\delta}\right)^2 + C_3 \left(\frac{y}{\delta}\right)^3 \qquad (12\text{-}108)$$

and determined the constants, adhering to the prescribed boundary conditions. The resulting profile was

$$\frac{u}{V_\infty} = \frac{3}{2}\frac{y}{\delta} - \frac{1}{2}\left(\frac{y}{\delta}\right)^3 \qquad (12\text{-}111)$$

Following the same procedure for the concentration boundary layer, i.e., assuming

$$\frac{\rho_A - \rho_{A_s}}{\rho_{A_\infty} - \rho_{A_s}} = C + C_1 \frac{y}{\delta_c} + C_2 \left(\frac{y}{\delta_c}\right)^2 + C_3 \left(\frac{y}{\delta_c}\right)^3 \qquad (14\text{-}34)$$

we get

$$\frac{\rho_A - \rho_{A_s}}{\rho_{A_\infty} - \rho_{A_s}} = \frac{3}{2}\frac{y}{\delta_c} - \frac{1}{2}\left(\frac{y}{\delta_c}\right)^3 \qquad (14\text{-}35)$$

Upon substituting Eqs. (12-111) and (14-35) into Eq. (14-33), the resulting Sherwood number is

$$\mathbf{Sh}_x = \frac{k_c x}{D_{AB}} = 0.36 \sqrt{\mathbf{Re}_x}\, \mathbf{Sc}^{\frac{1}{3}} \qquad (14\text{-}36)$$

which is approximately equivalent to the exact solution given by Eq. (14-18). The deviation occurs because of the slight difference between the assumed profile and the actual profile.

The degree of accuracy of this result suggests the versatility of this technique for other cases in which exact solutions are unknown.

Turbulent flow over a flat plate In the absence of exact solutions in turbulent flow, we can apply the von Kármán integral technique upon assuming turbulent profiles for the hydrodynamic and concentration boundary layers.

Except for a very thin layer next to the plate, the *one-seventh-power law*

$$\frac{u}{V_\infty} = \left(\frac{y}{\delta}\right)^{\frac{1}{7}} \qquad (13\text{-}82)$$

well represents the velocity distribution in the turbulent boundary layer. The concentration profile is assumed to be of the same form,

$$\frac{\rho_A - \rho_{A_s}}{\rho_{A_\infty} - \rho_{A_s}} = \left(\frac{y}{\delta_c}\right)^{\frac{1}{7}} \qquad (14\text{-}37)$$

Upon substituting Eqs. (13-82) and (14-37) into Eq. (14-33), the resulting expression for a Schmidt number of unity is

$$\mathbf{Sh}_x \equiv \frac{k_c x}{D_{AB}} = 0.0292 \mathbf{Re}_x^{\frac{4}{5}} \tag{14-38}$$

Since this results in

$$k_c = C x^{-\frac{1}{5}}$$

the average convective-mass-transfer coefficient, if the flow is turbulent throughout the entire length of the plate, is

$$\bar{k}_c = \frac{C \int_0^L x^{-\frac{1}{5}} dx}{\int_0^L dx} = \frac{5}{4} C L^{-\frac{1}{5}} \tag{14-39}$$

Therefore, the average Sherwood number is

$$\overline{\mathbf{Sh}}_L \equiv \frac{\bar{k}_c L}{D_{AB}} = 0.0365 \mathbf{Re}_L^{\frac{4}{5}} \tag{14-40}$$

If a significant length of the plate is covered by a laminar boundary layer, the average mass-transfer coefficient is found by integrating the laminar equation up to the critical length x_c and the turbulent equation from the critical length to the end of the plate. From Eqs. (14-14) and (14-38), the average mass-transfer coefficient for a flow which is partly laminar and partly turbulent is

$$\bar{k}_c = \frac{D_{AB}}{L} \left[0.332 \left(\frac{V_\infty}{\nu} \right)^{\frac{1}{2}} \int_0^{x_c} x^{-\frac{1}{2}} dx + 0.0292 \left(\frac{V_\infty}{\nu} \right)^{\frac{4}{5}} \int_{x_c}^L x^{-\frac{1}{5}} dx \right] \tag{14-41}$$

or

$$\bar{k}_c = \frac{D_{AB}}{L} \left[0.664 \sqrt{\mathbf{Re}_{x_c}} + 0.0365 (\mathbf{Re}_L^{\frac{4}{5}} - \mathbf{Re}_{x_c}^{\frac{4}{5}}) \right] \tag{14-42}$$

For a critical Reynolds number of 500,000 and $\mathbf{Sc} = 1.0$

$$\overline{\mathbf{Sh}}_L \equiv \frac{\bar{k}_c L}{D_{AB}} = 0.0365 \mathbf{Re}_L^{\frac{4}{5}} - 840 \tag{14-43}$$

Example Aviation fuel at 60°F, having properties

$\nu = 10^{-5}$ ft²/sec $\qquad R = 76$ ft-lb$_f$/lb$_m$-°R

$\rho = 54$ lb$_m$/ft³ $\qquad D_{AB} = 0.55$ ft²/hr

is spilled on a large flat surface and quickly spreads over a length of 6 ft to a depth of ¼ in. The vapor pressure of the fuel at 60°F is 2 psia. Estimate the evaporation rate if a gentle breeze at 70°F is blowing at 5 mph parallel to the fuel surface; the kinematic viscosity ν of air at the film temperature $T_f = 65°F$ is 1.62×10^{-4} ft²/sec.

CONVECTIVE MASS TRANSFER

Solution Since the evaporation rate is unknown at the outset, we shall assume that it is sufficiently low to have no influence upon the hydrodynamic boundary layer:

$$\mathbf{Re}_L = \frac{V_\infty L}{\nu} = \frac{(\frac{88}{12}\text{ ft/sec})(6\text{ ft})}{1.62 \times 10^{-4}\text{ ft}^2/\text{sec}} = 2.72 \times 10^6$$

It is obvious that both laminar and turbulent flow exist. Therefore, Eq. (14-43) gives the average Sherwood number, since

$$\mathbf{Sc} = \frac{(1.62 \times 10^{-4}\text{ ft}^2/\text{sec})(3{,}600\text{ sec/hr})}{0.55\text{ ft}^2/\text{hr}} = 1.05$$

we have

$$\overline{\mathbf{Sh}}_L \equiv \frac{\bar{k}_c L}{D_{AB}} = 0.0365(2.72 \times 10^6)^{\frac{4}{5}} - 840$$

$$\bar{k}_c = \frac{0.55\text{ ft}^2/\text{hr}}{6\text{ ft}}(5{,}100 - 840) = 391\text{ ft/hr}$$

The mass-transfer rate is given by Eq. (14-1). In terms of partial pressures

$$N_A = \frac{\bar{k}_c}{R_A \bar{T}_A}(p_{A_s} - p_{A_\infty})$$

where $\bar{T}_A$ is the film temperature, since the process is not isothermal; therefore,

$$N_A = \frac{391\text{ ft/hr}}{(76\text{ ft-lb}_f/\text{lb}_m\text{-°R})(460 + 65)°\text{R}}(2 - 0)\text{ lb}_f/\text{in}^2\,\frac{144\text{ in}^2}{\text{ft}^2} = 2.82\text{ lb}_m/\text{hr-ft}^2$$

or

$$\rho v_s = 2.82\text{ lb}_m/\text{hr-ft}^2$$

$$v_s = \frac{2.82\text{ lb}_m/\text{hr-ft}^2}{54\text{ lb}_m/\text{ft}^3}\,\frac{\text{hr}}{3{,}600\text{ sec}} = 1.45 \times 10^{-3}\text{ ft/sec}$$

This evaporation velocity permits us to check the initial assumption (low mass transfer), particularly for the laminar portion of the boundary layer.

The blowing parameter is

$$\frac{v_s}{V_\infty}\sqrt{\mathbf{Re}_L} = \frac{1.45 \times 10^{-3}}{\frac{88}{12}}\sqrt{2.72 \times 10^6} = 0.326$$

which is negligible when compared with the parameter of Fig. 14-2. Hence, the assumption of low mass-transfer rate was valid, and

$$N_A = 2.82\text{ lb}_m/\text{hr-ft}^2$$

14-3 ANALOGY BETWEEN MOMENTUM, HEAT, AND MASS TRANSFER

In the preceding sections we have noted the striking similarities among the transport processes. These similarities permit us to predict the behavior of one transport process from our experience with another.

Reynolds presented the first formal concept of the analogy between heat and momentum transfer in 1874 [16]. His results, introduced in Sec. 13-7 for the case of turbulent flow over a plane surface, are limited to flow systems having a Prandtl number of unity. Since then, many investigators [3, 15, 17, 19] have extended and refined the analogy to apply to a wide range of Prandtl numbers and to all types of flow situations.

Reynolds analogy The Reynolds analogy can readily be extended to include the case of mass transfer if the Schmidt number is also unity. Upon comparing Eq. (12-111) with Eq. (14-35), we see that

$$\frac{\partial}{\partial y}\left(\frac{u}{V_\infty}\right)\bigg|_{y=0} = \frac{\partial}{\partial y}\left(\frac{\rho_A - \rho_{A_s}}{\rho_{A\infty} - \rho_{A_s}}\right)\bigg|_{y=0} \qquad (14\text{-}44)$$

when $\delta = \delta_c$, which is valid when $\mathbf{Sc} = 1$; and it is valid for either laminar or turbulent flow. Replacing the gradients at the surface by their equivalents from Newton's law of viscosity and Fick's first law, respectively,

$$\frac{\partial u}{\partial y}\bigg|_{y=0} = \frac{\tau_s}{\mu} \equiv \frac{c_f\,\rho V_\infty^2}{\mu\,2} \qquad (14\text{-}45)$$

$$\frac{\partial}{\partial y}(\rho_A - \rho_{A_s})\bigg|_{y=0} = -\frac{N_{A_s}}{D_{AB}} \equiv -\frac{k_c(\rho_{A_s} - \rho_{A\infty})}{D_{AB}} \qquad (14\text{-}46)$$

we get

$$\frac{1}{V_\infty}\frac{c_f}{\mu}\frac{\rho V_\infty^2}{2} = -\frac{1}{\rho_{A\infty} - \rho_{A_s}}\frac{k_c(\rho_{A_s} - \rho_{A\infty})}{D_{AB}} \qquad (14\text{-}47a)$$

or

$$\frac{c_f}{2} = \frac{\mu}{\rho D_{AB}}\frac{k_c}{V_\infty} \qquad (14\text{-}47b)$$

But $\mu/\rho D_{AB} \equiv \mathbf{Sc} = 1$; therefore,

$$\frac{c_f}{2} = \frac{k_c}{V_\infty} \qquad (14\text{-}48)$$

Combining with Eq. (13-102), we have the Reynolds analogy for all three of the transport mechanisms

$$\blacksquare \qquad \frac{c_f}{2} = \frac{k_c}{V_\infty} = \frac{h_x}{c_p \rho V_\infty} \qquad (14\text{-}49a)$$

applicable when the Prandtl and Schmidt numbers are unity. This same relationship is valid for average values of the respective coefficients, i.e.,

$$\frac{C_f}{2} = \frac{\bar{k}_c}{V_\infty} = \frac{\bar{h}}{c_p \rho V_\infty} \qquad (14\text{-}49b)$$

Chilton-Colburn analogy Combining the exact expression for mass transfer in flow over a flat plate

$$\frac{k_c x}{D_{AB}} = 0.332 \sqrt{\mathbf{Re}_x}\, \mathbf{Sc}^{\frac{1}{3}} \tag{14-18}$$

with the Blasius result for laminar flow

$$c_f = \frac{0.664}{\sqrt{\mathbf{Re}_x}} \tag{12-94}$$

we get

$$\frac{k_c x}{D_{AB}} = \frac{c_f}{2} \mathbf{Re}_x \mathbf{Sc}^{\frac{1}{3}} \tag{14-50}$$

or

$$\frac{c_f}{2} = \frac{k_c}{V_\infty} \mathbf{Sc}^{\frac{2}{3}} \equiv j_M \tag{14-51}$$

where j_M is the j factor for mass transfer. This relationship was developed empirically by Chilton and Colburn [4], who verified its validity for $0.6 < \mathbf{Sc} < 2{,}500$. Referring to Eq. (13-104), the complete analogy is then

■ $$\frac{c_f}{2} = \frac{k_c}{V_\infty} \mathbf{Sc}^{\frac{2}{3}} = \frac{h_x}{c_p \rho V_\infty} \mathbf{Pr}^{\frac{2}{3}} = j_M = j_H \tag{14-52a}$$

or

$$\frac{C_f}{2} = \frac{\bar{k}_c}{V_\infty} \mathbf{Sc}^{\frac{2}{3}} = \frac{\bar{h}}{c_p \rho V_\infty} \mathbf{Pr}^{\frac{2}{3}} = \bar{j}_M = \bar{j}_H \tag{14-52b}$$

This relation is valid for gases and liquids in the range $0.6 < \mathbf{Sc} < 2{,}500$ and $0.6 < \mathbf{Pr} < 100$ [22]. When form drag is present, the analogy does not hold for momentum although it continues to be valid for the frictional-drag portion and for heat and mass, i.e.,

$$\frac{c_D}{2} \neq \frac{c_f}{2} = j_M = j_H \tag{14-53}$$

Eddy transport theory The eddy transport theory can be used to clarify the effects of deviating from Prandtl and Schmidt numbers of unity in making the analogies among the transport mechanisms.

In Sec 13-2 the eddy momentum diffusivity (eddy viscosity) ϵ was defined in describing the apparent shear stress in turbulent flow

$$\tau_{\text{turb}} \equiv \rho \epsilon \frac{d\bar{u}}{dy} = -\rho \overline{u'v'} \tag{13-24}$$

parallel with the definition of ordinary momentum diffusivity (kinematic viscosity) ν in laminar flow

$$\tau_{\text{lam}} \equiv \rho \nu \frac{d\bar{u}}{dy} \tag{14-54}$$

In the latter case the average value of velocity $\bar{u}$ is identically equal to the local velocity u, since $u' = 0$ in the equation $u = \bar{u} + u'$.

The total shear stress τ is given by

$$\tau \equiv \tau_{\text{lam}} + \tau_{\text{turb}} = \rho(\nu + \epsilon)\frac{d\bar{u}}{dy} \tag{14-55}$$

We have previously noted that the kinematic viscosity ν is a state property. The eddy viscosity ϵ, on the other hand, is not a function of state only but depends strongly upon the motion. In an effort to assign a value to ϵ, the Prandtl mixing length ℓ

$$\epsilon = \ell^2 \left|\frac{d\bar{u}}{dy}\right| = \ell \overline{v'} \tag{14-56}$$

was introduced in Sec. 13-2. Although somewhat nebulous, the magnitude of the Prandtl mixing length ℓ can be bracketed; it must lie between zero and the maximum dimension of the flow channel. The Prandtl mixing length ℓ was eliminated, however, by assuming it to be proportional to the distance from the surface

$$\ell = ky \tag{13-32}$$

and, by using the universal velocity profile, an expression for ϵ was obtained.

When temperature and concentration gradients exist in the fluid, heat and mass are transported by the turbulent fluctuations, or eddies, in a similar manner, giving

$$q'' = -\rho c_p (\alpha + \epsilon_H)\frac{d\bar{T}}{dy} \tag{14-57}$$

$$N_A = -(D_{AB} + \epsilon_M)\frac{d\bar{\rho}_A}{dy} \tag{14-58}$$

where

$$q''_{\text{turb}} = -\rho c_p \overline{u'T'} \tag{14-59}$$

$$N_{A_{\text{turb}}} = -\overline{u'\rho'_A} \tag{14-60}$$

The temporal average quantities came from the governing turbulent equations.

Prandtl's mixing-length theory is equally valid for the transport of heat and mass,

CONVECTIVE MASS TRANSFER

resulting in

$$\epsilon_H = \underline{\ell}^2 \left| \frac{d\bar{T}}{dy} \right| = \underline{\ell}\overline{v'} \tag{14-61}$$

$$\epsilon_M = \underline{\ell}^2 \left| \frac{d\bar{\rho}_A}{dy} \right| = \underline{\ell}\overline{v'} \tag{14-62}$$

where

$$T = \bar{T} + T' \tag{14-63}$$

$$\rho_A = \bar{\rho}_A + \rho'_A \tag{14-64}$$

These results assume that the Prandtl mixing length $\underline{\ell}$ is identical for each of the transport processes, giving

$$\epsilon = \epsilon_H = \epsilon_M \tag{14-65}$$

from Eqs. (14-46), (14-51), and (14-52). Figure 14-6 indicates that this is a good assumption for gases, vapors, and some liquids.

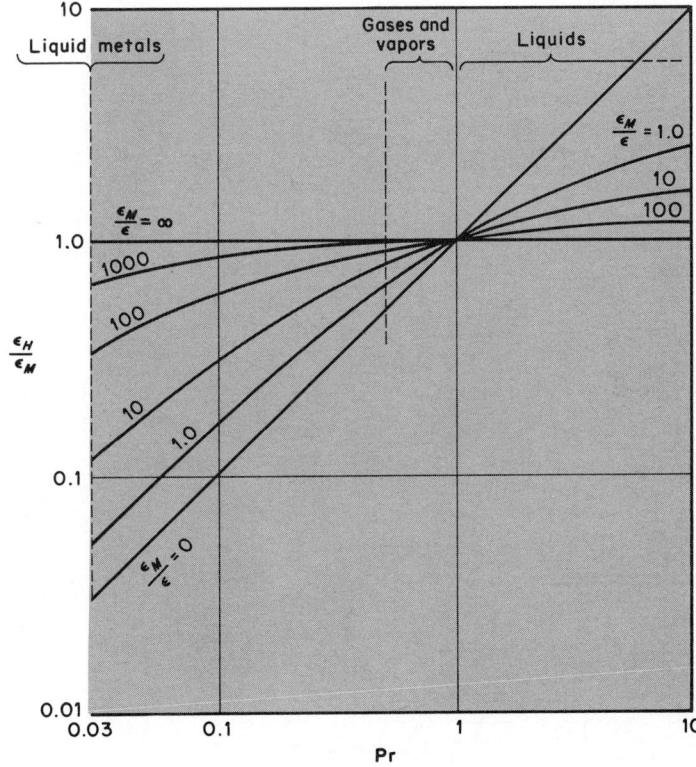

Fig. 14-6 The effect of Prandtl number on the ratio of diffusivities. (*Adapted from R. Jenkins, Heat Transfer Fluid Mech. Inst., Stanford, 1951.*)

For completely turbulent flow $v \ll \epsilon$, $\alpha \ll \epsilon_H$, and $D_{AB} \ll \epsilon_M$; therefore, upon dividing Eq. (14-55) by Eq. (14-57), we get

$$\tau c_p \, dT = q'' \, du \tag{14-66}$$

or

$$c_f \frac{\rho V_\infty^2}{2} c_p \, dT = h_x(T_s - T_\infty) \, du$$

which, integrated from the surface to the free stream, gives

$$\frac{c_f}{2} = \frac{h_x}{c_p \rho V_\infty} \tag{14-67}$$

identical with Eq. (13-102). In a similar manner, the simultaneous solution of Eqs. (14-55) and (14-58) results in

$$\frac{c_f}{2} = \frac{k_c}{V_\infty} \tag{14-48}$$

These results were obtained without assuming the Prandtl and Schmidt numbers to be unity; we did assume, however, completely turbulent flow. We might note that for **Pr** $= 1$ $(v = \alpha)$ and **Sc** $= 1$ $(v = D_{AB})$, it is unnecessary to assume completely turbulent flow, since $v + \epsilon = \alpha + \epsilon_H = D_{AB} + \epsilon_M$, giving the same results.

Prandtl-Taylor analogy In an effort to be more realistic, Prandtl [15] and Taylor [20] accounted for the laminar sublayer (Fig. 14-7) by writing a diffusion equation for it while using a Reynolds analogy equation for the turbulent core. The resulting relations [22], which reduce to the simple Reynolds analogy for **Pr** $= 1$ and **Sc** $= 1$, are

Heat-momentum:
$$\frac{h_x}{c_p \rho V_\infty} = \frac{1}{1 + 5\sqrt{c_f/2} \, (\mathbf{Pr} - 1)} \frac{c_f}{2} \tag{14-68}$$

Mass-momentum:
$$\frac{k_c}{V_\infty} = \frac{1}{1 + 5\sqrt{c_f/2} \, (\mathbf{Sc} - 1)} \frac{c_f}{2} \tag{14-69}$$

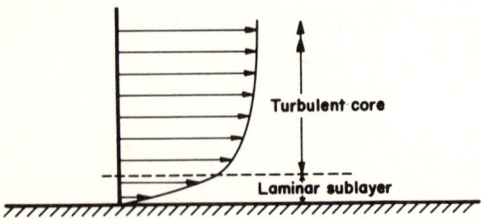

Fig. 14-7 Prandtl-Taylor flow model.

CONVECTIVE MASS TRANSFER

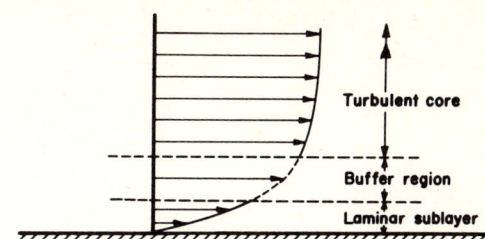

Fig. 14-8 von Kármán flow model.

von Kármán analogy Considering a buffer zone to lie between the turbulent core and the laminar sublayer, as illustrated in Fig. 14-8, von Kármán made a major improvement in the Prandtl-Taylor results [1]. His results

Heat-momentum:
$$\frac{h_x}{c_p \rho V_\infty} = \frac{1}{1 + 5\sqrt{c_f/2}\,\{\mathbf{Pr} - 1 + \ln[(1 + 5\mathbf{Pr})/6]\}} \frac{c_f}{2} \quad (14\text{-}70)$$

Mass-momentum:
$$\frac{k_c}{V_\infty} = \frac{1}{1 + 5\sqrt{c_f/2}\,\{\mathbf{Sc} - 1 + \ln[(1 + 5\mathbf{Sc})/6]\}} \frac{c_f}{2} \quad (14\text{-}71)$$

also reduce to the simple Reynolds analogy for fluids with $\mathbf{Pr} = 1$ and $\mathbf{Sc} = 1$.

14-4 DIMENSIONAL ANALYSIS

We have gained enough insight into the convective-mass-transfer process to know that the coefficient k_c depends upon (1) tube diameter D (or plate length L), (2) average fluid velocity V, (3) mass diffusivity D_{AB}, (4) fluid viscosity μ, and (5) fluid density ρ. Expressed functionally, the relationship is

$$k_c = F(D, V, D_{AB}, \mu, \rho) \quad (14\text{-}72)$$

Using the Buckingham pi theorem of Appendix C, these variables can be arranged in dimensionless groups to give

$$\frac{k_c D}{D_{AB}} = F_1\left(\frac{DV\rho}{\mu}, \frac{\mu}{\rho D_{AB}}\right) \quad (14\text{-}73a)$$

or

$$\mathbf{Sh}_D = F_1(\mathbf{Re}_D, \mathbf{Sc}) \quad (14\text{-}73b)$$

For mass transfer in external flows

$$\frac{k_c x}{D_{AB}} = F_2\left(\frac{Vx\rho}{\mu}, \frac{\mu}{\rho D_{AB}}\right) \quad (14\text{-}74a)$$

or

$$\mathbf{Sh}_x = F_2(\mathbf{Re}_x, \mathbf{Sc}) \quad (14\text{-}74b)$$

Dimensional analysis yields only the form of the relation; it does not give the pertinent constants, which must be determined by experiment.

The remainder of this chapter is devoted to design equations for some of the most common cases of convective mass transfer.

14-5 DESIGN EQUATIONS

Unless otherwise specified, best results are obtained when properties are evaluated at the film condition

$$\rho_{A_f} \equiv \frac{\rho_{A_s} + \rho_{A\infty}}{2} \tag{14-22}$$

in external flows and at the bulk condition

$$\rho_{A|b} \equiv \frac{\rho_{A_i} + \rho_{A_e}}{2} \tag{14-28}$$

for internal flows. For convenience, most of the correlation equations of this section are in terms of the j factors, $j_H = j_M$,

$$j_H \equiv \frac{h}{c_p \rho V_\infty} \mathbf{Pr}^{\frac{2}{3}} \tag{13-104}$$

$$j_M \equiv \frac{k_c}{V_\infty} \mathbf{Sc}^{\frac{2}{3}} \tag{14-51}$$

The equations which best represent the experimental data are shown on the correlation curves of Fig. 14-9.

The correlation of Fig. 14-10 is applicable to point conditions in the region of developed flow; if the tube length is greater than 50 diameters, it is valid for the entire pipe surface. It may be assumed that fully developed turbulent flow is attained in a length of 10 pipe diameters.

The correlations of Fig. 14-11 may be used for packed beds of cylinders with

$$d_p \equiv \sqrt{d_c L + d_c^2/2} \tag{14-75}$$

where d_c is the cylinder diameter and L is its length. Fluidized-bed data are correlated by the equation [1]

$$\bar{j}_M = 1.90 \mathbf{Re}_p^{-0.5} \tag{14-76}$$

Bennett and Myers [1] summarize data for other types of packing and for two-phase flow.

Superimposed on Fig. 14-12 are the friction-drag coefficient C_f and the form-drag coefficient C_D, illustrating the influence which form drag has on the result as discussed in Sec. 14-3.

The equations of this section are valid only when the mass-transfer rate does not

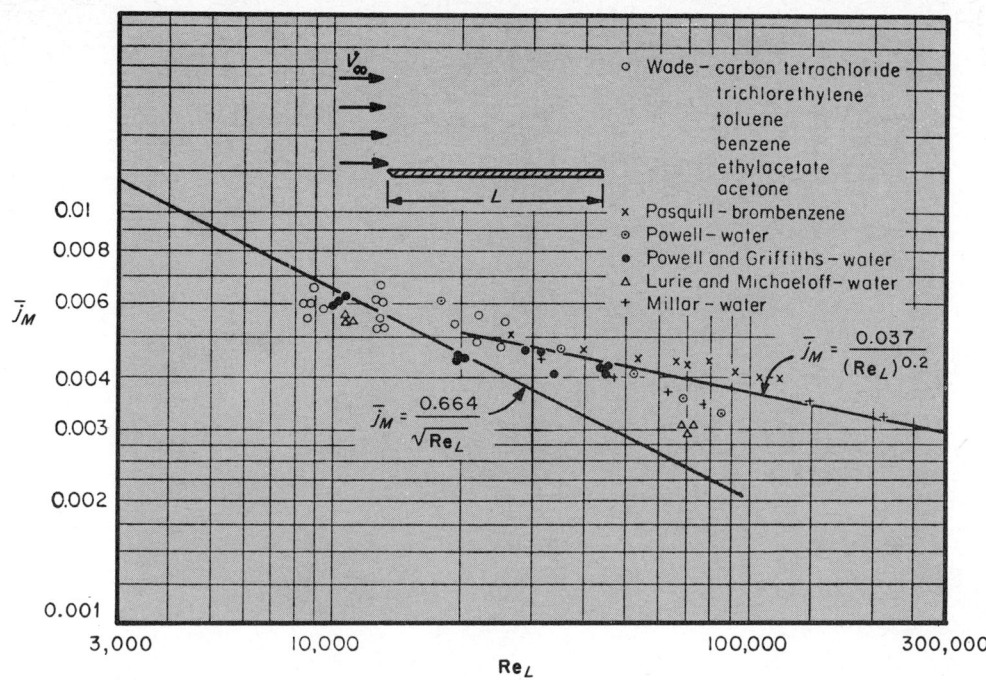

Fig. 14-9 Evaporation from flat plate, $Sc > 0.6$. (*Modified from T. K. Sherwood and R. L. Pigford, "Absorption and Extraction." Copyright 1952. McGraw-Hill Book Company. Used by permission.*)

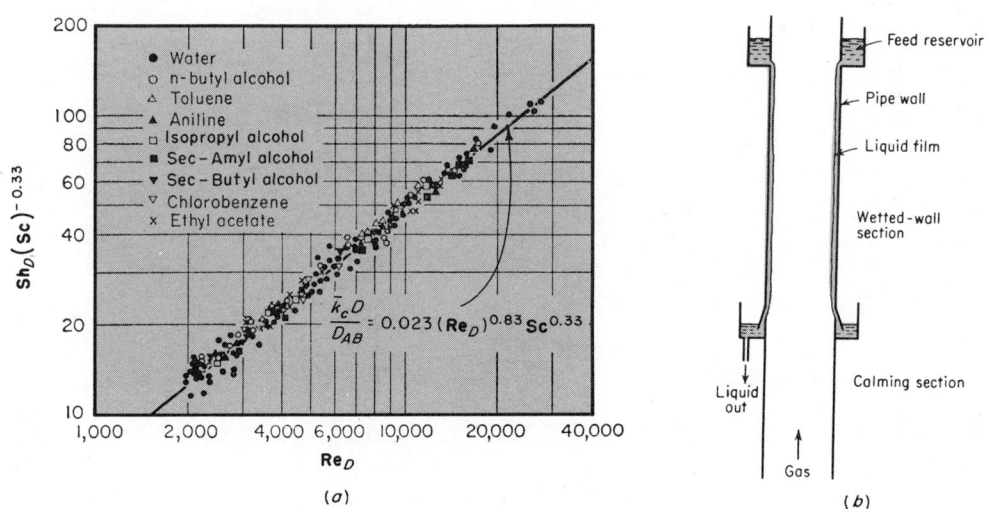

Fig. 14-10 Vaporization of liquids with gas flow inside circular tubes: (*a*) experimental data (wetted-wall column), $0.6 < Sc < 2{,}200$. (*Data from Gilliland and Sherwood* [8] *and Linton and Sherwood* [14].) (*b*) Wetted-wall tower.

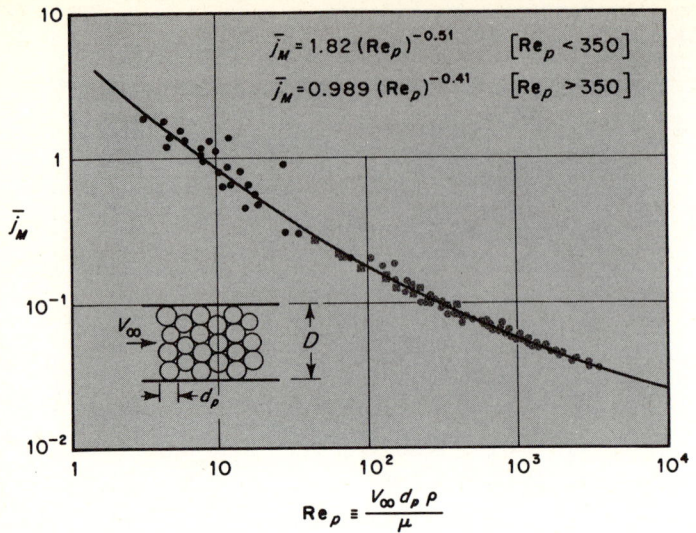

Fig. 14-11 Single-phase flow in packed beds of spherical granular solids. [*From M. Hobson and G. Thodos, Chem. Eng. Progr.,* **45**:*247–254 (1947).*]

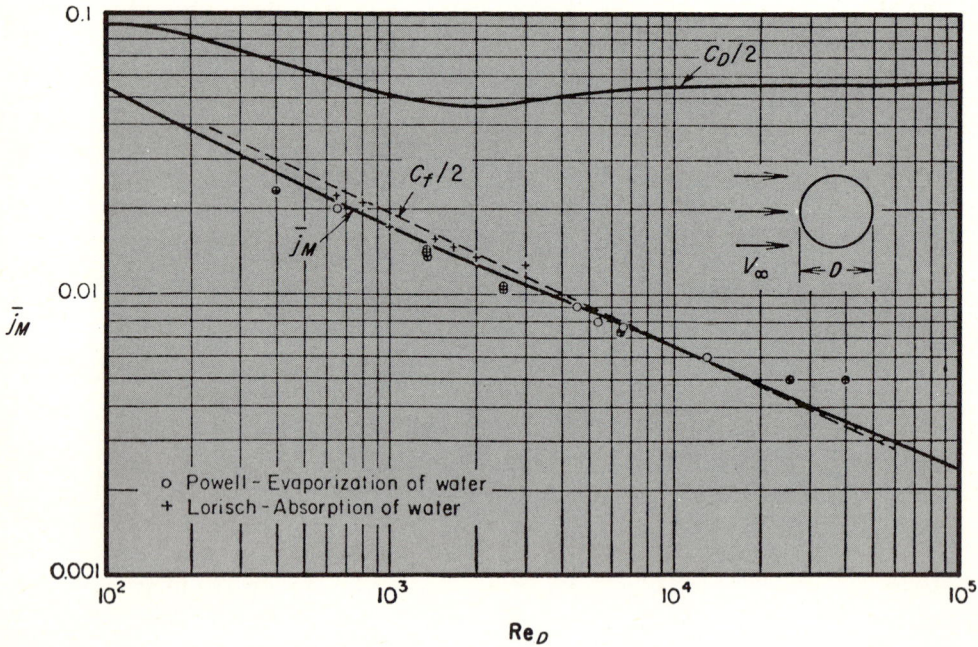

Fig. 14-12 Mass transfer in flow past single cylinder. (*Modified from T. K. Sherwood and R. L. Pigford, "Absorption and Extraction." Copyright 1952. McGraw-Hill Book Company. Used by permission.*)

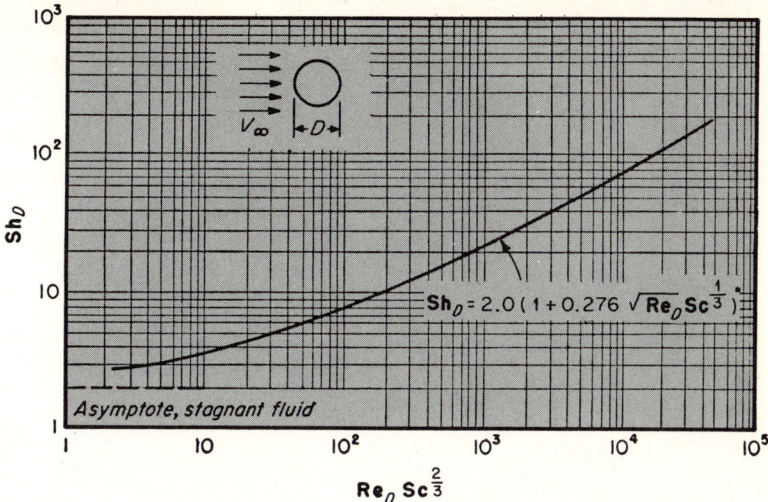

Fig. 14-13 Mass transfer in flow past single sphere. (*Modified from T. K. Sherwood and R. L. Pigford, "Absorption and Extraction." Copyright 1952. McGraw-Hill Book Company. Used by permission.*)

significantly alter the hydrodynamic boundary layer. For high mass-transfer rates, correction factors must be applied to account for the mass-transfer momentum-transfer interaction [2].

Example A 6-in-diam porous cylinder is placed in a wind tunnel. Dry air at 100°F flows normal to the cylinder at a velocity of 30 ft/sec. If the cylinder is continuously saturated with water, estimate the convective-mass-transfer coefficient; $D_{aw} = D_{wa} = 0.90$ ft²/hr.

Solution The Reynolds number based on the cylinder diameter is

$$\text{Re}_D = \frac{V_\infty D}{\nu} = \frac{(30 \text{ ft/sec})(0.5 \text{ ft})}{0.00018 \text{ ft}^2/\text{sec}} = 8.33 \times 10^4$$

where the kinematic viscosity ν is taken from Table A-3. Using Fig. 14-12, the average j factor for mass transfer is

$$\bar{j}_M \equiv \frac{\bar{k}_c}{V_\infty} \text{Sc}^{\frac{2}{3}} = 0.0026$$

Assuming the process to be isothermal at 100°F, the Schmidt number is

$$\text{Sc} \equiv \frac{\nu}{D_{aw}} = \frac{0.00018 \text{ ft}^2/\text{sec}}{(0.90/3,600) \text{ ft}^2/\text{sec}} = 0.72$$

Hence,

$$\bar{k}_c = \frac{0.0026}{0.72^{\frac{2}{3}}} (30 \text{ ft/sec}) = 0.097 \text{ ft/sec}$$

Example Air passes through a packed bed of 1-in-diam napththalene spheres. If the bed is at 50°F, determine the average convective-mass-transfer coefficient $\bar{k}_c$ for sublimation of the naphthalene when the average air velocity is 2 ft/sec. $D_{na} = 0.2$ ft²/hr.

Solution The Reynolds number based on the diameter of the spheres is

$$\mathbf{Re}_p = \frac{V_\infty d_p}{\nu} = \frac{(2 \text{ ft/sec})(\frac{1}{12} \text{ ft})}{0.156 \times 10^{-3} \text{ ft}^2/\text{sec}} = 1{,}070$$

where ν is taken from Table A-3. Figure 14-11 gives

$$j_M \equiv \frac{\bar{k}_c}{V_\infty} \mathbf{Sc}^{\frac{2}{3}} = 0.989(1{,}070^{-0.41})$$

The Schmidt number is

$$\mathbf{Sc} \equiv \frac{\nu}{D_{na}} = \frac{0.156 \times 10^{-3} \text{ ft}^2/\text{sec}}{(0.2/3{,}600) \text{ ft}^2/\text{sec}} = 2.81$$

Therefore,

$$\bar{k}_c = \frac{0.989(2 \text{ ft/sec})}{1{,}070^{0.41}(2.81^{\frac{2}{3}})} = 0.0572 \text{ ft/sec}$$

14-6 A NOTE ON TRANSPORT BETWEEN PHASES

This chapter has emphasized the transport process within a given phase. There are many important processes, however, such as distillation, absorption, and extraction, where it is necessary to consider the transport process between two contacting phases. In this case, referred to as *interphase mass transfer*, the mass-transfer rate is affected by the relative equilibrium between the phases in addition to being affected by the concentration gradients of both touching phases. The design correlations are largely empirical because of the complexity of the mechanism; the convective-mass-transfer coefficient itself depends upon the mass-transfer rate.

Readers who intend to study the unit operations processes in detail are referred to Welty, Wicks, and Wilson [22] and Bird, Stewart, and Lightfoot [2] for discussion of interphase mass transfer and some design equations. Details of such processes can then be found, for example, in Treybal [21], written primarily for the chemical engineer.

PROBLEMS

14-1. Mercury at 68°F flowing steadily at 0.15 ft/sec over a lead plate 2 ft long by 3 ft wide is subjected to irradiation. During a period of 36 weeks the mercury diffuses into the plate to a depth of 0.12 in. For a mass fraction of 0.15 at the plate surface, estimate the mass of mercury which diffuses into the plate during the 36-week period, assuming that the coefficient of diffusivity remains constant throughout the process. Is the process diffusion- or convection-controlled?

14-2. A shallow 2-ft-long pan of ether at 32°F is exposed to a 45°F airstream flowing at 8 ft/sec. Assume ether has thermodynamic properties very close to those of carbon monoxide. Assuming that the air is saturated when it contains 15 percent ether by weight, estimate the mass-transfer rate per unit area.

CONVECTIVE MASS TRANSFER

14-3. Helium at 70°F and 1 atm flows through a rectangular Pyrex duct at 10 ft/sec. Estimate the mass-transfer rate if the helium sees the duct as four large flat plates 7.0 ft long.

14-4. Fuel oil at 60°F flows through a 12-in-ID porous pipe with an average velocity of 2 ft/sec. The coefficient of diffusivity of the oil into the pipe is 0.8 ft²/hr. Estimate the mass-transfer rate after the exterior of the pipe reaches a mass concentration of 12 percent. The pipe is 200 ft long and serves as a filter for the oil diffusing through it. The wall thickness of the pipe is 0.5 in.

14-5. Dry air at 1 atm pressure enters a 20 ft-long 6-in-diam tube at 100°F and 0.5 ft/sec. The inner surface of the tube is lined with a felt material, which is continuously saturated with water at 60°F. Assuming constant temperature of the air and the pipe, determine the amount of water required per hour.

14-6. Gas from a supply line enters a porous 3-in-diam bulb 6 in long. Convective air currents flow over the bulb at 2 ft/sec. For an air temperature of 50°F, estimate the mass-transfer rate from the bulb if the gas has the following properties: $D_{AB} = 8$ ft²/hr; $R = 80$ ft-lb$_f$/lb$_m$-°R; vapor pressure $= 1$ lb$_f$/in².

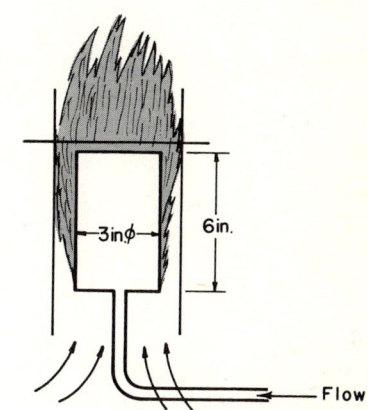

Fig. P14-6

14-7. (a) Air at 110°F and a relative humidity of 20 percent flows over a smooth water surface at an average velocity of 4 mph. The length of the water surface in the direction of the airflow is 6 ft. Assuming the relative humidity to be 100 percent at the liquid–air interface and the average water surface temperature to be 90°F and using Eq. (7-8) to calculate the coefficient of diffusivity of water vapor into air, calculate the water evaporated per hour per square foot of surface area.

(b) Repeat part (a) for an air velocity of 20 mph, assuming a critical Reynolds number of 500,000. Assume $Sc = 1$.

14-8. Using the expressions for the local Sherwood moduli in both laminar and turbulent regions resulting from integral analyses [Eqs. (14-36) and (14-38)], obtain an expression for the average Sherwood modulus for critical Reynolds numbers of 300,000, 500,000 and 600,000. Assume that the Schmidt number is unity.

14-9. A chemist in a laboratory accidentally upsets a beaker of alcohol on a smooth table top. If a fan blows 65°F air over the alcohol at the same temperature in such a way that the "effective" length of the puddle is 4 ft and the air moves at an average velocity of 30 ft/sec, how much alcohol will evaporate in 1 min? The vapor pressure of the alcohol is 0.182 psia, and $D_{AB} = 0.595$ ft²/hr.

14-10. Hydrogen at 0°F is forced over a bed of naphthalene. The average velocity of the hydrogen is 5 ft/sec, and the 200-ft long bed is at 212°F. How much naphthalene is transferred to the hydrogen in one hour? The vapor pressure of the naphthalene is 0.124 psia, and $D_{AB} = 0.199$ ft²/hr.

14-11. A very wide 8-ft-long flat porous plate is continually saturated with ammonia at 100°F. It is subjected to an airflow in the 8-ft direction, the air velocity and temperature being 20 ft/sec and 100°F, respectively. If the coefficient of diffusivity of ammonia into air is 0.982 ft²/hr, estimate the average convective-mass-transfer coefficient $\bar{k}_c$.

14-12. Crude oil near a well is stored in a large open square tank having 100-ft sides. A 7-mph wind blows over the top of the tank on a hot day (100°F) perpendicular to one of the sides. Estimate the mass transfer to the air per hour. The vapor pressure of the crude oil is 0.19 psia and $D_{AB} = 0.65$ ft²/hr.

14-13. A freshly washed bed sheet is hung on a clothesline in 20°F weather. The sheet quickly freezes, and a short time later a strong 20-mph wind arises. The sheet is elevated almost parallel to the ground, and the wind blows along the surface of the sheet in such a way that it is 5 ft from leading edge to trailing edge. If the coefficient of diffusivity of the water (ice) vapor to the air is 0.5 ft²/hr and the concentration at the surface of the sheet is 100 percent, estimate the initial mass-transfer rate to the air for 0 percent relative humidity. Estimate the time required for the sheet to dry if it initially contains 0.03 lb_m water/ft².

14-14. Excess urea is removed from a plastic sheet after manufacture by passing a stream of water over the sheet after it has cooled to a room temperature of 77°F. If the sheet is 2 ft long and water flows over it at 1 ft/sec, estimate the initial mass-transfer rate for a free urea concentration of 0.5 lb_m/ft³.

14-15. In an industrial process air at 75°F and 14.7 psia flows at 4 ft/sec over an 8-ft-long vat of acetone at the same temperature. The vapor pressure of the acetone is 3.87 psia. Estimate the mass-transfer rate.

14-16. A contractor realizes at the last minute that he has neglected to spray a second coat of paint on a wall on a house he is building. The wall is 50 ft long and 12 ft high. A gentle breeze at 7 mph blows along the wall on a 60°F day with 30 percent relative humidity. If the contractor uses a linseed-oil-base paint which is 40 percent linseed oil and it takes $1\frac{1}{2}$ gal to paint the wall at $8\frac{1}{2}$ lb_m/gal, what is the initial mass-transfer rate to the air? The vapor pressure of linseed oil at 60°F is 1.76 psia, and $D_{AB} = 0.48$ ft²/hr. Assume the molecular weight of linseed oil is 67.8.

14-17. In a solubility study, carbon tetrachloride is forced to flow over a plate with a hard sulfur surface. The plate is 4 ft long, and the fluid velocity is 2 ft/sec. If the kinematic viscosity of carbon tetrachloride is 3.9×10^{-6} ft²/sec, and if 20 lb_m flows over the 6-in-wide plate in 30 sec, what is the mass-transfer rate to the carbon tetrachloride on the second pass of the fluid? The mass fraction at the surface is 0.07, and $D_{AB} = 5.2 \times 10^{-5}$ ft²/hr.

14-18. If a rectangular container of chloroform is left uncovered in a cold room where a draft at 1.5 ft/sec blows along a 6-in side at 32°F, how much time is required for 4 oz of the chloroform to evaporate from the shallow 4- by 6-in dish? Chloroform has a saturation vapor pressure of 1.4 psia at 32°F, and the gas constant is 63 ft-lb_f/lb_m-°F.

14-19. A plant layout engineer is designing a floor plan for a complex plant. Part of the plant must have a controlled air supply with a high humidity. In an adjoining section there is to be a process using a hot-water bath. He decides that since there is a convenient supply of water for the bath, this might be a good source for humidifying the air. The bath consists of a 12-ft square tank with the water kept at 200°F. If outside air is allowed to flow directly over the bath with a velocity of 8 ft/sec, estimate the mass transfer per hour on a day when the temperature is 70°F and the relative humidity is 40 percent.

14-20. (a) Using the Prandtl-Taylor analogy, determine the mass-transfer coefficient for oxygen flowing at 5 ft/sec under a tank of hydrogen (both at 32°F) as shown. What is the mass flow rate of hydrogen into the tank at the top to maintain constant pressure of 20 psia in the 4-ft-wide tank?

(b) How do the results of part (a) compare with those obtained from the von Kármán analogy?

14-21. Compare the results of the von Kármán analogy to the eddy-transport theory in determining the mass-transfer coefficient for the flow of glucose over a flat plate 15 ft long at 3 ft/sec if the plate

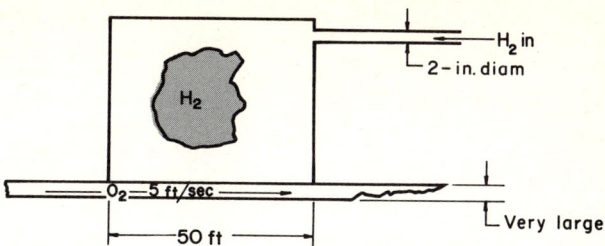

Fig. P14-20

is continually saturated with water at 77°F. Assume the glucose has a kinematic viscosity similar to aniline.

14-22. How long must a 3-ft-wide flat plate be to have an evaporation rate for benzene of 200 lb_m/hr to air at a temperature of 70°F if the air moves along the plate at 2 ft/sec? The vapor pressure of benzene at 70°F is 1.55 psia.

14-23. Hydrogen is transported at a velocity of 2 ft/sec through a large rectangular duct from an electrolytic vessel to a storage tank. The average hydrogen temperature is 85°F, and the duct walls are kept at 0°F. They are made of rubber with $D_{H_2-Rubber} = 0.7 \times 10^{-5}$ ft²/hr. If the duct is 30 ft long, estimate the mass-transfer rate to the duct, assuming that the rubber is thin enough for the process to be convection-controlled.

14-24. Water at 70°F evaporates to dry air blowing over a wetted flat plate 12 ft long at 2 ft/sec. Compare the result of Eq. (14-51) to the Reynolds analogy if for the Reynolds analogy a Schmidt number of unity is assumed.

14-25. Carbon monoxide flows over a flat plate which forms one side of a square duct. The process is isothermal at 32°F, and the carbon monoxide moves at 2 ft/sec. Oxygen is allowed to seep through small holes in the plate and is convected away. The holes are so spaced that the concentration of oxygen at the surface is 50 percent by volume. How long must the plate be to transfer 150 ft³/hr of oxygen to the carbon monoxide per foot of plate width?

14-26. What length of wetted-wall column is required to transfer 30 ft³/hr of saturated benzene to carbon dioxide at 32°F if the carbon dioxide flows through the 8-in-diam column at 4.0 ft/sec?

14-27. Carbon dioxide flows at 4 ft/sec through a wetted-wall column at 32°F. At the same time benzene is pumped from the lower reservoir to the feed reservoir. If in addition benzene is supplied to the feed reservoir from another source, what rate of flow from the other source is required to maintain a constant supply of benzene in the 6-in-diam column if it is 3 ft long? Assume the saturation vapor pressure of benzene is 0.484 psia.

14-28. If a bed of 2-in-diam CaCl spheres is 1 ft square and 34.68 in thick with the spheres in a hexagonal-close-pack structure (any four mutually adjacent spheres form a regular tetrahedron), estimate the mass transfer per hour from hydrogen at 100°F flowing through the bed at 1.0 ft/sec with an average water concentration of 0.01 lb_m/ft³.

14-29. Urea is passed through a bed of 1-ft-diam 1-ft-long packed porous cylinders; it has a kinematic viscosity of 4.4×10^{-4} ft²/sec. The cylinders are continually saturated with ethanol at 77°F. Determine the mass-transfer coefficient for a fluid velocity of 1.0 ft/sec.

14-30. Determine the average mass-transfer coefficient for air flowing at 2 ft/sec over a long porous 6-in-diam cylinder continuously saturated with ether at 77°F.

14-31. How long must a section of 3-in-ID porous tubing be to give the exiting fluid a hydrogen concentration of 0.0008 lb_m/ft³ if the tube wall has a hydrogen concentration of 0.0009 lb_m/ft³ and oxygen flows through the tube at 2 ft/sec? The process is isothermal at 32°F and 1 atm pressure.

14-32. If a 3-in-diam dry calcium chloride cylinder is placed in an air duct with air flowing at 60°F and 3 ft/sec with a relative humidity of 80 percent, estimate the initial mass-transfer rate of water to the calcium chloride cylinder; $D_{AB} = 0.87$ ft²/hr.

14-33. A cylindrical piece of ice is subjected to a normal cross flow of 32°F hydrogen at 6 ft/sec. The piece of ice is 1 ft in diameter. Assuming the ice to "wear" evenly, how much time is required for the outside 0.2-in layer to disappear?

14-34. The wall of a 6-in-diam cloth hose is continually saturated with benzene at 50°F. Hydrogen at 16.3 psia flows through the hose at an average bulk temperature of 100°F and at a velocity of 7 ft/sec. Determine $\bar{k}_c$ and the mass-transfer rate to the hydrogen, taking the vapor pressure of benzene at 50°F to be 1.35 psia and the average vapor pressure in the hose to be 0.675 psia.

14-35. A spherical sponge is used to mop up a quantity of chloroform from a shallow dish. The 4-in-diam sponge becomes completely saturated with the chloroform. If it is suspended in an open window with a gentle 70°F breeze at 1.0 ft/sec blowing over it, how much chloroform is transferred to the air in 1 sec? The vapor pressure of chloroform at 70°F is 3.68 psia.

14-36. A porous 3-in-diam sphere is saturated with glycerol. It is placed in a 1 ft/sec stream of ethanol while being supplied with enough glycerol to maintain saturation. The entire process takes place at 77°F. The amount of glycerol necessary to maintain saturation is found to be 4.58×10^{-3} ft³/hr. Determine the diffusivity. The kinematic viscosity of the ethanol may be taken as 4.48×10^{-5} ft²/sec.

14-37. (a) If a 6-in-diam mothball is suspended in an air duct to a cold-storage room, estimate the mass-transfer coefficient for airflow at 3 ft/sec and 40°F. $D_{AB} = 0.2$ ft²/hr.

(b) Suppose the air duct of part (a) contains a packed bed of 1-in-diam mothballs and there are just enough for the surface area of the large ball of part (a) to be equivalent to their combined surface areas. For identical flow conditions, compare the mass-transfer coefficients.

REFERENCES

1. Bennett, C. O., and J. E. Myers: "Momentum, Heat, and Mass Transfer," McGraw-Hill, New York, 1962.
2. Bird, R. B., W. E. Stewart, and E. N. Lightfoot: "Transport Phenomena," Wiley, New York, 1960.
3. Boelter, L. M. K., R. C. Martinelli, and F. Jonassen: Remarks on the Analogy between Heat and Momentum Transfer, *Trans. ASME*, **63**: 447–455 (1941).
4. Chilton, T. H., and A. P. Colburn: *Ind. Eng. Chem.*, **26**: 1183 (1934).
5. Christian, W. J., and S. P. Kezios: *AIChE J.*, **5**: 61 (1959).
6. Deissler, R. G.: Analysis of Turbulent Heat Transfer, Mass Transfer and Friction in Smooth Tubes at High Prandtl and Schmidt Numbers, *NACA Tech. Note* 3145, May 1945.
7. Eckert, E. R. G., and R. M. Drake, Jr.: "Heat and Mass Transfer," 2d ed., McGraw-Hill, New York, 1959.
8. Gilliland, E. R., and T. K. Sherwood: *Ind. Eng. Chem.*, **26**: 516 (1934).
9. Graetz, L.: Über die Wärmeleitungsfähigkeit von Flüssigkeiten, *Ann. Physik. Chem.*, **25**: 337–357 (1885).
10. Hartnett, J. P., and E. R. G. Eckert: *Trans. ASME*, **79**: 247–254 (1957).
11. Hobson, M., and G. Thodos: *Chem. Eng. Progr.*, **45**: 517–524 (1949).
12. Jenkins, R.: *Heat Transfer Fluid Mech. Inst. Stanford*, 1951, p. 147.
13. Kays, W. M.: "Convective Heat and Mass Transfer," McGraw-Hill, New York, 1966.
14. Linton, W. H., Jr., and T. K. Sherwood: *Chem. Engr. Progr.*, **46**: 258 (1950).
15. Prandtl, L.: Eine Beziehung zwischen Wärmeaustausch und Strömungswiederstand der Flüssigkeiten, *Phys. Z.*, **11**: 1072 (1910).
16. Reynolds, O.: *Proc. Manchester Lit. Phil. Soc.*, **8** (1874).

17. Rohsenow, W. M., and H. Y. Choi: "Heat, Mass, and Momentum Transfer," Prentice-Hall, Englewood Cliffs, N.J., 1961.
18. Sherwood, T. K., and R. L. Pigford: "Absorption and Extraction," McGraw-Hill, New York, 1952.
19. Spalding, D. B.: "Convective Mass Transfer," McGraw-Hill, New York, 1963.
20. Taylor, G. I.: *Rept. Mem. Brit. Aeron. Comm.* 272, 1916, p. 423.
21. Treybal, R. E.: "Mass Transfer Operations," 2d ed., McGraw-Hill, New York, 1968.
22. Welty, J. R., C. E. Wicks, and R. E. Wilson: "Fundamentals of Momentum, Heat and Mass Transfer," Wiley, New York, 1969.

CHAPTER 15
ONE-DIMENSIONAL COMPRESSIBLE FLOW

Except for the notions of density variations with change in pressure and temperature introduced as fundamentals, we have thus far dealt with fluids which behave as if they were incompressible. Of paramount importance in the study of compressible flow is the rate of change of density with pressure, which is very closely related to the velocity of propagation of a small pressure wave. Since temperature changes normally accompany density variations, thermodynamic laws, centered around the equation of state, are basic to compressible flow.

The analysis of this chapter will be restricted to one-dimensional flow; i.e., flow parameters are approximately constant over each cross section. Recall, however, that streamlines need not be parallel in one-dimensional flow.

15-1 THE SPEED OF SOUND

In an incompressible fluid the movement of a fluid particle at a point requires the simultaneous movement of all fluid particles in the flow field in order for the density

ONE-DIMENSIONAL COMPRESSIBLE FLOW

to remain constant; i.e., propagation velocities are infinite. Because of the greater distance between molecules in a compressible fluid, however, the small movement of a fluid element induces small disturbances in adjacent elements. This disturbance, referred to as an *elastic acoustic wave*, propagates throughout the fluid field at a relatively high velocity, the *sonic velocity*. We now seek a relationship between the velocity of propagation of this acoustic wave and fluid properties.

For an idealized model, let a stationary compressible fluid be contained in a piston–infinite-cylinder combination. As shown in Fig. 15-1a, a small impulse applied to the piston causes an elastic wave to propagate steadily with a velocity c. The fluid through which the wave passes has a pressure of $p + dp$ resulting from the impulse and a density of $\rho + d\rho$ and moves with a velocity dV. The fluid to the right of the wavefront, however, remains motionless. An observer moving with the wavefront sees the steady flow process as shown in Fig. 15-1b. Fluid having velocity c enters the control volume at the right-hand face, and its velocity is reduced to $c - dV$, resulting in a simultaneous increase in pressure and density.

The continuity equation (10-15) applied to the control volume gives

$$c\rho A = (c - dV)(\rho + d\rho)A$$

or, dropping the second-order differential and dividing out the common area A,

$$c \, d\rho = \rho \, dV \tag{15-1}$$

Neglecting the shear force on the wall (it reduces to zero as the control volume becomes infinitesimally small, of the order as the thickness of the wavefront), the momentum equation (10-50) yields

$$(p + dp)A - pA = \rho c A[-(c - dV)] - \rho c A(-c)$$

which upon simplification gives

$$dp = \rho c \, dV \tag{15-2}$$

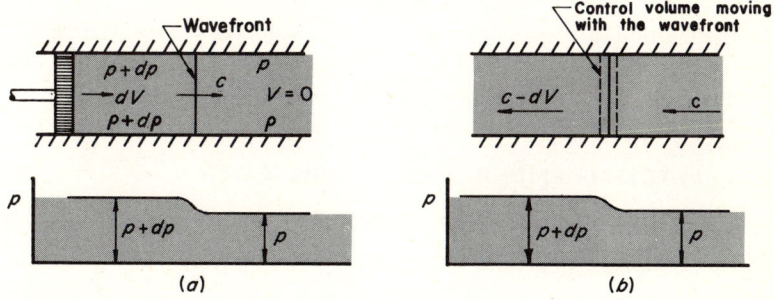

Fig. 15-1 Propagation of an elastic wave.

Combining Eqs. (15-1) and (15-2),

$$c^2 = \frac{dp}{d\rho} = \left.\frac{\partial p}{\partial \rho}\right|_s \tag{15-3}$$

where the partial derivative is valid since effects due to friction and heat transfer are negligible, giving an isentropic process. Frictional effects are of second order, and heat transfer is infinitesimal since temperature changes are minute during the propagation. These comments are valid with respect to the control volume containing the wave even in processes which are not isentropic.

For a perfect gas undergoing an isentropic process, the pressure-density relation is

$$\frac{p}{\rho^k} = \text{const}$$

Taking the logarithmic differential, we get

$$\ln p - k \ln \rho = \ln C$$

$$\frac{dp}{p} = k \frac{d\rho}{\rho}$$

or

$$\left.\frac{\partial p}{\partial \rho}\right|_s = k \frac{p}{\rho} \tag{15-4}$$

Replacing p by its equivalent ρRT from the equation of state, the speed of sound in a perfect gas is given by

$$c = \sqrt{kRT} \tag{15-5}$$

For air at normal pressure and temperature

$$c = 49.02\sqrt{T} \tag{15-6}$$

where T is in degrees Rankine, giving the speed of sound in feet per second.

While this analysis is valid for spherical and cylindrical waves as well, it should be remembered that the acoustic velocity was assumed constant (an inertial control volume with steady flow relative to it was used), and the compression wave resulted from a small pressure variation. An expansion wave (leftward movement of the piston) gives the same result. Large pressure variations produce shock waves, discussed in Sec. 15-7, which are not isentropic.

15-2 MACH NUMBER AND MACH CONE

The ratio of the local stream velocity V to the velocity of sound c in a compressible fluid is called the *Mach number* **M**, that is,

$$\mathbf{M} \equiv \frac{V}{c} \tag{15-7}$$

ONE-DIMENSIONAL COMPRESSIBLE FLOW

or, for a perfect gas,

$$\mathbf{M} = \frac{V}{\sqrt{kRT}} \tag{15-8}$$

which, after logarithmic differentiation, can be expressed in terms of the fluid velocity and temperature,

$$\frac{d\mathbf{M}}{\mathbf{M}} = \frac{dV}{V} - \frac{1}{2}\frac{dT}{T} \tag{15-9}$$

since k and R are constants for a perfect gas.

Because of its definition, the magnitude of the Mach number can be used to characterize expansion and compression waves in compressible flow. Consider a stationary source emitting small pressure waves as a fluid moves past it at different velocities, as indicated in Fig. 15-2. Note that the same patterns would result from a moving source in a stationary medium. Analogous patterns might be demonstrated by dropping pebbles in a quiescent lake at equal intervals of time and with equal spacing.

In Fig. 15-2a the spherical wavefront propagates at sonic velocity c in accordance with the analysis of the preceding section. In Fig. 15-2b the pressure intensity is not equivalent in all directions due to the simultaneous effects of waves emitted at different times.

In Fig. 15-2c the effects of pressure waves cannot move upstream against the fluid since the stream velocity is equal to the sonic velocity. The same thing is true in Fig. 15-2d, except that the zone of silence and zone of action are separated by a cone, the *Mach cone*, rather than by a plane as in c. The angle of the cone depends

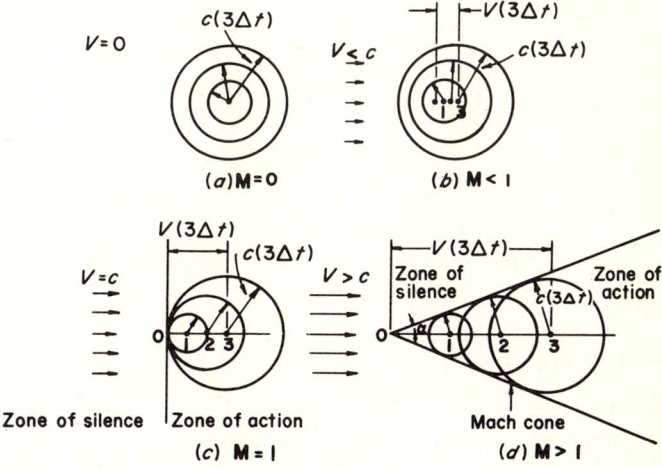

Fig. 15-2 Pressure field produced by wavefronts of different speeds.

upon the magnitude of the Mach number. The half-angle of the cone, termed the *Mach angle* α, can be directly related to the Mach number by trigonometry, giving

$$\sin \alpha = \frac{c}{V} = \frac{1}{\mathbf{M}} \tag{15-10}$$

Because of the unique characteristics of the flow fields represented in Fig. 15-2, it will be convenient to classify the flows depending upon the magnitude of the Mach number. If the medium is incompressible or approximately stationary, a pressure pulse spreads uniformly in all directions, as shown in Fig. 15-2a. If incompressible, the pulse is felt everywhere simultaneously because of the infinite acoustic velocity.

Figure 15-2b characterizes *subsonic* ($\mathbf{M} < 1$) flow. *Transonic* flow occurs when the Mach number is in the neighborhood of unity, typified in Fig. 15-2c. *Supersonic* flow ($\mathbf{M} > 1$), is characterized by Fig. 15-2d, which illustrates why a stationary observer on the ground cannot hear a supersonic plane until it has passed over him. Because of the real-gas effects of dissociation and ionization at high Mach numbers, the supersonic regime is sometimes referred to as *hypersonic* at Mach numbers above 5 [1, 6].

More detailed discussions of these flows and their distinctive characteristics will be found in texts, such as Shapiro [9] and John [3], devoted entirely to compressible flow.

Example The total included angle of the Mach cone produced by a bullet in flight is 58.4°. The temperature of the undisturbed air is 69°F, and its pressure is 14.6 psia. What is the velocity of the bullet relative to the undisturbed air?

Solution   From Eq. (15-10)

$$V = \frac{c}{\sin \alpha}$$

where α is the half-angle of the Mach cone, in this case 29.2°, and $c = \sqrt{kRT}$; therefore,

$$V = \frac{\sqrt{kRT}}{\sin \alpha} = \frac{\sqrt{1.4(53.3 \text{ ft-lb}_f/\text{lb}_m\text{-}°R)(529°R)(32.2 \text{ lb}_m\text{-ft/lb}_f\text{-sec}^2)}}{\sin 29.2°}$$

$$= \frac{1{,}128}{0.488} = 2{,}310 \text{ ft/sec}$$

Note that it was necessary to introduce the gravitational constant g_c for dimensional homogeneity.

15-3 GOVERNING EQUATIONS

In formulating the basic equations governing compressible flow, we must simultaneously satisfy all the conservation laws, the equation of state for the fluid medium, and the second law of thermodynamics. Throughout the analysis we shall assume steady flow. For a perfect gas the equations will be expressed in terms of the Mach

ONE-DIMENSIONAL COMPRESSIBLE FLOW

number corresponding to the local thermodynamic conditions in order to reduce the number of variables and facilitate integration.

Conservation of mass The continuity equation for steady flow is

$$\rho A V = \dot{m} = \text{const} \tag{15-11}$$

which can be logarithmically differentiated to give the *differential continuity equation:*

$$\frac{d\rho}{\rho} + \frac{dA}{A} + \frac{dV}{V} = 0 \tag{15-12}$$

For a perfect gas, Eq. (15-11) can be transformed into differential form in terms of the Mach number by algebraic manipulation and logarithmic differentiation, namely,

$$\dot{m} = \rho A V \frac{\sqrt{kRT}}{\sqrt{kRT}} = p A \mathbf{M} \sqrt{\frac{k}{RT}} \tag{15-13}$$

$$\frac{dp}{p} + \frac{dA}{A} + \frac{d\mathbf{M}}{\mathbf{M}} - \frac{1}{2}\frac{dT}{T} = 0 \tag{15-14}$$

since k and R are constants.

Conservation of momentum For a control volume as shown in Fig. 15-3 the steady-flow momentum equation for a compressible fluid is

$$pA + \left(p + \frac{dp}{2}\right) dA - (p + dp)(A + dA) - dF_\tau$$
$$= \rho A V(V + dV) - \rho A V^2 \tag{15-15}$$

where dF_τ is the shear force resulting from friction by the channel wall and $p + dp/2$ is the average pressure on the circumferential area of the control volume. Omitting second-order terms and rearranging, we get the *differential momentum equation*

$$\frac{dp}{\rho} + \frac{dF_\tau}{\rho A} + V\, dV = 0 \tag{15-16}$$

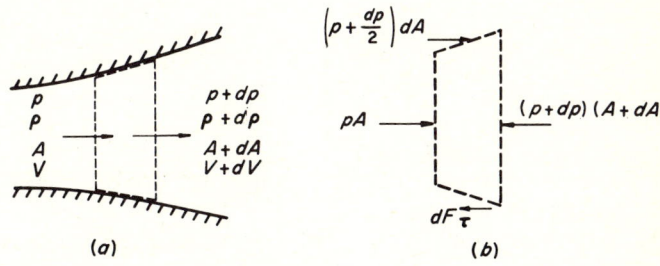

Fig. 15-3 Compressible flow through a variable-area duct.

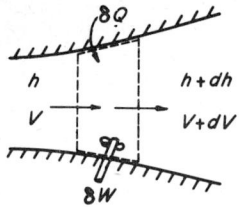

Fig. 15-4 Conservation of energy in compressible flow.

For a perfect gas, in terms of the Mach number, the equation becomes

$$\frac{1}{k\mathbf{M}^2}\frac{dp}{p} + \frac{dF_r}{\rho A V^2} + \frac{d\mathbf{M}}{\mathbf{M}} + \frac{1}{2}\frac{dT}{T} = 0 \tag{15-17}$$

with the aid of Eq. (15-9) and the definition of Mach number.

Conservation of energy Upon neglecting potential energy, which is normally insignificant in compressible fluids, the steady-flow *energy equation* (10-117) in *differential form* becomes

$$\frac{\delta Q}{\delta t} - \frac{\delta W_{\text{shaft}}}{\delta t} = \dot{m}\left(dh + d\frac{V^2}{2}\right) \tag{15-18}$$

which may be readily related to Fig. 15-4.

Recalling that $dh = c_p\,dT$ for a perfect gas, and noting that

$$c_p = \frac{k}{k-1}R \tag{15-19}$$

from Eq. (3-52), we see that Eq. (15-18) becomes

$$\frac{\delta Q/\delta t}{\dot{m}c_p T} - \frac{\delta W_{\text{shaft}}/\delta t}{\dot{m}c_p T} = \frac{dT}{T} + \frac{(V\,dV)V}{[k/(k-1)]RTV} = \frac{dT}{T} + (k-1)\mathbf{M}^2\frac{dV}{V} \tag{15-20}$$

Combining it with Eq. (15-9) gives

$$\frac{\delta Q/\delta t}{\dot{m}c_p T} - \frac{\delta W_{\text{shaft}}/\delta t}{\dot{m}c_p T} = \left(1 + \frac{k-1}{2}\mathbf{M}^2\right)\frac{dT}{T} + (k-1)\mathbf{M}\,d\mathbf{M} \tag{15-21}$$

which is the differential energy equation for a perfect gas.

Equation of state In general, an equation of the form

$$\rho = \rho(p,T) \tag{15-22}$$

is valid for a pure substance, as introduced in Chap. 1. In keeping with the format followed thus far, the equation of state for a perfect gas is

$$\rho = \frac{p}{RT} \tag{15-23}$$

although it seems trite to state it here.

Second law of thermodynamics For any known process $ds \geq 0$ and

$$ds = c_p \frac{dT}{T} - R \frac{dp}{p} \tag{15-24}$$

which was introduced in Chap. 3 by combining the first and second laws of thermodynamics.

15-4 ISENTROPIC FLOW OF A PERFECT GAS

A large class of one-dimensional compressible-flow problems are reversible (frictionless) and adiabatic (negligible heat transfer), considerably simplifying Eqs. (15-17) and (15-21). Further simplification results from the absence of shaft work.

In this section we seek to relate thermodynamic properties at different stations in the flow field to the Mach numbers at the respective stations under the conditions of isentropic flow in the absence of shaft work.

The energy equation (15-21) gives

$$\frac{dT}{T} = -\frac{2\mathbf{M}\, d\mathbf{M}}{2/(k-1) + \mathbf{M}} \tag{15-25}$$

which permits temperature to be determined in terms of Mach number only. Integrating between stations x_1 and x_2, we get a temperature ratio in terms of the respective local Mach numbers:

$$\frac{T_2}{T_1} = \frac{1 + [(k-1)/2]\mathbf{M}_1{}^2}{1 + [(k-1)/2]\mathbf{M}_2{}^2} \tag{15-26}$$

Substituting Eq. (15-25) into Eq. (15-17), a differential expression for pressure in terms of Mach number results with $F_r = 0$, that is,

$$\frac{dp}{p} = k\mathbf{M}^2 \left[\frac{\mathbf{M}\, d\mathbf{M}}{2/(k-1) + \mathbf{M}^2} - \frac{d\mathbf{M}}{\mathbf{M}} \right] \tag{15-27}$$

$$= -\frac{k}{k-1} \left[\frac{2\mathbf{M}\, d\mathbf{M}}{2/(k-1) + \mathbf{M}^2} \right]$$

giving

$$\frac{p_2}{p_1} = \left\{ \frac{1 + [(k-1)/2]\mathbf{M}_1{}^2}{1 + [(k-1)/2]\mathbf{M}_2{}^2} \right\}^{k/(k-1)} \tag{15-28}$$

The velocity ratio can be determined by combining Eqs. (15-9) and (15-25) and integrating as follows:

$$\frac{dV}{V} = \frac{d\mathbf{M}}{\mathbf{M}} - \frac{\mathbf{M}\,d\mathbf{M}}{2/(k-1) + \mathbf{M}^2}$$

$$= \left[1 - \frac{\mathbf{M}^2}{2/(k-1) + \mathbf{M}^2}\right] \frac{d\mathbf{M}}{\mathbf{M}} \tag{15-29}$$

$$\frac{V_2}{V_1} = \frac{\mathbf{M}_2}{\mathbf{M}_1} \left\{\frac{1 + [(k-1)/2]\mathbf{M}_1^2}{1 + [(k-1)/2]\mathbf{M}_2^2}\right\}^{\frac{1}{2}} \tag{15-30}$$

By combining the continuity equation (15-14) with Eqs. (15-25) and (15-27), the area variation results:

$$\frac{dA}{A} = \frac{k}{k-1} \frac{2\mathbf{M}\,d\mathbf{M}}{2/(k-1) + \mathbf{M}^2} - \frac{d\mathbf{M}}{\mathbf{M}} - \frac{\mathbf{M}\,d\mathbf{M}}{2/(k-1) + \mathbf{M}^2}$$

$$= \left[\frac{k+1}{k-1} \frac{\mathbf{M}^2}{2/(k-1) + \mathbf{M}^2} - 1\right] \frac{d\mathbf{M}}{\mathbf{M}} \tag{15-31}$$

$$\frac{A_2}{A_1} = \frac{\mathbf{M}_1}{\mathbf{M}_2} \left\{\frac{1 + [(k-1)/2]\mathbf{M}_1^2}{1 + [(k-1)/2]\mathbf{M}_2^2}\right\}^{(k+1)/[2(k-1)]} \tag{15-32}$$

Finally, by using the isentropic relation

$$\frac{\rho_2}{\rho_1} = \left(\frac{p_2}{p_1}\right)^{1/k}$$

the density ratio is

$$\frac{\rho_2}{\rho_1} = \left\{\frac{1 + [(k-1)/2]\mathbf{M}_1^2}{1 + [(k-1)/2]\mathbf{M}_2^2}\right\}^{1/(k-1)} \tag{15-33}$$

when Eq. (15-28) is used for the pressure ratio.

The preceding integrations might have been carried out between some reference state and a general station, simplifying the equations and facilitating the tabulation of the respective ratios in terms of their common parameter, the Mach number. This method, which is more conducive to engineering calculations, is detailed in the following discussion.

Reference states Let us choose as a reference state a station where the velocity approaches zero ($\mathbf{M} \simeq 0$), as might be the case in a very large chamber upstream from a small nozzle or at the inlet of a simple pitot tube (Fig. 15-5). The condition of the fluid when the velocity approaches zero is termed the *stagnation state*, and the properties of the fluid with zero velocity are stagnation properties. The zero subscript will be used to designate stagnation properties.

The fluid history in Fig. 15-5a and b obviously differs in that the fluid must be isentropically decelerated to reach the stagnation condition at the pitot tube whereas

ONE-DIMENSIONAL COMPRESSIBLE FLOW

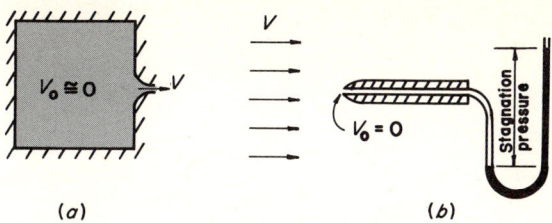

Fig. 15-5 Isentropic stagnation conditions.

the fluid in the tank might have been undisturbed for a week. This suggests the concept of localization—what would occur at a point if the flow were retarded to zero velocity isentropically, actually or hypothetically.

Writing the steady-flow energy equation (15-18) for the isentropic process of the fluid in either Fig. 15-5a or b gives

$$h_0 = h + \frac{V^2}{2} \tag{15-34}$$

which is illustrated in the hs diagram of Fig. 15-6a. All possible states for the process lie on the line of constant entropy s_0. The stagnation state is reached from any other state by proceeding along the line of constant entropy to the stagnation enthalpy.

For a perfect gas, Eq. (15-34) becomes

$$T_0 = T + \frac{V^2}{2c_p} \tag{15-35}$$

which is illustrated in Fig. 15-6b. Since $c_p = kR/(k-1)$,

$$\frac{T_0}{T} = 1 + \frac{k-1}{2} \frac{V^2}{kRT}$$

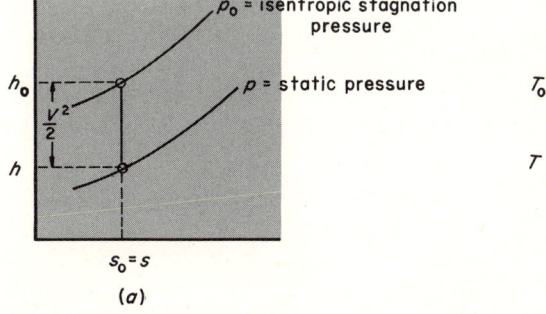

Fig. 15-6 Stagnation state.

or

$$\frac{T_0}{T} = 1 + \frac{k-1}{2} M^2 \qquad (15\text{-}36)$$

Note that this same result could have been obtained by directly integrating Eq. (15-25) between a general station and the station where the Mach number is zero (the reference state). This is equivalent to setting $M_2 = 0$ in Eq. (15-26).

At this point it is appropriate to point out the advantages in using this approach. Consider the *hs* diagram of Fig. 15-7 which depicts two states in addition to the stagnation state. The enthalpy difference between the two states can be ascertained directly or by referring both conditions to the reference state, i.e.,

$$h_1 - h_2 = (h_0 - h_2) - (h_0 - h_1) \qquad (15\text{-}37)$$

Similarly, we can write Eq. (15-36) for each state; hence

$$\frac{T_0}{T_1} = 1 + \frac{k-1}{2} M_1^2 \qquad (15\text{-}38a)$$

$$\frac{T_0}{T_2} = 1 + \frac{k-1}{2} M_2^2 \qquad (15\text{-}38b)$$

But for a perfect gas T_0 is the same for both cases, as illustrated in Fig. 15-6*b*. Therefore, dividing Eq. (15-38*a*) by Eq. (15-38*b*) gives

$$\frac{T_2}{T_1} = \frac{1 + [(k-1)/2]M_1^2}{1 + [(k-1)/2]M_2^2}$$

which is identical to Eq. (15-26).

The pressure and density ratios, analogous to the temperature ratio of Eq.

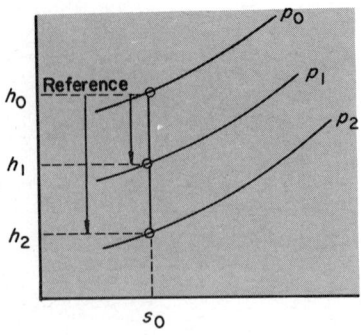

Fig. 15-7 The stagnation state as a reference.

(15-36), can be determined directly from Eqs. (15-28) and (15-33), respectively, giving

$$\frac{p_0}{p} = \left(1 + \frac{k-1}{2}\mathbf{M}^2\right)^{k/(k-1)} \tag{15-39}$$

$$\frac{\rho_0}{\rho} = \left(1 + \frac{k-1}{2}\mathbf{M}^2\right)^{1/(k-1)} \tag{15-40}$$

It is convenient to tabulate these ratios as a function of their common variable, the Mach number. Appendix Table A-12 gives these ratios for air ($k = 1.4$) for $0 \leq \mathbf{M} \leq 3$.

Example For a Mach number of 2.00 determine the temperature at station 1 if the temperature in the flow field is 680°R at station 2, where the Mach number is 1.60; $k = 1.4$.
Solution From the isentropic Table A-12

$$\left.\frac{T}{T_0}\right|_1 = 0.556 \quad \text{and} \quad \left.\frac{T}{T_0}\right|_2 = 0.661$$

But $T_{0_1} = T_{0_2}$; therefore,

$$\frac{T_1}{T_2} = \frac{0.556}{0.661}$$

and

$$T_1 = \frac{0.556}{0.661}(680) = 572°R$$

A second reference state, called the *critical state*, refers properties to the condition with $\mathbf{M} = 1$. An asterisk is used to designate critical properties. The critical state is reached by isentropically bringing the fluid, actually or hypothetically, to a Mach number of unity. All states having the same entropy and same stagnation temperature have the same isentropic stagnation state and the same critical state. Temperature, pressure, and density can be referred to the critical state by direct application of Eqs. (15-26), (15-28), and (15-33), respectively. Taking temperature [Eq. (15-26)] as an example, let $\mathbf{M}_2 = 1$, $T_2 = T^*$, and let station 1 be any station; we get

$$\frac{T^*}{T} = \frac{1 + [(k-1)/2]\mathbf{M}^2}{1 + (k-1)/2} = \frac{2 + (k-1)\mathbf{M}}{k+1} \tag{15-41}$$

It is often advantageous to relate the isentropic stagnation state and the critical state. These ratios, which are constant for a perfect gas, can be determined from Eqs. (15-36), (15-39), and (15-40) by setting $\mathbf{M} = 1$, giving the critical state $(\cdot)^*$. For temperature

$$\frac{T_0}{T^*} = 1 + \frac{k-1}{2} = \frac{k+1}{2} \tag{15-42}$$

or (since it will be more convenient to use the inverse later in setting limiting conditions on flow in nozzles)

$$\frac{T^*}{T_0} = \frac{2}{k+1} \tag{15-43}$$

Similarly,

$$\frac{p^*}{p_0} = \left(\frac{2}{k+1}\right)^{k/(k-1)} \tag{15-44}$$

$$\frac{\rho^*}{\rho_0} = \left(\frac{2}{k+1}\right)^{1/(k-1)} \tag{15-45}$$

For air ($k = 1.4$) these ratios are

$$\frac{T^*}{T_0} = 0.833 \qquad \frac{p^*}{p_0} = 0.528 \qquad \frac{\rho^*}{\rho_0} = 0.634 \tag{15-46}$$

We conclude from this that for the isentropic flow of a perfect gas, the temperature where **M** = 1 is 83.3 percent of the stagnation temperature; however, the pressure where **M** = 1 is only 52.8 percent of the stagnation pressure.

In isentropic flow the stagnation temperature and stagnation pressure are conserved by virtue of the absence of heat transfer and friction. It will be shown later that when heat transfer occurs, the stagnation temperature will change. Similarly, stagnation pressure varies when friction is present.

15-5 ISENTROPIC FLOW IN A VARYING-AREA DUCT

The dramatic distinction between subsonic and supersonic flow made in this section is contrary to intuition: we are acquainted primarily with subsonic flows in everyday life.

Rewriting the momentum equation (15-16), we get

$$V\,dV = -\frac{dp}{\rho}\frac{d\rho}{d\rho} = -c^2\frac{d\rho}{\rho} \tag{15-47a}$$

by making use of Eq. (15-3). In terms of the Mach number this becomes

$$\frac{d\rho}{\rho} = -\frac{V^2}{c^2}\frac{dV}{V} = -\mathbf{M}^2\frac{dV}{V} \tag{15-47b}$$

which can be substituted into the continuity equation (15-12) to give

$$\blacksquare \qquad \frac{dA}{A} = (\mathbf{M}^2 - 1)\frac{dV}{V} \tag{15-48}$$

Eliminating velocity between the continuity and momentum equations gives an important relation between pressure and area. Substituting

$$dV = -\frac{dp}{\rho V}$$

from the momentum equation into the continuity equations yields

$$\frac{d\rho}{\rho} + \frac{dA}{A} = \frac{dp}{\rho V^2} \tag{15-49}$$

But for the isentropic process $dp/d\rho = c^2$; hence,

$$\frac{1}{\rho c^2} + \frac{dA}{A\,dp} = \frac{1}{\rho V^2}$$

or

$$\blacksquare \quad \frac{dA}{A} = \frac{1}{\rho V^2}(1 - \mathbf{M}^2)\,dp \tag{15-50}$$

It is now obvious that the magnitude of the Mach number in Eqs. (15-48) and (15-50) determines how area varies with velocity and pressure, respectively. Note that the effects are opposite for velocity and pressure.

For *subsonic flow*, $\mathbf{M} < 1$, the two sides of Eq. (15-48) have opposite signs. Therefore, an increase in area causes a decrease in velocity. From Eq. (15-50), an increase in area causes an increase in pressure. The equations behave in accordance with normal experience, which is conditioned by low-speed or incompressible flows.

In the case of *supersonic flow*, $\mathbf{M} > 1$, we note surprisingly from Eq. (15-48) that an increase in area causes an increase in velocity. This phenomenon occurs because of the changes in density required by Eq. (15-47b). From Eq. (15-50), an increase in area causes a decrease in pressure.

These differences in behavior between subsonic and supersonic flows are illustrated in Fig. 15-8. A singular point in the behavior of the equations occurs when $\mathbf{M} = 1$, giving $dA = 0$. Since the area can neither increase nor decrease, sonic velocity can occur only at the minimum area, called the *throat*, of a nozzle or diffuser. Contrariwise *the velocity at a throat does not always have to be sonic*.

Mass flow rate of a perfect gas An equation convenient for calculating the mass flow rate per unit area can be obtained by simple rearrangement of the continuity equation (15-13) as follows:

$$\frac{\dot{m}}{A} = p\mathbf{M}\sqrt{\frac{k}{RT}}\sqrt{\frac{T_0}{T_0}\frac{p_0}{p_0}} = p_0\mathbf{M}\sqrt{\frac{k}{RT_0}}\sqrt{\frac{T_0}{T}}\frac{p}{p_0} \tag{15-51}$$

Expressing the static temperature and pressure in terms of the Mach number by using Eqs. (15-36) and (15-39), the mass flow rate is given in terms of stagnation properties

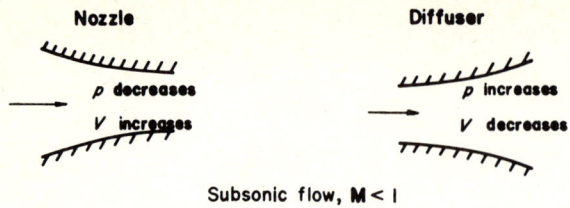

Subsonic flow, M < 1

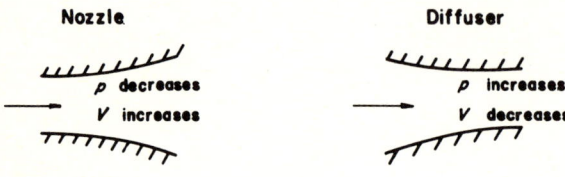

Supersonic flow, M > 1

Fig. 15-8 Effects of area change.

and the local Mach number:

$$\frac{\dot{m}}{A} = p_0 \mathbf{M} \sqrt{\frac{k}{RT_0}} \left(1 + \frac{k-1}{2} \mathbf{M}^2\right)^{-(k+1)/[2(k-1)]} \tag{15-52}$$

Since the cross-sectional area is a minimum where $\mathbf{M} = 1$, the maximum flow that can pass through a duct is that which occurs at the throat, found by setting $\mathbf{M} = 1$ in Eq. (15-52), i.e.,

$$\left.\frac{\dot{m}}{A}\right|_{\max} = \frac{\dot{m}}{A^*} = p_0 \sqrt{\frac{k}{RT_0}} \left(\frac{2}{k+1}\right)^{(k+1)/(k-1)} \tag{15-53}$$

Dividing Eq. (15-53) by Eq. (15-52) gives an area ratio which permits referring the area at any station to that at the throat:

$$\blacksquare \quad \frac{A}{A^*} = \frac{1}{\mathbf{M}} \left[\frac{2}{k+1} \left(1 + \frac{k-1}{2} \mathbf{M}^2\right)\right]^{k+1/[2(k-1)]} \tag{15-54}$$

This ratio is given in Table A-12 for $k = 1.4$. Note that for any area ratio, which is always greater than unity, there are always two values of $\mathbf{M}$, corresponding to subsonic flow and supersonic flow. This concept leads directly into the study of nozzles presented in the following section.

Example Air flows through the reducer shown in Fig. 15-9. Assuming isentropic flow, find p_2, T_2, and $\mathbf{M}_2$.

Solution The Mach number at the inlet is

$$\mathbf{M}_1 = \frac{V_1}{\sqrt{kRT_1}} = \frac{300 \text{ ft/sec}}{\sqrt{1.4(53.3 \text{ ft-lb}_f/\text{lb}_m\text{-}°R)(600°R)(32.2 \text{ lb}_m\text{-ft/lb}_f\text{-sec}^2)}} = 0.25$$

ONE-DIMENSIONAL COMPRESSIBLE FLOW

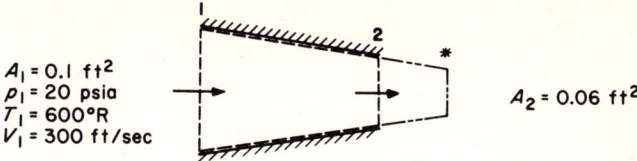

Fig. 15-9 Reducer with extension to critical condition.

We can now make use of the isentropic Table A-12 by referring conditions to the critical state ($M = 1$ where $A = A^*$). At $M_1 = 0.25$

$$\frac{A_1}{A^*} = 2.41$$

The outlet- to inlet-area ratio is

$$\frac{A_2}{A_1} = \frac{0.06}{0.1} = 0.6$$

Therefore,

$$\frac{A_2}{A^*} = \frac{A_1}{A^*}\frac{A_2}{A_1} = 2.41(0.6) = 1.446$$

and from the isentropic table

$M_2 = 0.4496$ (say 0.45)

Since $p_{0_1} = p_{0_2}$ and $T_{0_1} = T_{0_2}$ for isentropic flow, the tables may conveniently be used to find the exit conditions. From Table A-12

$$\left.\frac{p}{p_0}\right|_1 = 0.9575 \qquad \left.\frac{p}{p_0}\right|_2 = 0.8705$$

$$\left.\frac{T}{T_0}\right|_1 = 0.988 \qquad \left.\frac{T}{T_0}\right|_2 = 0.961$$

Therefore, the required properties are readily found by forming the appropriate ratios, i.e.

$$\frac{p_2}{p_1} = \frac{(p/p_0)_2}{(p/p_0)_1} = \frac{0.8705}{0.9575} = 0.919$$

$$p_2 = 0.919(20 \text{ psia}) = 18.38 \text{ psia}$$

$$\frac{T_2}{T_1} = \frac{(T/T_0)_2}{(T/T_0)_1} = \frac{0.961}{0.988} = 0.973$$

$$T_2 = 0.973(600°\text{R}) = 583°\text{R}$$

15-6 FLOW IN NOZZLES

In this section the conditions under which maximum flow occurs in a passage, as given by Eq. (15-53), will be considered.

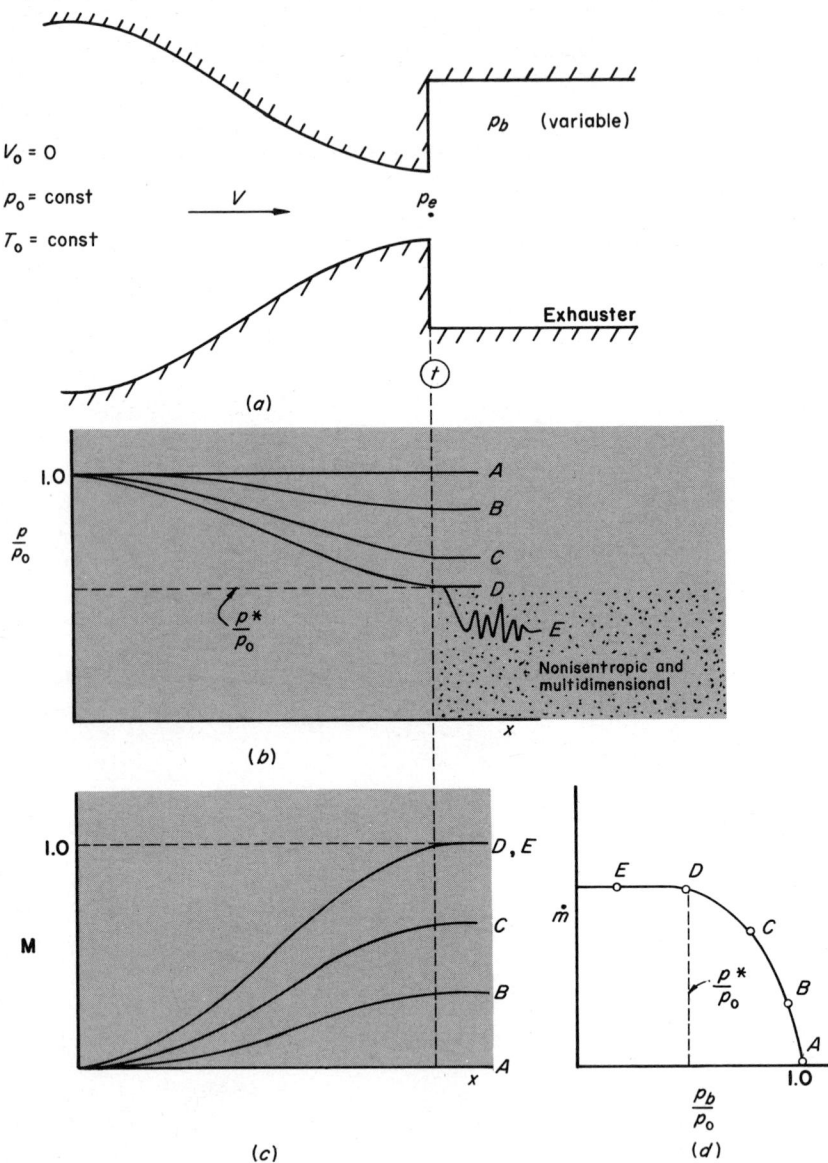

Fig. 15-10 Compressible flow in a convergent nozzle.

ONE-DIMENSIONAL COMPRESSIBLE FLOW

Convergent nozzle Figure 15-10 depicts a convergent nozzle with plots of the attendant variables for a variety of flow conditions. The nozzle is fed by a reservoir with a compressible fluid at a pressure p_0 and temperature T_0, stagnation conditions. The pressure p_b in the exhauster downstream of the nozzle may be varied, producing the different flow conditions shown by curves A, B, C, D, and E. The pressure in the exit plane of the nozzle is denoted by p_e and for the convergent nozzle coincides with the throat pressure p_t.

When $p_b = p_0$, no flow occurs, as indicated by curve A. As p_b is reduced lower and lower, the flow conditions typified by curves B and C, which are qualitatively the same, are reached. Of course the mass flow rate $\dot{m}$ and Mach number $\mathbf{M}$ become larger as the back pressure is reduced. Upon reducing the back pressure to the point where the Mach number is unity at the throat, $p = p^*$ (curve D), the mass flow rate cannot be increased by a further reduction in the back pressure (curve E); hence, the flow is *choked*. Reducing the back pressure p_b to any value below the critical p^* produces no changes throughout the nozzle, and complex pressure adjustments occur in the exhauster which are neither one-dimensional nor isentropic [7, 8, 9]. It is impossible to have a Mach number greater than unity in a convergent nozzle when the inlet condition is subsonic.

Example Air discharges from a 100-ft³ storage tank through a converging nozzle of exit area 1.0 in². Assuming the temperature inside the tank remains constant at 500°R, estimate the time required for the tank pressure to drop from 600 to 30 psia if the discharge is to the atmosphere, where $p_a = 14.6$ psia.

Solution Assuming that the steady-flow equations are applicable at each instant of time (quasi-steady), Eq. (15-52) gives the mass flow rate (hence the time, since the total discharge can be determined) in terms of the stagnation properties, area, and Mach number at the area selected. Taking the area to be that at the nozzle exit, we note that the Mach number is unity for the

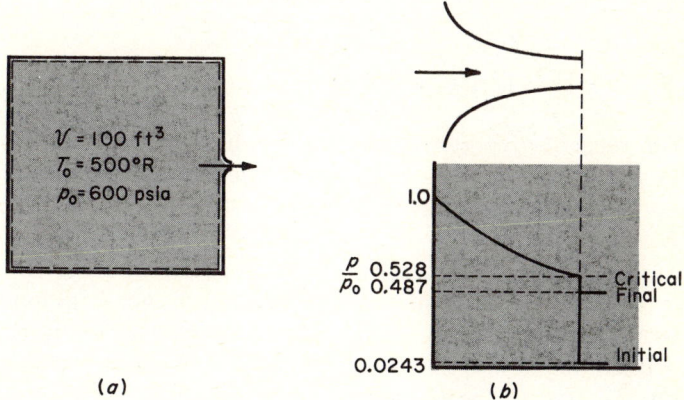

Fig. 15-11 Tank discharge through a converging nozzle.

pressure range considered since the pressure ratio at the exit is always less than the critical, as shown in Fig. 15-11b. Therefore, Eq. (15-53) is valid at any instant of time.

$$\dot{m} = p_0 A^* \sqrt{\frac{k}{RT_0}\left(\frac{2}{k+1}\right)^{(k+1)/(k-1)}}$$

$$= (p_0 \text{ lb}_f/\text{in}^2)(1.0 \text{ in}^2) \sqrt{\frac{1.4(32.2 \text{ lb}_m\text{-ft/lb -sec}^2)}{(53.3 \text{ ft-lb}_f/\text{lb}_m\text{-°R})(500°\text{R})}\left(\frac{2}{2.4}\right)^{2.4/0.4}}$$

$$= 0.02385 p_0 \text{ lb}_m/\text{sec}$$

where p_0 is expressed in psia.

The continuity equation (10-14) will be used since the mass contained in the tank varies with time, i.e.,

$$\frac{d}{dt}\int_{\text{cv}} \rho \, d\mathscr{V} = -\int_{\text{cs}} \rho \mathbf{V} \cdot d\mathbf{A} \qquad (10\text{-}14)$$

or

$$\frac{d}{dt} m = -(\rho V A)_{\text{out}} = -\dot{m}$$

But m is the mass inside the control volume at any time t, i.e.,

$$m = \frac{p_0 \mathscr{V}}{RT_0}$$

$$\frac{dm}{dt} = \frac{\mathscr{V}}{RT_0} \frac{dp_0}{dt}$$

Therefore,

$$\frac{\mathscr{V}}{RT_0}\frac{dp_0}{dt} + 0.02385 p_0 = 0$$

$$\frac{(144 \text{ in}^2/\text{ft}^2)(100 \text{ ft}^3)(\text{lb}_f/\text{in}^2)}{(0.02385 \text{ lb}_m/\text{sec})(53.3 \text{ ft-lb}_f/\text{lb}_m\text{-°R})(500°\text{R})}\int_{600}^{30}\frac{dp_0}{p_0} + \int_0^t dt = 0$$

$$22.65 \ln \frac{30}{600} + t = 0$$

$$t = 22.65 \ln \frac{600}{30} = 67.8 \text{ sec}$$

Convergent-divergent nozzle By adding a divergent section to a converging nozzle, as shown in Fig. 15-12, the flow can be accelerated to supersonic speeds in the divergent portion, as indicated by Eq. (15-48). Curves A to I illustrate the various possible flow conditions.

No flow occurs when $p_b = p_0$, as shown in curve A. As the back pressure p_b is reduced, the flow is subsonic throughout the nozzle as long as the pressure is greater than that shown by curve C, giving accelerating flow in the convergent section and decelerating flow in the divergent section. When p_b is reduced to a value corresponding to curve C, the Mach number at the throat is unity, $\mathbf{M}_t = 1$, and the pressure increases in the divergent section.

ONE-DIMENSIONAL COMPRESSIBLE FLOW

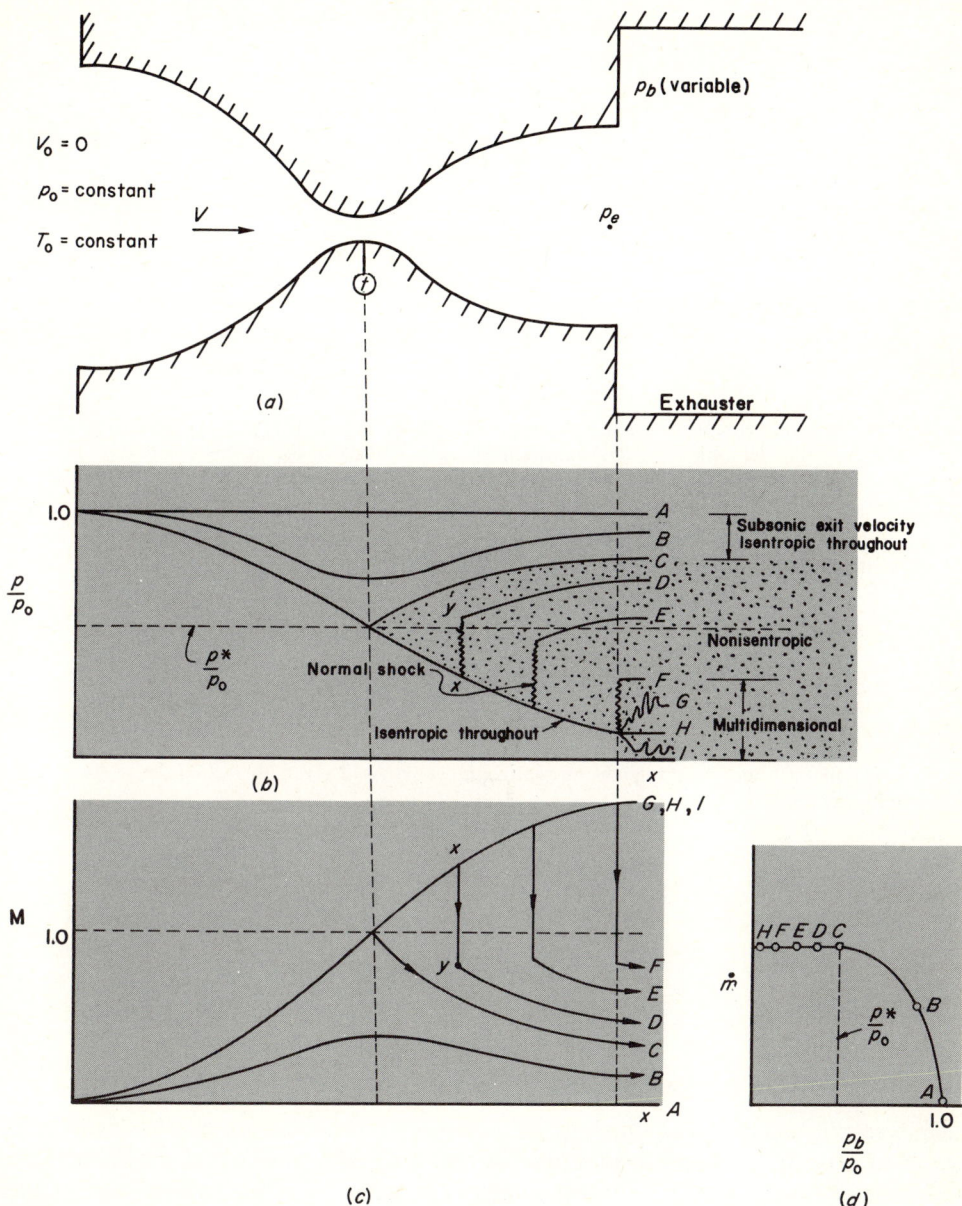

Fig. 15-12 Compressible flow in convergent-divergent nozzle.

Curve H represents flow which is isentropic throughout the nozzle and which accelerates throughout, being subsonic in the convergent section and supersonic in the divergent portion. In this case the pressure at the throat $p_t = p^*$ is identical to that of the subsonic flow of curve C. In both cases the flow is choked, passing the maximum amount of fluid.

For back pressures between those of curves C and H, one-dimensional isentropic solutions are not possible. These flows involve *shock fronts* which, idealized to be planar and very thin, are irreversible. These irreversible discontinuities will be considered in the following section. Curve F represents the lowest back pressure at which a normal shock will exist in the nozzle, standing in the exit plane. Curve G is the result of compression waves caused by a further reduction of the back pressure; curve I results from expansion waves. Compression waves and expansion waves are not one-dimensional, being oblique. The mechanism and analysis of oblique shocks are detailed by Shapiro [9], Courant and Friedricks [2], and Owczarek [6].

Example For a convergent-divergent nozzle of area ratio $A_e/A_t = 2.17$, $p_0 = 100$ psia, $T_0 = 500°$R, and $k = 1.4$, determine (*a*) the minimum pressure that can exist at the nozzle exit for isentropic *subsonic* flow throughout the nozzle and, (*b*) the pressure that will exist at the nozzle exit for isentropic *supersonic* flow throughout the nozzle.

Solution (*a*) From the isentropic Table A-12 in the subsonic regime $\mathbf{M} = 0.28$ for $A/A^* = 2.17$. At that Mach number

$$\frac{p}{p_0} = 0.947$$

Therefore,

$p = 94.7$ psia

(*b*) In the supersonic regime the isentropic Table A-12 gives $\mathbf{M} = 2.29$ for $A/A^* = 2.17$; the pressure ratio is

$$\frac{p}{p_0} = 0.0815$$

and

$p = 8.15$ psia

The pressure distributions for these cases are shown qualitatively in Fig. 15-13.

15-7 NORMAL-SHOCK WAVES

Schlieren photographs (which show flow patterns resulting from changes in density) taken of flow in nozzles or around high-speed projectiles indicate that at supersonic speeds sudden changes occur in the velocity and pressure of a fluid within distances of the order of 10^{-5} in. For engineering calculations it is advantageous to idealize the discontinuity to be a plane, considering only those conditions upstream and downstream from the discontinuity rather than considering the shock structure itself.

ONE-DIMENSIONAL COMPRESSIBLE FLOW

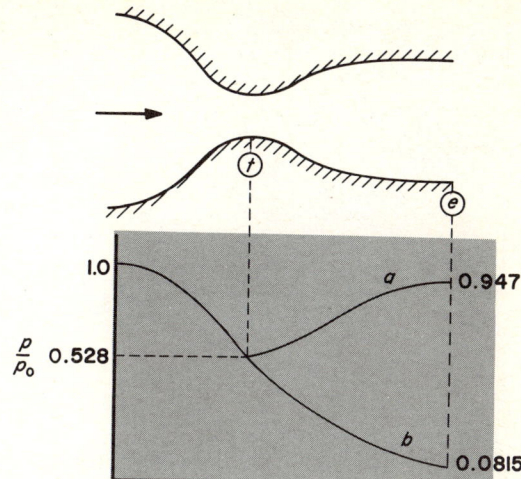

Fig. 15-13 Convergent-divergent nozzle.

Complex processes occurring within the shock are irreversible, but it is possible to ascertain the net changes in fluid properties across the shock without considering its structure.

Letting the normal shock shown in Fig. 15-14 be enclosed by a control volume of infinitesimal thickness, we seek relationships between the properties at stations x and y, upstream and downstream from the shock, respectively. Since the control volume is infinitesimally thick, area change is negligible. Therefore, the steady-flow governing equations, resulting from Eqs. (15-12), (15-16), (15-18), and (15-22), are

Continuity: $$\frac{\dot{m}}{A} = \rho_x V_x = \rho_y V_y \tag{15-55}$$

Momentum: $$p_x - p_y - \frac{F_r}{A} = \frac{\dot{m}}{A}(V_y - V_x) \tag{15-56}$$

Energy: $$h_x - h_y + \frac{\delta Q/\delta t}{\dot{m}} = \frac{V_y^2 - V_x^2}{2} \tag{15-57}$$

State: $$\rho = \rho(p, T)$$

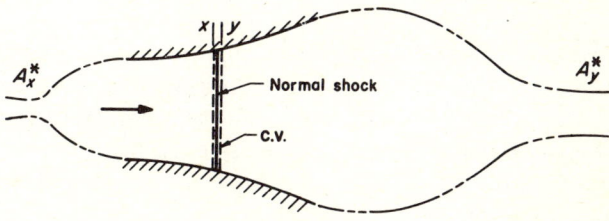

Fig. 15-14 Stationary normal shock.

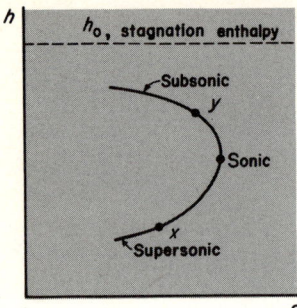

Fig. 15-15 The Fanno line.

or
$$h = h(s,\rho) \quad \text{and} \quad s = s(p,\rho) \tag{15-58}$$

Because of the infinitesimally thick discontinuity, both heat transfer and friction can be disregarded for the control volume. It will be instructive, however, to consider the effects of ignoring these phenomena one at a time.

The Fanno line Considering the effects of friction only and ignoring heat transfer, the simultaneous solution of the resulting equations (with $\delta Q/\delta t = 0$) can be represented on an hs diagram as shown in Fig. 15-15. The resulting curve, called the *Fanno line*, can be constructed by the following procedure if the conditions at x are completely known: (1) choose a velocity V_y; (2) determine ρ_y from Eq. (15-55) and h_y from Eq. (15-57); (3) using the equation of state (charts or tables may be required), find s_y and p_y which locate a point on the curve corresponding to the value of V_y. Then get the frictional force F_r from Eq. (15-56). The Fanno line is constructed by repeating this procedure. This is equivalent to changing the thickness of the control volume since F_r varies accordingly. The Fanno line represents the locus of states that can be reached, starting from the conditions at x, by changing the amount of friction in an adiabatic flow.

The point of maximum entropy corresponds to a Mach number of unity (see Prob. 15-38). The upper branch of the curve represents subsonic flow, approaching stagnation conditions. The lower branch represents supersonic flow.

The Rayleigh line The *Rayleigh line* represents a locus of points which results when heat transfer occurs but friction is neglected in steady flow. Figure 15-16 shows a typical Rayleigh line obtained by repeating the following procedure: (1) choose a velocity V_y and determine ρ_y from the continuity equation upon knowing the conditions at x; (2) determine p_y from the momentum equation (15-56); use the equation of state to get s_y and h_y. Determine the heat transfer by Eq. (15-57). Since heat is transferred, the stagnation enthalpy will be different for the different states represented on the curve.

It should be emphasized that the normal-shock process does not follow either

ONE-DIMENSIONAL COMPRESSIBLE FLOW

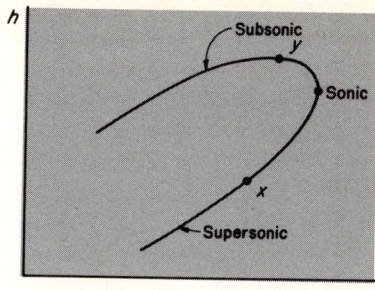

Fig. 15-16 The Rayleigh line.

the Fanno curve or the Rayleigh curve, since they result from neglecting heat transfer and friction, respectively, and the idealized normal-shock model neglects both. However, since the shock process simultaneously satisfies all the governing equations, when $F_r = \delta Q/\delta t = 0$, simple superposition of the Fanno and Rayleigh lines graphically represents the shock process. The points of intersection simultaneously satisfy the equations. Shown in Fig. 15-17, the shock process proceeds from x to y and never in the opposite direction, as limited by the second law of thermodynamics, being irreversible ($ds > 0$). Using an explicit equation of state, such as that for a perfect gas, property changes can be computed without using the enthalpy-entropy diagram.

Normal shock in a perfect gas Separating the variables and integrating Eq. (15-21) gives

$$\frac{T_y}{T_x} = \frac{1 + [(k-1)/2]\mathbf{M}_x^2}{1 + [(k-1)/2]\mathbf{M}_y^2} \tag{15-59}$$

which is identical to Eq. (15-26). It is more advantageous, however, to express property ratios in terms of the Mach number at a single station, say upstream from the shock. Rewriting the continuity equation (15-55) for a perfect gas in terms of the Mach number,

$$\frac{p_x}{RT_x} c_x \mathbf{M}_x = \frac{p_y}{RT_y} c_y \mathbf{M}_y \tag{15-60}$$

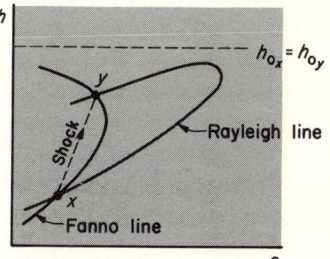

Fig. 15-17 The normal-shock process.

and replacing c by its equivalent $\sqrt{kRT}$ gives

$$\frac{p_y}{p_x} = \sqrt{\frac{T_y}{T_x} \frac{\mathbf{M}_x}{\mathbf{M}_y}} \tag{15-61}$$

which can be combined with Eq. (15-59) to give the pressure ratio in terms of the upstream and downstream Mach numbers. By expressing the momentum equation for a perfect gas in terms of Mach number we get an independent equation for the pressure ratio

$$p_x - p_y = \rho_y V_y^2 - \rho_x V_x^2 \tag{15-62}$$

$$p_x - p_y = \frac{p_y}{RT_y} \mathbf{M}_y^2 (kRT_y) - \frac{p_x}{RT_x} \mathbf{M}_x^2 (kRT_x)$$

$$p_x(1 + k\mathbf{M}_x^2) = p_y(1 + k\mathbf{M}_y^2)$$

$$\frac{p_y}{p_x} = \frac{1 + k\mathbf{M}_x^2}{1 + k\mathbf{M}_y^2} \tag{15-63}$$

Equating Eqs. (15-61) and (15-63) and using Eq. (15-59) gives

■ $$\mathbf{M}_y^2 = \frac{\mathbf{M}_x^2 + 2/(k-1)}{[2k/(k-1)]\mathbf{M}_x^2 - 1} \tag{15-64}$$

which permits the temperature and pressure ratios of Eqs. (15-59) and (15-63) to be expressed in terms of $\mathbf{M}_x$ only:

■ $$\frac{T_y}{T_x} = \frac{\left(1 + \frac{k-1}{2} \mathbf{M}_x^2\right)\left(\frac{2k}{k-1} \mathbf{M}_x^2 - 1\right)}{\frac{(k+1)^2}{2(k-1)} \mathbf{M}_x^2} \tag{15-65}$$

■ $$\frac{p_y}{p_x} = \frac{2k}{k+1} \mathbf{M}_x^2 - \frac{k-1}{k+1} \tag{15-66}$$

For a perfect gas the energy equation (15-57) is

$$c_p T_x + \frac{V_x^2}{2} = c_p T_y + \frac{V_y^2}{2} \tag{15-67a}$$

or

$$c_p (T_0)_x = c_p (T_0)_y \tag{15-67b}$$

since the gas may be isentropically decelerated either upstream or downstream to the stagnation condition, giving

■ $$T_{0_x} = T_{0_y} \tag{15-67c}$$

This result was anticipated since no heat was added or extracted.

ONE-DIMENSIONAL COMPRESSIBLE FLOW

The stagnation-pressure ratio can be determined by use of the preceding results. By algebraic manipulation

$$\frac{p_{0_y}}{p_{0_x}} = \left(\frac{p_0}{p}\right)_y \frac{p_y}{p_x} \left(\frac{p}{p_0}\right)_x \tag{15-68}$$

Since the ratios $(p_0/p)_y$ and $(p/p_0)_x$ are downstream and upstream from the shock, respectively, and do not relate pressures across the shock, isentropic relations may be used for them. Using Eqs. (15-39) and (15-66), the stagnation-pressure ratio becomes

$$\frac{p_{0_y}}{p_{0_x}} = \left\{\frac{1 + [(k-1)/2]\mathbf{M}_y^2}{1 + [(k-1)/2]\mathbf{M}_x^2}\right\}^{k/k-1} \left(\frac{2k}{k+1}\mathbf{M}_x^2 - \frac{k-1}{k+1}\right) \tag{15-68a}$$

or, with the aid of Eq. (15-64), in terms of $\mathbf{M}_x$

$$\frac{p_{0_y}}{p_{0_x}} = \frac{\left\{\dfrac{[(k+1)/2]\mathbf{M}_x^2}{1 + [(k-1)/2]\mathbf{M}_x^2}\right\}^{k/(k-1)}}{\left(\dfrac{2k}{k+1}\mathbf{M}_x^2 - \dfrac{k-1}{k+1}\right)^{1/(k-1)}} \tag{15-68b}$$

Another useful (and perhaps more meaningful) expression for the drop in stagnation pressure can be determined by noting that the reference critical areas upstream and downstream from the shock are different, as illustrated by phantom lines in Fig. 15-14. Since the flow is steady, the continuity equation (15-52) gives

$$\frac{\dot{m}}{A_x^*} = \sqrt{\frac{k}{R}\left(\frac{2}{k+1}\right)^{(k+1)/(k-1)}}\frac{p_{0_x}}{\sqrt{T_{0_x}}} \quad \text{and} \quad \frac{\dot{m}}{A_y^*} = \sqrt{\frac{k}{R}\left(\frac{2}{k+1}\right)^{(k+1)/(k-1)}}\frac{p_{0_y}}{\sqrt{T_{0_y}}}$$

Combining these relations,

$$\frac{A_x^*}{A_y^*} = \frac{p_{0_y}}{p_{0_x}} \tag{15-69}$$

since $T_{0_x} = T_{0_y}$. Having assumed the flow to be everywhere isentropic except across the shock, this equation is valid between any two points in the flow field when the only nonisentropic condition between the two points is a normal shock. This concept can be generalized to apply between any two stations, regardless of the flow conditions between the two stations, by introducing a discharge coefficient C, defined by

$$C \equiv \frac{\text{actual flow rate}}{\text{ideal (isentropic) flow rate}} \tag{15-70}$$

The following section illustrates the procedure for the common case of adiabatic flow in a duct with friction.

The change in entropy is readily determined by integrating Eq. (3-73), valid for an irreversible process, giving

$$s_y - s_x = c_p \ln \frac{T_y}{T_x} - R \ln \frac{p_y}{p_x} \tag{15-70a}$$

Further manipulation gives

$$s_y - s_x = -R \ln \frac{p_{0_y}}{p_{0_x}} \tag{15-70b}$$

which can be expressed in terms of $\mathbf{M}_x$ by using Eq. (15-68b).

The parameters of direct interest in engineering calculations are presented for $k = 1.4$ as a function of the upstream Mach number $\mathbf{M}_x$ in Table A-13.

For the formation, reflection, and translation of normal-shock waves the reader should consult John [3] or Shapiro [9].

Example The ratio of exit area to throat area of a convergent-divergent rocket exhaust nozzle is 1.63. In the combustion chamber, where the exhaust gases are generated, the stagnation conditions are $p_0 = 400$ psia and $T_0 = 3000°R$. Assuming the combustion products to behave as a perfect gas with $k = 1.4$ and molecular weight $\mathcal{M} = 20$, what is the exhaust velocity (a) for isentropic flow throughout the nozzle and (b) when a normal shock stands just inside the exit plane?

Solution (a) For $A/A^* = 1.63$ the exit Mach number from Table A-12 is

$$\mathbf{M} = 1.96$$

and

$$\frac{T}{T_0} = 0.566$$

Therefore,

$$T = 0.566(3000) = 1698°R$$

The exit velocity can be obtained from the definition of Mach number

$$\mathbf{M} = \frac{V}{\sqrt{kRT}}$$

upon knowing the gas constant, given by the ratio of the universal gas constant $\mathcal{R}$ to the molecular weight $\mathcal{M}$, that is,

$$R = \frac{\mathcal{R}}{\mathcal{M}} = \frac{1,545}{20} = 77.3 \text{ ft-lb}_f/\text{lb}_m\text{-°R}$$

Therefore

$$V = \mathbf{M}\sqrt{kRT}$$
$$= 1.96 \sqrt{1.4(77.3 \text{ ft-lb}_f/\text{lb}_m\text{-°R})(32.2 \text{ lb}_m\text{-ft/lb}_f\text{-sec}^2)(1698°R)}$$
$$= 2,430 \text{ ft/sec}$$

ONE-DIMENSIONAL COMPRESSIBLE FLOW

(b) From the shock Table A-13 at $M_x = 1.96$

$M_y = 0.584$

Since there is no loss in stagnation temperature across the shock,

$T_{0_x} = T_{0_y} = 3000°R$

but from the isentropic Table A-12

$\left.\dfrac{T}{T_0}\right|_{M_y=0.584} = 0.935$

hence,

$T = 0.935(3000) = 2805°R$

and the velocity may be determined as before:

$V = M\sqrt{kRT}$

$ = 0.584\sqrt{1.4(77.3)(32.2)(2805)}$

$ = 1,820 \text{ ft/sec}$

15-8 ADIABATIC FLOW WITH FRICTION

For the flow condition illustrated in Fig. 15-18 manipulation of the continuity equation (15-13) gives

$$\dfrac{\dot{m}}{A} = pM\sqrt{\dfrac{k}{RT}}\sqrt{\dfrac{T_i}{T_i}}$$

$$\dfrac{\dot{m}}{A} = \dfrac{pM}{\sqrt{T_i}}\sqrt{\dfrac{k}{R}\left(1 + \dfrac{k-1}{2}M^2\right)} \qquad (15\text{-}71)$$

when combined with Eq. (15-36), valid for the isentropic convergent portion of the nozzle. Combining Eqs. (15-53) and (15-70).

$$\dfrac{\dot{m}}{A_t} = C\dfrac{p_i}{\sqrt{T_i}}\sqrt{\dfrac{k}{R}\left(\dfrac{2}{k+1}\right)^{(k+1)/(k-1)}} \qquad (15\text{-}72)$$

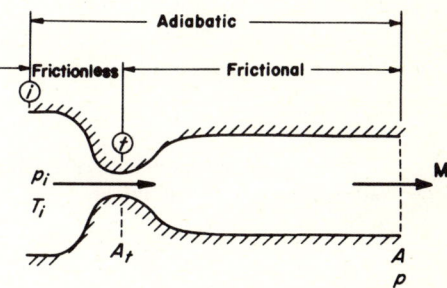

Fig. 15-18 Adiabatic flow.

Dividing Eq. (15-72) by Eq. (15-71) and algebraically manipulating, we get

$$\frac{1}{C}\left(\frac{A}{A_t}\frac{p}{p_i}\right) = \frac{1}{\mathbf{M}}\sqrt{\frac{[2/(k+1)]^{(k+1)/(k-1)}}{1 + [(k-1)/2]\mathbf{M}^2}\frac{\{1 + [(k-1)/2]\mathbf{M}^2\}^{k/(k-1)}}{\{1 + [(k-1)/2]\mathbf{M}^2\}^{k/(k-1)}}}$$

$$= \frac{1}{\mathbf{M}}\left[\frac{2}{k+1}\left(1 + \frac{k-1}{2}\mathbf{M}^2\right)\right]^{(k+1)/[2(k-1)]}\left(1 + \frac{k-1}{2}\mathbf{M}^2\right)^{-k(k-1)}$$

Since the right-hand side of this equation is the product of the area ratio and pressure ratio given by Eqs. (15-54) and (15-39), respectively,

$$\frac{1}{C}\left(\frac{A}{A_t}\frac{p}{p_i}\right) = \left(\frac{A}{A^*}\frac{p}{p_0}\right)_{\text{isentropic}} \tag{15-73}$$

This equation is extremely useful because it directly relates isentropic and non-isentropic conditions. For relatively high Reynolds numbers ($\mathbf{Re} > 10^6$) and/or thin boundary layers the discharge coefficient is of the order of 0.99 [9]. The thin-boundary-layer criterion applies when the nozzle is large with respect to the boundary layer. Therefore, for a large class of engineering problems the assumption of a discharge coefficient of the order of unity introduces only negligible error. For convenience in performing calculations the product $\dfrac{A}{A^*}\dfrac{p}{p_0}$ is tabulated for $k = 1.4$ in Table A-12. The following example will illustrate the usefulness of this equation.

Example Air flows through a convergent-divergent nozzle with stagnation properties $p_0 = 100$ psia and $T_0 = 1000°R$; $A_e/A_t = 2.84$. (a) For a back pressure $p_b = 80$ psia (i) find the exit Mach number; (ii) what is the nozzle area where the shock occurs? (b) At what back pressure will a normal shock stand in the exit plane of the nozzle?

Solution (a) The minimum pressure ratio for which subsonic flow exists throughout the nozzle is 0.97 (for $A/A^* = 2.84$) as found in Table A-12 and shown in Fig. 15-19; therefore, a normal shock stands in the nozzle, illustrated by pressure curve a.

(i) Assuming the air enters the nozzle at the stagnation conditions, the exit Mach number can be found from Eq. (15-73) upon assuming a discharge coefficient (taken as 0.99 in this case). Hence

$$\left(\frac{A}{A^*}\frac{p}{p_0}\right)_e = \frac{1}{C}\left(\frac{A_e}{A_t}\frac{p_e}{p_0}\right) = \frac{1}{0.99}[2.84(0.80)] = 2.30$$

From Table A-12

$$\mathbf{M}_e = 0.25$$

(ii) Reference properties downstream from the shock are found with the aid of the isentropic tables at the exit Mach number:

$$\left.\frac{p}{p_0}\right|_e = 0.96$$

ONE-DIMENSIONAL COMPRESSIBLE FLOW

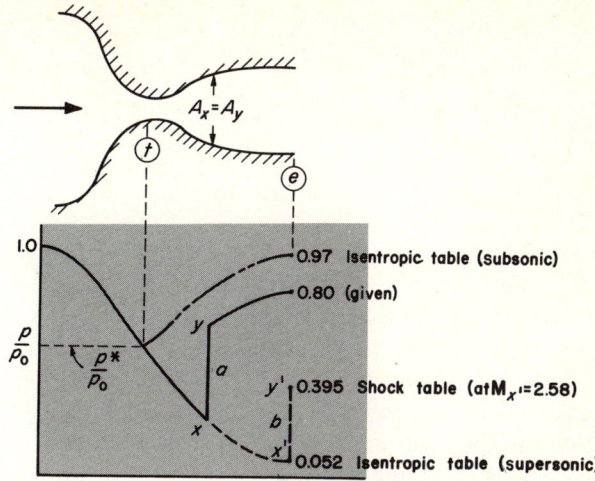

Fig. 15-19 Convergent-divergent nozzle with shock in diffuser.

Therefore,

$$p_{0_e} = p_{0_y} = \frac{80}{0.96} = 83.3 \text{ psia}$$

$$\frac{p_{0_y}}{p_{0_x}} = \frac{83.3}{100} = 0.833$$

From the shock Table A-13 at this stagnation-pressure ratio

$$\mathbf{M}_x \simeq 1.76 \quad \text{and} \quad \mathbf{M}_y \simeq 0.63$$

and from the isentropic Table A-12

$$\left.\frac{A}{A^*}\right|_x = 1.40$$

Therefore,

$$A_x = 1.40 A_t$$

(b) For a shock to stand in the exit plane of the nozzle, supersonic flow exists throughout the divergent portion, as shown by curve b in Fig. 15-19. The upstream Mach number $\mathbf{M}_{x'}$, found from the isentropic Table A-12, is

$$\mathbf{M}_{x'} = 2.58$$

Also

$$\left.\frac{p}{p_0}\right|_{x'} = 0.052$$

$$p_{x'} = 5.2 \text{ psia}$$

and from the shock table

$$\frac{p_{y'}}{p_{x'}} = 7.60$$

hence,

$$p_{y'} = 7.60(5.2) = 39.5 \text{ psia}$$

Flow in constant-area adiabatic ducts with friction (Fanno flow) In Chaps. 12 and 13 we noted that frictional effects produced velocity gradients and shear stresses at the flow boundaries. Because of the boundary-layer growth in a constant-area duct with friction, the effective area is reduced as the displacement thickness increases with length, as illustrated in Fig. 15-20a. Friction at the wall causes the flow properties to change, following a Fanno line as typified in Fig. 15-20b for a given mass flow rate per unit area. It should be obvious that the nature of the boundary and length will be significant parameters in this flow.

Various flow possibilities can be illustrated by referring to the figure. If the flow is subsonic initially, say point *a* on the curve, the second law, which requires an increase in entropy, requires the flow to proceed along the curve in the direction *ayz*. If the duct is long enough or rough enough, the flow becomes sonic. It cannot continue past the sonic condition at *z*, however, because of the second-law limitations.

If the flow is initially supersonic, it may remain supersonic throughout the duct length, represented on the curve by *bxz*. If the duct is sufficiently long, however, a shock will occur and the conditions vary in accordance with the curve represented by the path *bxyz*, always increasing in entropy. The flow is choked when sonic conditions occur at the duct exit.

If sonic flow exists at the duct exit and the duct length is increased, a readjustment in the flow occurs, resulting in a reduced mass flow rate; the flow shifts from one Fanno line to another, corresponding to the new mass flow rate. Sonic flow will recur at the exit.

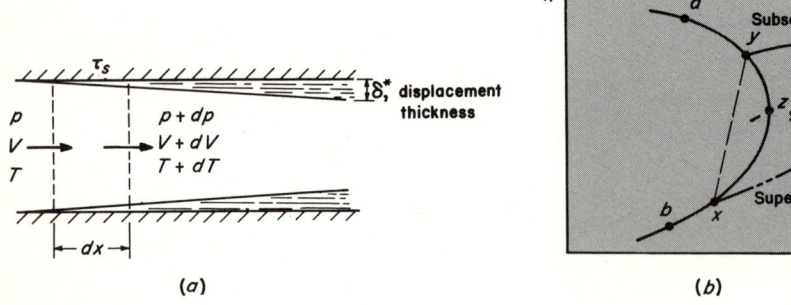

Fig. 15-20 Fanno flow.

ONE-DIMENSIONAL COMPRESSIBLE FLOW

Fanno flow in a perfect gas Considering the infinitesimal control volume shown in Fig. 15-20a and assuming that the flow is steady and that elevation changes, heat transfer, and shaft work are negligible, the governing equations are

Continuity: $$\frac{d\rho}{\rho} + \frac{dV}{V} = 0 \tag{15-74}$$

Momentum: $$dp + \frac{dF_\tau}{A} + \rho V\, dV = 0 \tag{15-75}$$

Energy: $$c_p\, dT + V\, dV = 0 \tag{15-76}$$

State: $$\frac{dp}{p} = \frac{d\rho}{\rho} + \frac{dT}{T} \tag{15-77}$$

Defining a friction factor f as 4 times the ratio of shear stress to dynamic pressure,

$$f \equiv \frac{4\tau_s}{\rho V^2/2} \tag{15-78}$$

the frictional force dF_τ can be written in terms of the friction factor and flow parameters:

$$dF_\tau = \tau_s\, dA_s = \frac{f}{8}\rho V^2 P\, dx$$

where dA_s is the wetted area and P is the wetted perimeter of the duct. Introducing the hydraulic diameter D_h as given by Eq. (12-49), we get

$$dF_\tau = f\frac{A}{D_h}\frac{\rho V^2}{2}\, dx \tag{15-79}$$

The momentum equation then becomes

$$dp + \frac{f}{D_h}\frac{\rho V^2}{2}\, dx + \rho V\, dV = 0 \tag{15-80}$$

The factor of 4 in the definition of friction factor was included for expediency rather than necessity. It makes the friction factors of Nikuradse [5] and Moody [4] directly applicable to Fanno flow. Since Mach number effects were ignored in the Moody diagram, data from it should be used only for low subsonic flow.

Following the procedure utilized in isentropic and normal-shock flows, it is convenient to form property ratios based on a reference state, in this case the critical state, with an asterisk being used to designate those quantities where $\mathbf{M} = 1$. To illustrate the procedure, consider the temperature ratio. Expressing Eq. (15-76) in terms of the variables $\mathbf{M}$ and T only, which is equivalent to setting the left-hand side of Eq. (15-21) equal to zero, the resulting equation is

$$\int_{T^*}^{T} \frac{dT}{T} = -\int_{\mathbf{M}=1}^{\mathbf{M}} \frac{(k-1)\mathbf{M}\, d\mathbf{M}}{1 + [(k-1)/2]\mathbf{M}^2} \tag{15-81}$$

giving

■ $$\frac{T}{T^*} = \frac{k+1}{2+(k-1)\mathbf{M}^2} \tag{15-82}$$

For completeness, other critical-property relations, determined in a similar manner, are given by the following equations [9]:

■ $$\frac{p}{p^*} = \frac{1}{\mathbf{M}}\sqrt{\frac{k+1}{2+(k-1)\mathbf{M}^2}} \tag{15-83}$$

■ $$\frac{V}{V^*} = \frac{\rho^*}{\rho} = \mathbf{M}\sqrt{\frac{k+1}{2+(k-1)\mathbf{M}^2}} \tag{15-84}$$

■ $$\frac{p_0}{p_0^*} = \frac{1}{\mathbf{M}}\sqrt{\left[\frac{2+(k-1)\mathbf{M}^2}{k+1}\right]^{(k+1)/(k-1)}} \tag{15-85}$$

$$s - s^* = \frac{kR}{k-1}\ln \mathbf{M}^2 \sqrt{\left[\frac{k+1}{\mathbf{M}^2[2+(k-1)\mathbf{M}^2]}\right]^{(k+1)/k}} \tag{15-86}$$

Equations (15-82) to (15-85) are tabulated as a function of $\mathbf{M}$ for $k = 1.4$ in Table A-14.

Of great use in making engineering calculations is the relation between the friction factor, duct length, and Mach number, obtained by combining Eqs. (15-74), (15-76), (15-77), and (15-80). The resulting equation in terms of Mach number is

$$\int \frac{f}{D_h} dx = \int \frac{1-\mathbf{M}^2}{1+[(k-1)/2]\mathbf{M}^2} \frac{d\mathbf{M}^2}{\mathbf{M}^2} \tag{15-87}$$

For an average value of f this can be integrated from station to station; however, in the interest of being able to tabulate properties as a function of Mach number, it is more desirably integrated between a local station and the station where the Mach number is unity, as illustrated in Fig. 15-21. Integrating with the critical condition ($\mathbf{M} = 1$) as a reference, we get

■ $$\frac{fL^*}{D_h} = \frac{1-\mathbf{M}^2}{k\mathbf{M}^2} + \frac{k+1}{2k}\ln\frac{(k+1)\mathbf{M}^2}{2+(k-1)\mathbf{M}^2} \tag{15-88}$$

This equation gives the maximum value of fL/D_h corresponding to any initial Mach number $\mathbf{M}$. Values of fL^*/D_h are tabulated as a function of $\mathbf{M}$ for $k = 1.4$ in Table A-14.

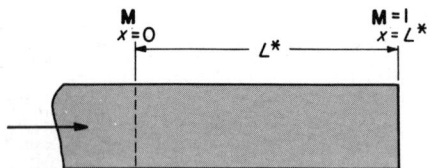

Fig. 15-21 Limits of integration for Eq. (15-87).

ONE-DIMENSIONAL COMPRESSIBLE FLOW

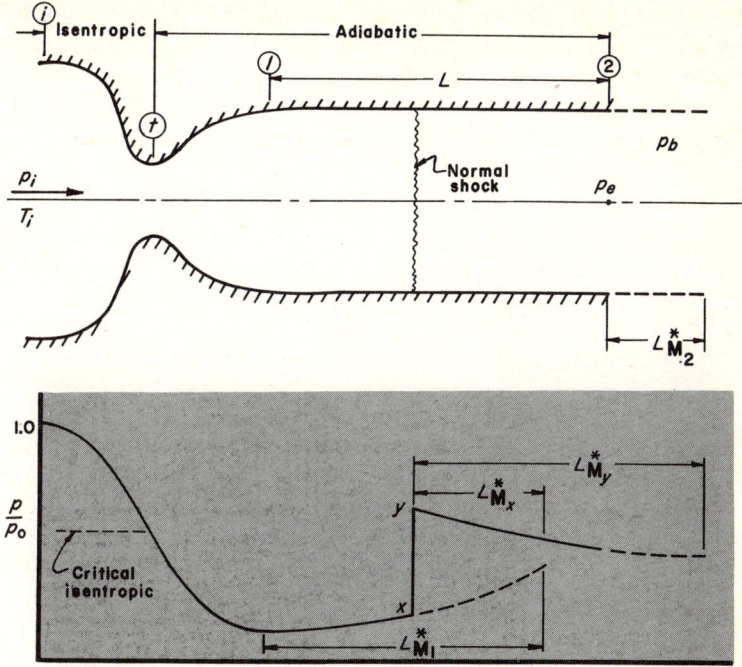

Fig. 15-22 Normal shock in constant-area adiabatic duct with friction.

The duct length required for the flow to change from an initial Mach number M_1 to another Mach number M_2 can conveniently be found from the table by use of the equation

$$\frac{fL}{D_h} = \left(\frac{fL^*}{D_h}\right)_{M_1} - \left(\frac{fL^*}{D_h}\right)_{M_2} \tag{15-89}$$

when no discontinuity exists between stations 1 and 2.

If a shock exists between station 1 and station 2, the length relationship becomes [10]

$$\left(\frac{fL^*}{D_h}\right)_{M_x} - \left(\frac{fL^*}{D_h}\right)_{M_y} = \left(\frac{fL^*}{D_h}\right)_{M_1} - \left(\frac{fL^*}{D_h}\right)_{M_2} - \frac{fL}{D_h} \tag{15-90}$$

This length relationship can be visualized by referring to Fig. 15-22, which shows a constant-area duct being fed by a convergent-divergent nozzle with the attendant pressure-ratio curve for the existence of a normal shock in the duct. When no shock exists in the duct, that is, $M_x = M_y$, Eq. (15-90) reduces to Eq. (15-89).

The left side of Eq. (15-90) may be thought of as the length that would be necessary to change fluid properties from values just upstream of the shock to those just downstream in an adiabatic duct with no shock. In other words, it is the length required for bringing about the same property changes in continuous Fanno flow that would occur in an infinitesimal length in the case of a discontinuity in the form of a normal shock.

For a given duct length and inlet conditions it is now possible to match the upstream and downstream properties by a methodical trial-and-error solution.

1. Obtain M_1 from the isentropic relations (or tables) if the nozzle is isentropic throughout or from Eq. (15-73) if isentropic only to the throat.
2. Knowing the back pressure p_b, obtain M_2 using Eq. (15-73).
3. Assume M_x and get M_y from shock relations (or tables).
4. Determine for each of these Mach numbers the corresponding (fL^*/D_h)'s from Fanno equations (or tables).
5. Use these results in Eq. (15-90) and continue this procedure until the left and right sides of the equation are equal. This procedure locates the position of the shock; i.e., all properties before and aft of the shock can now be determined.

The following procedure can be utilized for quick determination of local stream properties for either subsonic or supersonic flows. For the supersonic case the procedure makes use of the preceding analysis.

A. Subsonic flow
  1. Obtain M_1 from the isentropic relations (or tables) if the nozzle is isentropic throughout or from Eq. (15-73) if adiabatic throughout but isentropic only to the throat.
  2. If given fL/D_h, use Eq. (15-90) and the Fanno equations (or tables) to get the exit properties. The left side of the equation will be zero since no shock occurs.
  3. Or, if given p_b, check p^* using the duct inlet properties ($p_1^* = p_2^*$).
      a. If $p_b \leq p^*$, $M_2 = 1$.
      b. If $p_b > p^*$, $p_b = p_2$ and exit properties can be determined from the Fanno tables.
B. Supersonic flow with fL/D_h and p_b known
  Case I. $fL/D_h > fL^*/D_h$. A shock is always present. Calculate p^* using the duct inlet conditions.
  1. If $p_b \leq p^*$, $M_2 = 1$.
  2. If $p_b > p^*$, use the preceding shock-location analysis.
  Case II. $fL/D_h < fL^*/D_h$. If a shock occurs, $p_b = p_2$.
  1. Determine the exit-plane pressure assuming no shock occurs $p_{e'}$, using the nozzle exit properties, Eq. (15-87), and the Fanno tables.
  2. If $p_b \leq p_{e'}$ (no shock occurs and $p_e = p_{e'}$), get $M_2 > 1$ from part 1.
  3. If $p_b > p_{e'}$ (assuming a shock in the exit plane):
      a. Calculate M_x from the preceding shock-location analysis.
      b. Get $p_1^* = p_x^*$ from the Fanno tables.
      c. Get $(p/p^*)_x$, hence p_x, from the Fanno tables.

ONE-DIMENSIONAL COMPRESSIBLE FLOW

d. Obtain $p_y = p_G$, where p_G is the pressure just downstream from the assumed shock, from the shock tables.

  i. If $p_b \leq p_G$, no shock exists; duct exit properties are the same as in part 2.
  ii. If $p_b > p_G$, shock exists; use shock-location analysis.

Example Air at $p_0 = 100$ psia and $T_0 = 1000°R$ is fed into an adiabatic constant-area duct by an isentropic convergent-divergent nozzle having an exit-to-throat area ratio of 1.69. A normal shock occurs in the duct at a point where the Mach number is 1.30. For a duct of hydraulic diameter $D_h = 1.0$ ft and friction factor $f = 0.001$ (a) what is the maximum duct length? (b) Determine the maximum back pressure p_b for this length.

Solution (a) Since the flow is supersonic at the duct inlet, it is supersonic throughout the diffuser. For the area ratio $A_1/A^* = 1.69$ the Mach number at the nozzle exit is

$$\mathbf{M}_1 = 2.0$$

from the isentropic Table A-12, which also gives the attendant pressure ratio, shown in Fig. 15-23, i.e.,

$$\left.\frac{p}{p_0}\right|_1 = 0.128$$

For the maximum duct length, the exit Mach number must be unity; hence

$$\mathbf{M}_2 = 1.0$$

and from the shock Table A-13 at $\mathbf{M}_x = 1.3$ the Mach number just downstream from the shock is

$$\mathbf{M}_y = 0.786$$

When the relevant Mach numbers are known, Eq. (15-90) readily gives the required length with

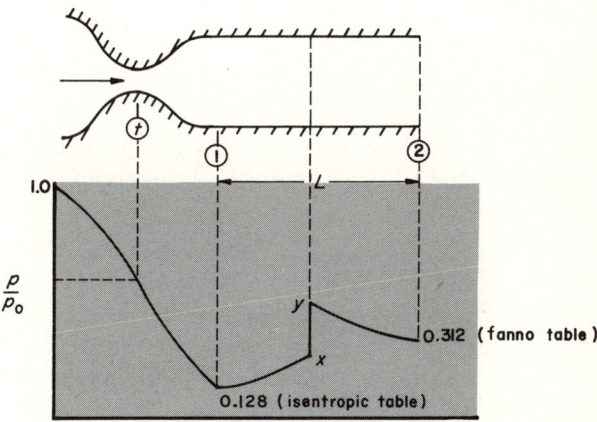

Fig. 15-23 Adiabatic constant-area duct with normal shock.

values of the (fL^*/D_h)'s taken from the Fanno Table A-14, i.e.,

$$\left(\frac{fL^*}{D_h}\right)_{M_x=1.3} - \left(\frac{fL^*}{D_h}\right)_{M_y=0.786} = \left[\left(\frac{fL^*}{D_h}\right)_{M_1=2} - \left(\frac{fL^*}{D_h}\right)_{M_2=1} - \frac{fL_{\max}}{D_h}\right]$$

$$0.0162 - 0.0219 = 0.0762 - 0 - \frac{fL_{\max}}{D_h}$$

$$\frac{fL_{\max}}{D_h} = 0.0819$$

Therefore,

$$L_{\max} = \frac{0.0819(1.0)}{0.001} = 81.9 \text{ ft}$$

(b) From the Fanno Table A-14

$$\left.\frac{p}{p^*}\right|_1 = 0.41 \quad \text{and} \quad \left.\frac{p}{p^*}\right|_2 = 1.00$$

and since $p_1^* = p_2^*$,

$$\frac{p_1}{p_2} = 0.41$$

But

$$p_1 = 0.128(100) = 12.8 \text{ psia}$$

and therefore,

$$p_2 = \frac{12.8}{0.41} = 31.2 \text{ psia}$$

This is the maximum back pressure for the given flow conditions. A higher pressure would cause the flow to readjust, shifting the shock upstream.

15-9 FLOW IN CONSTANT-AREA FRICTIONLESS DUCTS WITH HEAT TRANSFER (Rayleigh Flow)

Another simple flow case results when all changes in state occur because of heat transfer (Fig. 15-24). The Rayleigh line typifies all possible states of the fluid for a given mass flow.

For a flow which is initially subsonic, say point a, heat addition causes the Mach number to increase along the path $aybz$ until the flow is sonic at point z. Further heat addition causes the flow to readjust to a different Rayleigh line since the entropy cannot decrease. Heating of a flow which is initially supersonic, point m, reduces the Mach number, with changes of state occurring along the path $mxnz$, choking at z; or sonic flow may be reached by a normal shock with subsequent heating, path $mxybz$. In both cases, heat, like friction, always tends to make the Mach number approach unity.

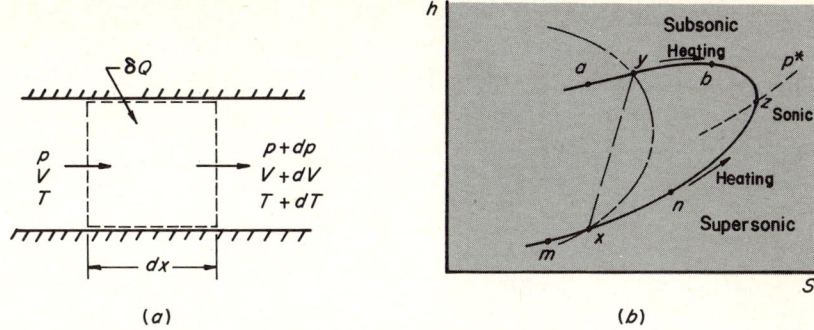

Fig. 15-24 Rayleigh flow.

Rayleigh flow in a perfect gas For steady flow through a constant-area frictionless duct, neglecting elevation changes and shaft work, the governing equations pertaining to the control volume of Fig. 15-24a are

Continuity: $\quad \dfrac{d\rho}{\rho} + \dfrac{dV}{V} = 0 \qquad$ (15-91)

Momentum: $\quad dp + \rho V\, dV = 0 \qquad$ (15-92)

Energy: $\quad \dfrac{\delta Q}{dm} = c_p\, dT + V\, dV = 0 \qquad$ (15-93)

State: $\quad \dfrac{dp}{p} = \dfrac{d\rho}{\rho} + \dfrac{dT}{T} \qquad$ (15-94)

Combining the continuity and momentum equations, expressing in terms of Mach number, and integrating between two stations gives

$$\frac{p_1}{p_2} = \frac{1 + k\mathbf{M}_2^2}{1 + k\mathbf{M}_1^2} \qquad (15\text{-}95)$$

Normalizing by setting the Mach number at station 2 equal to unity with the critical properties being designated with an asterisk leads to

■ $\quad \dfrac{p}{p^*} = \dfrac{k+1}{1 + k\mathbf{M}^2} \qquad$ (15-96)

Replacing p by ρRT and eliminating the density by using the continuity equation, we get

■ $\quad \dfrac{T}{T^*} = \left[\dfrac{\mathbf{M}(1+k)}{1 + k\mathbf{M}^2}\right]^2 \qquad$ (15-97)

The stagnation-temperature, stagnation-pressure, and density ratios, obtained directly from Eqs. (15-36), (15-39), and (15-40), respectively, are

$$\frac{T_0}{T_0^*} = \frac{(k+1)M^2[2+(k-1)M^2]}{(1+kM^2)^2} \qquad (15\text{-}98)$$

$$\frac{p_0}{p_0^*} = \frac{k+1}{1+kM^2}\left[\frac{2+(k-1)M^2}{k+1}\right]^{k/(k-1)} \qquad (15\text{-}99)$$

$$\frac{\rho^*}{\rho} = \frac{V}{V^*} = \frac{M^2(1+k)}{1+kM^2} \qquad (15\text{-}100)$$

These ratios are tabulated for $k = 1.4$ as a function of Mach number **M** in Table A-15.

The energy equation (15-93) can be written in terms of the stagnation temperature, that temperature the fluid would assume if adiabatically decelerated to zero velocity, i.e.,

$$\frac{\delta Q}{dm} = c_p(T_2 - T_1) + \frac{V_2^2 - V_1^2}{2} = c_p(T_{0_2} - T_{0_1}) \qquad (15\text{-}101)$$

Therefore, the change in stagnation temperature is a direct measure of the heat transferred.

The change in entropy which determines the direction of the process—heating or cooling—can be found from Eq. (3-73), expressed in terms of Mach number:

$$s - s^* = \frac{kR}{k-1}\ln\left[M^2\left(\frac{k+1}{1+kM^2}\right)^{(k+1)/k}\right] \qquad (15\text{-}102)$$

An anomaly in Rayleigh flow occurs at the point b in Fig. 15-24b. T/T^* is a maximum at that point, giving $M = 1/\sqrt{k}$. Heat addition beyond this point reduces the fluid temperature, contrary to intuition. Conversely, heat extraction increases the stream temperature.

Example Air enters the combustion chamber of a ramjet with a velocity of 400 ft/sec. The temperature of the undisturbed air is 600°R. For a fuel-air ratio of 0.02, what should the maximum heating value of the fuel be in order to maintain the flow rate?

Solution The inlet Mach number is

$$M_1 = \frac{V_1}{\sqrt{kRT_1}} = \frac{400\text{ ft/sec}}{\sqrt{(1.4 \times 32.2\text{ lb}_m\text{-ft/lb}_f\text{-sec}^2)(53.3\text{ ft-lb}_f/\text{lb}_m\text{-°R})(600\text{°R})}} = 0.33$$

giving

$$\left.\frac{T}{T_0}\right|_1 = 0.9785$$

from the isentropic Table A-12; therefore,

$$T_{0_1} = \frac{600}{0.9785} = 613\text{°R}$$

ONE-DIMENSIONAL COMPRESSIBLE FLOW

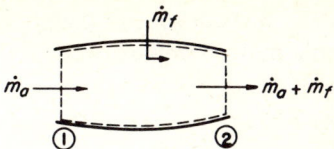

Fig. 15-25 Ramjet in flight.

Assuming the combustion products to behave like air, the reference temperature inside the combustion chamber is found from the Rayleigh Table A-15 at the inlet condition, i.e.,

$$\left.\frac{T_0}{T_0^*}\right|_1 = 0.4025$$

$$T_{0_1}^* = T_{0_2}^* = \frac{613}{0.4025} = 1520°R$$

An energy balance on the fluid in the combustion chamber, neglecting the enthalpy of the entering fuel, gives

$$\dot{m}_a c_{p_a} T_{0_1} + \dot{m}_f H = (\dot{m}_a + \dot{m}_f) c_{p_{\text{mix}}} T_{0_2}$$

where H is the heating value of the fuel. Dividing by the mass flow rate of the air and assuming that the specific heat of the mixture is equal to that of air, we have

$$\frac{\dot{m}_f}{\dot{m}_a} H = \left(1 + \frac{\dot{m}_f}{\dot{m}_a}\right) c_p (T_{0_2} - T_{0_1})$$

The heating value is a maximum when the exit stagnation temperature is a maximum. This occurs when the exit Mach number is unity, noting that $T_0/T_0^* = 1.000$ is a maximum in the Rayleigh Table A-15 when $\mathbf{M} = 1.00$. Therefore, the energy equation becomes

$$H_{\max} = \frac{1 + \dot{m}_f/\dot{m}_a}{\dot{m}_f/\dot{m}_a} c_p (T_{0_2}^* - T_{0_1})$$

$$= \frac{1 + 0.02}{0.02} (0.24 \text{ Btu/lb}_m\text{-}°R)(1520 - 613 R°)$$

$$= 111.0 \text{ Btu/lb}_m$$

14-10 CLOSURE

It should be emphasized that the Fanno and Rayleigh tables may be used for flows which are not choked. The critical properties pertain to hypothetical extensions of the duct to the point where $\mathbf{M} = 1$. In any flow situation the isentropic tables may be used to relate properties at a single station.

In most compressible-flow cases the state of the fluid is determined by a combination of the mechanisms discussed in this chapter. Many times, however, one mechanism is predominant. If the mechanisms are of equal order of magnitude, the resulting equations must retain the governing terms. For example, if heat transfer and friction are equally important, the energy and momentum equations must retain $\delta Q/\delta t$ and dF_r, respectively. Detailed analyses are found in more advanced textbooks or in texts devoted entirely to the flow of compressible fluids [1, 6, 9]. Integration of the more complex equations usually requires numerical techniques and computer solution.

PROBLEMS

15-1. Calculate the velocity of sound at 70°F in (a) hydrogen, (b) water vapor.

15-2. What is the speed of sound at standard atmospheric conditions (14.7 psia, 72°F)?

15-3. Calculate the speed of sound for NACA standard atmosphere at (a) sea level, (b) 10,000 ft, (c) 100,000 ft, (d) 200,000 ft.

15-4. Plot a curve of the speed of sound vs. temperature for air from 0 to 1000°F.

15-5. Air flows steadily, reversibly, and isothermally through a convergent nozzle. At a point in the flow the properties of the air are $V_1 = 1,000$ ft/sec, $T_1 = 50°F$. What is the Mach number at that point?

15-6. An airplane is traveling at a Mach number of 0.7 at an altitude of 10,000 ft. What is the velocity of the aircraft if we assume standard atmosphere?

15-7. A bullet travels at a Mach number of 2.0 through air at standard conditions ($T = 72°F$, $p = 14.7$ psia). What is the angle of the Mach cone produced by the bullet in flight? What is the velocity of the bullet?

15-8. A photograph of a meteorite taken on a space flight at an altitude of 31 miles (undisturbed air: $p = 1.95$ psfa; $T = 165°F$) shows that at a great distance from the meteorite the total included angle of its conical-shaped shock front is 60°.

  (a) What is the velocity of the meteorite in feet per second?

  (b) What is the maximum temperature (Fahrenheit) of the meteorite? Where does it occur? Explain.

15-9. Standard air ($p = 14.7$ psia, $T = 72°F$) at rest is accelerated isentropically until the velocity is 800 ft/sec.

  (a) What is the Mach number at this velocity?

  (b) What is the velocity when the airspeed becomes sonic?

15-10. Beginning with the continuity equation for steady flow, derive Eq. (15-14).

15-11. Using the steady-flow energy equation, develop the relation between the velocity at a point and the speed of sound. Plot the relationship for the flow regimes: (a) incompressible flow, $\mathbf{M} < 0.3$; (b) subsonic flow, $0.3 \leq \mathbf{M} \leq 1$; (c) transonic flow, $\mathbf{M} \approx 1$; (d) supersonic flow, $1 < \mathbf{M} < 5$; and (e) hypersonic flow, $\mathbf{M} > 5$. This plot is commonly known as the *steady-flow ellipse*.

15-12. Air flows steadily, reversibly, and isothermally through a variable-area duct. At a point in the flow the properties are $V_1 = 1,400$ ft/sec, $T_1 = 40°F$, $p_1 = 0.01$ psia, $s_1 = 0.475$ Btu/lb$_m$-°R.

  (a) What is the Mach number at the point?

  (b) What is the isentropic stagnation entropy?

  (c) What is the isentropic stagnation enthalpy?

15-13. Air, assumed to be a perfect gas, flows through a variable-area duct. At one point in the duct $\mathbf{M} = 0.7$, $p = 20$ psia, and the cross-sectional area is 2.05 in². What is the stagnation pressure for this flow, and what is the critical area?

15-14. Air at 50 psia in a large reservoir is heated to 200°F. The air from the reservoir is accelerated reversibly and adiabatically to a Mach number of 0.8.

  (a) What is the resulting temperature?

  (b) What is the velocity?

15-15. At a given location in an airflow field the Mach number is 1.5, and the pressure is 50 psia.

ONE-DIMENSIONAL COMPRESSIBLE FLOW

At another station in the airflow field the Mach number is 2.5. What is the pressure at the station where the Mach number is 2.5?

15-16. The pressure at station 1 in an airflow field is 100 psia, and the pressure and Mach number at station 2 are 50 psia and 2.0, respectively. Determine the Mach number at station 1.

15-17. Nitrogen gas flows through a duct, and the stagnation pressure and temperature are 20 psia and 40°F, respectively. The Mach number at a particular station in the duct is 0.8. What are the pressure, temperature, and velocity at this station?

15-18. Find the Mach numbers at which the following conditions are satisfied:

  (a) The temperature at some point in the flow of air through a duct is 85 percent of the stagnation temperature.

  (b) The stagnation pressure is 50 percent greater than the static pressure at some point in the flow of air through a duct.

15-19. The temperature at station 1 in an airflow field is 400°R, and the temperature at station 2, where the Mach number is 1.2, is 600°R. What is the Mach number at station 1?

15-20. The stagnation pressure on the nose of an aircraft in low-level flight is measured at 25 psia, and the ambient pressure is 14.5 psia. The air temperature is 35°F.

  (a) What is the free-flight Mach number of the aircraft?
  (b) What is its speed?
  (c) What is the stagnation temperature?

15-21. A rocket nozzle with an area ratio (exit area to throat area) of 4.00 is designed to exhaust into a region where the pressure is 14.6 psia. The fluid is assumed to be a perfect gas with specific-heat ratio of 1.4. What is the combustion-chamber pressure?

15-22. Air is flowing through a duct at a speed of 1,000 ft/sec with a pressure of 100 psia and a temperature of 1000°R. The pressure is to be isentropically increased to 150 psia. The mass flow rate through the duct is 100 lb_m/sec. What change in area is required to meet these conditions?

15-23. A perfect gas with $k = 1.4$ flows steadily, adiabatically, and frictionlessly from a large reservoir through the convergent duct shown. The temperature and pressure in the supply reservoir are 800°R and 100 psia, respectively, and the nozzle exhausts to a region in which the pressure is constant and equal to 25 psia. Determine the exit-plane velocity.

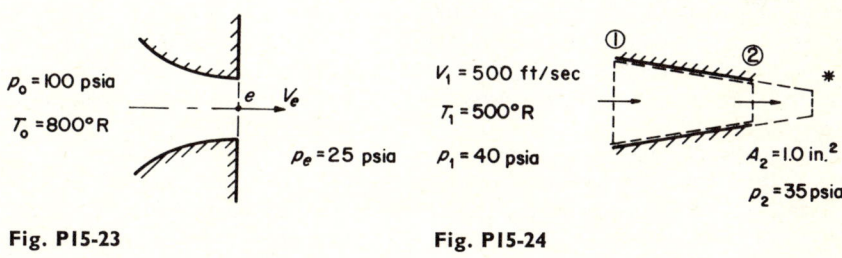

Fig. P15-23   Fig. P15-24

15-24. Air flows through a reducer as shown. Assume isentropic flow. What is the flow rate?

15-25. Air flows steadily and isentropically in the duct shown.
  (a) What is the isentropic stagnation pressure at section 1?
  (b) What is the isentropic stagnation temperature at section 1?
  (c) What is the Mach number at section 2?
  (d) What is the temperature at section 2?
  (e) What is the mass flow rate?

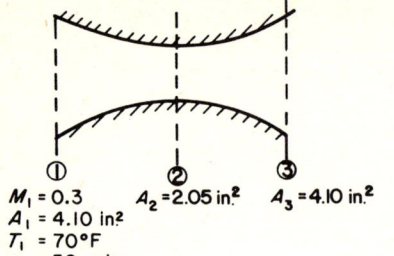

$M_1 = 0.3$
$A_1 = 4.10$ in.^{2}
$T_1 = 70°F$
$p_1 = 50$ psia
$A_2 = 2.05$ in.^{2} $A_3 = 4.10$ in.^{2}

Fig. P15-25

15-26. Air flows steadily through the system shown. What is the minimum diffuser throat area A_4 for the maximum flow rate?

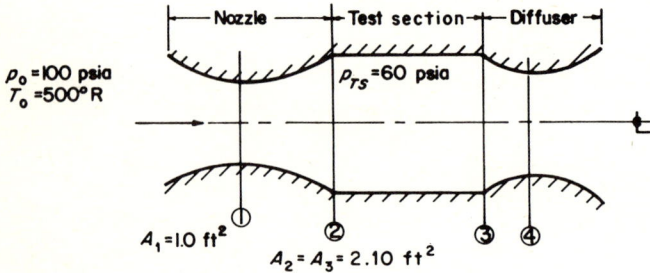

$p_0 = 100$ psia
$T_0 = 500°R$
$p_{TS} = 60$ psia
$A_1 = 1.0$ ft^2
$A_2 = A_3 = 2.10$ ft^2

Fig. P15-26

15-27. Air in a tank is at a pressure of 100 psia and a temperature of 75°F. The air will issue out of a small duct as a one-dimensional isentropic flow. What is the maximum possible flow per unit area at the exit of a convergent nozzle at the end of the duct?

15-28. Air at 100 psia and 200°F flows from a tank through a converging nozzle into the atmosphere ($p = 14.7$ psia).
  (a) What is the exit pressure?
  (b) What is the exit temperature?

15-29. Air is fed through a convergent nozzle from a reservoir as shown.
  (a) Plot the pressure distribution vs. x for varying back pressures, that is, $0 < p_b < 100$ psia.
  (b) Determine the mass flow rate.

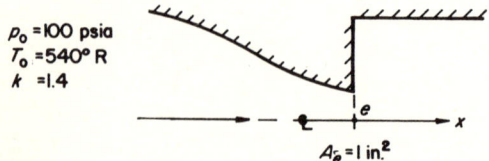

$p_0 = 100$ psia
$T_0 = 540°R$
$k = 1.4$
$A_e = 1$ in.^{2}

Fig. P15-29

15–30. Air at 100 psia and 200°F flows from a tank through a converging nozzle and discharges into a receiver whose pressure is 80 psia. What is the exit pressure and exit temperature?

15–31. Consider the flow of a gas from the tank shown. The flow at station 1 is sonic. Is the flow at station 2 sonic, supersonic, or subsonic? Discuss.

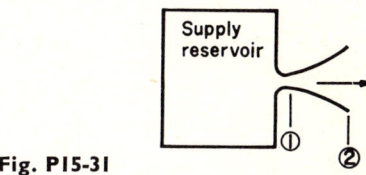

Fig. P15-31

15–32. Air flows through a converging-diverging nozzle as shown. Find p_e and p_t for $0 \leq p_e \leq 100$ psia.

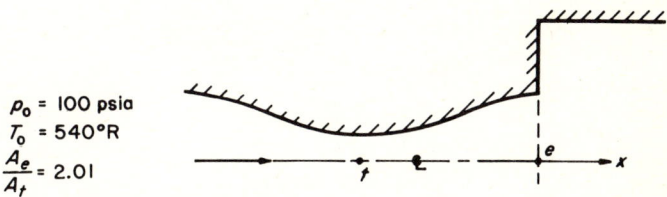

Fig. P15-32

15–33. For a convergent-divergent nozzle with $p_0 = 100$ psia, $A_e/A_t = 2.01$, $k = 1.4$ and exit pressure $p_e = 46.8$ psia, find the Mach number at the exit.

15–34. Air flows steadily through the duct shown. The duct is fed from a plenum where stagnation pressure and temperature are 100 psia and 70°F. What is the maximum mass flow rate?

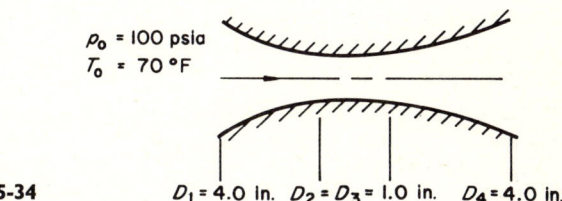

Fig. P15-34

15–35. A supersonic wind tunnel is to be designed for a Mach number of 2.5 with a 4-ft² test section. Air will be taken from the atmosphere (14.7 psia and 70°F) and will be accelerated to $M = 2.5$ in a converging-diverging nozzle. Assuming that the nozzle is frictionless to the throat, what is the pressure and temperature at the test section, and what is the throat diameter of the nozzle?

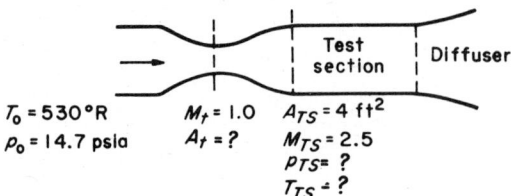

$T_0 = 530°R$ $M_t = 1.0$ $A_{TS} = 4 \text{ ft}^2$
$p_0 = 14.7 \text{ psia}$ $A_t = ?$ $M_{TS} = 2.5$
                                          $p_{TS} = ?$
                                          $T_{TS} = ?$ Fig. P15-35

15-36. Air flows through a converging-diverging nozzle, and a shock exists in the diverging portion of the nozzle, where the pressure is 35 psia. The stagnation conditions are 100 psia and 500°R.

(a) Determine the Mach number just upstream and just downstream from the shock.

(b) Determine the pressure and temperature just downstream from the shock.

15-37. A stream of air with Mach number of 1.5, pressure of 100 psia, and a temperature of 70°F passes through a normal shock. What is the stagnation pressure and stagnation temperature downstream from the shock?

15-38. Show that the point of maximum entropy on the Fanno line corresponds to a Mach number of unity.

15-39. Consider a frictionless steady adiabatic flow of air through the choked convergent-divergent duct shown. The stagnation conditions are 100 psia and 1000°R, respectively. A normal shock occurs somewhere in the duct. Where will this shock occur if the back pressure is 70 psia?

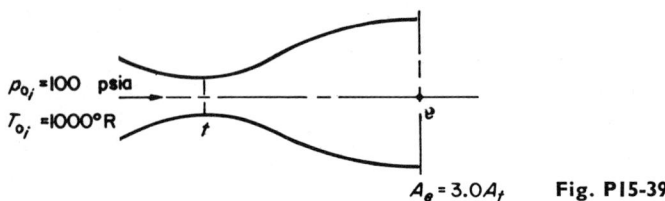

$p_{0i} = 100 \text{ psia}$
$T_{0i} = 1000°R$

$A_e = 3.0 A_t$ Fig. P15-39

15-40. Show that the point of maximum entropy on the Rayleigh line corresponds to a Mach number of unity.

15-41. Air assumed to be a perfect gas flows steadily, frictionlessly, and adiabatically through a converging-diverging section. The stagnation temperature and pressure are 600°R and 100 psia; $A_e/A_t = 2.50$. What back pressure is required for a normal shock to occur in the exit plane?

15-42. A steady frictionless adiabatic airflow exists in the choked convergent-divergent duct shown. The initial stagnation pressure and temperature are 100 psia and 1000°R, respectively. A shock stands in the divergent portion of the duct at a position where the cross-sectional area is $2.01 A_t$; $A_e/A_t = 3.0$.

(a) What is the Mach number just before the shock?

(b) What is the temperature just after the shock?

(c) What is the back pressure p_b?

(d) What back pressure would be required to move the normal shock to the exit plane?

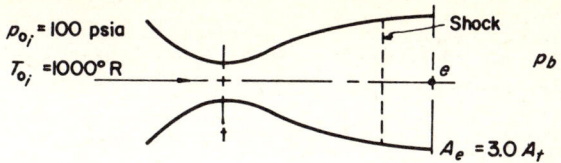

Fig. P15-42

15-43. Air flows through a convergent-divergent nozzle, and a normal shock exists in the divergent portion of the nozzle. The stagnation conditions are 100 psia and 600°R. The shock occurs where the cross-sectional area is 3 in²; $A_t = 1.5$ in², $A_e = 3.5$ in².
   (a) What is the loss in stagnation pressure during the flow through the nozzle?
   (b) Where would the shock occur if the back pressure were 60 psia?

15-44. Air at $p_0 = 100$ psia and $T_0 = 1000°R$ is fed into an adiabatic constant-area duct by an isentropic convergent-divergent nozzle having an exit-to-throat-area ratio of 1.18. No shock occurs anywhere in the duct. The hydraulic diameter $D_h = 1.0$ ft, and the friction factor $f = 0.001$.
   (a) What is the maximum length of the duct?
   (b) What are the exit temperature and pressure for this length?

15-45. Derive an expression, in terms of k, c_p, dp_0, and p_0, for the differential change in entropy ds of a perfect gas which flows adiabatically and with friction through a constant-area duct.

15-46. Air flows steadily through a choked convergent-divergent nozzle. The stagnation properties are $p_0 = 1,000$ psia and $T_0 = 1000°R$; $A_e/A_t = 2.0$. Where will a shock stand if the back pressure is 600 psia? What is the exit-plane Mach number for this back pressure?

15-47. Air at $p_0 = 100$ psia and $T_0 = 1000°R$ discharges into an insulated pipe (adiabatic constant-area duct) from an isentropic convergent-divergent nozzle having an area ratio of 1.51. The hydraulic diameter $D_h = 1$ ft, and the friction factor $f = 0.001$. A shock stands in the nozzle throat, and the back pressure is 40 psia. What is the duct length for these flow conditions?

15-48. Air at $p_0 = 100$ psia and $T_0 = 600°R$ is fed into an adiabatic constant-area duct by an isentropic convergent-divergent nozzle having an area ratio A_e/A_t of 2.5. The hydraulic diameter $D_h = 1$ ft and the friction factor $f = 0.001$. A shock stands in the nozzle exit plane. The duct back pressure is 30 psia. What is the duct length for these flow conditions?

15-49. An adiabatic constant-area frictional duct is fed by an isentropic nozzle with an area ratio $A_e/A_t = 1.6875$. A normal shock occurs in the duct at a point where the Mach number is 1.30. Find the maximum length of the duct and the maximum back pressure p_b for this length.

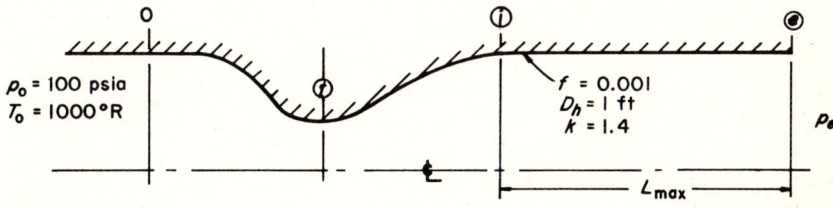

Fig. P15-49

15-50. Air at $p_0 = 100$ psia and $T_0 = 1000°R$ flows through an isentropic convergent-divergent nozzle into an adiabatic constant-area duct having an area ratio of 1.51. The hydraulic diameter $D_h = 1$ ft, and the friction factor $f = 0.001$. A normal shock occurs in the exit plane of the duct. What is the duct length for these conditions?

15-51. Fuel is burned in a combustion chamber with air, which is moving at a velocity of 400 ft/sec. The temperature of the air is 560°R. Determine the maximum possible rise in stagnation temperature.

15-52. A gaseous mixture of air and fuel enters a ramjet combustion chamber with a velocity of 150 ft/sec and at a temperature of 600°R. The heat of reaction H of the mixture for the particular fuel-air ratio employed is 600 Btu per pound of mixture. What is the Mach number at the exit of the combustion chamber?

15-53. Air flows steadily through a duct having a constant cross-sectional area. The velocity of the air is 300 ft/sec. At a position 10 ft from the end of the duct the pressure is 20 psia, and the temperature is 600°R. The fluid leaves the duct subsonically at a pressure of 14.7 psia. Determine the heat transfer per pound mass of fluid between these two sections; $c_p = 0.24$ Btu/lb$_m$-°F.

15-54. For Prob. 15-53, determine the heat transfer per unit mass between section 1 and the end that will cause the flow to choke.

REFERENCES

1. Cambel, A. B., and B. H. Jennings: "Gas Dynamics," McGraw-Hill, New York, 1958.
2. Courant, R., and K. O. Friedricks: "Supersonic Flow and Shock Waves," Interscience, New York, 1948.
3. John, J. E. A.: "Gas Dynamics," Allyn and Bacon, Boston, 1969.
4. Moody, L. F.: Friction Factors for Pipe Flow, *Trans. ASME*, November, 1944.
5. Nikuradse, J.: *VDI Forschungsh.* 356, 1932; *VDI Forschungsh.* 361, 1933.
6. Owczarek, J. A.: "Gas Dynamics," International Textbook, Scranton, Pa., 1964.
7. Pao, R. H. F.: "Fluid Dynamics," Merrill, Columbus, Ohio, 1967.
8. Shames, I. H.: "Mechanics of Fluids," McGraw-Hill, New York, 1962.
9. Shapiro, A. H.: "The Dynamics and Thermodynamics of Compressible Fluid Flow," vol. I, Ronald, New York, 1953.
10. Sissom, L. E.: Normal Shock Location in Adiabatic Constant-area Ducts with Friction, *AIAA J. Spacecraft Rockets*, **1967**: 810–811.

CHAPTER 16

FREE CONVECTIVE HEAT TRANSFER

Fluid motion due to buoyancy effects was briefly discussed in Chap. 12. In the present chapter heat transfer resulting from this type of *natural*, or *free*, convective flow will be treated. In this class of problem, the driving force for fluid motion is the body force of the Navier-Stokes equation, which arises from the temperature gradient. For the simple case of a body with constant wall temperature exposed to a quiescent ambient fluid, this force per unit volume is (as will be shown later)

$$\rho g \beta (T_s - T_\infty)$$

where β is the volume coefficient of expansion.

The ratio of this force to the viscous forces in the momentum equation can be written as a nondimensional parameter

$$\mathbf{Gr} = \frac{\rho g \beta (T_s - T_\infty) L^3}{\rho \nu^2} \propto \frac{\text{buoyancy force}}{\text{viscous force}}$$

called the *Grashof number*. *This parameter occupies a role in free convection corresponding to that of the Reynolds number in forced convection.* Recalling that the Reynolds number can be expressed as the ratio of inertia to viscous forces,

$$\mathbf{Re}_L{}^2 = \left(\frac{V_\infty L}{\nu}\right)^2 \propto \frac{\text{inertia force}}{\text{viscous force}}$$

the ratio of Grashof number to the square of the Reynolds number is an indication of the ratio of buoyancy to inertia forces

$$\frac{\mathbf{Gr}}{\mathbf{Re}_L{}^2} \propto \frac{\text{buoyancy force}}{\text{inertia force}}$$

In most free-convective problems the forced velocity V_∞ is zero, and the buoyancy term is the only significant driving force. The three possible cases are

1. $\mathbf{Gr} \gg \mathbf{Re}_L{}^2$, free convection
2. $\mathbf{Gr} \ll \mathbf{Re}_L{}^2$, forced convection (Chaps. 12 and 13)
3. $\mathbf{Gr} \simeq \mathbf{Re}_L{}^2$, mixed forced and free convection

Cases 1 and 3 are treated in the present chapter.

16-1 LAMINAR FREE CONVECTION ON A VERTICAL PLATE

The simplest free-convective-heat-transfer problem arises when a vertical wall is subjected to a cooler (or warmer) surrounding fluid. Consider the case of a heated wall at constant temperature, as shown in Fig. 16-1. The fluid adjacent to the plate is heated and consequently expands; this reduces its density and causes a vertical flow. The no-slip boundary condition is applicable at the wall, where $y = 0$, and since the ambient fluid velocity is zero, a velocity profile of the type depicted in Fig. 16-1 develops.

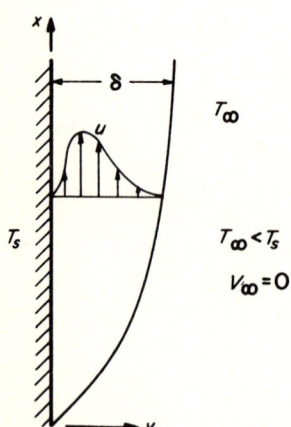

Fig. 16-1 Flow boundary layer on a vertical plate.

FREE CONVECTIVE HEAT TRANSFER

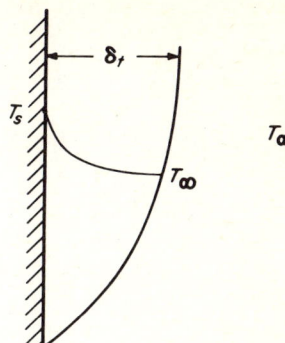

Fig. 16-2 Thermal boundary layer on a vertical plate.

For a plate cooler than the surroundings, the physical problem is simply inverted. Thus, the mathematical treatment of the case of a heated wall is applicable to a cooled wall.

A more complete description of the boundary layer involves a similar development of a thermal boundary layer, as shown in Fig. 16-2. Here the plate is at $T_s > T_\infty$. The thermal-boundary-layer thickness δ_t may or may not be identical with the flow-boundary-layer thickness δ at a given vertical position. This depends upon the fluid Prandtl number, as it did for forced-convective flow.

Also, it should be noted that these two figures imply a laminar boundary-layer flow. If the plane is sufficiently long, the flow will become turbulent. Experiments show that such flows become turbulent for a product[1] of the Grashof and Prandtl numbers between 10^8 and 10^9. For an environment of very low turbulence, the recommended rule of thumb for determination of the flow regime is

Laminar flow regime: **PrGr** $< 10^9$
Turbulent flow regime: **PrGr** $> 10^9$

The boundary-layer assumptions of Sec. 12-5 are equally applicable to the free-convection boundary-layer flow. Thus, the applicable momentum equation including the body force (which was omitted in Sec. 12-5) is

$$\rho\left(u\frac{\partial u}{\partial x} + v\frac{\partial u}{\partial y}\right) = -\frac{\partial p}{\partial x} - \rho g_x + \mu\frac{\partial^2 u}{\partial y^2} \tag{16-1}$$

for steady flow.

Recall from Sec. 12-5 that the pressure gradient is impressed through the boundary layer from the outer region. For the present case, the pressure gradient outside the boundary layer, where $v \equiv 0$, can be obtained directly from the Euler equation (10-82) and is

$$\frac{\partial p}{\partial x} = -\rho_\infty g_x \tag{16-2}$$

---

[1] The product **GrPr** is sometimes called the Rayleigh number in heat-transfer literature.

Substituting this into the momentum equation yields

$$\rho\left(u\frac{\partial u}{\partial x} + v\frac{\partial u}{\partial y}\right) = g_x(\rho_\infty - \rho) + \mu\frac{\partial^2 u}{\partial y^2} \tag{16-3}$$

Introducing the coefficient of volume expansion defined by

$$\beta \equiv \frac{1}{\mathscr{V}}\left(\frac{\partial \mathscr{V}}{\partial T}\right)_p \tag{16-4}$$

the density change across the boundary layer can be expressed as

$$\beta = \frac{1}{\mathscr{V}_\infty}\frac{\mathscr{V} - \mathscr{V}_\infty}{T - T_\infty} = \rho_\infty \frac{1/\rho - 1/\rho_\infty}{T - T_\infty}$$

or

$$\rho\beta(T - T_\infty) = \rho_\infty - \rho \tag{16-5}$$

When we substitute this last expression into the momentum equation, noting that the energy and continuity equations are unchanged from the forced-convective boundary-layer problem, the complete equations are

Momentum:
$$\rho\left(u\frac{\partial u}{\partial x} + v\frac{\partial u}{\partial y}\right) = g_x\rho\beta(T - T_\infty) + \mu\frac{\partial^2 u}{\partial y^2} \tag{16-6}$$

Energy:
$$u\frac{\partial T}{\partial x} + v\frac{\partial T}{\partial y} = \alpha\frac{\partial^2 T}{\partial y^2} \tag{16-7}$$

Continuity:
$$\frac{\partial u}{\partial x} + \frac{\partial v}{\partial y} = 0 \tag{16-8}$$

The boundary conditions are

At $y = 0$: $u = 0$ $v = 0$ $T = T_s$

At $y = \infty$: $u = 0$ $T = T_\infty$ $\frac{\partial u}{\partial y} = 0$ $\frac{\partial T}{\partial y} = 0$

Since there is one boundary condition for each order of the highest-order derivative in each spatial direction, the problem is completely specified. The reader should note that independent solution of the momentum equation is not possible; this is coupled to the energy equation through the buoyancy term, and the coupled equations must be solved simultaneously.

16-2 SOLUTION OF THE DIFFERENTIAL EQUATIONS

Pohlhausen, in collaboration with Schmidt and Beckmann [20], first solved this set of boundary-layer differential equations for a vertical flat plate. From Schmidt and

FREE CONVECTIVE HEAT TRANSFER

Beckmann's experimental data, Pohlhausen noted that the velocity profiles along the plate were *similar*, as Blasius had found for the forced-convection problem (see Sec. 12-5). He also observed that the temperature profiles were *similar*. Using a similarity parameter

$$\eta = \frac{y}{x}\left(\frac{\mathrm{Gr}_x}{4}\right)^{\frac{1}{4}} \tag{16-9}$$

he expressed the velocity components u and v in terms of a stream function (see Sec. 11-3) defined by[1]

$$\psi(x,y) = f(\eta)\left[4\nu\left(\frac{\mathrm{Gr}_x}{4}\right)^{\frac{1}{4}}\right] \tag{16-10}$$

Using these together with a dimensionless temperature

$$\theta = \frac{T - T_\infty}{T_s - T_\infty} \tag{16-11}$$

he reduced the set of three partial differential equations to two ordinary differential equations

Momentum: $\quad f''' + 3ff'' - 2(f')^2 + \theta = 0 \tag{16-12}$

Energy: $\quad \theta'' + 3\mathrm{Pr}f\theta' = 0 \tag{16-13}$

In these transformed equations f and θ are functions of η only, and consequently the prime refers to differentiation with respect to η. The boundary conditions are

At $\eta = 0$: $\quad f = 0 \quad f' = 0 \quad \theta = 1$
As $\eta \to \infty$: $\quad\quad\quad\quad f' \to 0 \quad \theta \to 0$

Notice that the energy equation (16-13) still contains the Prandtl number. As a consequence, a separate solution of the coupled equations is required for each Prandtl number. Pohlhausen effected a numerical solution for a Prandtl number of 0.733. Schuh [21] later solved these equations for Prandtl numbers of 10, 100, and 1,000.

Ostrach [18] conducted a thorough investigation of this problem in 1953. From a study of the more complete momentum and energy equations, he established the conditions for which Pohlhausen's transformed boundary-layer equations were valid. He then solved these equations with the aid of a digital computer for Prandtl numbers from 0.01 to 1,000. The velocity and temperature profiles resulting from

---

[1] By definition of the stream function, $u = \partial\psi/\partial y$. Hence, applying the chain rule for differentiation,

$$u = \frac{\partial\psi}{\partial\eta}\frac{\partial\eta}{\partial y} = 4\nu\left(\frac{\mathrm{Gr}_x}{4}\right)^{\frac{1}{4}}f'(\eta)\frac{1}{x}\left(\frac{\mathrm{Gr}_x}{4}\right)^{\frac{1}{4}} = \frac{2\nu}{x}\mathrm{Gr}_x^{\frac{1}{2}}f'$$

The velocity component v and the partial derivatives are found in a similar fashion, the details of which are left as an exercise.

his study are presented in Figs. 16-3 and 16-4, from which we note that for **Pr** > 1.0, the thermal boundary layer is thinner than the velocity boundary layer. This is especially evident for Prandtl numbers of 100 and 1,000 and is a result of the relative influence of viscous and conduction effects in the fluid.

The temperature profile permits determination of the local heat flux. Since the fluid velocity at the wall is zero, the heat flux is

$$\frac{q(x)}{A} = -k \left.\frac{\partial T}{\partial y}\right|_{y=0} = -\frac{k}{x}(T_s - T_\infty)\left(\frac{\mathrm{Gr}_x}{4}\right)^{\frac{1}{4}} \left.\frac{d\theta}{d\eta}\right|_{\eta=0} \tag{16-14}$$

By the definition of the convective-heat-transfer coefficient

$$\frac{q}{A} = h(T_s - T_\infty)$$

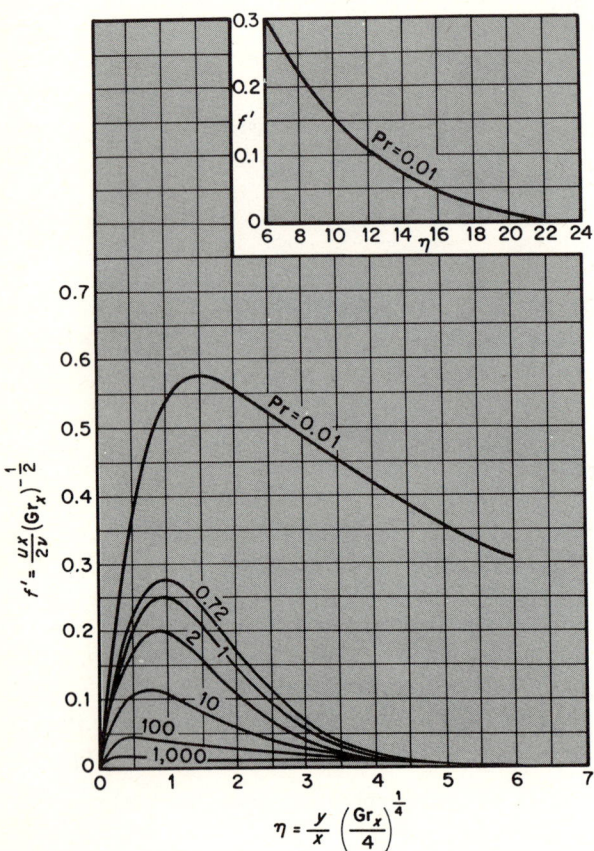

Fig. 16-3 Velocity profiles in laminar free convection along a vertical isothermal plate. (*Adapted from S. Ostrach, NACA Rept. 1111, 1953.*)

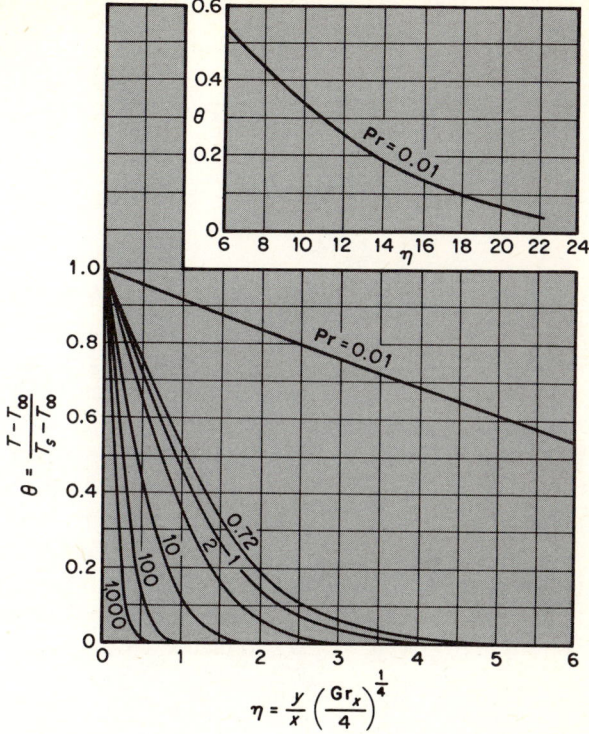

Fig. 16-4 Temperature profiles in laminar free convection along a vertical isothermal plate. (*Adapted from S. Ostrach, NACA Rept. 1111, 1953.*)

and then

$$\mathbf{Nu}_x = \frac{hx}{k} = -\left(\frac{\mathbf{Gr}_x}{4}\right)^{\frac{1}{4}} \frac{d\theta}{d\eta}\bigg|_{\eta=0} \tag{16-15}$$

Recalling that each solution of the coupled equations is for a single Prandtl number, this last expression can be written

$$\mathbf{Nu}_x = \left(\frac{\mathbf{Gr}_x}{4}\right)^{\frac{1}{4}} F_1(\mathbf{Pr}) \tag{16-16}$$

Values of $F_1(\mathbf{Pr})$ from Ostrach are:

Pr	0.01	0.72	0.733	1	2	10	100	1,000
$F_1(\mathbf{Pr})$	0.0812	0.5046	0.5080	0.5671	0.7165	1.1694	2.191	3.966

To obtain the total heat transfer from a plate of finite length the average heat-transfer coefficient is

$$\bar{h} = \frac{1}{L}\int_0^L h(x)\,dx = \frac{k}{L} F_1(\text{Pr})\left[\frac{g\beta(T_s - T_\infty)}{4\nu^2}\right]^{\frac{1}{4}} \int_0^L \frac{dx}{x^{\frac{1}{4}}}$$

Thus,

$$\overline{\text{Nu}} = \frac{\bar{h}L}{k} = \frac{4}{3}\left(\frac{\text{Gr}_L}{4}\right)^{\frac{1}{4}} F_1(\text{Pr}) \tag{16-17}$$

In obtaining the results of Eqs. (16-16) and (16-17), all fluid properties except the density were assumed constant. Consequently, application to problems involving temperature differences resulting in very large changes in μ and k should be undertaken with caution. For air, temperature differences of a few hundred degrees Fahrenheit are usually permissible. Also, the results should not be applied to problems where $\text{GrPr} < 10^4$, since the boundary-layer approximations are invalid below this value.

Frequently in free-convective problems, the uniform-heat-flux condition is more realistic than the uniform-wall-temperature approximation. Sparrow and Gregg [22] have presented a solution for the vertical plate in laminar free convection which either absorbs or dissipates heat at a uniform rate over its entire height. The differential equations (16-12) and (16-13) are applicable to this problem, but the boundary condition $\theta = 1$ at $\eta = 0$ is replaced by the heat-flux condition (Fourier's equation)

$$\frac{q}{A} = -k\frac{\partial T}{\partial y}\bigg|_{y=0}$$

Basing the similarity parameter on a modified Grashof number Gr_x^*, where $T_s - T_\infty$ is replaced with

$$\frac{q}{A}\frac{x}{k}$$

they used

$$\zeta = \frac{y}{x}\left(\frac{\text{Gr}_x^*}{5}\right)^{\frac{1}{5}} \tag{16-18}$$

and a stream function

$$\psi(x,y) = f(\zeta)\left[5\nu\left(\frac{\text{Gr}_x^*}{5}\right)^{\frac{1}{5}}\right] \tag{16-19}$$

together with a dimensionless temperature

$$\theta = \frac{T_\infty - T}{qx/Ak}\left(\frac{\text{Gr}_x^*}{5}\right)^{\frac{1}{5}} \tag{16-20}$$

FREE CONVECTIVE HEAT TRANSFER

to transform the momentum and energy equations to

Momentum: $\quad f''' - 3(f')^2 + 4ff'' - \theta = 0 \quad$ (16-21)

Energy: $\quad \theta'' + \mathbf{Pr}(4f\theta' - \theta f') = 0 \quad$ (16-22)

The boundary conditions are

At $\eta = 0$: $\quad f = 0 \quad f' = 0 \quad \theta' = 1$
As $\eta \to \infty$: $\quad \quad \quad f' \to 0 \quad \theta \to 0$

This set of equations contains the Prandtl number of the fluid, as did the set for the isothermal wall. Sparrow and Gregg obtained solutions for Prandtl numbers of 0.1, 1.0, 10, and 100. From the definition of θ we have

$$\frac{q(x)}{A} = -k\frac{\partial T}{\partial y}\bigg|_{y=0} = -\frac{k}{x}(T_s - T_\infty)\left(\frac{\mathbf{Gr}^*_x}{5}\right)^{\frac{1}{5}}\theta(0) \quad (16\text{-}23)$$

where $\theta(0)$ is evaluated at $\eta = 0$. Introducing the heat-transfer coefficient and Nusselt modulus, we obtain

$$\mathbf{Nu}_x = \frac{hx}{k} = -\left(\frac{\mathbf{Gr}^*_x}{5}\right)^{\frac{1}{5}}\theta(0) \quad (16\text{-}24)$$

This can be averaged by integrating over the plate length to yield

$$\overline{\mathbf{Nu}} = \frac{\bar{h}L}{k} = -\frac{6}{5}\left(\frac{\mathbf{Gr}^*_x}{5}\right)^{\frac{1}{5}}\theta(0) \quad (16\text{-}25)$$

Converting this to a conventional Grashof number with the average value of $T_s - T_\infty$ used as the temperature difference yields

$$\overline{\mathbf{Nu}} = \frac{\bar{h}L}{k} = \left\{\frac{6(4)^{\frac{1}{4}}}{5(5)^{\frac{1}{5}}[-\theta(0)]}\right\}^{\frac{5}{4}}\left(\frac{\mathbf{Gr}_x}{4}\right)^{\frac{1}{4}} \quad (16\text{-}26)$$

or, since θ is a function of the fluid Prandtl number,

$$\overline{\mathbf{Nu}} = F_2(\mathbf{Pr})\left(\frac{\mathbf{Gr}_x}{4}\right)^{\frac{1}{4}} \quad (16\text{-}27)$$

Values of $F_2(\mathbf{Pr})$ resulting from Sparrow and Gregg's investigation are:

Pr	0.1	1.0	10	100
$F_2(\mathbf{Pr})$	0.335	0.811	1.656	3.083

We recall that the average Nusselt modulus for the isothermal plate was given by Eq. (16-17). Comparing Ostrach's results with those of Sparrow and Gregg, we find that $\frac{4}{3}F_1(\mathbf{Pr})$ is approximately 94 to 95 percent of $F_2(\mathbf{Pr})$. This suggests that a single correlation equation may be suitable for several different cases of free convection.

Other exact solutions, i.e., solutions of the differential momentum and energy equations, include those of Sparrow and Gregg for variable surface temperature along the length of the plate [23] and for vertical cylinders with a small radius of curvature and constant wall temperature [24]. Millsaps and Pohlhausen [15, 16] treat the case of a vertical cylinder with small radius of curvature and a temperature difference between the cylinder and the surrounding fluid which varies linearly with height. *For vertical cylinders with radius of curvature an order of magnitude greater than the boundary-layer thickness, the flat-plate analyses are reasonable approximations.*

Another important geometrical problem is the horizontal cylinder. Hermann [7] solved the differential equations for this problem numerically, and it was treated later by Chiang and Kaye [1].

16-3 INTEGRAL METHODS OF SOLUTION

The integral forms of the momentum and energy equations for a vertical flat plate in laminar free convection can be obtained in a manner similar to that employed in Chap. 12 for the laminar forced-convective problem. These are

$$\text{Momentum:} \quad \frac{d}{dx}\left(\int_0^\delta \rho u^2 \, dy\right) = -\mu \left(\frac{\partial u}{\partial y}\right)_{y=0} + \int_0^\delta \rho g \beta (T - T_\infty) \, dy \quad (16\text{-}28)$$

$$\text{Energy:} \quad \frac{d}{dx}\left[\int_0^\delta u(T - T_\infty) \, dy\right] = -\alpha \left(\frac{dT}{dy}\right)_{y=0} \quad (16\text{-}29)$$

where it is assumed that the velocity and thermal boundary layers have the same local thickness δ.

The boundary conditions on u and T are as given in Sec. 16-1 for the differential equations for the isothermal-plate problem, with one additional boundary condition on u. From the velocity profile of Fig. 16-1 it is apparent that the derivative du/dy is zero at a point within the boundary layer. A velocity profile satisfying the boundary conditions (see Sec. 12-8) is

$$\frac{u}{V_x} = \frac{y}{\delta}\left(1 - \frac{y}{\delta}\right)^2 \quad (16\text{-}30)$$

This has a maximum value of u at $y/\delta = \frac{1}{3}$, which is seen to be slightly high by inspection of Ostrach's velocity profiles. In this expression V_x is an arbitrary function of the position along the plate x and has the dimensions of velocity.

A temperature profile satisfying the three thermal-boundary conditions is

$$\theta = \frac{T - T_\infty}{T_s - T_\infty} = \left(1 - \frac{y}{\delta}\right)^2 \quad (16\text{-}31)$$

Introducing Eqs. (16-30) and (16-31) into the momentum and energy equations and

FREE CONVECTIVE HEAT TRANSFER

integrating yields

Momentum: $$\frac{1}{105}\frac{d}{dx}V_x^2\delta = \tfrac{1}{3}g\beta(T_s - T_\infty)\delta - \nu\frac{V_x}{g} \tag{16-32}$$

Energy: $$\tfrac{1}{30}(T_s - T_\infty)\frac{d}{dx}V_x\delta = 2\alpha\frac{T_s - T_\infty}{\delta} \tag{16-33}$$

Following Squire [27] and Eckert [2], we assume that V_x and δ can be expressed as exponential functions of x. Trying

$$V_x = C_1 x^a \quad \text{and} \quad \delta = C_2 x^b$$

we obtain upon substituting these into Eqs. (16-32) and (16-33) and then differentiating

Momentum: $$\frac{2a+b}{105}(C_1^2 C_2 x^{2a+b-1}) = g\beta(T_s - T_\infty)\frac{C_2 x^b}{3} - \frac{C_1}{C_2}\nu x^{a-b} \tag{16-34}$$

Energy: $$\frac{a+b}{30}C_1 C_2 x^{a+b-1} = \frac{2\alpha}{C_2}x^{-b} \tag{16-35}$$

Since these must be valid for all x, we can equate exponents of x. This yields the set of algebraic equations

$$2a + b - 1 = b = a - b$$
$$a + b - 1 = -b$$

which has the solution

$$a = \tfrac{1}{2} \quad b = \tfrac{1}{4}$$

With these exponents the momentum and energy equations become

Momentum: $$\frac{C_1 C_2}{84}x^{\frac{1}{4}} = g\beta(T_s - T_\infty)\frac{C_2}{3}x^{\frac{1}{4}} - \frac{C_1}{C_2}\nu x^{\frac{1}{4}} \tag{16-36}$$

Energy: $$\frac{C_1 C_2}{40}x^{-\frac{1}{4}} = \frac{2\alpha}{C_2}x^{-\frac{1}{4}} \tag{16-37}$$

Solving these two equations for the constants, we obtain

$$C_1 = 5.17\nu\left(\frac{20}{21} + \frac{\nu}{\alpha}\right)^{-\frac{1}{2}}\left[\frac{g\beta(T_s - T_\infty)}{\nu^2}\right]^{\frac{1}{2}} \tag{16-38}$$

$$C_2 = 3.93\left(\frac{20}{21} + \frac{\nu}{\alpha}\right)^{\frac{1}{4}}\left[\frac{g\beta(T_s - T_\infty)}{\nu^2}\right]^{-\frac{1}{4}}\left(\frac{\nu}{\alpha}\right)^{-\frac{1}{2}} \tag{16-39}$$

Noting that ν/α is the Prandtl number, we have by the assumed form of the expression for δ,

$$\delta = 3.93 \mathbf{Pr}^{-\frac{1}{2}}(\tfrac{20}{21} + \mathbf{Pr})^{\frac{1}{4}}\left[\frac{g\beta(T_s - T_\infty)x^3}{\nu^2}\right]^{-\frac{1}{4}} x$$

or

$$\frac{\delta}{x} = 3.93 \mathrm{Pr}^{-\frac{1}{2}}(0.952 + \mathrm{Pr})^{\frac{1}{4}} \mathrm{Gr}_x^{-\frac{1}{4}} \tag{16-40}$$

The heat transfer at the wall is calculated from

$$\frac{q}{A} = -k\left.\frac{dT}{dy}\right|_{y=0} = h(T_s - T_\infty)$$

With the temperature profile of Eq. (16-31) this becomes

$$h = \frac{2k}{\delta}$$

or, in dimensionless form,

$$\frac{hx}{k} = \mathrm{Nu}_x = \frac{2x}{\delta} \tag{16-41}$$

Combining Eqs. (16-40) and (16-41) yields

$$\mathrm{Nu}_x = 0.508 \mathrm{Pr}^{\frac{1}{2}}(0.952 + \mathrm{Pr})^{-\frac{1}{4}} \mathrm{Gr}_x^{\frac{1}{4}} \tag{16-42}$$

It is interesting to compare the results of the integral analysis with those of Ostrach. For air with a Prandtl number of 0.72, Ostrach's results indicate

$$\mathrm{Nu}_x \approx 0.356 \mathrm{Gr}_x^{\frac{1}{4}}$$

whereas the integral analysis yields

$$\mathrm{Nu}_x \approx 0.379 \mathrm{Gr}_x$$

Eckert and Jackson [3] have presented an integral analysis for the portion of a vertical plate subjected to a turbulent free-convective flow. Basing assumed velocity- and temperature-profile forms on the one-seventh-power law, they obtained

$$\mathrm{Nu}_x = 0.0295 \mathrm{Gr}_x^{\frac{2}{5}} \mathrm{Pr}^{\frac{7}{15}} (1 + 0.494 \mathrm{Pr}^{\frac{2}{3}})^{-\frac{2}{5}} \tag{16-43}$$

Levy [12] has formulated a general integral analysis applicable to several geometrical configurations including the vertical flat plate and the horizontal cylinder.

Fluid property variations In all the analyses, the fluid properties other than density are assumed constant. For large differences between the free-stream and surface temperatures, this can result in significant error. Sparrow and Gregg [25] have shown that for laminar heat transfer from vertical isothermal plates to gases the properties other than β should be evaluated at a reference temperature

$$T_{\mathrm{ref}} = T_s + 0.38(T_\infty - T_s) \tag{16-44}$$

FREE CONVECTIVE HEAT TRANSFER

β should be taken as $1/T_\infty$, where T_∞ is in degrees Rankine. This is valid over the temperature range $0.5 < T_s/T_\infty < 3.0$, where again the temperatures are in degrees Rankine.

16-4 EMPIRICAL CORRELATIONS FOR NATURAL CONVECTION

The preceding analyses reveal that the Nusselt number is a function of the Grashof and Prandtl numbers. It has been found that the experimental data in both the laminar and turbulent regimes for many geometrical configurations investigated can be correlated by an equation of the form

$$\overline{\mathbf{Nu}} = \frac{\bar{h}L}{k_f} = C(\mathbf{Gr}_f \mathbf{Pr}_f)^a \tag{16-45}$$

The exponent a is usually $\frac{1}{4}$ for laminar and $\frac{1}{3}$ for turbulent flow. The subscript f denotes that properties are to be evaluated at the film temperature of Eq. (12-172),

$$T_f = \frac{T_s + T_\infty}{2}$$

Free convection from vertical plates and cylinders of large diameter The Grashof number is formed with the height of the surface as the significant distance, and the appropriate constants for use with Eq. (16-45) as recommended by McAdams [13] are given in Table 16-1. For $\mathbf{Gr}_f \mathbf{Pr}_f$ ranging from 10^{-1} to 10^4, the data of Fig. 16-5 are to be used.

Free convection from horizontal square plates The Grashof number is formed with the length of one side. The constants in Table 16-2 are recommended. There is some question about extending these correlations to rectangular plates. For small length-to-width ratios, the use of the longer side in both **Nu** and **Gr** is recommended. Notice that this results in a lower average heat-transfer coefficient than the use of the shorter side for laminar flow and has no effect on $\bar{h}$ for turbulent flow.

Free convection from long horizontal cylinders For horizontal cylinders which are sufficiently long to render end effects negligible, the values of C and a for use in Eq. (16-45) recommended by McAdams [13] are given in Table 16-3. The Grashof

Table 16-1 Vertical plate constants

Physical condition	$\mathbf{Gr}_f \mathbf{Pr}_f$	C	a
Laminar	10^4–10^9	0.59	$\frac{1}{4}$
Turbulent	10^9–10^{12}	0.13	$\frac{1}{3}$

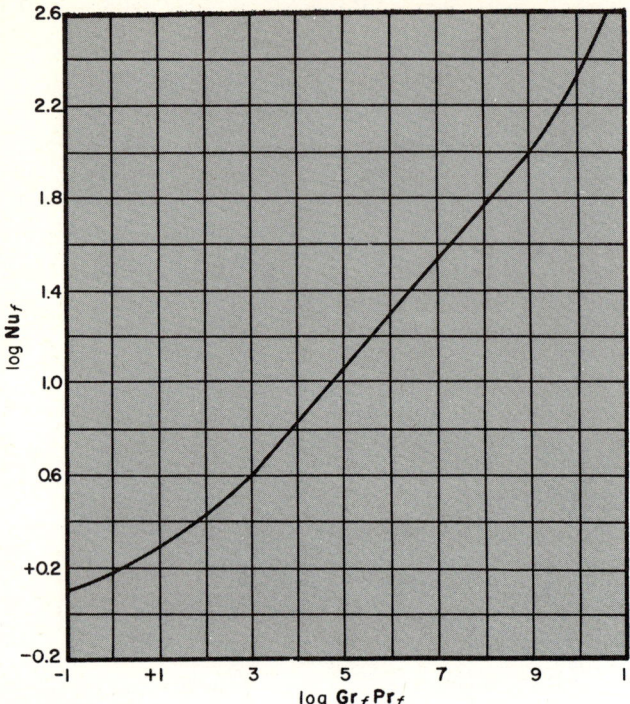

Fig. 16-5 Free-convective heat transfer from heated vertical plates. (*Adapted from W. H. McAdams, "Heat Transmission," 3d ed., p. 173. Copyright 1954. McGraw-Hill Book Company. Used by permission.*)

Table 16-2 Horizontal plate constants

Physical condition	$Gr_f Pr_f$	C	a
Laminar-heated surface up or cooled surface down [5]	$10^5 - 2 \times 10^7$	0.54	$\frac{1}{4}$
Turbulent-heated surface up or cooled surface down [5]	$2 \times 10^7 - 3 \times 10^{10}$	0.14	$\frac{1}{3}$
Laminar-heated surface down or cooled surface up [13]	$3 \times 10^5 - 3 \times 10^{10}$	0.27	$\frac{1}{4}$

Table 16-3 Horizontal cylinder constants

Physical condition	$Gr_f Pr_f$	C	a
Laminar	$10^4 - 10^9$	0.53	$\frac{1}{4}$
Turbulent	$10^9 - 10^{12}$	0.13	$\frac{1}{3}$

FREE CONVECTIVE HEAT TRANSFER

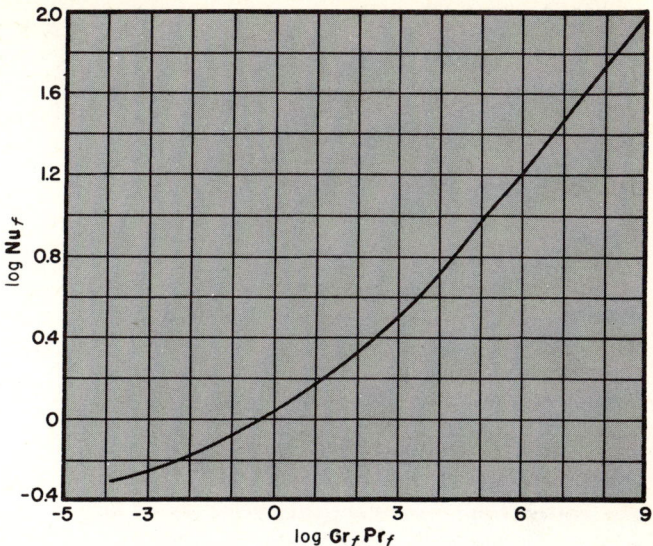

Fig. 16-6 Free-convective heat transfer from heated horizontal cylinders. (*Adapted from W. H. McAdams, "Heat Transmission," 3d ed., p. 176. Copyright 1954. McGraw-Hill Book Company. Used by permission.*)

and Nusselt moduli are formed with the cylinder diameter as the significant length. For $10^{-5} < \mathbf{Gr}_f \mathbf{Pr}_f < 10^4$, Fig. 16-6 should be used. If the cylinder is short, it is also advisable to evaluate the heat transfer by the method described below for miscellaneous shapes.

Example The front panel of a home dishwasher is 3 ft high, and it is at 105°F during the drying cycle. The room temperature is 70°F. Estimate the heat gain by the room from this panel if it is 2.5 ft wide.

Solution We first must examine the product of the Grashof and Prandtl numbers with all properties evaluated at the film temperature

$$T_f = \frac{T_s + T_\infty}{2} = \frac{105 + 70}{2} = 87.5°F$$

From Table A-3 by interpolation

$c_p = 0.24$ Btu/lb$_m$-°F $k = 0.151$ Btu/hr-ft-°F
$\nu = 0.174 \times 10^{-3}$ ft²/sec $\beta = 1.83 \times 10^{-3}$ °F^{-1}
$\mathbf{Pr} = 0.72$

Then the Grashof number is

$$\mathbf{Gr}_f = \frac{(32.2 \text{ ft/sec}^2)(1.83 \times 10^{-3} \text{ °F}^{-1})(105 - 70)\text{°F}(3 \text{ ft})^3}{(0.174 \times 10^{-3} \text{ ft}^2/\text{sec})^2} = 1.85 \times 10^9$$

The product $Gr_f Pr_f$ is

$(1.85 \times 10^9)(0.72) = 1.32 \times 10^9$

and the flow is turbulent. The correlation equation for this vertical plate is

$\overline{Nu} = 0.13 Gr_f Pr_f^{\frac{1}{3}}$

and

$\dfrac{\bar{h}L}{k_f} = 0.13(1.32 \times 10^9)^{\frac{1}{3}} = 142.7$

$\bar{h} = \dfrac{142.7(0.151 \text{ Btu/hr-ft-}°\text{F})}{3 \text{ ft}} \approx 7.2 \text{ Btu/hr-ft}^2\text{-}°\text{F}$

The heat-transfer rate to the room is

$q = \bar{h} A \, \Delta T = (7.2 \text{ Btu/hr-ft}^2\text{-}°\text{F})(3 \times 2.5) \text{ ft}^2 (105 - 70)°\text{F}$
$= 1890 \text{ Btu/hr}$

Free convection from vertical cylinders of small diameter For small-diameter vertical cylinders such as wires, the boundary-layer thickness may become large when compared with the radius of curvature. Then the flat-plate approximation becomes questionable. According to an analysis by Elenbaas [4], the Nusselt modulus is a function of the product of the Rayleigh number based on the diameter of the wire and the dimensionless diameter-height ratio D/L. This is supported by the experimental measurements of Kyte et al. [11], whose results are presented in Fig. 16-7.

Example A 0.01-in-diam vertical wire 6 in long is heated electrically to 130°F and exposed to ambient air at 70°F and 14.7 psia. Determine the heat-transfer rate.

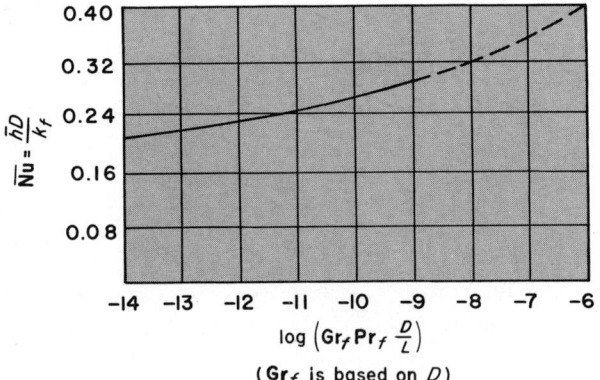

Fig. 16-7 Free-convective heat transfer from small vertical cylinders. [*Adapted from J. R. Kyte, A. J. Madden, and E. L. Piret, Chem. Eng. Progr.,* **49**: *657 (1953), by permission.*]

FREE CONVECTIVE HEAT TRANSFER

Solution The fluid properties evaluated at the film temperature

$$T_f = \frac{T_s + T_\infty}{2} = \frac{130 + 70}{2} = 100°F$$

are

$c_p = 0.240$ Btu/lb$_m$-°F $\quad$ Pr $= 0.72$

$k = 0.0154$ Btu/hr-ft-°F $\quad \dfrac{g\beta\rho^2}{\mu^2} = 1.76 \times 10^6 (°F\text{-}ft^3)^{-1}$

The Grashof number formed with the wire diameter is

$$\text{Gr}_f = \frac{g\beta}{\nu^2}(T_s - T_\infty)D^3 = \frac{1.76 \times 10^6}{°F\text{-}ft^3}(130 - 70)°F\left(\frac{0.01\ ft}{12}\right)^3 = 0.0612$$

Then

$$\text{Gr}_f \text{Pr}_f \frac{D}{L} = 0.0612(0.72)\frac{0.01}{6.0} = 0.0000735$$

From Fig. 16-7 at log $(7.35 \times 10^{-5}) = -5.866$

$$\overline{\text{Nu}} = \frac{hD}{k_f} \simeq 0.4$$

$$\frac{0.4(0.0154\ \text{Btu/hr-ft-}°F)}{(0.01/12)\ ft} = 7.40\ \text{Btu/hr-ft}^2\text{-}°F$$

Thus,

$$q = hA\,\Delta T = \frac{7.4\ \text{Btu}}{\text{hr-ft}^2\text{-}°F}\left[\frac{0.01(\pi)(6)}{144}\ ft^2\right](130 - 70)°F = 0.582\ \text{Btu/hr}$$

Free convection from inclined plates Experiments with inclined plates indicate that the heat-transfer coefficient is less than for a vertical plate. This is due to the reduction in the buoyant body force in the direction parallel to the plate, which in turn reduces the fluid velocity in that direction. Rich [19] indicates that the vertical-plate correlations may be used if the local acceleration due to gravity is replaced in the Grashof number by $g\cos\theta$, where θ is the angle of inclination from the vertical. This correlates experimental results to ± 10 percent for small angles of inclination.

Free convection from miscellaneous shapes The preceding correlation equations are for well-defined configurations. Physical shapes are frequently encountered, however, which do not conform to any of these cases. A particular example would be a short horizontal cylinder. King's [10] experiments with cylinders, blocks, and spheres indicate that Eq. (16-45) is applicable with

$$C = 0.60 \qquad a = \tfrac{1}{4}$$

for $10^4 < \text{Gr}_f\text{Pr}_f < 10^9$. The length for forming the Grashof number is

$$L = \frac{1}{1/L_h + 1/L_v} \tag{16-46}$$

where L_h and L_v are the predominant horizontal and vertical dimensions of the object.

16-5 FREE CONVECTION IN ENCLOSED LAYERS

Natural-convective heat transport occurs in a body of fluid enclosed by surfaces when the surfaces and the fluid are not at a uniform temperature. Two common cases are *horizontal* and *vertical* fluid layers. In either, we define an average heat-transfer coefficient $\bar{h}$ by

$$\frac{q}{A} = \bar{h}(T_1 - T_2) \tag{16-47}$$

where the temperatures are those of the opposing surfaces, as shown in Fig. 16-8.

Horizontal air layers There are two general cases: (1) upper plate warmer and (2) lower plate warmer. For the first, there is no fluid motion, and the heat transfer is purely by conduction. Then the heat-transfer coefficient of Eq. (16-47) can be evaluated by

$$\frac{q}{A} = \bar{h}(T_1 - T_2) = k\frac{T_1 - T_2}{b}$$

and

$$\frac{\bar{h}b}{k} = \overline{\text{Nu}} = 1.0 \tag{16-48}$$

For a warmer lower surface, Jakob [9] recommends

$$\overline{\text{Nu}} = 0.195\text{Gr}_b^{\frac{1}{4}} \qquad 10^4 < \text{Gr}_b < 4 \times 10^5 \tag{16-49}$$

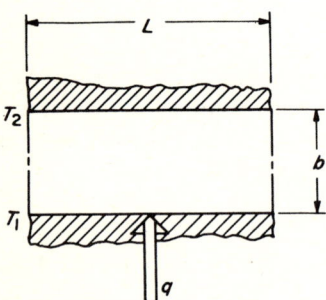

Fig. 16-8 General configuration for an enclosed free-convection system.

Fig. 16-9 Cellular motion in an enclosed fluid layer heated from the bottom.

and

$$\overline{\mathbf{Nu}} = 0.068 \mathbf{Gr}_b^{\frac{1}{4}} \qquad 4 \times 10^5 < \mathbf{Gr}_b \tag{16-50}$$

Notice that the subscript b refers to the significant dimension for forming the Grashof number; this is the distance between the two plates. The fluid properties are evaluated at the average of the two plate temperatures. For the lower range of $\mathbf{Gr}_b$, the fluid motion is a well-ordered, cellular process, as shown in Fig. 16-9. For the higher range, the flow is turbulent and disordered.

For Grashof numbers less than approximately 2,000, the convective velocities are very low. The heat transfer is then primarily due to conduction, and Eq. (16-48) applies. Between this range and the lower limit for Eq. (16-49), engineering estimates are based on the more conservative value obtained with Eqs. (16-48) and (16-49).

Horizontal liquid layers Globe and Dropkin [6] measured heat-transfer rates in enclosed horizontal layers of water, oils, and liquid metals. Their experiments indicate that

$$\overline{\mathbf{Nu}} = 0.069 \mathbf{Gr}_b^{\frac{1}{3}} \mathbf{Pr}^{0.407} \tag{16-51}$$

is applicable for $3 \times 10^5 < \mathbf{Gr}_b \mathbf{Pr} < 7 \times 10^9$, where all fluid properties are evaluated at the arithmetic average of the two plate temperatures.

Vertical air layers Jakob [8] has considered the results of several investigators and recommends

$$\overline{\mathbf{Nu}} = 0.18 \mathbf{Gr}_b^{\frac{1}{4}} \left(\frac{L}{b}\right)^{-\frac{1}{9}} \qquad 2 \times 10^4 < \mathbf{Gr}_b < 2 \times 10^5 \tag{16-52}$$

and

$$\overline{\mathbf{Nu}} = 0.065 \mathbf{Gr}_b^{\frac{1}{3}} \left(\frac{L}{b}\right)^{-\frac{1}{9}} \qquad 2 \times 10^5 < \mathbf{Gr}_b < 11 \times 10^6 \tag{16-53}$$

These expressions are applicable for $L/b > 3$, where fluid properties are evaluated at $(T_1 + T_2)/2$. For smaller values of this geometrical parameter, the empirical expressions for heat transfer to or from a single vertical plate may be applied to each surface.

Again, for a Grashof number less than approximately 2,000, the convective effects are small. The problem may be treated as pure conduction, and Eq. (16-48) is applicable.

16-6 COMBINED FORCED AND FREE CONVECTION

In a forced, nonisothermal flow, density changes invariably occur due to heating or cooling of the fluid. These changes in turn result in added body forces in the momentum equations, and the forced-convection momentum and energy equations are coupled in the sense that they must be solved simultaneously. The solutions are quite difficult due to the number of parameters involved.

To analyze mathematically any complicated process involving convective transport, we have found it expedient to eliminate terms of lesser importance; the reader will recall that the solution of the laminar Navier-Stokes equations for flow over a flat plate involved an order-of-magnitude analysis to simplify the equations. For the present problem, it will be helpful if we can determine regimes for which either buoyant terms or forced-convective terms are negligible. If, for example, buoyant effects are negligible in a problem, the solution for the similar forced-flow problem is applicable.

Qualitatively we would expect very large forced velocities to overshadow free-convective effects. Several configurations have been extensively investigated, and summary guidelines for them are available.

Vertical isothermal plate Sparrow and Gregg [26] investigated the effects of buoyancy on laminar forced boundary-layer flow along a vertical isothermal plate. They found the buoyancy effects on the *local* heat-transfer coefficient to be less than 10 percent when

$$\frac{\text{Gr}_x}{\text{Re}_x^2} \leq 0.15 \tag{16-54}$$

Note that the significant parameters involve the Grashof number and the square of the Reynolds number. This is in agreement with the discussion at the beginning of this chapter.

The effect of free convection on the *average* heat-transfer coefficient is less than 5 percent for

$$\frac{\text{Gr}_L}{\text{Re}_L^2} \leq 0.225 \tag{16-55}$$

These estimations are applicable for the free-convective flow either in the same or opposite direction to the forced flow and for Prandtl numbers ranging from 0.01 to 10.0.

Horizontal isothermal plate Mori [17] has shown the effects of natural convection on the *local* heat-transfer coefficient for forced flow over the upper or lower surface of a horizontal isothermal plate to be less than 10 percent when

$$\frac{\text{Gr}_x}{\text{Re}_x^2} \leq 0.083 \tag{16-56}$$

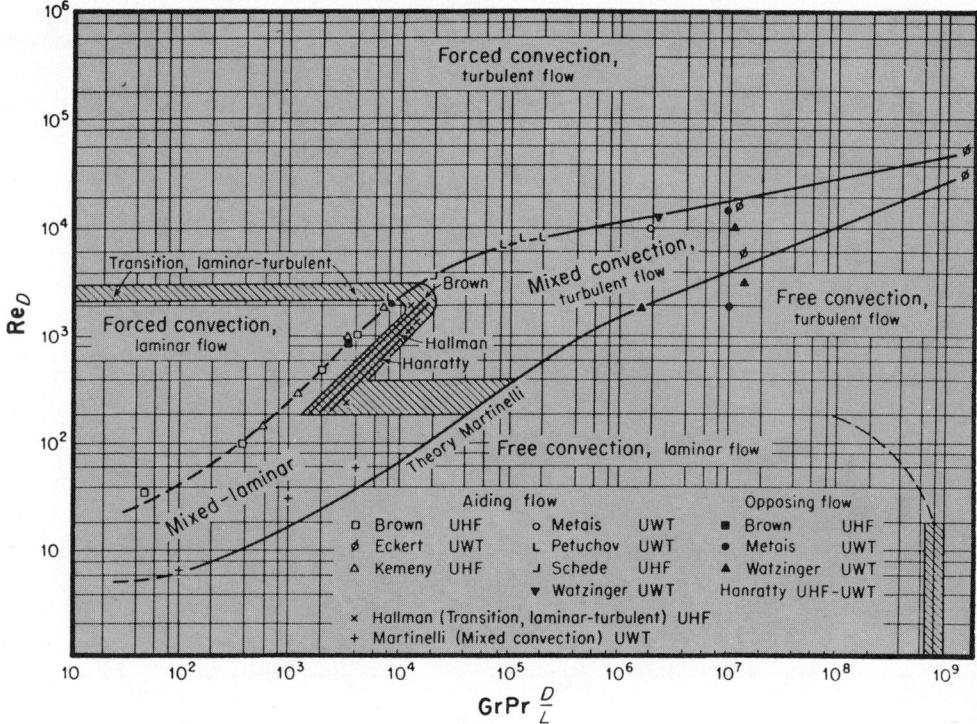

Fig. 16-10 Regimes of forced, free, and mixed convection for flow in vertical tubes. [*Adapted from B. Metais and E. R. G. Eckert, Trans. ASME, Ser. C, J. Heat Transfer,* **86:** *295 (1964).*]

Inside vertical tubes Mixed convective effects for this configuration have been reported by several investigators. Metais and Eckert [14] have summarized these results and have made tentative recommendations, as shown in Fig. 16-10. The abbreviations UHF and UWT stand for uniform heat flux and uniform wall temperature, respectively. Here the Grashof number is formed with (1) the tube diameter as the significant length dimension and (2) the difference between the tube wall and fluid bulk temperatures. These recommendations are applicable for

$$10^{-2} < \text{Pr} \frac{D}{L} < 1.0$$

The divisions between the regimes are so established that the actual heat flux in a forced- or free-convective regime does not deviate more than 10 percent from that caused by external forces alone or body forces alone, respectively. Notice that there are six regimes; forced laminar, forced turbulent, free laminar, free turbulent, mixed laminar, and mixed turbulent. Heat transfer in either of the mixed or free-convective regimes is the subject of current research.

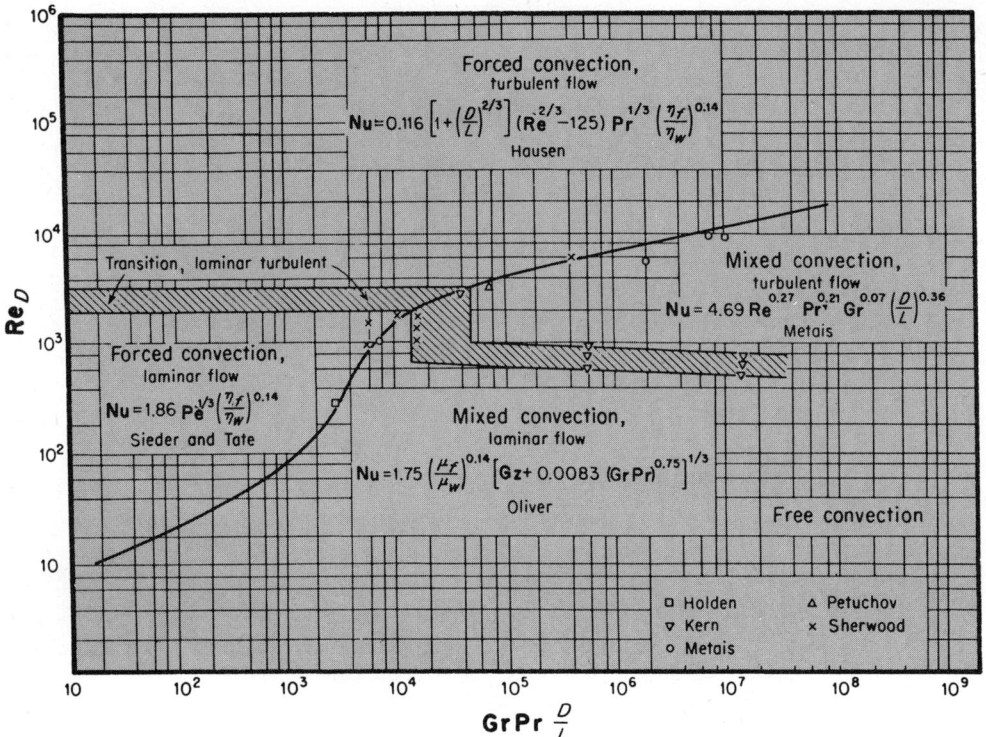

Fig. 16-11 Regimes of forced, free, and mixed convection for flow in horizontal tubes. [*Adapted from B. Metais and E. R. G. Eckert, Trans. ASME, Ser. C, J. Heat Transfer*, **86**: *295 (1964)*.]

Inside horizontal tubes Experimental results of several investigators with this configuration have also been summarized by Metais and Eckert. The results available to them did not permit establishment of limiting conditions between free and mixed convection; a limit between mixed and forced convection, however, was possible. Their tentative recommendations are presented in Fig. 16-11, which includes recommended correlation equations for the forced laminar, forced turbulent, mixed laminar, and mixed turbulent regimes. In these equations **Gz** is the Graetz number, **Pe** is the Peclet number,

$$\mathbf{Gz} = \mathbf{RePr}\frac{D}{L} \qquad \mathbf{Pe} = \mathbf{RePr}$$

and η_f/η_w is the product of μ_f/μ_w and D/L.

These recommendations are again applicable for

$$10^{-2} < \mathbf{Pr}\frac{D}{L} < 1.0$$

FREE CONVECTIVE HEAT TRANSFER

and the limiting condition between mixed- and forced-convective regimes was established to result in heat-flux errors of 10 percent or less due to neglecting body forces in the forced-flow regime.

PROBLEMS

16–1. For a plane vertical wall 1 ft high held at 140°F and exposed to air at 60°F, determine whether the problem is free, forced, or mixed convection if the forced-approach or free-stream velocity is:
    (a) 10 ft/sec   (b) 2 ft/sec   (c) 100 ft/sec   (d) 0.1 ft/sec   (e) 0.01 ft/sec   (f) 1 ft/sec

16–2. For a vertical wall at 140°F exposed to still air at 60°F, determine the maximum height for laminar free-convective flow.

16–3. Verify Eqs. (16-3) and (16-5), and obtain the resulting momentum equation (16-6).

16–4. Find the velocity component v in terms of the similarity parameter and the stream function as was done for u in Sec. 16-2.

16–5. For the situation of Prob. 16-2, determine the heat transfer per foot of wall width by (a) the results of Ostrach's solution of the differential equations and (b) the results of the integral analysis.

16–6. A 1- by 1-ft vertical square plate is maintained at a temperature of 275°F and is exposed to still air at 75°F. Using Ostrach's results, determine the heat loss from the plate.

16–7. A vertical cylinder which is 1 ft in diameter and 2 ft high is maintained at a temperature of 170°F and is exposed to air at 75°F. Use the results of Ostrach to determine the heat transfer from the cylinder.

16–8. A vertical 3-in-diam cylinder 1 ft high is maintained at a temperature of 118°F and is exposed to air at 70°F and atmospheric pressure. Using the results of Sparrow and Gregg, determine the rate of heat transfer from the cylinder.

16–9. A vertical flat rectangular plate which is 5 ft wide and 2 ft high is maintained at a temperature of 193°F and is exposed to nitrogen at 80°F and atmospheric pressure. Determine the heat loss from the plate, using the results of Sparrow and Gregg.

16–10. Repeat Prob. 16-9 using the results of Ostrach's analysis to determine the heat loss from the plate.

16–11. For the vertical wall of Prob. 16-2, determine the heat transfer per foot of wall width by (a) the empirical correlation equation for laminar flow and (b) the empirical correlation equation for turbulent flow.

16–12. Using the integral-analysis assumptions, determine an expression for the velocity in the laminar boundary layer as a function of distance along a plate x and distance from the plate y. Determine the y location of the maximum velocity at a given x position, and compare it with the location given by Ostrach for $\mathbf{Pr} = 1.0$.

16–13. Using the results of the integral analysis, plot the free-convective boundary-layer thickness as a function of length along the plate for a vertical plate maintained at 200°F and exposed to air at 100°F. Consider the range $10^4 < \mathbf{GrPr} < 10^9$.

16–14. A vertical flat plate maintained at 225°F is exposed to still air at 90°F. Determine the boundary-layer thickness at a distance of 15 in from the lower edge of the plate.

16–15. Using the expression for the Nusselt number obtained from the integral method of solution [Eq. (16-42)], show that the average Nusselt number is $\frac{4}{3}$ times the Nusselt number evaluated at the end of the plate; that is, $\overline{\mathbf{Nu}} = \frac{4}{3}\mathbf{Nu}_L$.

16–16. A vertical flat plate 2 ft high is maintained at 120°F and is exposed to still air at 80°F. Determine the heat transfer per foot of width using the results of the integral analysis.

16–17. A vertical plate 2 ft wide and 1 ft high is exposed to oxygen at 90°F. Calculate the heat loss from the plate (one side) if it is maintained at 110°F.

16-18. Estimate the initial rate of heat transfer to a cylinder 3 ft in diameter and 6 ft long if it is submerged vertically in light oil at a temperature of 80°F. The initial temperature of the cylinder is 40°F.

16-19. On a summer day the air temperature reaches 105°F, and the air is relatively still. A milkshake in a 9-in-diam cup 8 in tall is exposed to this air but shaded from the sun. Estimate the heat gained by the milkshake if its surface temperature is 35°F. Compare the results using the miscellaneous-shapes approach with those using the cylinder approach.

16-20. A vertical thin 4-in-diam cylinder 18 in tall is maintained at a temperature of 125°F in an atmospheric environment of 75°F. Determine the total heat loss from the cylinder.

16-21. A 20-in-high vertical plate is at a constant surface temperature of 250°F and exposed to still air at 50°F. Determine the heat transfer per foot of width by the (*a*) exact solution of the differential equations (Ostrach's results), (*b*) result of the integral analysis, and (*c*) suggested empirical correlation equation.

16-22. Calculate the heat loss from a horizontal 2- by 2-ft plate with an upper-surface temperature of 330°F to air at 70°F.

16-23. A horizontal square plate with its lower surface heated to 125°F is losing heat to still air at a temperature of 75°F; the length of one side of the plate is 1.5 ft. Determine the rate of heat transfer. (The upper surface is insulated.)

16-24. A horizontal 3- by 3-ft square plate with its upper surface heated to 120°F is exposed to carbon dioxide at 80°F. Estimate the heat-transfer rate.

16-25. A horizontal square plate with its upper surface maintained at 95°F is exposed to water at 65°F. The length of one side is 1 ft. Calculate the heat-transfer rate from the upper surface.

16-26. A 6-in-diam horizontal cylinder (very long) is maintained at 120°F. It is in a large tank of glycerin, which is at 50°F.
    (*a*) Determine the heat transfer to the glycerin per foot of cylinder length.
    (*b*) Repeat part (*a*), assuming vertical-flat-plate theory, with the height equal to the half circumference beginning at the bottom of the cylinder and extending to the top.

16-27. A 4-in-diam horizontal cylinder is 12 ft long. The wall is at 300°F, and the surrounding fluid is still air at 100°F (far from the cylinder). Estimate the average Nusselt modulus for free-convective heat transfer.

16-28. A 0.1-in-diam horizontal wire 10 in long is heated electrically so that the surface temperature is maintained at 115°F. The wire is in a vessel of glycerin at 45°F. Determine the heat lost by the wire.

16-29. A vertical heated 1-in-diam rod 1 ft long is at a uniform temperature of 350°F and exposed to air at 50°F and atmospheric pressure. Determine the heat transfer from the rod. *Note:* Examine the maximum boundary-layer thickness by flat-plate theory as a guide to selecting the appropriate type of thermal correlation.

16-30. A 0.01-in-diam vertical wire at 400°F is exposed to air at 70°F. Determine the rate of heat transfer for a 12-in length of the wire.

16-31. Estimate the rate of heat transfer from a 0.003-in-diam wire 1 ft long placed vertically in light oil at 65°F. The surface temperature of the wire is maintained at 275°F.

16-32. The surface temperature of a 6-in-OD 10-ft-long pipe carrying steam in a large room is 200°F. The air in the room is at 14.7 psia and 100°F. Determine the free-convective heat loss from the pipe when it is (*a*) vertical and (*b*) horizontal.

16-33. A 1-ft flat plate is inclined from the vertical at an angle of 20°. The surface temperature of the plate is 140°F, and the temperature of the surrounding air is 60°F.
    (*a*) Determine the rate of heat transfer from the plate per foot of width.
    (*b*) Repeat part (*a*) with water instead of air.

16-34. A 9-in flat plate 1 ft wide is inclined from the vertical at an angle of 15°. The plate is maintained at a temperature of 150°F and is exposed to carbon dioxide at 50°F and 14.7 psia. Determine the rate of heat transfer from the plate.

16-35. A smooth rock which approximates a 3-in-diam sphere is dropped into water at 60°F. Estimate the initial rate of heat transfer if the initial temperature of the rock is 100°F.

16-36. Two horizontal surfaces separated by a 5-in layer of air have upper and lower temperatures of 80 and 120°F, respectively. Determine the rate of heat loss per square foot.

16-37. Two horizontal surfaces with air between them are separated by a distance of 2 in. Calculate the heat flux (Btu/hr-ft^2) if (*a*) the upper surface is at a temperature of 110°F and the lower is at 90°F and (*b*) the upper surface is at a temperature of 90°F and the lower is at 110°F.

16-38. Two horizontal surfaces which are separated by a 1-in layer of water have upper and lower temperatures of 75 and 105°F, respectively. Determine the heat flux in Btu/hr-ft^2.

16-39. The walls of a building have an air space which is 3 in thick and 4 ft high. If the temperature of one surface is 105°F and that of the other is 95°F, determine the convective heat loss per unit area in Btu/hr-ft^2.

16-40. Saturated liquid ammonia at 30°F is forced through a vertical tube with a uniform wall temperature of 70°F. The 2.8-in-diam tube is 1 ft long. Estimate the average convective heat-transfer coefficient if the velocity (*a*) is 0.01 ft/sec and (*b*) is 5 ft/sec.

16-41. Air at 14.7 psia and 100°F is forced through a horizontal 1-in-diam tube at an average velocity of 1 ft/sec. The tube is 1 ft long, and the tube wall is maintained at 500°F. Calculate the average heat-transfer coefficient.

16-42. Steam at 14.7 psia and 800°F is forced at a velocity of 5 ft/sec through a 6-in-diam horizontal tube. The length of the tube is 2 ft, and the temperature of the tube wall is 200°F. Determine the heat-transfer coefficient.

REFERENCES

1. Chiang, T., and J. Kaye: On Laminar Free Convection from a Horizontal Cylinder, *Proc. 4th U.S. Natl. Congr. Appl. Mech.*, 1962, vol. 2, p. 1213.
2. Eckert, E. R. G., and R. M. Drake: "Heat and Mass Transfer," 2d ed., pp. 315–317, McGraw-Hill, New York, 1959.
3. Eckert, E. R. G., and T. W. Jackson: Analysis of Turbulent Free Convection Boundary Layer on a Flat Plate, *NACA Rept.* 1015, 1951.
4. Elenbaas, W.: Dissipation of Heat by Free Convection from Vertical and Horizontal Cylinders, *J. Appl. Phys.*, **19:** 1148 (1948).
5. Fishenden, M., and O. A. Saunders: "Heat Transfer," Oxford University Press, New York, 1950.
6. Globe, S., and D. Dropkin, *Trans. ASME, ser. C*, **81:** 24 (1959).
7. Hermann, R.: Heat Transfer by Free Convection from Horizontal Cylinders in Diatomic Gases, *NACA Tech. Mem.* 1366, 1954.
8. Jakob, M.: Free Convection through Enclosed Plane Gas Layers, *Trans. ASME*, **68:** 189 (1946).
9. Jakob, M.: "Heat Transfer," vol. 1, p. 535, Wiley, New York, 1949.
10. King, W. J.: "Mechanical Engineering," **54:** 347 (1932).
11. Kyte, J. R., A. J. Madden, and E. L. Piret: *Chem. Eng. Progr.*, **49:** 653 (1953).
12. Levy, S.: Integral Methods in Natural Convection Flow, *J. Appl. Mech.*, **22:** 515 (1955).
13. McAdams, W. H.: "Heat Transmission," 3d ed., pp. 172–180, McGraw-Hill, New York, 1954.
14. Metais, B., and E. R. G. Eckert: Forced, Mixed, and Free Convective Regimes, *J. Heat Transfer*, **86:** 295 (1964).
15. Millsaps, K., and K. Pohlhausen: *J. Aeron. Sci.*, **23:** 381 (1956).

16. Millsaps, K., and K. Pohlhausen: *J. Aeron. Sci.*, **25**: 357 (1958).
17. Mori, Y.: *ASME Paper* 60-WA-220, 1960.
18. Ostrach, S.: An Analysis of Laminar Free Convection Flow and Heat Transfer about a Flat Plate Parallel to the Direction of the Generating Body Force, *NACA Rept.* 1111, 1953.
19. Rich, B. R.: *Trans. ASME*, **75**: 489 (1953).
20. Schmidt, E., and W. Beckmann: Das Temperatur und Geschwindigkeitsfeld vor einer wärmeabgebenden senkrechten Platte bei natürlicher Konvection, *Tech. Mech. Thermodyn.*, **1**: 341, 391 (1930).
21. Schuh, H.: Boundary Layers of Temperature, in W. Tollmien (ed.), "Boundary Layers," British Ministry of Supply, German Document Center, Ref. 3220T, 1948.
22. Sparrow, E. M., and J. L. Gregg: Laminar Free Convection from a Vertical Flat Plate, *Trans. ASME*, **78**: 435 (1956).
23. Sparrow, E. M., and J. L. Gregg: Similar Solutions for Free Convection from a Nonisothermal Vertical Plate, *Trans. ASME*, **80**: 379 (1958).
24. Sparrow, E. M., and J. L. Gregg: *Trans. ASME*, **78**: 1823 (1956).
25. Sparrow, E. M., and J. L. Gregg: The Variable Fluid Property Problem in Free Convection, *Trans. ASME*, **80**: 879 (1958).
26. Sparrow, E. M., and J. L. Gregg: *Trans. ASME*, **81**: 133 (1959).
27. Squire, H. B.: as reported by S. Goldstein in "Modern Developments in Fluid Mechanics," vol. 2, p. 641, Oxford University Press, London, 1938.

CHAPTER 17
MULTIPHASE PHENOMENA

This chapter introduces the class of problems in which more than one phase of a substance is present. Industrially important cases include cavitation, two-phase flow with and without heat transfer, pool boiling, condensation, and conduction with melting or freezing. Here, as in the treatment of single-phase flow, isothermal multiphase problems will be considered first, followed by cases involving heat transfer. Finally, nonconvective problems of heat transfer with phase change, i.e., conduction with melting or freezing, will be discussed.

17-1 BUBBLE DYNAMICS

A common characteristic of cavitation, two-phase flow, and pool boiling is the presence of bubbles of gas or vapor. An understanding of the growth and collapse of bubbles is fundamental to the study of these phenomena, for this growth and collapse—*bubble dynamics*—separates these topics from all others in transport phenomena.

There are two general classifications of bubbles, *gas bubbles* or *vapor bubbles*, depending upon whether the substance inside the bubble is the same as the surrounding liquid or a different fluid altogether. A gas bubble is a single quantity of gas submerged in a liquid of different chemical composition, e.g., an air bubble in water. It is usually at a temperature well above its thermodynamic critical point and is essentially insoluble in the surrounding liquid.

A vapor bubble, on the other hand, is a single quantity of a gaseous phase of a substance submerged in a liquid of the same chemical composition. The most common example is a steam bubble in liquid water. A vapor bubble is at a temperature below its thermodynamic critical point. Since this chapter is devoted to multiphase phenomena, we shall be concerned with the behavior of vapor bubbles rather than gas bubbles. Liquid-gas bubble flows may be treated by the methods of Chap. 20. We may note in passing that growth or collapse is usually far more significant with vapor than with gas bubbles. Ordinary water at room temperature, for example, contains only about 2 percent air by volume if saturated. Hence gas bubbles can grow only very slowly (and very little) compared with the growth of vapor bubbles that occurs when the pressure is reduced to the saturation value corresponding to the fluid temperature.

Static bubble In Sec. 4-4 it was shown by a force balance on a static vapor bubble that the relationship between the vapor pressure inside the bubble and the liquid pressure outside is

$$p_i - p_o = \frac{2\sigma}{R} \tag{4-37}$$

where p_i = vapor pressure inside bubble
p_o = liquid pressure outside bubble
R = radius of the bubble
σ = surface tension of liquid in contact with its vapor

The surface tension of liquid water in contact with its vapor is presented for the entire range of liquid-phase temperatures in Fig. 17-1, and surface-tension data for several fluids of engineering importance are given in Table 17-1.

The nucleation process for the origin of a vapor bubble is not completely understood. The nucleus of a vapor bubble is believed to originate on a solid or gas particle suspended in the liquid or the container surface. This is well supported by experiments with pure, degassed liquids which have resulted in very high superheats with no change in phase.

Momentum equation for a spherical bubble The momentum equation for the liquid surrounding a spherical bubble is the primary equation which describes the bubble growth or collapse. It was first obtained by Besant [5] and applied to calculate pressures in a liquid undergoing cavitation by Rayleigh [48]. The primary

MULTIPHASE PHENOMENA

Table 17-1 Surface tension of liquid in contact with its vapor†

Liquid	Temperature °C	Temperature °F	Surface tension Dynes/cm	Surface tension lb_f/ft
Acetone	0	32	26.2	0.00179
	20	68	23.7	0.00162
	60	140	18.6	0.00127
Ammonia	11.1	52	23.0	0.00157
	34.1	93.4	18.0	0.00123
	59	138.2	13.0	0.00089
Benzene	0	32	31.7	0.00217
	20	68	29.02	0.00199
	100	212	18.8	0.00129
Bromine	0	32	45.0	0.00308
	20	68	41.5	0.00284
	50	122	36.2	0.00248
n-Butyl alcohol	0	32	26.2	0.00179
	20	68	24.6	0.00168
	100	212	17.8	0.00122
Carbon dioxide	0	32	4.5	0.000308
	20	68	1.16	0.0000794
	30	86	0.06	
Carbon tetrachloride	10	50	28.2	0.00193
	20	68	27.0	0.00185
	100	212	17.3	0.00118
Ethyl alcohol	20	68	22.75	0.00154
	50	122	20.14	0.00138
	100	212	15.47	0.00106
Glycol	0	32	49.0	0.00335
	20	68	47.7	0.00326
	80	176	42.3	0.00289
Hydrogen	−258.4	−433.1	2.88	0.000197
	−255.1	−427.2	2.32	0.000159
	−252.7	−422.9	1.91	0.000131
Methyl alcohol‡	0	32	24.5	0.00168
	20	68	22.6	0.00155
	100	212	15.7	0.00107
Nitrogen	−203	−333.4	10.5	0.000719
	−193	−315.4	8.3	0.000568
	−183	−297.4	6.2	0.000424
Oxygen	−203	−333.4	18.3	0.00125
	−193	−315.4	15.7	0.00107
	−183	−297.4	13.2	0.000904
Toluene	20	68	28.5	0.00195
	50	122	25.0	0.00171
	130	266	16.3	0.00112

† Adapted from 10th ed., N. A. Lange, "*Handbook of Chemistry*," Copyright 1967. McGraw-Hill Book Company. Used by permission.
‡ In contact with air rather than vapor.

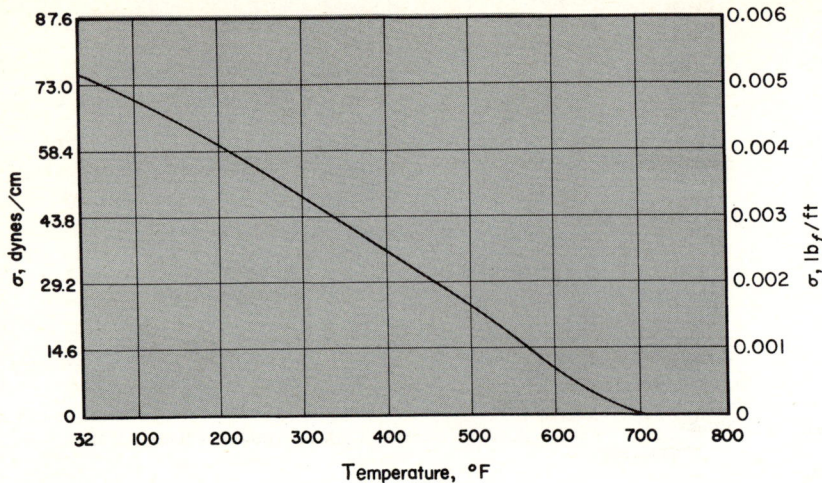

Fig. 17-1 Surface tension of saturated water.

assumptions in deriving this equation are these:

1. The bubble consists of vapor phase of the surrounding liquid.
2. The vapor pressure in the bubble is independent of radius.
3. The liquid phase is incompressible.
4. The motion is purely radial for both vapor and liquid.
5. Viscous forces are negligible.

For these conditions the continuity equation for the liquid in spherical coordinates reduces to

$$\frac{\partial v_r}{\partial r} + \frac{2v_r}{r} = 0 \tag{17-1}$$

and the Navier-Stokes equation becomes

$$\frac{\partial v_r}{\partial t} + v_r \frac{\partial v_r}{\partial r} = -\frac{1}{\rho_L}\frac{dp}{dr} \tag{17-2}$$

The boundary conditions are

$$p(\infty,t) = p_\infty \tag{17-3}$$

and

$$p(R,t) = p_v - \frac{2\sigma}{R} \tag{17-4}$$

where the latter condition is taken from Eq. (4-37) with the inner bubble pressure being the vapor pressure p_v and the pressure just outside the bubble being $p(R,t)$.

The radial velocity at any position in the liquid is obtained by integration of the continuity equation. This yields

$$r^2 v_r = F_1(t) \tag{17-5}$$

since the velocity must be a function of time for a growing or collapsing bubble. It is customary in the literature to denote the velocity of the bubble wall, v_r at $r = R$, by $\dot{R}$. Then at the wall Eq. (17-5) can be written

$$R^2 \dot{R} = F_1(t) \tag{17-6}$$

and combining this with Eq. (17-5) the velocity at any radial position in the liquid ($r > R$) can be expressed in terms of the wall velocity by

$$v_r = \left(\frac{R}{r}\right)^2 \dot{R} \tag{17-7}$$

Since R is a function of time for a growing or collapsing bubble, it is clear that the radial velocity of the liquid is a function of both r and t.

Substituting Eq. (17-7) into (17-2) and performing the indicated differentiation yields

$$\left[\frac{2R}{r^2}\dot{R}^2 + \left(\frac{R}{r}\right)^2 \ddot{R}\right] + \left[\left(\frac{R}{r}\right)^2 R^2 \dot{R}^2 - \frac{2}{r^3}\right] = -\frac{1}{\rho_L}\frac{dp}{dr} \tag{17-8}$$

Integrating with respect to r while holding t constant gives

$$-\frac{1}{r}(2R\dot{R}^2 + R^2\ddot{R}) + \frac{R^4\dot{R}^2}{2r^4} = -\frac{1}{\rho_L}p(r,t) + F_2(t) \tag{17-9}$$

The unknown function $F_2(t)$ can readily be eliminated by the boundary condition of Eq. (17-3); thus

$$F_2(t) = \frac{p_\infty}{\rho_L}$$

Applying the remaining boundary condition gives the final result

■ $$R\ddot{R} + \tfrac{3}{2}\dot{R}^2 = \frac{p_v - p_\infty}{\rho_L} - \frac{2\sigma}{\rho_L R} \tag{17-10}$$

This is the momentum equation for a growing or collapsing vapor bubble and is often referred to as the *Rayleigh equation*. When the bubble also contains a gas, the vapor pressure p_v is replaced by the sum of the gas and vapor pressure inside the bubble, $p_g + p_v$.

The Rayleigh equation is an ordinary, nonlinear, second-order differential equation for $R(t)$, whereas the beginning Navier-Stokes equation (17-2) was a nonlinear partial differential equation. Solutions of this equation have contributed significantly to our understanding of nucleate boiling and cavitation problems, and this is an area of recent and current research.

17-2 CAVITATION

The growth of vapor pockets in a liquid flow is known as *cavitation*, and it can occur with flow over external surfaces, through fixed passages, and through moving or rotating passages, e.g., in turbines, pumps, and propellors. Cavitation is of considerable importance: it is usually accompanied by a reduction in performance of fluid machinery and by rapid deterioration of surfaces exposed to a cavitating flow. Cavitation is easily detected by its associated noise. A flow passage experiencing cavitation sounds as though large solid particles were in the fluid and banging on the surfaces. Much is still unknown about cavitation, and this section is intended to serve only as an introductory treatment.

Nuclei The growth of a vapor bubble in a liquid that is not highly superheated requires a nucleus. In most cases, this is provided by gas (or air) bubbles or pockets. Air bubbles as large as 20 μm in diameter are present in ordinary water in very large numbers immediately after discharge from a tap. Such bubbles, however, disappear after approximately 1 hr even if the water is saturated with air. If the water is not saturated, the life of bubbles of this size is much shorter, frequently measurable in seconds. Even so, many flow situations provide sufficient air entrainment to supply nuclei for cavitation. An example is the intake flow of pumps and compressors.

It is also true that very clear water and even distilled clear water cavitates readily. Distilled water certainly contains no solid nuclei, and, if clear, any air bubbles must be small indeed. Since small bubbles are rapidly absorbed by the liquid, there must be another source of nuclei. Harvey et al. [24] proposed the existence of gas nuclei in surface cavities, as shown in Fig. 17-2. A very small air bubble lodged in a small crevice of a surface is on the convex side of the interface, and thus by Eq. (4-37) the air is at a lower pressure than the water. For a very small bubble, it is at a much lower pressure, and there is no pressure diffusion of gas into the liquid. Thus, the bubble in such a crevice can persist for a very long time. The number of crevices suitable for microscopic gas pockets of this type is very large on most commercial material surfaces. The existence of large numbers of nuclei, either from entrained gas or from surface crevice pockets, is assured for most flow

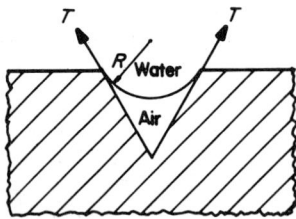

Fig. 17-2 A nucleus of air in a surface cavity.

MULTIPHASE PHENOMENA

situations, and cavitation can occur if the fluid static pressure is reduced to a sufficiently low value.

Critical pressure The critical pressure for *incipient cavitation* depends upon the radius of the bubble and the nucleating gas pressure. Since the volume of a spherical bubble is proportional to the cube of its radius and the mass of gas (not vapor) is constant in a growing bubble, the partial pressure of the gas is $p_g = C/r^3$. Then by Eq. (4-37) the equilibrium relationship between the pressure in the liquid and the vapor pressure in the bubble is

$$p_v + p_g - p_l = \frac{2\sigma}{r} \tag{17-11}$$

or

$$p_l - p_v = \frac{C}{r^3} - \frac{2\sigma}{r} \tag{17-11a}$$

From this the minimum value of $p_l - p_v$ can be found by differentiating with respect to r and equating to zero. This yields a critical radius

$$r_c = \left(\frac{3C}{2\sigma}\right)^{\frac{1}{2}} \tag{17-12}$$

or

$$r_c = \frac{-4\sigma}{3(p_l - p_v)} \tag{17-13}$$

According to Daily and Johnson [15], this negative sign requires the fluid pressure p_l to be less than the vapor pressure since the bubble critical radius and the surface tension are both positive. Accordingly the bubble will assume an equilibrium radius satisfying Eq. (17-11) for liquid pressures greater than the value required by Eq. (17-13). If the liquid pressure is less than this value, the bubble is unstable and will grow.

The form of Eq. (17-13) masks the fact that we do not know a priori—or rather we are unable to predict—the size of the original nuclei. Hence this result is not readily usable in elementary engineering calculations. It is quite useful in laboratory investigations, where the size of gas nuclei prior to cavitation is measurable. It should also be noted that this development was for a spherical bubble with constant gas mass and constant temperature and without consideration of dynamic effects. Deviations from these conditions render Eq. (17-13) useful as a first-order approximation to the critical pressure for incipient cavitation when the gas mass in a bubble is known.

Engineering approximations of the minimum pressure to prevent bubble growth (and hence cavitation) are frequently based on the simpler relationship

$$p_l \geq p_v(T) \tag{17-14}$$

While this is admittedly inaccurate, it is frequently a sufficient guide for predicting whether cavitation will occur.

Example Gasoline with a vapor pressure of 2 psia and a specific gravity of 0.68 is pumped through a 3-in-diam line having a venturi type of flowmeter with a $1\frac{1}{2}$-in-diam throat. If the pressure just upstream from the venturi is 10 psig and the atmospheric pressure is 14.7 psia, what flow rate in gallons per minute can the system sustain without the onset of cavitation?

Solution For an engineering approximation we shall assume incipient cavitation if the static fluid pressure drops to the saturation vapor pressure, 2 psia. Treating the flow as one-dimensional and applying Bernouilli's equation along the centerline (streamline) from the point just upstream from the venturi to the throat

$$p_1 + \rho \frac{V_1^2}{2g_c} = p_2 + \rho \frac{V_2^2}{2g_c}$$

For one-dimensional flow (constant velocity at a given cross section) the continuity equation is

$$A_1 V_1 = A_2 V_2$$

and hence

$$V_2 = \frac{A_1}{A_2} V_1 = \left(\frac{D_1}{D_2}\right)^2 V_1$$

$$V_2 = \left(\frac{3.0}{1.5}\right)^2 V_1 = 4 V_1$$

Thus, the Bernoulli equation becomes

$$16 V_1^2 - V_1^2 = (p_1 - p_2) \frac{2g_c}{\rho}$$

$$15 V_1^2 = (24.7 - 2)(144 \text{ lb}_f/\text{ft}^2) \frac{64.4 \text{ ft-lb}_m/\text{lb}_f\text{-sec}^2}{0.68 \times 62.4 \text{ lb}_m/\text{ft}^3}$$

$$V_1^2 = \frac{22.7(144)(64.4)}{15(0.68)(62.4)} \text{ ft}^2/\text{sec}^2$$

$$V_1 = 18.2 \text{ ft/sec}$$

The volumetric flow rate is

$$Q = AV = \frac{\pi}{4}\left(\frac{3}{12} \text{ ft}\right)^2 (18.2 \text{ ft/sec})(7.48 \text{ gal/ft}^3)(60 \text{ sec/min}) = 400 \text{ gal/min}$$

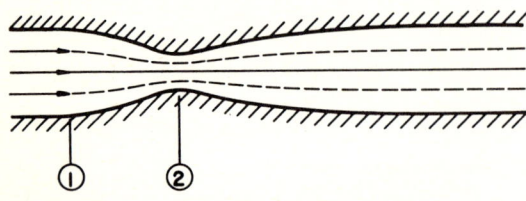

Fig. 17-3 Flow through a venturi flowmeter.

MULTIPHASE PHENOMENA

The reader is again warned that the method of treatment in the preceding problem affords only a first approximation to the correct answer, and such calculations should serve only as a guide to indicate the possibility of cavitation. This example illustrates the fact that the significant flow parameters influencing the onset of cavitation include the fluid dynamic head and the difference between the local fluid static pressure and the saturation vapor pressure. The dimensionless ratio

$$\mathbf{C}_v = \frac{p_l - p_v(T)}{\rho V^2 / 2 g_c} \tag{17-15}$$

is called the *cavitation number*. It is useful in scaling cavitation effects for a particular model configuration. The *critical cavitation number* is the value of $\mathbf{C}_v$ for incipient cavitation, but this is difficult to predict because of its dependence upon many factors including the Reynolds number, Weber number, surface-roughness ratio, and gas content.

Mechanism of damage There is considerable discussion in the literature concerning the cause of damage during cavitation. A widely held theory is that whenever a vapor bubble encounters a region of high fluid pressure, a rapid collapse occurs. This results in a local pressure pulse of large magnitude, and if the collapse occurs near a surface, damage could result from this pressure fluctuation. It is interesting to note that Rayleigh [48] in 1917 gave a numerical example showing that the internal pressure in the latter stages of collapse of a water-vapor bubble (radius equal to one-twentieth the initial value) is 1,260 times the surrounding fluid pressure far from the bubble wall.

An example of agreement between experimental bubble-collapse rate and that predicted by the Rayleigh equation is given by Knapp [27]. The collapse history of bubbles with an 0.14-in original radius is presented in Fig. 17-4.

There has been much debate concerning the direct mechanical damage of pressure pulses resulting from cavitation. In many cases of technical interest it can be shown that bubble collapse is too far from surfaces to result in extensive mechanical damage, but serious damage results nevertheless! At low flow velocity this is probably due to the agitation which increases the corrosive action over that under static conditions by an order of magnitude or more. At high flow velocity, the scouring mechanical action is dominant and results in extensive surface-material loss. Knapp [28] has shown that the threshold velocity for mechanical damage to become important in a flow of water is approximately 50 ft/sec. Below this value, the damage is due mainly to increased corrosion.

Cavitation in flow about immersed bodies is frequently encountered but will not be discussed in detail here. For a summary the reader should consult Csanady [14]. It is interesting to note that there is a regime of cavitation about a submerged nonenclosed body where the actual cavitation is in a downstream region, detached from the surface. This is known as *supercavitation*, and propellors frequently

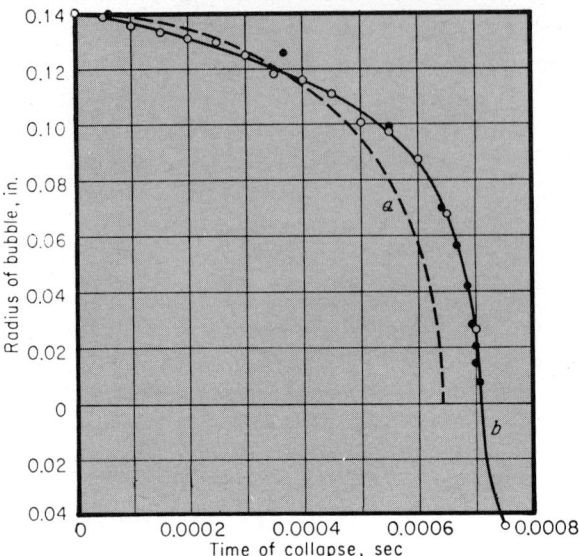

Fig. 17-4 Collapse history of a bubble with a 0.14-in. maximum radius compared with Rayleigh theory: (*a*) theory with equal maximum diameter and pressure; (*b*) theory with equal maximum diameter and time. [*Adapted from R. T. Knapp, Proc. Inst. Mech. Engrs.* (*London*), **166**: *150* (*1952*).]

operate in this regime with no resulting damage. Also, since the flow downstream from a propellor is unrestrained, there is very little choking effect.

17-3 FUNDAMENTALS OF TWO-PHASE FLOW

Flow of two phases of a single substance occurs in many situations, including flow in parts of vapor-compression refrigeration systems, flow in boiler tubes, and flow in heat exchangers with phase change. This is referred to as *two-phase single-component flow*. Equally important is the problem of *two-phase two-component flow*, such as the flow of oil and gas in a pipeline. The physical descriptions and definitions of the present section apply to both these cases, though the former is a liquid-vapor flow and the latter is a liquid-gas flow.[1]

The most important parameter characterizing a two-phase flow is the ratio of vapor- or gaseous-phase flow rate to liquid-phase flow rate. This ratio determines the type of flow pattern, which markedly influences such macroscopic design parameters as pressure loss and heat transfer.

---

[1] Theoretical analyses of either type can be undertaken with the equations developed in Chap. 20 for a fluid-solid flow so long as the fluids are considered to be incompressible.

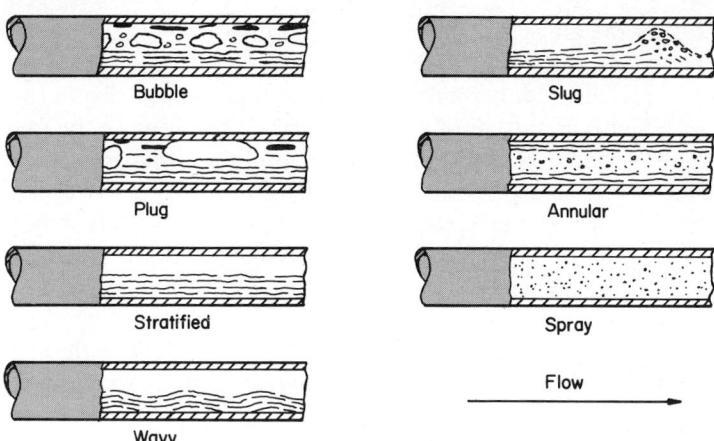

Fig. 17-5 Typical two-phase flow patterns.

Typical flow patterns and a correlation plot for determination of flow regimes corresponding to these patterns are given in Figs. 17-5 and 17-6, respectively. The flow patterns of Fig. 17-5 are in order of increasing ratio of vapor to liquid flow rate. At very small values of G_v/G_l, only small bubbles of vapor are present, and they tend to collect in the upper part of a horizontal tube. The term G represents the mass flow ρV per unit area in pounds mass per hour per square foot based on the total pipe cross-sectional area, and the subscripts v and l denote vapor and liquid phases, respectively. As this ratio increases, the small bubbles coalesce into plugs of vapor, which also tend to remain near the top of the tube. If the ratio is increased still

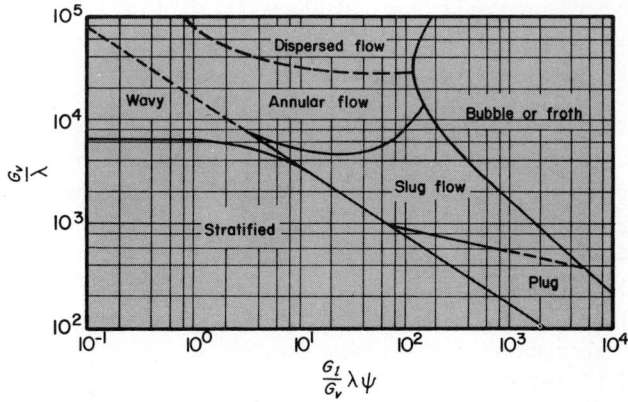

Fig. 17-6 Flow regimes. [*From O. Baker, Oil Gas J.*, **53**: 185 (1954).]

further, the plugs grow until a continuous vapor stream occurs in the top part of the tube, as shown in the stratified-flow sketch.

The next discernible pattern change results from the vapor phase velocity being higher than that of the liquid phase, resulting in the formation of waves on the liquid surface. Continuing to increase the vapor-phase velocity over the liquid velocity can result in liquid slugs distributed along the tube or in an annular liquid flow with the vapor in the core. Here, liquid droplets may be dispersed in the core flow. Finally the vapor-phase velocity may be great enough to disperse the liquid completely so that a mist or spray type of flow with droplets more or less uniformly dispersed in the vapor results.

A reversed annular flow can occur in flow with film-boiling heat transfer. In this case the annular flow can be vapor with a core flow of liquid. This does not occur, however, in adiabatic or isothermal flows.

There have been several attempts by investigators to establish criteria for predicting the type of flow pattern which will exist under given flow conditions. The most widely accepted is that due to Baker [2], shown in Fig. 17-6. This plot resulted from a systematic study of adiabatic air-water-mixture flows in horizontal pipes. The parameters are G_v/λ and $(G_l/G_v)\lambda\psi$, where G_l is as previously defined and G_v is the mass flow rate of air per unit area. Baker defines

$$\lambda \equiv \left(\frac{\rho_v}{0.075} \frac{\rho_l}{62.3}\right)^{\frac{1}{2}} \tag{17-16}$$

and

$$\psi \equiv \left(\frac{73}{\sigma_l} \mu_l \frac{62.3}{\rho_l}\right)^{\frac{1}{3}} \tag{17-17}$$

where ρ_v = density of gas phase, lb_m/ft^3
ρ_l = density of liquid phase, lb_m/ft^3
μ_l = dynamic viscosity of liquid, centipoises
σ_l = surface tension of liquid, dynes/cm

Caution should be exercised in applying Fig. 17-6: the lines on this plot do not represent sharp boundaries between flow patterns but are intended to indicate the area of transition. Further, the flow patterns in an inclined or vertical tube are likely to be quite different from those in a horizontal tube, and the Baker plot would be less applicable to these situations. Kozlov [30] as reported by Tong [59] has studied flow patterns with an air-water two-phase flow in vertical tubes, and the interested reader should consult these references.

Slip ratio and void fraction The fact that the vapor- and liquid-phase velocities are usually not the same is obvious from the discussion of flow patterns. The difference in velocities is sometimes referred to as the *slip*, and the *slip ratio* is defined

MULTIPHASE PHENOMENA

as

$$S \equiv \frac{\text{velocity of vapor}}{\text{velocity of liquid}} \qquad (17\text{-}18)$$

The *void fraction* is defined as the ratio of the local vapor volume per unit length to the total local flow volume per unit length. Thus

$$\epsilon \equiv \frac{\text{volume of vapor}}{\text{volume of vapor} + \text{volume of liquid}} \qquad (17\text{-}19)$$

or, in terms of the cross-sectional areas experiencing vapor and liquid flow at a given section,

$$\epsilon = \frac{A_v}{A_v + A_l} \qquad (17\text{-}19a)$$

An additional useful term is the *flow quality* x defined by

$$x \equiv \frac{\text{mass flow rate of vapor}}{\text{total mass flow rate}} \qquad (17\text{-}20)$$

Using this, the flow average density is

$$\bar{\rho}_F = x\rho_v + (1-x)\rho_l \qquad (17\text{-}21)$$

and the slip ratio is

$$S = \frac{V_v}{V_l} \equiv \frac{x}{1-x} \frac{1-\epsilon}{\epsilon} \frac{\rho_l}{\rho_v} \qquad (17\text{-}22)$$

Example A saturated steam-water mixture at 100 psia with a flow quality of 2 percent flows in a horizontal insulated pipe. The local void fraction as determined by a nuclear-source densitometer is 65 percent. Determine the ratio of the vapor velocity to that of the liquid and the probable flow pattern if the total flow rate per unit area of pipe is 50,000 lb_m/hr-ft².

Solution The ratio of vapor to liquid velocities is the slip ratio, which can be obtained with Eq. (17-22). From steam tables,

$$\rho_l = \frac{1}{0.01774} \; lb_m/ft^3 \quad \text{and} \quad \rho_v = \frac{1}{4.432} \; lb_m/ft^3$$

Thus,

$$S = \frac{0.02}{1.00 - 0.02} \frac{1.00 - 0.65}{0.65} \frac{4.432}{0.01774} = 2.74$$

and the vapor velocity is 2.74 times that of the liquid. Applying the Baker plot to predict the flow pattern,

$$\lambda = \left(\frac{1}{4.432 \times 0.075} \frac{1}{0.01774 \times 62.3}\right)^{\frac{1}{2}}$$

$$= [3.01(0.905)]^{\frac{1}{2}} = 1.65$$

$$\psi = \left\{\frac{73}{44}\left[(0.112 \times 10^{-3} \; lb_m/ft\text{-sec}) \frac{1{,}488 \text{ centipoises}}{lb_m/ft\text{-sec}}\right] \frac{62.3}{1/0.01774}\right\}^{\frac{1}{3}}$$

$$= [1.66(0.1667)(1.106)]^{\frac{1}{3}} = 0.674$$

where the surface tension and viscosity are evaluated at the saturation temperature of the liquid.

By Eq. (17-20), the vapor flow rate per unit area is

$$G_v = 0.02(50,000) = 1,000 \text{ lb}_m/\text{hr-ft}^2$$

and clearly $G_l = 49,000 \text{ lb}_m/\text{hr-ft}^2$. Thus

$$\frac{G_v}{\lambda} = \frac{1,000}{1.65} = 606$$

and

$$\frac{G_l}{G_v} \lambda \psi = 49(1.65)(0.674) = 54.5$$

From the Baker plot of Fig. 17-6 we see that the flow pattern is most probably *stratified* (see Fig. 17-5).

17-4 PRESSURE DROP IN ISOTHERMAL FLOW

The frictional pressure loss in two-phase flow is frequently much larger than that for either phase flowing alone at the total mass flow rate, and the methods of calculating the pressure loss in single-phase flow (Chaps. 12 and 13) are inadequate for engineering design purposes. For isothermal[1] two-phase flow, the pattern remains essentially unchanged over extensive lengths, and the pressure-loss analysis can be based upon a single flow pattern. For very long pipes the change in density due to large frictional static-pressure changes may necessitate a piecewise solution with a particular flow pattern applicable only to a portion of the system. Large static-pressure changes also occur, of course, with significant change in flow area or elevation, and these must be considered where applicable.

The total pressure loss in an adiabatic two-phase pipe flow is

$$\left(\frac{\Delta p}{\Delta L}\right)_{tot} = \left(\frac{\Delta p}{\Delta L}\right)_{fric} + \left(\frac{\Delta p}{\Delta L}\right)_{mom} + \left(\frac{\Delta p}{\Delta L}\right)_{elev} \tag{17-23}$$

Direct measurement of the frictional loss is difficult in two-phase flow. The loss due to elevation changes can be readily eliminated or accounted for, but any frictional loss results in some momentum change in a two-phase flow, which complicates experimental verification of analyses.

Tong [59] recommends the use of two principal types of model for analyses of two-phase flow pressure drop, the *homogeneous* model of Owens and the *slip* model of Lockhart and Martinelli. The homogeneous model is applicable for the *spray* pattern of Fig. 17-5 and is perhaps the better approach for the *bubble* flow of that figure.

---

[1] Most adiabatic two-phase flows in pipes are essentially isothermal, and these two descriptive terms are interchangeable in this section.

Homogeneous model Owens [45] developed an analytical model for the total pressure loss under the assumptions of (1) equal velocities of vapor and liquid and (2) thermodynamic equilibrium between the phases. Defining a mean specific volume of the flowing fluid by

$$\bar{v} = v_l\left[1 + \frac{x}{v_l}(v_v - v_l)\right] \tag{17-24}$$

the three components of Eq. (17-23) can be expressed as

$$\left(\frac{\Delta p}{\Delta L}\right)_{\text{fric}} = \frac{f_{TP}G^2\bar{v}}{2g_c D_h} \tag{17-25}$$

$$\left(\frac{\Delta p}{\Delta L}\right)_{\text{mom}} = \frac{G^2}{g_c}\left(\frac{\Delta \bar{v}}{\Delta L}\right) \tag{17-26}$$

$$\left(\frac{\Delta p}{\Delta L}\right)_{\text{elev}} = \frac{1}{\bar{v}}\frac{g}{g_c} \tag{17-27}$$

With these Eq. (17-23) becomes

$$\left(\frac{\Delta p}{\Delta L}\right)_{\text{tot}} = \frac{\dfrac{f_{TP}G^2 v_l}{2g_c D_h}\left[1 + x\left(\dfrac{v_v}{v_l} - 1\right)\right] + \dfrac{G^2 v_l}{g_c}\left(\dfrac{v_v}{v_l} - 1\right)\dfrac{\Delta x}{\Delta L} + \dfrac{g}{g_c v_l\{1 + x[(v_v/v_l) - 1]\}}}{1 + x\dfrac{G^2}{g_c}\left(\dfrac{\Delta v_v}{\Delta p}\right)} \tag{17-28}$$

where G = total mass flow rate per unit area, $lb_m/sec\text{-}ft^2$
v = specific volume, ft^3/lb_m
D_h = channel hydraulic diameter, ft

Owens suggested the use of the single-phase friction factor for the liquid flow f_l for f_{TP}. This factor f_l is obtained at a superficial Reynolds number based on a superficial liquid velocity

$$V_{ls} = \frac{\text{mass flow rate of liquid}}{(\text{liquid density})(\text{area of tube})} \tag{17-29}$$

This approach has been shown to yield satisfactory results in the spray and bubble flow-pattern regimes.

Slip model Lockhart and Martinelli [35] developed an empirical correlation for adiabatic two-phase annular flow which predicts the *frictional pressure loss only*. It is recommended for all flow patterns of Fig. 17-5 except the bubble and spray regimes.

The essence of their approach was to correlate the actual measured frictional loss as a function of superficial calculated frictional losses of both phases. These

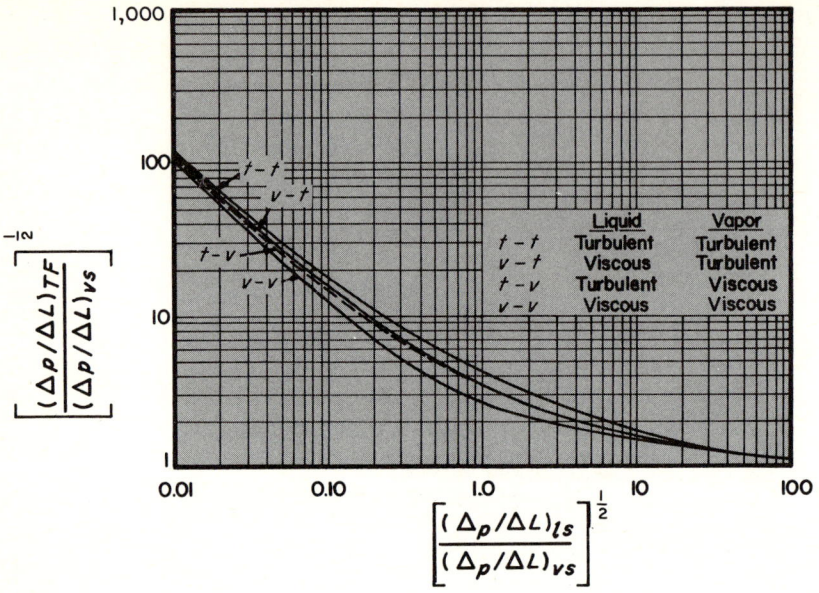

Fig. 17-7 Slip-model frictional-pressure-loss correlation. TF = actual two-phase condition, ls = superficial liquid-phase flow, vs = superficial vapor-phase flow. [*Adapted from R. W. Lockhart and R. C. Martinelli, Chem. Eng. Progr.*, **45**: *39 (1949), by permission.*]

calculated losses are based on the friction factors of single-phase flow using superficial velocities and Reynolds numbers. The superficial liquid-phase velocity is calculated by Eq. (17-29), and the superficial vapor-phase velocity is defined similarly. The superficial Reynolds number of each phase is based on the properties of that phase, the superficial velocity, and the pipe diameter.

They defined four two-phase flow regimes depending upon whether each superficial Reynolds number indicates turbulent or laminar flow, i.e., above or below 2,000. Their results are presented in Fig. 17-7.

Example Calculate the pressure drop due to friction for the last example, i.e., a saturated steam-water-mixture flow at 100 psia, flow quality of 2 percent, and total flow rate per unit area of 50,000 lb_m/hr-ft², if the pipe is 2-in-ID commercial steel, and 1,000 ft long.

Solution In the previous example we found the flow pattern to be stratified, and hence we would expect the *slip-flow* model to be best for estimation of the pressure loss. The properties of the two phases are

$\rho_l = 56.4 \text{ lb}_m/\text{ft}^3$ $\nu_l = 0.208 \times 10^{-5} \text{ ft}^2/\text{sec}$

$\rho_v = 0.2255 \text{ lb}_m/\text{ft}^3$ $\nu_v = 0.333 \times 10^{-3} \text{ ft}^2/\text{sec}$

MULTIPHASE PHENOMENA

The mass flow rate of vapor per unit area is

$$G_v = xG = 0.02(50{,}000 \text{ lb}_m/\text{hr-ft}^2)$$

$$= \left(1{,}000 \frac{\text{lb}_m}{\text{hr-ft}^2}\right) \frac{\text{hr}}{3{,}600 \text{ sec}} = 0.278 \text{ lb}_m/\text{sec-ft}^2$$

and the mass flow rate of liquid per unit area is

$$G_l = 50{,}000 - 1{,}000 = 49{,}000 \text{ lb}_m/\text{hr-ft}^2$$

$$= \left(49{,}000 \frac{\text{lb}_m}{\text{hr-ft}^2}\right) \frac{\text{hr}}{3{,}600 \text{ sec}} = 13.6 \text{ lb}_m/\text{sec-ft}^2$$

The superficial velocities are

$$V_{vs} = \frac{G_v}{\rho_v} = \frac{0.278 \text{ lb}_m/\text{sec-ft}^2}{0.2255 \text{ lb}_m/\text{ft}^3} = 1.23 \text{ ft/sec}$$

$$V_{ls} = \frac{G_l}{\rho_l} = \frac{13.6 \text{ lb}_m/\text{sec-ft}^2}{56.4 \text{ lb}_m/\text{ft}^3} = 0.242 \text{ ft/sec}$$

The superficial Reynolds numbers are

$$\mathbf{Re}_{vs} = \frac{(1.23 \text{ ft/sec})(\tfrac{2}{12} \text{ ft})}{0.333 \times 10^{-3} \text{ ft}^2/\text{sec}} = 616 \qquad \text{laminar}$$

$$\mathbf{Re}_{ls} = \frac{(0.242 \text{ ft/sec})(\tfrac{2}{12} \text{ ft})}{0.208 \times 10^{-5} \text{ ft}^2/\text{sec}} = 19{,}400 \qquad \text{turbulent}$$

and the flow is turbulent viscous.

The friction factor for the superficial vapor flow is, by Eq. (12-38),

$$f_{vs} = \frac{64}{\mathbf{Re}_D} = \frac{64}{616} = 0.1038$$

The friction factor for the liquid flow from Figs. 13-10 and 13-11 is

$$f_{ls} = 0.028$$

The superficial pressure changes are

$$\left(\frac{\Delta p}{\Delta L}\right)_{vs} = \frac{-\rho_v V_{vs}^2 f_{vs}}{2 g_c D}$$

$$= \frac{-(0.2255 \text{ lb}_m/\text{ft}^3)(1.23 \text{ ft/sec})^2(0.1038)}{(64.4 \text{ ft-lb}_m/\text{lb}_f\text{-sec}^2)(\tfrac{2}{12} \text{ ft})}$$

$$= -3.3 \times 10^{-4} \text{ lb}_f/\text{ft}^2\text{-ft}$$

$$\left(\frac{\Delta p}{\Delta L}\right)_{ls} = \frac{-\rho_l V_{ls}^2 f_{ls}}{2 g_c D}$$

$$= \frac{-(56.4 \text{ lb}_m/\text{ft}^3)(0.242 \text{ ft/sec})^2(0.028)}{(64.4 \text{ ft-lb}_m/\text{lb}_f\text{-sec}^2)(\tfrac{2}{12} \text{ ft})}$$

$$= -8.62 \times 10^{-3} \text{ lb}_f/\text{ft}^2\text{-ft}$$

Thus,

$$\left[\frac{(\Delta p/\Delta L)_{ls}}{(\Delta p/\Delta L)_{vs}}\right]^{\frac{1}{2}} = \left(\frac{8.62 \times 10^{-3}}{3.3 \times 10^{-4}}\right)^{\frac{1}{2}} = 5.11$$

and from Fig. 17-7,

$$\left[\frac{(\Delta p/\Delta L)_{TP}}{(\Delta p/\Delta L)_{ls}}\right]^{\frac{1}{2}} \simeq 1.9$$

and thus,

$$\left(\frac{\Delta p}{\Delta L}\right)_{TP} = 3.61(-8.62 \times 10^{-3}) \text{ lb}_f/\text{ft}^2\text{-ft}$$

$$\simeq 31.1 \times 10^{-3} \text{ lb}_f/\text{ft}^2\text{-ft}$$

The total frictional pressure drop is

$$\Delta p = (31.1 \times 10^{-3} \text{ lb}_f/\text{ft}^2\text{-ft})(1{,}000 \text{ ft})\frac{\text{ft}^2}{144 \text{ in}^2} = 0.216 \text{ psi}$$

17-5 FUNDAMENTALS OF BOILING HEAT TRANSFER

Boiling heat transfer occurs whenever there is a change in phase from liquid to vapor. The interest in this mode of heat exchange was first due to widespread commercial applications—in processing kettles, steam boiler tubes, refrigeration and air-conditioning evaporators, etc. The heat flux per unit area is quite large compared with that obtainable without phase change, and this fact has generated interest in applying phase-change techniques to such areas as transient cooling of rocket nozzles and steady-state heat pipes.

The two general boiling situations are known as *pool boiling* and *flow* or *forced-convective boiling*. A thorough understanding of the first is necessary before undertaking a study of the second.

Pool boiling Modern investigations into this mode of heat transfer began in 1934, when Nukiyama [43] aroused academic curiosity by his then highly unusual findings relative to heat transfer during boiling of water from a submerged resistance wire. Leidenfrost [33] in 1756 and Lang [32] in 1888 had previously found that maximum and minimum heat-transfer rates exist in boiling. Nukiyama's publication marked the beginning of intensive efforts by researchers throughout the world to understand and explain the various phenomena associated with boiling heat transfer.

Pool boiling in its simplest form can occur whenever a surface hotter than the fluid saturation temperature is exposed to a liquid at or near saturation conditions with no external agitation or forced-convective currents. This is perhaps the most frequently encountered form of boiling and is typified by the process which occurs in the common teakettle on a stove. The primary correlation for steady-state pool-boiling heat transfer is usually expressed in the form of the *boiling curve*, as shown in Fig. 17-8. The rather odd shape of this curve is a result of the fact that it is actually composed of four main regimes, each representing a totally different mechanism of heat transfer.

In regime I, there is no phase change, and the heat transfer is due solely to natural convection. In this region the customary free-convective-heat-transfer relationships are quite adequate for predicting heat-transfer rates. As the wall temperature is increased such that the *excess temperature* (this term is customarily used to represent the difference between the wall and the fluid saturation temperatures) approaches point b, phase change begins to occur in the form of bubbles at favored nucleation sites along the surface. Nucleation sites for boiling are generally believed to be quite similar to those for cavitation, as shown in Fig. 17-2. The number of such active sites depends upon several variables; generally, nucleation begins with only one site or at most a few. For a given surface and fluid, the density of active nucleating sites increases with increasing excess temperature.

The mechanism of heat transport in regime II, the *nucleate boiling regime*, is very complex; heat is being removed from the surface both by the vaporization process and by high-velocity convective currents associated with breakaway of the bubbles. In a majority of nucleate pool-boiling situations, the energy transport by vaporization is considerably less than the total heat flux. In a typical experiment, Gunther and Kreith [23] measured a total heat flux of 1.04×10^6 Btu/hr-ft². The latent heat transport calculated by measuring the average bubble departure size and departure rate was 20,000 Btu/hr-ft². This is only 2 percent of the total heat flux. The major part of the heat transfer in nucleate boiling must be due to some other transport

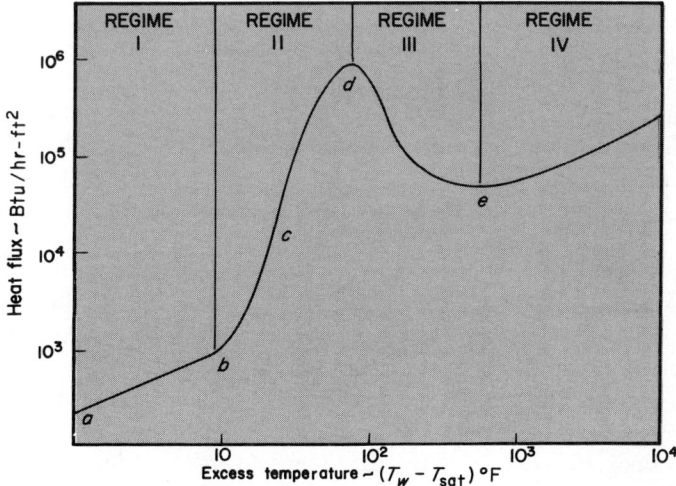

Fig. 17-8 Typical pool-boiling curve for heat transfer from wire to water at atmospheric pressure.

mechanism. Forster and Grief [20] attribute this to intense fluid agitation close to the surface, while Moore and Mesler [41] and Snyder and Robin [55] have postulated a high energy flux due to vaporization of a microlayer of water underneath a bubble with subsequent phase change to liquid at the bubble upper surface. Whichever model is dominant does not alter the fact that the primary energy-transport mode (except perhaps very near the surface) must be convective and must depend upon the inrush of cooler liquid to replace exiting bubbles.

As the excess temperature passes through point c, an inflection occurs in the curve as a result of increased nucleation, with the bubbles coalescing to form continuous columns of vapor and thereby causing a reduction in individual site effectiveness. As point d is approached, nucleating sites become so numerous that the interference of columns from individual sites causes a further loss in individual effectiveness. At point d, the site density is so great that it prevents adequate motion of liquid to the heated surface to replenish that leaving by vaporization and being forced away due to buoyancy forces. Here, the increase in number of sites with increase in excess temperature exactly balances the loss of effectiveness at each site. This results in a maximum local value of the heat flux, which is usually referred to as the *critical heat flux*, the *departure from nucleate boiling* (DNB), the *boiling crisis*, or the *burnout point*. All four terms have been used interchangeably throughout the boiling-heat-transfer literature.

Increase in the excess temperature beyond point d results in *transition boiling* (also called *partial film boiling* and *unstable film boiling*), regime III of Fig. 17-8. This is perhaps the most unusual part of the boiling curve since an increase in excess temperature results in a decrease in the heat flux. The actual mechanism, like that of nucleate boiling, is very complex. The phenomenon appears to oscillate between film and nucleate boiling, the surface at times being blanketed by a film of vapor and then returning to a nucleating condition. In the case of a heated wire, the film appears to move along the length and then return. There is considerable disagreement among researchers concerning the mechanism(s) of transition boiling. While this is undoubtedly the least understood of the boiling regimes, it is also the least useful, which probably helps explain why it is so poorly understood.

As the excess temperature is increased to point e, the mechanism of heat transfer changes to one of *stable film boiling* (regime IV). Here the heated surface is completely separated from the liquid by a vapor film. An example occurs wherever a droplet of water is placed on a very hot surface (such as a hot skillet). The droplet immediately vaporizes at its base and appears to float on the resulting vapor film. This has been referred to as the *Leidenfrost phenomenon* in the literature. In this regime, it is generally agreed that the heat transport through the vapor film is by conduction and radiation to the vapor-liquid interface where liquid is being evaporated. At moderate values of excess temperature, the heat flux in film boiling is quite small due to the thermal resistance of this vapor film. The effect of radiation becomes increasingly significant at high surface temperatures, especially when accompanied by high surface emissivity.

MULTIPHASE PHENOMENA

Both nucleate and film boiling can occur simultaneously on a heated wire element. Figure 17-9 illustrates this, with film boiling evident on the left portion and nucleate boiling on the right side. Here the temperature of the wire on the right side is slightly less than on the left due to conduction along the wire. A slight increase in the wire temperature would cause the entire surface to experience film boiling.

Excess temperature Experimental correlations for boiling heat transfer are usually based on the temperature above the boiling point, simply called the *excess temperature*. For a fixed surface temperature and fluid saturation condition this is

$$\Delta T_x = T_w - T_{\text{sat}} \tag{17-30}$$

That the actual fluid temperature is not used in determining the thermal potential or driving temperature difference is a peculiarity of boiling. To illustrate this point, the heat transfer from a plate of given size at, say, 215°F to a pool of liquid water at atmospheric pressure is essentially the same whether the water is at a bulk temperature of 100°F, 200°F, or 212°F. In reality, there is a slight effect due to subcooling, but this is omitted in engineering calculations.

The term *bulk fluid boiling* is applied to the situation where the fluid is at the saturation temperature and vapor bubbles originate and grow in the fluid itself. The more commonly encountered situation is *subcooled boiling*. Here, the nucleation and resulting bubble growth can occur only at the heated surface. In subcooled pool boiling, the bubbles leaving the surface experience a decrease in size due to

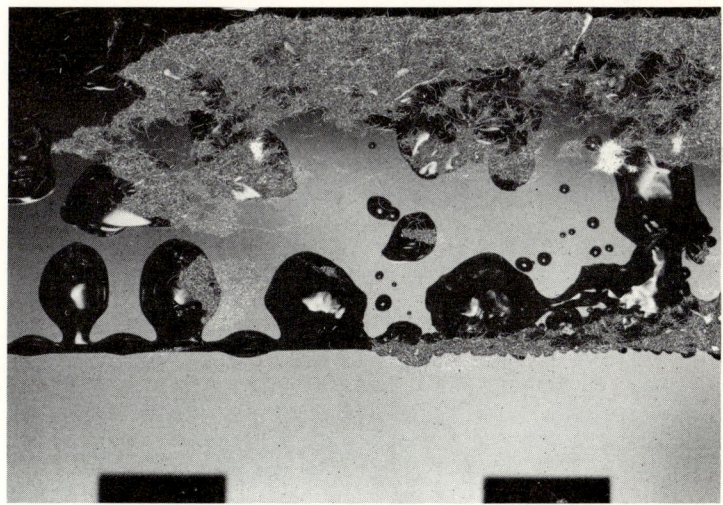

Fig. 17-9 Simultaneous film and nucleate boiling of isopropanol on a heated 0.008-in.-diam wire ($q = 187{,}000$ Btu/hr-ft²). (*Photograph courtesy of J. Lienhard.*)

condensation as they travel through the subcooled liquid. This is discussed in more detail in the following section.

Nucleate boiling Theoretical predictions of the superheat (excess temperature) required for initiation of nucleation in a pure degassed liquid are much higher than that actually observed, except when extreme precautions are taken to ensure the absence of surface gas pockets as shown in Fig. 17-2. A widely held theory is that most commercial surfaces contain many microscopic conelike cavities which provide vapor or gas nuclei.

Bubbles grow in nucleate boiling at these nucleating sites (cavities) on a heated surface. Actually, the base of a single bubble can cover more than one such site. The growth rate and life of a bubble depend on both the surface excess temperature and the bulk fluid temperature. If the fluid is highly subcooled, the bubble collapses without leaving the surface due to heat transfer around the upper portion of the bubble. If the fluid is moderately subcooled, say not more than 5°F below the saturation temperature, the bubble may break away from the surface and rise due to buoyancy effects. In the liquid it experiences a heat-transfer-controlled collapse due to condensation at its vapor-liquid interface.

If the liquid is superheated, bubbles leaving the surface will grow as they move through the liquid, and bubbles may also form in the body of the liquid. Under this condition and that of very small subcooling, bubbles usually rise to the free surface in pool boiling.

Analyses of growth or collapse rates of spherical vapor bubbles in boiling based on solutions to the Rayleigh equation (17-10) have been undertaken by numerous investigators, including Plesset and Zwick [47], Forster and Zuber [18], Florschuetz and Chao [17], Wittke and Chao [60], and Cho and Seban [11]. The details of these solutions are beyond the scope of the present book. It is informative, however, to consider the equilibrium requirement for a spherical bubble in a body of liquid at uniform pressure and temperature, p_l and T_l. The objective is to determine for given bubble radius and fluid properties the temperature requirement for the vapor bubble to exist in thermodynamic equilibrium.

For a spherical bubble containing a noncondensable gas having a partial pressure p_g, the equilibrium pressure difference, as given by Eq. (17-11), is

$$p_v - p_l = \frac{2\sigma}{r} - p_g \tag{17-11}$$

Along the saturation line a relationship between temperature and pressure is the Clausius-Clapeyron equation, which together with the perfect-gas approximation is, for $v_{fg} \simeq v_v$,

$$\frac{dp}{dT} = \frac{h_{fg}}{v_v T_v} = \frac{p_v h_{fg}}{R_v T_v^2} \tag{17-31}$$

MULTIPHASE PHENOMENA

Now

$$\frac{p_v - p_l}{T_v - T_{\text{sat}}} \simeq \frac{dp}{dT}$$

and using this equation (17-31) can be approximated by

$$T_v - T_{\text{sat}} \simeq \frac{R_v T_v^2}{p_v h_{fg}} (p_v - p_l) \tag{17-32}$$

Combining Eqs. (17-11) and (17-32) yields

$$T_v - T_{\text{sat}} \simeq \frac{R_v T_v^2}{p_v h_{fg}} \left(\frac{2\sigma}{r} - p_g \right) \tag{17-33}$$

which is the equilibrium relationship between bubble radius and amount of superheat. If the liquid temperature surrounding the bubble results in $T_l - T_{\text{sat}}$ greater than $T_v - T_{\text{sat}}$ of Eq. (17-33), a bubble of radius r containing a noncondensable gas exerting pressure p_g will grow; if $T_l - T_{\text{sat}}$ is less than this $T_v - T_{\text{sat}}$, the bubble will collapse. It should also be evident that the effect of the noncondensable gas is to reduce the amount of superheat necessary for bubble growth.

Effects of various parameters
1. *Aging.* A boiling surface that has been in service for an extensive time usually requires increasing excess temperature with time for the same heat flux. This is probably due to a decrease in cavity size or effectiveness by oxidation and an increased thermal resistance due to oxidation and mineral deposits.
2. *Gravity.* Merte and Clark [39] have shown that in regime I and the lower part of regime II of Fig. 17-8, the curve shifts to the left with increasing gravity. In these regions free convection is very significant, and their results are as would be expected. At zero gravity, experiments have shown that there is no nucleate boiling.
3. *Dissolved gases.* Dissolved gases are released by the liquid at the hot surface during boiling. This activity increases the liquid agitation, and experiments by McAdams show that there is a significant increase in heat flux over that obtainable with a degassed liquid.
4. *Geometry.* In the nucleate boiling regime, the heat-transfer rate is rather insensitive to the geometry of the heater. According to Rohsenow and Choi [52], the same correlation is applicable to horizontal, vertical, or inclined flat surfaces and to wires with diameters above 0.004 in. They point out that this is not true in the lower part of the nucleate regime, where free-convective effects are significant.
5. *Pressure.* Numerous investigations of the effect of pressure indicate that the heat flux increases with increasing pressure. McAdams [37] presents a rather extensive summary of this work. The data of Addoms [1] presented in Fig. 17-10 show very large increases in q/A with increasing p. This is in keeping with the equilibrium requirement for excess temperature of Eq. (17-33). For a given

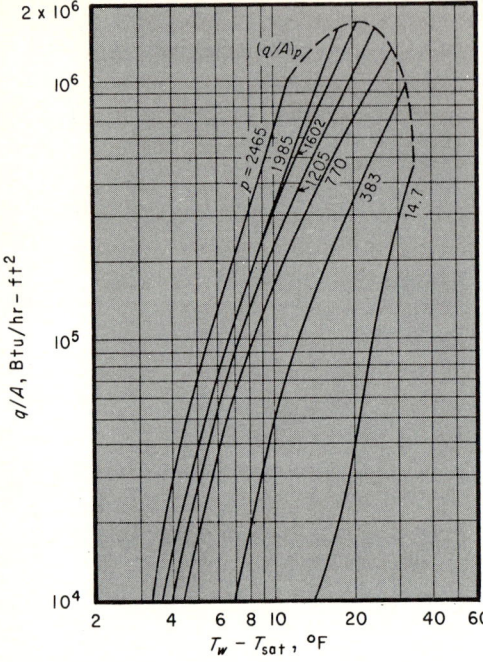

Fig. 17-10 Effect of pressure on heat flux for water on a 0.024-in. platinum wire. (*Data from Addoms* [1].)

cavity size, a larger surrounding pressure (and hence a larger vapor pressure) requires a smaller value of $T_v - T_{sat}$. Thus, a bubble can exist at a lower excess temperature, and the heat flux for a high pressure is greater than that for a low pressure at the same excess temperature.

6. *Roughness.* The effect of surface roughness is intimately associated with nucleating cavity size. Since more superheat is required for smaller cavities [by Eq. (17-33)], rougher surfaces with larger cavities exhibit nucleation at smaller values of the wall superheat (excess temperature). Experiments show that the heat flux is directly related to the number of cavities and the cavity size. Berenson [3] has shown that the required superheat for a given heat flux increases with decreasing artificial roughness obtained by finishing the surface with graded emery cloth, lapping, and mirror finishing. Indeed, the key to advances in predicting the nucleate-boiling heat flux from a surface with a given fluid lies in a better understanding and description of the surface roughness.

17-6 BOILING-HEAT-TRANSFER CORRELATIONS

From the preceding discussion of boiling mechanisms, it is evident that no single correlation equation could apply to the entire range of pool boiling, as shown in Fig. 17-8. The most important regimes are nucleate boiling and stable film boiling.

MULTIPHASE PHENOMENA

The purpose of this section is to introduce the methods of predicting heat transfer in nucleate or film boiling, both for pool boiling and with forced convection.

Nucleate pool boiling Several investigators [19, 20, 21] have attempted to formulate a general correlation suitable for this boiling regime. The most widely accepted correlation is that due to Rohsenow [50], whose semiempirical result can be expressed as

$$\left(\frac{q}{A}\right)_b = \mu_l h_{fg} \sqrt{\frac{g(\rho_l - \rho_v)}{g_c \sigma}} \left[\frac{c_l(T_w - T_{\text{sat}})}{h_{fg} \text{Pr}_l^{1.7} C_{sf}}\right]^3 \tag{17-34}$$

where
- c_l = specific heat of saturated liquid, Btu/lb$_m$-°F
- C_{sf} = experimental constant (Table 17-2)
- g = local gravitational acceleration, ft/sec²
- g_c = constant of proportionality, 32.17 ft-lb$_m$/lb$_f$-sec²
- h_{fg} = enthalpy of vaporization, Btu/lb$_m$
- Pr_l = Prandtl number of saturated liquid
- q/A = heat flux per unit area, Btu/hr-ft²
- $T_w - T_{\text{sat}}$ = excess temperature, °F
- μ_l = liquid viscosity, lb$_m$/ft-hr
- σ = surface tension, lb$_f$/ft
- ρ_l = density of saturated liquid, lb$_m$/ft³
- ρ_v = density of saturated vapor, lb$_m$/ft³

The experimental constant C_{sf} depends upon the liquid-surface combination. In particular, it is a function of the number of nucleating sites (or surface roughness) and the angle of contact between the vapor bubble and the heated surface. Little progress has been made toward the prediction of C_{sf}, and consequently resort is made to purely experimental values. Table 17-2 reports some widely accepted values.

Table 17-2 Values of C_{sf}

Fluid-surface combination	C_{sf}	Reference
Water–brass	0.006	13
Water–copper	0.013	46
Water–nickel	0.006	53
Water–platinum	0.013	1
CCl$_4$–copper	0.013	46
Benzene–chromium	0.010	12
n-Pentane–chromium	0.015	12
Ethyl alcohol–chromium	0.0027	12
Isopropyl alcohol–copper	0.0025	46
35% K$_2$CO$_3$–copper	0.0054	46
50% K$_2$CO$_3$–copper	0.0027	46
n-Butyl alcohol–copper	0.0030	46

Example Estimate the heat flux for water boiling on a 10-mil-diam platinum wire at 1 atm pressure, 30°F excess temperature, and standard gravity.

Solution Since Table 17-2 contains a value of C_{sf} for this fluid-surface combination, that is, $C_{sf} = 0.013$, we can use the Rohsenow equation. (Note from Fig. 17-8 that at the stated surface excess temperature, the regime is nucleate boiling.) The appropriate fluid properties are

$c_l = 1.0$ Btu/lb$_m$-°F $\quad h_{fg} = 970.3$ Btu/lb$_m$
$\mathbf{Pr}_l = 1.78$ $\quad\quad\quad\quad\quad \mu_l = 0.699$ lb$_m$/ft-hr
$\rho_l = 59.85$ lb$_m$/ft^3 $\quad\quad \rho_v = 0.0373$ lb$_m$/ft^3

Then by Eq. (17-34)

$$\frac{q}{A} = \left(0.699 \frac{\text{lb}_m}{\text{ft-hr}}\right)\left(970.3 \frac{\text{Btu}}{\text{lb}_m}\right) \sqrt{\frac{\left(32.17 \frac{\text{ft}}{\text{sec}^2}\right)\left(59.85 - 0.037\right)\frac{\text{lb}_m}{\text{ft}^3}}{\left(32.17 \frac{\text{ft-lb}_m}{\text{lb}_f\text{-sec}^2}\right)\left(0.00403 \frac{\text{lb}_f}{\text{ft}}\right)}} \times \left[\frac{\left(1.0 \frac{\text{Btu}}{\text{lb}_m\text{-°F}}\right)(30°\text{F})}{\left(970 \frac{\text{Btu}}{\text{lb}_m}\right)(1.78^{1.7})(0.013)}\right]^3$$

$= 0.699(970.3)(101.986)(0.7168) = 49{,}500$ Btu/hr-ft^2

An important observation is that *the heat flux in nucleate boiling is proportional to the cube of the excess temperature.*

Peak heat flux in nucleate pool boiling The prediction of the maximum heat flux is important for two reasons: (1) this value indicates the maximum performance for many systems, and (2) the reduction in heat flux beyond this point can result in actual physical failure—burnout. Consequently the prediction of the maximum flux has been the goal of several investigations [6, 31, 40, 54, 58, 61, 62].

Zuber [61] developed an expression for the maximum heat flux in nucleate boiling based on hydrodynamic or Helmholtz instability theory. From photographic evidence depicting a regular spacing of bubble columns rising from the vapor blanket at the heated surface, he proposed a model with round, pulsating vapor jets flowing away from the surface and liquid flow to the surface. The volume flow rate of vapor from the surface is postulated to be equal to the volume rate of liquid flow toward the surface. The resulting relative velocity at the vapor-liquid interface of the jets causes an instability which results in breakup of the jet. His result is

$$\left(\frac{q}{A}\right)_{\max} = \frac{\pi}{24} \rho_v h_{fg} \left[\frac{\sigma(\rho_l - \rho_v)gg_c}{\rho_v^2}\right]^{\frac{1}{4}} \left(\frac{\rho_l}{\rho_l + \rho_v}\right)^{\frac{1}{2}} \quad (17\text{-}35)$$

Experimental results are better correlated by Eq. (17-35) if the constant is increased to 0.18. All terms and units of Eq. (17-35) are the same as for Eq. (17-34).

Other widely recognized correlation equations for the peak heat flux in nucleate pool boiling are the Rohsenow and Griffith [54] equation based on energy transport by the vapor bubbles

$$\left.\frac{q}{A}\right|_{\max} = 143\rho_v h_{fg}\left(\frac{g}{g_0}\right)^{\frac{1}{4}}\left(\frac{\rho_l - \rho_v}{\rho_v}\right)^{0.6} \tag{17-36}$$

and the Kutateladze [31] equation obtained through dimensional analysis

$$\left.\frac{q}{A}\right|_{\max} = C(\rho_v)^{\frac{1}{2}}h_{fg}[\sigma(\rho_l - \rho_v)gg_c]^{\frac{1}{4}} \tag{17-37}$$

In Eq. (17-36), g_0 is the earth's standard gravitation, 32.17 ft/sec². All other terms in both equations are as previously defined. The constant C recommended by Kutateladze is 0.14. It is interesting to note that the Kutateladze equation and the Zuber equation differ only in the value of this constant and the last term of Eq. (17-35) (which is usually close to unity), though they were obtained by vastly different approaches.

Film pool boiling Bromley [7] investigated stable film boiling on horizontal heated tubes both experimentally and analytically. His experiments included several fluids and 0.19-, 0.24-, 0.35-, and 0.47-in-diam tubes. His recommended equation for a boiling-heat-transfer coefficient is

$$h_c = 0.62\left[\frac{k_v^3 \rho_v(\rho_l - \rho_v)g(h_{fg} + 0.4Cp_v\,\Delta T_x)}{D\mu_v\,\Delta T_x}\right]^{\frac{1}{4}} \tag{17-38}$$

where D = outside diameter of the cylinder
ΔT_x = excess temperature, $T_w - T_{\text{sat}}$

and subscripts l and v refer to saturated liquid and vapor, respectively.

The constant 0.62 is the average between the theoretical values of 0.512 for a stagnant liquid surrounding the vapor and 0.724 for the liquid moving at the same velocity as the vapor. This equation agrees with Bromley's experiments with carbon tetrachloride, nitrogen, and pentane. It does not correlate his results with water, however, and should be used with caution. The equation is also not applicable for very small or very large diameters, as the heat flux approaches infinity and zero as the diameter approaches zero and infinity, respectively.

The heat-transfer coefficient of Eq. (17-38) is that due only to conduction through the vapor film formed around the wire in stable film boiling. It is evident from Fig. 17-8 that this regime occurs only at high values of the excess temperature. For water at atmospheric pressure the minimum surface temperature is around 1000°R. Consequently, the heat transfer by radiation becomes significant in film boiling. Bromley recommends the empirical relation for the total boiling-heat-transfer coefficient

$$h_b = h_c\left(\frac{h_c}{h_b}\right)^{\frac{1}{3}} + h_r \tag{17-39}$$

Assuming the heated surface to be surrounded by a liquid of infinite extent (this makes the effective liquid emissivity unity), the radiative-heat-transfer coefficient is

$$h_r = \frac{\sigma\epsilon(T_w^4 - T_{\text{sat}}^4)}{T_w - T_{\text{sat}}} \tag{17-40}$$

Here σ is the Stefan-Boltzmann constant and ϵ is the surface emissivity.

Nucleate boiling with forced convection The method of estimating the heat flux in forced-convective nucleate-boiling flow in tubes and channels depends upon the regime, as defined in Fig. 17-11. Below the incipience of boiling, the single-phase correlation equations of Chaps. 12 and 13 are applicable. In the *partial-nucleate-boiling regime* in tubes and channels, Bergles and Rohsenow [4] suggest

$$q'' = q''_{\text{conv}}\left\{1 + \left[\frac{q''_B}{q''_{\text{conv}}}\left(1 - \frac{q''_{Bi}}{q''_B}\right)\right]^2\right\}^{\frac{1}{2}} \tag{17-41}$$

where q'' is the heat flux per unit area. The value of q''_B is found by the non-flow-boiling correlation [Eq. (17-34)]. The value of q''_{Bi}, the nonflow nucleate-boiling heat flux at the incipience of boiling, depends upon the ΔT_x at which boiling begins. This can be estimated by a complete construction of the non-flow-boiling curve.

In the fully developed nucleate-flow boiling regime of Fig. 17-11, the heat flux

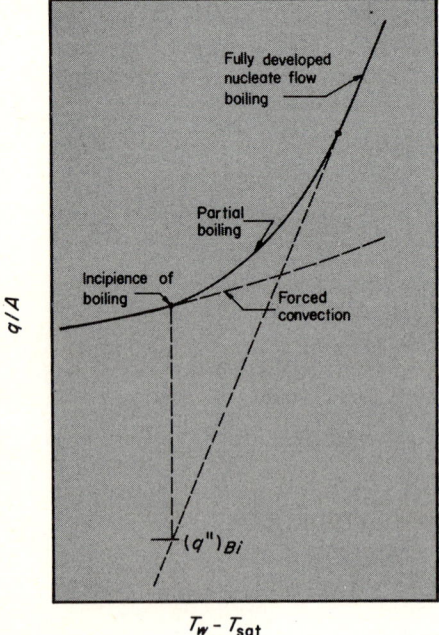

Fig. 17-11 Nucleate-flow boiling regimes. (*After A. E. Bergles and W. M. Rohsenow, ASME Paper 63-HT-22, 1963.*)

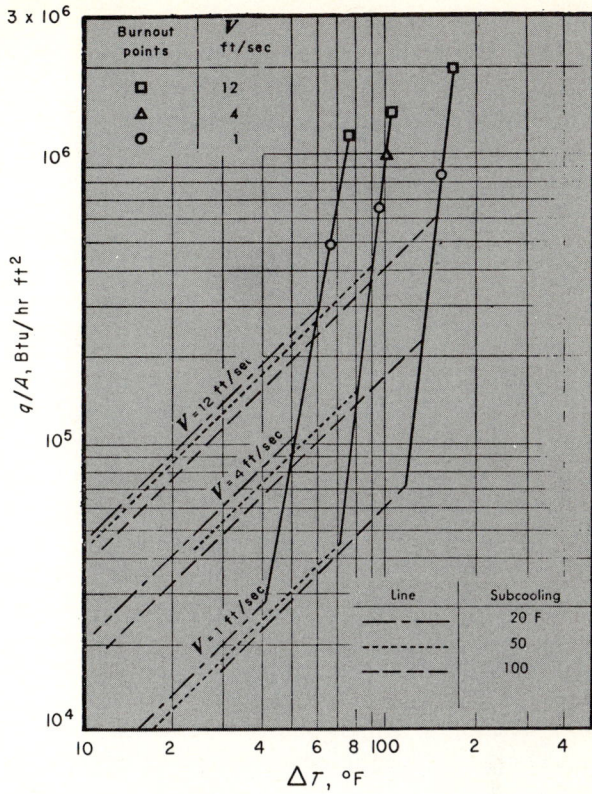

Fig. 17-12 Typical forced-convective subcooled boiling. (*Reprinted from W. H. McAdams et al., Ind. Eng. Chem.,* **41**: *1945. Copyright 1949 by the American Chemical Society. Reprinted by permission of the copyright owner.*)

is independent of velocity, but it depends upon pressure and temperature, as illustrated in Fig. 17-12. Here the heat flux for subcooled water at low pressures can be estimated by the equation from McAdams et al. [38]

$$\frac{q}{A} = 0.074 \Delta T_x^{3.86} \qquad 30 \leq p \leq 90 \text{ psia} \tag{17-42}$$

For higher pressures, the generalized correlation of Levy [34] can be approximated by

$$\frac{q}{A} = \frac{p^{\frac{4}{3}}}{495} \Delta T_x^3 \qquad 100 \leq p \leq 2{,}000 \text{ psia} \tag{17-43}$$

The peak heat flux in nucleate forced-convective boiling of water is given by the empirical equation of Mirshak et al., as reported by Rohsenow and Choi [52]. This is

$$\left(\frac{q}{A}\right)_{\max} = 480{,}000(1 + 0.0365V)(1 + 0.00508\Delta T_{sc})(1 + 0.0131p) \qquad (17\text{-}44)$$

where the terms, units, and recommended ranges of parameters are

$\left(\dfrac{q}{A}\right)_{\max}$ = burnout flux, Btu/hr ft^2
V = mass average velocity, 5 to 45 ft/sec
p = pressure, 25 to 85 psia
ΔT_{sc} = subcooling, 9 to 135°F

Their equation was found to correlate experimental burnout for flow inside tubes and external flow over heated metal strips with equivalent diameters from 0.21 to 0.46 in to within ±16 percent. For a correlation of the maximum heat flux obtainable with other fluids, the reader should consult Griffith [22], and for a complete treatment of this highly complicated subject the chapter on flow-boiling crisis in Tong [59] is recommended.

Forced convective vaporization Nucleate boiling with a high void fraction can progress into an annular flow with a very high velocity vapor core. The vapor velocity can be so great that the entire heat-transfer mechanism changes to one of conduction through the liquid along the wall, with vaporization occurring at the interface between this liquid annulus and the vapor core.

The many complications which must be considered with this flow-heat-transfer model are discussed in detail by Tong [59].

Film boiling with forced convection Bromley et al. [8] investigated laminar film boiling with forced flow on the outside of horizontal carbon tubes. Their experiments included benzene, carbon tetrachloride, ethanol, and n-heptane, tube diameters from 0.387 to 0.637 in, and free-stream velocities normal to the tubes from 0 to 14 ft/sec.

Their results show that the pool film-boiling equation (17-38) is applicable for $V_\infty \leq \sqrt{gD}$ and that

$$h_c = 2.7\left[\frac{V_\infty k_v \rho_v (h_{fg} + 0.4 C p_v \Delta T_x)}{D\,\Delta T_x}\right]^{\frac{1}{2}} \qquad (17\text{-}45)$$

for $V_\infty \leq 2\sqrt{gD}$. The effects of high pressure, subcooling, and high velocity (turbulent film boiling) are discussed by Tong [59].

17-7 CONDENSATION HEAT TRANSFER

In many industrial applications involving boiling, the inverse process, *condensation*, also occurs. Certainly in all closed-loop designs in which a liquid is vaporized,

MULTIPHASE PHENOMENA

it must be condensed. Such is the case in steam power plants and vapor-compression-cycle refrigeration equipment. Condensation is sometimes accomplished by bulk mixing of a vapor with its subcooled liquid but more frequently by exposing a vapor to a subcooled solid surface.

Whenever a superheated or saturated vapor touches a cooled surface (below the saturation temperature), condensate forms on the surface and flows in the direction dictated by the gravitational field. If the condensate does not wet the surface, individual droplets grow by coalescence and run down the surface. This is called *dropwise condensation*. If the liquid wets the surface, a smooth, continuous liquid film is formed which flows downward. This is called *film condensation*.

In dropwise condensation, much of the condenser surface is directly exposed to the vapor, whereas the liquid film in film condensation effectively blankets the surface. The film serves as a thermal resistance and thereby reduces the heat flux and the rate of condensation. Heat fluxes of the order of 250,000 Btu/hr-ft² have been obtained with dropwise condensation. Film-condensation heat fluxes are much lower and may be as low as one-tenth the value obtained with dropwise condensation under otherwise identical conditions.

Unfortunately most surfaces become wetted with use, and film condensation is the commoner of the two modes. Considerable effort has been directed toward surface treatments and fluid additives to promote dropwise condensation, but they have not had great success.

Laminar film condensation on plates This is an important practical case since most condensing surfaces are short and the film velocity is too low for transition to turbulent flow. Consequently, it has attracted numerous investigators.

Nusselt [44] presented an analysis for laminar condensation on a vertical flat plate in 1916. He made four major assumptions:

1. A linear temperature gradient in the liquid film
2. A uniform wall temperature T_w
3. Pure vapor at the saturation temperature T_{sat}
4. Negligible shear at the liquid-vapor interface, i.e., low velocity

Under these conditions a force balance on an element of fluid of unit depth in the film shown in Fig. 17-13 is, neglecting inertia terms,

$$\mu_l \frac{du}{dy} dx = g(\rho_l - \rho_v)(\delta - y) \, dx \tag{17-46}$$

The term on the left is the viscous shear at y, and the term on the right is simply the gravitational force acting downward. Integrating at a particular x location from $y = 0$ to $y = \delta$ and applying the no-slip boundary condition $u = 0$ at $y = 0$ yields

$$u = \frac{g(\rho_l - \rho_v)}{\mu_l}\left(\delta_y - \frac{y^2}{2}\right) \tag{17-47}$$

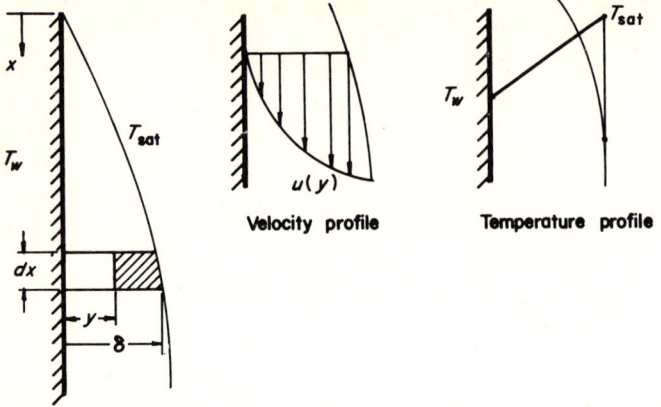

Fig. 17-13 Laminar film condensation on a vertical plate.

Using this velocity profile, the mass flow rate per unit width (perpendicular to the xy plane) is

$$\frac{\dot{m}}{W} = \int_0^\delta \rho_l u \, dy = \int_0^\delta \frac{\rho_l g (\rho_l - \rho_v)}{\mu_l} \left(\delta_y - \frac{y^2}{2} \right) dy = \frac{\rho_l (\rho_l - \rho_v) g \delta^3}{3 \mu_l} \quad (17\text{-}48)$$

From this, the rate of change of $\dot{m}/W$ with δ is

$$\frac{d(\dot{m}/W)}{d\delta} = \frac{g \rho_l (\rho_l - \rho_v)}{\mu_l} \delta^2 \quad (17\text{-}49)$$

The average enthalpy change which occurs as the vapor condenses and cools to the average temperature of the liquid film (which is subcooled) is

$$h'_{fg} = h_{fg} + \frac{1}{\dot{m}/W} \int_0^\delta \rho_l u c_p (T_{\text{sat}} - T) \, dy \quad (17\text{-}50)$$

Since a linear temperature profile was assumed,

$$T = T_w + \frac{y}{\delta} (T_{\text{sat}} - T_w)$$

and Eq. (17-50) integrates to

$$h'_{fg} = h_{fg} + \tfrac{3}{8} C_p (T_{\text{sat}} - T_w) \quad (17\text{-}51)$$

Assuming heat to be transferred through the condensate only by conduction and equating this to the enthalpy change of the condensing vapor per unit width yields

$$\frac{q}{A} = k_l \frac{T_{\text{sat}} - T_w}{\delta} = h'_{fg} \frac{d(\dot{m}/W)}{dx} \quad (17\text{-}52)$$

MULTIPHASE PHENOMENA

where h'_{fg} is defined by Eq. (17-51). Combining Eqs. (17-49) and (17-52) to eliminate $d(\dot{m}/W)$ results in

$$\delta^3 \, d\delta = \frac{k_l(T_{\text{sat}} - T_w)\mu_l}{g\rho_l(\rho_l - \rho_v)h'_{fg}} \, dx$$

Integration and application of the boundary condition $\delta = 0$ at $x = 0$ to eliminate the constant of integration gives the following expression for the film thickness:

$$\delta = \left[\frac{4k_l\mu_l x(T_{\text{sat}} - T_w)}{g\rho_l(\rho_l - \rho_v)h'_{fg}}\right]^{\frac{1}{4}} \tag{17-53}$$

The local heat-transfer coefficient is obtained from

$$h_x A(T_{\text{sat}} - T_w) = k_l A \frac{T_{\text{sat}} - T_w}{\delta}$$

and hence $h_x = k_l/\delta$ (as in laminar free-convective single-phase heat transfer). Thus,

$$h_x = \left[\frac{k_l^3 g \rho_l(\rho_l - \rho_v)h'_{fg}}{4\mu_l x(T_{\text{sat}} - T_w)}\right]^{\frac{1}{4}} \tag{17-54}$$

By the usual technique of averaging

$$\bar{h} = \frac{1}{L}\int_0^L h_x \, dx$$

which, since $h_x \propto x^{-\frac{1}{4}}$, gives $\bar{h} = \frac{4}{3}h_{x=L}$, or

$$\bar{h} = 0.943\left[\frac{k_l^3 g \rho_l(\rho_l - \rho_v)h'_{fg}}{\mu_l L(T_{\text{sat}} - T_w)}\right]^{\frac{1}{4}} \tag{17-55}$$

The Nusselt type of analysis can be readily extended to the case of a plate inclined at the angle ϕ to the horizontal. The result is

$$\bar{h} = 0.943\left[\frac{k_l^3 g \rho_l(\rho_l - \rho_v)h'_{fg} \sin\phi}{\mu_l L(T_{\text{sat}} - T_w)}\right]^{\frac{1}{4}} \tag{17-56}$$

In most cases with laminar film condensation on a plate, the experimental results are higher than predicted by Eqs. (17-55) or (17-56). This led McAdams [38] to suggest that the values be increased by 20 percent. Thus the recommended correlation equation is

$$\bar{h} = 1.13\left[\frac{k_l^3 g \rho_l(\rho_l - \rho_v)h'_{fg} \sin\phi}{\mu_l L(T_{\text{sat}} - T_w)}\right]^{\frac{1}{4}} \tag{17-57}$$

This result can also be applied to both the inside and outside surfaces of vertical cylinders, provided the diameter is large compared with the film thickness. It does not apply, however, if the cylinder is inclined.

An improved integral analysis of film condensation on a flat plate has been carried out by Rohsenow [51]. Boundary-layer analyses of this problem have been presented by Sparrow and Gregg [56] and Koh, Sparrow, and Hartnett [29]. For an outline of this work the reader should consult Rohsenow and Choi [52]. None of the theoretical analyses can be used to predict condensation rates for liquid metals with reliability.

Laminar film condensation on horizontal tubes The Nusselt type of analysis can be carried out for a horizontal circular tube, the details of which are presented by Jakob [25]. The resulting expression for the average coefficient is

$$\bar{h} = 0.725 \left[\frac{k_l^3 g \rho_l (\rho_l - \rho_v) h'_{fg}}{D \mu_l (T_{\text{sat}} - T_w)} \right]^{\frac{1}{4}} \tag{17-58}$$

In a vertical tier of n horizontal tubes, as frequently used in designs, the condensate from the upper horizontal tube flows directly onto the next tube, etc. This effectively increases the average liquid-film thickness and decreases the average heat-transfer coefficient. In this case Eq. (17-58) is modified by replacing D with nD, where n is the number of tubes:

$$\bar{h} = 0.725 \left[\frac{k_l^3 g \rho_l (\rho_l - \rho_v) h'_{fg}}{n D \mu_l (T_{\text{sat}} - T_w)} \right]^{\frac{1}{4}} \tag{17-59}$$

This should be considered as a preliminary estimate only, since splashing and increased velocity may render each tube just as effective as the upper one. In practice, empirical correlations are frequently used for such designs.

An improved analysis for a single cylinder which accounts for subcooling of the liquid film is presented by Chen [10].

Turbulent condensation Turbulent flow of the liquid film occurs on vertical or inclined plates and vertical cylinders if the Reynolds number exceeds a critical value of 1,800. This occurs very infrequently with horizontal cylinders due to the short path of the flowing condensate. If the length is sufficient, the film thickness and instabilities result in turbulent motion quite analogous to that experienced in single-phase boundary-layer flow. The turbulent motion provides an additional mechanism of energy transport through the film and results in an improved heat-transfer coefficient.

The customary practice in condensation is to define the Reynolds number of the condensate film using the equivalent hydraulic diameter,

$$\mathbf{Re}_f = \frac{D_h \rho_l V}{\mu_l} = \frac{4 A \rho_l V}{P \mu_l}$$

where D_h is defined by Eq. (12-49). In terms of the mass flow rate $\dot{m}$ this is

$$\mathbf{Re}_f = \frac{4\dot{m}}{P \mu_l} \tag{17-60}$$

where $\dot{m}$ is $\rho_l A V$, A being the area of condensate flow and V the average condensate velocity over the area. This is frequently expressed for a flat plate in terms of the mass flow per unit width at the bottom of the surface, $\dot{m}/W = \Gamma$, as

$$\mathbf{Re}_f = \frac{4\Gamma}{\mu_l} \tag{17-61}$$

since the wetted surface perimeter is the width W for a plate. For a vertical cylinder, the film Reynolds number is computed with the wetted perimeter equal to πD, the inside tube diameter being used for condensation on the inside surface, etc.

The average heat-transfer coefficient for turbulent film condensation on vertical surfaces is given by Kirkbride [26] as

$$\bar{h} = 0.0076 \, \mathbf{Re}_f^{0.4} \left[\frac{k_l^3 \rho_l (\rho_l - \rho_v) g}{\mu_l^2} \right]^{\frac{1}{3}} \tag{17-62}$$

for $\mathbf{Re}_f > 1{,}800$.

Determination of film Reynolds number The first step in determining the heat flux in film condensation is the determination of the film flow regime. Since the condensate velocity depends upon the average heat-transfer coefficient, a trial-and-error approach is required. The approach to a problem is then to assume either laminar or turbulent flow, calculate the heat-transfer coefficient with the appropriate correlation equation, calculate the resulting flow rate, and then check the film Reynolds number to ascertain that the correct flow regime was selected initially.

For convenience in carrying out the calculations, the Reynolds number may be expressed as a function of the average heat-transfer coefficient as follows. The total heat transfer from a surface of area A is

$$q = \dot{m} h'_{fg} = \bar{h} A (T_{\text{sat}} - T_w)$$

which can be arranged to give

$$\dot{m} = \frac{\bar{h} A (T_{\text{sat}} - T_w)}{h'_{fg}}$$

Combining this with Eq. (17-60) results in

$$\mathbf{Re}_f = \frac{4 \bar{h} A (T_{\text{sat}} - T_w)}{P \mu_l h'_{fg}} \tag{17-63}$$

where P is the wetted perimeter (which is simply the width for a flat plate) and A is the total area of the surface.

Example A vertical plate 1 ft high is maintained at 188°F and exposed to saturated steam at atmospheric pressure. Calculate the heat-transfer and condensation rate per hour per foot of plate width.

Solution The liquid properties are evaluated at the film temperature, $T_f = (212 + 188)/2 = 200°F$. Thus,

$\rho_l = 60.1 \text{ lb}_m/\text{ft}^3$ \qquad $k_l = 0.394 \text{ Btu/hr-ft-°F}$

$\mu_l = 0.205 \times 10^{-3} \text{ lb}_m/\text{ft-sec}$ \qquad $c_{p_l} = 1.00 \text{ Btu/lb}_m\text{-°F}$

The vapor density is $\rho_v = 0.0372 \text{ lb}_m/\text{ft}^3$ at saturation conditions, and $h_{fg} = 970.3 \text{ Btu/lb}_m$. We shall assume the condensate flow to be laminar and use Eq. (17-57) with $\sin \phi = 1.0$. For this problem

$$\rho_l(\rho_l - \rho_v) \simeq \rho_l^2 = 60.1^2 = 3{,}615 \; (\text{lb}_m/\text{ft}^3)^2$$

and

$$h'_{fg} = 970.3 + \tfrac{3}{8}(1.00)(212 - 188) = 979.3 \text{ Btu/lb}_m$$

Then

$$h = 1.13 \left[\frac{\left(0.394 \dfrac{\text{Btu}}{\text{hr-ft-°F}}\right)^3 \left(32.17 \dfrac{\text{ft}}{\text{sec}^2}\right)(3{,}615)\left(\dfrac{\text{lb}_m}{\text{ft}^3}\right)^2 \left(979.3 \dfrac{\text{Btu}}{\text{lb}_m}\right)}{\left(0.205 \times 10^{-3} \dfrac{\text{lb}_m}{\text{ft-sec}}\right)(1 \text{ ft})(24°\text{F})\left(\dfrac{\text{hr}}{3{,}600 \text{ sec}}\right)} \right]^{\frac{1}{4}}$$

$$= (5.16 \times 10^{12})^{\frac{1}{4}} = 1508 \text{ Btu/hr-ft}^2\text{-°F}$$

Checking the Reynolds number by Eq. (17-63) with $A = LW = 1 \text{ ft} \times W$ and $P = W$,

$$\text{Re}_f = \frac{4(1508 \text{ Btu/hr-ft}^2\text{-°F})(1 \text{ ft} \times W)(24°\text{F})}{W(0.205 \times 10^{-3} \text{ lb}_m/\text{ft-sec})(3{,}600 \text{ sec/hr})(979.3 \text{ Btu/lb}_m)} = 200$$

and the laminar-flow assumption was correct. The heat-transfer rate per unit width of the plate is

$$q = hA(T_{\text{sat}} - T_w) = (1508 \text{ Btu/hr-ft}^2\text{-°F}^2)(1 \times 1 \text{ ft}^2)(24°\text{F})$$

$$= 36{,}200 \text{ Btu/hr}$$

and the rate of mass condensation per unit width is

$$\dot{m} = \frac{q}{h'_{fg}} = \frac{36{,}200 \text{ Btu/hr}}{979.3 \text{ Btu/lb}_m} = 37.4 \text{ lb}_m/\text{hr}$$

In the preceding analyses and correlation equations, negligible vapor velocity at the film-vapor interface is assumed. There are many applications where this is not so. Of particular importance is the case of high vapor flow inside vertical tubes, which introduces a shear stress at the interface. Downward flow increases the film velocity, thins the film, and increases the heat-transfer coefficient. Upward flow retards the film velocity and has the opposite effect of downward flow on film thickness and heat-transfer coefficient. Jakob [25] compares analytical solutions with experimental results for a range of velocity and ΔT_x.

Another complication arises whenever the vapor is above the saturation temperature. The mechanism for film condensation of superheated vapor is not the same as for saturated vapor; nevertheless, the previously presented equations can be used with reasonable accuracy. McAdams [37] cites an experimental example where the

MULTIPHASE PHENOMENA

measured q/A for condensation of steam vapor superheated 180°F was only 3 percent more than that for saturated steam at the same pressure and wall temperature. It is emphasized that the actual temperature of the vapor is not to be used in such calculations; the appropriate temperature difference is still $T_{\text{sat}} - T_w$.

The effect of a noncondensable gas in the vapor can be to reduce the rate of condensation significantly. For more information on this subject the reader should consult Jakob [25] or McAdams [37].

Dropwise condensation As previously stated, the heat flux in dropwise condensation may be as high as 10 times that in film condensation under otherwise identical conditions. It rarely occurs, however, unless *promoters* are used to prevent wetting of the surface. Drew, Nagle, and Smith [16] conclude that film-type condensation always results when clean steam condenses on uncontaminated surfaces, both smooth and rough, regardless of the presence of noncondensing gases.

Many contaminants can promote nonwetting and result in dropwise condensation, but only those absorbed by the surface are effective for a significant length of time. Some promoters are oleic acid for copper, brass, nickel, and chromium; benzyl mercaptan and octyl thiocyanate for copper; fats and waxes for most metals (but these last two are not very long-lasting). Fats and waxes, as well as silicones and Teflon coatings, are especially good when the fluid is water.

17-8 CONDUCTION WITH PHASE CHANGE

Change of phase occurs in many practical problems, including freezing and thawing of foods and soils, solidification of metal castings, welding, and ice manufacture, to name a few. The object of analyses of these problems is to predict the rate of phase change, i.e., the rate of solidification, freezing, or melting, and to determine the heat flux necessary to accomplish the resulting phase-change rate.

For the general case of transient conduction in a solid with constant properties and no internal energy generation, Eq. (5-9) reduces to

$$\frac{\partial T}{\partial t} = \alpha \, \nabla^2 T \qquad (17\text{-}64)$$

Stefan [57] first discussed solving the one-dimensional form of this equation with a moving boundary. He was concerned with the rate of ice formation in polar seas, and the general conduction problem with a moving boundary is referred to as *Stefan's problem* in the literature.

Actually the first exact solution was determined by Franz Neumann and presented in his lectures in the 1860s. They were not published, however, until 1912 [49]. Neumann's solution was for the case illustrated in Fig. 17-14, where the water is a semi-infinite body initially at a uniform temperature T_i. T_i is above the freezing temperature T_{fr}, and the transient problem is assumed to begin with the

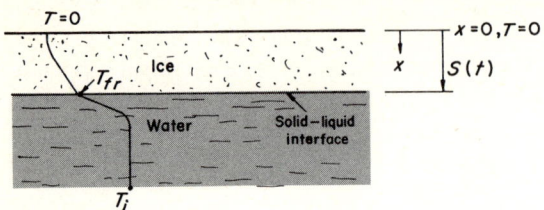

Fig. 17-14 Model for ice formation in a semi-infinite liquid.

surface temperature at $x = 0$ suddenly being lowered to zero temperature[1] and maintained at that value for all $t > 0$.

The solid-liquid interface is at the melting-freezing temperature T_{fr}. This is higher than the temperature in the ice and lower than that in the liquid. Consequently an energy balance at the solid-liquid interface with change of phase occurring is

Energy conducted away through ice = energy removal for solidification + energy conducted in through water

or per unit area this is

$$k_s \frac{\partial T_s}{\partial x} = \rho L \frac{dS}{dx} + k_l \frac{\partial T_l}{\partial x} \tag{17-65}$$

where the subscripts s and l refer to the solid and liquid phases, respectively. L is the latent heat of solidification.

The differential equations for the temperature distributions in the solid and the liquid are

Solid: $$\frac{\partial T_s}{\partial t} = \alpha_s \frac{\partial^2 T_s}{\partial x^2} \tag{17-66}$$

Liquid: $$\frac{\partial T_l}{\partial t} = \alpha_l \frac{\partial^2 T_l}{\partial x^2} \tag{17-67}$$

The complete Neumann problem then is to determine the temperature distributions $T_s(x)$ and $T_l(x)$ which satisfy Eqs. (17-65) to (17-67) subject to the boundary conditions

$$T_s(0,t) = 0 \tag{17-68}$$
$$T_l(\infty,t) = T_i \tag{17-69}$$
$$T_s = T_l = T_{fr} \quad \text{at } x = S(t) \tag{17-70}$$

---

[1] The value of zero can be replaced by any suitable temperature T_{ref} below the freezing point by simply defining $T = T(°F) - T_{\text{ref}}(°F)$.

and the initial condition $T(x) = T_i$ for all $t < 0$. The details of the mathematical solution are given by Carslaw and Jaeger [9]. For the solution to satisfy the conditions for all time, the interface location must be given by

$$S(t) = 2\lambda\sqrt{\alpha_s t} \tag{17-71}$$

where the proportionality constant λ is determined from the transcendental equation

$$\frac{e^{-\lambda^2}}{\operatorname{erf}\lambda} - \frac{k_l\sqrt{\alpha_s}(T_i - T_{fr})e^{-(\alpha_s/\alpha_l)\lambda^2}}{k\sqrt{\alpha_l}T_{fr}(\operatorname{erfc}\lambda)(\alpha_s/\alpha_l)^{\frac{1}{2}}} = \frac{\lambda L\sqrt{\pi}}{(c_p)_s T_{fr}} \tag{17-72}$$

Here the complementary error function is

$$\operatorname{erfc} x = 1 - \operatorname{erf} x$$

(the error function is discussed in Chap. 6). Equation (17-71), with λ as given by (17-72), is the solution for the freezing rate $S(t)$.

The temperature profiles are given by

$$T_l = T_i - \frac{T_i - T_{fr}}{(\operatorname{erfc}\lambda)(\alpha_s/\alpha_l)^{\frac{1}{2}}}\operatorname{erfc}\frac{x}{2(\alpha_l t)^{\frac{1}{2}}} \tag{17-73}$$

and

$$T_s = \frac{T_{fr}}{\operatorname{erf}\lambda}\operatorname{erf}\frac{x}{2(\alpha_s t)^{\frac{1}{2}}} \tag{17-74}$$

Other "exact" mathematical solutions for this class of problem include melting of a semi-infinite solid, phase change between a semi-infinite solid at zero temperature and a semi-infinite liquid at a constant temperature other than zero, and the extension of these to melting or solidification of alloys where the phase change occurs over a finite temperature range. They are discussed by Carslaw and Jaeger [9].

The problem illustrated in Fig. 17-14 utilized a boundary condition which is quite unrealistic; how can the surface of a semi-infinite body of water be brought to an arbitrary temperature of zero and held at that value? The more realistic problem involves a convective heat transfer at the upper surface of the solid, but this boundary condition has prevented an exact solution to the partial differential equations for the temperature.

An approximate solution for the freezing problem as shown in Fig. 17-15 has been presented by London and Seban [36]. The simplest case is when the liquid is initially at the freezing temperature T_{fr} and there is no resulting conduction in the liquid. Neglecting the heat capacity of the ice (which is the major approximation) and treating $\bar{h}_0$, T_∞, and the ice properties as constants, the heat flux is

$$\frac{q}{A} = \frac{T_{fr} - T_\infty}{1/\bar{h}_0 + S/k_s} \tag{17-75}$$

Under these conditions there is no heat removal due to subcooling of the ice (and

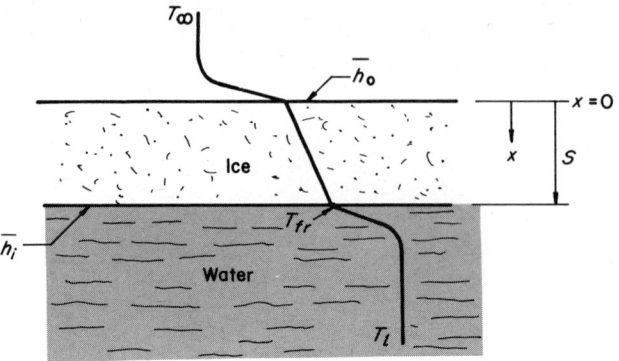

Fig. 17-15 Freezing of water with a convective boundary condition.

clearly none conducted through the liquid) so that this heat flux must originate due to freezing at the interface. This is

$$\frac{q}{A} = \rho_s L \frac{dS}{dt} \qquad (17\text{-}76)$$

where dS/dt is the volume rate of solidification per unit area. Equating (17-75) and (17-76) and separating variables yields

$$\left(\frac{1}{\bar{h}_0} + \frac{S}{k_s}\right) dS = \frac{T_{fr} - T_\infty}{\rho_s L} dt \qquad (17\text{-}77)$$

The parameters of this equation can be nondimensionalized by setting

$$S^+ = \frac{\bar{h}_0}{k_s} S \qquad (17\text{-}78)$$

and

$$t^+ = \bar{h}_0{}^2 \left(\frac{T_{fr} - T_\infty}{\rho_s L k_s}\right) t \qquad (17\text{-}79)$$

Substituting these into Eq. (17-27) gives

$$(1 + S^+)\, dS^+ = dt^+ \qquad (17\text{-}80)$$

Integrating and applying the condition that $S = S^+ = 0$ at $t = t^+ = 0$ results in

$$S^+ = (1 + 2t^+)^{\frac{1}{2}} - 1 \qquad (17\text{-}81)$$

which gives the location of the interface as a function of time.

The commonest problem involves freezing of a liquid initially at a temperature $T_l > T_{fr}$. In this case there is heat transfer from the liquid to the phase-change

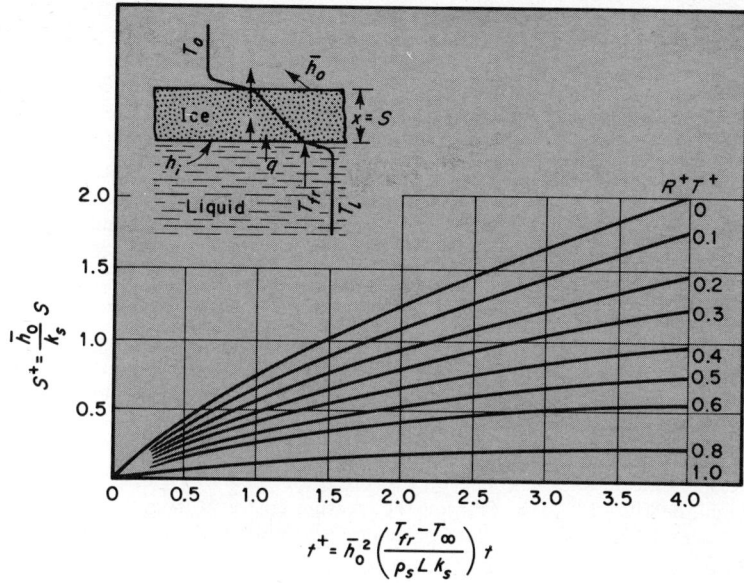

Fig. 17-16 Solidification of an ice slab. [*Adapted from A. L. London and R. A. Seban, Trans. ASME*, **64**:777 (*1943*).]

interface at $x = S$. For the case where the heat-transfer coefficient at the solid-liquid interface is $\bar{h}_i$, London and Seban [36] obtained the solution

$$t^+ = \frac{S^+}{R^+T^+} - \frac{1}{(R^+T^+)^2} \ln\left(1 + \frac{R^+T^+S^+}{1 + R^+T^+}\right) \tag{17-82}$$

where

$$R^+ = \frac{\bar{h}_i}{\bar{h}_0} \tag{17-83}$$

$$T^+ = \frac{T_l - T_{fr}}{T_{fr} - T_\infty} \tag{17-84}$$

and t^+ and S^+ are as defined by Eqs. (17-78) and (17-79). This result is plotted for a range of the appropriate parameters in Fig. 17-16. London and Seban also present approximate solutions in graphical form for ice formation inside and outside of cylinders.

Several other methods of solution have been applied to phase-change problems, including variational calculus, heat-balance integral, and contour-integral techniques, in addition to numerical and analog methods. A thorough review of the literature in this field through 1965 is presented by Muehlbauer and Sunderland [42].

PROBLEMS

17-1. Using the differential form of the continuity equation (10-25) and the assumptions given in Sec. 17-1, obtain the continuity equation in spherical coordinates [Eq. (17-1)] for a bubble in a liquid.

17-2. In an isothermal laboratory experiment utilizing water at 60°F in an open container, spherical vapor bubbles of approximately $\frac{1}{16}$-in diameter are observed at a fluid depth of 1 ft. The laboratory barometer reading is 29.25 ± 0.01 cm Hg. Determine the vapor pressure inside the bubble, assuming static conditions and no substance other than water vapor in the bubble. Is the effect of the surface tension negligible? Is the fluid head of 1 ft negligible? (Consider the accuracy of the barometer.)

17-3. Determine the critical radius of a spherical vapor bubble in n-butyl alcohol at 68°F for a liquid pressure of 1.2 psia. The vapor pressure is 1.94 psia.

17-4. A petroleum product with specific gravity of 0.55 and a vapor pressure of 3 psia is pumped through a 6-in-diam line. A venturi flowmeter with a 4-in-diam throat is located in the line. The pressure just upstream from the venturi is 30 psia. Estimate the flow rate in gallons per minute that the system can sustain without the onset of cavitation.

17-5. An in-flight aircraft-refueling system incorporates a venturi-type pressure regulator with a 1.5-in-diam throat in a 3-in-diam fuel line. For flow of JP-4 fuel with a vapor pressure of 2.0 psia and a density of 47.5 lb_m/ft^3, plot the pressure upstream from the venturi to prevent cavitation as a function of flow rate in gallons per minute. Consider the flow range from 200 to 600 gal/min.

17-6. A saturated mixture of steam and water at 1,000 psia is flowing through a pipe at a flow rate per unit area of 100,000 lb_m/hr-ft^2. The flow quality is 10 percent. Determine the probable flow regime.

17-7. A mixture of air and water at 90°F and 14.7 psia flows through a horizontal pipe. The liquid flow rate is 2,000 gal/hr, and the gas flow rate is 250 ft^3/hr (measured at flow conditions). Using Baker's data, determine the flow pattern which will probably exist for (*a*) a 3-in-ID pipe and (*b*) a 6-in-ID pipe.

17-8. A saturated mixture of steam and water at 30 psia is flowing in an insulated horizontal pipe. The flow quality is 4 percent, and the void fraction is 50 percent. If the total flow rate per unit area of pipe is 35,000 lb_m/hr-ft^2, determine the slip ratio and the probable flow pattern.

17-9. A 2,000-ft length of a 4-in-ID commercial steel piping is carrying a saturated steam-water mixture which is at 450 psia and has a flow quality of 6 percent. The total flow rate per unit area is 50,000 lb_m/hr-ft^2. Calculate the pressure drop.

17-10. A saturated steam-water mixture at 300 psia and with flow quality of 5 percent is flowing in a 4-in-ID commercial steel pipe 2,500 ft long. The total flow rate per unit area is 75,000 lb_m/hr-ft^2. Calculate the pressure drop due to friction.

17-11. A saturated steam-water mixture at 2,500 psia is flowing through a 1,500-ft length of 3-in-ID wrought iron pipe. If the flow quality is 4 percent and the total flow rate per unit area is 2,000,000 lb_m/hr-ft^2, determine the pressure drop.

17-12. In an experiment with bulk-fluid boiling of water at standard atmospheric pressure, spherical vapor bubbles of $\frac{1}{8}$-in diameter are observed. Assuming pure water vapor in the bubble and vapor pressure that is the same as the atmospheric pressure, 14.696 psia, calculate the temperature of the vapor by Eq. (17-33).

17-13. For the water-vapor bubble of Prob. 17-12, use the external-flow correlation Eq. (13-127) to estimate the average heat-transfer coefficient $\bar{h}$. The bubble rise velocity is 10 ft/sec. Assuming the bubble diameter to remain constant, how much will the vapor temperature drop while the bubble moves 1 ft?

17-14. Estimate the heat flux which would occur in nucleate boiling of saturated water at 400°F with a platinum heater at 425°F.

17-15. Estimate the surface temperature of a brass heater required to maintain a heat flux of 650,000 Btu/hr-ft² in nucleate boiling of water at 700 psia.

17-16. Estimate the heat flux to water boiling on a very small copper wire, assuming nucleate pool boiling. The conditions are atmospheric pressure and an excess temperature of 40°F.

17-17 Water is boiling on a 0.005-in-diam nickel wire at 1 atm pressure and 20°F excess temperature. Estimate the heat flux.

17-18. Determine the peak heat flux for nucleate boiling of water at 1 standard atmosphere pressure and standard gravity by the Zuber correlation.

17-19. Repeat Prob. 17-18 using (*a*) the Kutateladze equation; (*b*) the Rohsenow and Griffith equation.

17-20. Approximate the heat flux in film boiling from a 0.40-in-diam tube in saturated water at atmospheric pressure. The surface temperature of the tube is 500°F. Use Bromley's equation, neglecting radiation.

17-21. A vertical plate, 18 in high and maintained at 88°F, is exposed to saturated steam at atmospheric pressure. Determine the rate of heat transfer and the condensate rate per hour per foot of plate width for film condensation.

17-22. For the situation of Prob. 17-21, estimate the height of wall necessary for turbulent condensate flow.

17-23. Saturated ammonia at 10°F condenses on a plate 14 in long inclined 75° to the horizontal and maintained at −10°F. Determine the rate of heat transfer to the wall and the rate of condensate per foot of plate width. The following properties of ammonia at 0°F may be used:

$h_{fg} = 611.8$ Btu/lb$_m$ $v_l = 0.02419$ ft³/lb$_m$

$c_{pl} = 1.085$ Btu/lb$_m$-°F $v_v = 9.116$ ft³/lb$_m$

17-24. Determine the effective heat-transfer coefficient for saturated Freon-12 at 140°F experiencing film condensation on a ½-in-OD tube held at 80°F by an internal flow of cool air. The Freon properties at the film temperature are

$h_{fg} = 54.313$ Btu/lb$_m$ $v_l = 0.012924$ ft³/lb$_m$

$c_{pl} = 0.2456$ Btu/lb$_m$-°F $v_v = 0.26769$ ft³/lb$_m$

17-25. Repeat Prob. 17-24 for saturated Freon-12 vapor flowing over a bank of ½-in-OD tubes, 10 in each column and 12 in each row. If the tubes are 2 ft long, estimate the total rate of condensation.

REFERENCES

1. Addoms, J. N.: "Heat Transfer at High Rates to Water Boiling outside Cylinders," D.Sc. thesis, Chemical Engineering Department, Massachusetts Institute of Technology, June, 1948; as reported in Rohsenow and Choi [52].
2. Baker, O.: Simultaneous Flow of Oil and Gas, *Oil Gas J.*, **53**: 185 (1954).
3. Berenson, P.: Transition Boiling Heat Transfer from a Horizontal Surface, *J. Heat Transfer*, **83**: 351 (1961).
4. Bergles, A. E., and W. M. Rohsenow: The Determination of Forced-convection Surface-boiling Heat Transfer, *ASME Paper* 63-HT-22, 1963.
5. Besant, W.: "Hydrostatics and Hydrodynamics," Cambridge, 1859.
6. Borishanski, V. M.: An Equation Generalizing Experimental Data on the Cessation of Bubble Boiling in a Large Volume of Liquid, *Zh. Tekhn. Fiz.*, **25**: 252 (1956).
7. Bromley, L. A.: Heat Transfer in Stable Film Boiling, *Chem. Eng. Progr.*, **46**: 221 (1950).

8. Bromley, L. A., N. LeRoy, and J. A. Robbers: Heat Transfer in Forced Convective Film Boiling, *Ind. Eng. Chem.*, **45**: 2639 (1953).
9. Carslaw, H. S., and J. C. Jaeger: "Conduction of Heat in Solids," 2d ed., Oxford University Press, London, 1959.
10. Chen, M. M.: An Analytical Study of Laminar Film Condensation, II, Single and Multiple Horizontal Tubes, *J. Heat Transfer*, **83**: 48 (1961).
11. Cho, S. M., and R. A. Seban: On Some Aspects of Steam Bubble Collapse, *J. Heat Transfer*, **91**: 537 (1969).
12. Cichelli, M. T., and C. F. Bonilla: *Trans. AIChE*, **41**: 755 (1945).
13. Cryder, D. S., and A. C. Finalborgo: *Trans. AIChE*, **33**: 346 (1947).
14. Csanady, G. T.: "Theory of Turbomachines," McGraw-Hill, New York, 1964.
15. Daily, J. W., and V. E. Johnson: *Trans. ASME*, **78**: 1695 (1956).
16. Drew, T. B., W. M. Nagle, and W. Q. Smith: The Conditions for Dropwise Condensation of Steam, *Trans. AIChE*, **31**: 605 (1935).
17. Florschuetz, L. W., and B. T. Chao: On the Mechanics of Vapor Bubble Collapse, *J. Heat Transfer*, **87**: 209 (1965).
18. Forster, H. K., and N. Zuber: Growth of a Vapor Bubble in a Superheated Liquid, *J. Appl. Phys.*, **25**: 474 (1954).
19. Forster, H. K., and N. Zuber: Dynamics of Vapor Bubbles and Boiling Heat Transfer, *AIChE J.*, **1**: 531 (1955).
20. Forster, K., and R. Grief: Heat Transfer to a Boiling Liquid: Mechanism and Correlations, *J. Heat Transfer*, **81**: 43 (1959).
21. Gilmour, C. H.: *Chem. Eng. Progr.*, **54**: 77 (1958).
22. Griffith, P.: Correlation of Nucleate Boiling Burnout Data, *ASME Paper* 57-HT-21, 1957.
23. Gunther, F., and F. Kreith: Photographic Study of Bubble Formation in Heat Transfer to Subcooled Water, *Calif. Inst. Tech., JPL Progr. Rept.* 4-120, 1950.
24. Harvey, E. N., W. D. McElroy, and A. H. Whiteley: *J. Appl. Phys.*, **18**: 162 (1947).
25. Jakob, M.: "Heat Transfer," vol. I, Wiley, New York, 1949.
26. Kirkbride, C. G.: Heat Transfer by Condensing Vapors on Vertical Tubes, *Trans. AIChE*, **30**: 170 (1934).
27. Knapp, R. T.: *Proc. Inst. Mech. Engrs. (London)*, **166**: 150 (1952).
28. Knapp, R. T.: *Trans. ASME*, **77**: 1045 (1955).
29. Koh, J. C. Y., E. M. Sparrow, and J. P. Hartnett: *Intern. J. Heat Mass Transfer*, **2**: 69 (1961).
30. Kozlov, B. K.: Forms of Flow of Gas-Liquid Mixtures and Their Stability Limits in Vertical Tubes, *Ass. Tech. Serv. Trans.* 136R of *Zh. Tekh. Fiz.*, **24**: 2285 (1954).
31. Kutateladze, S. S.: Heat Transfer in Condensation and Boiling, *USAEC Rept.* AEC-tr-3770, 1952.
32. Lang, C.: *Trans. Inst. Engrs. Shipbuilders (Scotland)*, **32**: 279 (1888).
33. Leidenfrost, J. G.: "De aquae communis nonnullis qualitatibus tractatus," Duisburg, 1756.
34. Levy, S.: Generalized Correlation of Boiling Heat Transfer, *J. Heat Transfer*, **81**: 37 (1959).
35. Lockhart, R. W., and R. C. Martinelli: Proposed Correlation of Data for Isothermal Two-phase Two-component Flow in Pipes, *Chem. Eng. Progr.*, **45**: 39 (1949).
36. London, A. L., and R. A. Seban: Rate of Ice Formation, *Trans. ASME*, **64**: 771 (1943).
37. McAdams, W. H.: "Heat Transmission," 3d ed., McGraw-Hill, New York, 1954.
38. McAdams, W. H., W. E. Kennel, C. S. Minden, C. Rudolf, C. Picornell, and J. E. Dow: Heat Transfer at High Rates to Water with Surface Boiling, *Ind. Eng. Chem.*, **41**: 1945 (1949).
39. Merte, H., and J. A. Clark: A Study of Pool Boiling in an Accelerating System, *J. Heat Transfer*, **83**: 233 (1961).
40. Moissis, R., and P. J. Berenson: On the Hydrodynamic Transitions in Nucleate Boiling, *J. Heat Transfer*, **85**: 221 (1963).

41. Moore, F. D., and R. B. Mesler: The Measurement of Rapid Surface Temperature Fluctuations during Nucleate Boiling of Water, *AIChE J.*, **7**: 620 (1961).
42. Muehlbauer, J. C., and J. E. Sunderland: Heat Conduction with Freezing or Melting, *Appl. Mech. Rev.*, **18**: 951 (1965).
43. Nukiyama, S.: Maximum and Minimum Values of Heat Transmitted from Metal to Boiling Water under Atmospheric Pressure, *J. Soc. Mech. Engrs. Japan*, **37**: 367 (1934).
44. Nusselt, W.: Die Oberflachenkondensation des Wasserdampfes, *Z. VDI*, **B60**: 541 (1916).
45. Owens, W. L.: Two Phase Pressure Gradient, *Intern. Devel. Heat Transfer*, Pt. II, p. 363, 1961.
46. Piret, E. L., and H. S. Isbin: Two-phase Heat Transfer in Natural Circulation Evaporators, *AIChE Heat Transfer Symp. St. Louis, Dec. 13–16, 1953.*
47. Plesset, M. S., and S. A. Zwick: On the Dynamics of Small Vapor Bubbles in Liquids, *J. Math. Phys.*, **33**: 308 (1955).
48. Rayleigh, Lord: On the Pressure Developed in a Liquid during the Collapse of a Spherical Cavity, *Phil. Mag.*, **34**: 94 (1917).
49. Riemann, G. F. B., and H. Weber: "Die partiellen Differentialgleichungen der mathematischen Physik," vol. 2, p. 121, 1912.
50. Rohsenow, W. M.: A Method of Correlating Heat Transfer Data for Surface Boiling Liquids, *Trans. ASME*, **74**: 969 (1952).
51. Rohsenow, W. M.: *Trans. ASME*, **78**: 1645 (1956).
52. Rohsenow, W. M., and H. Y. Choi: "Heat, Mass, and Momentum Transfer," Prentice-Hall, Englewood Cliffs, N.J., 1961.
53. Rohsenow, W. M. and J. A. Clark: Heat Transfer and Pressure Drop Data for High Heat Flux Densities to Water at High Sub-critical Pressures, *Fluid Mech. Heat Transfer Inst.*, Stanford, 1951.
54. Rohsenow, W. M., and P. Griffith: Correlation of Maximum Heat Flux Data for Boiling of Saturated Liquids, *AIChE-ASME Heat Trans. Symp. Louisville, Ky.*, 1955.
55. Snyder, N. W., and T. T. Robin: Mass-transfer Model in Subcooled Nucleate Boiling, *ASME Paper* 68-HT-51, *10th Natl. Heat Transfer Conf. Philadelphia, Aug.* 11, *1968*.
56. Sparrow, E. M., and J. L. Gregg: *J. Heat Transfer*, **81**: 13 (1959).
57. Stefan, J.: Über die Theorie der Eisbildung, insbesondere über der Eisbildung in Polarmäre, *Ann. Phys. Chem.*, **42**: 269 (1891).
58. Sun, K., and J. H. Lienhard: The Peak Pool Boiling Heat Flux on Horizontal Cylinders, *Univ. Kentucky Off. Res. Engrg. Serv. Bull.* 88, 1969.
59. Tong, L. S.: "Boiling Heat Transfer and Two-phase Flow," Wiley, New York (1965).
60. Wittke, D. D., and B. T. Chao: Collapse of Vapor Bubbles with Translatory Motion, *J. Heat Transfer*, **89**: 17 (1967).
61. Zuber, N.: On Stability of Boiling Heat Transfer, *Trans. ASME*, **80**: 711 (1958).
62. Zuber, N., and M. Tribus: Further Remarks on the Stability of Boiling Heat Transfer, *Univ. California, Los Angeles, Dept. Engrg. Rept.* 58-5, 1958.

PART IV

SPECIAL TOPICS AND APPLICATIONS

CHAPTER 18

THERMAL ANALYSIS OF HEAT EXCHANGERS

18-1 INTRODUCTION

A prime objective in the study of heat transfer is the thermal analysis of heat exchangers. These devices are widely encountered, and their importance justifies an extensive treatment.

The simplest type of heat exchanger is a device in which a hot and a cold fluid are directly mixed. An example is the *open* feedwater heater of a steam power plant. Here, a simple thermodynamic first-law analysis suffices since the exit fluid is at a uniform condition; the heat loss by the hotter fluid is equal to the gain by the cooler fluid.

Perhaps the most frequently used heat exchanger is the type where the hot and cold fluids are separated by a partition through which heat, but not matter, flows. Called *recuperators*, they range in design from simple plate-type recuperators, with the two fluids separated by a flat plate or tube-within-a-tube designs, to very complex systems involving multiple passes and baffles. The purpose of multiple passes and baffles is to increase the efficiency or the heat-transfer rate for a given package size.

A complete treatment of heat exchanger design would include

1. Thermal analysis
2. Structural analysis
3. Manufacturing considerations
4. Cost analysis
5. Size and pressure-drop considerations

Thermal analysis is the subject of the present chapter, but every engineer should be aware of these other important design factors. For many large recuperators, structural design must comply with the ASME Code for Unfired Pressure Vessels. For most designs this is a suitable guide whether it is a requirement of the customer or not. Frequently the thermal analyst is at odds, so to speak, with the manufacturing engineer. Every design must be practical with regard to manufacturing dimensional tolerances, processes, and tooling.

The other major design topics, cost analysis and size and pressure-drop considerations, are frequently inseparable. From our study of convective heat transfer, we should immediately recognize that an effective means of increasing the heat-transfer coefficient is to increase the Reynolds number. For fixed size this requires an increase in velocity, which, of course, results in an increase in the pressure drop through the heat exchanger. This means that more pump work is required, thereby affecting

Table 18-1 Approximate overall heat-transfer coefficients

Condition	U, $Btu/hr\text{-}ft^2\text{-}°F$
Oil to oil	30–55
Organics to organics	10–60
Steam to:	
Aqueous solutions	100–600
Fuel oil, heavy	10–30
Light	30–60
Gases	5–50
Water	175–600
Water to:	
Alcohol	50–150
Brine	100–200
Compressed air	10–30
Condensing alcohol	45–120
Condensing ammonia	150–250
Condensing Freon-12	80–150
Condensing oil	40–100
Gasoline	60–90
Lubricating oil	20–60
Organic solvents	50–150
Water	150–300

the cost of initial equipment as well as operating costs. In aircraft and space applications, size (or weight) of the complete package is usually the most important factor; designs are optimized with regard to the reduction in payload of the vehicle due to the weight of the component considered. On the other hand, designs for large stationary power plants are optimized purely on economic grounds. Such a cost study involves initial plus operating costs, anticipated life, maintenance, and interest rates.

Before studying methods of detailed thermal analysis, it is appropriate to consider approximate estimates of the overall heat-transfer coefficients for some commonly encountered conditions. These are most convenient for preliminary design purposes. Extensive tabulations of this sort are presented in Refs. 7 and 9, and a list of some is given in Table 18-1.

18-2 CLASSIFICATIONS OF HEAT EXCHANGERS

Common heat exchangers include the *shell-and-tube*, *cross-flow tube-bank*, and the *flat-plate* types. The simplest form of the first is a double pipe arrangement, as shown in Fig. 18-1. This configuration has found wide acceptance because of its ease of manufacture and low cost. A particular application is in cooling compressed air from reciprocating compressors. In this case, fluid a is the warm compressed air and fluid b is usually cold water.

When both fluids flow in the same direction, as in Fig. 18-1, the exchanger is called a *parallel-flow type*. If they move in opposite directions, it is called a *counter-flow exchanger*.

To increase the heat-transfer coefficient, multiple tubes, multiple passes, and shell-side baffles are frequently employed in shell-and-tube exchangers. An exchanger with several tubes, two passes, and shell-side baffles is illustrated in Fig. 18-2. Equipment manufacturers offer many variations of the shell-and-tube heat exchanger in stock designs. For very long units or for high-temperature applications, thermal-expansion problems are sometimes encountered. One solution is to use a floating tube head. Other ramifications include removable tubes and a variety of baffle designs.

If the two fluids flowing through a heat exchanger move at right angles to each

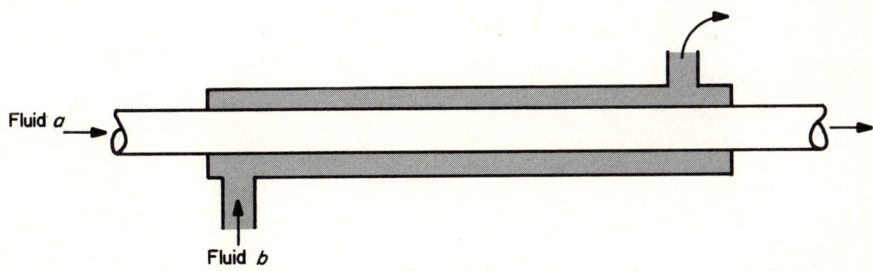

Fig. 18-1 Double-pipe heat exchanger.

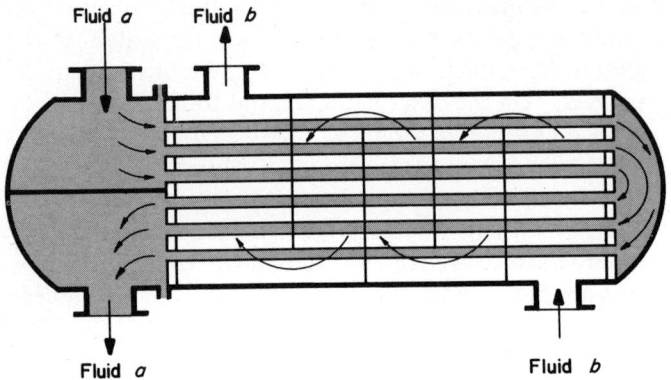

Fig. 18-2 Schematic of a one-shell-pass, two-tube-pass heat exchanger.

other, the unit is further classified as a *cross-flow exchanger*, frequently in both tube-bank and flat-plate heat exchangers. Cross-flow heat exchangers with both fluids unmixed and with one fluid mixed and one unmixed are illustrated in Fig. 18-3.

Plate-type heat exchangers are simple to construct and are employed for low-pressure applications. These may be parallel-flow, counterflow, or cross-flow. A cross-flow unit with both fluids unmixed is illustrated in Fig. 18-4, together with the temperature variation of the fluids. Note that the temperature for either fluid forms a curved plane on a temperature-vs.-space plot. This is to be expected since both fluids are changing temperature in their direction of flow, and the fluid in a given flow path is subjected to a temperature difference unlike that experienced by the fluid

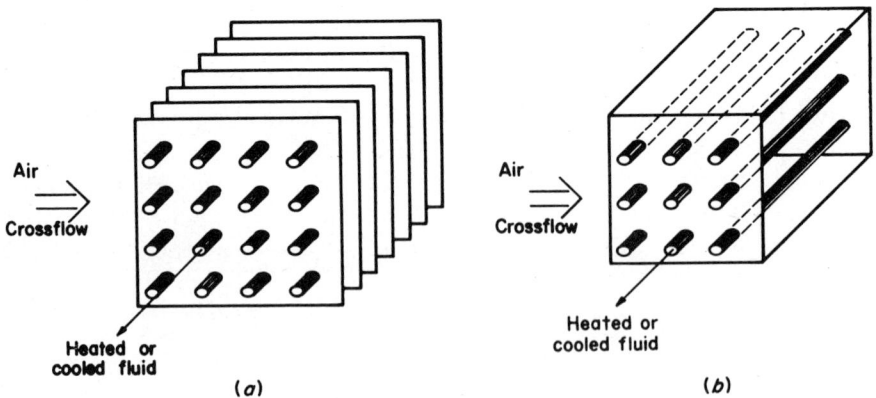

Fig. 18-3 Cross-flow heat exchangers: (*a*) both fluids unmixed; (*b*) one fluid mixed, one fluid unmixed.

THERMAL ANALYSIS OF HEAT EXCHANGERS

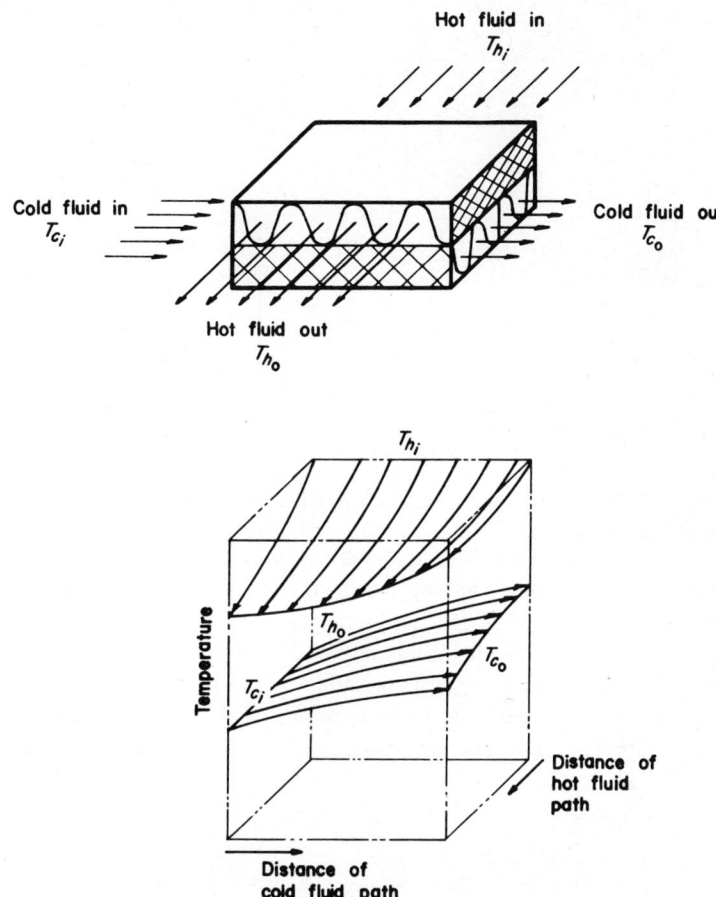

Fig. 18-4 Cross-flow plate heat exchanger with both fluids unmixed.

in any other path at the same distance from the inlet. For example, consider the hot-fluid flow path closest to the cold-fluid exit; this hot fluid exchanges heat along its entire path with a cold fluid at a higher average T_c than any other hot-fluid path experiences; consequently this path has the lowest heat transfer in the exchanger and the highest T_h at the exit.

18-3 HEAT-TRANSFER CALCULATIONS

The case of a flat-plate heat exchanger (or a double-pipe unit) with either parallel or counterflow is shown in Figs. 18-5 and 18-6, which illustrate the advantages of a counterflow exchanger. In parallel flow, the discharge temperature of the colder

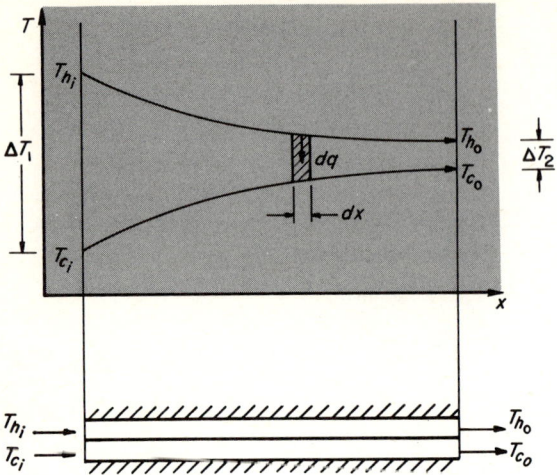

Fig. 18-5 Temperature profiles for a parallel-flow flat-plate heat exchanger.

fluid cannot exceed that of the hotter; in counterflow, however, this restriction does not exist. Also, the surface area required to accomplish a prescribed heat transfer with given fluid conditions is less in a counterflow than a parallel-flow design.

Clearly, the temperature difference between the hot and cold fluid is dependent upon the location along the flow path. This is also true of single-pass evaporators or condensers. Temperature profiles for a condenser are shown in Fig. 18-7. The

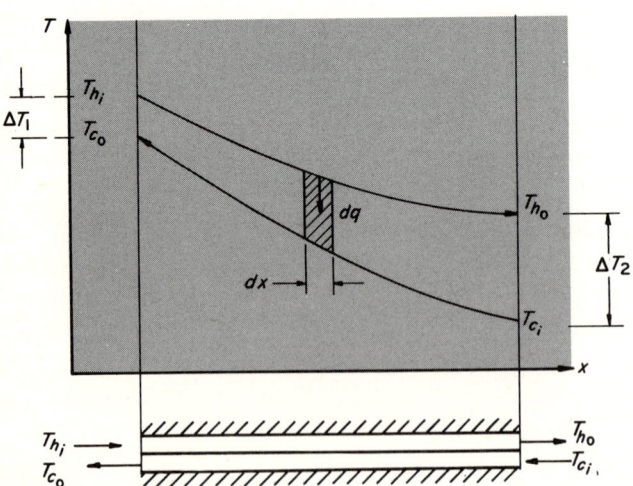

Fig. 18-6 Temperature profiles for a counterflow flat-plate heat exchanger.

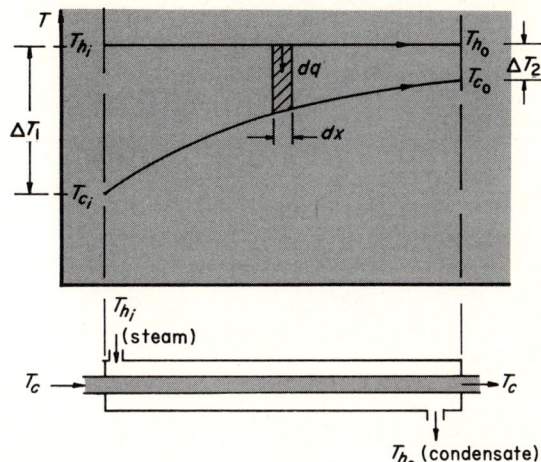

Fig. 18-7 Temperature profiles for a single-pass condenser.

primary difference between this case and that of Fig. 18-6 is that one fluid temperature is constant. This case arises when, for example, a flow of cooling water is provided to condense steam at constant pressure, the steam-condensate temperature being essentially unaffected by the heat-transfer rate.

In any of these exchangers, it is advantageous to employ a simplified equation of the form

$$q = UA(\Delta T)_{\text{mean}} \tag{18-1}$$

where
U = overall heat-transfer coefficient (Sec. 5-4)
A = area for heat transfer used in determining U
$(\Delta T)_{\text{mean}}$ = appropriate temperature difference between hot and cold fluids

Determination of the heat-transfer coefficient U presents no new obstacle. For convenience, appropriate expressions for single-wall plane and cylindrical cases are restated here:

Plane wall:
$$U = \frac{1}{1/h_o + L/k + 1/h_i} \tag{18-2}$$

Cylindrical wall:
$$U_0 = \frac{1}{r_o/r_i h_i + [r_o \ln (r_o/r_i)]/k + 1/h_o} \tag{18-3}$$

These equations are written for a single homogeneous material of thermal conductivity k separating the two fluids. In the plane-wall case it has thickness L, in the cylindrical-wall case it has thickness $r_o - r_i$. The cylindrical-case expression for U is formulated

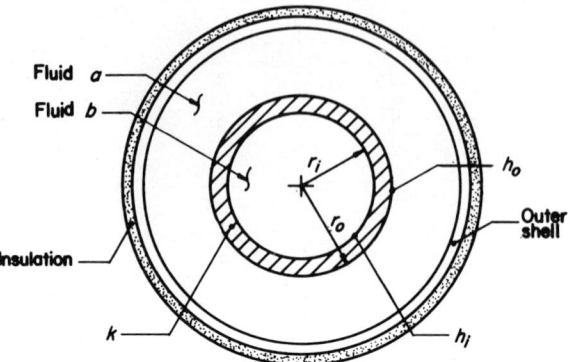

Fig. 18-8 Nomenclature for evaluating U_o by Eq. (18-3).

for the area in Eq. (18-1) based on the wall outer radius,

$$A = \pi r_o^2 \times \text{length}$$

These comments are further clarified by Fig. 18-8. Calculation of numerical values for h_o and h_i can be accomplished by the methods of Sec. 13-8; the determination of a suitable $(\Delta T)_{\text{mean}}$ for the entire heat exchanger is undertaken in the next subsection.

Log mean temperature difference In the following analysis we shall assume (1) that the overall heat-transfer coefficient U is constant throughout the exchanger, (2) that the temperature of either fluid is constant at a given cross section and can be represented by a bulk temperature, (3) that there is no heat exchange except between the two fluids, and (4) that the fluid specific heats are constant. This last statement means that the temperature of a condensing or evaporating fluid remains constant in the exchanger; mixed latent- and sensible-heat changes in one fluid are not considered in the analysis.

The energy exchange between the hot and cold fluids for a differential length of the parallel-flow heat exchanger (Fig. 18-5) may be expressed as

$$dq = U(T_h - T_c)\, dA \tag{18-4}$$

where dA is clearly the product of a constant and dx. An equivalent expression for the heat transfer in length dx is

$$dq = \dot{m}_c c_c\, dT_c = -\dot{m}_h c_h\, dT_h \tag{18-5}$$

where $\dot{m}$ is the mass flow rate and c is the fluid specific heat. Thus[1]

$$dT_c = \frac{dq}{\dot{m}_c c_c} \qquad dT_h = -\frac{dq}{\dot{m}_h c_h}$$

[1] For the case of the counterflow heat exchanger of Fig. 18-6, both $\dot{m}_c$ and dT_c are negative, and the development of the log mean temperature difference is the same as for parallel flow.

THERMAL ANALYSIS OF HEAT EXCHANGERS

and

$$d(T_h - T_c) = -dq\left(\frac{1}{\dot{m}_h c_h} + \frac{1}{\dot{m}_c c_c}\right) \tag{18-6}$$

Substituting dq from (18-4) into (18-6),

$$d(T_h - T_c) = -U(T_h - T_c)\,dA\left(\frac{1}{\dot{m}_h c_h} + \frac{1}{\dot{m}_c c_c}\right)$$

and regrouping yields

$$\frac{d(T_h - T_c)}{T_h - T_c} = -U\left(\frac{1}{\dot{m}_h c_h} + \frac{1}{\dot{m}_c c_c}\right)dA$$

Integrating over the length of the exchanger we obtain

$$\ln\frac{\Delta T_2}{\Delta T_1} = -UA\left(\frac{1}{\dot{m}_h c_h} + \frac{1}{\dot{m}_c c_c}\right) \tag{18-7}$$

where the ΔT terms are as shown in Fig. 18-5. Now from an energy balance on each fluid

$$\dot{m}_h c_h = \frac{q}{T_{h_i} - T_{h_o}} \quad \text{and} \quad \dot{m}_c c_c = \frac{q}{T_{c_o} - T_{c_i}}$$

and substituting these expressions into Eq. (18-7) yields

$$\ln\frac{\Delta T_2}{\Delta T_1} = -UA\frac{(T_{h_i} - T_{h_o}) + (T_{c_o} - T_{c_i})}{q}$$

Rearranging gives

$$q = UA\frac{\Delta T_2 - \Delta T_1}{\ln(\Delta T_2/\Delta T_1)} \tag{18-8}$$

and by comparison with Eq. (18-1)

$$(\Delta T)_{\text{mean}} = \text{LMTD} = \frac{\Delta T_2 - \Delta T_1}{\ln(\Delta T_2/\Delta T_1)} \tag{18-9}$$

The form of the *log mean temperature difference* (LMTD) can easily be remembered: it is simply the difference in temperatures of the two streams at end 2 minus the difference at end 1 divided by the natural logarithm of the ratio of the difference at end 2 to that at end 1. The reader can readily show that the subscripts in Eq. (18-9) can be interchanged without changing the numerical value (or sign) of the LMTD. Hence the designation of ends for use in Eqs. (18-8) and (18-9) is quite arbitrary.

Example A double-pipe heat exchanger is used to cool a compressed-air flow of 50 lb_m/min from 120 to 100°F. The cooling fluid is water, which enters the heat exchanger at 70°F and leaves

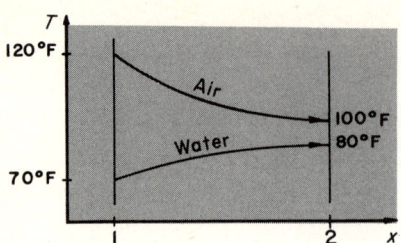

Fig. 18-9 Temperature plots for air-water parallel-flow heat exchanger.

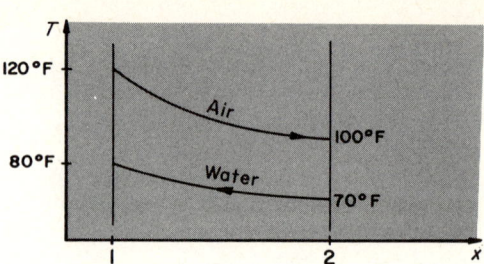

Fig. 18-10 Temperature profiles for air-water counterflow heat exchanger.

at 80°F. If the overall heat-transfer coefficient is 25 Btu/hr-ft²-°F, determine the heat-exchanger area for (*a*) parallel flow and (*b*) counterflow.

Solution (*a*) The temperature-distance plots for parallel flow are shown in Fig. 18-9. The total heat transfer from the air is

$$q = \dot{m}_a c_{pa} \Delta T_a$$

$$= (50 \text{ lb/min})(60 \text{ min/hr})(0.240 \text{ Btu/lb}_m\text{-°F})(120 - 100)°F = 14{,}400 \text{ Btu/hr}$$

$$\text{LMTD} = \frac{(100 - 80) - (120 - 70)}{\ln\left[(100 - 80)/(120 - 70)\right]}$$

$$= \frac{20 - 50}{\ln \frac{20}{50}} = \frac{-30}{\ln 0.4} = 32.8°F$$

Thus,

$$A = \frac{q}{U(\text{LMTD})} = \frac{14{,}400 \text{ Btu/hr}}{(25 \text{ Btu/hr-ft}^2\text{-°F})(32.8°F)} = 17.56 \text{ ft}^2$$

(*b*) The temperature-distance plots for counterflow are shown in Fig. 18-10, and the LMTD is

$$\text{LMTD} = \frac{(100 - 70) - (120 - 80)}{\ln\left[(100 - 70)/(120 - 80)\right]}$$

$$= \frac{30 - 40}{\ln \frac{30}{40}} = \frac{-10}{\ln \frac{3}{4}} = 34.7°F$$

Thus,

$$A = \frac{q}{U(\text{LMTD})} = \frac{14{,}400 \text{ Btu/hr}}{(25 \text{ Btu/hr-ft}^2\text{-°F})(34.7°F)} = 16.6 \text{ ft}^2$$

Determination of an appropriate mean temperature difference for more complex heat exchangers such as those utilizing multiple tube passes, several shell passes, or cross-flow designs is considerably more difficult than that for the single-pass shell-and-tube exchanger. The resulting $(\Delta T)_{\text{mean}}$ for a simple design such as one shell pass and two tube passes is so complex that the usual practice has been to modify the LMTD

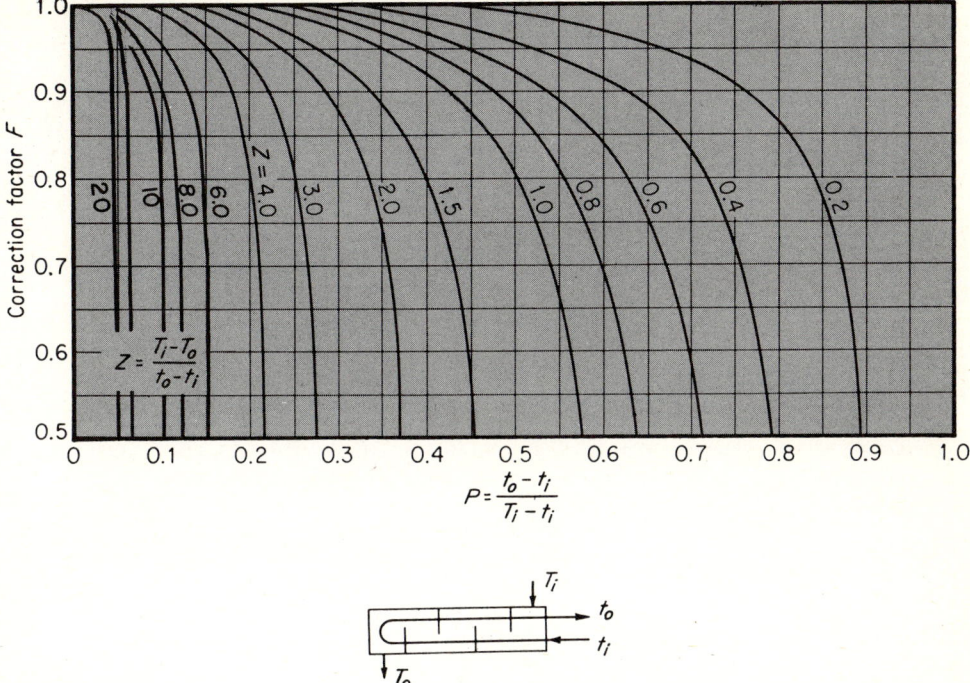

Fig. 18-11 Correction factors for heat exchangers with one shell pass and two (or multiples of two) tube passes.

by appropriate correction factors for more complex designs. Bowman, Mueller, and Nagle [1] published correction charts for this purpose. Figures 18-11 to 18-14 present correction factors for several common configurations. These, together with

$$q = UAF(\text{LMTD}) \tag{18-10}$$

provide a practical method of calculating the heat transfer. *The LMTD to be used with the correction factor of these figures is for a counterflow double-pipe heat exchanger with the same fluid inlet and exit temperatures as in the more complex design.* In these four figures, the simplified notation of T, t to denote temperatures of the two fluid streams has been introduced since it is immaterial whether the colder fluid flows through the shell or the tubes.

Example Commercial aniline at a rate of 70 lb_m/min is heated from 60 to 100°F with oil having an inlet temperature of 180°F and an exit temperature of 135°F. A shell-and-tube exchanger is used, with the aniline making two tube passes and the oil making one shell pass. Determine the area required for the exchanger if the overall heat-transfer coefficient is 45 Btu/hr-ft²-°F.

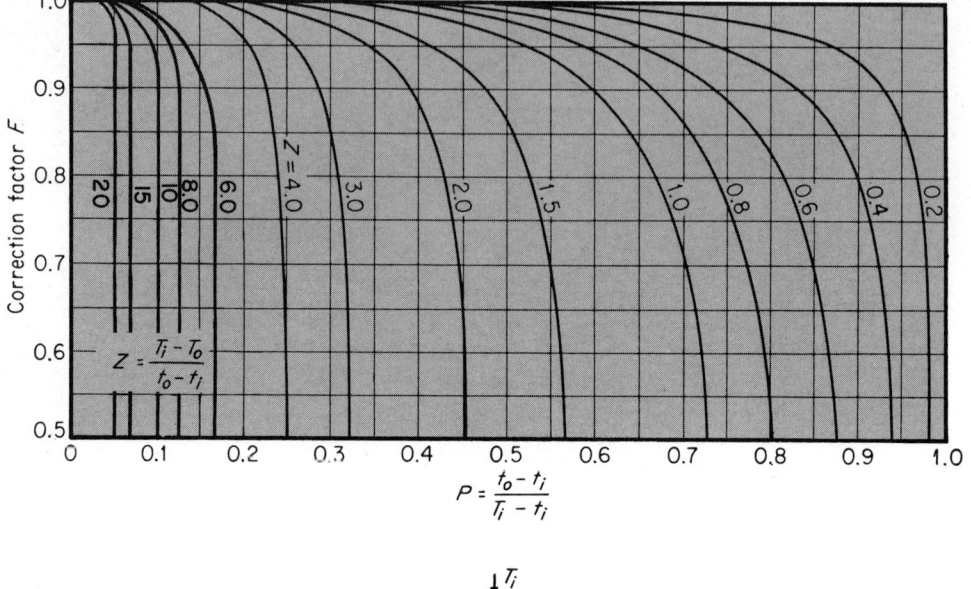

Fig. 18-12 Correction factors for heat exchangers with two shell passes and four (or multiple) tube passes.

Solution Let T be used for oil (shell) and t for aniline (tube). Then the LMTD is (for a counterflow double-pipe exchanger)

$$\text{LMTD} = \frac{(T_i - t_o) - (T_o - t_i)}{\ln[(T_i - t_o)/(T_o - t_i)]}$$

$$= \frac{(180 - 100) - (135 - 60)}{\ln[(180 - 100)/(135 - 60)]} = 77.2°F$$

From Fig. 18-11 with

$$P = \frac{100 - 60}{180 - 60} = 0.333 \quad \text{and} \quad Z = \frac{180 - 135}{100 - 60} = 1.125$$

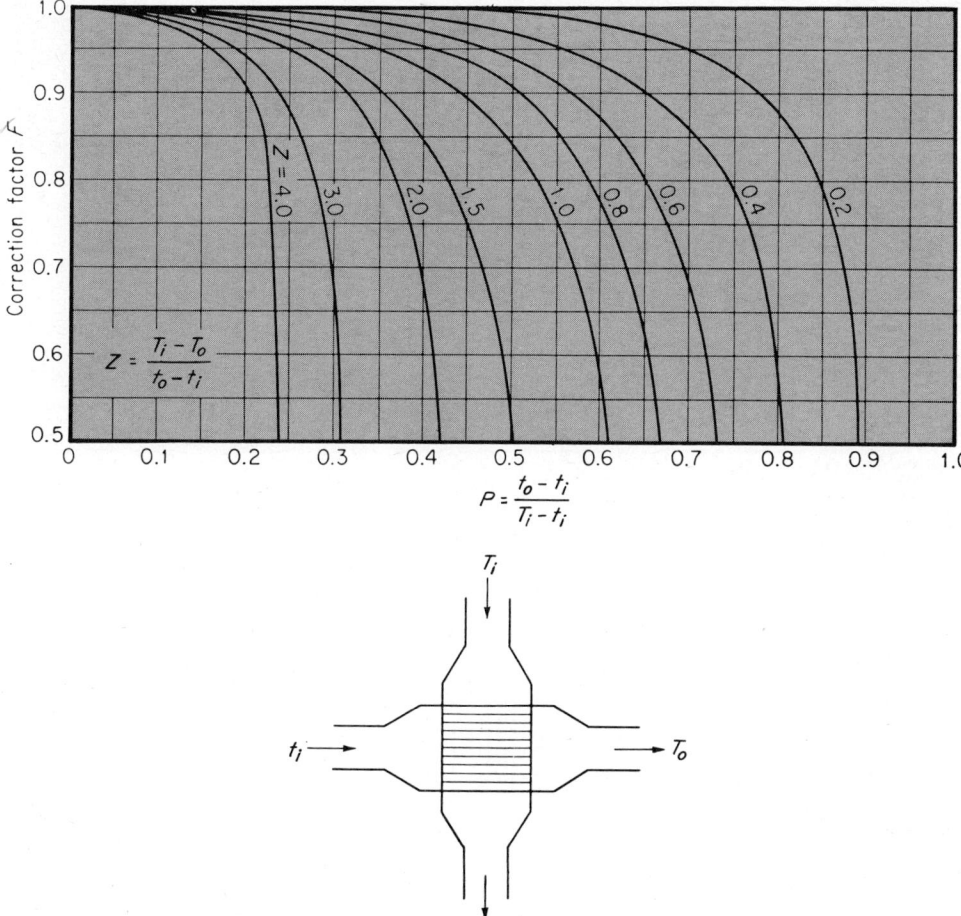

Fig. 18-13 Correction factors for cross-flow heat exchanger with one fluid mixed and one fluid unmixed.

we obtain $F = 0.95$. For an average c_p of aniline of 0.485 Btu/lb$_m$-°F,

$$q = \dot{m}c(t_o - t_i) = (70 \text{ lb}_m/\text{min})(60 \text{ min/hr})(0.485 \text{ Btu/lb}_m\text{-°F})(40°\text{F})$$

$$= 81{,}500 \text{ Btu/hr}$$

and by Eq. (18-10)

$$A = \frac{q}{UF(\text{LMTD})} = \frac{81{,}500 \text{ Btu/hr}}{(45 \text{ Btu/hr-ft}^2\text{-°F})(0.95)(77.2°\text{F})} = 24.7 \text{ ft}^2$$

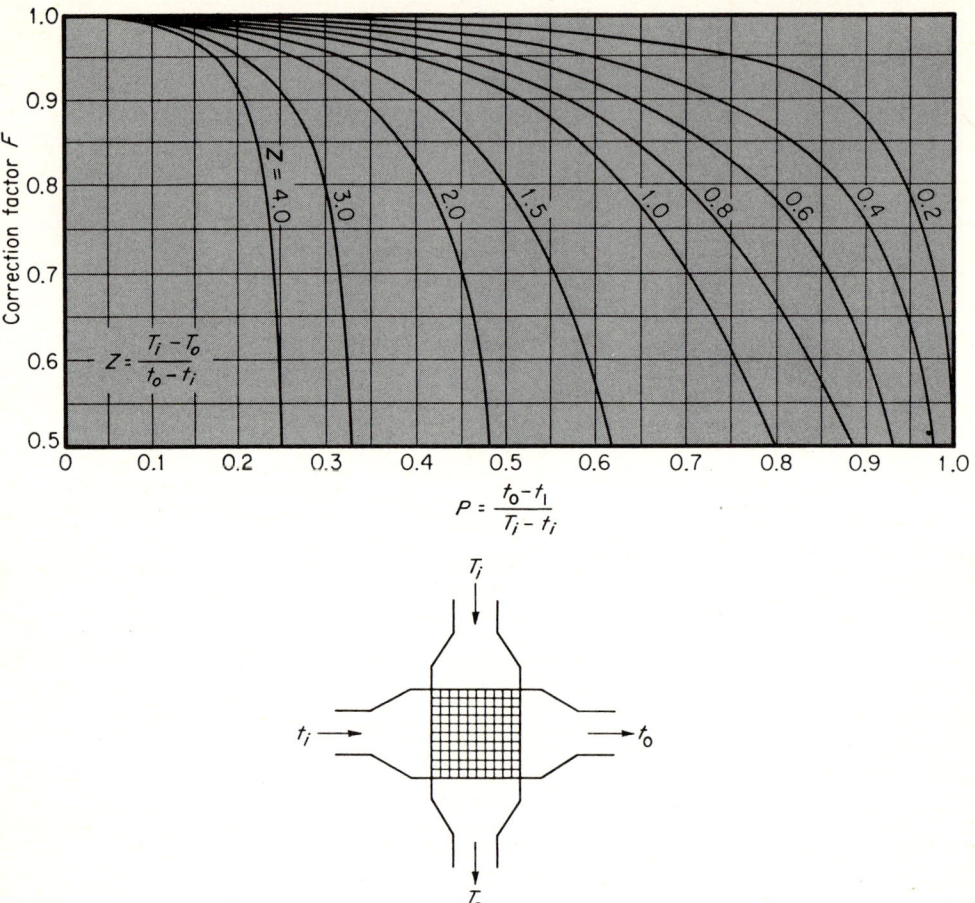

Fig. 18-14 Correction factors for cross-flow heat exchanger with both fluids unmixed.

Example A water flow of 25,000 lb$_m$/hr is to be heated from 70 to 150°F in a water-to-water heat exchanger. Hot water at 195°F is available, but the minimum discharge temperature of this fluid is to be 160°F. The overall heat-transfer coefficient based on the inside diameter of 1.0-in tubes in a shell-and-tube exchanger is 190 Btu/hr-ft²-°F. If the hot water makes one shell pass and the design water velocity in the tubes is 1.5 ft/sec, calculate the number of tubes and length for a two-tube-pass design.

Solution The number of tubes is given by the continuity equation

$$\dot{m}_c = \rho_c u_c n \frac{\pi d^2}{4}$$

$$n = \frac{25{,}000 \text{ lb}_m/\text{hr}}{(62.4 \text{ lb}_m/\text{ft}^3)[1.5(3{,}600 \text{ ft/hr})][\pi(1)/4(144)] \text{ ft}^2}$$

$$= 13.64 = 14 \text{ tubes}$$

THERMAL ANALYSIS OF HEAT EXCHANGERS

Letting T be used for the shell side and t for the tube side, the LMTD for a single-pass counterflow shell-and-tube exchanger is

$$\text{LMTD} = \frac{(T_i - t_o) - (T_o - t_i)}{\ln[(T_i - t_o)/(T_o - t_i)]} = \frac{(195 - 150) - (160 - 70)}{\ln[(195 - 150)/(160 - 70)]} = 64.9°F$$

From Fig. 18-11 with

$$P = \frac{150 - 70}{195 - 70} = 0.64 \quad \text{and} \quad Z = \frac{195 - 160}{150 - 70} = 0.438$$

we obtain $F = 0.87$. From an energy balance on the fluid being heated we have

$$q = (25{,}000 \text{ lb}_m/\text{hr})(1.0 \text{ Btu/lb}_m\text{-°F})(150 - 70)°F = 2{,}000{,}000 \text{ Btu/hr}$$

and by Eq. (18-10)

$$A = \frac{q}{UF(\text{LMTD})} = \frac{2{,}000{,}000 \text{ Btu/hr}}{(190 \text{ Btu/hr-ft}^2\text{-°F})(0.87)(64.9°F)}$$

$$= 186.7 \text{ ft}^2$$

Now the tube surface area per linear foot is

$$\frac{A}{\text{ft}} = 2(14)(\pi)(\tfrac{1}{12}\text{ ft}) = 7.33 \text{ ft}^2$$

and thus the required length is

$$A_{\text{tot}} = \frac{A}{\text{ft}} L$$

or

$$L = 24.5 \text{ ft}$$

Thus,

Number of tubes = 14

Number of passes = 2

Length of tubes = 24.5 ft

18-4 EFFECTIVENESS METHOD

In the design or selection of a typical heat exchanger, the inlet temperatures of both fluids are usually known; one or both discharge temperatures, however, are frequently not known. This situation arises when the overall heat-transfer coefficient U is known (or can be closely estimated) and the size of the unit is fixed, as in the selection of an off-the-shelf design to accomplish a specified temperature change in one fluid. This type of problem could be handled by a trial-and-error approach using the conventional LMTD with appropriate correction factor, but this is a rather laborious and inefficient approach.

Nusselt [8] and Ten Broeck [11] have developed a method which does not require iterative procedures. It introduces a definition of heat-exchanger *effectiveness*, which can be used to eliminate the unknown discharge temperature, and a solution for this effectiveness in terms of other known parameters (flow rates, specific heats, area, and U). The heat-exchanger effectiveness is

$$\epsilon \equiv \frac{\text{actual heat transfer}}{\text{maximum possible heat transfer}} \tag{18-11}$$

where the maximum possible heat transfer is that which would result if one fluid underwent a temperature change equal to the maximum temperature difference available, the temperature of the entering hot fluid minus that of the entering cold fluid. Letting the product of the mass flow rate and the specific heat be C, we find

$$q_{\max} = (\dot{m}c)_{\min}(T_{h_i} - T_{c_i}) = C_{\min}(T_{h_i} - T_{c_i}) \tag{18-12}$$

Clearly the fluid with minimum C must be used to determine the maximum possible heat transfer, for if the other fluid were allowed to undergo the maximum temperature change available, an energy balance would then require the fluid with minimum C to undergo a greater temperature change, an obvious impossibility.

By combining Eqs. (18-11) and (18-12) we obtain

$$q_{\text{actual}} = \epsilon C_{\min}(T_{h_i} - T_{c_i}) \tag{18-13}$$

and this is the basic equation for calculating the heat transfer with unknown discharge temperatures. The problem then is to determine appropriate values of the heat-exchanger effectiveness ϵ. Note that either the hot or the cold fluid may have the minimum value of C, and consequently there are two possible values of effectiveness

$$C_h < C_c: \quad \epsilon_h = \frac{(\dot{m}c)_h(T_{h_i} - T_{h_o})}{(\dot{m}c)_h(T_{h_i} - T_{c_i})} = \frac{T_{h_i} - T_{h_o}}{T_{h_i} - T_{c_i}} \tag{18-14a}$$

$$C_c < C_h: \quad \epsilon_c = \frac{(\dot{m}c)_c(T_{c_o} - T_{c_i})}{(\dot{m}c)_c(T_{h_i} - T_{c_i})} = \frac{T_{c_o} - T_{c_i}}{T_{h_i} - T_{c_i}} \tag{18-14b}$$

These two possibilities exist for either a parallel or a counterflow exchanger.

To obtain one expression for effectiveness, let us direct attention to the parallel-flow exchanger of Fig. 18-5 and consider the case when the hotter fluid has the minimum value of $\dot{m}c$. The development from Eqs. (18-4) to (18-7) in Sec. 18-3 is valid for any parallel-flow exchanger (under the restrictions listed for the development of the LMTD). Thus, from Eq. (18-7) we have

$$\ln \frac{T_{h_o} - T_{c_o}}{T_{h_i} - T_{c_i}} = -UA\left(\frac{1}{C_h} + \frac{1}{C_c}\right) \tag{18-15}$$

or

$$\frac{T_{h_o} - T_{c_o}}{T_{h_i} - T_{c_i}} = \exp\left[-\frac{UA}{C_h}\left(1 + \frac{C_h}{C_c}\right)\right] \tag{18-16}$$

THERMAL ANALYSIS OF HEAT EXCHANGERS

An energy balance on the two fluids yields

$$C_h(T_{h_i} - T_{h_o}) = C_c(T_{c_o} - T_{c_i})$$

and thus

$$T_{c_o} = T_{c_i} + \frac{C_h}{C_c}(T_{h_i} - T_{h_o}) \tag{18-17}$$

Using this, we obtain

$$\frac{T_{h_o} - T_{c_o}}{T_{h_i} - T_{c_i}} = \frac{(T_{h_o} - T_{h_i}) - (C_h/C_c)(T_{h_i} - T_{h_o}) + (T_{h_i} - T_{c_i})}{T_{h_i} - T_{c_i}}$$

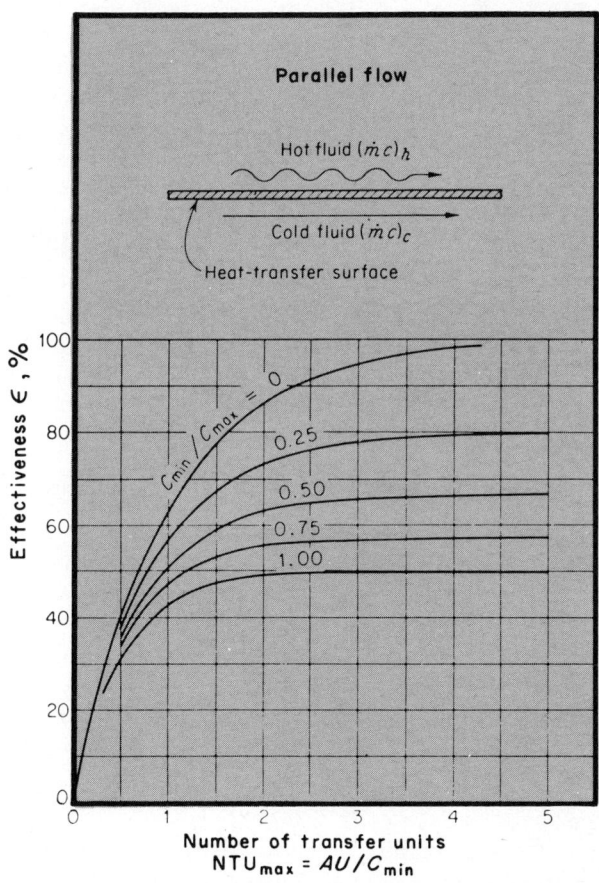

Fig. 18-15 Effectiveness for parallel-flow heat exchanger.

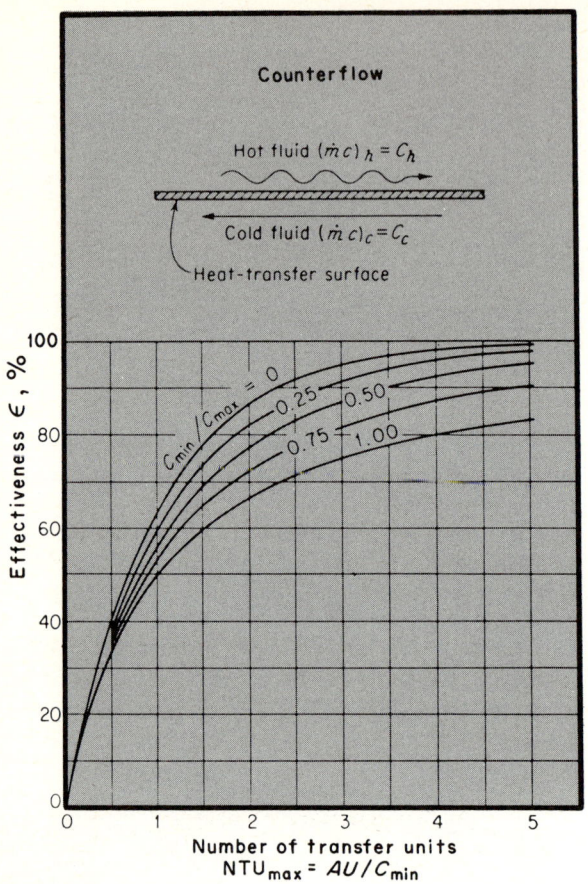

Fig. 18-16 Effectiveness for counterflow heat exchanger.

which together with Eq. (18-14a) is

$$\frac{T_{h_o} - T_{c_o}}{T_{h_i} - T_{c_i}} = 1 - \epsilon_h \left(1 + \frac{C_h}{C_c}\right) \tag{18-18}$$

Substitution of (18-18) into (18-16) and rearrangement yields

$$\epsilon_h = \frac{1 - \exp\left[-(UA/C_h)(1 + C_h/C_c)\right]}{1 + C_h/C_c} \tag{18-19}$$

It can readily be shown (see, for example, Kreith [5]) that the same result is obtained for a parallel-flow exchanger with C of the cold fluid smaller than that of the hot fluid except that C_h and C_c are interchanged. Thus, the general expression for

THERMAL ANALYSIS OF HEAT EXCHANGERS

the effectiveness of a parallel-flow heat exchanger is usually written as

$$\epsilon = \frac{1 - \exp\left[-\mathrm{NTU}(1 + C_{\min}/C_{\max})\right]}{1 + C_{\min}/C_{\max}} \tag{18-20}$$

where the number of heat transfer units $\mathrm{NTU} = UA/C_{\min}$ has been introduced. This expression for ϵ contains only U, area, fluid properties, and flow rates. Again it is emphasized that it was developed under the restrictions of constant U, constant fluid specific heats, no heat exchange except between the two fluids, and temperature of each fluid constant at a given cross section.

Expressions for effectiveness for counterflow, cross flow, and shell-and-tube exchangers have been determined, and Kays and London [4] present convenient graphical values for numerous configurations. Curves for some common designs are presented in Figs. 18-15 to 18-20.

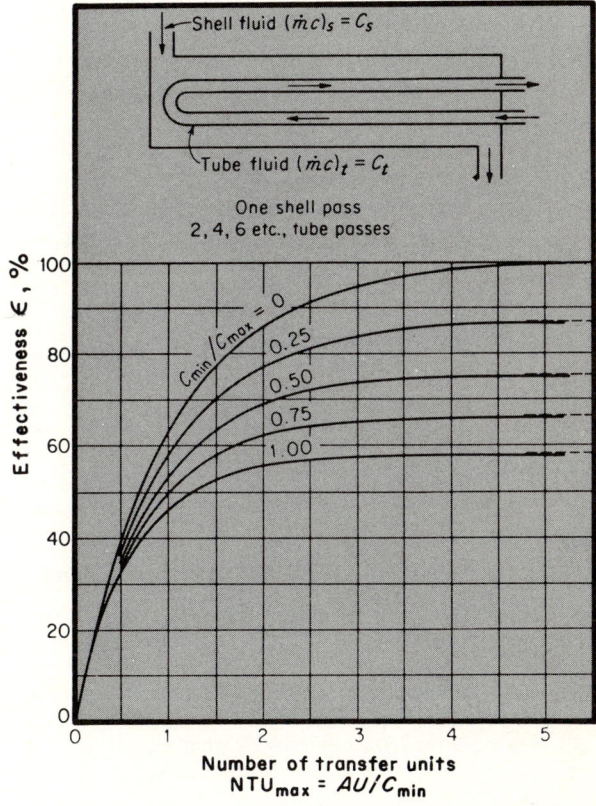

Fig. 18-17 Effectiveness for shell-and-tube heat exchanger with one shell pass.

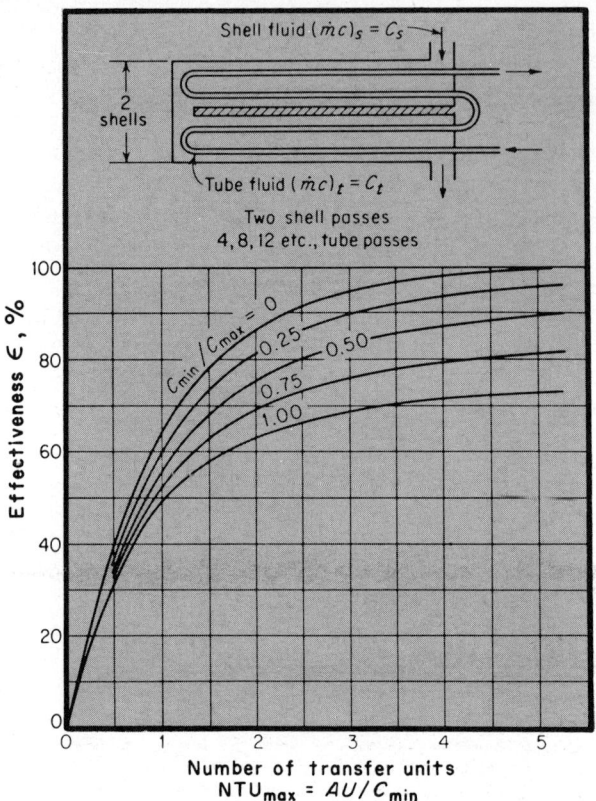

Fig. 18-18 Effectiveness for shell-and-tube heat exchanger with two shell passes.

The term NTU can be considered as a heat-exchanger size factor. From Fig. 18-15 it is apparent that for given fluid properties and flow rates the effectiveness increases with increasing values of NTU. Such increases can be accomplished by increasing the exchanger size or the overall heat-transfer coefficient U. U, however, is directly related to the fluid velocities through the film heat-transfer coefficients; consequently, increases in NTU are obtained at additional fabrication costs or operating expense.

Example A parallel-flow double-pipe heat exchanger uses hot water at 180°F, 100 lb_m/min flow rate, to heat a 50 lb_m/min flow of cold water from 80 to 120°F. If the overall heat-transfer coefficient is 200 Btu/hr-ft²-°F, what area of heat exchanger is required?

Solution The minimum fluid is clearly the cold water, since the specific heat does not vary greatly with temperature. Thus

$$C_{\min} = \dot{m}_c c_c = (50 \text{ lb}_m/\text{min})(0.998 \text{ Btu/lb}_m\text{-°F}) = 49.9 \text{ Btu/min-°F}$$

$$C_{\max} = \dot{m}_h c_h = (100 \text{ lb}_m/\text{min})(1.00 \text{ Btu/lb}_m\text{-°F}) = 100 \text{ Btu/min-°F}$$

THERMAL ANALYSIS OF HEAT EXCHANGERS

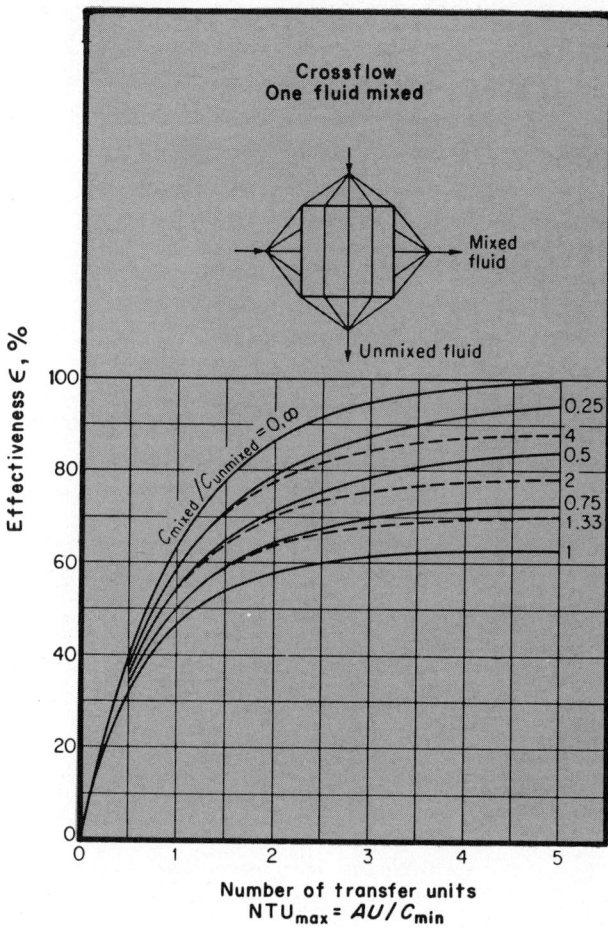

Fig. 18-19 Effectiveness for cross-flow heat exchanger with one fluid mixed, one fluid unmixed.

where the specific heat of the cold fluid is evaluated at the average temperature and the hot-side average temperature is assumed to be 150°F for property evaluation. Thus

$$\frac{C_{\min}}{C_{\max}} = \frac{49.9}{100} \simeq 0.5$$

The effectiveness is given by Eq. (18-14b)

$$\epsilon_c = \frac{T_{c_o} - T_{c_i}}{T_{h_i} - T_{c_i}} = \frac{120 - 80}{180 - 80} = 0.40$$

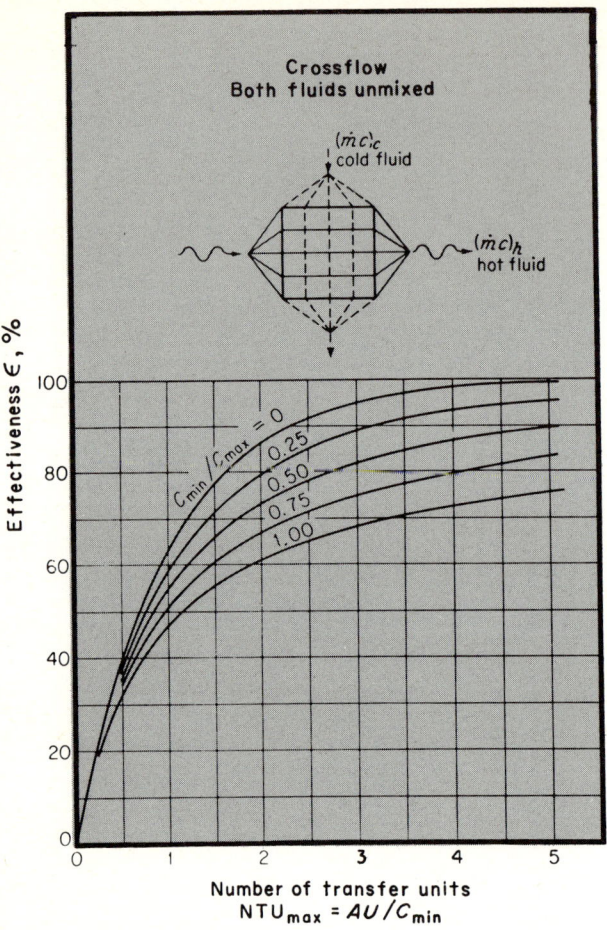

Fig. 18-20 Effectiveness for cross-flow heat exchanger with both fluids unmixed.

Then from Fig. 18-15,

$$\text{NTU} = \frac{AU}{C_{\min}} = 0.62$$

$$A = \frac{0.62(49.9 \text{ Btu/min°F})}{200 \text{ Btu/hr-ft}^2\text{-°F}} (60 \text{ min/hr})$$

$$= 9.27 \text{ ft}^2$$

Example An exchanger with oil flowing through tubes subjected to an air cross flow as shown in Fig. 18-3a is used to cool 60 lb_m/min of oil, which enters at 260°F. The design exit oil temperature is 180°F, and the oil specific heat is 0.50 Btu/lb_m-°F. The cooling air enters at 85°F, and

THERMAL ANALYSIS OF HEAT EXCHANGERS

Table 18-2

$\dfrac{C_{min}}{C_{max}}$	C_{min}	ΔT_c	NTU_{max}	ϵ Calc.	ϵ From Fig. 18-20
0.5	900	160	3.0	0.915	0.82
0.75	1,350	59.3	2.0	0.34	0.67
0.60	1,180	133	2.29	0.76	0.75

the overall heat-transfer coefficient is 30 Btu/hr-ft²-°F. For an exchanger area of 90 ft² determine the required airflow rate and the exit air temperature.

Solution The problem statement does not permit direct determination of the minimum fluid. If the oil is the minimum fluid, we can solve directly for ϵ and NTU_{max} and then use Fig. 18-20 to determine C_{min}/C_{max}, which would permit determination of the airflow rate. Assuming oil to be the minimum fluid,

$$\epsilon_h = \frac{T_{h_i} - T_{h_o}}{T_{h_i} - T_{c_i}} = \frac{260 - 180}{260 - 85} = 0.457$$

$$C_{min} = (60 \text{ lb/min})(0.5 \text{ Btu/lb-°F}) = 30 \text{ Btu/min-°F}$$

$$NTU_{max} = \frac{AU}{C_{min}} = \frac{(90 \text{ ft}^2)(30 \text{ Btu/hr-ft}^2\text{-°F})}{30(60) \text{ Btu/hr-°F}} = 1.50$$

From Fig. 18-20 it is apparent that C_{min}/C_{max} would be greater than 1.0—an impossibility. Clearly the minimum fluid is the air. The solution requires the assumption of an airflow rate and a check by Fig. 18-20. Assumptions and calculations are given in Table 18-2. Note that $c_c = c_p = 0.240$ and is constant for air in this temperature range. The exit air temperature is

$$T_{c_o} = T_{c_i} + \Delta T_c = 218°F$$

and the air mass flow rate is

$$\dot{m} = \frac{C_{min}}{c_c} = \frac{1,180}{0.24} = 4,920 \text{ lb}_m/\text{hr}$$

18-5 FOULING FACTORS

Heat-exchanger surfaces are subject to deposition of a film of foreign material. This film may be quite thin or, in the case of scale, quite thick. In either case, the thermal resistance offered by the film can be appreciable and should be included in the formulation of the overall heat-transfer coefficient U. For either a plane or a cylindrical wall without fins, the expression for U_{fouled} is

$$U_f = \frac{1}{R_{c_o} + R_{f_o} + R_k + R_{f_i}(A_o/A_i) + R_{c_i}(A_o/A_i)} \tag{18-21}$$

where R_{c_o} = outside convective resistance, $1/h_o$
R_{f_o} = outside resistance due to fouling

R_k = conductive resistance of the metal tubing or plate [Eqs. (18-2) and (18-3)]
R_{f_i} = inside resistance due to fouling
R_{c_i} = inside convective resistance, $1/h_i$
A_o/A_i = ratio of outside to inside area

Values of the fouling thermal resistance are usually called *fouling factors*. They depend upon the cleanliness of the fluid and the length of service and also (but less

Table 18-3 Fouling factors for heat-transfer equipment†

Fluid	Fouling resistance, hr-ft²-°F/Btu‡	
	Below 125°F	Above 125°F
Water:		
Sea	0.0005	0.001
Distilled	0.0005	0.0005
Treated boiler feedwater	0.001–0.0005	0.001
City or well	0.001	0.002
Great Lakes	0.001	0.002
Cooling tower, treated makeup	0.001	0.002
Untreated	0.003	0.004–0.005
River, minimum	0.001–0.002	0.002–0.003
Mississippi, Delaware, Schuylkill, East River, New York Bay	0.002–0.003	0.003–0.004
Muddy or silty	0.002–0.003	0.003–0.004
Hard (above 15 grains/gal)	0.003	0.005

Fluid	All temperatures
Industrial liquids:	
Brine (cooling equipment)	0.001
Organic	0.001
Refrigeration liquids	0.001
Industrial gases and vapors:	
Alcohol vapors	0.0005
Air	0.002
Coke over gas (other manufactured gas)	0.01
Organic vapor	0.0005
Refrigerating vapors (condensing)	0.002
Steam, oil-free	0.0005
Oil-bearing from reciprocating compressors	0.001
Industrial oils:	
Fuel oil	0.005
Machinery and transformer oils	0.001
Vegetable oil	0.003
Quenching oil	0.004

† Abstracted from "Standards of the Tubular Exchanger Manufacturers Association" [10] by permission.
‡ Ratings for service up to 125°F are based on heating-medium temperatures up to 240°F; for service over 125°F on heating-medium temperatures up to 400°F.

THERMAL ANALYSIS OF HEAT EXCHANGERS 691

obviously) upon operating temperatures and flow velocities. A rather extensive list of recommended values is presented in Ref. 10, a selected number of which are given in Table 18-3. Where a range of values is indicated in this table for water, the lower value is applicable if the velocity is over 3 ft/sec.

18-6 CLOSURE

In all the preceding sections we have assumed that the overall heat-transfer coefficient U is known. Actually the most difficult part of any heat-exchanger analysis is an accurate prediction of U. This requires determination of the convective coefficient h for each fluid, which is accomplished by the methods of Chap. 13.

Variable-property analysis An additional complication is frequently encountered in heat exchangers with fluids whose properties are strongly temperature-dependent. This can result in significant changes in h and consequently in U through the exchanger, necessitating a variable-property analysis. A suggested approach is to divide the exchanger into small sections over which the fluid properties do not change significantly and perform a numerical stepwise solution. The reader should note that this requires iteration for most designs, since if we are to begin with a small area at the hot-fluid inlet of a simple counterflow design, the cold-side fluid temperature for this incremental area is unknown. Clearly a digital-computer program is required for variable-property analyses.

Compact exchangers A compact heat exchanger as defined by Macklin [6] is one that has 200 ft² of surface area or more per cubic foot of exchanger volume. These are widely used in aircraft and space-vehicle applications and where the convective-heat-transfer coefficient on one side is much smaller than the other. This situation is frequently encountered in gas-to-liquid or gas–to–condensing-vapor applications. By increasing the effective area on the side with the lower value of h by the use of extended surfaces (such as fins), the total package size and weight can be reduced.

Other techniques used to increase the heat-transfer coefficient include high gas velocities and artificially induced turbulence. Either of these also causes high pressure drop and consequently results in high operation costs.

Predictions of heat transfer or pressure drop in compact exchangers by analytical means is very difficult due to the complexity of the flow. Resort is usually made to experiment, and Kays and London [4] present extensive results for tests on a variety of compact-exchanger designs.

Regenerators The emphasis of this chapter has been on steady-flow heat exchangers with two fluids at different temperatures separated by a wall. These are frequently termed recuperators and are the most common type of exchanger. A heat exchanger in which the hot and cold fluids alternately pass over the same surface is known as a *regenerator*.

Cyclic regenerators are commonly used only with gases. They usually consist of a flow passage containing pellets of high thermal capacity. During part of the cycle warm fluid passes through the exchanger and energy is stored in the pellet material. During the other part the cold fluid passes through the same passage and is heated by the pellets. For design and theory of such exchangers Coppage and London [2] and Jakob [3] should be consulted.

PROBLEMS

18-1. Air is cooled from 140 to 100°F by passing it through the center pipe of a double-pipe counterflow heat exchanger, as shown in Fig. 18-1. The average air velocity is 108 ft/sec, and the center-pipe diameter is 1.0 in. The tube wall is held at 80°F by a cooling water flow in the annulus; the average bulk temperature of the water is 60°F. Using the methods of Chap. 13, determine an appropriate $\bar{h}_a$ for the air side. By an energy balance determine $\bar{h}_w$ for the water side and then calculate U for the exchanger, neglecting the thermal resistance of the pipe (pipe wall is thin). Compare the value of U thus obtained with the range given for this fluid pair in Table 18-1.

18-2. Water is heated from 60 to 190°F in a shell-and-tube heat exchanger with condensing steam at 100 psia. The water flows through 1-in-ID mild steel tubes with 1.18-in OD. The maximum velocity of the water established by pressure-loss considerations is 4 ft/sec. The heat-transfer coefficient for the condensate side may be taken as 800 Btu/hr-ft²-°F. Determine the maximum value of U obtainable and compare with the range in Table 18-1. *Hint:* Use the Dittus-Boelter equation for the water-side coefficient.

18-3. Lubricating oil for a large industrial machine is cooled from 240 to 120°F with a single-pass shell-and-tube counterflow heat exchanger. The shell-side fluid is water, which enters at 70°F and leaves at 100°F. If the overall heat-transfer coefficient is 35 Btu/hr-ft²-°F, (*a*) calculate by the LMTD method the heat-exchanger surface area for an oil flow of 6 gal/min and (*b*) determine the number of 0.5-in-OD tubes required if the heat-exchanger length is 6 ft. The oil specific heat is 0.48 Btu/lb$_m$-°F, and the density is 54 lb$_m$/ft³.

18-4. Repeat Prob. 18-3 for two tube passes and one shell-side pass. The oil is the tube-side fluid.

18-5. An aircraft oil cooler (heat exchanger) is of cross-flow design with both fluids unmixed. The oil enters at 260°F and leaves at 135°F while the coolant air enters at −40°F and leaves at 60°F. For an oil flow rate of 10 lb$_m$/min, calculate the heat-exchanger area by the LMTD method. The overall heat-transfer coefficient is 50 Btu/hr-ft²-°F.

18-6. Cooling water for an internal combustion engine is cooled from 180 to 120°F in a radiator type of heat exchanger by cross flow of forced air. The flow rate of the water is 20 lb$_m$/min, and the air enters at a temperature of 75°F and leaves at 115°F. The overall heat-transfer coefficient is 30 Btu/hr-ft²-°F. Determine the required heat-transfer area if both fluids are unmixed.

18-7. Benzene at a flow rate of 15,000 lb$_m$/hr is to be heated from 80 to 150°F in a shell-and-tube exchanger. Water enters the shell side at 10,000 lb$_m$/hr and 200°F and makes one pass. The overall heat-transfer coefficient may be taken as 50 Btu/hr-ft²-°F. The average velocity of the oil in the tubes cannot exceed 2 ft/sec, and the heat exchanger can be no longer than 30 ft because of space limitations. Calculate the number of tube passes, the number of tubes per pass, and the length of the tubes using $\frac{1}{2}$-in-ID thin-walled tubes.

18-8. A feedwater heater for a steam boiler is to heat 60,000 lb$_m$/hr of water from 60 to 190°F. The design is to be a shell-and-tube type using condensing steam at 100 psia and 1.0-in-ID 1.18-in-OD mild steel tubes. The maximum water velocity is 4 ft/sec, and the overall heat-transfer coefficient based on tube outside surface area is 376 Btu/hr-ft²-°F (see Prob. 18-2). The maximum tube length is 30 ft. Determine the number of tubes and the number of tube passes by the LMTD method. Assume one shell pass.

THERMAL ANALYSIS OF HEAT EXCHANGERS

18-9. Light oil at a flow rate of 40 lb_m/min is to be cooled from 150 to 100°F by means of a double-pipe heat exchanger with cooling water in the annular section. The inner pipe is 2.0 in OD, and U based on the OD of this pipe is 50 Btu/hr-ft²-°F. The cooling water enters at 65°F and leaves at 90°F. Calculate the length required for (a) parallel flow and (b) counterflow.

18-10. Liquid sodium is heated from 200 to 700°F with superheated steam in a shell-and-tube exchanger. The steam enters at 1600°F, makes two shell-side passes, and leaves at 1000°F. The liquid sodium undergoes four tube passes. The overall heat-transfer coefficient may be taken as 300 Btu/hr-ft²-°F. Calculate the area required.

18-11. Oil is to be cooled from 200 to 150°F in a shell-and-tube heat exchanger. The oil makes two tube passes while water makes a single shell pass. The water is heated from 60 to 80°F. The tubes are 1 in OD, and each tube pass is 5 ft long. If the overall heat-transfer coefficient based on tube OD is 60 Btu/hr-ft²-°F, what mass flow rate of oil can be cooled per tube? Assume $c_p = 0.5$ Btu/lb_m-°F.

18-12. A room air heater consists of a finned-tube heat exchanger (cross flow, both fluids unmixed). The airflow rate is 2,000 ft³/min, and it is heated from 65 to 85°F. The tube flow is hot water, which enters at 175°F. The heat-exchanger surface area is 60 ft², and the overall heat-transfer coefficient is 32 Btu/hr-ft²-°F. Calculate the exit water temperature and the water flow rate by the effectiveness method.

18-13. For the situation of Prob. 18-11, except for tube length and flow rate, determine the length per tube pass by the effectiveness method if the oil flow rate per tube is 15 lb_m/min.

18-14. Begin with Eq. (18-7) and develop an expression for ϵ_c analogous to Eq. (18-19).

18-15. An economizer is to preheat 40,000 lb_m/hr of pressurized water from 170 to 370°F. (An economizer is a heat exchanger which utilizes energy from boiler exhaust gas which would otherwise be lost.) A flue gas flow of 90,000 lb_m/hr at 750°F is available. The gas specific heat may be taken as 0.25 Btu/lb_m-°F. Determine the exchanger area and the exit air temperature for a single-pass counterflow design with an overall heat-transfer coefficient of 12 Btu/hr-ft²-°F by (a) the LMTD method and (b) the effectiveness method.

18-16. An intercooler is used between two stages of an air compressor. Air leaves the first stage at 180°F and is cooled to 120°F before entering the second stage. The design is a double tube pass, single shell pass with air in the tubes and cooling water in the shell. Water from a cooling tower is available at 90°F. For an airflow rate of 300 lb_m/min, an overall heat-transfer coefficient of 20 Btu/hr-ft²-°F, and an area of 200 ft², determine water flow rate and exit temperature of the water.

18-17. An air-cooled Freon-12 condenser (both fluids unmixed) for an air-conditioning application is designed to remove 48,000 Btu/hr with an airflow velocity of 40 ft/sec. Under design conditions with inlet air at 95°F, Freon pressure of 221 psia corresponding to a saturation temperature of 140°F, and an exit air temperature of 110°F, the overall heat-transfer coefficient is found by experiment to be 32 Btu/hr-ft²-°F. The condenser is to be provided with a two-speed blower for reduced-noise operation at night; the lower velocity is 20 ft/sec. Assume that U varies with the $\frac{3}{4}$ power of air velocity and that the mass flow of air varies directly with velocity to determine the cooling capacity under the reduced-noise condition. Use the same inlet air temperature. Assume that C is much greater for the condensing Freon than for the air.

18-18. A counterflow heat exchanger is used to cool 10,000 lb_m/hr of oil having a specific heat of 0.5 Btu/lb_m-°F with a 7,500 lb_m/hr water flow. The area is 125 ft², and U is 60 Btu/hr-ft²-°F. The oil enters at 140°F, and the water enters at 70°F. Determine the rate of heat transfer and the exit temperatures of both fluids.

18-19. Repeat Prob. 18-18 for (a) parallel flow and (b) cross flow with both fluids unmixed. Compare results.

18-20. After extensive use in heating boiler feedwater, the exchanger of Prob. 18-2 is reported to be operating with U_0 about 65 percent of the original value of 376 Btu/hr-ft²-°F. Determine whether this is due to tube internal blockage by large foreign particles or to normal fouling of the surfaces.

18-21. Vegetable oil in a food-processing plant is heated from 80 to 100°F by means of a hot-water double-pipe heat exchanger. The oil flows with a velocity of 3 ft/sec through a $\frac{1}{2}$-in steel pipe (ID = 0.622 in, OD = 0.840 in) while the hot water flows through the annular section formed by a 1-in steel pipe (ID = 1.049 in, OD = 1.315 in). The water inlet temperature is 150°F, and the water-side Reynolds number based on hydraulic diameter may be taken as 50,000.

(a) Use the Dittus-Boelter equation to determine an outside heat-transfer coefficient assuming an average water temperature of 140°F, and use the constant-wall-temperature, infinite-length laminar-flow equation to obtain an inside heat-transfer coefficient, assuming the vegetable oil to have the same properties as those tabulated for light oil. Use these results in obtaining an appropriate "clean" U.

(b) What would be the expected U after extended use, assuming average city water?

(c) What length counterflow exchanger should be used?

REFERENCES

1. Bowman, R. A., A. C. Mueller, and W. M. Nagle: Mean Temperature Difference in Design, *Trans. ASME*, **62**: 283–294 (1940).
2. Coppage, J. E., and A. L. London: The Periodic-flow Regenerator: A Summary of Design Theory, *Trans. ASME*, **75**: 779–787 (1953).
3. Jakob, J.: "Heat Transfer," vol. 2, chap. 35, Wiley, New York, 1957.
4. Kays, W. M., and A. L. London: "Compact Heat Exchangers," National Press, Palo Alto, Calif., 1955.
5. Kreith, F.: "Principles of Heat Transfer," 2d ed., pp. 497–498, International Textbook, Scranton, Pa., 1965.
6. Macklin, M.: A Guide to Optimum Design of Compact Heat Exchangers, *Machine Design*, April 12, 1962, p. 132.
7. Mueller, A. C.: Thermal Design of Shell-and-tube Heat Exchangers for Liquid-to-liquid Heat Transfer, *Purdue Univ. Eng. Exp. Sta. Eng. Bull. Res. Ser.* 121, 1954.
8. Nusselt, W.: A New Heat Transfer Formula for Cross-flow, *Tech. Mech. Thermodyn.* **12** (1930).
9. Perry, J. H. (ed.): "Chemical Engineers' Handbook," 4th ed., McGraw-Hill, New York, 1963.
10. "Standards of Tubular Exchanger Manufacturers Association," 4th ed., New York, 1959.
11. Ten Broeck, H.: Multipass Exchanger Calculations, *Ind. Eng. Chem.*, **30**: 1041–1042 (1938).

CHAPTER 19
OPEN-CHANNEL FLOW

Man's first contact with fluid transport came as he observed the flow of water in rivers and streams. He was alternately blessed by the flow of water under control, in irrigation and drainage, and plagued as flood waters raced out of control, destroying his crops.

The Roman aqueducts stand as testimony to man's early accomplishments using open-channel flow. These early successes came with very little fundamental understanding of the flow mechanism. It was not until the eighteenth century that the relationship between flow rate and channel characteristics began to evolve. Since then, man has learned much about the regulation and control of natural channel flow and has devised a multitude of artificial channels for his convenience.

In order to overcome the shearing forces in closed-conduit flow, the energy is supplied by the gradient of piezometric head, the sum of the gradients of pressure head and elevation head

$$\frac{d}{dx}\frac{p}{\gamma} + \frac{dz}{dx}$$

In open channels there is no gradient of pressure head; changes occur as a result of gravitational influence. In this respect, open-channel flow is simpler than flow in pipes that are full; however, the geometry of the free surface is not known a priori, giving a "floating" boundary condition. As one might suspect after these comments, the mechanism of open-channel flow pertains to closed conduits which are flowing partially full, such as storm sewers and sanitary sewers, characterized by their free surface.

With increasing depth of flow in closed conduits flowing partially full, a transition point is reached between open-channel and closed-conduit flow. Due to the continuous nature of natural phenomena, we should not expect a significant discontinuity in the flow process; hence, we should expect to find a single function which might be used to analyze both types of flow. The Darcy-Weisbach equation is this function, being applicable to both types of flow.

19-1 THE SPEED OF AN ELEMENTARY WAVE

The velocity of a single simple wave on the surface of a liquid is found in much the same way as the speed of sound was determined in Chap. 15. The unsteady propagation of a wave of celerity c from a point of disturbance in a nonaccelerating fluid is shown in Fig. 19-1a. Figure 19-1b is the same flow transformed to steady flow by superimposing the negative of the wave velocity on the control volume chosen. This is directly analogous to the propagation of an elastic wave in a compressible fluid depicted in Fig. 15-1.

For unit depth, the continuity equation applied to the control volume of Fig. 19-1b gives

$$cy = (c - dV)(y + dy)$$

or, neglecting higher-order terms,

$$\frac{dV}{c} - \frac{dy}{y} = 0 \tag{19-1}$$

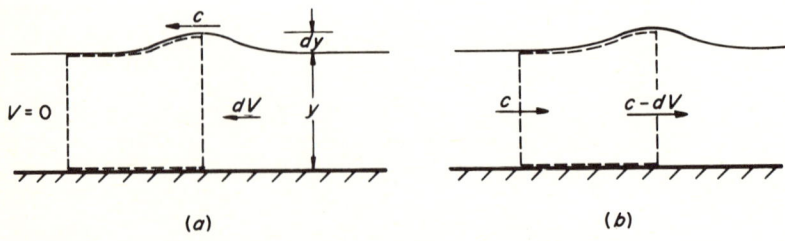

Fig. 19-1 Elementary surface wave: (a) observer at rest; (b) observer moving with wave.

OPEN-CHANNEL FLOW

Assuming the energy loss in the wave to be negligible, the Bernoulli equation (10-91) applied between the two control surfaces across which the fluid flows yields

$$y + \frac{c^2}{2g} = (y + dy) + \frac{(c - dV)^2}{2g}$$

Again neglecting higher-order terms, this becomes

$$c\,dV - g\,dy = 0 \tag{19-2}$$

Combining Eqs. (19-1) and (19-2), the speed of a wave whose height is small compared with the channel depth, $dy \ll y$, is

$$c = \sqrt{gy} \tag{19-3}$$

For nonrectangular channels the result is

$$c = \sqrt{gy_h} \tag{19-4}$$

where y_h is the hydraulic depth defined as the area of the flow cross section divided by the surface width; for a rectangular channel $y = y_h$.

19-2 TYPES OF MOTION

Four flow classifications are needed to describe the flow in an open channel completely:

1. Steady or unsteady
2. Uniform or nonuniform
3. Laminar or turbulent
4. Tranquil or rapid

One from each of these four types must exist simultaneously. In Chap. 9 we considered the first three types. It should be noted, however, that for the Reynolds number which classifies the flow as laminar or turbulent in an open channel the

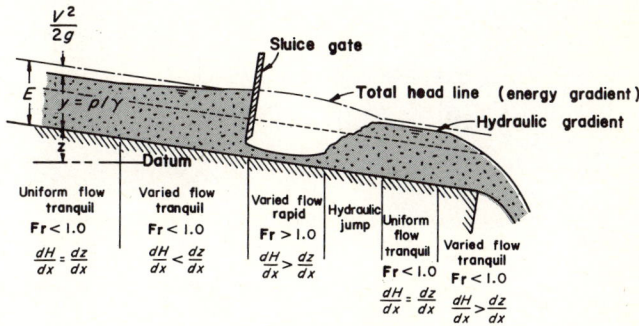

Fig. 19-2 Common types of open-channel flow.

Table 19-1

Open-channel flow		Compressible flow	
Fr < 1	tranquil (subcritical)	M < 1	subsonic
Fr = 1	critical	M = 1	sonic
Fr > 1	rapid (supercritical)	M > 1	supersonic

characteristic length is the hydraulic radius R_h; most open-channel flows are turbulent [1].

A flow is classified as tranquil or rapid depending upon the magnitude of the Froude number

$$\mathbf{Fr} \equiv \frac{V}{\sqrt{gy_h}} \tag{19-5}$$

Note the analogy between the definition of the Froude number and that of the Mach number [Eq. (15-8)]. Table 19-1, which characterizes open-channel flow, extends the analogy. Figure 19-2 illustrates the most common types of open-channel flow;

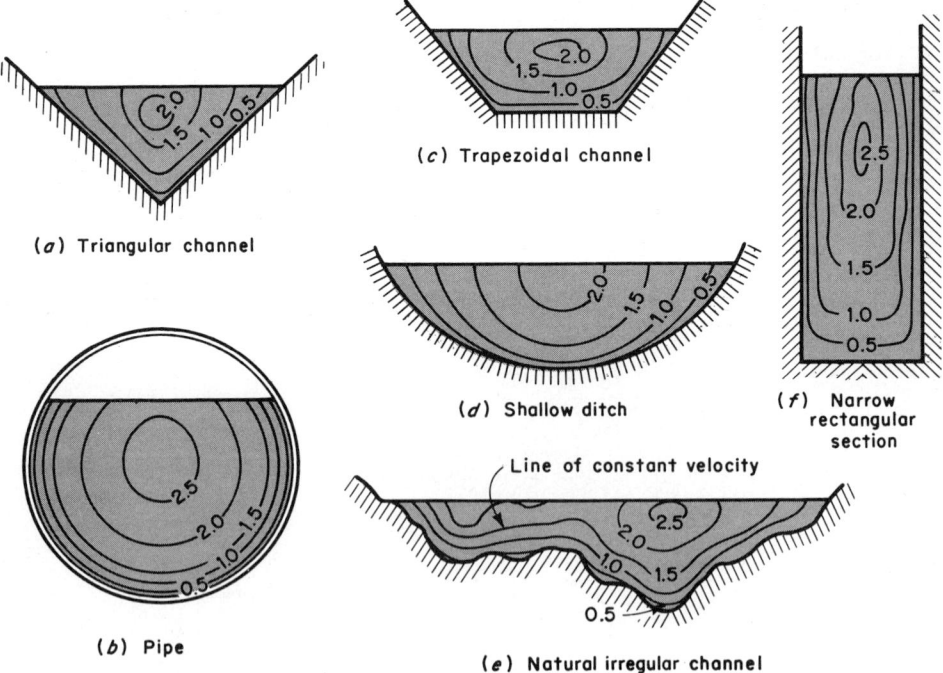

Fig. 19-3 Typical velocity variations. (*From V. J. Chow, "Open-channel Hydraulics." Copyright 1959. McGraw-Hill Book Company. Used by permission.*)

OPEN-CHANNEL FLOW

$H = E + z$. The hydraulic jump and flow under a sluice gate are examples of nonuniform flow, discussed in Sec. 19-5 when energy is qualitatively considered.

In open-channel flow the velocity varies from zero along the wetted perimeter to a maximum at or near the free surface, depending upon the shape of the cross section. Figure 19-3 illustrates typical velocity variations in open channels.

19-3 STEADY, UNIFORM, TURBULENT FLOW

Since the same shear-resistance mechanism pertains to tranquil and rapid flows, no distinction will be made between them in this analysis. The restriction of this analysis to uniform flow is much more stringent than the assumption of steady turbulent flow since most open-channel flows are nonuniform, especially in natural channels such as river beds. Approximate analyses can be made of most open-channel flows, however, by assuming uniform flow.

Consider the flow in a prismatic open channel of uniform slope S, depicted in Fig. 19-4, where S is the drop per unit length of the channel bed. Applying the momentum equation to the control volume, we get

$$\tau_s P L = \gamma A L \sin \theta$$

where P is the wetted perimeter. But

$$\sin \theta = \frac{\Delta y}{L} \simeq \tan \theta = S$$

and

$$\frac{A}{P} \equiv R_h$$

therefore

$$\tau_s \cong \gamma R_h S \tag{19-6}$$

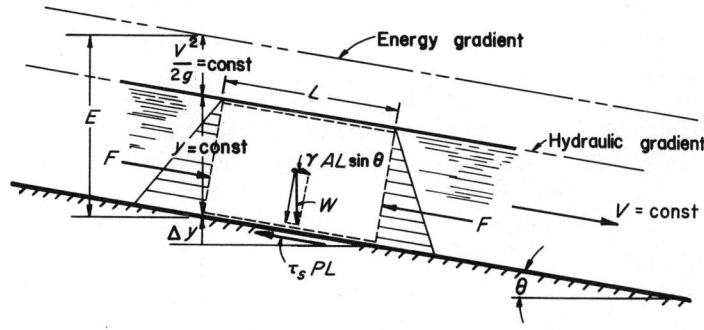

Fig. 19-4 Segment of open-channel flow.

Measurements made on open channels show that the shear mechanism relates the friction factor to the Reynolds number in the same manner as in closed conduits [4]. From the development of Eq. (12-46), the shear on a cylindrical fluid element (Fig. 12-5) in a pipe is

$$\tau = \frac{\Delta p}{L} \frac{r}{2}$$

This equation, which is valid for both laminar and turbulent flow, gives the shear stress at the wall τ_s

$$\tau_s = \frac{\Delta p}{L} \frac{D}{4}$$

Using the Darcy-Weisbach equation (12-37a), this becomes

$$\tau_s = f \frac{\rho V^2}{8} \tag{19-7}$$

Eliminating τ_s between Eqs. (19-6) and (19-7) gives the celebrated *Chézy equation* [6]

$$V = C\sqrt{R_h S} \tag{19-8}$$

where

$$C \equiv \sqrt{\frac{8g}{f}} \tag{19-9}$$

is the *resistance factor*.

The Chézy resistance factor depends upon many things, including the relative roughness of the channel, obstructions, bends, marine and plant growths, and cross-sectional changes. The large number of parameters which affect its magnitude make it impossible to express the resistance factor explicitly. The most popular empirical relation was proposed by the Irish engineer Robert Manning in 1891 [10], namely,

$$C = \frac{1.486}{n} R_h^{\frac{1}{6}} \tag{19-10}$$

where n is a roughness factor, published widely for a variety of surface materials. Appendix Table A-16 gives average values for a select spectrum of materials. Using this formulation for the Chézy resistance factor, the velocity is conveniently expressed by the *Manning equation*

$$\blacksquare \quad V = \frac{1.486}{n} R_h^{\frac{2}{3}} S^{\frac{1}{2}} \tag{19-11}$$

in British engineering units, $V \sim$ feet per second. Note that n is not a dimensionless factor but has dimensions of $L^{\frac{1}{6}}$. Also, the constant 1.486 has dimensions $\sqrt{L}/T$.

OPEN-CHANNEL FLOW

Example A 2-ft-diam storm drain of unfinished concrete flows one-half full of water. For a slope of 0.002 (*a*) determine the flow rate. (*b*) What is the flow regime?

Solution (*a*) The hydraulic radius is

$$R_h = \frac{A}{P} = \frac{\pi R^2/2}{\pi R} = \frac{R}{2} = \frac{1}{2} \text{ ft}$$

From Table A-16 the Manning roughness factor is

$$n = 0.014 \text{ ft}^{\frac{1}{6}}$$

Using Eq. (19-11), the velocity is

$$V = \frac{1.486 \sqrt{\text{ft/sec}^2}}{0.014 \text{ ft}^{\frac{1}{6}}} (0.5 \text{ ft})^{\frac{2}{3}} (0.002)^{\frac{1}{2}} = 2.98 \text{ ft/sec}$$

giving a flow rate of

$$Q = VA = (2.98 \text{ ft/sec}) \left(\frac{\pi}{2} \text{ ft}^2\right) = 4.68 \text{ ft}^3/\text{sec}$$

(*b*) The Froude number determines the flow regime; hence, from Eq. (19-5)

$$\mathbf{Fr} = \frac{V}{\sqrt{gy_h}} = \frac{2.98 \text{ ft/sec}}{\sqrt{32.2 \text{ ft/sec}^2 (\pi/4 \text{ ft})}}$$

since $y_h = A/w$ where w is the cross-sectional width, or

$$y_h = \frac{\pi R^2/2 \text{ ft}^2}{2 \text{ ft}}$$

Therefore,

$$\mathbf{Fr} = 0.593$$

and the flow is tranquil.

19-4 OPTIMUM SHAPE OF CROSS SECTION

For the maximum flow rate in an open channel of a given slope and surface material, it is obvious from Eq. (19-11) that the hydraulic radius must be maximized. From the elementary calculus, the cross section with the minimum wetted perimeter for a given area is a circle or semicircle. Although hydraulically optimum, it is not normally feasible from the construction and erosion standpoints to use semicircular channels; trapezoidal channels are more feasible.

The optimum trapezoidal shape, shown in Fig. 19-5, can be determined by solving for the wetted perimeter, differentiating with respect to depth while holding the area constant, and setting the result equal to zero for a minimum value of wetted perimeter. The resulting hydraulically optimum trapezoid is one in which a semicircle can be inscribed with its center at the surface as shown in Fig. 19-5*b*. This shape is feasible only when the surface is lined with a stabilizing material which holds the shape against erosion. In practice, such factors as bank stability, excavation costs, and ease of maintenance require that channels be much wider than those which are hydraulically optimum.

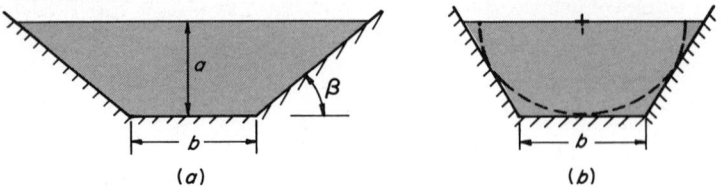

Fig. 19-5 Trapezoidal cross section: (*a*) general; (*b*) optimum.

19-5 TRANSITIONS IN OPEN CHANNELS

If the bed of a channel of constant width is raised within a short reach (length) of the channel, as shown in Fig. 19-6, the velocity increases as a result of the reduction in cross section, decreasing the depth of the fluid. Since the frictional resistance is small in a short distance, the Bernoulli equation (10-91) is applicable, giving

$$H = E_1 + z_1 = E_2 + z_2 \tag{19-12}$$

where

$$E \equiv y + \frac{V^2}{2g} \tag{19-13}$$

is the *specific energy*, the head above the floor of the channel. For a rectangular channel the discharge per unit width is $q = Vy$; therefore,

$$E = y + \frac{q^2}{2gy^2} \tag{19-14}$$

which plots as a specific-energy diagram when q is held constant. Figure 19-7 shows a family of curves for different q's. As the depth approaches zero, the kinetic energy approaches infinity, becoming asymptotic to the abscissa. As y increases without limit, $V \to 0$ since the flow rate is fixed and $A \to \infty$, and the upper branch of the curve is asymptotic to the 45° line shown.

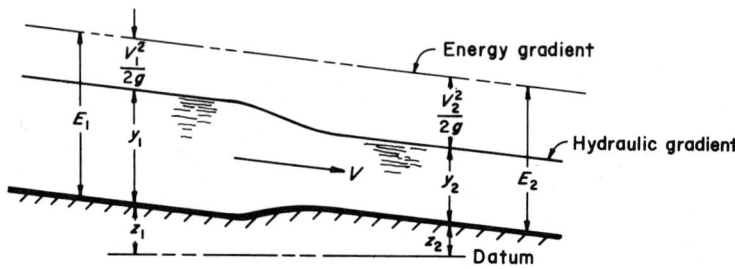

Fig. 19-6 Flow in transition.

OPEN-CHANNEL FLOW

At the point of minimum specific energy the depth y_c is called the *critical depth* and occurs when the Froude number is unity. When the specific energy is greater than this minimum, the flow can exist at two different or *alternate depths*. Alternate depths, qualitatively representing those of Fig. 19-6, are shown on the specific-energy diagram in Fig. 19-7. The line of **Fr** = 1 divides the tranquil and rapid flow regimes. The specific-energy curve is analogous to the Fanno curve of Fig. 15-15.

Flow in transition is nonuniform. Important examples of nonuniform flow are the hydraulic jump and flow under a sluice gate.

The hydraulic jump Often used to dissipate energy when rapid flow is encountered in open channels, the hydraulic jump is closely akin to the sudden expansion in closed-conduit flow. It can be utilized to reduce the water velocity downstream from the spillway of a dam in order to minimize erosion of the downstream channel.

A hydraulic jump occurs only when the flow upstream is rapid and the flow downstream is tranquil, as illustrated in Fig. 19-8. Assuming the bed of the channel to be horizontal, we may neglect gravity forces in the direction of flow. Also, since the jump occurs in a very short reach of channel, we shall neglect shear resistance. Writing the continuity and momentum equations for the control volume shown, under these conditions for steady flow, we get

$$V_1 y_1 = V_2 y_2 \tag{19-15}$$

$$\frac{\gamma y_1^2}{2} - \frac{\gamma y_2^2}{2} = \frac{V_1 y_1 \gamma}{g}(V_2 - V_1) \tag{19-16}$$

Combining and rearranging,

$$\mathbf{Fr_1} \equiv \frac{V_1}{\sqrt{gy_1}} = \left[\frac{1}{2}\frac{y_2}{y_1}\left(\frac{y_2}{y_1} + 1\right)\right]^{\frac{1}{2}} \tag{19-17}$$

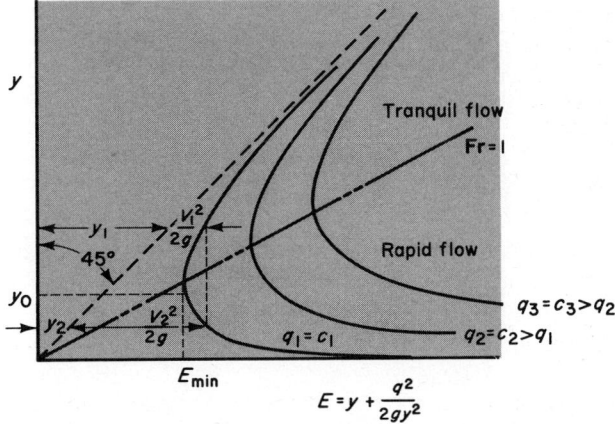

Fig. 19-7 Specific-energy diagram.

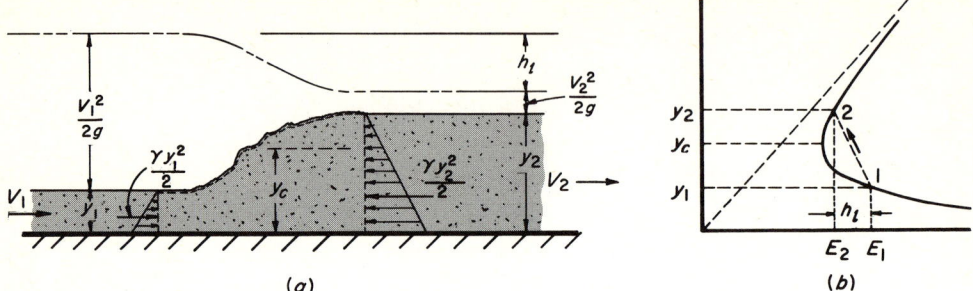

Fig. 19-8 Hydraulic jump.

which relates the alternate depths to the approach velocity. Solving for the depth ratio in terms of the Froude number,

$$\frac{y_2}{y_1} = \frac{1}{2}(\sqrt{1 + 8\mathbf{Fr}_1^2} - 1) \tag{19-18}$$

When the Froude number is unity, the upstream and downstream depths are equal to the critical depth y_c. Experimental measurements indicate that Eq. (19-18) relates the upstream and downstream depths within 1 percent [15].

The energy which might be dissipated by a hydraulic jump h_l is found from the energy equation, i.e.,

$$h_l = E_1 - E_2 \tag{19-19a}$$

$$h_l = (y_1 - y_2) + \left(\frac{V_1^2}{2g} - \frac{V_2^2}{2g}\right) \tag{19-19b}$$

Using continuity to eliminate V_2,

$$h_l = (y_1 - y_2) + \frac{V_1^2}{2g}\left[1 - \left(\frac{y_1}{y_2}\right)^2\right] \tag{19-20}$$

Eliminating V_1^2/g between Eqs. (19-17) and (19-20), the energy loss in a hydraulic jump is

$$h_l = \frac{(y_2 - y_1)^3}{4y_1 y_2} \tag{19-21}$$

where the units are ft. We see from this equation that the jump must occur from a depth below y_c, that is, $y_2 > y_1$, in order for a loss of energy to occur even though Eq. (19-18) is symmetrical in y_1 and y_2. This again reminds us of compressible flow, with the hydraulic jump being analogous to the normal shock [5, 12].

Example Water at 60 ft/sec flows down a spillway into a horizontal channel. For a flow rate of 180 ft³/sec per foot of channel width (a) what depth of tail water will produce a hydraulic jump? (b) How much energy is dissipated in the jump?

OPEN-CHANNEL FLOW

Solution (a) The upstream depth is

$$y_1 = \tfrac{180}{60} = 3 \text{ ft}$$

which can be substituted into Eq. (19-18) to get the downstream depth y_2

$$\frac{y_2}{3 \text{ ft}} = \frac{1}{2}\left[\sqrt{1 + \frac{8(60 \text{ ft/sec})^2}{(32.2 \text{ ft/sec}^2)(3 \text{ ft})}} - 1\right]$$

$$y_2 = 24.45 \text{ ft}$$

(b) The energy dissipated may be found from Eq. (19-21), i.e.,

$$h_l = \frac{(24.45 - 3)^3}{4(3)(24.45)} = 33.6 \text{ ft}$$

or, in terms of energy,

$$h_l = \frac{(33.6 \text{ ft})(180 \text{ ft}^3/\text{sec})(62.4 \text{ lb}_f/\text{ft}^3)}{550 \text{ ft-lb}_f/\text{hp-sec}}$$

$$= 686 \text{ hp/ft of channel width}$$

19-6 THE ANALOGY BETWEEN LIQUID AND GAS FLOW

Throughout this chapter we have pointed out the similarities between open-channel flow and the flow of a compressible fluid. In this section we present a simple and economical means of experimentally studying compressible flow by using a free-surface-flow device called a *water table*, since water is the liquid commonly used.

The similarity between the surface waves produced by ships and waves formed by a body moving through air at supersonic speeds was noted as early as 1887 by Mach [9]. The general theory relating the liquid and gas flows was developed by Jouquet in 1920 [7]. Black and Mediratta extended the generalized treatment to include the analogy between hydraulic jump and normal shock [2].

For two-dimensional frictionless steady flow Table 19-2 summarizes the governing equations for liquid and gas flows. From the continuity equations we note that the vertical depth of the liquid y is analogous to the density ρ of the compressible fluid, or $y/y_0 \sim \rho/\rho_0$. For complete analogy it is necessary, of course, to have analogous boundary conditions. This can be satisfied by keeping the geometry of the open channel identical to that of the compressible-flow conduit. Obviously, boundary-layer effects must be negligible in accordance with our assumption of frictionless flow.

Integrating the energy equations between a stagnation station ($V_0 = 0$) and a general station, we get from the liquid and gas equations, respectively,

$$V^2 = 2g(y_0 - y) \qquad (19\text{-}22)$$

$$V_{\max}^2 = 2gy_0 \qquad (19\text{-}23)$$

Table 19-2 Governing equations for two-dimensional steady flow

	Frictionless liquid	Perfect gas
Continuity	$\dfrac{\partial}{\partial \xi} yu + \dfrac{\partial}{\partial \eta} yv = 0$	$\dfrac{\partial}{\partial \xi} \rho u + \dfrac{\partial}{\partial \eta} \rho v = 0$
Energy	$g\,dy + \mathbf{V}\,d\mathbf{V} = 0$	$\dfrac{kR}{k-1} dT + \mathbf{V}\,d\mathbf{V} = 0$
Velocity potential [11, 8]	$\dfrac{\partial^2 \phi}{\partial \xi^2}\left[gy - \left(\dfrac{\partial \phi}{\partial \xi}\right)^2\right]$ $+ \dfrac{\partial^2 \phi}{\partial \eta^2}\left[gy - \left(\dfrac{\partial \phi}{\partial \eta}\right)^2\right]$ $- 2\dfrac{\partial^2 \phi}{\partial \xi\,\partial \eta}\dfrac{\partial \phi}{\partial \xi}\dfrac{\partial \phi}{\partial \eta} = 0$	$\dfrac{\partial^2 \phi}{\partial \xi^2}\left[c^2 - \left(\dfrac{\partial \phi}{\partial \xi}\right)^2\right]$ $- \dfrac{\partial^2 \phi}{\partial \eta^2}\left[c^2 - \left(\dfrac{\partial \phi}{\partial \eta}\right)^2\right]$ $- 2\dfrac{\partial^2 \phi}{\partial \xi\,\partial \eta}\dfrac{\partial \phi}{\partial \xi}\dfrac{\partial \phi}{\partial \eta} = 0$

where y = vertical depth of liquid
u, v = velocity components in the ξ, η orthogonal directions
$\mathbf{V} = \mathbf{i}u + \mathbf{j}v$
ϕ = velocity potential
c = acoustic velocity

or

$$\left(\frac{V}{V_{\max}}\right)^2 = 1 - \frac{y}{y_0} \quad \text{liquid} \tag{19-24}$$

$$V^2 = \frac{2kR}{k-1}(T_0 - T) \tag{19-25}$$

$$V^2_{\max} = \frac{2kR}{k-1} T_0 \tag{19-26}$$

or

$$\left(\frac{V}{V_{\max}}\right)^2 = 1 - \frac{T}{T_0} \quad \text{gas} \tag{19-27}$$

where y_0 is the liquid depth where the velocity is zero. From Eqs. (19-24) and (19-27) we observe that the depth ratio y/y_0 for the liquid is analogous to the temperature ratio T/T_0 for the gas; that is, $y/y_0 \sim T/T_0$.

OPEN-CHANNEL FLOW

Table 19-3

Frictionless liquid	Perfect-gas analog ($k = 2$)
$\dfrac{y}{y_0}$	$\dfrac{\rho}{\rho_0}$
$\dfrac{y}{y_0}$	$\dfrac{T}{T_0}$
$\left(\dfrac{y}{y_0}\right)^2$	$\dfrac{p}{p_0}$
$\mathbf{Fr} \equiv \dfrac{V}{\sqrt{gy}}$	$\mathbf{M} \equiv \dfrac{V}{\sqrt{kRT}}$

We now have two quantities for the gas, ρ and T, which are analogous to the same quantity y/y_0 of the liquid, and we must ask how this is possible. Consider the isentropic relations for a perfect gas:

$$\frac{\rho}{\rho_0} = \left(\frac{T}{T_0}\right)^{1/(k-1)} \qquad (19\text{-}28)$$

$$\frac{p}{p_0} = \left(\frac{\rho}{\rho_0}\right)^k = \left(\frac{T}{T_0}\right)^{k/(k-1)} \qquad (19\text{-}29)$$

From Eq. (19-28), $\rho/\rho_0 = T/T_0$ only when $k = 2$, giving $p/p_0 = (T/T_0)^2$ from Eq. (19-29). Therefore, the analogy holds exactly only when the specific-heat ratio $k = 2$—for a *fictitious gas*, since k for most gases is in the range 1.2 to 1.5. Fortunately, the error introduced with this assumed fictitious gas is small in both the subsonic and supersonic regimes [3, 14]. The parameters can be corrected, however, by the correction factors discussed below. From the velocity-potential equations of Table 19-2, we note that the acoustic velocity c in a perfect gas is analogous to the velocity of an elementary wave $\sqrt{gy}$ in a liquid, that is, $c \sim \sqrt{gy}$, as mentioned earlier in the development of the Froude number equation (19-5).

The analogous quantities are summarized in Table 19-3. Expressing Eq. (19-22) in terms of the Froude number,

$$\frac{y_0}{y} = 1 + \tfrac{1}{2}\mathbf{Fr}^2 \qquad (19\text{-}30)$$

which, for $k = 2$, gives

$$\frac{T_0}{T} = 1 + \tfrac{1}{2}\mathbf{Fr}^2 \qquad (19\text{-}31)$$

But, for $k = 2$ Eq. (15-36) gives

$$\frac{T_0}{T} = 1 + \tfrac{1}{2}\mathbf{M}^2 \tag{19-32}$$

hence,

$$\mathbf{M} = \mathbf{Fr} = \frac{V}{\sqrt{gy}} = \frac{\sqrt{2g(y_0 - y)}}{\sqrt{gy}}$$

$$\mathbf{M}^2 = \frac{2(y_0 - y)}{y} \tag{19-33a}$$

or

$$\frac{y}{y_0} = \frac{2}{\mathbf{M}^2 + 2} \tag{19-33b}$$

The geometry of the flow channel can be established by using Eq. (15-54), giving

$$\frac{w}{w^*} = \frac{1}{\mathbf{M}} \left(\frac{2 + \mathbf{M}^2}{3} \right)^{\frac{3}{2}} \tag{19-34}$$

where w is the width of the channel having a uniform depth of y and w^* is the critical width.

Correction factors for the fictitious gas ($k = 2$) By noting that the stagnation temperature will be the same for either a real or a fictitious gas, we can get a temperature correction factor by making use of Eq. (15-36):

$$\frac{T_{k=1.4}}{T_{k=2}} = \frac{(T_0/T)_{k=2}}{(T_0/T)_{k=1.4}} = \frac{1 + 0.5\mathbf{M}^2}{1 + 0.2\mathbf{M}^2} \tag{19-35}$$

Combining with Eq. (19-33a),

$$\frac{T_{k=1.4}}{T_{k=2}} = \frac{(T_0/T)_{k=2}}{(T_0/T)_{k=1.4}} = \frac{1}{0.4 + 0.6(y/y_0)} \tag{19-36}$$

Similarly, the pressure and density correction factors for air ($k = 1.4$) are

$$\frac{p_{k=1.4}}{p_{k=2}} = \frac{(p_0/p)_{k=2}}{(p_0/p)_{k=1.4}} = \frac{(y/y_0)^{1.5}}{[0.4 + 0.6(y/y_0)]^{3.5}} \tag{19-37}$$

$$\frac{\rho_{k=1.4}}{\rho_{k=2}} = \frac{(\rho_0/\rho)_{k=2}}{(\rho_0/\rho)_{k=1.4}} = \frac{(y/y_0)^{1.5}}{[0.4 + 0.6(y/y_0)]^{2.5}} \tag{19-38}$$

In order for the analogy to hold in the presence of shock phenomena, the isentropic gas relations must be replaced by shock equations while the hydraulic relations are replaced by the equations for a hydraulic jump [2].

Other flow phenomena, such as high-speed combustion [13], can be studied by use of the hydraulic analog.

OPEN-CHANNEL FLOW

Example A small closed-loop water table, used in an undergraduate laboratory to study the operating characteristics of a convergent-divergent nozzle in supersonic flow, is shown in Fig. 19-9. Depth measurements taken along the channel centerline with the aid of the micrometer attachment were:

x (in)	y/y_0	x (in)	y/y_0	x (in)	y/y_0
1.00	1.00	5.00	0.7612	8.00	0.2067
2.00	0.9700	6.00	0.4436	9.00	0.1710
3.00	0.9718	7.00	0.2800	9.50	0.1691
4.00	0.9304				

The depth above the horizontal table for no flow was $y_0 = 0.532$ in. From the given data, plot $\mathbf{M}$, T/T_0, ρ/ρ_0, and p/p_0 vs. the distance along the nozzle centerline.

Solution The results are plotted in Fig. 19-10. To illustrate the procedure, calculations at the station $x = 9.50$ in follow.

From Eq. (19-33a)

$$\mathbf{M}^2 = 2\left(\frac{y_0}{y} - 1\right)$$

$$= 2\left(\frac{1}{0.1691} - 1\right)$$

$$\mathbf{M} = 3.134$$

The temperature ratio comes from Eq. (19-36) combined with Eq. (19-32):

$$\left.\frac{T}{T_0}\right|_{k=1.4} = \left(\frac{T}{T_0}\right)_{k=2} \frac{1}{0.4 + 0.6(y/y_0)}$$

$$= \frac{y/y_0}{0.4 + 0.6(y/y_0)}$$

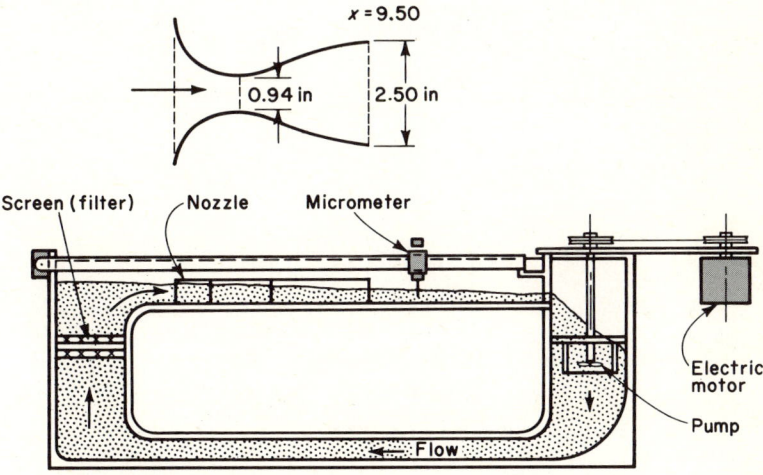

Fig. 19-9 Water table.

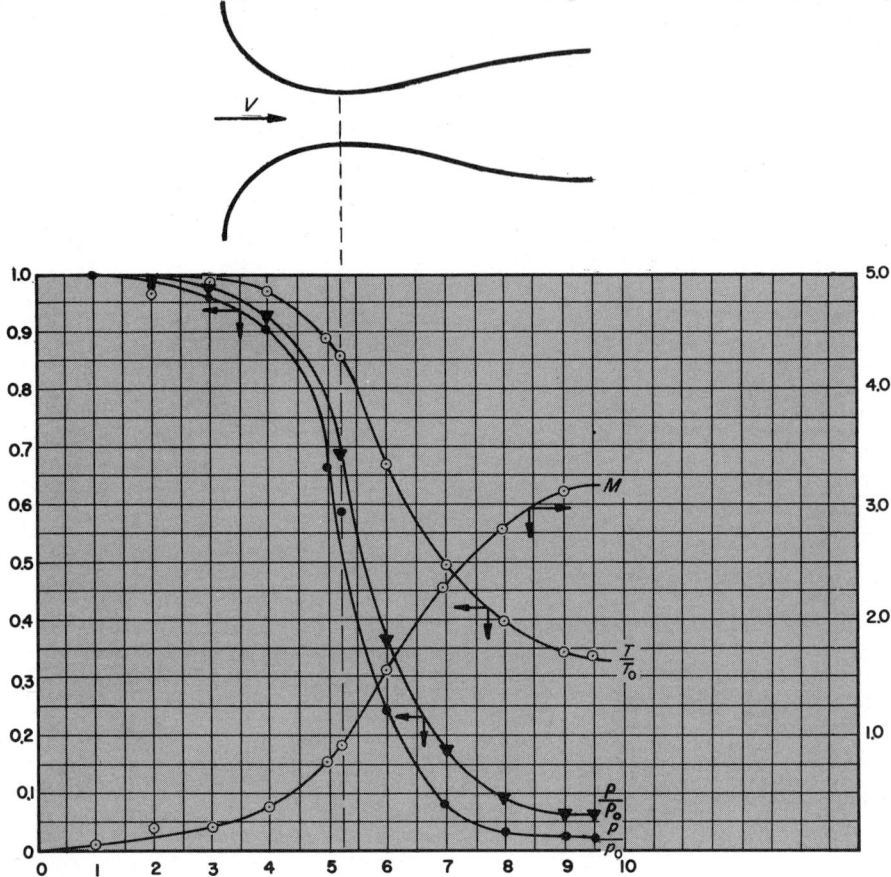

Fig. 19-10 Characteristics of air flowing in a convergent-divergent nozzle, from the hydraulic analog.

Therefore,

$$\left.\frac{T}{T_0}\right|_{k=1.4} = \frac{0.1691}{0.4 + 0.6(0.1691)} = 0.337$$

It is interesting to note that this is the same result from Eq. (15-36) at $\mathbf{M} = 3.134$, validating the use of the analogy.

Using Eq. (19-37), the pressure ratio is

$$\left.\frac{p}{p_0}\right|_{k=1.4} = \left(\frac{p}{p_0}\right)_{k=2} \frac{(y/y_0)^{1.5}}{[0.4 + 0.6(y/y_0)]^{3.5}}$$

$$= \left[\frac{y/y_0}{0.4 + 0.6(y/y_0)}\right]^{3.5}$$

OPEN-CHANNEL FLOW

since $p/p_0 = (y/y_0)^2$ for $k = 2$; therefore,

$$\left.\frac{p}{p_0}\right|_{k=1.4} = \left[\frac{0.1691}{0.4 + 0.6(0.1691)}\right]^{3.5} = 0.0222$$

In a similar manner the density ratio is determined from Eq. (19-38):

$$\left.\frac{\rho}{\rho_0}\right|_{k=1.4} = \left(\frac{\rho}{\rho_0}\right)_{k=2} \frac{(y/y_0)^{1.5}}{[0.4 + 0.6(y/y_0)]^{2.5}}$$

$$= \left[\frac{y/y_0}{0.4 + 0.6(y/y_0)}\right]^{2.5}$$

using $(\rho/\rho_0)_{k=2} = y/y_0$; hence,

$$\left.\frac{\rho}{\rho_0}\right|_{k=1.4} = 0.0660$$

The slight deviations of the critical conditions (throat) shown in Fig. 19-10 are primarily due to boundary-layer effects, since the water table was relatively small and no additives were used to reduce surface tension or friction.

The water table is a convenient, accurate, and economical method of simulating the flow of compressible fluids. It can be used to investigate flow through orifices, flow stability, the effect of model size on wind-tunnel choking, shock waves, and a variety of other phenomena which could not be studied economically otherwise.

PROBLEMS

19-1. At what speed will a small disturbance move in the channel shown?

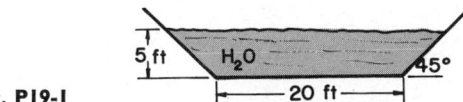

Fig. P19-1

19-2. For the three channels shown, having equal surface width and depth but different cross sections, compare the speed of an elementary wave.

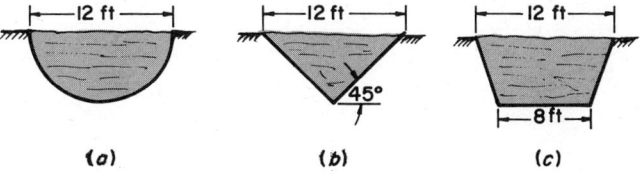

Fig. P19-2

19-3. The trapezoidal channel shown has a discharge rate of 1,695 ft³/sec. Is the flow tranquil or rapid?

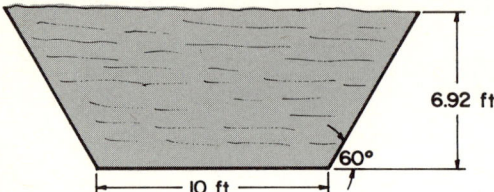

Fig. P19-3

19-4. An equilateral-triangular channel having 4-ft sides has a flow rate of 60 ft³/sec and a slope of 0.004.
  (a) Determine the average velocity.
  (b) What classification describes the flow?

19-5. A 6-ft-diam corrugated-metal cylindrical duct is to carry 100 ft³/sec when it is half full. What slope must it have?

19-6. A 20-ft-wide natural stream runs through a 6-in-deep rectangular channel of smooth stone ($n = 0.030$) before it cascades over a waterfall. The channel is long enough for the flow to be uniform, and the slope is 0.004. Classify the flow.

19-7. A rectangular flume of trowel-finished concrete has a slope of 0.012, a depth of 12 ft, and a width of 15 ft. Determine the flow rate and classify the flow if the flume is flowing (a) half full and (b) full.

19-8. A 15-ft-wide by 10-ft-deep rectangular channel has a slope of 0.0081 and a trowel-finished concrete surface. For depths of 1, 1½, 3, 5, and 7 ft (a) classify the flow, (b) determine the average shear stress, and (c) find the flow rate.

19-9. A circular corrugated-metal storm drain is to carry 50 ft³/sec of water at a slope of 0.008 when flowing half full. What size should it be?

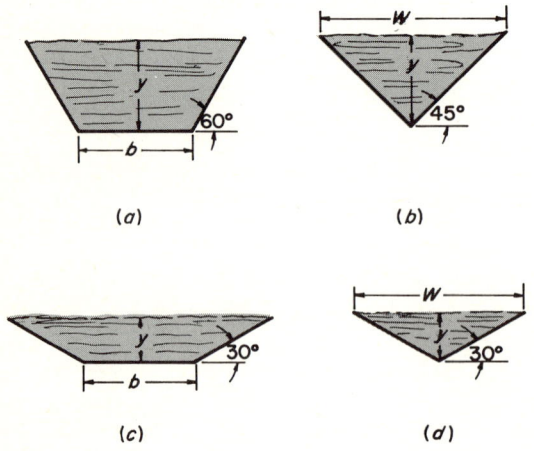

Fig. P19-12

19-10. A rectangular-shaped trowel-finished-concrete ditch with a bottom slope of 0.002 has a width of 8 ft and a depth of 4 ft and is flowing three-fourths full. Estimate the average velocity in the ditch.

19-11. A rectangular channel of trowel-finished concrete is to carry 60 ft³/sec at a slope of 0.007. Determine the minimum channel dimensions.

19-12. For the channel cross sections shown derive expressions for their hydraulic radii in terms of their depth.

19-13. A rectangular flume of trowel-finished concrete has a slope of 0.0007 and a discharge of 2,000 ft³/sec. What are the optimum dimensions of the channel?

19-14. A rectangular drainage canal is to be constructed of unfinished concrete. It will have a 1.5-ft drop per mile, and the maximum flow rate is to be 4,000 ft³/sec. Determine its optimum width and depth.

19-15. A trapezoidal-shaped channel of unfinished concrete has a discharge of 800 ft³/sec when flowing full with a slope of 0.0006.
  (*a*) Determine the optimum dimensions.
  (*b*) What is the average flow velocity?

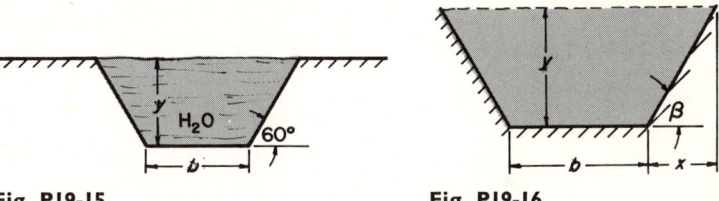

Fig. P19-15   Fig. P19-16

19-16. For the trapezoidal section shown, prove that the minimum wetted perimeter is $P_{min} = 2y(2 - \cos \beta)/\sin \beta$.

19-17. A trapezoidal channel with side slopes of 3:1 is to carry water at 300 ft³/sec with a bottom slope of 0.0016. Determine the bottom width, the depth, and the average velocity for the optimum design.

19-18. If 160 ft³/sec of wash water flows over an 8-ft-wide spillway at a depth of 3 in, is a hydraulic jump possible? If so, what is its downstream depth, and how much energy is dissipated?

19-19. A spillway channel has a flow rate of 140 ft³/sec per foot of width and a velocity of 35 ft/sec.
  (*a*) What downstream depth will produce a hydraulic jump?
  (*b*) How much energy is lost per pound of water as it flows through the jump?

19-20. Water flows 2 ft deep in a rectangular concrete flume 12 ft wide. The flow rate is 384 ft³/sec.
  (*a*) Is it possible for a hydraulic jump to occur?
  (*b*) If so, at what rate does the jump dissipate energy?

19-21. The tail water below a dam is 10 ft deep downstream from a hydraulic jump which occurs at the base of the spillway. The flow rate is 80 ft³/sec per foot of width; $Fr_1 = 2.5$. What is the upstream depth y_1, and how much energy is dissipated per foot of width?

19-22. A hydraulic jump occurs at the base of an 80-ft-wide spillway. The downstream depth is four times the upstream depth of 1.5 ft. What is the flow rate over the spillway?

19-23. Water flows at the rate of 1,300 ft³/sec in a horizontal rectangular channel which is 12 ft wide and has a water depth of 4 ft. Is a hydraulic jump possible for these conditions? If so, what is the downstream depth, and how much energy is dissipated?

19-24. If the variable area ducts shown were modeled on a water table, determine by inspection whether the water depth y is increasing or decreasing.

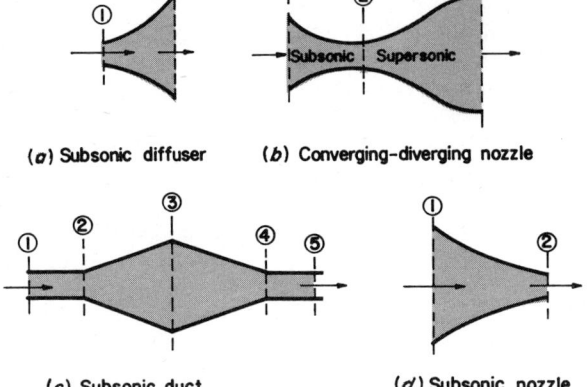

(a) Subsonic diffuser   (b) Converging-diverging nozzle

(c) Subsonic duct   (d) Subsonic nozzle

Fig. P19-24

19-25. For the subsonic nozzle shown, the following data were obtained from a water-table experiment:

x, in	1.0	2.0	3.0	4.0	5.0
y/y_0	1.0	0.96	0.82	0.71	0.667

The value for y_0 was 0.72 in. Plot **M**, T/T_0, and ρ/ρ_0 vs. the distance x along the nozzle.

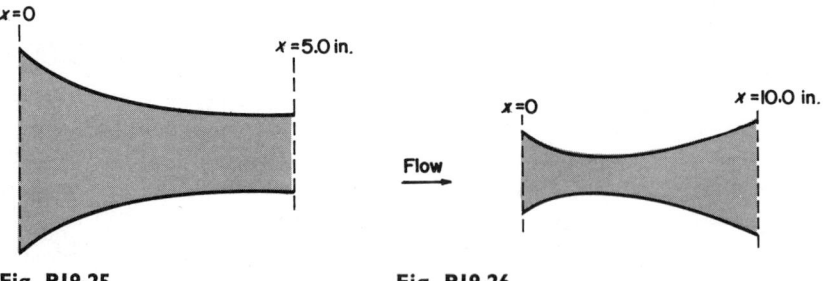

Fig. P19-25   **Fig. P19-26**

19-26. A water-table analogy of the converging-diverging duct shown yields the following data for $y_0 = 0.55$ in:

x	0.0	1.0	2.0	3.0	4.0	5.0	6.0	7.0	8.0	9.0	10.0
y/y_0	0.98	0.91	0.86	0.77	0.71	0.78	0.88	0.95	0.97	0.99	0.995

Plot **M**, T/T_0, ρ/ρ_0, and p/p_0 vs. the distance along the duct x.

OPEN-CHANNEL FLOW

19-27. A closed-loop water table is used to study a supersonic nozzle. The following data were obtained for the water depth along the centerline:

x, in	1.00	2.00	3.00	4.00	5.00
y/y_0	1.00	0.85	0.54	0.20	0.15

The water depth of the horizontal table for $V = 0$ was $y_0 = 0.45$ in.
  (a) Plot M and T/T_0 vs. x.
  (b) Locate the position of the nozzle throat.

19-28. The variable-area duct shown was modeled on a water table. For $y_0 = 1.0$ in the following data were obtained:

x, in	1.0	2.0	3.0	4.0	5.0	6.0	7.0	8.0	9.0	10.0
y/y_0	1.0	0.89	0.77	0.667	0.52	0.52	0.52	0.56	0.61	0.64

Plot M and T/T_0 vs. x.

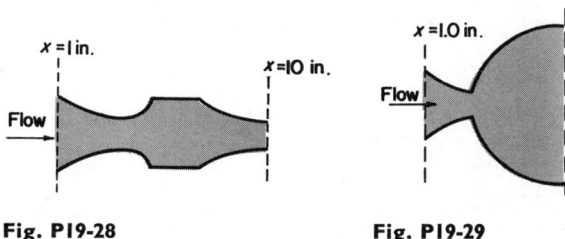

Fig. P19-28   **Fig. P19-29**

19-29. The converging-diverging duct shown resulted in the following data when it was modeled on a water table; $y_0 = 1.25$ in.

x, in	1.0	2.0	3.0	4.0	5.0	6.0
y/y_0	0.92	0.84	0.667	0.35	0.27	0.18

Plot the Mach number vs. the distance along the duct centerline.

REFERENCES

1. Albertson, M. L., J. R. Barton, and D. B. Simons: "Fluid Mechanics for Engineers," Prentice-Hall, Englewood Cliffs, N.J., 1960.
2. Black, J., and O. P. Mediratta: Supersonic Flow Investigations with a "Hydraulic Analogy" Water Channel, *Aeron. Quart.*, **2**: 227–253 (1951).
3. Cambel, A. B., and B. H. Jennings: "Gas Dynamics," McGraw-Hill, New York, 1958.
4. Chow, V. T.: "Open-channel Hydraulics," McGraw-Hill, New York, 1959.
5. Henderson, F. M.: "Open Channel Flow," Macmillan, New York, 1966.

6. Herschel, C.: On the Origin of the Chézy Formula, *J. Assoc. Eng. Soc.*, **17**: 363, 369 (1887).
7. Jouquet, E.: Some Problems in General Hydrodynamics, *J. Math. Pures Appl.*, (8) **3**(1): 3–13 (1920).
8. Kinslow, R.: The Hydraulic Analogy as a Teaching Aid, *J. Engrg. Educ.*, **49**(8): 731–734 (1959).
9. Mach, E.: Photography of Projectile Phenomena in Air, *Sitzber. Wiener Akad.*, **95**: 164 (1887).
10. Manning, R.: On the Flow of Water in Open Channels and Pipes, *Trans. Inst. Civil Engrs. Ireland*, **20**: 161–207 (1891).
11. Murphy, G., D. J. Shippy, and H. L. Luo: "Engineering Analogies," Iowa State University Press, Ames, 1963.
12. Olson, R. M.: "Engineering Fluid Mechanics," 2d ed., International Textbook, Scranton, Pa., 1966.
13. Oppenheim, A. K.: Water-channel Analog to High-velocity Combustion, *J. Appl. Mech.*, March, 1953, p. 115.
14. Shapiro, A. H.: Analogue Methods, in B. Lewis, R. N. Pease, and H. S. Taylor (eds.), "Physical Measurements in Gas Dynamics and Combustion," Princeton University Press, Princeton, N.J., 1954.
15. Streeter, V. L.: "Fluid Mechanics," 4th ed., McGraw-Hill, New York, 1966.

CHAPTER 20

FLOW THROUGH PERMEABLE MEDIA

The fundamentals of multiphase flow were developed in Chap. 17. In this chapter we shall consider the flow of a single-phase fluid through a permeable and sometimes discontinuous medium, making use of the ideas presented in Chap. 17.

The discontinuous medium may be another phase of the same fluid, e.g., steam bubbling through water in a teakettle, or it may be a completely different chemical substance, e.g., water flowing through a bed of gravel. The term *two-component flow* is valid for either case. Although incorrect for the latter case, the term *two-phase flow* is sometimes used for both since the mathematical models which describe two-phase or two-component flows are identical.

We are surrounded in nature by some prime examples of two-component flows: rain, snow, fog, smog, quicksand, and the movement of water, gas, and oil through the earth and the blood in our body. Engineering processes requiring multi-component flows are numerous, e.g., refrigeration, air-conditioning, food-processing, and combustion processes. For simplicity in this chapter, we shall refer to the two components as solids and fluid.

The flow of solid particles within a fluid has been of interest since man first observed dust, twigs, and other small particles being lifted and carried about by the wind. A study of this phenomenon falls logically into three categories: (1) fixed bed, (2) fluidized bed, and (3) solids transport. These categories, depicted graphically in Fig. 20-1 for vertical cogravity solids flow and countercurrent fluid flow, have the following distinguishing characteristics:

1. *Fixed bed:* the solid particles remain fixed, being restrained by porous screens, while the fluid is passed through the bed of particles.
2. *Fluidized bed:* the solid particles are unrestrained at the top, and the bed expands as the fluid velocity increases. There is no net solids flow.
3. *Solids transport:* both solids and fluid flow; obviously, then, the relative velocity is of importance, and the effects produced by this relative motion lead to complexities which cannot be predicted by analysis alone.

To cite only a limited number of disciplines and examples, flow through a fixed bed is of great importance to the petroleum engineer in extracting crude oil or gas from an underground reservoir, to the civil engineer in water-purification processes, to the chemical engineer in distillation and sedimentation processes, and to the mechanical engineer in power generation and filtration.

The fluidized bed is used when an intimate, uniform contact between a fluid and solid particles is required. It is advantageous in catalytic reactions, such as the

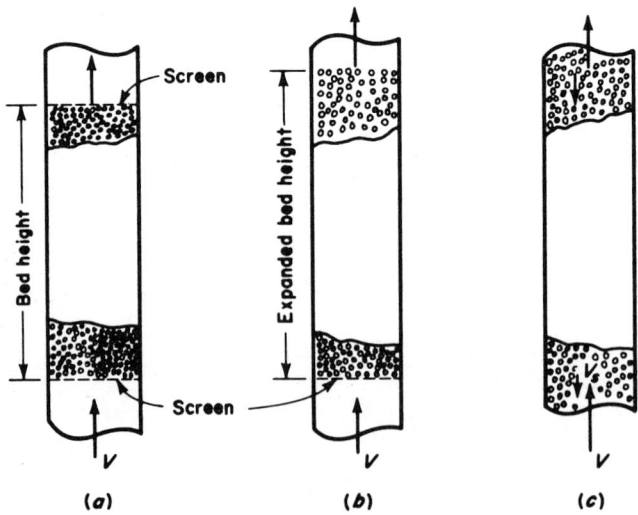

Fig. 20-1 Two-component flow systems: (*a*) fixed bed; (*b*) fluidized bed; (*c*) solids transport.

FLOW THROUGH PERMEABLE MEDIA

cracking of crude oil to make gasoline, for reactions between a fluid and solids, and for the drying of solid particles.

Solids-transport techniques have been successfully applied in moving coal, grain, chemicals, and various slurries in the chemical, mining, and food-processing industries. Gravity flow of granular solids (dense phase) is commonly encountered in solids handling and storage [7]. Grain (dilute phase) has been transported pneumatically for several decades [4]. For many years a coal-water mixture (high ratio of solids to fluid) has been pumped through a 10-in pipeline from Cadiz, Ohio, to the Cleveland Electric Company's Eastlake Station, a distance of 108 miles [15].

20-1 FORCES ON PARTICLES

Whether or not the solids move in the same direction as the fluid depends upon the relative balance of forces; for equilibrium

$$\text{Drag force} + \text{buoyant force} + \text{gravity force} = 0 \tag{20-1}$$

For a *single spherical particle* of diameter D_p moving counter to gravity

$$\frac{\pi D_p^2}{4} C_D \rho \frac{(V - V_s)^2}{2} + \frac{\pi D_p^3}{6} g\rho - \frac{\pi D_p^3}{6} g\rho_s = 0 \tag{20-2}$$

where C_D = drag coefficient given by Fig. 13-28 or 20-2
ρ = fluid density
ρ_s = solids density

Simplifying, we get

$$V - V_s = \left[\frac{4}{3} \frac{g D_p (\rho_s - \rho)}{\rho C_D}\right]^{\frac{1}{2}} \tag{20-3}$$

For a spherical particle to be transported counter to gravity, the relative velocity $V - V_s$ must exceed the *terminal settling velocity* V_t

$$V_t = \left[\frac{4}{3} \frac{g D_p (\rho_s - \rho)}{\rho C_D}\right]^{\frac{1}{2}} \tag{20-4}$$

obtained by settling $V = 0$ in Eq. (20-3). The terminal settling velocity is the velocity at which a particle will fall in a stagnant fluid. For a Reynolds number less than 1

$$C_D = \frac{24}{\text{Re}} \tag{20-5}$$

and

$$V_t = \frac{D_p^2 (\rho_s - \rho) g}{18 \mu} \qquad \text{Stokes' law} \tag{20-6}$$

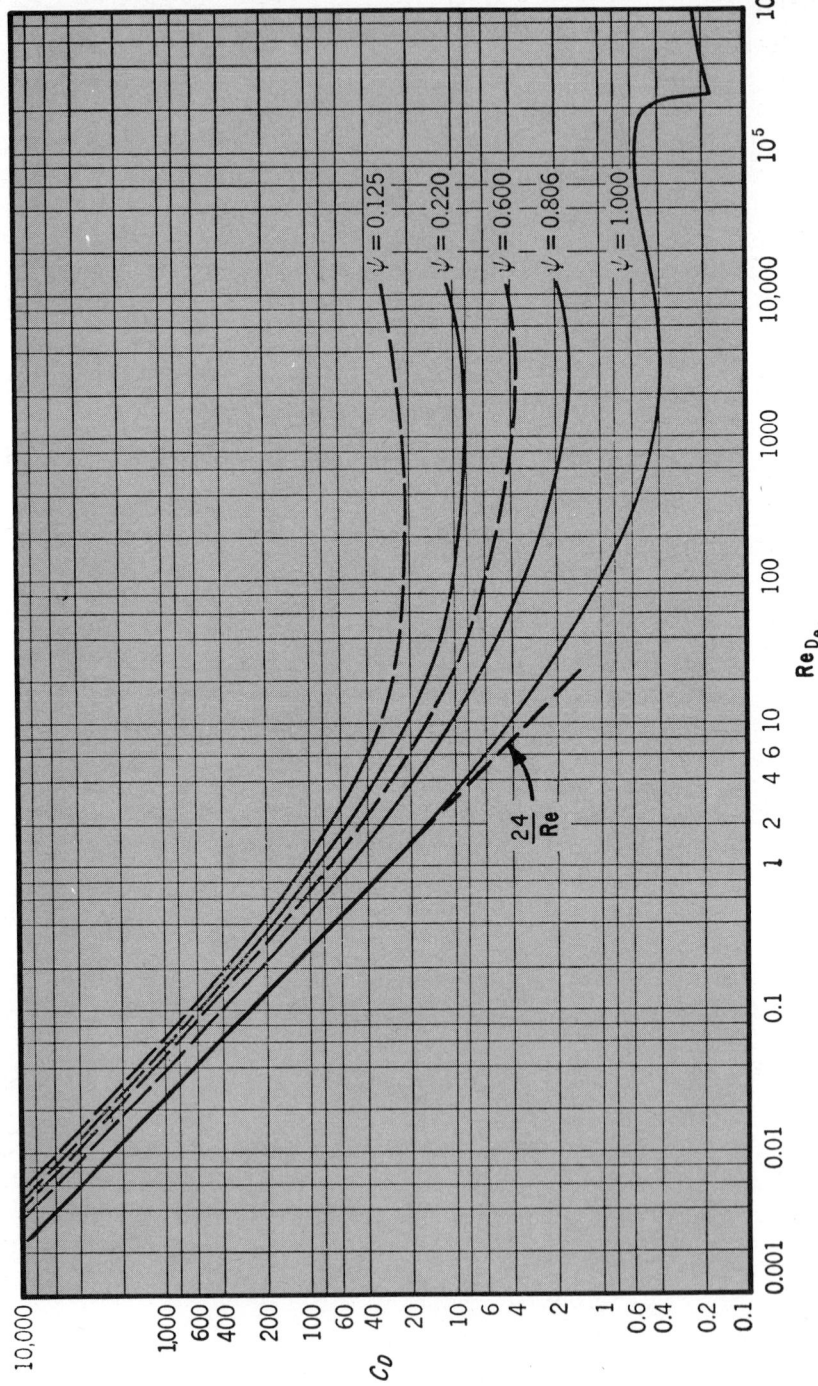

Fig. 20-2 Drag coefficients for single particles. [*Adapted from H. Waddel, J. Franklin Inst.*, **217**:459–490 (1934).]

FLOW THROUGH PERMEABLE MEDIA

These results can be applied to nonspherical particles using the *sphericity* ψ, defined as

$$\psi = \frac{\text{surface area of sphere having volume equal to that of the particle } S_e}{\text{surface area of the particle } S_p}$$

and the corresponding drag-coefficient correction factors as given in Fig. 20-2. To facilitate the use of Fig. 20-2, Table 20-1 gives the sphericity of some common geometrical shapes. Also given are the ratios of the diameter D_e of an equivalent-volume sphere (a sphere having the same volume as the particle) to the average diameter D_{av} determined by screen analysis.

To apply the preceding results to large numbers of particles moving together requires further modifications in the drag coefficient to account for the interaction between particles. It is more convenient, however, to consider the flow system to be made up of a bundle of tubes of unusual cross section and then to modify the theory for straight uniform tubes to apply. With the proper evaluation of effective parameters, either approach is satisfactory.

Table 20-1 Sphericity and the value of D_e related to screen size†

Shape	Sphericity ψ	$\dfrac{D_e}{D_{av}}$ ‡
Sphere	1.00	1.00
Octahedron	0.847	0.965
Cube	0.806	1.24
Prism:		
$a \times a \times 2a$	0.767	1.564
$a \times 2a \times 2a$	0.761	0.985
$a \times 2a \times 3a$	0.725	1.127
Cylinder:		
$h = 3r$	0.860	1.31
$h = 10r$	0.691	1.96
$h = 20r$	0.580	2.592
Disk:		
$h = r$	0.827	0.909
$h = r/3$	0.594	0.630
$h = r/10$	0.323	0.422
$h = r/15$	0.254	0.368

† Reprinted by permission from G. G. Brown et al., "Unit Operations," p. 77, John Wiley & Sons, Inc., New York, 1950.

‡ Multiply screen size D_{av} by the factor indicated to get correct value for D_e to be used in the equations.

20-2 ONE-DIMENSIONAL CAPILLARY-TUBE MODEL

Considering the general case of a moving bed as shown in Fig. 20-3a, where both solids (dense) and fluid are assumed to move upward with average velocities V_s and V, respectively, the two-component system is made up of a heterogeneous network of pores and solid particles.

To simplify the analysis, we replace the actual fluid-particle system by an idealized model, shown in Fig. 20-3b, containing uniformly dispersed parallel cylindrical passages which will produce the same momentum, heat, and mass-transfer effects. The volume of the "capillary tubes" will be taken equal to the void volume of the actual system, valid since no size or cross-sectional limitations have been made on the cylindrical passages.

Defining the void fraction or porosity ϵ as

$$\epsilon = \frac{\text{void volume}}{\text{total volume}} \tag{20-7}$$

the flow area per unit total area for the fluid is $A_f = \epsilon$, and the solid area is $A_s = 1 - \epsilon$ per unit total area. The remaining analysis of this section pertains to cases where $\epsilon = \epsilon_{\text{fixed bed}} = \epsilon_{\text{moving bed}}$.

Continuity equation Applying the integral control-volume continuity equation

$$\int_{\text{cv}} \frac{\partial \rho}{\partial t} d\mathscr{V} = -\int_{\text{cs}} \rho \mathbf{V} \cdot d\mathbf{A} \tag{10-15}$$

to the fluid, we get (per unit total area)

$$\left.\frac{\partial \rho}{\partial t}\right|_{\text{av}} \epsilon \, \Delta y = -(\rho V \epsilon)_{y+\Delta y} - (\rho V \epsilon)_y \tag{20-8}$$

When we divide by $\epsilon \, \Delta y$ and take the limit as Δy approaches zero, the resulting continuity equation is

$$\frac{\partial \rho}{\partial t} = -\frac{\partial}{\partial y} \rho V \tag{20-9}$$

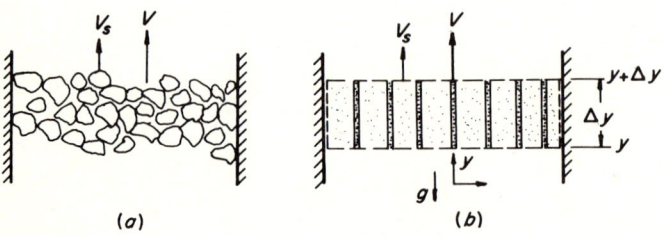

Fig. 20-3 Flow through a porous medium: (a) moving bed; (b) capillary-tube model.

FLOW THROUGH PERMEABLE MEDIA

It is often more convenient to use the *superficial fluid velocity* V_∞, that velocity which would exist if the fluid occupied the duct alone, given by

$$V_\infty = \epsilon V \tag{20-10}$$

Therefore, in terms of the superficial fluid velocity, the *continuity equation* becomes

$$\epsilon \frac{\partial \rho}{\partial t} + \frac{\partial}{\partial y} \rho V_\infty = 0 \tag{20-11}$$

Momentum equation Recalling the integral momentum equation for an inertial control volume

$$\sum \mathbf{F} = \frac{d}{dt} \int_{cv} \rho \mathbf{V} \, d\mathscr{V} + \int_{cs} \mathbf{V}(\rho \mathbf{V} \cdot d\mathbf{A}) \tag{10-50}$$

the respective terms *for the fluid* of the capillary-tube model of Fig. 20-3b are (per unit total area)

$$\sum \mathbf{F} = \epsilon(p|_y - p|_{y+\Delta y}) - F_D - \rho g \epsilon \, \Delta y \tag{20-12}$$

where

F_D = frictional drag in the capillary tubes

In terms of the relative velocity $V_{\text{rel}} = V - V_s$

$$\frac{d}{dt} \int_{cv} \rho \mathbf{V} \, d\mathscr{V} = \frac{\partial}{\partial t} \rho V_{\text{rel}} \bigg|_{\text{av}} \epsilon \, \Delta y \tag{20-13}$$

$$\int_{cs} \mathbf{V}(\rho \mathbf{V} \cdot d\mathbf{A}) \big|_y = -[V_{\text{rel}}(\rho V_{\text{rel}})\epsilon]_y \tag{20-14}$$

$$\int_{cs} \mathbf{V}(\rho \mathbf{V} \cdot d\mathbf{A}) \big|_{y+\Delta y} = [V_{\text{rel}}(\rho V_{\text{rel}})\epsilon]_{y+\Delta y} \tag{20-15}$$

Substituting Eqs. (20-12) to (20-15) into Eq. (10-50), we get

$$-\frac{p|_{y+\Delta y} - p|_y}{\Delta y} - \frac{F_D}{\epsilon \, \Delta y} - \rho g = \frac{\partial}{\partial t}(\rho V_{\text{rel}})\bigg|_{\text{av}} + \frac{[V_{\text{rel}}(\rho V_{\text{rel}})]_{y+\Delta y} - [V_{\text{rel}}(\rho V_{\text{rel}})]_y}{\Delta y} \tag{20-16}$$

after dividing by $\epsilon \, \Delta y$. Taking the limit as Δy approaches zero, we have the momentum equation

$$-\frac{\partial p}{\partial y} - F_D''' - \rho g = \frac{\partial}{\partial t} \rho V_{\text{rel}} + \frac{\partial}{\partial y} \rho V_{\text{rel}}^2 \tag{20-17}$$

if it is agreed that the equation gives average values rather than point values, since the flow system is not a continuum in the strictest sense. F_D''' is the frictional drag per unit volume. Differentiating the products on the right-hand side of Eq. (20-17) and using the continuity equation to simplify, we get

$$\rho \frac{D(V_{\text{rel}})}{Dt} = -\rho g - \frac{\partial p}{\partial y} - F_D''' \qquad (20\text{-}18)$$

which may be compared with the y-direction Navier-Stokes equation (12-17a).

To get an expression for the drag force we may use the fundamental Hagen-Poiseuille flow of Sec. 12-3, but this requires knowing (or assuming) the size and number of capillary tubes. A more convenient method is to use the empirical relation from Darcy's classical experiment [5], which for laminar flow results in

$$F_D''' = \frac{\mu}{K}(V_\infty)_{\text{rel}} \qquad (20\text{-}19)$$

where $(V_\infty)_{\text{rel}} = V_\infty - V_s =$ superficial relative fluid velocity (that relative velocity which would exist if the fluid occupied the duct alone, $V_\infty = \epsilon V$)
$\mu =$ fluid viscosity
$K =$ permeability of porous medium

The superficial relative fluid velocity can then be expressed by *Darcy's law*

$$\blacksquare \quad (V_\infty)_{\text{rel}} = -\frac{K}{\mu}\left(\frac{dp}{dy} + \rho g\right) \qquad (20\text{-}20)$$

which was deduced for the steady state [12]. Table 20-2 gives some typical values of permeability for incompressible porous media. Permeability data are often given in terms of the *darcy*, named in honor of the pioneer investigator; the conversion factor is

$$1 \text{ ft}^2 = 1.0623 \times 10^{-11} \text{ darcy}$$

Energy equation; steady state Following these procedures, the integral energy equation (10-113) for a control volume with y as the vertical coordinate

$$\frac{\delta Q}{\delta t} - \frac{\delta W_{\text{shaft}}}{\delta t} = \frac{d}{dt}\int_{\text{cv}} \rho e \, d\mathcal{V} + \int_{\text{cs}} \left(h + \frac{V^2}{2} + gy\right)(\rho \mathbf{V} \cdot d\mathbf{A})$$

can be simplified for steady state, negligible kinetic- and potential-energy changes, and no shaft work to give

$$\frac{\delta Q}{\delta t} = \int_{\text{cs}} h(\rho \mathbf{V} \cdot d\mathbf{A}) \qquad (20\text{-}21)$$

Or *for the solids and fluid* per unit total area

$$(1-\epsilon)(q_s''|_{y+\Delta y} - q_s''|_y) + \epsilon(q''|_{y+\Delta y} - q''|_y) = (1-\epsilon)[(\rho_s V_s h_s)_y \\ - (\rho_s V_s h_s)_{y+\Delta y}] + (\rho V h)_y - (\rho V h)_{y+\Delta y} \qquad (20\text{-}22)$$

FLOW THROUGH PERMEABLE MEDIA

and *for the solids only*

$$(1 - \epsilon)(q_s''|_{y+\Delta y} - q_s''|_y) - h_c S(T - T_s) = (1 - \epsilon)[(\rho_s V_s h_s)_y - (\rho_s V_s h_s)_{y+\Delta y}] \quad (20\text{-}23)$$

where S is the effective surface area of the particles in height Δy and h_c is the local convective-heat-transfer coefficient.

Using Fourier's law, $q'' = -k\, \partial T/\partial y$, dividing by Δy, and taking the limit as Δy approaches zero, the equations simplify for constant fluid properties to [11, 13, 14]

Solids and fluid:
$$\frac{d^2 T_s}{dy^2} + \frac{\epsilon k}{(1-\epsilon)k_{se}} \frac{d^2 T}{dy^2} - \frac{\rho_s V_s c_s}{k_{se}} \frac{dT_s}{dy} - \frac{\epsilon \rho V c_p}{(1-\epsilon)k_{se}} \frac{dT}{dy} = 0 \quad (20\text{-}24)$$

Solids:
$$\frac{d^2 T_s}{dy^2} - \frac{\rho_s V_s c_s}{k_{se}} \frac{dT_s}{dy} + \frac{h_c S''''}{(1-\epsilon)k_{se}}(T - T_s) = 0 \quad (20\text{-}25)$$

where k = thermal conductivity of fluid
k_{se} = effective thermal conductivity of solids
c = specific heat
S'''' = effective surface area per unit volume

Table 20-2 Typical values of permeability for incompressible porous media

Medium	Porosity	Permeability, ft^2
Silica powder[†]	0.37–0.49	1.4×10^{-13}–5.4×10^{-13}
Sand, mesh:[‡]		
30–40	0.40	3.7×10^{-9}
40–50	0.40	7.0×10^{-10}
50–60	0.40	4.6×10^{-10}
60–70	0.40	3.3×10^{-10}
70–80	0.40	2.8×10^{-10}
80–100	0.40	1.1×10^{-10}
Fine heterogeneous	0.30–0.35	10^{-11}–10^{-10}
Soils[†]	0.43–0.54	3.1×10^{-10}–1.5×10^{-9}
Sandstone[†]	0.08–0.38	5.3×10^{-15}–3.2×10^{-11}
Limestone[†]	0.04–0.10	2.1×10^{-15}–4.8×10^{-13}
Brick[†]	0.12–0.34	5.1×10^{-14}–2.3×10^{-12}
Leather[†]	0.56–0.59	10^{-12}–1.3×10^{-12}
Fiber glass[†]	0.88–0.93	2.5×10^{-10}–5.4×10^{-10}

[†] From R. E. Collins, "Flow of Fluids Through Porous Materials," copyright © 1961 by Litton Educational Publishing, Inc., by permission of Van Nostrand Reinhold Company.

[‡] From M. Muskat, "The Flow of Homogeneous Fluids through Porous Media." Copyright 1937. McGraw-Hill Book Company. Used by permission.

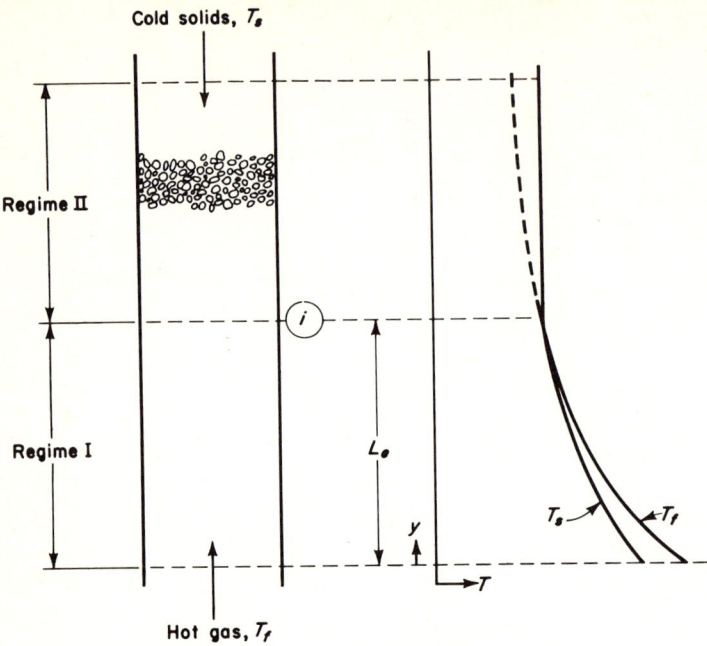

Fig. 20-4 Temperature distributions in gas-solids flow.

These two equations can be solved simultaneously for the temperature distribution in the fluid and solids.

For a gas flowing through small particles the temperature distributions in both solids and gas are exponential, as indicated in Fig. 20-4 for the boundary conditions

At $y = 0$: $\quad T = T|_{in}$

At $y = L_e$ (1): $\quad T_s = T_s|_{in} \quad \dfrac{dT_s}{dy} = 0 \quad \dfrac{dT}{dy} = 0$

(2): $\quad T|_{out} = T_s|_{in}$

The effective length L_e is not fixed until the flow conditions and temperature differentials are specified; this requires the additional boundary condition. Boundary condition 2 defines the position where heat transfer ends (regime I); no heat is transferred in regime II. For small particles of high thermal conductivity the effective length is small, since the temperature of the solids will approach that of the gas very rapidly [13, 14].

20-3 PERCOLATION: FIXED BED

The governing equations for flow through a fixed bed of porous media can be obtained from the preceding section by setting $V_s = 0$. It is more convenient, however, to

FLOW THROUGH PERMEABLE MEDIA

take advantage of empirical data and express design equations in the same form as equations for flow through empty pipes. In Sec. 12-3 the pressure loss in a pipe was given by

$$\Delta p = f \frac{L}{D} \frac{\rho V^2}{2} \tag{12-37a}$$

We can write an analogous equation for the pressure drop Δp in a packed bed:

$$\Delta p = f_p \frac{L}{D_p} \frac{\rho V_\infty^2}{2} \tag{20-26}$$

where f_p = packed-bed friction factor
L = length of packed column
D_p = particle diameter
ρ = fluid density
V_∞ = superficial fluid velocity

This expression may be considered as the definition of the packed-bed friction factor f_p. It is valid for both laminar and turbulent flow.

For particles which are not spherical D_p is taken as the diameter of a sphere having the same volume as that of the particle. It is convenient to express it in terms of the packing area a_p (interstitial area per unit total volume, in square feet per cubic foot), since these data are available from manufacturers of packing materials. The surface area per unit volume of the particles is $a_p/(1-\epsilon)$, which can be equated to the respective factors for a sphere in order to get a mean particle diameter, i.e.,

$$\frac{a_p}{1-\epsilon} = \frac{\pi D_p^2}{(\pi/6) D_p^3}$$

or

$$\blacksquare \quad D_p = \frac{6(1-\epsilon)}{a_p} \tag{20-27}$$

The concept of packing area a_p is also advantageous in expressing the hydraulic radius R_h for flow through porous media. The hydraulic radius is defined by

$$R_h \equiv \frac{\text{cross-sectional flow area } A}{\text{wetted perimeter } P} \tag{20-28}$$

Multiplying numerator and denominator by the length L of the packed column, we get

$$R_h = \frac{AL}{PL} = \frac{\text{flow volume}}{\text{total wetted surface}} \tag{20-29}$$

where the flow volume is that volume which is available for flow, i.e., the void volume; therefore,

$$R_h = \frac{\text{void volume/total volume}}{\text{total wetted surface/total volume}}$$

or
$$R_h = \frac{\epsilon}{a_p} \tag{20-30}$$
and in terms of the particle diameter we get
$$R_h = \frac{D_p}{6} \frac{\epsilon}{1-\epsilon} \tag{20-31}$$
Recalling the *Hagen-Poiseuille equation* for laminar flow in single-component flow through a duct
$$\Delta p = \frac{8\mu VL}{R^2} \tag{11-43}$$
we can envision a packed bed as being made up of a series of parallel capillary tubes and express the pressure drop in a packed bed by
$$\Delta p = \frac{2\mu VL}{R_h^2} \tag{20-32}$$
where the radius R of a round duct has been expressed in terms of the more general hydraulic radius $R_h = R/2$. Using the superficial fluid velocity
$$V_\infty = V\epsilon \tag{20-10}$$
and the expression for the hydraulic radius from Eq. (20-31), the pressure drop for laminar flow through a packed bed is
$$\Delta p = 72 \frac{V_\infty \mu L}{D_p^2} \frac{(1-\epsilon)^2}{\epsilon^3} \tag{20-33}$$

So far we have tacitly taken the capillary tubes to be equal in length to the bed depth. This is, of course, incorrect since the fluid must actually pass through a tortuous path of greater length than L. An abundance of experimental data have shown that the constant should be multiplied by a factor of $\frac{25}{12}$, giving

■ $$\Delta p = 150 \frac{V_\infty \mu L}{D_p^2} \frac{(1-\epsilon)^2}{\epsilon^3} \tag{20-34}$$

which is known as the *Blake-Kozeny equation* [2]. This result is valid only in the *laminar* regime given by **Re**$'(1-\epsilon) < 10$, where

$$\mathbf{Re'} \equiv \frac{V_\infty D_p \rho}{\mu} \qquad \text{modified Reynolds number} \tag{20-35}$$

Equating the pressure drop in Eqs. (20-26) and (20-34), the *packed-bed friction factor for laminar flow* is

■ $$f_p = \frac{300}{\mathbf{Re'}} \frac{(1-\epsilon)^2}{\epsilon^3} \tag{20-36}$$

The Blake-Kozeny equation is plotted in Fig. 20-5.

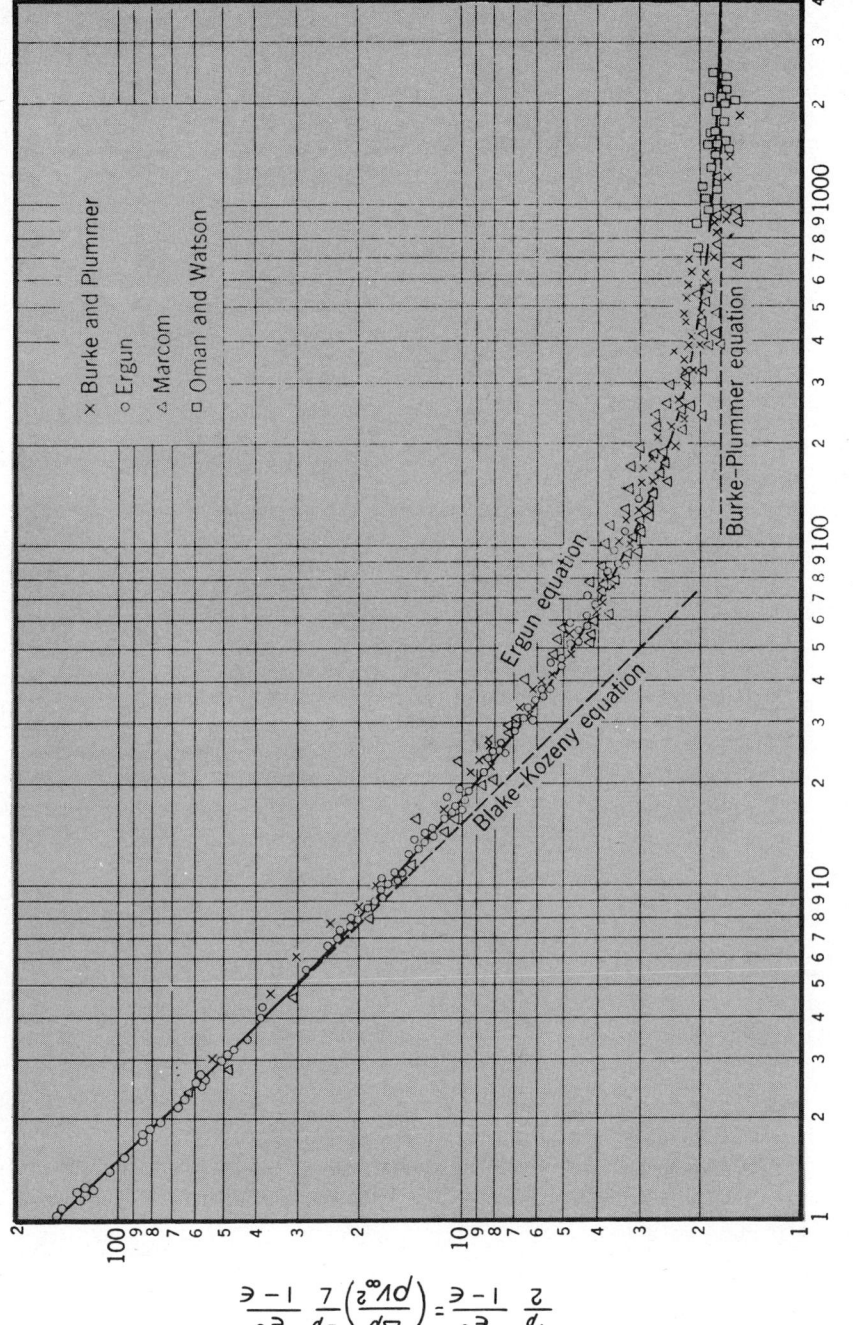

Fig. 20-5 Friction factors for flow through packed beds. [*Adapted from S. Ergun, Chem. Eng. Progr.,* **48**(2):89 (1952).]

Following the same procedure for turbulent flow, we define a turbulent friction factor; for single-component flow through a duct

$$\Delta p = f \frac{L}{D} \frac{\rho V^2}{2} \tag{13-60}$$

Using the preceding relations

$$D = 4R_h = 4\frac{\epsilon}{a_p} = \frac{4D_p}{6}\frac{\epsilon}{1-\epsilon}$$

and

$$V^2 = \frac{V_\infty^2}{\epsilon^2}$$

Eq. (13-60) can be modified to give

$$\Delta p = \tfrac{3}{2} f \frac{L}{D_p} \frac{\rho V_\infty^2}{2} \frac{1-\epsilon}{\epsilon^3} \tag{20-37}$$

for flow through a packed bed. For *highly turbulent flow* experimental data show that $\tfrac{3}{2} f = 3.50$; therefore, we get

$$\blacksquare \quad \Delta p = 3.50 \frac{L}{D_p} \frac{\rho V_\infty^2}{2} \frac{1-\epsilon}{\epsilon^3} \tag{20-38}$$

which is valid for $\mathbf{Re}' (1 - \epsilon) > 1{,}000$. This result, known as the *Burke-Plummer equation* [2], is plotted in Fig. 20-5. Equating the pressure drop in Eqs. (20-26) and (20-38), the *packed-bed friction factor for turbulent flow* is

$$\blacksquare \quad f_p = 3.50 \frac{1-\epsilon}{\epsilon^3} \tag{20-39}$$

For flow in the transition regime, $10 < \mathbf{Re}'(1-\epsilon) < 1{,}000$, Ergun [6] showed that the data of several investigators correlate well by simply adding the laminar and turbulent expressions, giving

$$\frac{\Delta p}{L} = 150 \frac{V_\infty \mu}{D_p^2} \frac{(1-\epsilon)^2}{\epsilon^3} + 1.75 \frac{\rho V_\infty^2}{D_p} \frac{1-\epsilon}{\epsilon^3} \tag{20-40}$$

For compressible flow the density of the fluid may be taken as the arithmetic average at the end pressures. In extreme cases it may be necessary to treat the equation in differential form, i.e., letting

$$\frac{dp}{dy} = \frac{\Delta p}{L} \tag{20-41}$$

and integrating to get the effects of compressibility. Since V_∞ changes throughout the bed for a compressible fluid, Eq. (20-40) can be expressed in terms of the superficial

FLOW THROUGH PERMEABLE MEDIA

Table 20-3 Characteristics of random packings†

Packing	Nominal size, in‡	n''', pieces/ft^3	γ, lb$_f$/ft^3	a_p, ft^2/ft^3	ϵ
Raschig rings, carbon	¼	85,000	46	212	0.55
	½	10,600	27	114	0.74
	¾	3,140	34	75	0.67
	1	1,325	27	57	0.74
	1½	392	34	38	0.67
	2	166	27	28	0.74
Berl saddles, stoneware	¼	107,000	56	274	0.60
	½	16,700	54	142	0.62
	¾	4,950	49	87	0.66
	1	2,180	45	76	0.68
	1½	645	40	46	0.71
	2	250	39	32	0.72
Intalox saddles, stoneware	½	20,700	45	190	0.78
	¾	6,500	44	102	0.77
	1	2,385	44	78	0.77
	1½	709	42	59	0.80
	2	265	42	36	0.79
Pall rings, metal	⅝	5,930	37	104	0.93
	1	1,405	30	63	0.94
	1½	377	26	39	0.95
	2	171	24	31	0.96

† From the U.S. Stoneware Company.
‡ Outside diameter and length for Raschig rings and Pall rings.

mass velocity $G_\infty \equiv \rho V_\infty = \rho \epsilon V$, which is constant throughout the bed, resulting in

$$\frac{\rho \Delta p}{G_\infty^2} \frac{D_p}{L} \frac{\epsilon^3}{1-\epsilon} = 150 \frac{1-\epsilon}{\text{Re}'} + 1.75 \tag{20-42}$$

This result, known as the *Ergun equation*, well represents the flow in all flow regimes, as shown in Fig. 20-5.

The "stretching" factors used in plotting Fig. 20-5, that is, $1/(1-\epsilon)$ for the abscissa and $\epsilon^3/(1-\epsilon)$ for the ordinate, are readily apparent from the form of the Ergun equation. The similarity between this figure and Fig. 13-9 or 13-10 should be noted.

Table 20-3 gives some characteristics of packing materials commonly used in columns for continuous-contacting processes, where the aim is to expose the fluid to a very large area per unit volume. Since solids pack more loosely adjacent to a wall

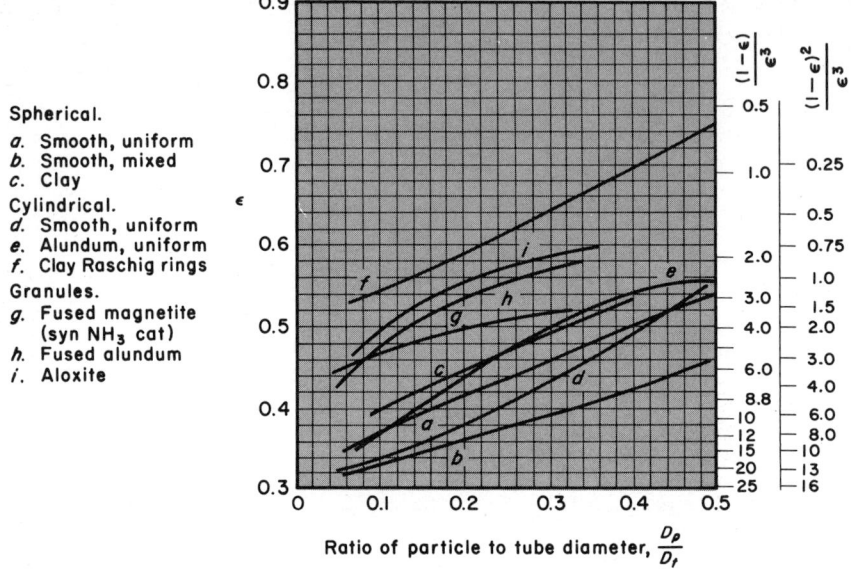

Fig. 20-6 Void fractions in beds of typical solids. (*Adapted from M. Leva, "Fluidization." Copyright 1959. McGraw-Hill Book Company. Used by permission.*)

than farther out in the bed, Fig. 20-6 can be used to account for this wall effect on the porosities of some typical solids. Superimposed on the figure are the geometrical parameters, $(1 - \epsilon)/\epsilon^3$ and $(1 - \epsilon)^2/\epsilon^3$, required in Eqs. (20-34) and (20-38).

Convective-heat- and mass-transfer coefficients for flow through packed beds of granular solids are presented in Fig. 14-11.

Example Air at 100°F flows upward through a 2-ft-diam column 4 ft long packed with $\tfrac{1}{2}$-in carbon Raschig rings. Estimate the flow rate for a pressure differential of 100 lb_f/ft^2.

Solution The necessary characteristics are found in Table 20-3, i.e.,

$$\epsilon = 0.74 \quad \text{and} \quad a_p = 114$$

The mean particle diameter is then given by Eq. (20-27):

$$D_p = \frac{6(1 - \epsilon)}{a_p} = \frac{6(1 - 0.74)}{114} = 0.0137 \text{ ft}$$

From Table A-3 the thermodynamic properties of the air are

$$\rho = 0.071 \text{ lb}_m/\text{ft}^3 \quad \text{and} \quad \mu = 1.285 \times 10^{-5} \text{ lb}_m/\text{ft-sec}$$

Since it is not known a priori whether the flow is laminar or turbulent, the velocity V_∞ cannot be determined explicitly. Both the abscissa and ordinate of Fig. 20-5, however, can be expressed in terms of the unknown velocity V_∞, and the result can be determined by trial intersections on the curve. This is equivalent to solving the Ergun equation by trial. The latter method will be used.

FLOW THROUGH PERMEABLE MEDIA

Expressing Eq. (20-42) in terms of the superficial velocity, we get

$$\frac{\Delta p}{\rho V_\infty^2} \frac{D_p}{L} \frac{\epsilon^3}{1-\epsilon} = 150 \frac{(1-\epsilon)\mu}{V_\infty D_p \rho} + 1.75$$

All quantities are known except V_∞; therefore,

$$\frac{(100 \text{ lb}_f/\text{ft}^2)(32.2 \text{ lb}_m\text{-ft/lb}_f\text{-sec}^2)}{(0.071 \text{ lb}_m/\text{ft})V_\infty^2} \frac{0.0137 \text{ ft}}{4 \text{ ft}} \frac{0.74^3}{0.26} = 150 \frac{0.26(1.285 \times 10^{-5} \text{ lb}_m/\text{ft-sec})}{V_\infty(0.0137 \text{ ft})(0.071 \text{ lb}_m/\text{ft}^3)} + 1.75$$

where it was necessary to introduce the gravitational constant g_c for dimensional homogeneity. Simplifying, the resulting equation is

$$\frac{247}{V_\infty^2} = \frac{0.00343}{V_\infty} + 1.75$$

or

$$V_\infty^2 + 0.00196 V_\infty - 141 = 0$$

which can be solved by the quadratic equation to give

$$V_\infty = \frac{-0.00196 + \sqrt{0.00196^2 + 4(141)}}{2}$$

where only the positive root has physical meaning; hence,

$V_\infty = 11.87$ ft/sec

giving a flow rate of

$Q = V_\infty A = (11.87 \text{ ft/sec})(\pi \text{ ft}^2) = 37.3 \text{ ft}^3/\text{sec}$

or

$\dot{m} = \rho V_\infty A = (0.071 \text{ lb}_m/\text{ft}^3)(37.3 \text{ ft}^3/\text{sec}) = 2.65 \text{ lb}_m/\text{sec}$

Filtration An important application of flow through permeable media is the separation of solids from liquids or slurries. The mechanism is to pump the liquid or slurry through a porous filter cloth so that the solid matter adheres to the cloth and builds up to form a *filter cake*. The idealized process is shown schematically in Fig. 20-7.

Two complicating factors enter the filtration mechanism: (1) the filter-cake thickness changes with time as the buildup of solids occurs, and (2) most filter cakes are compressible, causing the porosity to vary with pressure differential, which obviously changes with filter-cake thickness. Because of these factors filter design is largely empirical; however, the mechanism can be understood by a simple analysis.

Since the particles making up the filter cake are relatively small, the flow is usually laminar; hence, we can use either Darcy's law [Eq. (20-20)] or the Blake-Kozeny equation. Neglecting the body force ρg, Eq. (20-20) can be integrated for constant viscosity and permeability to give

$$\Delta p = \frac{V_\infty \mu L}{K} \tag{20-43}$$

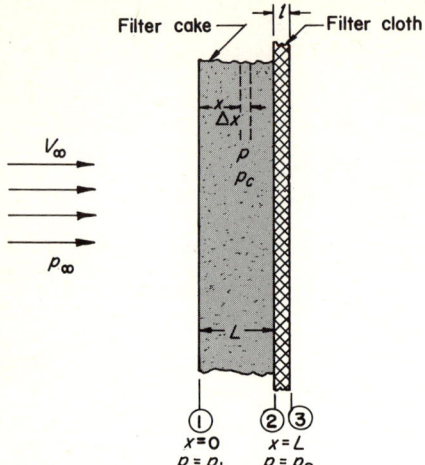

Fig. 20-7 Cross section of a filter.

By equating this result to the pressure differential given by Eq. (20-34) the permeability is seen to be a function of geometrical properties only, i.e.,

$$K = \frac{D_p^2}{150} \frac{\epsilon^3}{(1-\epsilon)^2} \tag{20-44}$$

Expressing the Blake-Kozeny equation in differential form, we get

$$dp = R_c V_\infty \mu \, dx \tag{20-45}$$

where R_c is the specific resistance of the filter cake defined by

$$R_c \equiv \frac{150}{D_p^2} \frac{(1-\epsilon)^2}{\epsilon^3} \tag{20-46}$$

For flow through a compressible filter cake the local specific resistance varies with the mechanical compressive stress p_c existing in the solid; it is given by

$$p_c = p_\infty - p \tag{20-47}$$

in which p_∞ is the upstream pressure and p is the local fluid pressure. Since p_∞ is usually constant,

$$dp_c = -dp \tag{20-48}$$

and from Eq. (20-45)

$$-\int \frac{dp_c}{R_c} = \int_0^L V_\infty \mu \, dx = \int_{p_1}^{p_2} \frac{dp}{R_c} \tag{20-49}$$

This gives the relation *for flow through a compressible filter cake*

■ $\quad p_2 - p_1 = \bar{R}_c V_\infty \mu L \tag{20-50}$

where

$$\bar{R}_c = \frac{p_2 - p_1}{-\int (dp_c/R_c)} \qquad (20\text{-}51)$$

For flow through an incompressible filter cake the average specific resistance $\bar{R}_c$ is equal to the local specific resistance R_c, and the same relation applies.

During buildup of the filter cake the pressure drop across the filter cloth is greater than that across the filter cake; hence, it must be accounted for in the overall pressure differential. For the filter cloth the Blake-Kozeny equation gives

$$p_3 - p_2 = R_{cl} V_\infty \mu l \qquad (20\text{-}52)$$

in which the specific resistance R_{cl} of the filter cloth is assumed to remain constant at

$$R_{cl} = \left[\frac{150}{D_p^2} \frac{(1-\epsilon)^2}{\epsilon^3}\right]_{cl} \qquad (20\text{-}53)$$

The total pressure drop is therefore given by the sum of the pressure differentials in Eqs. (20-50) and (20-52):

$$p_3 - p_1 = \mu V_\infty (\bar{R}_c L + R_{cl} l) \qquad (20\text{-}54)$$

For flow through either an incompressible or a compressible filter cake the rate of accumulation of the cake is given by

$$\frac{dL}{dt} = \chi V_\infty \qquad (20\text{-}55)$$

where χ is the volume fraction or concentration of solid matter in the fluid. Eliminating the velocity between Eqs. (20-54) and (20-55), we get

$$\frac{dL}{dt} = \frac{\chi(p_3 - p_1)}{\mu(\bar{R}_c L + R_{cl} l)} \qquad (20\text{-}56)$$

If the filtration process is carried out with a *constant pressure differential*, Eq. (20-56) can be integrated to give the time t

$$t = \frac{\mu}{\chi(p_3 - p_1)} \left(\bar{R}_c \frac{L^2}{2} + R_{cl} l L\right) \qquad (20\text{-}57)$$

Since $L = \int \chi V_\infty \, dt = \chi \mathscr{V}''$, where $\mathscr{V}''$ is the total volume of fluid passed through the filter per unit area, the time t is also given by

$$t = \frac{\mu}{p_3 - p_1} \left[\bar{R}_c \frac{\chi(\mathscr{V}'')^2}{2} + R_{cl} \mathscr{V}''\right] \qquad (20\text{-}58)$$

If the filtration is carried out with a constant volumetric flow rate, the time required to filter a volume $\mathscr{V}''$ per unit area of filter is

$$t = \frac{\mathscr{V}''}{V_\infty} \qquad (20\text{-}59)$$

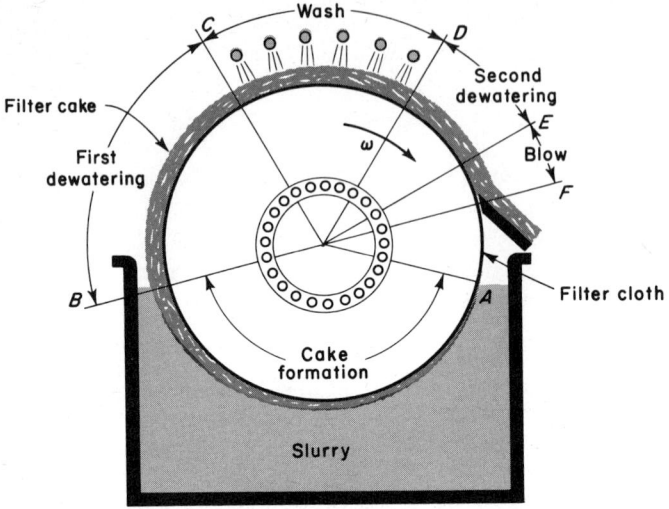

Fig. 20-8 Continuous rotary filter.

and after time t the pressure drop is given by

$$p_3 - p_1 = \mu V_\infty (\bar{R}_c \chi \mathscr{V}'' + R_{cl} l) \tag{20-60}$$

or

$$p_3 - p_1 = \frac{\mu}{t} [\bar{R}_c \chi (\mathscr{V}'')^2 + R_{cl} l \mathscr{V}''] \tag{20-61}$$

Figure 20-8 illustrates the operation of a continuous rotary vacuum filter [1, 9]. Filter cloth is stretched over the outer surface of the rotary drum. The drum is divided internally into segments, which are connected to the ports of a rotary valve. As the drum rotates slowly, suction is applied between positions A and B. The filter cake builds up, and its buildup can be determined with Eq. (20-56). As the cake leaves the slurry, it partially dries from the filtrate between B and C, a spray wash is applied from C to D to facilitate its later removal, and it is dried (sometimes by sucking air through it) from the washing process between D and E. Finally, a positive pressure is applied between E and F, and the cake is removed with a scraper or other cleaning devices.

20-4 FLUIDIZATION: EXPANDED BED

For a bed of solids to expand upward, as shown in Fig. 20-1b, the superficial velocity of the fluid V_∞ must exceed the minimum fluidization velocity V_{∞_f}. To find this velocity we make a force balance on a bed of particles of area A and bed depth L. The net downward force due to the effects of gravity is

$$F_g = (\rho_s - \rho)gAL(1 - \epsilon) \tag{20-62}$$

FLOW THROUGH PERMEABLE MEDIA

The upward force due to the frictional action of the fluid on the solids is

$$F_{f_p} = A\,\Delta p = \left[\frac{150(1-\epsilon)}{\mathbf{Re'}} + 1.75\right]\frac{\rho V_\infty^2 L(1-\epsilon)A}{D_p \epsilon^3} \tag{20-63}$$

where Eq. (20-42) was used for Δp. At the moment of incipient fluidization, $F_g = F_{f_p}$. Equating (20-62) and (20-63) while replacing V with V_{∞_f}, we get

$$(\rho_s - \rho)g = \frac{150(1-\epsilon)\mu V_{\infty_f}}{D_p^2 \epsilon^3} + \frac{1.75\rho V_{\infty_f}^2}{D_p \epsilon^3} \tag{20-64}$$

If the void fractions and the fluid properties are known, the minimum fluidization velocity V_{∞_f} can be determined.

Equation (20-64) gives the onset of fluidization, the beginning of bed expansion, characterized by minimum void fraction. The other extreme is reached when the void fraction approaches unity, which corresponds to a single particle being suspended in the bed. For this case the maximum velocity is the same as the velocity at which the particle would fall through a large body of the stagnant fluid. This maximum velocity is the terminal velocity V_t, given in Sec. 20-1,

$$V_t = \frac{D_p^2}{18}\frac{(\rho_s - \rho)g}{\mu} \qquad \mathbf{Re} < 1 \tag{20-6}$$

and by Eq. (20-4)

$$V_t = \left[\frac{3D_p(\rho_s - \rho)g}{\rho}\right]^{\frac{1}{2}} \qquad 500 < \mathbf{Re} < 200{,}000 \tag{20-65}$$

since C_D is essentially constant at 0.44 in this regime, as observed from Fig. 13-28. For other Reynolds numbers the drag coefficient C_D can be obtained from Fig. 13-28 and used in Eq. (20-4) to get the terminal velocity.

The bed is expanded for all velocities between V_{∞_f} and V_t. The relation between void fraction and the superficial fluid velocity at these intermediate velocities is satisfactorily represented by a log-log plot like that illustrated in Fig. 20-9.

The variation of pressure drop with velocity in a fixed bed was predicted by both Darcy's law and the Ergun equation. Above the minimum fluidization velocity V_{∞_f} the pressure differential remains almost constant until the solids begin to separate. These conditions are depicted typically in Fig. 20-10.

Heat- and mass-transfer coefficients for fluidized beds are given by Eq. (14-76) and the Chilton-Colburn analogy expressed by Eq. (14-52b).

For more extensive analyses and design equations the student should consult Zenz and Othmer [19] and Leva [10].

Example In a batch fluidization process 80°F water, flowing at the rate of 22.8 lb_m/sec, must contact 400 lb_m of smooth, solid cylindrical pellets (diameter $\frac{1}{4}$ in and length $\frac{1}{4}$ in) having a density of 90 lb_m/ft^3. Estimate the size of the cylindrical bed if it is to be expanded to 1.5 times the volume formed by a fixed bed.

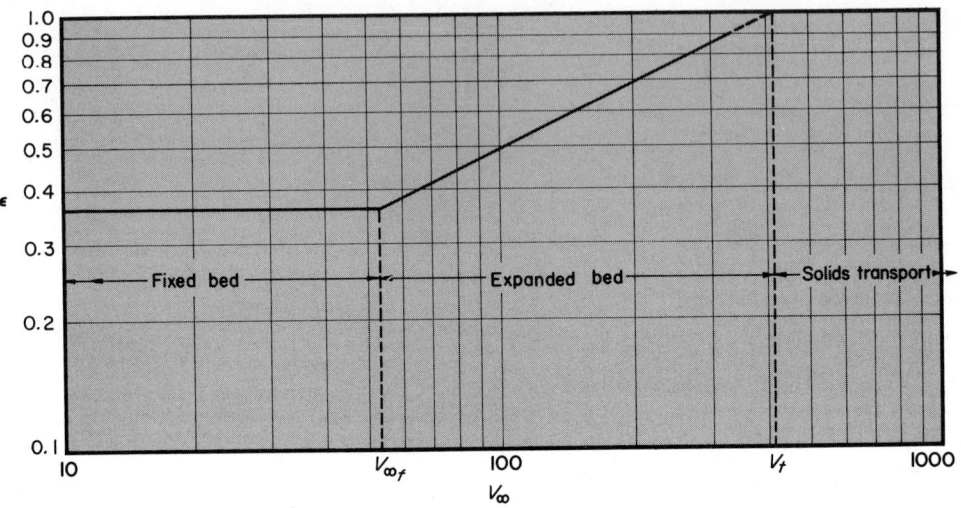

Fig. 20-9 Variation of void fraction with superficial fluid velocity.

Solution At the outset we shall determine the mean particle diameter D_p given by Eq. (20-27)

$$D_p = \frac{6(1-\epsilon)}{a_p}$$

The void fraction can be estimated from Fig. 20-6 by assuming that the bed diameter D_t is much larger than the mean particle diameter D_p, giving

$$\epsilon \simeq 0.32$$

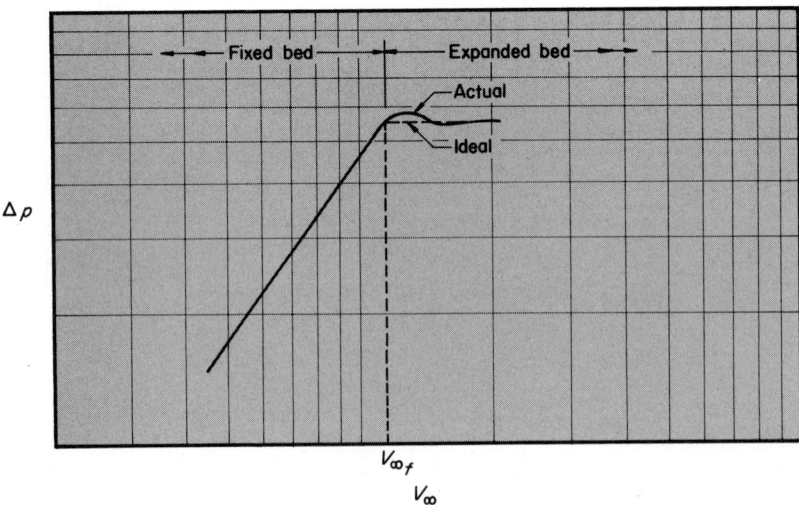

Fig. 20-10 Variation of pressure drop with superficial fluid velocity.

FLOW THROUGH PERMEABLE MEDIA

Each particle has a surface area S_p given by

$$S_p = \pi DL + 2\frac{\pi D^2}{4}$$

or, since $D = L$,

$$S_p = \tfrac{3}{2}\pi D^2 = \tfrac{3}{2}\pi \left(\frac{0.25}{12}\right)^2 \simeq 0.002 \text{ ft}^2$$

It is reasonable to assume that the number of pellets which will pack into a given volume will be approximately equal to the number of Raschig rings of the same size which will pack into the same volume; therefore, from Table 20-3

$n''' = 85{,}000$ pieces/ft^3

The packing area a_p is then given by

$a_p = n''' S_p = (85{,}000 \text{ pieces/ft}^3)(0.002 \text{ ft}^2/\text{piece}) = 170 \text{ ft}^2/\text{ft}^3$

Then

$$D_p = \frac{6(1 - 0.32)}{170} = 0.024 \text{ ft}$$

From Table A-3

$\mu = 5.78 \times 10^{-4}$ lb$_m$/ft-sec   and $\rho = 62.2$ lb$_m$/ft^3

Substituting these parameters and the given data into equation (20-64), we get

$$(90 - 62.2) \text{ lb}_m/\text{ft}^3 (32.2 \text{ ft/sec}^2) = \frac{150(1 - 0.32)(5.78 \times 10^{-4} \text{ lb}_m/\text{ft-sec})V_{\infty_f}}{(0.024 \text{ ft})^2 (0.32^3)}$$
$$+ 1.75 \frac{(62.2 \text{ lb}_m/\text{ft}^3)V_{\infty_f}^2}{(0.024 \text{ ft})(0.32^3)}$$

or

$895 = 3{,}120 V_{\infty_f} + 138{,}500 V_{\infty_f}^2$

from which the *minimum fluidization velocity* $V_{\infty_f} = 0.07$ ft/sec.

A first estimate of the terminal velocity V_t is given by Eq. (20-65):

$$V_t = \left[\frac{3(0.024 \text{ ft})(90 - 62.2) \text{ lb}_m/\text{ft}^3 (32.2 \text{ ft/sec}^2)}{62.2 \text{ lb}_m/\text{ft}^3}\right]^{\frac{1}{2}}$$
$$= 1.035^{\frac{1}{2}} = 1.017 \text{ ft/sec}$$

This can now be used to get the Reynolds number to check the validity of using Eq. (20-65); hence,

$$\text{Re} = \frac{V_t D_p \rho}{\mu} = \frac{(1.017 \text{ ft/sec})(0.024 \text{ ft})(62.2 \text{ lb}_m/\text{ft}^3)}{5.78 \times 10^{-4} \text{ lb}_m/\text{ft-sec}} = 4{,}220$$

which is within the range for which Eq. (20-65) is valid; however, the drag coefficient used to get that equation was for a sphere. This can be refined by using Fig. 20-2, requiring the sphericity, which we now obtain:

$$\psi = \frac{S_e}{S_p} = \frac{\pi D_e^2}{S_p}$$

But the volume of the pellet is

$$\mathscr{V} = \frac{\pi D^2}{4} L = \frac{\pi D_{pellet}^3}{4}$$

since $D = L$, and

$$\frac{\pi D_{pellet}^3}{4} = \frac{\pi D_e^3}{6}$$

$D_e = (\tfrac{6}{4})^{\frac{1}{3}} D_{pellet} = 0.0238$ ft

giving

$$\psi = \frac{\pi (0.0238^2)}{0.002} = 0.89$$

From Fig. 20-2 at **Re** $\simeq 4{,}220$,

$C_D \approx 0.6$

Using this value in Eq. (20-4), a corrected value of the terminal velocity is obtained:

$$V_t = \left[\frac{4(32.2 \text{ ft/sec}^2)(0.024 \text{ ft})(90 - 62.2) \text{ lb}_m/\text{ft}^3}{(62.2 \text{ lb}_m/\text{ft}^3)(0.6)} \right]^{\frac{1}{2}} = 0.88 \text{ ft/sec}$$

We can now plot a diagram similar to that of Fig. 20-9 from which we can read the fluidizing velocity for the expanded bed. For a 50 percent increase in bed expansion,

$$\epsilon_{ex} = \frac{1 - \epsilon}{1.50} = \frac{0.68}{1.50} = 0.453$$

For the expanded-bed void fraction we get $V_\infty = 0.152$ ft/sec from the plot. The required cross-sectional area can now be found from the continuity equation $\dot{m} = \rho A_t V_\infty$; thus,

$$A_t = \frac{22.8 \text{ lb}_m/\text{sec}}{(0.152 \text{ ft/sec})(62.2 \text{ lb}_m/\text{ft}^3)} = 2.41 \text{ ft}^2$$

from which the diameter is

$$D_t = \left(\frac{4 A_t}{\pi} \right)^{\frac{1}{2}} = 1.75 \text{ ft}$$

Hence

$$\frac{D_p}{D_t} = \frac{0.024}{1.75} = 0.0137$$

and our initial assumption that the bed is much larger than the mean particle diameter is seen to be valid upon consulting Fig. 20-6.

The bed volume is

$$\mathscr{V} = \frac{400 \text{ lb}_m}{90 \text{ lb}_m/\text{ft}^3} = 4.45 \text{ ft}^3$$

from which the expanded bed height is

$$L_{ex} = \frac{\mathscr{V}}{A} = \frac{4.45 \text{ ft}^3}{2.41 \text{ ft}^2} = 1.84 \text{ ft}$$

FLOW THROUGH PERMEABLE MEDIA

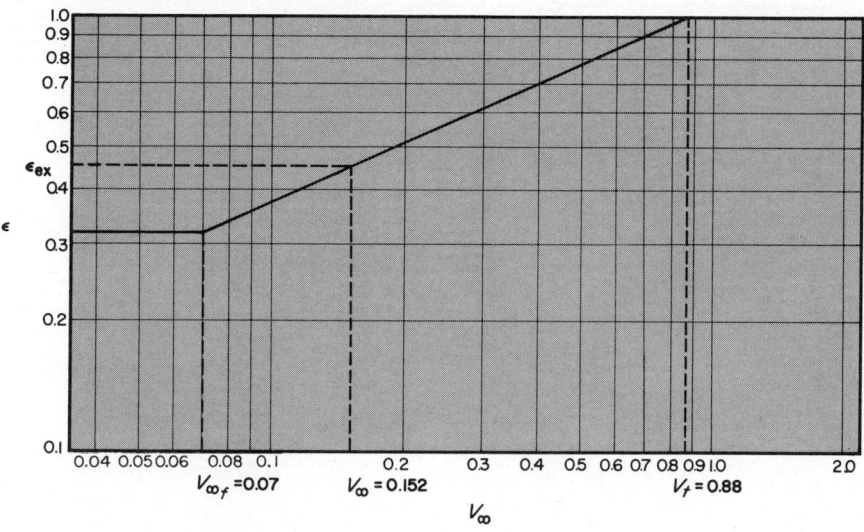

Fig. 20-11 Void fraction for bed expansion.

20-5 SOLIDS TRANSPORT

The flow systems of this section are characterized by a net movement of the solids relative to the containing vessel. The physical nature of the *dense phase* (moving bed) is approximately the same as that of a fixed bed (low voidage); however, the *dilute phase* is distinguished by high voidages.

To better understand the physics of solids transport let us consider an extension of Fig. 20-10 to include solids movement [18]. This is shown schematically in Fig. 20-12 for a vertical bed with upward fluid flow. Three typical cases, designated by the mass flow rate of solids $\dot{m}_s$, are shown: $\dot{m}_s = 0$ (fixed bed for curve A and suspended particle for curve tt'); $\dot{m}_{s_1} < \dot{m}_{s_2}$.

The pressure drop becomes a maximum at the onset of fluidization, point A for the fixed-bed case. Solids which are moving downward under the influence of gravity (curves C' and D') are stopped as the pressure differential reaches points C and D, respectively. As the bed expands, the pressure differential decreases while the solids separate, reducing their resistance to the flow of the fluid. In the vicinity of point A the bed is quiescent, but as the fluid flow rate is increased, it becomes agitated (particularly when a gas or vapor is the fluidizing medium) and slugging, illustrated in the inset, occurs at point S. Because of instabilities the curve breaks off at this point.

With sufficiently high fluid flow rates the unstable zone is passed over, becoming stable again at point s, where the solids concentration has been reduced from dense at point S to dilute at point s. From this point, a reduction in fluid velocity permits the bed to move downward under the influence of gravity (curves cc' and dd'); an increase

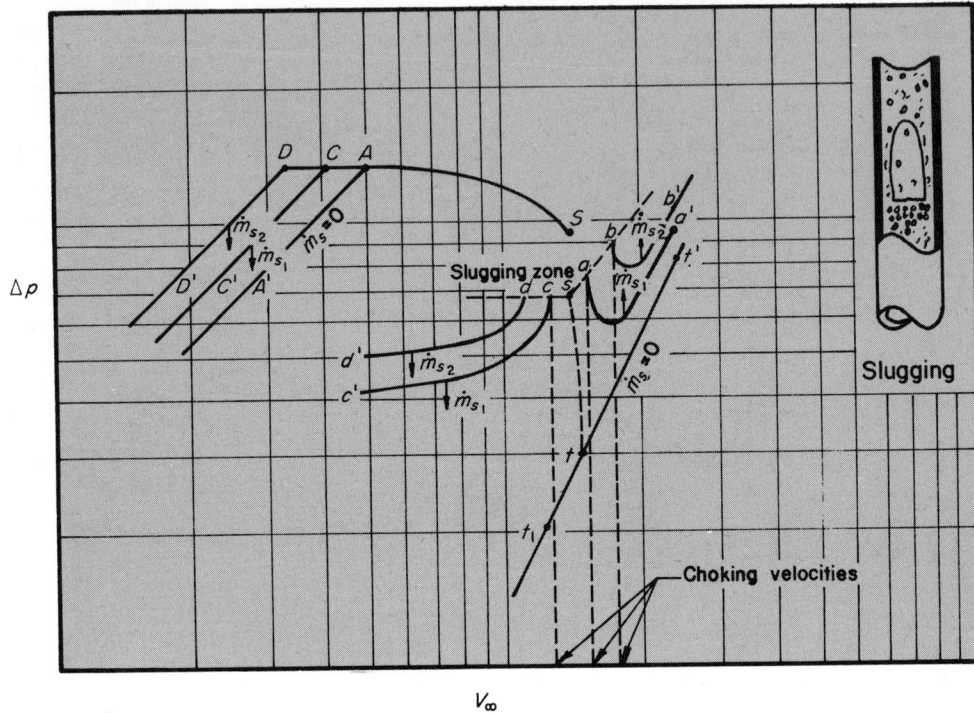

Fig. 20-12 Phase diagram for upward fluid flow.

in fluid velocity causes a substantial decrease in pressure drop (the solids are being separated farther) limited by the terminal velocity of a single particle (point t), where the fluid velocity just equals the terminal velocity. The pressure drop is now that which occurs in an empty conduit. An increase in fluid velocity causes the single particle to move upward. As the velocity is increased from V_t, the solid particle moves upward with a velocity $V_\infty - V_t$. If instead of reducing the pressure drop from point s the pressure is increased while increasing the fluid velocity, the solids will move upward, as indicated by curves aa' and bb'.

The *choking velocity* is that velocity below which solids will precipitate out of the flow system. In horizontal flow the limiting velocity marking the onset of precipitation is known as the *saltation velocity*.

Slugging rarely occurs when the fluidizing medium is a liquid. It can be expected in gas-fluidized systems, however, in the voidage range $0.75 < \epsilon < 0.95$. In the following paragraphs design equations are presented for the most common cases of solids transport.

Dense phase (moving bed) *For gravity flow of solids with concurrent or countercurrent gas or vapor flow* Happel [7] correlated data for void fractions ranging from 0.33 to 0.49 and $D_p = 0.015$ to 0.276 in by the use of a modified friction factor f_m

FLOW THROUGH PERMEABLE MEDIA

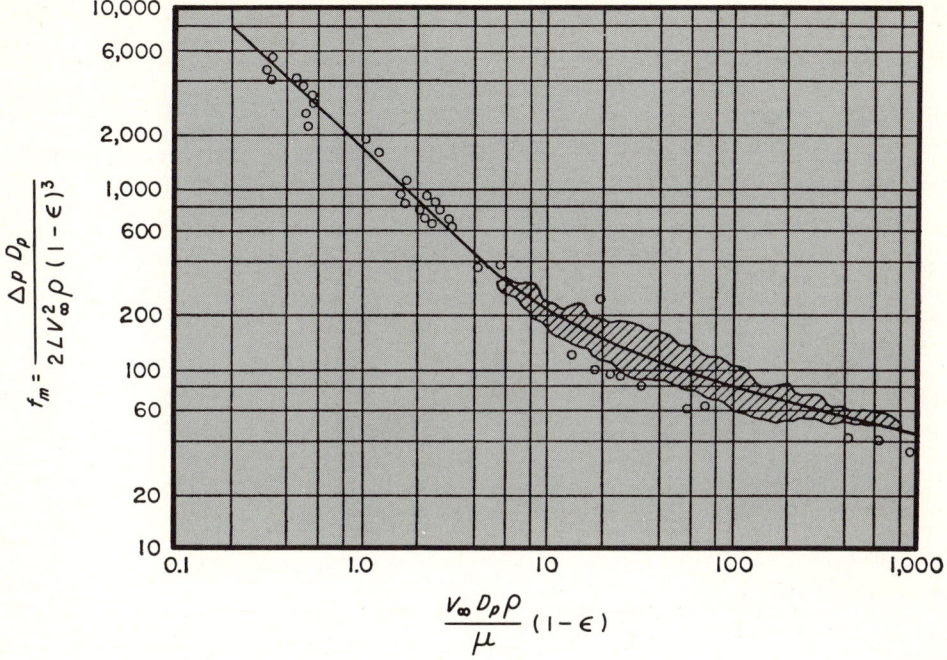

Fig. 20-13 Dense-phase vertical moving-bed data.

related to a modified Reynolds number

$$f_m = \frac{\Delta p\, D_p}{2LV_\infty^2 \rho (1-\epsilon)^3} = \phi(\mathbf{Re'}) \tag{20-66}$$

where

$$D_p = \frac{1}{\sum \dfrac{\rho_{si}/\rho_s}{D_i}} \tag{20-67}$$

in which the subscript i refers to particles of a given size. The relation is shown in Fig. 20-13. The superficial fluid velocity V_∞ is relative to the wall for a solids velocity less than 20 percent of the gas or vapor velocity. For higher solids velocities, V_∞ should be replaced by the fluid velocity relative to the moving particles.

Intermediate phase (high ratio solids to fluid) For solids-to-gas ratios from 25 to 850 the data of Wen [17] for *flow in horizontal pipes* are correlated by a solid-phase friction factor f_s defined by

$$f_s = \frac{g_c}{2} \frac{D_t}{L} \frac{\Delta p_t}{\rho_s V_s^2} \tag{20-68}$$

and the solids velocity V_s as presented in Fig. 20-14. The total pressure drop Δp_t is that due to the combined gas and solids flow. D_t is the diameter of the confining tube.

Solids density for the correlation ranged from 81 to 156 lb$_m$/ft^3. It should be noted that the data correlate with V_s only, indicating that the flow is independent of the particle size.

Available data for flow for a high ratio of solids to gas indicate that the *gas velocity is approximately double the solids velocity*.

Dilute phase *Pneumatic transport* has long been important in the process industries, beginning with the conveying of grains and seeds. As in other solids-transport mechanisms, the slip velocity V_{slip}, defined by

$$V_{\text{slip}} = V_\infty - V_p^* \tag{20-69}$$

is an important parameter. The asterisk is used to indicate the fully accelerated condition.

The fully accelerated particle velocity V_p^*, given by Hinkle [8], is

$$V_p^* = V_\infty \left(1 - 1.41 D_p^{0.3} \frac{\rho_s}{62.3}\right)^{0.5} \tag{20-70}$$

where D_p is the particle diameter in feet and ρ_s is the solids density in pounds mass per cubic foot. The variables from which the correlation resulted are

Air velocity = 66 to 119 ft/sec

Particle diameter = 0.014 to 0.33 in

Solids density = 65.5 to 113 lb$_m$/ft^3

Solids loading ratio α = up to 5 lb$_m$ solids/lb$_m$ air

The total pressure drop Δp_t due to combined gas and solids flow in a *horizontal circular conduit* can be obtained from the equation

$$\Delta p_t = \frac{fV_\infty^2 \rho L}{2g_c D_t}\left(1 + \frac{f_p^* V_p^*}{fV_\infty}\alpha\right) \tag{20-71}$$

in which f is the Fanning friction factor given in Fig. 13-10 and f_p^* is the solids flow friction factor, existing under fully accelerated flow conditions, given by

$$f_p^* = \frac{3\rho C_D D_t}{2 D_p \rho_s}\left(\frac{V_\infty - V_p^*}{V_p^*}\right)^2 \tag{20-72}$$

C_D is the drag coefficient obtained from Fig. 20-2 or from the appropriate equation.

For pneumatic transport in a *vertical circular conduit* a term for solids head must be added to the correlation, giving [8]

$$\Delta p_t = \frac{fV_\infty^2 \rho L}{2g_c D_t}\left(1 + \frac{f_p^* V_p^*}{fV_\infty}\alpha + \frac{2g D_t}{fV_\infty V_p^*}\alpha\right) \tag{20-73}$$

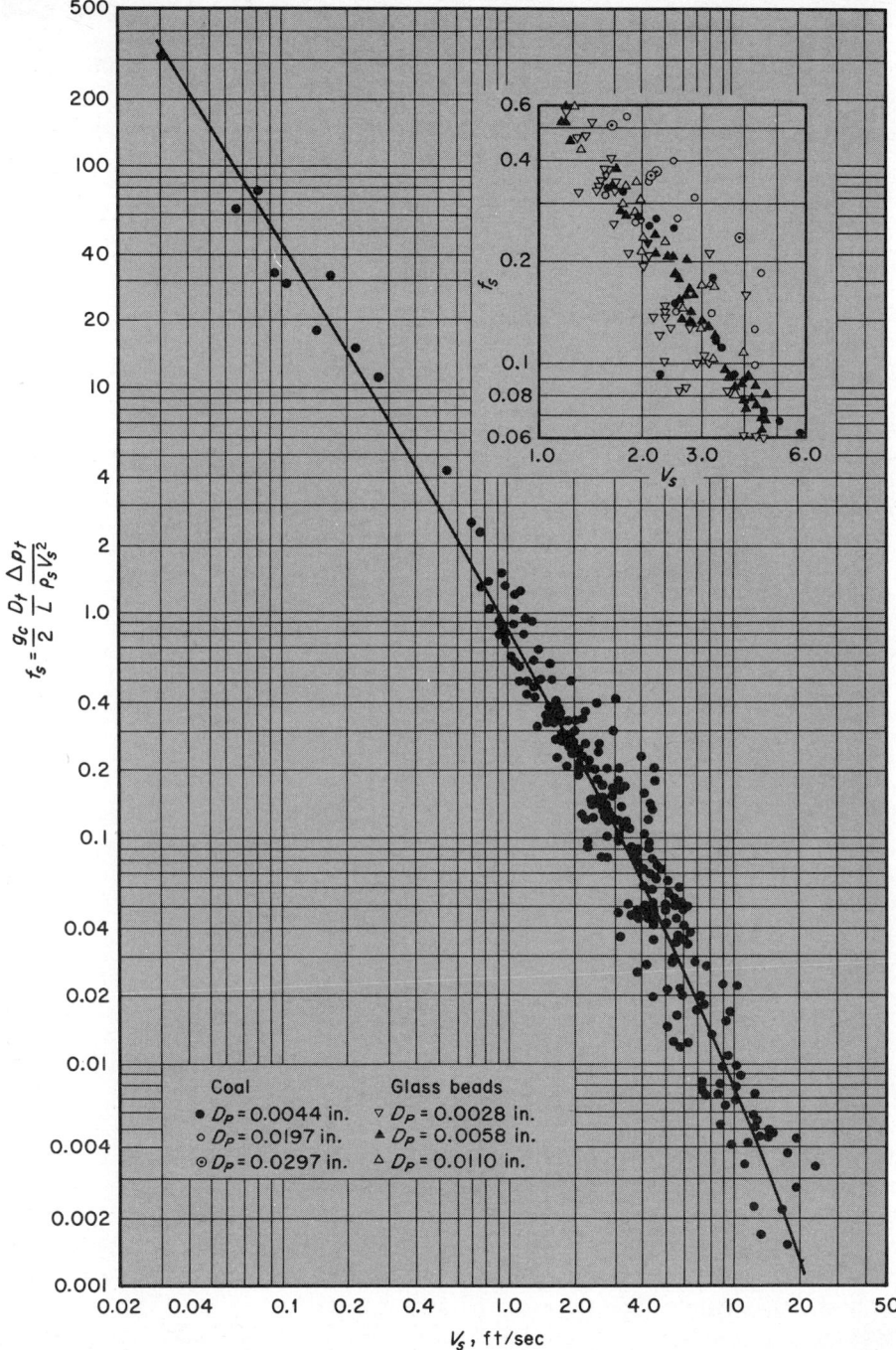

Fig. 20-14 Data for transport in horizontal pipes for a high solids-air ratio. (*From M. Leva, "Fluidization." Copyright 1959. McGraw-Hill Book Company. Used by permission.*)

The maximum particle velocities are independent of the solids loading ratio α, but the rate of acceleration to the maximum particle velocity is slower with greater solids loads. Lengths to achieve fully accelerated conditions may be appreciable [10].

For more detailed analyses and design equations, the reader should consult Zenz and Othmer [19] and Leva [10].

PROBLEMS

20-1. A 1-in-diam rubber sphere is released in a column of fluid ($\rho = 70.4$ lb$_m$/ft^3). It settles at a velocity of 0.2 ft/sec. What is the dynamic viscosity μ of the fluid?

20-2. What velocity will a $\frac{1}{2}$-in-diam glass marble reach when dropped in a deep pool of water at 60°F?

20-3. Adsorption tests on an odd-shaped particle of density $\rho_s = 96$ lb$_m$/ft^3 and a sphericity of 0.7 give its surface area as $S_p = 0.0042$ ft^2. At what velocity will it move when released in a column of 100°F air moving upward at a velocity of 40 ft/sec?

20-4. Air at 200°F is forced through a 2-ft bed of 80- to 100-mesh sand at a superficial velocity of 4 ft/sec.
    (a) What is the pressure drop across the bed (psi)?
    (b) What is the average interstitial fluid velocity?

20-5. A very thin porous leather diaphragm permits 0.004 ft^3/hr-ft^2 of n-butyl alcohol at 60°F to pass through it with a pressure drop of 18 in H$_2$O. What flow rate of 80°F ammonia can be expected through the diaphragm for the same pressure drop?

20-6. Crude oil percolates through a large bed of very porous limestone into a 6-in-diam oil well. What pressure must be exerted at a radial distance of 50 ft to cause a flow of 100 ft^3/day through a 1-ft slab, assuming the oil to flow radially inward? The oil has density $\rho = 54$ lb$_m$/ft^3 and dynamic viscosity $\mu = 8 \times 10^{-5}$ lb$_f$-sec/ft^2, and the temperature is 56°F.

20-7. Air at 125°F is forced through a 4-ft bed of grain to prevent rot. The equivalent particle diameter is 0.08 in, and the void fraction is 0.24. What pressure drop is required for a superficial velocity of (a) 2 ft/sec; (b) 0.01 ft/sec?

20-8. What delivery pressure is required to force 100 ft^3/min of air at 32°F through a bed of clay spheres contained in a 6-in-ID pipe? The exit pressure from the 6-ft bed must be 15.0 psia, and the size distribution of the spheres is as follows:

Mesh	Weight fraction $\mathscr{W}$	Size d_p, in
6–8	0.15	0.110
8–10	0.75	0.078
10–14	0.10	0.055

The average particle diameter D_p is given by

$$D_p = \frac{1}{\Sigma(\mathscr{W}/d_p)}$$

where $\mathscr{W}$ is the weight fraction of particles having diameter d_p.

20-9. Benzene vapor mixed with an inert gas is absorbed in a 5-ft tower packed with $\frac{1}{4}$-in berl saddles. The density of the mixture is 0.089 lb$_m$/ft^3, and it has a dynamic viscosity $\mu = 2.0$ centipoises. Estimate the superficial fluid velocity V_∞ for a pressure drop of 340 lb$_f$/ft^2.

FLOW THROUGH PERMEABLE MEDIA

20-10. Aloxite granules are packed in a 2.27-in-ID tube. For laminar flow estimate the permeability of the bed if $D_p/D_t = 0.22$.

20-11. To filter 0.3 gal/ft² of a slurry, having 5 percent solids and a viscosity of 2×10^{-4} lb$_f$-sec/ft², in an initially clean bed of fibers at a constant pressure differential of 72 lb$_f$/ft² requires 20 min. What pressure drop can be expected in a 2-in clean bed of the same fibrous medium in filtering the slurry flowing at a superficial velocity of 0.0002 ft/sec?

20-12. A mixture of glass beads, having density $\rho_s = 150$ lb$_m$/ft³ and the size distribution given below, is to be fluidized in a 6-in-ID column by water at 80°F. What is the minimum fluidization velocity?

d_p, in	0.172	0.228	0.388	0.508
Weight fraction W	0.1655	0.1515	0.1985	0.4845

20-13. A 1-ft-ID bed of ¼-in carbon Raschig rings is to be expanded to a void fraction of 0.87 by a fluid having properties $\rho = 34.0$ lb$_m$/ft³ and $\mu = 1.20 \times 10^{-3}$ lb$_m$/ft-sec. What flow rate is required?

20-14. A prototype moving-bed compact regenerative heat exchanger for use in a turbine-powered truck is tested with aluminum granules ($\rho_s = 92.0$ lb$_m$/ft³) having the following size distribution:

Mean sieve diameter d_p	0.0470	0.0272	0.0176	0.0130
Weight fraction W	0.101	0.670	0.225	0.004

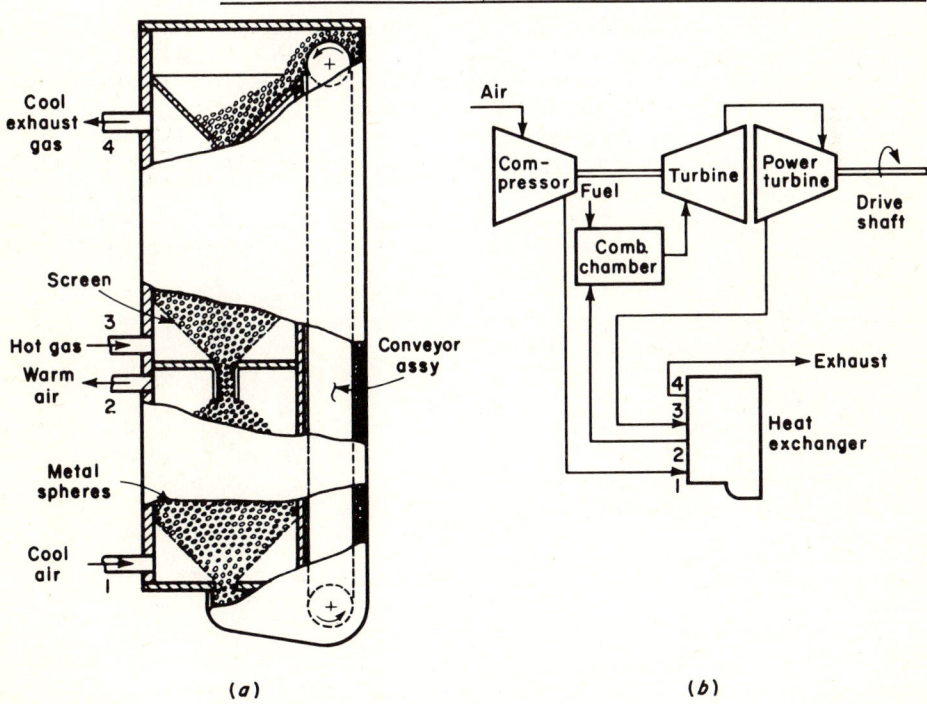

Fig. P20-14

The void fraction under static conditions is 0.46. Solids flow with gravity at the superficial rate of 15,400 $lb_m/hr\text{-}ft^2$ while air at 100°F flows countercurrently, as shown schematically, at the superficial rate of 940 $lb_m/hr\text{-}ft^2$. Assuming that the void fraction varies only slightly from its static value:

(a) What is the effective heat-transfer surface area per unit volume?

(b) Estimate the pressure drop in a 1-in-ID tube 6 in long connecting the hot and cold chambers.

20-15. Coal having a density of 81.0 lb_m/ft^3 and pulverized to a mean particle diameter of 0.03 in is transported to a furnace in a horizontal 2-in-ID pipe by air at 100°F. Estimate the total pressure drop for a mixture in a 100-ft length of pipe for an airflow rate of 25 ft^3/min.

20-16. Clover seed of $\rho_s = 40$ lb_m/ft^3 and $D_p = 0.05$ in is to be pneumatically transported through a 1-in-ID smooth tube. The air-seed mixture leaves the tube at 14.7 psia and 70°F. The exit air velocity is 50 ft/sec. Assuming that the seeds have been fully accelerated, find the pressure drop in a 20-ft tube for a solids-to-gas ratio of 2 lb_m seed per pound mass of air when the tube is (a) horizontal, (b) vertical.

REFERENCES

1. Bennett, C. O., and J. E. Myers: "Momentum, Heat, and Mass Transfer," McGraw-Hill, New York, 1962.
2. Bird, R. B., W. E. Stewart, and E. N. Lightfoot: "Transport Phenomena," Wiley, New York, 1960.
3. Brown, G. G., et al.: "Unit Operations," Wiley, New York, 1951.
4. Cramp, W., and A. Priestley: Pneumatic Grain Elevators, *Engineer*, **1924**: 139.
5. Darcy, L.: Les Fontaines publiques de la ville de Dijon, Victor Delmont, Paris, 1856.
6. Ergun, S.: Fluid Flow through Packed Columns, *Chem. Engrg. Progr.*, **48**(2): 89 (1952).
7. Happel, J.: Pressure Drop Due to Vapor Flow through Moving Beds, *Ind. Eng. Chem.*, **41**: 1161 (1949).
8. Hinkle, B. L.: Ph.D. thesis, Georgia Institute of Technology, 1953.
9. Kay, J. M.: "An Introduction to Fluid Mechanics and Heat Transfer," 2d ed., Cambridge University Press, London, 1963.
10. Leva, M.: "Fluidization," McGraw-Hill, New York, 1959.
11. Parker, J. D., J. H. Boggs, and E. F. Blick: "Introduction to Fluid Mechanics and Heat Transfer," Addison-Wesley, Reading, Mass., 1969.
12. Scheidegger, A. E.: "The Physics of Flow through Porous Media," Macmillan, New York, 1957.
13. Sissom, L. E., and T. W. Jackson: Temperature Distribution of Solids and Fluid in Fluid-to-particle Heat Exchange, *ASME* 65-HT-17, 1965.
14. Sissom, L. E., and T. W. Jackson: Heat Exchange in Fluid-Dense Particle Moving Beds, *J. Heat Transfer*, February, 1967.
15. Stepanoff, A. J.: Pumping Solid-Liquid Mixtures, *Mech. Engrg.*, September, 1964, p. 29.
16. Waddel, H.: The Coefficient of Resistance as a Function of Reynolds' Number for Solids of Various Shapes, *J. Franklin Inst.*, **217**: 459–490 (1934).
17. Wen, C. Y.: Ph.D. thesis, West Virginia, 1956.
18. Zenz, F. A.: Two-phase Fluid-Solid Flow, *Ind. Eng. Chem.*, **41**(12): 2801 (1949).
19. Zenz, F. A., and D. F. Othmer: "Fluidization and Fluid-Particle Systems," Reinhold, New York, 1960.

APPENDIXES

APPENDIX A

TABLES OF PROPERTIES AND FUNCTIONS

Table A-1 Kinematic-viscosity conversion table†
(Centistokes to Engler, Saybolt, and Redwood units)

Centistokes	Engler degrees	Saybolt seconds at 130°F	Redwood seconds at 140°F	Centistokes	Engler degrees	Saybolt seconds at 130°F	Redwood seconds at 140°F
2.0	1.140	32.66	30.95	29.0	3.945	136.8	120.4
2.5	1.182	34.46	32.20	30.0	4.070	141.2	124.4
3.0	1.224	36.07	33.45	31.0	4.195	145.6	128.3
3.5	1.266	37.67	34.70	32.0	4.320	150.0	132.3
4.0	1.308	39.17	35.95	33.0	4.445	154.5	136.3
4.5	1.350	40.78	37.20	34.0	4.570	159.0	140.2
5.0	1.400	42.38	38.45	35.0	4.695	163.5	144.2
5.5	1.441	43.98	39 80	36.0	4.825	168.0	148.2
6.0	1.481	45.59	41.05	37.0	4.955	172.5	152.2
6.5	1.521	47.19	42.40	38.0	5.080	177.0	156.2
7.0	1.563	48.79	43.70	39.0	5.205	181.5	160.3
7.5	1.605	50.44	45.00	40.0	5.335	186.0	164.3
8.0	1.653	52.10	46.35	41.0	5.465	190.6	168.3
8.5	1.700	53.80	47.75	42.0	5.590	195.1	172.3
9.0	1.746	55.51	49.10	43.0	5.720	199.6	176.4
9.5	1.791	57.21	50.55	44.0	5.845	204.2	180.4
10.0	1.837	58.91	52.00	45.0	5.975	208.8	184.5
11.0	1.928	62.42	55.00	46.0	6.105	213.4	188.5
12.0	2.020	66.03	58.10	47.0	6.235	218.0	192.6
13.0	2.120	69.73	61.30	48.0	6.365	222.6	196.6
14.0	2.219	73.54	64.55	49.0	6.495	227.2	200.7
15.0	2.323	77.35	67.95	50.0	6.630	231.8	204.7
16.0	2.434	81.25	71.40	52.0	6.890	241.1	212.8
17.0	2.540	85.26	74.85	54.0	7.106	250.3	221.0
18.0	2.644	89.37	78.45	56.0	7.370	259.5	229.1
19.0	2.755	93.48	82.10	58.0	7.633	268.7	237.2
20.0	2.870	97.69	85.75	60.0	7.896	277.9	245.3
21.0	2.984	101.9	89.50	62.0	8.159	287.2	253.5
22.0	3.100	106.2	93.25	64.0	8.422	296.4	261.6
23.0	3.215	110.5	97.05	66.0	8.686	305.6	269.8
24.0	3.335	114.8	100.9	68.0	8.949	314.8	277.9
25.0	3.455	119.1	104.7	70.0	9.212	324.0	286.0
26.0	3.575	123.5	108.6	72.0	9.475	333.3	294.1
27.0	3.695	127.9	112.5	74.0	9.738	342.5	302.2
28.0	3.820	132.4	116.5	75.0	9.870	347.2	306.3

† From T. Baumeister (ed.), "Marks' Mechanical Engineers' Handbook," 6th ed.   Copyright 1958. McGraw-Hill Book Company.   Used by permission.

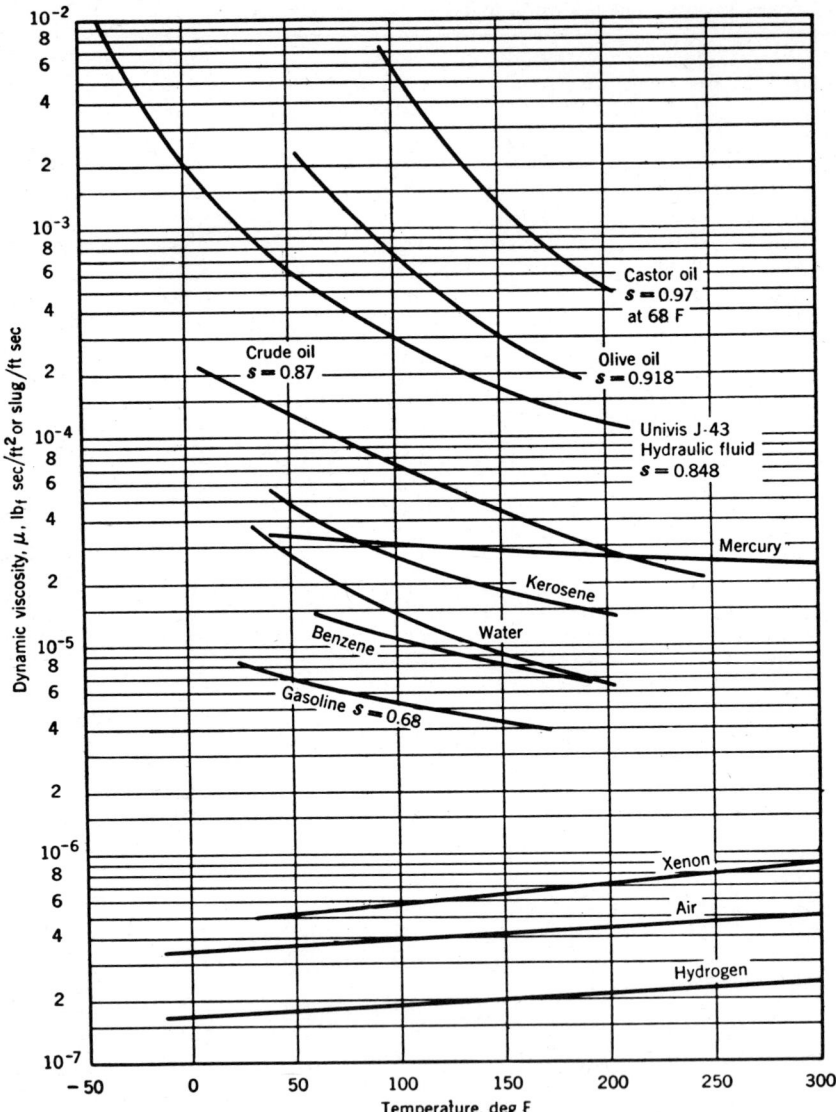

Fig. A-1 Dynamic (absolute) viscosity of fluids. Specific-gravity values apply at 70°F. (*By permission from R. M. Olson, "Essentials of Engineering Fluid Mechanics," International Textbook Company, Scranton, Pa., 1961.*)

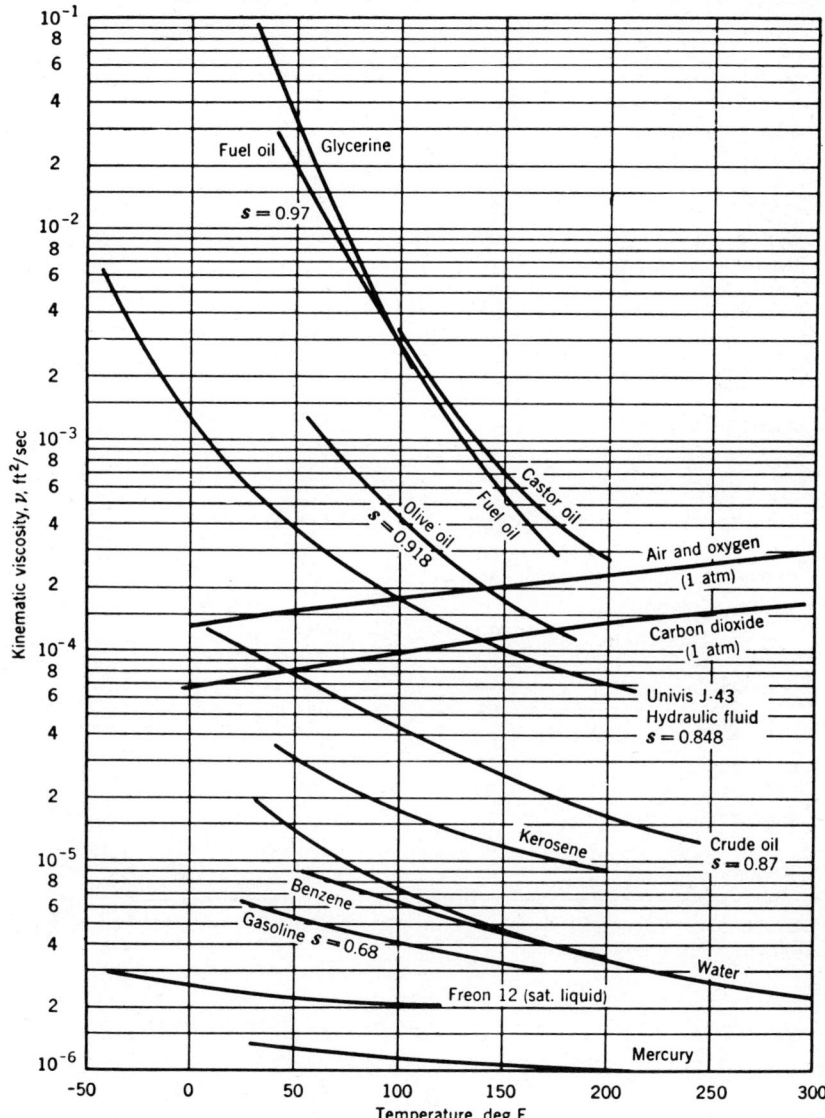

Fig. A-2 Kinematic viscosity of fluids. Specific-gravity values apply at 70°F. (*By permission from R. M. Olson, "Essentials of Engineering Fluid Mechanics," 2d ed., International Textbook Company, Scranton, Pa., 1961.*)

Table A-2 Physical properties of some nonmetals†

Material	Average temperature, °F	k, Btu/hr-ft-°F	c, Btu/lb$_m$-°F	ρ, lb$_m$/ft^3	α, ft^2/hr
Insulating materials:					
Asbestos	32	0.087	0.25	36	∼0.01
	392	0.12	....	36	∼0.01
Cork	86	0.025	0.04	10	∼0.006
Cotton, fabric	200	0.046			
Diatomaceous earth,					
powdered	100	0.030	0.21	14	∼0.01
	300	0.036			
	600	0.046			
Molded pipe covering	400	0.051	....	26	
	1600	0.088			
Glass wool, fine	20	0.022			
	100	0.031	....	1.5	
	200	0.043			
Packed	20	0.016			
	100	0.022	....	6.0	
	200	0.029			
Hair felt	100	0.027	....	8.2	
Kaolin insulating					
brick	932	0.15	....	27	
	2102	0.26			
Kaolin insulating					
firebrick	392	0.05	....	19	
	1400	0.11			
85% magnesia	32	0.032	....	17	
	200	0.037	....	17	
Rock wool	20	0.017		8	
	200	0.030			
Rubber	32	0.087	0.48	75	0.0024

† By permission from F. Kreith, "Principles of Heat Transfer," 2d ed. International Textbook Company, Scranton, Pa., 1965.

TABLE A-2 (Continued)

Material	Average temperature, °F	k, Btu/hr-ft-°F	c, Btu/lb_m-°F	ρ, lb_m/ft^3	α, ft^2/hr
Building materials:					
Brick, fireclay	392	0.58	0.20	144	0.02
	1832	0.95			
Masonry	70	0.38	0.20	106	0.018
Zirconia	392	0.84	....	304	
	1832	1.13			
Chrome brick	392	0.82	....	246	
	1832	0.96			
Concrete, stone	~70	0.54	0.20	144	0.019
10% moisture	~70	0.70	....	140	~0.025
Glass, window	~70	~0.45	0.2	170	0.013
Limestone, dry	70	0.40	0.22	105	0.017
Sand, dry	68	0.20	....	95	
10% H$_2$O	68	0.60	....	100	
Soil, dry	70	~0.20	0.44	....	~0.01
wet	70	~1.5	....	....	~0.03
Wood:					
Oak, ⊥ to grain	70	0.12	0.57	51	0.0041
∥ to grain	70	0.20	0.57	51	0.0069
Pine, ⊥ to grain	70	0.06	0.67	31	0.0029
∥ to grain	70	0.14	0.67	31	0.0067
Ice	32	1.28	0.46	57	0.048

Table A-3 Physical properties of gases, liquids, and liquid metals†
All gas properties are for atmospheric pressure

T, °F	ρ, lb_m/ft^3	c_p, $Btu/lb_m\text{-}°F$	$\mu \times 10^5$, $lb_m/ft\text{-}sec$	$\nu \times 10^3$, ft^2/sec	k, $Btu/hr\text{-}ft\text{-}°F$	Pr	α, ft^2/hr	$\beta \times 10^3$, $°F^{-1}$	$g\beta\rho^2/\mu^2$, $(°F\text{-}ft^3)^{-1}$
				A. GASES					
Air									
0	0.086	0.239	1.110	0.130	0.0133	0.73	0.646	2.18	4.2×10^4
32	0.081	0.240	1.165	0.145	0.0140	0.72	0.720	2.03	3.16
100	0.071	0.240	1.285	0.180	0.0154	0.72	0.905	1.79	1.76
200	0.060	0.241	1.440	0.239	0.0174	0.72	1.20	1.52	0.850
300	0.052	0.243	1.610	0.306	0.0193	0.71	1.53	1.32	0.444
400	0.046	0.245	1.750	0.378	0.0212	0.689	1.88	1.16	0.258
500	0.0412	0.247	1.890	0.455	0.0231	0.683	2.27	1.04	0.159
600	0.0373	0.250	2.000	0.540	0.0250	0.685	2.68	0.943	0.106
700	0.0341	0.253	2.14	0.625	0.0268	0.690	3.10	0.862	70.4×10^3
800	0.0314	0.256	2.25	0.717	0.0286	0.697	3.56	0.794	49.8
900	0.0291	0.259	2.36	0.815	0.0303	0.705	4.02	0.735	36.0
1000	0.0271	0.262	2.47	0.917	0.0319	0.713	4.50	0.685	26.5
1500	0.0202	0.276	3.00	1.47	0.0400	0.739	7.19	0.510	7.45
2000	0.0161	0.286	3.45	2.14	0.0471	0.753	10.2	0.406	2.84
2500	0.0133	0.292	3.69	2.80	0.051	0.763	13.1	0.338	1.41
3000	0.0114	0.297	3.86	3.39	0.054	0.765	16.0	0.289	0.815
Steam									
212	0.0372	0.451	0.870	0.234	0.0145	0.96	0.864	1.49	0.877×10^6
300	0.0328	0.456	1.000	0.303	0.0171	0.95	1.14	1.32	0.459
400	0.0288	0.462	1.130	0.395	0.0200	0.94	1.50	1.16	0.243
500	0.0258	0.470	1.265	0.490	0.0228	0.94	1.88	1.04	0.139
600	0.0233	0.477	1.420	0.610	0.0257	0.94	2.31	0.943	82×10^3
700	0.0213	0.485	1.555	0.725	0.0288	0.93	2.79	0.862	52.1
800	0.0196	0.494	1.700	0.855	0.0321	0.92	3.32	0.794	34.0
900	0.0181	0.50	1.810	0.987	0.0355	0.91	3.93	0.735	23.6
1000	0.0169	0.51	1.920	1.13	0.0388	0.91	4.50	0.685	17.1
1200	0.0149	0.53	2.14	1.44	0.0457	0.88	5.80	0.603	9.4
1400	0.0133	0.55	2.36	1.78	0.053	0.87	7.25	0.537	5.49
1600	0.0120	0.56	2.58	2.14	0.061	0.87	9.07	0.485	3.38
1800	0.0109	0.58	2.81	2.58	0.068	0.87	10.8	0.442	2.14
2000	0.0100	0.60	3.03	3.03	0.076	0.86	12.7	0.406	1.43
2500	0.0083	0.64	3.58	4.30	0.096	0.86	18.1	0.338	0.603
3000	0.0071	0.67	4.00	5.75	0.114	0.86	24.0	0.289	0.293

† By permission from F. Krieth, "Principles of Heat Transfer," 2d ed., International Textbook Company, Scranton, Pa., 1965.

APPENDIXES

TABLE A-3 (continued)

T, °F	ρ, lb_m/ft^3	c_p, Btu/lb_m-°F	$\mu \times 10^5$ lb_m/ft-sec	$\nu \times 10^3$ ft^2/sec	k, Btu/hr-ft-°F	Pr	α, ft^2/hr	$\beta \times 10^3$, °F^{-1}	$g\beta\rho^2/\mu^2$, (°F-ft^3)$^{-1}$
Oxygen									
0	0.0955	0.2185	1.215	0.127	0.0131	0.73	0.627	2.18	4.33×10^4
100	0.0785	0.2200	1.420	0.181	0.0159	0.71	0.880	1.79	1.76
200	0.0666	0.2228	1.610	0.242	0.0179	0.722	1.20	1.52	0.84
400	0.0511	0.2305	1.955	0.382	0.0228	0.710	1.94	1.16	0.256
600	0.0415	0.2390	2.26	0.545	0.0277	0.704	2.79	0.943	0.103
800	0.0349	0.2465	2.53	0.725	0.0324	0.695	3.76	0.794	48.5×10^3
1000	0.0301	0.2528	2.78	0.924	0.0366	0.690	4.80	0.685	25.8
1500	0.0224	0.2635	3.32	1.480	0.0465	0.677	7.88	0.510	7.50
Nitrogen									
0	0.0840	0.2478	1.055	0.125	0.0132	0.713	0.635	2.18	4.55×10^6
100	0.0690	0.2484	1.222	0.177	0.0154	0.71	0.898	1.79	1.84
200	0.0585	0.2490	1.380	0.236	0.0174	0.71	1.20	1.52	0.876
400	0.0449	0.2515	1.660	0.370	0.0212	0.71	1.88	1.16	0.272
600	0.0364	0.2564	1.915	0.526	0.0252	0.70	2.70	0.943	0.110
800	0.0306	0.2583	2.145	0.702	0.0291	0.70	3.62	0.794	52.0×10^3
1000	0.0264	0.2689	2.355	0.891	0.0330	0.69	4.65	0.685	27.7
1500	0.0197	0.2835	2.800	1.420	0.0423	0.676	7.58	0.510	8.12
Carbon monoxide									
0	0.0835	0.2482	1.065	0.128	0.0129	0.75	0.621	2.18	4.32×10^6
200	0.0582	0.2496	1.390	0.239	0.0169	0.74	1.16	1.52	0.860
400	0.0446	0.2532	1.670	0.374	0.0208	0.73	1.84	1.16	0.268
600	0.0362	0.2592	1.910	0.527	0.0246	0.725	2.62	0.943	0.109
800	0.0305	0.2662	2.134	0.700	0.0285	0.72	3.50	0.794	52.1×10^3
1000	0.0263	0.2730	2.336	0.887	0.0322	0.71	4.50	0.685	28.0
1500	0.0196	0.2878	2.783	1.420	0.0414	0.70	7.33	0.510	8.13
Helium									
0	0.012	1.24	1.140	0.950	0.078	0.67	5.25	2.18	77,800
200	0.00835	1.24	1.480	1.77	0.097	0.686	9.36	1.52	15,600
400	0.0064	1.24	1.780	2.78	0.115	0.70	14.5	1.16	4,840
600	0.0052	1.24	2.02	3.89	0.129	0.715	20.0	0.943	2,010
800	0.00436	1.24	2.285	5.24	0.138	0.73	25.5	0.794	932
1000	0.00377	1.24	2.520	6.69	...	...	...	0.685	494
1500	0.0028	1.24	3.160	11.30	...	...	...	0.510	129

TABLE A-3 (continued)

T, °F	ρ, lb_m/ft^3	c_p, Btu/lb_m-°F	$\mu \times 10^5$, lb_m/ft-sec	$\nu \times 10^3$, ft^2/sec	k, Btu/hr-ft-°F	Pr	α, ft^2/hr	$\beta \times 10^3$, °F^{-1}	$g\beta\rho^2/\mu^2$, (°F-ft^3)$^{-1}$
Hydrogen									
0	0.0060	3.39	0.540	0.89	0.094	0.70	4.62	2.18	86,600
100	0.0049	3.42	0.620	1.26	0.110	0.695	6.56	1.79	36,600
200	0.0042	3.44	0.692	1.65	0.122	0.69	8.45	1.52	18,000
500	0.0028	3.47	0.884	3.12	0.160	0.69	16.5	1.04	3,360
1000	0.0019	3.51	1.160	6.2	0.208	0.705	31.2	0.685	591
1500	0.0014	3.62	1.415	10.2	0.260	0.71	51.4	0.510	161
2000	0.0011	3.76	1.64	14.4	0.307	0.72	74.2	0.406	59
3000	0.0008	4.02	1.72	24.2	0.380	0.66	118.0	0.289	20
Carbon dioxide									
0	0.132	0.184	0.88	0.067	0.0076	0.77	0.313	2.18	15.8×10^6
100	0.108	0.203	1.05	0.098	0.0100	0.77	0.455	1.79	6.10
200	0.092	0.216	1.22	0.133	0.0125	0.76	0.63	1.52	2.78
500	0.063	0.247	1.67	0.266	0.0198	0.75	1.27	1.04	0.476
1000	0.0414	0.280	2.30	0.558	0.0318	0.73	2.75	0.685	71.4×10^3
1500	0.0308	0.298	2.86	0.925	0.0420	0.73	4.58	0.510	19.0
2000	0.0247	0.309	3.30	1.34	0.050	0.735	6.55	0.406	7.34
3000	0.0175	0.322	3.92	2.25	0.061	0.745	10.8	0.289	1.85

T, °F	ρ, lb_m/ft^3	c_p, Btu/lb_m-°F	$\mu \times 10^5$, lb_m/ft-sec	$\nu \times 10^5$, ft^2/sec	k, Btu/hr-ft-°F	Pr	$\alpha \times 10^3$, ft^2/hr	$\beta \times 10^3$, °F^{-1}	$g\beta\rho^2/\mu^2$, (°F-ft^3)$^{-1}$
B. LIQUIDS									
Commercial aniline									
60	64.0	0.48	325.0	5.08	0.10	56.0	3.25		
100	63.0	0.49	170.0	2.70	0.10	30.0	3.24	0.49	21.6×10^6
150	61.5	0.505	96.5	1.57	0.098	18.0	3.16	0.492	64.5
200	60.0	0.515	61.1	1.02	0.096	11.8	3.11		
300	57.5	0.54	32.5	0.565	0.093	6.8	3.00		
Ammonia (saturated liquid)									
−20	42.4	1.07	17.6	0.417	0.317	2.15	6.94		
0	41.6	1.08	17.1	0.410	0.316	2.09	7.04		
10	40.8	1.09	16.6	0.407	0.314	2.07	7.08		
32	40.0	1.11	16.1	0.402	0.312	2.05	7.03	1.2	238×10^6
50	39.1	1.13	15.5	0.396	0.307	2.04	6.95	1.3	266
80	37.2	1.17	14.5	0.386	0.293	2.01	6.73		
120	35.2	1.22	13.0	0.355	0.275	1.99	6.40		

TABLE A-3 (continued)

T, °F	ρ, lb_m/ft^3	c_p, Btu/lb_m-°F	$\mu \times 10^5$, lb_m/ft-sec	$\nu \times 10^5$, ft^2/sec	k, Btu/hr-ft-°F	Pr	$\alpha \times 10^3$, ft^2/hr	$\beta \times 10^3$, °F^{-1}	$g\beta\rho^2/\mu^2$, (°F-ft^3)$^{-1}$
Freon–12, CCl_2F_2 (saturated liquid)									
−40	94.8	0.211	28.4	0.300	0.040	5.4	2.00		
−20	93.0	0.214	25.0	0.272	0.040	4.8	2.01	1.03	4.6×10^9
0	91.2	0.217	23.1	0.253	0.041	4.4	2.07	1.05	5.27
20	89.2	0.220	21.0	0.238	0.042	4.0	2.14	1.34	7.80
32	87.2	0.223	20.0	0.230	0.042	3.8	2.16	1.72	10.5
60	83.0	0.231	18.0	0.213	0.042	3.5	2.19	2.1	14.4
100	78.5	0.240	16.0	0.206	0.040	3.5	2.12	2.5	19.4
120	75.9	0.244	15.5	0.204	0.039	3.5	2.12		
n–Butyl alcohol									
60	50.5	0.55	226	4.48	0.097	46.6	3.49		
100	49.7	0.61	129	2.60	0.096	29.5	3.16	0.45	21.5×10^6
150	48.5	0.68	67.5	1.39	0.095	17.4	2.88	0.48	80
200	47.2	0.77	38.6	0.815	0.094	11.3	2.58		
300	...	...	19.0						
Benzene									
60	55.1	0.40	46.0	0.835	0.093	7.2	4.22	0.60	0.3×10^9
80	54.6	0.42	39.6	0.725	0.092	6.5	4.01		
100	54.0	0.44	35.1	0.650	0.087	5.1	3.53		
150	53.5	0.46	26.0	0.480	...	4.5			
200	...	...	20.3	...	...	4.0			
Light oil									
60	57.0	0.43	5,820	102	0.077	1,170	3.14	0.38	1.17×10^4
80	56.8	0.44	2,780	49	0.077	570	3.09	0.38	5.1
100	56.0	0.46	1,530	27.4	0.076	340	2.95	0.39	16.7
150	54.3	0.48	530	9.8	0.075	122	2.88	0.40	1.34×10^6
200	54.0	0.51	250	4.6	0.074	62	2.69	0.42	6.4
250	53.0	0.52	139	2.6	0.074	35	2.67	0.44	21.0
300	51.8	0.54	83	1.6	0.073	22	2.62	0.45	56.5
Glycerin									
50	79.3	0.554	256×10^3	3,230	0.165	31×10^3	3.76		
70	78.9	0.570	100	1,270	0.165	12.5	3.67	0.28	56
85	78.5	0.584	42.4	540	0.164	5.4	3.58	0.30	332
100	78.2	0.600	18.8	240	0.163	2.5	3.45		
120	77.7	0.617	12.4	160	...	$\simeq$1.6			

TABLE A-3 (Continued)

T, °F	ρ, lb_m/ft^3	c_p, Btu/lb_m-°F	$\mu \times 10^3$, lb_m/ft-sec	$\nu \times 10^6$, ft^2/sec	k, Btu/hr-ft-°F	Pr	$\alpha \times 10^3$, ft^2/hr	$\beta \times 10^3$, °F^{-1}	$g\beta\rho^2/\mu^2$, (°F-ft^3)$^{-1}$
Water									
32	62.4	1.01	1.20	19.3	0.319	13.7	5.07	−0.037	
40	62.4	1.00	1.04	16.7	0.325	11.6	5.21	0.020	2.3×10^6
50	62.4	1.00	0.88	14.0	0.332	9.55	5.33	0.049	8.0
60	62.3	0.999	0.76	12.2	0.340	8.03	5.47	0.085	18.4
70	62.3	0.998	0.658	10.6	0.347	6.82	5.57	0.12	34.6
80	62.2	0.998	0.578	9.3	0.353	5.89	5.68	0.15	56.0 0
90	62.1	0.997	0.514	8.25	0.359	5.13	5.79	0.18	85.0
100	62.0	0.998	0.458	7.40	0.364	4.52	5.88	0.20	118×10^6
150	61.2	1.00	0.292	4.77	0.384	2.74	6.27	0.31	440.0
200	60.1	1.00	0.205	3.41	0.394	1.88	6.55	0.40	1.11×10^9
250	58.8	1.01	0.158	2.69	0 396	1.45	6.69	0.48	2.14
300	57.3	1.03	0.126	2.20	0.395	1.18	6.70	0.60	4.00
350	55.6	1.05	0.105	1.89	0.391	1.02	6.69	0.69	6.24
400	53.6	1.08	0.091	1.70	0.381	0.927	6.57	0.80	8.95
450	51.6	1.12	0.080	1.55	0.367	0.876	6.34	0.90	12.1
500	49.0	1.19	0.071	1.45	0.349	0.87	5.99	1.00	15.3
550	45.9	1.31	0.064	1.39	0.325	0.93	5.05	1.10	17.8
600	42.4	1.51	0.058	1.37	0.292	1.09	4.57	1.20	20.6
C. LIQUID METALS									
Bismuth									
600	625	0.0345	1.09	1.74	9.5	0.014	440	0.065	0.687×10^9
800	616	0.0357	0.90	1.5	9.0	0.013	410	0.068	
1000	608	0.0369	0.74	1.2	9.0	0.011	400	0.070	
1200	600	0.0381	0.62	1.0	9.0	0.009	390		
1400	591	0.0393	0.53	0.9	9.0	0.008	390		
Mercury									
50	847	0.033	1.07	1.2	4.7	0.027	170	0.1	2.02×10^9
200	834	0.033	0.84	1.0	6.0	0.016	220	0.1	2.02
300	826	0.033	0.74	0.9	6.7	0.012	250		
400	817	0.032	0.67	0.8	7.2	0.011	270		
600	802	0.032	0.58	0.7	8.1	0.008	310		
Sodium									
200	58.0	0.33	0.47	8.1	49.8	0.011	2,600	0.150	73.5×10^6
400	56.3	0.32	0.29	5.1	46.4	0.007	2,600	0.20	243
700	53.7	0.31	0.19	3.5	41.8	0.005	2,500		
1000	51.2	0.30	0.14	2.7	37.8	0.004	2,400		
1300	48.6	0.30	0.12	2.5	34.5	0.004	2,400		

Table A-4 Thermal conductivity k, specific heat c, density ρ, and thermal diffusivity α of metals and alloys[†]

Material	k, Btu/hr-ft-°F				c, Btu/lb_m-°F	ρ, lb_m/ft^3	α, ft^2/hr
	32°F	212°F	372°F	932°F	32°F	32°F	32°F
Metals:							
Aluminum	117	119	133	155	0.208	169	3.33
Bismuth	4.9	3.9	...	...	0.029	612	0.28
Copper, pure	224	218	212	207	0.091	558	4.42
Gold	169	170	...	...	0.030	1,203	4.68
Iron, pure	35.8	36.6	...	...	0.104	491	0.70
Lead	20.1	19	18	...	0.030	705	0.95
Magnesium	91	92	...	...	0.232	109	3.60
Mercury	4.8	...	...	...	0.033	849	0.17
Nickel	34.5	34	32	...	0.103	555	0.60
Silver	242	238	...	...	0.056	655	6.6
Tin	36	34	...	...	0.054	456	1.46
Zinc	65	64	59	...	0.091	446	1.60
Alloys:							
Admiralty metal	65	64					
Brass, 70% Cu, 30% Zn	56	60	66	...	0.092	532	1.14
Bronze, 75% Cu, 25% Sn	15	...	...	...	0.082	540	0.34
Cast iron, Plain	33	31.8	27.7	24.8	0.11	474	0.63
Alloy	30	28.3	27	...	0.10	455	0.66
Constantan, 60% Cu, 40% Ni	12.4	12.8	...	...	0.10	557	0.22
18–8 stainless steel, Type 304	8.0	9.4	10.9	12.4	0.11	488	0.15
Type 347	8.0	9.3	11.0	12.8	0.11	488	0.15
Steel, mild, 1% C	26.5	26	25	22	0.11	490	0.49

[†] By permission from F. Kreith, "Principles of Heat Transfer," 2d ed., International Textbook Company, Scranton, Pa., 1965.

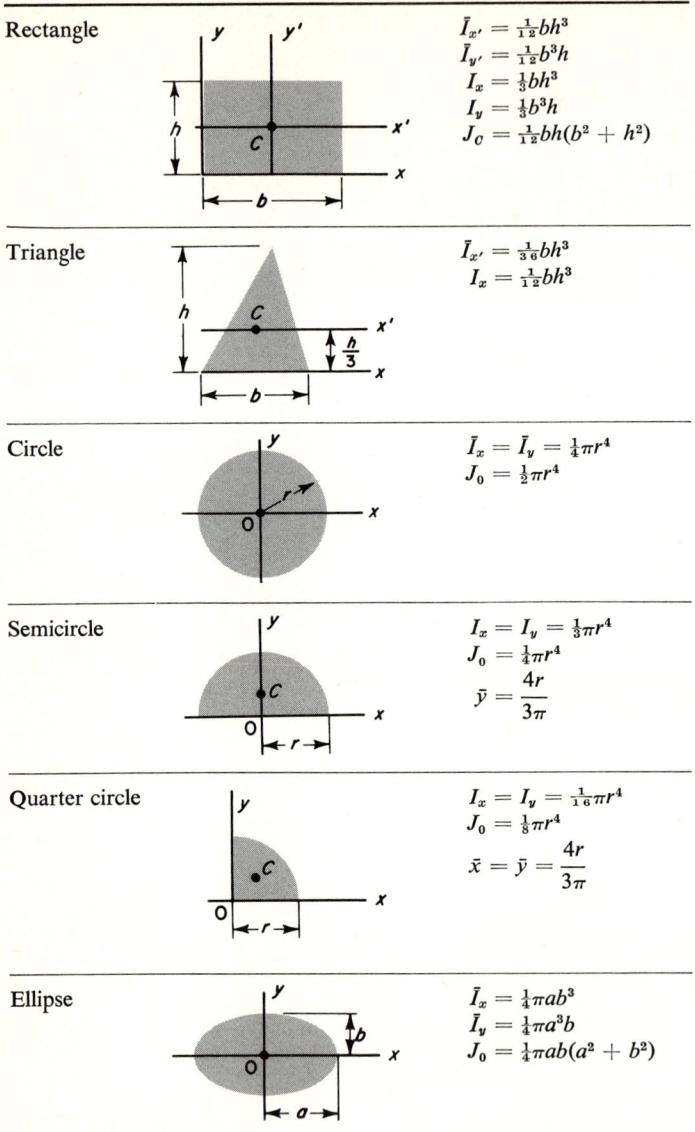

Fig. A-3 Centroids and moments of inertia of common geometrical shapes.

Table A-5 Experimental diffusivities of pairs of gases and vapors at 1 atm, D_{AB}, ft²/hr†

Substance A	Substance B						Temp., °R
	Air	H_2	O_2	N_2	CO_2	CH_4	
Argon				0.753			528
Acetone	0.423	1.4					492
Benzene	0.299	1.19	0.309		0.136		492
Carbon dioxide	0.535	2.15	0.54			0.594	492
				0.632			528
		3.49					1392
Carbon monoxide		2.52	0.718		0.532		492
			3.88				1302
Chloroform	0.353						492
Ethanol		1.46			0.266		492
Ether, diethyl	0.302	1.16			0.212		492
Hydrogen	2.37		2.70	2.62	2.13	2.42	492
					2.51		537
			16.3				1392
Oxygen	0.69	2.71		0.702	0.54		492
Water	0.853	2.91			0.535		492
			5.04				1302

† Abstracted and modified from J. H. Perry (ed.), "Chemical Engineers' Handbook," 4th ed. Copyright 1963. McGraw-Hill Book Company. Used by permission.

Table A-6 Experimental diffusivities of liquids in dilute solutions at 1 atm and 537°R, D_{AB}, ft²/hr × 10⁵ †

Substance A, solute	Substance B, solvent					
	Water	Acetone	Benzene	Ethanol	Glycerol	Carbon tetrachloride
Acetic acid		12.82	8.18			
Caffeine	2.44					
Carbon dioxide	7.60			15.5		
Chloroform			9.70	5.36		
Ether, diethyl	3.30		10.6			
Ethanol	4.97					
Glucose	2.68					
Glycerol	3.65			2.17		
Iodine			7.68	5.04		5.63
Methanol	6.22					
Urea	5.32			2.84		
Water					0.0814	

† Abstracted and modified by permission from J. H. Perry (ed.), "Chemical Engineers' Handbook," 4th ed. Copyright 1963. McGraw-Hill Book Company.

Table A-7 Experimental diffusivities in solids, D_{AB}, ft²/hr†

Substance A	Temp., °R	Silicon dioxide	Pyrex	Nickel	Lead	Silver	Copper
Helium	528	$9.31-21.3 \times 10^{-10}$	1.74×10^{-10}				
	1392		7.76×10^{-8}				
Hydrogen	1392	$2.22-8.15 \times 10^{-8}$					
	645			4.5×10^{-8}			
	789			4.07×10^{-7}			
Bismuth	528				4.27×10^{-16}		
Mercury	528				9.7×10^{-15}		
Antimony	528					1.36×10^{-20}	
Aluminum	528						5.04×10^{-30}
Cadmium	528						1.05×10^{-14}

† Taken by permission from R. M. Barrer, "Diffusion in and through Solids," pp. 141, 222, and 275, The Macmillan Company, New York, 1941.

Table A-8 A short summary of Bessel functions

x	$J_0(x)$	$J_1(x)$	$Y_0(x)$	$Y_1(x)$
0.0	1.000000	0.000000	$-\infty$	$-\infty$
0.2	0.990025	0.099501	−1.081105	−3.323825
0.4	0.960398	0.196027	−0.606025	−1.780872
0.6	0.912005	0.286701	−0.308051	−1.260391
0.8	0.846287	0.368842	−0.086802	−0.978144
1.0	0.765198	0.440051	0.088257	−0.781213
1.2	0.671133	0.498289	0.228084	−0.621136
1.4	0.566855	0.541948	0.337895	−0.479147
1.6	0.455402	0.569896	0.420427	−0.347578
1.8	0.339986	0.581517	0.477432	−0.223665
2.0	0.223891	0.576725	0.510376	−0.107032
2.2	0.110362	0.555963	0.520784	0.001488
2.4	0.002508	0.520185	0.510415	0.100489
2.6	−0.096805	0.470818	0.481331	0.188364
2.8	−0.185036	0.409709	0.435916	0.263545
3.0	−0.260052	0.339059	0.376850	0.324674
3.2	−0.320188	0.261343	0.307053	0.370711
3.4	−0.364296	0.179226	0.229615	0.401015
3.6	−0.391769	0.095466	0.147710	0.415392
3.8	−0.402556	0.012821	0.064503	0.414115
4.0	−0.397150	−0.066043	−0.016941	0.397926
4.2	−0.376557	−0.138647	−0.093751	0.368013
4.4	−0.342257	−0.202776	−0.163336	0.325971
4.6	−0.296138	−0.256553	−0.223460	0.273745
4.8	−0.240425	−0.298500	−0.272304	0.213565
5.0	−0.177597	−0.327580	−0.308518	0.147863
5.2	−0.110290	−0.243223	−0.331251	0.079190
5.4	−0.041210	−0.345345	−0.340168	0.010127
5.6	0.026971	−0.334333	−0.335444	−0.056806
5.8	0.091703	−0.311028	−0.317746	−0.119234
6.0	0.150645	−0.276684	−0.288195	−0.175010
6.2	0.201747	−0.232917	−0.248310	−0.222284
6.4	0.243311	−0.181638	−0.199949	−0.259560
6.6	0.274043	−0.124980	−0.145226	−0.285747
6.8	0.293096	−0.065219	−0.086434	−0.300187
7.0	0.300079	−0.004683	−0.025950	−0.302667
7.2	0.295071	0.054327	0.033850	−0.293423
7.4	0.278596	0.109625	0.090681	−0.273115
7.6	0.251602	0.159214	0.142429	−0.242801
7.8	0.215408	0.201357	0.187227	−0.203885
8.0	0.171651	0.234636	0.223521	−0.158060
8.2	0.122215	0.257999	0.250118	−0.107241
8.4	0.069157	0.270786	0.266222	−0.053485
8.6	0.014623	0.272755	0.271458	0.001084
8.8	−0.039234	0.264074	0.265875	0.054356
9.0	−0.090334	0.245312	0.249937	0.104315
9.2	−0.136748	0.217409	0.224494	0.149113
9.4	−0.176772	0.181632	0.190744	0.187136
9.6	−0.208979	0.139525	0.150180	0.217059
9.8	−0.232276	0.092840	0.104527	0.237893
10.0	−0.245936	0.043473	0.055671	0.249015

Table A-9 A short summary of modified Bessel functions

x	$e^{-x}I_0(x)$	$e^{-x}I_1(x)$	$e^{x}K_0(x)$	$e^{x}K_1(x)$
0.0	1.000000	0.000000	∞	∞
0.2	0.826939	0.082283	2.140757	5.833386
0.4	0.697402	0.136763	1.662682	3.258674
0.6	0.599327	0.172164	1.416738	2.373920
0.8	0.524149	0.194499	1.258203	1.917930
1.0	0.465760	0.207910	1.144463	1.636153
1.2	0.419782	0.215257	1.057485	1.442898
1.4	0.383063	0.218508	0.988070	1.301054
1.6	0.353315	0.219019	0.930946	1.191868
1.8	0.328872	0.217726	0.882834	1.104805
2.0	0.308508	0.215269	0.841568	1.033477
2.2	0.291317	0.212088	0.805654	0.973770
2.4	0.276622	0.208481	0.774018	0.922914
2.6	0.263914	0.204652	0.745868	0.878967
2.8	0.252806	0.200737	0.720604	0.840530
3.0	0.243000	0.196827	0.697762	0.806563
3.2	0.234269	0.192979	0.676975	0.776280
3.4	0.226431	0.189230	0.657952	0.749072
3.6	0.219346	0.185602	0.640456	0.724461
3.8	0.212900	0.182108	0.624292	0.702065
4.0	0.207002	0.178751	0.609298	0.681576
4.2	0.201577	0.175533	0.595339	0.662742
4.4	0.196566	0.172450	0.582301	0.645356
4.6	0.191916	0.169500	0.570087	0.629243
4.8	0.187586	0.166676	0.558613	0.614257
5.0	0.183541	0.163972	0.547808	0.600274
5.2	0.179749	0.161383	0.537607	0.587189
5.4	0.176186	0.158903	0.527958	0.574910
5.6	0.172829	0.156524	0.518812	0.563359
5.8	0.169658	0.154242	0.510126	0.552468
6.0	0.166657	0.152051	0.501863	0.542176
6.2	0.163811	0.149946	0.493990	0.532431
6.4	0.161107	0.147922	0.486477	0.523187
6.6	0.158534	0.145974	0.479298	0.514403
6.8	0.156080	0.144098	0.472428	0.506042
7.0	0.153738	0.142289	0.465845	0.498072
7.2	0.151498	0.140545	0.459531	0.490462
7.4	0.149354	0.138861	0.453467	0.483188
7.6	0.147299	0.137235	0.447637	0.476225
7.8	0.145327	0.135663	0.442027	0.469552
8.0	0.143432	0.134142	0.436623	0.463149
8.2	0.141609	0.132671	0.431413	0.456999
8.4	0.139855	0.131245	0.426385	0.451086
8.6	0.138164	0.129864	0.421529	0.445395
8.8	0.136534	0.128525	0.416835	0.439912
9.0	0.134960	0.127225	0.412296	0.434625
9.2	0.133439	0.125963	0.407901	0.429523
9.4	0.131968	0.124738	0.403644	0.424596
9.6	0.130546	0.123547	0.399518	0.419833
9.8	0.129168	0.122389	0.395516	0.415226
10.0	0.127833	0.121263	0.391632	0.410767

APPENDIXES

Table A-10 Zero-pressure gas constants†

Substance	$\mathcal{M}$	R, ft-lb_f/lb_m-°R	c_p,‡ Btu/lb_m-°R	c_v,‡ Btu/lb_m-°R	$k = \dfrac{c_p}{c_v}$
Acetylene, C_2H_2	26.04	59.39	0.403	0.327	1.23
Air	28.97	53.36	0.240	0.172	1.40
Ammonia, NH_3	17.03	90.77	0.501	0.384	1.30
Argon, Ar	39.94	38.73	0.124	0.077	1.67
Benzene, C_6H_6	78.11	19.78	0.250	0.224	1.11
n-Butane, C_4H_{10}	58.12	26.61	0.400	0.366	1.09
Isobutane, C_4H_{10}	58.12	26.59	0.398	0.364	1.09
1-Butene, C_4H_8	56.11	27.54	0.365	0.328	1.11
Carbon dioxide, CO_2	44.01	35.12	0.202	0.156	1.29
Carbon monoxide, CO	28.01	55.19	0.248	0.178	1.40
Dodecane, $C_{12}H_{26}$	170.34	9.07	0.393	0.381	1.03
Ethane, C_2H_6	30.07	51.43	0.418	0.352	1.19
Ethylene, C_2H_4	28.05	55.13	0.371	0.300	1.24
Freon-12, CCl_2F_2	120.92	12.78	0.137	0.120	1.14
Helium, He	4.00	386.33	1.241	0.745	1.67
n-Heptane, C_7H_{16}	100.20	15.42	0.396	0.376	1.05
n-Hexane, C_6H_{14}	86.18	17.93	0.397	0.374	1.06
Hydrogen, H_2	2.02	766.53	3.416	2.431	1.40
Methane, CH_4	16.04	96.04	0.532	0.408	1.30
Neon, Ne	20.18	76.58	0.246	0.148	1.67
Nitric oxide, NO	30.01	51.49	0.238	0.172	1.39
Nitrogen, N_2	28.02	55.15	0.248	0.177	1.40
Octane, C_8H_{18}	114.23	13.54	0.395	0.378	1.05
Oxygen, O_2	32.00	48.29	0.219	0.157	1.40
n-Pentane, C_5H_{12}	72.15	21.42	0.398	0.370	1.07
Isopentane, C_5H_{12}	72.15	21.42	0.397	0.370	1.07
Propane, C_3H_8	44.10	35.07	0.398	0.353	1.13
Propylene, C_3H_6	42.08	36.72	0.363	0.306	1.19
Sulfur dioxide, SO_2	64.07	24.12	0.148	0.117	1.26
Water vapor, H_2O	18.02	85.80	0.445	0.335	1.33
Xenon, Xe	131.30	11.78	0.038	0.023	1.67

† Abstracted from J. F. Masi, *Trans. ASME*, **76:** 1067 (1954); *Natl. Bur. Std. (U.S.) Circ. 500*, February, 1952; *Natl. Bur. Std., (U.S.), API Res. Proj. 44*, 1952.
‡ At 77°F.

Table A-11 Normal total emissivity of various surfaces†

Surface	T, °F	Emissivity ε
Metals and their oxides		
Aluminum:		
Highly polished plate, 98.3% pure	440–1070	0.039–0.057
Commercial sheet	212	0.09
Heavily oxidized	299–940	0.20–0.31
Brass:		
Highly polished:		
73.2% Cu, 26.7% Zn	476–674	0.028–0.031
62.4% Cu, 36.8% Zn, 0.4% Pb, 0.3% Al	494–710	0.033–0.037
82.9% Cu, 17.0% Zn	530	0.030
Hard-rolled, polished, but direction of polishing visible	70	0.038
Dull plate	120–660	0.22
Copper:		
Polished	242	0.023
	212	0.052
Plate, heated long time, covered with thick oxide layer	77	0.78
Gold, pure, highly polished	440–1160	0.018–0.035
Iron and steel (not including stainless):		
Steel, polished	212	0.066
Iron, polished	800–1880	0.14–0.38
Cast iron, newly turned	72	0.44
Cast iron, turned and heated	1620–1810	0.60–0.70
Mild steel	450–1950	0.20–0.32
Oxidized surfaces:		
Iron plate, pickled, then rusted red	68	0.61
Iron, dark-gray surface	212	0.31
Rough ingot iron	1700–2040	0.87–0.95
Sheet steel with strong, rough oxide layer	75	0.80
Lead:		
Unoxidized, 99.96% pure	260–440	0.057–0.075
Gray oxidized	75	0.28
Oxidized at 300°F	390	0.63
Magnesium, magnesium oxide	530–1520	0.55–0.20
Molybdenum:		
Filament	1340–4700	0.096–0.202
Massive, polished	212	0.071
Monel metal, oxidized at 1110°F	390–1110	0.41–0.46
Nickel:		
Polished	212	0.072
Nickel oxide	1200–2290	0.59–0.86
Nickel alloys:		
Copper nickel, polished	212	0.059
Nichrome wire, bright	120–1830	0.65–0.79
Nichrome wire, oxidized	120–930	0.95–0.98

† Abstracted by permission from H. C. Hottel in W. H. McAdams (ed.), "Heat Transmission," 3d ed., pp. 472–478. Copyright 1954. McGraw-Hill Book Company.

Table A-II (Continued)

Surface	T, °F	Emissivity ϵ
Platinum; polished plate, pure	440–1160	0.054–0.104
Silver:		
Polished, pure	440–1160	0.020–0.032
Polished	100–700	0.022–0.031
Stainless steels:		
Polished	212	0.074
Type 301	450–1725	0.54–0.63
Tin, bright tinned iron	76	0.043 and 0.064
Tungsten, filament	6000	0.39
Zinc, galvanized sheet iron, fairly bright	82	0.23
Refractories, building materials, paints, and miscellaneous		
Alumina (85–99.5% Al_2O_3, 0–12% SiO_2, 0–1% Ge_2O_3); effect of mean grain size, μm		
10 μm		0.30–0.18
50 μm		0.39–0.28
100 μm		0.50–0.40
Asbestos, board	74	0.96
Brick:		
Red, rough, but no gross irregularities	70	0.93
Fireclay	1832	0.75
Carbon:		
T-carbon (Gebruder Siemens) 0.9% ash, started with emissivity of 0.72 at 260°F but on heating changed to values given	260–1160	0.81–0.79
Filament	1900–2560	0.526
Rough plate	212–608	0.77
Lampblack, rough deposit	212–932	0.84–0.78
Concrete tiles	1832	0.63
Enamel, white fused, on iron	66	0.90
Glass:		
Smooth	72	0.94
Pyrex, lead, and soda	500–1000	0.95–0.85
Paints, lacquers, varnishes:		
Snow-white enamel varnish on rough iron plate	73	0.906
Black shiny lacquer, sprayed on iron	76	0.875
Black shiny shellac on tinned iron sheet	70	0.821
Black matte shellac	170–295	0.91
Black or white lacquer	100–200	0.80–0.95
Flat black lacquer	100–200	0.96–0.98
Porcelain, glazed	72	0.92
Quartz, rough, fused	70	0.93
Roofing paper	69	0.91
Rubber, hard, glossy plate	74	0.94
Water	32–212	0.95–0.963

Table A-12 One-dimensional isentropic flow parameters for perfect gas, $k = 1.4$

M	$\dfrac{A}{A^*}$	$\dfrac{p}{p_0}$	$\dfrac{\rho}{\rho_0}$	$\dfrac{T}{T_0}$	$\left(\dfrac{A}{A^*}\right)\left(\dfrac{p}{p_0}\right)$	M	$\dfrac{A}{A^*}$	$\dfrac{p}{p_0}$	$\dfrac{\rho}{\rho_0}$	$\dfrac{T}{T_0}$	$\left(\dfrac{A}{A^*}\right)\left(\dfrac{p}{p_0}\right)$
0.00	...	1.000	1.000	1.000		0.78	1.05	0.669	0.750	0.891	0.701
0.01	57.87	0.9999	0.9999	0.9999	57.87	0.80	1.04	0.656	0.740	0.886	0.681
0.02	28.94	0.9997	0.9999	0.9999	28.93	0.82	1.03	0.643	0.729	0.881	0.663
0.04	14.48	0.999	0.999	0.9996	14.46	0.84	1.02	0.630	0.719	0.876	0.645
0.06	9.67	0.997	0.998	0.999	9.64	0.86	1.02	0.617	0.708	0.871	0.628
0.08	7.26	0.996	0.997	0.999	7.23	0.88	1.01	0.604	0.698	0.865	0.612
0.10	5.82	0.993	0.995	0.998	5.78	0.90	1.01	0.591	0.687	0.860	0.596
0.12	4.86	0.990	0.993	0.997	4.82	0.92	1.01	0.578	0.676	0.855	0.582
0.14	4.18	0.986	0.990	0.996	4.13	0.94	1.00	0.566	0.666	0.850	0.568
0.16	3.67	0.982	0.987	0.995	3.61	0.96	1.00	0.553	0.655	0.844	0.554
0.18	3.28	0.978	0.984	0.994	3.20	0.98	1.00	0.541	0.645	0.839	0.541
0.20	2.96	0.973	0.980	0.992	2.88	1.00	1.00	0.528	0.632	0.833	0.528
0.22	2.71	0.967	0.976	0.990	2.62	1.02	1.00	0.516	0.623	0.828	0.516
0.24	2.50	0.961	0.972	0.989	2.40	1.04	1.00	0.504	0.613	0.822	0.505
0.26	2.32	0.954	0.967	0.987	2.21	1.06	1.00	0.492	0.602	0.817	0.493
0.28	2.17	0.947	0.962	0.985	2.05	1.08	1.01	0.480	0.592	0.810	0.483
0.30	2.04	0.939	0.956	0.982	1.91	1.10	1.01	0.468	0.582	0.805	0.472
0.32	1.92	0.932	0.951	0.980	1.79	1.12	1.01	0.457	0.571	0.799	0.462
0.34	1.82	0.923	0.944	0.977	1.68	1.14	1.02	0.445	0.561	0.794	0.452
0.36	1.74	0.914	0.938	0.975	1.59	1.16	1.02	0.434	0.551	0.788	0.443
0.38	1.66	0.905	0.931	0.972	1.50	1.18	1.02	0.423	0.541	0.782	0.434
0.40	1.59	0.896	0.924	0.969	1.42	1.20	1.03	0.412	0.531	0.776	0.425
0.42	1.53	0.886	0.917	0.966	1.35	1.22	1.04	0.402	0.521	0.771	0.416
0.44	1.47	0.876	0.909	0.963	1.29	1.24	1.04	0.391	0.512	0.765	0.408
0.46	1.42	0.865	0.902	0.959	1.23	1.26	1.05	0.381	0.502	0.759	0.400
0.48	1.38	0.854	0.893	0.956	1.18	1.28	1.06	0.371	0.492	0.753	0.392
0.50	1.34	0.843	0.885	0.952	1.13	1.30	1.07	0.361	0.483	0.747	0.385
0.52	1.30	0.832	0.877	0.949	1.08	1.32	1.08	0.351	0.474	0.742	0.378
0.54	1.27	0.820	0.868	0.945	1.04	1.34	1.08	0.342	0.464	0.736	0.370
0.56	1.24	0.808	0.859	0.941	1.00	1.36	1.09	0.332	0.455	0.730	0.364
0.58	1.21	0.796	0.850	0.937	0.966	1.38	1.10	0.323	0.446	0.724	0.357
0.60	1.19	0.784	0.840	0.933	0.932	1.40	1.11	0.314	0.437	0.718	0.350
0.62	1.17	0.772	0.831	0.929	0.899	1.42	1.13	0.305	0.429	0.713	0.344
0.64	1.16	0.759	0.821	0.924	0.869	1.44	1.14	0.297	0.420	0.707	0.338
0.66	1.13	0.747	0.812	0.920	0.841	1.46	1.15	0.289	0.412	0.701	0.332
0.68	1.12	0.734	0.802	0.915	0.814	1.48	1.16	0.280	0.403	0.695	0.326
0.70	1.09	0.721	0.792	0.911	0.789	1.50	1.18	0.272	0.395	0.690	0.320
0.72	1.08	0.708	0.781	0.906	0.765	1.52	1.19	0.265	0.387	0.684	0.315
0.74	1.07	0.695	0.771	0.901	0.742	1.54	1.20	0.257	0.379	0.678	0.309
0.76	1.06	0.682	0.761	0.896	0.721	1.56	1.22	0.250	0.371	0.672	0.304

APPENDIXES

Table A-12 (Continued)

M	$\dfrac{A}{A^*}$	$\dfrac{p}{p_0}$	$\dfrac{\rho}{\rho_0}$	$\dfrac{T}{T_0}$	$\left(\dfrac{A}{A^*}\right)\left(\dfrac{p}{p_0}\right)$	M	$\dfrac{A}{A^*}$	$\dfrac{p}{p_0}$	$\dfrac{\rho}{\rho_0}$	$\dfrac{T}{T_0}$	$\left(\dfrac{A}{A^*}\right)\left(\dfrac{p}{p_0}\right)$
1.58	1.23	0.242	0.363	0.667	0.299	2.30	2.19	0.080	0.165	0.486	0.175
1.60	1.25	0.235	0.356	0.661	0.294	2.32	2.23	0.078	0.161	0.482	0.173
1.62	1.27	0.228	0.348	0.656	0.289	2.34	2.27	0.075	0.157	0.477	0.171
1.64	1.28	0.222	0.341	0.650	0.285	2.36	2.32	0.073	0.154	0.473	0.169
1.66	1.30	0.215	0.334	0.645	0.280	2.38	2.36	0.071	0.150	0.469	0.166
1.68	1.32	0.209	0.327	0.639	0.275	2.40	2.40	0.068	0.147	0.465	0.164
1.70	1.34	0.203	0.320	0.634	0.271	2.42	2.45	0.066	0.144	0.461	0.162
1.72	1.36	0.197	0.313	0.628	0.267	2.44	2.49	0.064	0.141	0.456	0.160
1.74	1.38	0.191	0.306	0.623	0.262	2.46	2.54	0.062	0.138	0.452	0.158
1.76	1.40	0.185	0.300	0.617	0.258	2.48	2.59	0.060	0.135	0.448	0.156
1.78	1.42	0.179	0.293	0.612	0.254	2.50	2.64	0.059	0.132	0.444	0.154
1.80	1.44	0.174	0.287	0.607	0.250	2.52	2.69	0.057	0.129	0.441	0.152
1.82	1.46	0.169	0.281	0.602	0.247	2.54	2.74	0.055	0.126	0.437	0.151
1.84	1.48	0.164	0.275	0.596	0.243	2.56	2.79	0.053	0.123	0.433	0.149
1.86	1.51	0.159	0.269	0.591	0.239	2.58	2.84	0.052	0.121	0.429	0.147
1.88	1.53	0.154	0.263	0.586	0.236	2.60	2.90	0.050	0.118	0.425	0.145
1.90	1.56	0.149	0.257	0.581	0.232	2.62	2.95	0.049	0.115	0.421	0.143
1.92	1.58	0.145	0.251	0.576	0.229	2.64	3.01	0.047	0.113	0.418	0.142
1.94	1.61	0.140	0.246	0.571	0.225	2.66	3.06	0.046	0.110	0.414	0.140
1.96	1.63	0.136	0.240	0.566	0.222	2.68	3.12	0.044	0.108	0.410	0.138
1.98	1.66	0.132	0.235	0.561	0.219	2.70	3.18	0.043	0.106	0.407	0.137
2.00	1.69	0.128	0.230	0.556	0.216	2.72	3.24	0.042	0.103	0.403	0.135
2.02	1.72	0.124	0.225	0.551	0.213	2.74	3.31	0.040	0.101	0.400	0.134
2.04	1.75	0.120	0.220	0.546	0.210	2.76	3.37	0.039	0.099	0.396	0 132
2.06	1.78	0.116	0.215	0.541	0.207	2.78	3.43	0.038	0.097	0.393	0.130
2.08	1.81	0.113	0.210	0.536	0.204	2.80	3.50	0.037	0.095	0.389	0.129
2.10	1.84	0.109	0.206	0.531	0.201	2.82	3.57	0.036	0.093	0.386	0.128
2.12	1.87	0.106	0.201	0.526	0.198	2.84	3.64	0.035	0.091	0.383	0.126
2.14	1.90	0.103	0.197	0.522	0.195	2.86	3.71	0.034	0.089	0.379	0.125
2.16	1.94	0.100	0.192	0.517	0.193	2.88	3.78	0.033	0.087	0.376	0.123
2.18	1.97	0.097	0.188	0.513	0.190	2.90	3.85	0.032	0.085	0.373	0.122
2.20	2.01	0.094	0.184	0.508	0.188	2.92	3.92	0.031	0.083	0.370	0.120
2.22	2.04	0.091	0.180	0.504	0.185	2.94	4.00	0.030	0.081	0.366	0.119
2.24	2.08	0.088	0.176	0.499	0.183	2.96	4.08	0.029	0.080	0.363	0.118
2.26	2.12	0.085	0.172	0.495	0.180	2.98	4.15	0.028	0.078	0.360	0.117
2.28	2.15	0.083	0.168	0.490	0.178	3.00	4.23	0.027	0.076	0.357	0.115

Table A-13 One-dimensional normal-shock parameters for perfect gas, $k = 1.4$

M_x	M_y	$\dfrac{p_y}{p_x}$	$\dfrac{T_y}{T_x}$	$\dfrac{(p_0)_y}{(p_0)_x}$	M_x	M_y	$\dfrac{p_y}{p_x}$	$\dfrac{T_y}{T_x}$	$\dfrac{(p_0)_y}{(p_0)_x}$
1.00	1.000	1.000	1.000	1.000	1.80	0.617	3.613	1.532	0.813
1.02	0.980	1.047	1.013	1.000	1.82	0.612	3.698	1.547	0.804
1.04	0.962	1.095	1.026	1.000	1.84	0.608	3.783	1.562	0.795
1.06	0.944	1.144	1.039	1.000	1.86	0.604	3.869	1.577	0.786
1.08	0.928	1.194	1.052	0.999	1.88	0.600	3.957	1.592	0.777
1.10	0.912	1.245	1.065	0.999	1.90	0.596	4.045	1.608	0.767
1.12	0.896	1.297	1.078	0.998	1.92	0.592	4.134	1.624	0.758
1.14	0.882	1.350	1.090	0.997	1.94	0.588	4.224	1.639	0.749
1.16	0.868	1.403	1.103	0.996	1.96	0.584	4.315	1.655	0.740
1.18	0.855	1.458	1.115	0.995	1.98	0.581	4.407	1.671	0.730
1.20	0.842	1.513	1.128	0.993	2.00	0.577	4.500	1.688	0.721
1.22	0.830	1.570	1.140	0.991	2.02	0.574	4.594	1.704	0.711
1.24	0.818	1.627	1.153	0.988	2.04	0.571	4.689	1.720	0.702
1.26	0.807	1.686	1.166	0.986	2.06	0.567	4.784	1.737	0.693
1.28	0.796	1.745	1.178	0.983	2.08	0.564	4.881	1.754	0.683
1.30	0.786	1.805	1.191	0.979	2.10	0.561	4.978	1.770	0.674
1.32	0.776	1.866	1.204	0.976	2.12	0.558	5.077	1.787	0.665
1.34	0.766	1.928	1.216	0.972	2.14	0.555	5.176	1.805	0.656
1.36	0.757	1.991	1.229	0.968	2.16	0.553	5.277	1.822	0.646
1.38	0.748	2.055	1.242	0.963	2.18	0.550	5.378	1.839	0.637
1.40	0.740	2.120	1.255	0.958	2.20	0.547	5.480	1.857	0.628
1.42	0.731	2.186	1.268	0.953	2.22	0.544	5.583	1.857	0.619
1.44	0.723	2.253	1.281	0.948	2.24	0.542	5.687	1.892	0.610
1.46	0.716	2.320	1.294	0.942	2.26	0.539	5.792	1.910	0.601
1.48	0.708	2.389	1.307	0.936	2.28	0.537	5.898	1.929	0.592
1.50	0.701	2.458	1.320	0.930	2.30	0.534	6.005	1.947	0.583
1.52	0.694	2.529	1.334	0.923	2.32	0.532	6.113	1.965	0.575
1.54	0.687	2.600	1.347	0.917	2.34	0.530	6.222	1.984	0.566
1.56	0.681	2.673	1.361	0.910	2.36	0.527	6.331	2.003	0.557
1.58	0.675	2.746	1.374	0.903	2.38	0.525	6.442	2.021	0.549
1.60	0.668	2.820	1.388	0.895	2.40	0.523	6.553	2.040	0.540
1.62	0.663	2.895	1.402	0.888	2.42	0.521	6.666	2.060	0.532
1.64	0.657	2.971	1.416	0.880	2.44	0.519	6.779	2.079	0.523
1.66	0.651	3.048	1.430	0.872	2.46	0.517	6.894	2.098	0.515
1.68	0.646	3.126	1.444	0.864	2.48	0.515	7.009	2.118	0.507
1.70	0.641	3.205	1.458	0.856	2.50	0.513	7.125	2.138	0.499
1.72	0.635	3.285	1.473	0.847	2.52	0.511	7.242	2.157	0.491
1.74	0.631	3.366	1.487	0.839	2.54	0.509	7.360	2.177	0.483
1.76	0.626	3.447	1.502	0.830	2.56	0.507	7.479	2.198	0.475
1.78	0.621	3.530	1.517	0.821	2.58	0.506	7.599	2.218	0.468

Table A-13 *(Continued)*

M_x	M_y	$\dfrac{p_y}{p_x}$	$\dfrac{T_y}{T_x}$	$\dfrac{(p_0)_y}{(p_0)_x}$	M_x	M_y	$\dfrac{p_y}{p_x}$	$\dfrac{T_y}{T_x}$	$\dfrac{(p_0)_y}{(p_0)_x}$
2.60	0.504	7.720	2.238	0.460	2.80	0.488	8.980	2.451	0.389
2.62	0.502	7.842	2.260	0.453	2.82	0.487	9.111	2.473	0.383
2.64	0.500	7.965	2.280	0.445	2.84	0.485	9.243	2.496	0.376
2.66	0.499	8.088	2.301	0.438	2.86	0.484	9.376	2.518	0.370
2.68	0.497	8.213	2.322	0.431	2.88	0.483	9.510	2.541	0.364
2.70	0.496	8.338	2.343	0.424	2.90	0.481	9.645	2.563	0.358
2.72	0.494	8.465	2.364	0.417	2.92	0.480	9.781	2.586	0.352
2.74	0.493	8.592	2.396	0.410	2.94	0.479	9.918	2.609	0.346
2.76	0.491	8.721	2.407	0.403	2.96	0.478	10.055	2.632	0.340
2.78	0.490	8.850	2.429	0.396	2.98	0.476	10.194	2.656	0.334
					3.00	0.475	10.333	2.679	0.328

Table A-14 Fanno flow parameters for perfect gas, $k = 1.4$

M	$\dfrac{T}{T^*}$	$\dfrac{p}{p^*}$	$\dfrac{p_0}{p_0^*}$	$\dfrac{V}{V^*}$	$\dfrac{fL^*}{D_h}$	M	$\dfrac{T}{T^*}$	$\dfrac{p}{p^*}$	$\dfrac{p_0}{p_0^*}$	$\dfrac{V}{V^*}$	$\dfrac{fL^*}{D_h}$
0	1.200	∞	∞	0	∞	0.78	1.070	1.33	1.05	0.807	0.02292
0.01	1.200	109.54	57.87	0.011	1783.60	0.80	1.064	1.29	1.04	0.825	0.01807
0.02	1.200	57.77	28.94	0.022	444.61	0.82	1.058	1.25	1.03	0.843	0.01398
0.04	1.200	27.38	14.48	0.044	110.09	0.84	1.052	1.22	1.02	0.861	0.01056
0.06	1.199	18.25	9.67	0.066	48.26	0.86	1.045	1.19	1.02	0.879	0.007742
0.08	1.199	13.68	7.26	0.088	26.68	0.88	1.039	1.16	1.01	0.897	0.005450
0.10	1.198	10.94	5.82	0.109	16.73	0.90	1.033	1.13	1.01	0.914	0.003628
0.12	1.197	9.12	4.86	0.131	11.35	0.92	1.026	1.10	1.01	0.932	0.002229
0.14	1.195	7.81	4.18	0.153	8.128	0.94	1.020	1.07	1.00	0.949	0.001204
0.16	1.194	6.83	3.67	0.175	6.049	0.96	1.013	1.05	1.00	0.966	0.000514
0.18	1.192	6.07	3.28	0.197	4.636	0.98	1.007	1.02	1.00	0.983	0.0001232
0.20	1.191	5.46	2.96	0.218	3.633	1.00	1.000	1.00	1.00	1.000	0
0.22	1.189	4.96	2.71	0.240	2.899	1.02	0.993	0.98	1.00	1.016	0.0001145
0.24	1.186	4.54	2.50	0.261	2.347	1.04	0.986	0.96	1.00	1.033	0.0004427
0.26	1.184	4.19	2.32	0.283	1.922	1.06	0.980	0.93	1.00	1.049	0.0009592
0.28	1.182	3.88	2.17	0.304	1.589	1.08	0.973	0.91	1.01	1.065	0.001645
0.30	1.179	3.62	2.04	0.326	1.325	1.10	0.966	0.89	1.01	1.081	0.002488
0.32	1.176	3.39	1.92	0.347	1.112	1.12	0.959	0.87	1.01	1.097	0.003456
0.34	1.173	3.19	1.82	0.368	0.9380	1.14	0.952	0.86	1.02	1.113	0.004547
0.36	1.170	3.00	1.74	0.389	0.7950	1.16	0.946	0.84	1.02	1.128	0.005745
0.38	1.166	2.84	1.66	0.410	0.6764	1.18	0.939	0.82	1.02	1.143	0.007035
0.40	1.163	2.70	1.59	0.431	0.5771	1.20	0.932	0.80	1.03	1.158	0.008410
0.42	1.159	2.56	1.53	0.452	0.4936	1.22	0.925	0.79	1.04	1.173	0.009855
0.44	1.155	2.44	1.47	0.473	0.4229	1.24	0.918	0.77	1.04	1.188	0.01137
0.46	1.151	2.33	1.42	0.494	0.3627	1.26	0.911	0.76	1.05	1.203	0.01293
0.48	1.147	2.23	1.38	0.514	0.3113	1.28	0.904	0.74	1.06	1.217	0.01455
0.50	1.143	2.14	1.34	0.535	0.2673	1.30	0.897	0.73	1.07	1.231	0.01621
0.52	1.138	2.05	1.30	0.555	0.2293	1.32	0.890	0.71	1.08	1.245	0.01790
0.54	1.134	1.97	1.27	0.575	0.1967	1.34	0.883	0.70	1.08	1.259	0.01962
0.56	1.129	1.90	1.24	0.595	0.1684	1.36	0.876	0.69	1.09	1.273	0.02137
0.58	1.124	1.83	1.21	0.615	0.1439	1.38	0.869	0.68	1.10	1 286	0.02315
0.60	1.119	1.76	1.19	0.635	0.1227	1.40	0.862	0.66	1.11	1.300	0.2493
0.62	1.114	1.70	1.17	0.654	0.1043	1.42	0.855	0.65	1.13	1.313	0.02673
0.64	1.109	1.65	1.15	0.674	0.08832	1.44	0.848	0.64	1.14	1.326	0.02855
0.66	1.104	1.59	1.13	0.693	0.07446	1.46	0.841	0.63	1.15	1.339	0.03036
0.68	1.098	1.54	1.11	0.713	0.06245	1.48	0.834	0.62	1.16	1.352	0.03219
0.70	1.093	1.49	1.09	0.732	0.05203	1.50	0.828	0.61	1.18	1.365	0.03401
0.72	1.087	1.45	1.08	0.751	0.04304	1.52	0.821	0.60	1.19	1.377	0.03584
0.74	1.082	1.41	1.07	0.770	0.03528	1.54	0.814	0.59	1.20	1.389	0.03766
0.76	1.076	1.36	1.06	0.788	0.02886	1.56	0.807	0.58	1.22	1.402	0.03947

Table A-14 (Continued)

M	$\frac{T}{T^*}$	$\frac{p}{p^*}$	$\frac{p_0}{p_0^*}$	$\frac{V}{V^*}$	$\frac{fL^*}{D_h}$	M	$\frac{T}{T^*}$	$\frac{p}{p^*}$	$\frac{p_0}{p_0^*}$	$\frac{V}{V^*}$	$\frac{fL^*}{D_h}$
1.58	0.800	0.57	1.23	1.414	0.04128	2.30	0.583	0.33	2.19	1.756	0.09656
1.60	0.794	0.56	1.25	1.425	0.04309	2.32	0.578	0.33	2.23	1.764	0.09777
1.62	0.787	0.55	1.27	1.437	0.04488	2.34	0.573	0.32	2.27	1.771	0.09897
1.64	0.780	0.54	1.28	1.449	0.04667	2.36	0.568	0.32	2.32	1.778	0.1002
1.66	0.774	0.53	1.30	1.460	0.04894	2.38	0.563	0.32	2.36	1.785	0.1013
1.68	0.767	0.52	1.32	1.471	0.05020	2.40	0.558	0.31	2.40	1.792	0.1025
1.70	0.760	0.51	1.34	1.483	0.05195	2.42	0.553	0.31	2.45	1.799	0.1036
1.72	0.754	0.50	1.36	1.494	0.05368	2.44	0.548	0.30	2.49	1.806	0.1047
1.74	0.747	0.50	1.38	1.504	0.05540	2.46	0.543	0.30	2.54	1.813	0.1058
1.76	0.741	0 49	1.40	1.515	0 05711	2.48	0.538	0.30	2.59	1.819	0.1069
1.78	0.735	0.48	1.42	1.526	0.05880	2.50	0.533	0.29	2.64	1.826	0.1080
1.80	0.728	0.47	1.44	1.536	0.06047	2.52	0.529	0.29	2.69	1.832	0.1091
1.82	0.722	0.47	1.46	1.546	0.06213	2.54	0.524	0.28	2.74	1.839	0.1101
1.84	0.716	0.46	1.48	1.556	0.06377	2.56	0.519	0.28	2.79	1.845	0.1111
1.86	0.709	0.45	1.51	1.566	0.06539	2.58	0.515	0.28	2.84	1.851	0.1121
1.88	0.703	0.45	1.53	1.576	0.06699	2.60	0.510	0.27	2.90	1.857	0.1131
1.90	0.697	0.44	1.56	1.586	0.06888	2.62	0.506	0.27	2.95	1.863	0.1141
1.92	0.691	0.43	1.58	1.596	0.07015	2.64	0.501	0.27	3.01	1.869	0.1151
1.94	0.685	0.43	1.61	1.605	0.07170	2.66	0.497	0.26	3.06	1.875	0.1161
1.96	0.679	0.42	1.63	1.615	0.07324	2.68	0.492	0.26	3.12	1.881	0.1170
1.98	0.673	0.41	1.66	1.624	0.07475	2.70	0.488	0.26	3.18	1.886	0.1180
2.00	0.667	0.41	1.69	1.633	0.07625	2.72	0.484	0.26	3.24	1.892	0.1189
2.02	0.661	0.40	1.72	1.642	0.07773	2.74	0.480	0.25	3.31	1.898	0.1198
2.04	0.655	0.40	1.75	1.651	0.07919	2.76	0.476	0.25	3.37	1.903	0.1207
2.06	0.649	0.39	1.78	1.660	0.08063	2.78	0.471	0.25	3.43	1.909	0.1216
2.08	0.643	0.38	1.81	1.668	0.08205	2.80	0.467	0.24	3.50	1.914	0.1224
2.10	0.638	0.38	1.84	1.677	0.08346	2.82	0.463	0.24	3.57	1.919	0.1233
2.12	0.632	0.37	1.87	1.685	0.08485	2.84	0.459	0.24	3.64	1.925	0.1241
2.14	0.626	0.37	1.90	1.694	0.08622	2.86	0.455	0.24	3.71	1.930	0.1250
2.16	0.621	0.36	1.94	1.702	0.08757	2.88	0.451	0.23	3.78	1.935	0.1258
2.18	0.615	0.36	1.97	1.710	0.08891	2.90	0.447	0.23	3.85	1.940	0.1266
2.20	0.610	0.35	2.00	1.718	0.09023	2.92	0.444	0.23	3.92	1.945	0.1274
2.22	0.604	0.35	2.04	1.726	0.09153	2.94	0.440	0.22	4.00	1.950	0.1282
2.24	0.599	0.34	2.08	1.734	0.09281	2.96	0.436	0.22	4.08	1.954	0.1290
2.26	0.594	0.34	2.12	1.741	0.09407	2.98	0.432	0.22	4.15	1.959	0.1298
2.28	0.588	0.34	2.15	1.749	0.09532	3.00	0.428	0.22	4.23	1.964	0.1305

Table A-15 Rayleigh flow parameters for perfect gas, $k = 1.4$

M	$\dfrac{T_0}{T_0^*}$	$\dfrac{T}{T^*}$	$\dfrac{p}{p^*}$	$\dfrac{p_0}{p_0^*}$	$\dfrac{V}{V^*}$	M	$\dfrac{T_0}{T_0^*}$	$\dfrac{T}{T^*}$	$\dfrac{p}{p^*}$	$\dfrac{p_0}{p_0^*}$	$\dfrac{V}{V^*}$
0	0	0	2.40	1.27	0	1.78	0.955	1.022	1.296	1.023	0.788
0.01	0.000	0.000	2.40	1.27	0.000	0.80	0.964	1.025	1.266	1.019	0.810
0.02	0.002	0.002	2.40	1.27	0.001	0.82	0.972	1.028	1.236	1.016	0.831
0.04	0.008	0.009	2.39	1.27	0.004	0.84	0.978	1.028	1.207	1.012	0.852
0.06	0.017	0.020	2.39	1.26	0.009	0.86	0.984	1.028	1.179	1.010	0.872
0.08	0.030	0.036	2.38	1.26	0.015	0.88	0.988	1.027	1.152	1.007	0.892
0.10	0.047	0.056	2.37	1.26	0.024	0.90	0.992	1.024	1.125	1.005	0.911
0.12	0.067	0.080	2.35	1.26	0.034	0.92	0.995	1.021	1.098	1.003	0.930
0.14	0.089	0.107	2.34	1.25	0.046	0.94	0.997	1.017	1.073	1.002	0.948
0.16	0.115	0.137	2.32	1.25	0.059	0.96	0.999	1.012	1.048	1.001	0.966
0.18	0.143	0.171	2.30	1.24	0.074	0.98	1.000	1.006	1.024	1.000	0.983
0.20	0.174	0.207	2.27	1.23	0.091	1.00	1.000	1.000	1.000	1.000	1.000
0.22	0.206	0.244	2.25	1.23	0.109	1.02	1.000	0.993	0.977	1.000	1.016
0.24	0.239	0.284	2.22	1.22	0.128	1.04	0.999	0.986	0.954	1.001	1.032
0.26	0.274	0.325	2.19	1.21	0.148	1.06	0.998	0.978	0.933	1.002	1.048
0.28	0.310	0.367	2.16	1.21	0.170	1.08	0.996	0.969	0.911	1.003	1.063
0.30	0.347	0.409	2.13	1.20	0.192	1.10	0.994	0.960	0.891	1.005	1.078
0.32	0.384	0.451	2.10	1.19	0.215	1.12	0.991	0.951	0.871	1.007	1.092
0.34	0.421	0.493	2.06	1.18	0.239	1.14	0.989	0.942	0.851	1.010	1.106
0.36	0.457	0.535	2.03	1.17	0.263	1.16	0.986	0.932	0.832	1.012	1.120
0.38	0.493	0.576	2.00	1.16	0.288	1.18	0.982	0.922	0.814	1.016	1.133
0.40	0.529	0.615	1.96	1.16	0.314	1.20	0.979	0.912	0.796	1.019	1.146
0.42	0.564	0.653	1.92	1.15	0.340	1.22	0.975	0.902	0.778	1.023	1.158
0.44	0.597	0.690	1.89	1.14	0.366	1.24	0.971	0.891	0.761	1.028	1.171
0.46	0.630	0.725	1.85	1.13	0.392	1.26	0.967	0.881	0.745	1.033	1.182
0.48	0.661	0.759	1.81	1.12	0.418	1.28	0.962	0.870	0.729	1.038	1.194
0.50	0.691	0.790	1.78	1.11	0.444	1.30	0.958	0.859	0.713	1.044	1.205
0.52	0.720	0.820	1.74	1.10	0.471	1.32	0.953	0.848	0.698	1.050	1.216
0.54	0.747	0.847	1.70	1.10	0.497	1.34	0.949	0.838	0.683	1.056	1.226
0.56	0.772	0.872	1.67	1.09	0.523	1.36	0.944	0.827	0.669	1.063	1.237
0.58	0.796	0.896	1.63	1.08	0.549	1.38	0.939	0.816	0.655	1.070	1.247
0.60	0.819	0.917	1.60	1.08	0.574	1.40	0.934	0.805	0.641	1.078	1.256
0.62	0.840	0.936	1.56	1.07	0.600	1.42	0.929	0.795	0.628	1.086	1.266
0.64	0.859	0.953	1.52	1.06	0.625	1.44	0.924	0.784	0.615	1.094	1.275
0.66	0.877	0.968	1.49	1.06	0.649	1.46	0.919	0.773	0.602	1.103	1.284
0.68	0.894	0.981	1.46	1.05	0.674	1.48	0.914	0.763	0.590	1.112	1.293
0.70	0.908	0.993	1.423	1.043	0.698	1.50	0.909	0.752	0.578	1.122	1.301
0.72	0.922	1.003	1.391	1.038	0.721	1.52	0.904	0.742	0.567	1.132	1.309
0.74	0.934	1.011	1.358	1.032	0.744	1.54	0.899	0.732	0.556	1.142	1.318
0.76	0.945	1.017	1.327	1.028	0.766	1.56	0.894	0.722	0.544	1.153	1.325

APPENDIXES

Table A-15 (Continued)

M	$\dfrac{T_0}{T_0^*}$	$\dfrac{T}{T^*}$	$\dfrac{p}{p^*}$	$\dfrac{p_0}{p_0^*}$	$\dfrac{V}{V^*}$	M	$\dfrac{T_0}{T_0^*}$	$\dfrac{T}{T^*}$	$\dfrac{p}{p^*}$	$\dfrac{p_0}{p_0^*}$	$\dfrac{V}{V^*}$
1.58	0.889	0.712	0.534	1.164	1.333	2.30	0.740	0.431	0.286	1.886	1.510
1.60	0.884	0.702	0.524	1.176	1.340	2.32	0.736	0.426	0.281	1.916	1.513
1.62	0.879	0.692	0.513	1.188	1.348	2.34	0.733	0.420	0.277	1.948	1.516
1.64	0.874	0.682	0.504	1.200	1.355	2.36	0.730	0.414	0.273	1.979	1.520
1.66	0.869	0.672	0.494	1.213	1.361	2.38	0.727	0.409	0.269	2.012	1.522
1.68	0.864	0.663	0.485	1.226	1.368	2.40	0.724	0.404	0.265	2.045	1.525
1.70	0.860	0.654	0.476	1.240	1.374	2.42	0.721	0.399	0.261	2.079	1.528
1.72	0.855	0.644	0.467	1.254	1.381	2.44	0.718	0.384	0.257	2.114	1.531
1.74	0.850	0.635	0.458	1.269	1.387	2.46	0.716	0.388	0.253	2.149	1.533
1.76	0.846	0.626	0.450	1.284	1.393	2.48	0.713	0.384	0.250	2.185	1.536
1.78	0.841	0.618	0.442	1.300	1.399	2.50	0.710	0.379	0.246	2.222	1.538
1.80	0.836	0.609	0.434	1.316	1.405	2.52	0.707	0.374	0.243	2.259	1.541
1.82	0.832	0.600	0.426	1.332	1.410	2.54	0.705	0.369	0.239	2.298	1.543
1.84	0.827	0.592	0.418	1.349	1.416	2.56	0.702	0.365	0.236	2.337	1.546
1.86	0.823	0.584	0.411	1.367	1.421	2.58	0.700	0.360	0.232	2.377	1.548
1.88	0.818	0.575	0.403	1.385	1.426	2.60	0.697	0.356	0.229	2.418	1.551
1.90	0.814	0.567	0.396	1.403	1.431	2.62	0.694	0.351	0.226	2.459	1.553
1.92	0.810	0.559	0.390	1.422	1.436	2.64	0.692	0.347	0.223	2.502	1.555
1.94	0.806	0.552	0.383	1.442	1.441	2.66	0.690	0.343	0.220	2.545	1.557
1.96	0.802	0.544	0.376	1.462	1.446	2.68	0.687	0.338	0.217	2.589	1.559
1.98	0.797	0.536	0.370	1.482	1.450	2.70	0.685	0.334	0.214	2.634	1.561
2.00	0.793	0.529	0.364	1.503	1.454	2.72	0.683	0.330	0.211	2.680	1.563
2.02	0.789	0.522	0.357	1.525	1.459	2.74	0.680	0.326	0.208	2.727	1.565
2.04	0.785	0.514	0.352	1.547	1.463	2.76	0.678	0.322	0.206	2.775	1.567
2.06	0.782	0.507	0.346	1.569	1.467	2.78	0.676	0.319	0.203	2.824	1.569
2.08	0.778	0.500	0.340	1.592	1.471	2.80	0.674	0.315	0.200	2.873	1.571
2.10	0.774	0.494	0.334	1.616	1.475	2.82	0.672	0.311	0.198	2.924	1.573
2.12	0.770	0.487	0.329	1.640	1.479	2.84	0.670	0.307	0.195	2.975	1.575
2.14	0.767	0.480	0.324	1.665	1.483	2.86	0.668	0.304	0.193	3.028	1.577
2.16	0.763	0.474	0.319	1.691	1.487	2.88	0.665	0.300	0.190	3.081	1.578
2.18	0.760	0.467	0.314	1.717	1.490	2.90	0.664	0.297	0.188	3.136	1.580
2.20	0.756	0.461	0.309	1.743	1.494	2.92	0.662	0.293	0.186	3.191	1.582
2.22	0.753	0.455	0.304	1.771	1.497	2.94	0.660	0.290	0.183	3.248	1.583
2.24	0.749	0.449	0.299	1.799	1.501	2.96	0.658	0.287	0.181	3.306	1.585
2.26	0.746	0.443	0.294	1.827	1.504	2.98	0.656	0.283	0.179	3.365	1.587
2.28	0.743	0.437	0.290	1.856	1.507	3.00	0.654	0.280	0.176	3.424	1.588

Table A-16 Average roughness factors for Manning's equation†

Surface	n, $ft^{\frac{1}{6}}$
Closed conduits flowing partly full:	
Smooth brass	0.010
Corrugated-metal storm drains	0.024
Concrete culvert with bends and connections	0.013
Unfinished concrete (smooth wood form)	0.014
Clay drain tile	0.013
Rubble masonry	0.025
Lined or built-up channels:	
Unpainted steel	0.012
Planed wood	0.012
Unplaned wood	0.013
Trowel-finished concrete	0.013
Unfinished concrete	0.017
Glazed brick	0.013
Brick in cement mortar	0.015
Excavated channels:	
Clean earth (straight channel)	0.022
Earth with weeds (winding channel)	0.030
Natural streams:	
Clean and straight	0.030
Weedy reaches, deep pools	0.100

† Abstracted by permission from V. T. Chow, "Open-channel Hydraulics," Copyright 1959. McGraw-Hill Book Company.

Table A-17 Properties of water, saturation pressures†

Pressure P, psia	Temp., °F	Specific volume			Specific enthalpy			Specific entropy		
		v_f	v_{fg}	v_g	h_f	h_{fg}	h_g	s_f	s_{fg}	s_g
0.4	72.87	0.016056	792.0	792.1	40.92	1,052.4	1,093.3	0.0799	1.9762	2.0562
0.6	85.22	0.016085	540.0	540.1	53.25	1,045.5	1,098.7	0.1028	1.9186	2.0215
0.8	94.38	0.016112	411.7	411.7	62.39	1,040.3	1,102.6	0.1195	1.8775	1.9970
1.0	101.74	0.016136	333.6	333.6	69.73	1,036.1	1,105.8	0.1326	1.8455	1.9781
2.0	126.07	0.016230	173.74	173.76	94.03	1,022.1	1,116.2	0.1750	1.7450	1.9200
3.0	141.47	0.016300	118.71	118.73	109.42	1,013.2	1,122.6	0.2009	1.6854	1.8864
4.0	152.96	0.016358	90.63	90.64	120.92	1,006.4	1,127.3	0.2199	1.6428	1.8626
5.0	162.24	0.016407	73.515	73.532	130.20	1,000.9	1,131.1	0.2349	1.6094	1.8443
6.0	170.05	0.016451	61.967	61.984	138.03	996.2	1,134.2	0.2474	1.5820	1.8294
8.0	182.86	0.016527	47.328	47.345	150.87	988.5	1,139.3	0.2676	1.5384	1.8060
10.0	193.21	0.016592	38.404	38.420	161.26	982.1	1,143.3	0.2836	1.5043	1.7879
12.0	201.96	0.016650	32.377	32.394	170.05	976.6	1,146.7	0.2970	1.4762	1.7731
14.696	212.00	0.016719	26.782	26.799	180.17	970.3	1,150.5	0.3121	1.4447	1.7568
15	213.03	0.016726	26.274	26.290	181.21	969.7	1,150.9	0.3137	1.4415	1.7552
20	227.96	0.016834	20.070	20.087	196.27	960.1	1,156.3	0.3358	1.3962	1.7320
25	240.07	0.016927	16.284	16.301	208.52	952.1	1,160.6	0.3535	1.3607	1.7141
30	250.34	0.017009	13.727	13.744	218.9	945.2	1,164.1	0.3682	1.3313	1.6995
35	259.29	0.017083	11.879	11.896	228.0	939.1	1,167.1	0.3809	1.3063	1.6872
40	267.25	0.017151	10.479	10.497	236.1	933.6	1,169.8	0.3921	1.2844	1.6765
45	274.44	0.017214	9.382	9.399	243.5	928.6	1,172.0	0.4021	1.2649	1.6671
50	281.02	0.017274	8.497	8.514	250.2	923.9	1,174.1	0.4112	1.2474	1.6586
55	287.08	0.017329	7.768	7.785	256.4	919.5	1,175.9	0.4196	1.2314	1.6510
60	292.71	0.017383	7.156	7.174	262.2	915.4	1,177.6	0.4273	1.2167	1.6440
65	297.98	0.017433	6.636	6.653	267.6	911.5	1,179.1	0.4344	1.2031	1.6375
70	302.93	0.017482	6.188	6.205	272.7	907.8	1,180.6	0.4411	1.1905	1.6316
75	307.61	0.017529	5.797	5.814	277.6	904.3	1,181.9	0.4474	1.1786	1.6260
80	312.04	0.017573	5.454	5.471	282.1	900.9	1,183.1	0.4534	1.1675	1.6208
90	320.28	0.017659	4.878	4.895	290.7	894.6	1,185.3	0.4643	1.1470	1.6113
100	327.82	0.017740	4.413	4.431	298.5	888.6	1,187.2	0.4743	1.1284	1.6027
110	334.79	0.01782	4.031	4.048	305.8	883.1	1,188.9	0.4834	1.1115	1.5950

† Abstracted by permission from the "ASME Steam Tables," American Society of Mechanical Engineers, New York, 1967.

Table A-17 (Continued)

Pressure p, psia	Temp., °F	Specific volume			Specific enthalpy			Specific entropy		
		v_f	v_{fg}	v_g	h_f	h_{fg}	h_g	s_f	s_{fg}	s_g
120	341.27	0.01789	3.71	3.728	312.6	877.8	1,190.4	0.4919	1.0960	1.5879
130	347.33	0.01796	3.436	3.454	319.0	872.8	1,191.7	0.4998	1.0815	1.5813
140	353.04	0.01803	3.201	3.219	325.0	868.0	1,193.0	0.5071	1.0681	1.5752
150	358.43	0.01809	2.996	3.014	330.6	863.4	1,194.1	0.5141	1.0554	1.5695
160	363.55	0.01815	2.816	2.834	336.1	859.0	1,195.1	0.5206	1.0435	1.5641
170	368.42	0.01821	2.656	2.674	341.2	854.8	1,196.0	0.5269	1.0322	1.5591
180	373.08	0.01827	2.513	2.531	346.2	850.7	1,196.9	0.5328	1.0215	1.5543
190	377.53	0.01833	2.385	2.403	350.9	846.7	1,197.6	0.5384	1.0113	1.5498
200	381.80	0.01839	2.269	2.287	355.5	842.8	1,198.3	0.5438	1.0016	1.5454
250	400.97	0.01865	1.8245	1.8432	376.1	825.0	1,201.1	0.5679	0.9585	1.5264
300	417.35	0.01889	1.5238	1.5427	394.0	808.9	1,202.9	0.5882	0.9223	1.5105
350	431.73	0.01912	1.3064	1.3255	409.8	794.2	1,204.0	0.6059	0.8909	1.4968
400	444.60	0.01934	1.1416	1.1610	424.2	780.4	1,204.6	0.6217	0.8630	1.4847
450	456.28	0.01954	1.0122	1.0318	437.3	767.5	1,204.8	0.6360	0.8378	1.4738
500	467.01	0.01975	0.9079	0.9276	449.5	755.1	1,204.7	0.6490	0.8148	1.4639
600	486.20	0.02013	0.7496	0.7698	471.7	732.0	1,203.7	0.6723	0.7738	1.4461
700	503.08	0.02050	0.6351	0.6556	491.6	710.2	1,201.8	0.6928	0.7377	1.4304
800	518.21	0.02087	0.5481	0.5690	509.8	689.6	1,199.4	0.7111	0.7051	1.4163
900	531.95	0.02123	0.4797	0.5009	526.7	669.7	1,196.4	0.7279	0.6753	1.4032
1,000	544.58	0.02159	0.4244	0.4460	542.6	650.4	1,192.9	0.7434	0.6476	1.3910
1,200	567.19	0.02232	0.3401	0.3625	571.9	613.0	1,184.8	0.7714	0.5969	1.3683
1,400	587.07	0.02307	0.2787	0.3018	598.8	576.5	1,175.3	0.7966	0.5507	1.3474
1,600	604.87	0.02387	0.2316	0.2555	624.2	540.3	1,164.5	0.8199	0.5076	1.3274
1,800	621.02	0.02472	0.1939	0.2186	648.5	503.8	1,152.3	0.8417	0.4662	1.3079
2,000	635.80	0.02565	0.1627	0.1883	672.1	466.2	1,138.3	0.8625	0.4256	1.2881
2,500	668.11	0.02859	0.1021	0.1307	731.7	361.6	1,093.3	0.9139	0.3026	1.2345
3,000	695.33	0.03428	0.0507	0.0850	801.8	218.4	1,020.3	0.9728	0.1891	1.1619
3,208.2	705.47	0.05078	0	0.0508	906.0	0	906.0	1.0612	0	1.0612

Table A-I8 Properties of water, saturation temperatures†

Temp. t, °F	Pressure P, psia	Specific volume			Specific enthalpy			Specific entropy		
		v_f	v_{fg}	v_g	h_f	h_{fg}	h_g	s_f	s_{fg}	s_g
32.018	0.08865	0.016022	3,302.4	3,302.4	0.0003	1,075.5	1,075.5	0	2.1872	2.1872
35	0.09991	0.016020	2,948.1	2,948.1	3.002	1,073.8	1,076.8	0.0061	2.1706	2.1767
40	0.12163	0.016019	2,445.8	2,445.8	8.027	1,071.0	1,079.0	0.0162	2.1432	2.1594
45	0.14744	0.016020	2,037.7	2,037.8	13.044	1,068.1	1,081.2	0.0262	2.1164	2.1426
50	0.17796	0.016023	1,704.8	1,704.8	18.054	1,065.3	1,083.4	0.0361	2.0901	2.1262
60	0.2561	0.016033	1,207.6	1,207.6	28.060	1,059.7	1,087.7	0.0555	2.0391	2.0946
70	0.3629	0.016050	868.3	868.4	38.052	1,054.0	1,092.1	0.0745	1.9900	2.0645
80	0.5068	0.016072	633.3	633.3	48.037	1,048.4	1,096.4	0.0932	1.9426	2.0359
90	0.6981	0.016099	468.1	468.1	58.018	1,042.7	1,100.8	0.1115	1.8970	2.0086
100	0.9492	0.016130	350.4	350.4	67.999	1,037.1	1,105.1	0.1295	1.8530	1.9825
110	1.2750	0.01617	265.37	265.39	77.98	1,031.4	1,109.3	0.1472	1.8105	1.9577
120	1.6927	0.01620	203.25	203.26	87.97	1,025.6	1,113.6	0.1646	1.7693	1.9339
130	2.223	0.01625	157.32	157.33	97.96	1,019.8	1,117.8	0.1817	1.7295	1.9112
140	2.889	0.01629	122.98	123.00	107.95	1,014.0	1,122	0.1985	1.6910	1.8895
150	3.718	0.01634	97.05	97.07	117.95	1,008.2	1,126.1	0.2150	1.6536	1.8686
160	4.741	0.0164	77.27	77.29	127.96	1,002.2	1,130.2	0.2313	1.6174	1.8487
170	5.993	0.01645	62.04	62.06	137.97	996.2	1,134.2	0.2473	1.5822	1.8295
180	7.511	0.01651	50.21	50.22	148.00	990.2	1,138.2	0.2631	1.5480	1.8111
190	9.340	0.01657	40.94	40.96	158.04	984.1	1,142.1	0.2787	1.5148	1.7934
200	11.526	0.01664	33.62	33.64	168.09	977.9	1,146.0	0.2940	1.4824	1.7764
210	14.123	0.01671	27.799	27.816	178.15	971.6	1,149.7	0.3091	1.4509	1.7600
212	14.696	0.01672	26.782	26.799	180.17	970.3	1,150.5	0.3121	1.4447	1.7568
220	17.19	0.01678	23.131	23.148	188.23	965.2	1,153.4	0.3241	1.4201	1.7442
230	20.78	0.01685	19.364	19.381	198.33	958.7	1,157.1	0.3388	1.3902	1.7290
240	24.97	0.01693	16.304	16.321	208.45	952.1	1,160.6	0.3533	1.3609	1.7142
260	35.43	0.01709	11.745	11.762	228.76	938.6	1,167.4	0.3819	1.3043	1.6862
280	49.20	0.01726	8.627	8.644	249.17	924.6	1,173.8	0.4098	1.2501	1.6599
300	67.01	0.01745	6.4483	6.4658	269.7	910.0	1,179.7	0.4372	1.1979	1.6351
320	89.64	0.01766	4.8961	4.9138	290.4	894.8	1,185.2	0.4640	1.1477	1.6116
340	117.99	0.01787	3.7699	3.7878	311.3	878.8	1,190.1	0.4902	1.0990	1.5892

† Abstracted by permission from the "ASME Steam Tables," American Society of Mechanical Engineers, New York, 1967.

Table A-18 (Continued)

Temp. t, °F	Pressure p, psia	Specific volume			Specific enthalpy			Specific entropy		
		v_f	v_{fg}	v_g	h_f	h_{fg}	h_g	s_f	s_{fg}	s_g
360	153.01	0.01811	2.9392	2.9573	332.3	862.1	1,194.4	0.5161	1.0517	1.5678
380	195.73	0.01836	2.3170	2.3353	353.6	844.5	1,198.0	0.5416	1.0057	1.5473
400	247.26	0.01864	1.8444	1.8630	375.1	825.9	1,201.0	0.5667	0.9607	1.5274
420	308.78	0.01894	1.4808	1.4997	396.9	806.2	1,203.1	0.5915	0.9165	1.5080
440	381.54	0.01926	1.1976	1.2169	419.0	785.4	1,204.4	0.6161	0.8729	1.4890
460	466.87	0.01961	0.97463	0.99424	441.5	763.2	1,204.8	0.6405	0.8299	1.4704
480	566.15	0.02000	0.79716	0.81717	464.5	739.6	1,204.1	0.6648	0.7871	1.4518
500	680.86	0.02043	0.65448	0.67492	487.9	714.3	1,202.2	0.6890	0.7443	1.4333
540	962.79	0.02146	0.44367	0.46513	536.8	657.5	1,194.3	0.7378	0.6577	1.3954
580	1,326.17	0.02279	0.29937	0.32216	589.1	589.9	1,179.0	0.7876	0.5673	1.3550
600	1,543.2	0.02364	0.24384	0.26747	617.1	550.6	1,167.7	0.8134	0.5196	1.3330
640	2,059.9	0.02595	0.15427	0.18021	679.1	454.6	1,133.7	0.8686	0.4134	1.2821
680	2,708.6	0.03037	0.08080	0.11117	758.5	310.1	1,068.5	0.9365	0.2720	1.2086
700	3,094.3	0.03662	0.03857	0.07519	822.4	172.7	995.2	0.9901	0.1490	1.1390
705.47	3,208.2	0.05078	0	0.05078	906.0	0	906.0	1.0612	0	1.0612

APPENDIX B

BASIC EQUATIONS FOR SELECTED COORDINATE SYSTEMS

VECTOR IDENTITIES AND OPERATIONS

Vector: $\quad \mathbf{A} \equiv \mathbf{e}_1 A_1 + \mathbf{e}_2 A_2 + \mathbf{e}_3 A_3$

$\mathbf{e}_1, \mathbf{e}_2, \mathbf{e}_3 \equiv$ unit vectors in the coordinate directions

$A_1, A_2, A_3 \equiv$ components of the vector

Scalar: $\quad \Phi$

VECTOR IDENTITIES

$\mathbf{A} \cdot \mathbf{B} = \mathbf{B} \cdot \mathbf{A} = A_1 B_1 + A_2 B_2 + A_3 B_3$

$\mathbf{A} \cdot (\mathbf{B} + \mathbf{C}) = \mathbf{A} \cdot \mathbf{B} + \mathbf{A} \cdot \mathbf{C}$

$\mathbf{A} \times \mathbf{B} = -\mathbf{B} \times \mathbf{A} = \begin{vmatrix} \mathbf{e}_1 & \mathbf{e}_2 & \mathbf{e}_3 \\ A_1 & A_2 & A_3 \\ B_1 & B_2 & B_3 \end{vmatrix}$

$$(\mathbf{A} + \mathbf{B}) \times \mathbf{C} = (\mathbf{A} \times \mathbf{C}) + (\mathbf{B} \times \mathbf{C})$$

$$\mathbf{A} \times (\mathbf{B} + \mathbf{C}) = (\mathbf{A} \times \mathbf{B}) + (\mathbf{A} \times \mathbf{C})$$

$$\mathbf{A} \times (\mathbf{B} \times \mathbf{C}) = \mathbf{B}(\mathbf{A} \cdot \mathbf{C}) - \mathbf{C}(\mathbf{A} \cdot \mathbf{B})$$

$$\mathbf{A} \cdot (\mathbf{B} \times \mathbf{C}) = (\mathbf{A} \times \mathbf{B}) \cdot \mathbf{C} = \mathbf{B} \cdot (\mathbf{C} \times \mathbf{A}) = \begin{vmatrix} A_1 & A_2 & A_3 \\ B_1 & B_2 & B_3 \\ C_1 & C_2 & C_3 \end{vmatrix}$$

$$(\mathbf{A} \times \mathbf{B}) \cdot (\mathbf{C} \times \mathbf{D}) = (\mathbf{A} \cdot \mathbf{C})(\mathbf{B} \cdot \mathbf{D}) - (\mathbf{A} \cdot \mathbf{D})(\mathbf{B} \cdot \mathbf{C})$$

$$(\mathbf{A} \times \mathbf{B}) \times (\mathbf{C} \times \mathbf{D}) = \mathbf{B}[\mathbf{A} \cdot (\mathbf{C} \times \mathbf{D})] - \mathbf{A}[\mathbf{B} \cdot (\mathbf{C} \times \mathbf{D})]$$
$$= \mathbf{C}[\mathbf{A} \cdot (\mathbf{B} \times \mathbf{D})] - \mathbf{D}[\mathbf{A} \cdot (\mathbf{B} \times \mathbf{C})]$$

$$\nabla^2 \Phi = \nabla \cdot \nabla \Phi$$

$$\nabla^2 \mathbf{A} = (\nabla \cdot \nabla) \mathbf{A}$$

$$\nabla \cdot \nabla \times \mathbf{A} = 0$$

$$\nabla \times \nabla \Phi = 0$$

$$\nabla \times (\nabla \times \mathbf{A}) = \nabla(\nabla \cdot \mathbf{A}) - \nabla^2 \mathbf{A}$$

$$(\mathbf{A} \cdot \nabla)\mathbf{A} = \nabla \frac{|\mathbf{A}|^2}{2} - \mathbf{A} \times (\nabla \times \mathbf{A})$$

$$\nabla \times (\mathbf{A} \times \mathbf{B}) = (\mathbf{B} \cdot \nabla)\mathbf{A} - \mathbf{B}(\nabla \cdot \mathbf{A}) - (\mathbf{A} \cdot \nabla)\mathbf{B} + \mathbf{A}(\nabla \cdot \mathbf{B})$$

$$\nabla \cdot (\mathbf{A} \times \mathbf{B}) = \mathbf{B} \cdot \nabla \times \mathbf{A} - \mathbf{A} \cdot \nabla \times \mathbf{B}$$

$$\nabla(\mathbf{A} \cdot \mathbf{B}) = (\mathbf{B} \cdot \nabla)\mathbf{A} + (\mathbf{A} \cdot \nabla)\mathbf{B} + \mathbf{B} \times (\nabla \times \mathbf{A}) + \mathbf{A} \times (\nabla \times \mathbf{B})$$

Table B-1 Vector operations

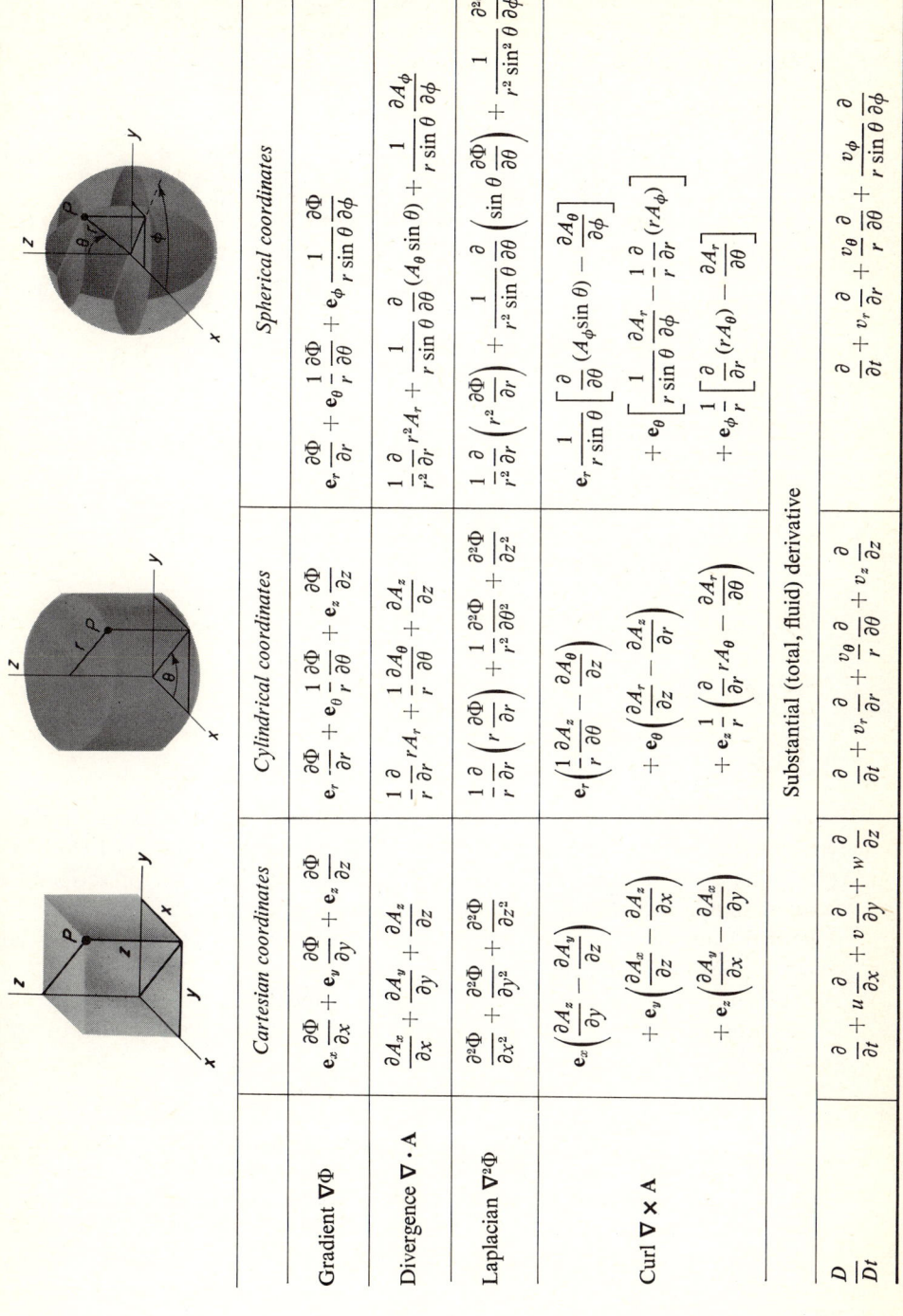

	Cartesian coordinates	Cylindrical coordinates	Spherical coordinates
Gradient $\nabla \Phi$	$\mathbf{e}_x \dfrac{\partial \Phi}{\partial x} + \mathbf{e}_y \dfrac{\partial \Phi}{\partial y} + \mathbf{e}_z \dfrac{\partial \Phi}{\partial z}$	$\mathbf{e}_r \dfrac{\partial \Phi}{\partial r} + \mathbf{e}_\theta \dfrac{1}{r}\dfrac{\partial \Phi}{\partial \theta} + \mathbf{e}_z \dfrac{\partial \Phi}{\partial z}$	$\mathbf{e}_r \dfrac{\partial \Phi}{\partial r} + \mathbf{e}_\theta \dfrac{1}{r}\dfrac{\partial \Phi}{\partial \theta} + \mathbf{e}_\phi \dfrac{1}{r \sin \theta}\dfrac{\partial \Phi}{\partial \phi}$
Divergence $\nabla \cdot \mathbf{A}$	$\dfrac{\partial A_x}{\partial x} + \dfrac{\partial A_y}{\partial y} + \dfrac{\partial A_z}{\partial z}$	$\dfrac{1}{r}\dfrac{\partial}{\partial r} r A_r + \dfrac{1}{r}\dfrac{\partial A_\theta}{\partial \theta} + \dfrac{\partial A_z}{\partial z}$	$\dfrac{1}{r^2}\dfrac{\partial}{\partial r} r^2 A_r + \dfrac{1}{r \sin \theta}\dfrac{\partial}{\partial \theta}(A_\theta \sin \theta) + \dfrac{1}{r \sin \theta}\dfrac{\partial A_\phi}{\partial \phi}$
Laplacian $\nabla^2 \Phi$	$\dfrac{\partial^2 \Phi}{\partial x^2} + \dfrac{\partial^2 \Phi}{\partial y^2} + \dfrac{\partial^2 \Phi}{\partial z^2}$	$\dfrac{1}{r}\dfrac{\partial}{\partial r}\left(r \dfrac{\partial \Phi}{\partial r}\right) + \dfrac{1}{r^2}\dfrac{\partial^2 \Phi}{\partial \theta^2} + \dfrac{\partial^2 \Phi}{\partial z^2}$	$\dfrac{1}{r^2}\dfrac{\partial}{\partial r}\left(r^2 \dfrac{\partial \Phi}{\partial r}\right) + \dfrac{1}{r^2 \sin \theta}\dfrac{\partial}{\partial \theta}\left(\sin \theta \dfrac{\partial \Phi}{\partial \theta}\right) + \dfrac{1}{r^2 \sin^2 \theta}\dfrac{\partial^2 \Phi}{\partial \phi^2}$
Curl $\nabla \times \mathbf{A}$	$\mathbf{e}_x \left(\dfrac{\partial A_z}{\partial y} - \dfrac{\partial A_y}{\partial z}\right)$ $+ \mathbf{e}_y \left(\dfrac{\partial A_x}{\partial z} - \dfrac{\partial A_z}{\partial x}\right)$ $+ \mathbf{e}_z \left(\dfrac{\partial A_y}{\partial x} - \dfrac{\partial A_x}{\partial y}\right)$	$\mathbf{e}_r \left(\dfrac{1}{r}\dfrac{\partial A_z}{\partial \theta} - \dfrac{\partial A_\theta}{\partial z}\right)$ $+ \mathbf{e}_\theta \left(\dfrac{\partial A_r}{\partial z} - \dfrac{\partial A_z}{\partial r}\right)$ $+ \mathbf{e}_z \dfrac{1}{r}\left(\dfrac{\partial}{\partial r} r A_\theta - \dfrac{\partial A_r}{\partial \theta}\right)$	$\mathbf{e}_r \dfrac{1}{r \sin \theta}\left[\dfrac{\partial}{\partial \theta}(A_\phi \sin \theta) - \dfrac{\partial A_\theta}{\partial \phi}\right]$ $+ \mathbf{e}_\theta \left[\dfrac{1}{r \sin \theta}\dfrac{\partial A_r}{\partial \phi} - \dfrac{1}{r}\dfrac{\partial}{\partial r}(r A_\phi)\right]$ $+ \mathbf{e}_\phi \dfrac{1}{r}\left[\dfrac{\partial}{\partial r}(r A_\theta) - \dfrac{\partial A_r}{\partial \theta}\right]$
		Substantial (total, fluid) derivative	
$\dfrac{D}{Dt}$	$\dfrac{\partial}{\partial t} + u \dfrac{\partial}{\partial x} + v \dfrac{\partial}{\partial y} + w \dfrac{\partial}{\partial z}$	$\dfrac{\partial}{\partial t} + v_r \dfrac{\partial}{\partial r} + \dfrac{v_\theta}{r}\dfrac{\partial}{\partial \theta} + v_z \dfrac{\partial}{\partial z}$	$\dfrac{\partial}{\partial t} + v_r \dfrac{\partial}{\partial r} + \dfrac{v_\theta}{r}\dfrac{\partial}{\partial \theta} + \dfrac{v_\phi}{r \sin \theta}\dfrac{\partial}{\partial \phi}$

APPENDIX C
DIMENSIONAL ANALYSIS

There are five fundamental entities which serve as a measure of physical systems. These entities—mass $[M]$, length $[L]$, time $[T]$, temperature $[\theta]$, and electric charge $[Q]$*—first called *dimensions* in 1871 by J. Clerk Maxwell [5], are prescribed by international standards. Mathematically, dimensions are a code for telling us how the numerical value of a quantity changes when the basic units of measurement are subjected to prescribed changes [4].

Often force $[F]$ is considered to be the fundamental entity, rather than mass $[M]$. This affords no complications, since force and mass are uniquely related by Newton's second law

$$F = \frac{ma}{g_c} \tag{C-1}$$

where g_c is a dimensional constant whose value depends upon the units chosen for mass. In some cases both force $[F]$ and mass $[M]$ are taken as basic entities, although

---
* Brackets will be used to denote dimensions.

Table C-1 Dimensions of entities

Quantity, Q	Dimension ($MLT\theta$ system)
Primary:	
Mass m	M
Length l	L
Time t	T
Temperature T	θ
Secondary:	
Force F	ML/T^2
Area A	L^2
Volume $\mathscr{V}$	L^3
Velocity V	L/T
Acceleration a (or g)	L/T^2
Density ρ	M/L^3
Specific weight γ	M/L^2T^2
Momentum mV	ML/T
Pressure and stress p, σ, τ	M/LT^2
Energy and work W	ML^2/T^2
Power $\dot{W}$	ML^2/T^3
Moment of inertia of area I	L^4
Moment of inertia of mass	ML^2
Volume coefficient of expansion β	$1/\theta$
Angle (radian measure) α	1 (dimensionless)
Torque (moment)	ML^2/T^2
Angular velocity ω	$1/T$
Angular acceleration α	$1/T^2$
Angular momentum M	ML^2/T
Modulus of elasticity E	M/LT^2
Surface tension σ	M/T^2
Viscosity, dynamic (absolute) μ	M/LT
Kinematic ν	L^2/T
Heat q	ML^2/T^2
Enthalpy h	L^2/T^2
Specific heat c	$L^2/T^2\theta$
Thermal conductivity k	$ML/T^3\theta$
Thermal diffusivity α	L^2/T
Mass diffusivity D_{AB}	L^2/T
Convective-heat-transfer coefficient h	$M/T^3\theta$
Convective-mass-transfer coefficient k_c	L/T

this requires the addition of g_c to the fundamental list of quantities. In keeping with the most common usage, we adopt the $MLT\theta$ system, electric charge [Q] being absent in our transport-phenomena relations; however, the theory and techniques presented herein apply equally well to the $FLT\theta$ system or to the $FMLT\theta g_c$ system. Table C-1 gives the dimensions of some of the most common physical quantities. The search for the correct dimensional form of an unknown equation is called *dimensional analysis*.

C-I BUCKINGHAM PI THEOREM

The Buckingham pi theorem[1] [1] provides a systematic technique for arranging a problem's governing parameters into dimensionless groups, facilitating experimentation and understanding of the physical problem.

Letting Q represent a physical quantity which controls the behavior of a phenomenon, the complete physical equation can be written as an unknown function of all of the Q's:

$$F(Q_1, Q_2, Q_3, \ldots, Q_n) = 0 \tag{C-2}$$

An equivalent form of this equation can be expressed in terms of a power series with each term containing all the Q's, giving

$$C_1 Q_1^{r_1} Q_2^{r_2} \cdots Q_n^{r_n} + C_2 Q_1^{s_1} Q_2^{s_2} \cdots Q_n^{s_n} + \cdots = 0 \tag{C-3}$$

where the C's are numerical coefficients. Dividing Eq. (C-3) by the first term, $C_1 Q_1^{r_1} Q_2^{r_2} \cdots Q_n^{r_n}$, we get

$$1 + \frac{C_2}{C_1} Q_1^{s_1 - r_1} Q_2^{s_2 - r_2} \cdots Q_n^{s_n - r_n} + \cdots = 0 \tag{C-4}$$

Assuming the equation is dimensionally homogeneous at the outset, each term, including the 1 and the zero, is dimensionless (principle of dimensional homogeneity). The numerical coefficients C_2/C_1, C_3/C_1, $\ldots$ are dimensionless; hence, the products of the powers of the Q's in each term are dimensionless. Therefore, Eq. (C-4) can be expressed dimensionally as

$$[Q_1^{x_1} Q_2^{x_2} \cdots Q_n^{x_n}] = [M^0 L^0 T^0 \theta^0] \tag{C-5}$$

where $x_1, x_2, \ldots, x_n$ are unknown exponents.

Since there may be a large number of Q's, as suggested by the partial listing in Table C-1, it is more convenient to work with a smaller number of products of Q's, for example,

$$\pi_1' = [Q_1^{a_1} Q_2^{b_1} Q_3^{c_1} Q_4^{d_1}] = [M^0 L^0 T^0 \theta^0] \tag{C-6}$$

$$\pi_2' = [Q_1^{a_2} Q_2^{b_2} Q_3^{c_2} Q_5^{d_2}] = [M^0 L^0 T^0 \theta^0] \tag{C-7}$$

$$\pi_3' = [Q_1^{a_3} Q_2^{b_3} Q_3^{c_3} Q_6^{d_3}] = [M^0 L^0 T^0 \theta^0] \tag{C-8}$$

$$\pi_{n-k}' = [Q_1^{a_{n-k}} Q_2^{b_{n-k}} Q_3^{c_{n-k}} Q_n^{d_{n-k}}] = [M^0 L^0 T^0 \theta^0] \tag{C-9}$$

where k is the number of primary dimensions and π_1', π_2', $\ldots$ are the independent nondimensional products which can be formed when using all the Q's. In this scheme, the repeating variables (Q_1, Q_2, and Q_3 in this case) which are chosen must not, when taken together, form a dimensionless product. Both sides of these equations can be raised to the $1/d_i$ power, which is equivalent to making the exponent of

---

[1] The symbol π is the conventional representation for a dimensionless product; it is in no way related to the number 3.14159.

APPENDIXES

one variable equal to unity, giving

$$\pi_1 = Q_1^{a_1/d_1} Q_2^{b_1/d_1} Q_3^{c_1/d_1} Q_4 = [M^0 L^0 T^0 \theta^0]$$

or

$$\pi_1 = [Q_1^a Q_2^b Q_3^c Q_4] = [M^0 L^0 T^0 \theta^0] \tag{C-6a}$$

Similarly,

$$\pi_2 = [Q_1^a Q_2^b Q_3^c Q_5] = [M^0 L^0 T^0 \theta^0] \tag{C-7a}$$

$$\pi_3 = [Q_1^a Q_2^b Q_3^c Q_6] = [M^0 L^0 T^0 \theta^0] \tag{C-8a}$$

$$\pi_{n-k} = [Q_1^a Q_2^b Q_3^c Q_n] = [M^0 L^0 T^0 \theta^0] \tag{C-9a}$$

For simplicity, the subscripts on the exponents have been dropped. Each set of a, b, c is unique to its respective π group. And in terms of the dimensionless parameter, Eq. (C-2) can therefore be expressed as

$$\pi_1 = f(\pi_2, \pi_3, \ldots, \pi_{n-k}) \tag{C-10}$$

The equivalence of Eqs. (C-2) and (C-10) constitutes the Buckingham pi theorem [2, 6]; stated verbally,

The number of dimensionless products used to describe a situation involving n variables is equal to the total number of variables minus the number of primary dimensions k in the problem, n − k.

Although applicable to a majority of cases, the pi theorem is limited to situations where the set of simultaneous equations formed by equating the exponents of each primary dimension to zero is linearly independent. The exceptional case, which can be treated by the more general Van Driest [7] method, is illustrated in the second of the following examples.

Example An observer watching the action of a simple pendulum "feels" that its period of vibration τ depends upon the length l of the pendulum, its mass m, and the gravitational acceleration g. Based upon his observations, what is the form of the governing equation?

Solution The functional equation is

$$F(l, m, g, \tau) = 0$$

or, in terms of dimensions,

$$F\left([L], [M], \left[\frac{L}{T^2}\right], [T]\right) = 0$$

Since there are four variables involving three primary dimensions, the number of dimensionless products is

$$n - k = 4 - 3 = 1$$

Therefore

$$\pi_1 = l^a m^b g^c \tau$$

where the exponent of τ has been taken as unity since the statement of the problem suggests that it is the dependent variable. In dimensional form

$$\pi_1 = [L^a M^b (LT^{-2})^c T] = [M^0 L^0 T^0]$$

Equating exponents on M, L, and T, respectively, we get a set of simultaneous equations for a, b, and c:

$M:$ $\quad b = 0$

$L:$ $\quad a + c = 0$

$T:$ $\quad -2c + 1 = 0$

Therefore,

$a = -c = -\frac{1}{2}$

$b = 0$

$c = \frac{1}{2}$

giving

$$\pi_1 = l^{-\frac{1}{2}} m^0 g^{\frac{1}{2}} \tau = \tau \sqrt{\frac{g}{l}}$$

or

$$\tau = \phi \sqrt{\frac{l}{g}}$$

It should be noted that the mass is not a factor in the equation as assumed by the observer, being eliminated in the process. The inclusion of irrelevant variables adds to the work but does not invalidate the method. On the other hand, the omission of a pertinent variable will give a wrong result. Therefore, the careful analyst will make sure that all possible variables are included from the outset.

Example In the heat treatment of a small sphere of steel by quenching in a constant-temperature bath, it is assumed that the temperature of the sphere above the bath temperature $(T - T_\infty)$ depends upon the following parameters:

1. The time t after immersion
2. The temperature of the sphere above the bath temperature at the time of immersion $T_i - T_\infty$
3. The thermal heat capacity of the sphere $C = \rho c V$
4. The thermal surface resistance of the sphere $R = \bar{h} A_s$

If these assumptions are correct, what is the dimensional form of the governing equation?

Solution The governing equation in functional form is

$$F(t, (T_i - T_\infty), C, R, (T - T_\infty)) = 0$$

Expressed in terms of dimensions, this becomes

$$F\left([T], [\theta], \left[\frac{ML^2}{T^2 \theta}\right], \left[\frac{ML^2}{T^3 \theta}\right], [\theta]\right) = 0$$

The number of dimensionless products is

$n - k = 5 - 4 = 1$

since there are five variables and four primary dimensions; hence,

$\pi_1 = t^a(T_i - T_\infty)^b C^c R^d (T - T_\infty)$

The exponent of the dependent variable is taken as unity. Written in dimensional form, this becomes

$\pi_1 = [T^a \theta^b (ML^2 T^{-2} \theta^{-1})^c (ML^2 T^{-3} \theta^{-1})^d \theta] = [M^0 L^0 T^0 \theta^0]$

To be dimensionless the exponents on the respective terms must be equal, giving

$M:$ $\quad c + d = 0$
$L:$ $\quad 2c + 2d = 0$
$T:$ $\quad a - 2c - 3d = 0$
$\theta:$ $\quad b - c - d + 1 = 0$

This set of equations is not linearly independent since the L equation is twice the M equation, making only three independent equations. In this case a more rigorous technique [4, 3] gives the number of dimensionless groups as

$n -$ number of independent equations $= 5 - 3 = 2$

Accordingly, two π terms should be formed; hence,

$\pi_1 = t^a(T_i - T_\infty)^b C^c (T - T_\infty) = [T^a \theta^b (ML^2 T^{-2} \theta^{-1})^c] = [M^0 L^0 T^0 \theta^0]$
$\pi_2 = t^a(T_i - T_\infty)^b C^c R = [T^a \theta^b (ML^2 T^{-2} \theta^{-1})^c (ML^2 T^{-3} \theta^{-1})] = [M^0 L^0 T^0 \theta^0]$

Equating the exponents for π_1,

$M:$ $\quad c = 0$
$L:$ $\quad 2c = 0$
$T:$ $\quad a - 2c = 0$
$\theta:$ $\quad b - c + 1 = 0$

from which

$a = c = 0 \quad$ and $\quad b = -1$

Therefore,

$\pi_1 = \dfrac{T - T_\infty}{T_i - T_\infty}$

This might have been deduced from the outset by noting that any variables having the same dimensions and occurring together will result in their ratio being dimensionless. For π_2,

$M:$ $\quad c + 1 = 0$
$L:$ $\quad 2c + 2 = 0$
$T:$ $\quad a - 2c - 3 = 0$
$\theta:$ $\quad b - c - 1 = 0$

giving

$c = -1 \quad a = 1 \quad b = 0$

Thus

$$\pi_2 = \frac{Rt}{C}$$

and the functional relation can be written

$$\frac{T - T_\infty}{T_i - T_\infty} = \phi\left(\frac{Rt}{C}\right)$$

Dimensional analysis does not reveal the nature of the function. For this case, however, the governing relation as derived analytically in Chap. 6 is given by

$$\frac{T - T_\infty}{T_i - T_\infty} = e^{-Rt/C} \tag{6-9}$$

and we note that π_2 is a negative exponential.

C-2 MOMENTUM-TRANSPORT PARAMETERS

The general fluid-dynamics problem involves at least one parameter from each of the three classes of variables—geometric, kinematic, and dynamic.

In order to limit the number of variables to keep the analysis tractable the minimum number of characteristic variables will be used. To illustrate, the use of velocity, density, and area (length squared) make the use of mass flow rate and/or volume flow rate redundant. In a similar manner, either pressure or force may be used, realizing that one can be expressed in terms of the other by using an area.

The minimum variables which will fully describe the behavior of a fluid-dynamics problem in a simple stationary system having steady flow are length l, diameter D, velocity V, density ρ, pressure drop Δp, specific weight γ, viscosity μ, modulus of elasticity E, and surface tension σ. Expressed in functional form,

$$F(l,D,V,\rho,\Delta p,\gamma,\mu,E,\sigma) = 0 \tag{C-11}$$

From this array of parameters one must be chosen as the dependent variable. For convenience in application to physical problems the performance variable Δp is chosen, and Eq. (C-11) can be rewritten explicitly in Δp to give

$$\Delta p = f(l,D,V,\rho,\gamma,\mu,E,\sigma) \tag{C-12}$$

Since these are nine variables involving three primary dimensions, there will be $n - k = 9 - 3 = 6$ dimensionless π terms. Choosing V, l, and ρ as the repeating variables, the π terms which result, when combining with the remaining variables in turn, are

$$\pi_1 = V^a l^b \rho^c \, \Delta p = [(LT^{-1})^a L^b (ML^{-3})^c (ML^{-1}T^{-2})] = [M^0 L^0 T^0] \tag{C-13}$$

$$\pi_2 = V^a l^b \rho^c D = [(LT^{-1})^a L^b (ML^{-3})^c L] = [M^0 L^0 T^0] \tag{C-14}$$

$$\pi_3 = V^a l^b \rho^c \gamma = [(LT^{-1})^a L^b (ML^{-3})^c (ML^{-2}T^{-2})] = [M^0 L^0 T^0] \tag{C-15}$$

$$\pi_4 = V^a l^b \rho^c \mu = [(LT^{-1})^a L^b (ML^{-3})^c (ML^{-1}T^{-1})] = [M^0 L^0 T^0] \tag{C-16}$$

$$\pi_5 = V^a l^b \rho^c E = [(LT^{-1})^a L^b (ML^{-3})^c (ML^{-1}T^{-2})] = [M^0 L^0 T^0] \tag{C-17}$$

$$\pi_6 = V^a l^b \rho^c \sigma = [(LT^{-1})^a L^b (ML^{-3})^c (MT^{-2})] = [M^0 L^0 T^0] \tag{C-18}$$

APPENDIXES

In selecting the repeating variables the following factors were taken into consideration.

1. One was chosen from each class of variables: $V \sim$ kinematic; $l \sim$ geometric; $\rho \sim$ dynamic.
2. Since Δp was chosen to be explicit, it could appear in only one π term.
3. The number of repeating variables was taken equal to the number of primary dimensions.

Equating exponents for π_1,

$M:$ $c + 1 = 0$

$L:$ $a + b - 3c - 1 = 0$

$T:$ $-a - 2 = 0$

we get $a = -2$, $b = 0$, and $c = -1$, giving

$$\pi_1 = \frac{\Delta p}{\rho V^2} \tag{C-19}$$

which represents the change in Euler number **Eu** and pressure coefficient $\mathbf{C}_p$ introduced in Chap. 9, namely,

$$\mathbf{Eu} \equiv \frac{p}{\rho V^2} \tag{9-34}$$

$$\mathbf{C}_p \equiv \frac{p}{\frac{1}{2}\rho V^2} \tag{9-35}$$

Solving for the exponents of π_2,

$M:$ $c = 0$

$L:$ $a + b - 3c + 1 = 0$

$T:$ $-a = 0$

from which $a = c = 0$, $b = -1$, and

$$\pi_2 = \frac{D}{l} \tag{C-20}$$

It should be pointed out here that this is a ratio of *characteristic lengths*, which might not specifically be length and diameter. For example, the same relationship holds for the aspect ratio (AR) of an aircraft wing defined by

$$\mathrm{AR} \equiv \frac{\mathrm{chord}^2}{\mathrm{area}} \tag{C-21}$$

For flow through a porous bed the relationship might involve the bed size and the particle diameter. The dimensionless characteristic would be retained in either case.

For π_3, $a = -2$, $b = 1$, and $c = -1$, yielding

$$\pi_3 = \frac{\gamma l}{\rho V^2} \tag{C-22}$$

which is the reciprocal of Froude number squared, noting that $\gamma = \rho g$, that is,

$$\mathbf{Fr} \equiv \frac{V}{\sqrt{gl}} \tag{9-33}$$

Following the same procedure, we get

$$\pi_4 = \frac{\mu}{\rho V l} \tag{C-23}$$

which is the reciprocal of the Reynolds number

$$\mathbf{Re} = \frac{Vl}{\nu} \tag{9-31}$$

since $\nu \equiv \mu/\rho$. Similarly,

$$\pi_5 = \frac{E}{\rho V^2} \tag{C-24}$$

Using Eqs. (1-10), (1-17), (15-3), and (15-7), this can readily be shown to be equal to the reciprocal of the Mach number

$$\mathbf{M} \equiv \frac{V}{c} \tag{9-32}$$

The final dimensionless product

$$\pi_6 = \frac{\sigma}{V^2 l \rho} \tag{C-25}$$

is the reciprocal of the Weber number

$$\mathbf{We} \equiv \frac{V^2 l \rho}{\sigma} \tag{9-36}$$

The general functional equation can now be written using the π terms

$$\frac{\Delta p}{\rho V^2} = f\left(\frac{D}{l}, \frac{\gamma l}{\rho V^2}, \frac{\mu}{\rho V l}, \frac{E}{\rho V^2}, \frac{\sigma}{V^2 l \rho}\right) \tag{C-26}$$

or in terms of the well-known dimensionless parameters of Chap. 9, i.e.,

$$\frac{\Delta p}{\rho V^2} = \phi\left(\frac{D}{l}, \mathbf{Fr}, \mathbf{Re}, \mathbf{M}, \mathbf{We}\right) \tag{C-27}$$

Note that the development of Eq. (13-60) makes use of some of these parameters.

C-3 HEAT-TRANSPORT PARAMETERS

The variables required to describe a general heat-transfer problem are characteristic length l, velocity V, density ρ, temperature increment ΔT, volume coefficient of expansion β, gravitational acceleration g, absolute viscosity μ, thermal conductivity k, heat-transfer coefficient h, and specific heat c_p. In functional form the relationship is

$$F(l,V,\rho,\Delta T,\beta,g,\mu,k,h,c_p) = 0 \tag{C-28}$$

There are ten variables involving four primary dimensions; hence, there will be $10 - 4 = 6$ π terms.

Choosing l, V, ρ, and k as the core variables, the dimensionless products can be written in terms of unknown exponents a, b, c, and d, giving

$$\pi_1 = l^a V^b \rho^c k^d \Delta T = [L^a(LT^{-1})^b(ML^{-3})^c(MLT^{-3}\theta^{-1})^d \theta] = [M^0 L^0 T^0 \theta^0] \tag{C-29}$$

$$\pi_2 = l^a V^b \rho^c k^d \beta = [L^a(LT^{-1})^b(ML^{-3})^c(MLT^{-3}\theta^{-1})^d \theta^{-1}] = [M^0 L^0 T^0 \theta^0] \tag{C-30}$$

$$\pi_3 = l^a V^b \rho^c k^d g = [L^a(LT^{-1})^b(ML^{-3})^c(MLT^{-3}\theta^{-1})^d(LT^{-2})] = [M^0 L^0 T^0 \theta^0] \tag{C-31}$$

$$\pi_4 = l^a V^b \rho^c k^d \mu = [L^a(LT^{-1})^b(ML^{-3})^c(MLT^{-3}\theta^{-1})^d(ML^{-1}T^{-1})] = [M^0 L^0 T^0 \theta^0] \tag{C-32}$$

$$\pi_5 = l^a V^b \rho^c k^d h = [L^a(LT^{-1})^b(ML^{-3})^c(MLT^{-3}\theta^{-1})^d(MT^{-3}\theta^{-1})] = [M^0 L^0 T^0 \theta^0] \tag{C-33}$$

$$\pi_6 = l^a V^b \rho^c k^d c_p = [L^a(LT^{-1})^b(ML^{-3})^c(MLT^{-3}\theta^{-1})^d(L^2 T^{-2}\theta^{-1})] = [M^0 L^0 T^0 \theta^0] \tag{C-34}$$

Following the procedures established in the preceding sections, we get

$$\pi_1 = \frac{k \Delta T}{\rho l V^3} \tag{C-35}$$

$$\pi_2 = \frac{\rho l V^3 \beta}{k} \tag{C-36}$$

$$\pi_3 = \frac{gl}{V^2} \tag{C-37}$$

$$\pi_4 = \frac{\mu}{\rho V l} \tag{C-38}$$

$$\pi_5 = \frac{hl}{k} \tag{C-39}$$

$$\pi_6 = \frac{\rho V l c_p}{k} \tag{C-40}$$

We recognize some of these dimensionless products; other familiar ones result from combinations of the π terms. The Reynolds number is the reciprocal of π_4

$$\mathbf{Re} = \frac{\rho V l}{\mu} \tag{9-31}$$

The Nusselt number, given by π_5, was first encountered in Chap. 12:

$$\mathbf{Nu} = \frac{hl}{k} \tag{12-149}$$

Multiplying π_4 by π_6, which retains the dimensionless nature of the product, gives the Prandtl number

$$\mathbf{Pr} = \frac{\mu c_p}{k} \tag{12-144}$$

The Grashof number, first used in Chap. 16, results from $\pi_1 \pi_2 \pi_3 (1/\pi_4^2)$;

$$\mathbf{Gr} = \frac{\beta g \rho^2 l^3 \, \Delta T}{\mu^2} \tag{C-41}$$

Different combinations of these parameters occur in heat-transfer relations, depending upon the predominant mode of transfer. In *forced convection*

$$\mathbf{Nu} = f(\mathbf{Re}, \mathbf{Pr}) \tag{C-42}$$

For example, in turbulent forced convection in tubes the governing equation is

$$\overline{\mathbf{Nu}}_D = 0.023 \mathbf{Pr}^{\frac{1}{3}} \mathbf{Re}_D^{0.8} \tag{13-119}$$

where the characteristic length is the tube diameter. In *free convection* the functional equation is

$$\mathbf{Nu} = \phi(\mathbf{Gr}, \mathbf{Pr}) \tag{C-43}$$

typified by

$$\overline{\mathbf{Nu}} = C(\mathbf{Gr}_f \mathbf{Pr}_f)^a \tag{16-45}$$

where C and a are constants which depend upon the physics of the problem and the subscript f denotes that the parameters are evaluated at the film temperature.

C-4 MASS-TRANSPORT PARAMETERS

For brevity, only mass transfer into a stream flowing under forced convection will be considered. The resulting equation will be analogous to the forced-convection heat-transfer relation of Eq. (C-42). The equation for mass transfer into a stream moving under free-convection conditions can be written down by direct analogy with Eq. (C-43).

In forced-convection mass transfer the pertinent variables are characteristic length l, density ρ, mass diffusivity D_{AB}, viscosity μ, velocity V, and the convective-mass-transfer coefficient k_c. In function form the equation is

$$F(l, \rho, D_{AB}, \mu, V, k_c) = 0 \tag{C-44}$$

having six variables with three primary dimensions; hence, there will be three π terms. Selecting l, ρ, and D_{AB} as the repeating variables, the π groups are

$$\pi_1 = l^a \rho^b D_{AB}^c \mu = [L^a(ML^{-3})^b(L^2T^{-1})^c(ML^{-1}T^{-1})] = [M^0L^0T^0] \tag{C-45}$$

$$\pi_2 = l^a \rho^b D_{AB}^c V = [L^a(ML^{-3})^b(L^2T^{-1})^c(LT^{-1})] = [M^0L^0T^0] \tag{C-46}$$

$$\pi_3 = l^a \rho^b D_{AB}^c k_c = [L^a(ML^{-3})^b(L^2T^{-1})^c(LT^{-1})] = [M^0L^0T^0] \tag{C-47}$$

giving

$$\pi_1 = \frac{\mu}{\rho D_{AB}} \tag{C-48}$$

$$\pi_2 = \frac{lV}{D_{AB}} \tag{C-49}$$

$$\pi_3 = \frac{k_c l}{D_{AB}} \tag{C-50}$$

Multiplying π_2 by the reciprocal of π_1, we get the Reynolds number

$$\mathbf{Re} = \frac{\rho V l}{\mu} \tag{9-31}$$

π_1 is identically equal to the Schmidt number

$$\mathbf{Sc} = \frac{\mu}{\rho D_{AB}} \tag{2-11}$$

and π_3 is the Sherwood number

$$\mathbf{Sh} = \frac{k_c l}{D_{AB}} \tag{14-21}$$

The resulting functional equation for forced-convection mass transfer is

$$\mathbf{Sh} = f(\mathbf{Re}, \mathbf{Sc}) \tag{C-51}$$

typified by

$$\overline{\mathbf{Sh}} = 0.664\sqrt{\mathbf{Re}}\mathbf{Sc}^{\frac{1}{3}} \tag{14-21}$$

REFERENCES

1. Buckingham, E.: *Phys. Rev.*, **2**: 345 (1914).
2. Hunsaker, J. C., and B. G. Rightmire: "Engineering Applications of Fluid Mechanics," McGraw-Hill, New York, 1947.

3. Kreith, F.: "Principles of Heat Transfer," 2d ed., International Textbook, Scranton, Pa., 1966.
4. Langhaar, H. L.: "Dimensional Analysis and Theory of Models," Wiley, New York, 1951.
5. Maxwell, J. C.: On the Mathematical Classification of Physical Quantities, *Proc. London Math. Soc.*, **3**(34): 224 (1871).
6. Shepherd, D. G.: "Principles of Turbomachinery," Macmillan, New York, 1957.
7. Van Driest, E. R.: On Dimensional Analysis and the Presentation of Data in Fluid Flow Problems, *J. Appl. Mech.*, **13**(1): A-34 (March, 1946).

ANSWERS TO ODD-NUMBERED PROBLEMS

CHAPTER 1

- **1-1.** 2.88×10^{-3} lb_f-sec/ft²
- **1-3.** 0.684 lb_m/ft³
- **1-5.** 1,600 lb_f/ft²-sec
- **1-7.** (a) 548°R
  (b) 14.72 ft³
- **1-9.** 47.4 atm
- **1-11.** 0.0497 in³/in³ psi
- **1-13.** p
- **1-15.** (a) 2 lb_m
  (b) 1.864 lb_f
- **1-17.** 1.316 lb_f

CHAPTER 2

- **2-1.** 1.5 slugs/ft³
- **2-3.** 11.1 ft-lb_f/ft
- **2-5.** 46.6 ft/sec
- **2-7.** 1.193×10^{-4} lb_f-sec/ft²
- **2-9.** 212.613°F
- **2-11.** 1.06 lb_m/hr-ft²
- **2-13.** 0.00768

CHAPTER 3

- **3-1.** Lagrangian analysis: observer moves with the fluid (system); eulerian analysis: observer concentrates on fixed region (control volume)
- **3-3.** (a) 0
  (b) 1
  (c) Undefined in this region
  (d) Undefined in this region
  (e) Undefined in this region
- **3-5.** 4.75 lb_m/ft³
- **3-7.** (ab) $n = 1 \to p\mathcal{V} = \text{const}$; (bc) constant-volume; (cd) adiabatic: if reversible $\to$ isentropic: $n = k$ in general $k > 1$; $p\mathcal{V}^k = \text{const}$ for $k > 1$, p de-

799

creases with $\mho$ more than for $n = 1$; (da) constant pressure

3-9. 20.7×10^3 ft-lb$_f$

3-11. $RT \ln\left(\dfrac{v_2 - b}{v_1 - b}\right) + a\left(\dfrac{1}{v_2} - \dfrac{1}{v_1}\right)$

3-13. 13.31×10^3 ft-lb$_f$

3-15. (a) < 0; $= 0$
(b) $= 0$; < 0
(c) > 0; $= 0$
(d) $= 0$; > 0
(e) $= 0$; < 0
(f) $= 0$; > 0
(g) $= 0$; $= 0$

3-17. (b) $du = 0$, since u is a property

3-19. (a) -60 Btu
(b) 60 Btu

3-21. (a) $140.46°$F
(b) 0.001376 ft^3/lb$_m$
(c) 144.93 Btu/lb$_m$
(d) 145.39 Btu

3-23. 610 Btu

3-25. (a) No
(b) None

3-27. Not valid

3-29. $16,550$ Btu

3-31. No

3-33. du

3-35. $1,238.42$ Btu
$1,259.73$ Btu

3-37. 0.10882 Btu/°R

3-39. 0.061 Btu/lb$_m$-°R

CHAPTER 4

4-1. 19.86 lb$_f$

4-3. $PA_{\text{base 2}} = F_{\text{total}} = W + F_1 + F_2 = PA_{\text{base 1}}$

4-5. (a) 47.6 psig; 43.3 psig
(b) 38.8 psig; 21.65 psig

4-9. 14.532 psia; -0.168 psig

4-11. -1.2 ft

4-13. 12.35 lb$_f$

4-15. 20.95 lb$_f$; $\tfrac{1}{8} x_1$

4-17. 1.1 in

4-19. -0.115 psig

4-21. 14.964 psia

4-23. (a) 13.501 psia
(b) 4.54 ft H$_2$O gauge

4-25. 13.22 psia

4-27. $\dfrac{12 \text{ in H}_2\text{O}}{S[1 + (a/A)]}$

4-29. 21.85 psia

4-31. 0.141 lb$_f$/ft

4-33. $\tfrac{3}{4}\sigma_2$, compound curvature

4-37. 9.326 in

4-39. 240 lb$_f$/in^2

4-41. (a) 9 ft
(b) $3,727.2$ lb$_f$

4-43. (a) $-4,018$ ft-lb$_f$ (clockwise)
(b) The maximum moment will occur when the benzene side is full and the oil side is empty; $-12,690$ ft-lb$_f$ (clockwise)

4-45. $-1,150$ ft-lb$_f$/ft (clockwise)

4-47. 11.3 ft

4-49. -247.95 ft-lb$_f$ (clockwise)

4-51. 3.84 ft

4-53. The square gate will open first at a depth of 18.65 ft

4-55. 0.75

4-57. $276,000$ ft^3

4-59. $\geq (32 - 3d^2)/6d^2$

4-61. 1.35 lb$_f$

4-63. 11

4-65. 50.2 lb$_f$; 748.8 lb$_f$

4-67. 221.18 in^3

4-69. 6.56 ft-lb$_f$

4-71. $319,488$ ft-lb$_f$

CHAPTER 5

5-1. (a) $86.36°$F
(b) $92.86°$F
(c) $75°$F

5-3. $319.02°$F; $179.85°$F

5-5. 232.39 Btu/hr-ft

5-7. 0.0103 ft

5-9. $2\pi k(T_1 - T_2)$
$\times \left[\dfrac{L}{\ln(r_2/r_1)} + \dfrac{2}{(1/r_1) - (1/r_2)}\right]$

5-11. 5.55 Btu/hr-ft-°F

5-13. $30,959.75$ Btu/hr-ft^2

5-15. $\dfrac{4\pi k_0(T_1^3 - T_2^3)}{3(1/r_1 - 1/r_2)}$

5-17. 21.4 hr-°F/Btu; 17.24 hr-°F/Btu

5-19. 21.12 Btu/hr-ft^2-°F

5-21. 39.9%

5-23. 198.34 Btu/hr-ft

5-25. $324.21°$F

5-27. (a) 6.61 Btu/hr
(b) 3.17 Btu/hr

5-29. 477.6 Btu/hr

ANSWERS TO ODD-NUMBERED PROBLEMS

5-31. 98.38 watts
5-33. The 0.032-in-thick fin
5-35. The tapered fin dissipates more heat per unit weight than the straight fin
5-37. $4\sum_{n=0}^{\infty}\sum_{m=0}^{\infty} \dfrac{(-1)^{n+m}\exp[-(\lambda_n^2+\mu_m^2)^{1/2}z]}{(\lambda_n L)(\mu_m l)} \times \cos\lambda_n x \cos\mu_m y$
5-39. The water is not boiling
5-41. (a) 0.342
(b) 51,350 Btu/hr
5-43. 1,965 Btu/hr
5-45. 12,428.49°F

CHAPTER 6

6-1. Yes
6-3. No
6-5. 80°F
6-7. (a) 0.0227
(b) 473.41°F
6-9. 2.592 in
6-11. 513.6°F
6-13. 578.4°F
6-15. 52.6 hr
6-17. 11.58 hr
6-19. 34.83°F
6-21. 1.884 min
6-23. 28.93 min
6-25. 4.57 hr
6-27. 53.8 min
6-29. 173.44°F
6-31. (a) 6.08 days
(b) Yes, 75.25°F
6-33. 267.42°F

CHAPTER 7

7-1. (a) Yes
(b) Yes
(c) Yes
(d) The gradient causing mass transport depends on what system is being investigated
(e) No, unless external forces are applied
(f) No, unless external forces are applied

7-3. $+5.17 \times 10^{-3}$ lb$_m$/ft²-hr
7-7. 0.51 ft²/hr
7-9. 0.715 ft²/hr
7-11. 0.884 ft²/hr; 0.974 ft²/hr
7-13. 6.48×10^{-5} ft²/hr
7-15. (a) $\frac{3}{4}$
(b) 41 lb$_m$/lb-mole; 32 lb$_m$/lb-mole; 44 lb$_m$/lb-mole
(c) 0.0051 lb-mole/ft³; 0.001275 lb-mole/ft³; 0.00382 lb-mole/ft³
(d) 0.196; 0.804
(e) 0.0408 lb$_m$/ft³; 0.1680 lb$_m$/ft³; 0.2088 lb$_m$/ft³
(f) 18 ft/hr; -6 ft/hr
(g) 19.3 ft/hr; -4.7 ft/hr
(h) 10.54×10^{-3} lb-mole/ft²-hr
25.5×10^{-3} lb-mole/ft²-hr
-15.3×10^{-3} lb-mole/ft²-hr
(i) 0.815 lb$_m$/ft²-hr; -0.672 lb$_m$/ft²-hr; 0.146 lb$_m$/ft²-hr
7-19. 0.0045 lb$_m$/ft²-hr
7-21. 3.68 min
7-23. 0.0216 lb$_m$/ft²-hr
7-25. 2.29 ft/hr
7-27. Mass loss insignificant compared with initial mass
7-29. 8,270 lb$_m$/hr-ft²
7-31. 49.2×10^{-4} lb$_m$/hr-ft²
7-33. $\dfrac{2\rho D_{AB}}{x_1}\dfrac{\omega_1-\omega_2}{\ln(x_2/x_1)}$
which from the example in Sec. 7-7 is twice that of the circular duct
7-35. 26.25 lb$_m$/hr
7-37. $0.1105\,\omega_A|_{3\text{in}}$ is the same as at 1 in
7-39. 3.69 days
7-41. 17.6×10^{-2} ft
7-43. 0.997 lb$_m$ fluid/lb$_m$ mixture
7-45. 3.44×10^{-3} lb$_m$/ft³

CHAPTER 8

8-1. 3.48×10^{20} sec^{-1}; 8.64×10^{-11} cm
8-3. 0.65
8-5. 94 Btu/hr-ft²
8-7. 643°F
8-9. 10,470°R
8-11. $\alpha = \epsilon$
8-13. $\frac{1}{2}$
8-17. 2.02×10^4 Btu/hr
8-19. 0.17; 0.28
8-21. 0.045

ANSWERS TO ODD-NUMBERED PROBLEMS

8-23. 1100°R
8-25. 7,260 Btu/hr
8-27. 364 Btu/hr
8-29. 5.83×10^5 Btu/hr
8-31. (a) 1.9×10^6 Btu/hr
 (b) 5.3×10^6 Btu/hr
8-33. 317°F
8-35. 144 Btu/hr-ft
8-37. -4.46 Btu/hr
8-39. (a) 8.5×10^4 Btu/hr
 (b) 2.48×10^5 Btu/hr
8-41. 0.042
8-43. 3.39×10^3 Btu/hr-ft^2
8-45. 9.795×10^4 Btu/hr-ft

CHAPTER 9

9-1. $x^2 - y^2 = 3$
9-3. $3(e^x - e) - y^3 = -9$
9-5. (a) $xy = c$
 (b) $25x\mathbf{i} + 25y\mathbf{j}$
9-7. $2\mathbf{i} + 12\mathbf{j} + 1\mathbf{k}; 4\mathbf{i} + 108\mathbf{j} + 6\mathbf{k}$
9-9. $2xyz\,dx + (x^2z + z)\,dy$
 $\qquad + (x^2y + y - 3)\,dz$

 $2xyz\dfrac{dx}{dt} + (x^2z + z)\dfrac{dy}{dt}$
 $\qquad + (x^2y + y - 3)\dfrac{dz}{dt}$

9-11. 1.768×10^{-4} ft^3/sec
9-13. (a) 2,010 ft/sec
 (b) No
9-15. (a) 25 ft/sec
 (b) 1,600 lb$_f$
9-17. 15 ft/sec
9-19. 25.8 ft/sec^2
9-23. 36.48 ft^3
9-25. -42.9 ft/sec^2
9-27. 15.5 psia
9-29. $\tan^{-1}\left(\dfrac{1}{1.732 - 2g/a}\right)$

CHAPTER 10

10-1. 40 ft/sec
10-3. $9\pi/64$
10-5. 1 ft/sec
10-7. 2 ft/sec
10-9. $42\rho\,(6 + 4t^2)$
10-11. 78.15 ft/sec
10-13. 106 ft/sec
10-15. 610 ft/sec
10-19. $\dfrac{d\rho}{\rho} + \dfrac{dA}{A} + \dfrac{dV}{V} = 0$
10-21. 7.07 ft/sec
10-23. 4,950 ft/sec
10-25. (a) 14.5 ft/sec
 (b) 10.52 psi
10-27. 2.58 lb$_f$/ft
10-31. (a) 14,900 lb$_f$
 (b) 14,900 lb$_f$
 (c) 16,110 lb$_f$
10-33. $-\frac{2}{3}(\rho V_\infty^2 d)$
10-35. $(p_2 - p_1)(2h) - \rho[V_1^2(2h) + 2V_1 h]$
10-37. 289 lb$_f$/ft
10-39. 124.52 lb$_f$
10-41. $+4,570$ lb$_f$
10-43. 4,920 lb$_f$
10-45. 12.45 lb$_f$
10-47. 9.07 lb$_m$/sec
10-49. 76.2 lb$_f$ (to the left)
10-51. (a) 10.1 lb$_f$ (to the left)
 (b) 3.255 ft/sec (to the left)
10-53. $V_R = \left(V_e + \dfrac{p_e A_e}{\dot{m}}\right)\ln\dfrac{m_o}{m_o - \dot{m}t}$
10-55. 164 ft^3/sec
10-57. 0.2185 ft^3/sec
10-59. 39.0 ft/sec
10-61. $V_1 = \sqrt{2gh_2(\gamma_b/\gamma_a)}$
10-63. 110.2 ft^3/sec
10-65. 31.8 ft/sec
10-67. 0.6175 ft^3/sec
10-69. 263 ft; 2.38 in
10-71. No
10-73. 25.6 lb$_m$/sec
10-77. (a) $\dfrac{d^2z}{dt^2} + \dfrac{2g}{L}z = \dfrac{p_1 - p_2}{\rho L}$
 (b) $z = 0;\ dz/dt = 0$ at $t = 0$
 (c) $\dfrac{h}{2}(1 - \cos\sqrt{2g/L}\,t)$
10-79. 330.2 lb$_f$
10-81. (a) 79.8 lb$_f$-ft
 (b) 14.5 hp
10-83. $-23,100$ ft-lb$_f$
10-85. $\rho Q[r_1 V_1 \cos\alpha_1$
 $\qquad + r_2(V_{r_2}\cos\beta_2 - r_2\omega r_2)]$
10-87. 1,865 hp
10-89. 7,909 hp
10-91. -1.18×10^6 Btu/hr
10-93. 1.134 hp (input)
10-95. $T_f = kT_i$

CHAPTER 11

11-1. The flow is irrotational; $\Omega = 0$

11-3. $\left(\dfrac{x}{z} - 2xy\right)\mathbf{i} + \left(1 - \dfrac{y}{z}\right)\mathbf{j} + (2yz - 3)\mathbf{k}$

11-7. $\Gamma = 0$

11-9. (a) $\mathbf{i} + 2\mathbf{j}$
(b) $\tfrac{4}{15}\mathbf{i} + 3\mathbf{j}$
(c) $6\mathbf{i} + 9\mathbf{j}$

11-11. (a) Yes
(b) No

11-13. 32.2 ft/sec^2

11-15. No

11-17. $\tfrac{3}{2}x^2y + \dfrac{y^3}{3} + \dfrac{x^2}{2} + C$

11-19. $r = 4$ ft
$r = +4.53; -3.53$

11-23. $\phi = k\theta + \text{const}$

11-25. $\phi = -x^2 + y^2 + 2y + \text{const}$

11-27. (a) $6y\mathbf{i} - 6x\mathbf{j}$
(b) -6
(c) The flow is rotational
(d) $x = 0, y = 0$

11-29. $\psi = -r\cos\theta + \text{const}$

11-33. (a) $u = 3y^2; v = -2x$
(b) No

11-35. $\psi = -e^y x + \text{const}$

11.39. $x = \tfrac{4}{3}$; (b) $\tfrac{1}{8}$

11-41. $\Lambda_\theta/2\pi + V_\infty r \sin\theta$

11-43. $15r(1 - 1/r^2)\sin\theta$

11-45. $\theta_s = 48.6°$
$-7{,}000\rho$ lb$_f$/ft^2

11-47. $\pm 14.48°$

CHAPTER 12

12-3. $\dfrac{dp}{dz} = \mu\left(\dfrac{d^2V_z}{dr^2} + \dfrac{1}{r}\dfrac{dV_r}{dr}\right)$

12-5. (a) $\dfrac{\partial V_r}{\partial r} + \dfrac{V_r}{r} = 0$

$\rho\left(V_r\dfrac{\partial V_r}{\partial r}\right) = -\dfrac{\partial p}{\partial r} + \mu\left(\dfrac{1}{r}\dfrac{\partial V_r}{\partial r} - \dfrac{V_r}{r^2}\right)$

$\dfrac{\partial p}{\partial \theta} = 0$

$\rho\left(V_r\dfrac{\partial V_r}{\partial r}\right) = -\dfrac{\partial p}{\partial z} + \mu\left(\dfrac{\partial^2 V_z}{\partial r^2} + \dfrac{1}{r}\dfrac{\partial V_z}{\partial r}\right)$

(b) $r = a$: $V_r = \dfrac{Q}{2\pi a(1)}$;
$V_\theta = 0, V_z = 0$

$r = b$: $V_r = \dfrac{Q}{2\pi b(1)}$;
$V_\theta = 0, V_z = 0$

(c) $V_r = \dfrac{Q}{2\pi r}$

12-7. 0.215 rad/sec

12-9. (a) $\dfrac{\partial v}{\partial y} = 0 \rightarrow v = c(x)$

(b) $\rho v \dfrac{\partial u}{\partial y} = -\dfrac{\partial p}{\partial x} + \mu\dfrac{\partial^2 u}{\partial y^2}$

$\rho u \dfrac{\partial v}{\partial x} = -\dfrac{\partial p}{\partial y} + \mu\dfrac{\partial^2 v}{\partial x^2}$

$p \neq p(z)$
$\mathbf{V} \neq \mathbf{V}(t,x) \neq \sqrt{u^2 + v^2}$
$u \neq u(x); v \neq v(x)$
$v = v_i; \dfrac{\partial p}{\partial y} = 0; p = p(x)$

(d) $v = v_i$ (injection velocity)

(e) $u = \dfrac{1}{\rho v_i}\dfrac{dp}{dx}$

$\times \left(\dfrac{2e^{Ay} - e^{Ah} - e^{-Ah}}{e^{Ah} - e^{-Ah}} - \dfrac{y}{h}\right)$

where $A = \rho v_i/\mu$

12-11. $h_l = 0.0484$ ft

12-13. $r = R/\sqrt{2}$

12-15. 0.00223 gal/min

12-17. (a) $\Delta P = 0$
(b) 11.95 lb$_f$/in^2
(c) 11.95 lb$_f$/in^2

12-19. 0.1 in

12-21. 1.15 lb$_f$/ft^2

12-23. 0.108 ft^3/sec

12-25. 0.0278 ft^3/sec

12-27. 2.65 ft^3/sec-ft

12-29. Yes

12-31. 0.0612

12-33. $\dfrac{1}{3}\dfrac{\delta}{x}$

12-35. (a) $6.01\mu V_\infty$
(b) $1.85\left(\dfrac{v}{V_\infty}\right)^{1/2}$

12-37. The solutions are comparable

12-39. $\dfrac{\delta^*}{x} = \dfrac{1.75}{\sqrt{\text{Re}_x}}$

12-41. 0.0709 lb$_f$

12-43. 0.383 lb$_f$

ANSWERS TO ODD-NUMBERED PROBLEMS

12-47. 0.3145 Btu/lb_m-hr
12-49. 0.0380 sec^{-2}
12-53. 2,910 Btu/hr-ft^2
12-55. $\bar{h} = 4.23$ Btu/hr-ft^2-°F
$h_x = 3.0$ Btu/hr-ft^2-°F
12-57. 447 Btu/hr
12-59. 12.53 Btu/hr-ft^2
12-61. 231 Btu/hr-ft^2
12-63. 1.095 ft
12-65. (a) 47.5 Btu/hr-ft^2-°F
(b) 1,120 Btu/hr-ft
(c) 2.32°F
(d) 2.275 lb_f/in^2
12-67. 35.5 ft; 3,980 ft
12-71. 189.5 Btu/hr

CHAPTER 13

13-3. 0.525ρ
13-5. 0.012
13-7. 22.4 ft/sec
13-9. No
13-11. 0.0293
13-13. 0.709 psi
13-15. (a) 91.5 ft-lb_f/lb_m
(b) 11.55 hp
13-17. (a) 64,200
(b) 15.65 psf/ft
(c) 25.85 psf/ft
13-19. (a) 0.0442 Moody; 0.0475 Nikuradse
(b) 0.033 Moody; 0.0275 Nikuradse
(c) 0.020 Moody; 0.0195 Nikuradse
13-21. 2.09 psi
13-23. (a) 17 ft^3/sec
(b) 255 hp
13-25. 10 in
13-27. 40-ft length
13-29. 92.4 psi
13-31. 49.2 in
13-33. 15.2 psi
13-35. 2.1 ft
13-37. 18.3 psi
13-39. 454.7 ft
13-41. $p_B = 73.6$ psi; $p_C = 72.69$ psi
13-43. 7.1 ft^3/sec; 6.96 ft^3/sec
13-45. 1,470 lb_f
13-47. 13.0 lb_f
13-49. 1.91 ft
13-51. (a) 0.553 in
(b) 0.64 lb_f
13-53. 0.412 lb_f/ft^2
13-55. (a) 0.143 ft
(b) 0.01788 ft
(c) 108.8 ft/sec
13-57. (a) 0.195 ft
(b) 1.7 ft on a side
13-59. 0.1124 lb_f
13-63. 1.45 in
13-65. 0.00196 by Eq. (13-92); 0.00188 by Eq. (12-95)
13-67. 940 lb_f/ft^2
13-69. 0.0395 lb_f; 2.545 ft/sec^2
13-71. 1.36×10^{-3} lb_f
13-73. (a) 25.2 in
13-75. 39.4 Btu/hr-ft^2-°F
13-77. 7,490 Btu/hr
13-79. (a) 0.0975 ft
(b) 98,000 Btu/hr
13-81. 135 kw
13-83. 40.0°F
13-85. 1,908 Btu/hr
13-87. 175 Btu/hr-ft^2-°F
13-89. 25,750 Btu/hr
13-91. 29.6 ft
13-93. Yes
13-95. (a) 2,256 Btu/hr-ft^2
(b) 2,185 Btu/hr-ft^2
13-97. (a) 822 Btu/hr-ft^2-°F
(b) 928 Btu/hr-ft^2-°F
13-99. (a) 0.867 lb_m/sec
(b) 164
(c) 50,800 Btu/hr; 86.3°F
(d) 85.15°F
(e) 51,800 Btu/hr; 87.55°F
13-101. 0.675 Btu/hr-ft^2-°R
13-103. 112.1°F
13-105. 49.2 Btu/hr-ft
13-107. 56,400 Btu/hr
13-109. 7.43 Btu/hr
13-111. 52,600 Btu/hr
13-113. 1,050 Btu/hr-ft^2-°F
13-115. 119°F; no

CHAPTER 14

14-1. (a) 16.79×10^{-2} lb_m/hr-ft^2
(b) 3.26×10^{-2} lb_m/hr-ft^2
14-3. 55.3×10^{-14} lb_m/sec
14-5. 0.795 lb_m/hr
14-7. (a) 6.45×10^{-2} lb_m/hr-ft^2
(b) 0.37 lb_m/hr-ft^2
14-9. 2.38×10^{-3} lb_m/ft^2

ANSWERS TO ODD-NUMBERED PROBLEMS

14-11. 155 ft/hr
14-13. 0.0276 lb_m/hr; 1.086 hr
14-15. 5.41 lb_m/hr
14-17. 31.0 lb_m/hr
14-19. 165 lb_m/hr
14-21. von Kármán: 0.0155 ft/hr; eddy transport: 16.4 ft/hr
14-23. 2.46×10^{-5} lb_m/hr-ft^2
14-25. 9.6 ft
14-27. 1.265 lb_m/hr
14-29. 0.096 ft/hr
14-31. 5.8 ft
14-33. 6.91 hr
14-35. 3.04×10^{-4} lb_m
14-37. (a) 40.6 ft/hr
(b) 264 ft/hr

CHAPTER 15

15-1. (a) 4,280 ft/sec
(b) 1,392 ft/sec
15-3. (a) 1,116 ft/sec
(b) 1,078 ft/sec
(c) 970 ft/sec
(d) 850 ft/sec
15-5. 0.895
15-7. 30°; 2,260.4 ft/sec
15-9. (a) 0.707
(b) 1,130.2 ft/sec
15-13. 27.75 psia; 1.88 in^2
15-15. 10.85 psia
15-17. 13.12 psia; 443°R; 837 ft/sec
15-19. 2.16
15-21. 487 psia
15-23. 1,263 ft/sec
15-25. (a) 53.25 psia
(b) 539°R
(c) 0.85
(d) 471°R
(e) 2.45 lb_m/sec
15-27. 626 lb_m/ft^2-sec
15-29. (b) 2.285 lb_m/sec
15-33. 0.595
15-35. 0.867 psia; 235°R; 1.387 ft
15-37. 342 psia; 758°R
15-39. $A_x = 1.64 A_t$
15-41. 43.4 psia
15-43. (a) 37.2 psi
(b) $A_x = 3.06$ in^2
15-45. $ds = -c_p \dfrac{k-1}{k} \dfrac{dp_0}{p_0}$

15-47. 452.8 ft
15-49. 81.3 ft; 31.3 psia
15-51. 755°R
15-53. 309 Btu/lb_m

CHAPTER 16

16-1. (a) Forced
(b) Mixed
(c) Forced
(d) Free
(e) Free
(f) Free
16-5. (a) 115 Btu/hr
(b) 116 Btu/hr
16-7. 420 Btu/hr
16-9. 806.0 Btu/hr
16-11. (a) 130 Btu/hr
(b) 160 Btu/hr
16-17. 26.6 Btu/hr
16-19. 66.35 Btu/hr; 49.4 Btu/hr (cylinder)
16-21. (a) 247 Btu/hr
(b) 315 Btu/hr
(c) 334 Btu/hr
16-23. 37.8 Btu/hr
16-25. 3,177 Btu/hr
16-27. 24.5
16-29. 93.9 Btu/hr
16-31. 19.8 Btu/hr
16-33. (a) 71.6 Btu/hr
(b) 13,000 Btu/hr
16-35. 608 Btu/hr
16-37. (a) 1.848 Btu/hr-ft^2
(b) 7.24 Btu/hr-ft^2
16-39. 0.675 Btu/hr-ft^2
16-41. 1.485 Btu/hr-ft^2-°F

CHAPTER 17

17-1. $\dfrac{\partial V_r}{\partial r} + \dfrac{2V_r}{r} = 0$
17-3. 2.525×10^{-4} in
17-7. Stratified
17-9. 0.569 psi
17-11. 0.134 psi
17-13. 0.13°F
17-15. 511°F
17-17. 177,500 Btu/hr-ft^2
17-19. (a) 373,000 Btu/hr-ft^2
(b) 435,000 Btu/hr-ft^2
17-21. 173,000 Btu/hr; 170 lb_m/hr

17-23. 26,400 Btu/hr; 42.5 lb_m/hr
17-25. 114.3 Btu/hr-ft²-°F; 3,595 lb_m/hr

CHAPTER 18

18-1. $\bar{h}_a = 21.85$ Btu/hr-ft²-°F; $\bar{h}_w = 43.7$ Btu/hr-ft²-°F
$U = 14.5$ Btu/hr-ft²-°F
18-3. (a) 49.2 ft²
(b) 63
18-5. 3.7 ft²
18-7. 2 passes; 29 tubes; 24.2 ft
18-9. (a) 61.5 ft
(b) 46.5 ft
18-11. 10.8 lb_m/min
18-13. 7.09 ft
18-15. (a) 2,333.5 ft²; 387.4°F
(b) 2,440 ft²
18-17. 27,360 Btu/hr
18-19. (a) 192,500 Btu/hr; $T_{h_o} = 101.5$°F; $T_{w_o} = 95.67$°F
(b) 219,000 Btu/hr; $T_{h_o} = 96.25$°F; $T_{w_o} = 99.20$°F
18-21. (a) $h_o = 4,040$ Btu/hr-ft²-°F; $h_i = 36.3$ Btu/hr-ft²-°F; $U = 39.0$ Btu/hr-ft²-°F
(b) 26.1 Btu/hr-ft²-°F
(c) 57.4 ft

CHAPTER 19

19-1. 11.6 ft/sec
19-3. Rapid
19-5. 0.006289
19-7. (a) 2,510 ft³/sec; rapid
(b) 6,260 ft³/sec; rapid
19-9. 4.6 ft diam
19-11. 3.70 by 1.85 ft
19-13. 10.5 by 21.0 ft
19-15. (a) $y = 8.95$ ft; $b = 10.3$ ft
(b) 5.81 ft/sec
19-17. $b = 9.05$ ft; $y = 6.30$ ft; 4.27 ft/sec
19-19. (a) 15.54 ft
(b) 6.19 ft-lb_f/lb_m
19-21. 3.27 ft; 20.95 hp per foot of width
19-23. Yes; 11.64 ft; 2.394 ft-lb_f/lb_m

CHAPTER 20

20-1. 0.285 lb_m/ft-sec
20-3. 15.8 ft/sec (upward)
20-5. 0.0624 ft³/hr
20-7. (a) 347 psi
(b) 46.74 lb_f/ft²
20-9. 1.1 ft/sec
20-11. 7.15×10^4 lb_f/ft²
20-13. 33.8 lb_m/sec
20-15. 2,480 lb_f/ft²

NAME INDEX

Acousta, A. J., 288
Addoms, J. N., 641, 642, 661
Albertson, M. L., 715
Arpaci, V. S., 151, 153, 169, 174, 185
Artley, J., 30

Baird, D. C., 194, 225
Baker, O., 629, 630, 661
Barrer, R. M., 764
Barton, J. R., 715
Baumeister, T., 751
Beckmann, W., 596, 618
Beckwith, T. G., 35, 61
Bennett, C. O., 225, 525, 536, 544, 748
Berenson, P. J., 642, 661, 662
Bergles, A. E., 646, 661
Besant, W., 620, 661
Bird, R. B., 30, 225, 228, 540, 544, 748
Black, J., 705, 715
Blasius, H., 438
Blick, E. F., 748
Boelter, L. M. K., 491, 515, 544
Boggs, J. H., 748
Bonilla, C. F., 662
Borishanski, V. M., 661
Boussinesq, J., 445, 515
Bowman, R. A., 677, 694
Brewley, L. V., 151
Bromley, L. A., 645, 648, 661
Brown, D. R., 125, 151
Brown, G. G., 721, 748
Buck, N. L., 35, 61
Buckingham, E., 788, 797

Callandar, H. L., 427, 438
Cambel, A. B., 592, 715
Carslaw, H. S., 151, 198, 225, 515, 657, 662
Cess, R. D., 248, 256, 265, 502, 515
Chaddock, J. B., 225
Chandrasekhar, S., 256, 265
Chao, B. T., 640, 662, 663
Chapin, J. H., 515
Chapman, A. J., 122, 123, 151
Chapman, W., 189, 225
Chen, M. M., 652, 662
Chiang, T., 602, 618
Chilton, T. H., 531, 544
Cho, S. M., 640, 662
Choi, H. Y., 30, 190, 225, 545, 641, 647, 652, 661, 663
Chow, V. T., 698, 715, 778
Christian, W. J., 523, 544

Cichelli, M. T., 662
Clark, J. A., 641, 662, 663
Colburn, A. P., 486, 515, 531, 544
Collins, R. E., 725
Coppage, J. E., 692, 694
Coral, M., 377
Courant, R., 566, 592
Cramp, W., 748
Crank, J., 198, 225
Cross, H., 473, 515
Cryder, D. S., 662
Csanady, G. T., 627, 662

Daily, J. W., 288, 625, 662
Darcy, L., 748
Diessler, R. G., 544
Dow, J. E., 662
Downey, G. L., 346
Dowty, E. L., 181, 185
Drake, R. M., 265, 544, 617
Drew, T. B., 655, 662
Dropkin, D., 503, 515, 611, 617
Dunkle, P. V., 234, 265
Dusinberre, G. M., 180, 185
Dwyer, O. E., 503, 515
Dyer, D. F., 225

Eckert, E. R. G., 229, 265, 521, 544, 603, 604, 613, 614, 617
Egbert, R. B., 257, 265
Elenbaas, W., 608, 617
Ergun, S., 729, 748
Erk, S., 167, 171, 173, 185, 225
Eskinazi, S., 14, 102

Fick, A., 225
Finalborgo, A. C., 662
Fishenden, M., 617
Florschuetz, L. W., 640, 662
Folley, K. W., 377
Forster, H. K., 638, 640, 662
Friedricks, K. O., 566, 592

Gaylord, E. W., 416, 438
Gibson, A. H., 466, 515
Giedt, W. H., 494, 495, 515
Gilliland, E. R., 189, 225, 537, 544
Gilmour, C. H., 662
Globe, S., 611, 617
Goldstein, S., 618
Graetz, L., 427, 438, 523, 544
Gregg, J. L., 600, 602, 604, 612, 618, 652, 663
Grief, R., 638, 662

Griffith, P., 644, 648, 662, 663
Grigull, U., 167, 171, 173, 185, 225
Grimson, E. D., 499, 500, 515
Grober, H., 167, 171, 173, 185, 217, 225
Grosh, R. J., 502, 515
Gubareff, G. G., 231, 265
Gunther, F., 637, 662

Hagen, G., 389, 438
Halliday, D., 30
Hamilton, D. C., 239, 241, 242, 265
Happel, J., 742, 748
Harleman, D. R. F., 288
Harper, W. B., 125, 151
Hartnett, J. P., 521, 544, 652, 662
Harvey, E. N., 624, 662
Hausen, H., 429, 438
Haworth, D. R., 181, 185
Heisler, M. P., 165–167, 170–173, 185, 225
Henderson, F. M., 715
Hermann, R., 602, 617
Herschel, C., 716
Hinkle, B. L., 744, 748
Hobson, M., 538, 544
Hoe, R. J., 503, 515
Holman, J. P., 30, 151, 158, 185
Hottel, H. C., 256–260, 265, 768
Howarth, L., 399, 400, 438
Hughes, W. F., 416, 438
Hunsaker, J. C., 797
Hutchison, T. S., 194, 225

Isbin, H. S., 663
Iverson, H. W., 515

Jackson, T. W., 604, 617, 748
Jaeger, J. C., 151, 198, 225, 515, 657, 662
Jakob, M., 30, 497, 515, 610, 611, 617, 652, 654, 655, 662, 692, 694
Janssen, J. E., 231, 265
Jeans, Sir James, 189, 225
Jenkins, R., 533, 544
Jennings, B. H., 592, 715
John, J. E. A., 550, 572, 592
Johnson, K. R., 137
Johnson, V. E., 625, 662
Johnston, H. F., 499, 515
Jonassen, F., 544
Jouquet, E., 705, 716

Karamchiet, K., 377
Katz, D. L., 494, 516
Kaufman, S. J., 503, 516
Kay, J. M., 748
Kayan, C. F., 138, 151
Kaye, J., 602, 617
Kays, W. M., 428, 429, 431, 438, 500, 501, 515, 525, 544, 685, 691, 694
Keenan, J. H., 59
Kennel, W. E., 662
Kestin, J., 439
Keyes, F. G., 59
Kezios, S. P., 523, 544
King, W. J., 609, 617
Kinslow, R., 716
Kirkbride, C. G., 653, 662
Knapp, R. T., 627, 628, 662
Knudsen, J. G., 494, 516
Koh, J. C. Y., 652, 662
Kozlov, B. K., 630, 662
Kreith, F., 158, 185, 637, 662, 684, 694, 754, 756, 761, 798
Kreysig, E., 185
Kuethe, A. M., 357, 377
Kutateladze, S. S., 645, 662
Kyte, J. R., 608, 617

Lang, C., 636, 662
Lange, N. A., 621
Langhaar, H. L., 393, 438, 798
Latzko, H., 457, 491, 516
Lay, J. E., 35, 61
Leidenfrost, J. G., 636, 662
Leppert, G., 498, 516
LeRoy, N., 661
Leva, M., 732, 737, 745, 746, 748
Levy, S., 604, 617, 647, 662
Lewis, B., 716
Lienhard, J., 639, 663
Lightfoot, E. N., 30, 255, 288, 540, 544, 748
Linton, W. H., Jr., 537, 544
Lo, R. K., 500, 501, 515
Lockhart, R. W., 633, 634, 662
London, A. L., 657, 659, 662, 685, 692, 694
Lubarsky, B., 503, 516
Lundgren, T. S., 393, 438
Luo, H. L., 716
Lyon, R. N., 503, 516

McAdams, W. H., 258–260, 265, 496, 498, 516, 605–607, 617, 641, 647, 651, 654, 655, 662, 768
McElroy, W. D., 662
Mach, E., 705, 716

Macklin, M., 691, 694
Madden, A. J., 608, 617
Manning, R., 700, 716
Martinelli, R. C., 544, 633, 634, 662
Masi, J. F., 767
Maxwell, J. C., 786, 798
Mediratta, O. P., 705, 715
Merte, H., 641, 662
Mesler, R. B., 638, 662
Metais, B., 613, 614, 617
Millsaps, K., 617, 618
Milne-Thomson, L. M., 357, 377
Minden, C. S., 662
Moissis, R., 662
Moody, L. F., 458–460, 516, 577, 592
Moore, F. D., 638, 662
Morgan, W. R., 239, 241, 242, 265
Mori, Y. Y., 612, 618
Muehlbauer, J. C., 659, 662
Mueller, A. C., 677, 694
Murphy, G., 716
Muskat, M., 725
Myers, J. E., 225, 525, 536, 544, 748

Nagle, W. N., 655, 662, 677, 694
Navier, M., 379, 438
Nelson, A. L., 377
Neumann, Franz, 655
Nikuradse, J., 393, 438, 448, 449, 451, 452, 516, 577, 592
Nukiyama, S., 636, 663
Nusselt, W., 427, 438, 649, 663, 682, 694

Olson, R. M., 716, 752, 753
Oppenheim, A. K., 716
Ostrach, S., 598, 599, 618
Othmer, D. F., 737, 746, 748
Owczarek, J. A., 566, 592
Owens, W. L., 633, 663

Pai, Shih-I, 383, 438, 447, 516
Pao, R. H. F., 346, 357, 377, 592
Parker, J. D., 748
Pease, R. N., 716
Perry, J. H., 694, 763
Picornall, C., 662
Pigford, R. L., 226, 515, 537–539, 545
Piret, E. L., 608, 617, 663
Planck, Max, 265
Plesset, M. S., 640, 663
Pohlhausen, E., 418, 438
Pohlhausen, K., 602, 617, 618
Poiseuille, J., 389, 738
Poisson, S. D., 379, 438

Pope, A., 377
Prandtl, L., 379, 438, 446, 452, 483, 516, 534, 544
Priestley, A., 748

Rauscher, M., 377
Rayleigh, Lord, 620, 627, 663
Redheffer, R. M., 346
Reichardt, H., 449, 516
Reynolds, O., 275, 288, 530, 544
Rich, B. R., 609, 618
Riemann, G. F. B., 663
Rightmire, B. G., 797
Robbers, J. A., 661
Robin, T. T., 638, 663
Rohsenow, W. M., 30, 190, 225, 545, 641, 643, 644, 646, 647, 652, 661, 663
Rudolf, C., 662

Sabersky, R. H., 288
St. Venant, B. de, 379, 438
Sarofim, A. G., 256, 265
Saunders, O. A., 617
Scheidegger, A. E., 748
Schenck, H., 180, 185
Schetzer, J. D., 357, 377
Schiller, L., 393, 439
Schlichting, H., 383, 398, 439, 447, 450–453, 456, 475, 481, 482, 516
Schmidt, E., 596, 618
Schneider, P. J., 30, 122, 151, 159, 176, 185
Schubauer, G. B., 475, 516
Schuh, H., 597, 618
Seban, R. A., 504, 516, 640, 657, 659, 662
Shames, I. H., 14, 102, 346, 592
Shapiro, A. H., 14, 550, 566, 572, 592, 716
Shepherd, D. G., 102, 797
Sherwood, T. K., 226, 537–539, 544, 545
Shimazaki, T. T., 504, 516
Shippy, D. J., 716
Sieder, E. N., 493, 516
Simons, D. B., 715
Sissom, L. E., 592, 748
Skramstad, H. K., 475, 516
Smith, G. M., 346, 655
Snyder, N. W., 638, 663
Sokolnikoff, I. S., 346
Sonntag, R. E., 14, 35, 61
Spalding, D. B., 545
Sparrow, E. M., 248, 256, 265, 438, 600, 602, 604, 612, 618, 652, 662, 663
Spiegel, M. R., 151

NAME INDEX

Squire, H. B., 603, 618
Starr, J. B., 438
Stefan, J., 655, 663
Stein, F. M., 377
Stepanoff, A. J., 748
Stewart, W. E., 30, 225, 288, 540, 544, 748
Stokes, G. G., 379, 439
Streeter, V. L., 102, 716
Sun, K., 663
Sunderland, J. E., 137, 181, 185, 659, 662
Sutherland, W., 189, 226

Tate, C. E., 493, 516
Taylor, H. S., 716
Taylor, G. I., 534, 545
Ten Broeck, H., 682, 694
Thodos, G., 538, 544

Thomas, G. B., 346
Thomson, L. M., 357
Tietjens, O., 438
Tong, L. S., 630, 632, 648, 663
Torborg, R. H., 231, 265
Treybal, R. E., 226, 540, 545
Tribus, M., 663

Van Driest, E. R., 789, 798
Van Wylen, G. J., 14, 35, 61
Vliet, G. C., 498, 516
Von Kármán, T., 403, 438, 535

Waddel, H., 720, 748
Wark, K., 61
Weber, H., 663
Weisbach, J., 463, 516

Welty, J. R., 226, 540, 545
Wen, C. Y., 743, 748
Whiteley, A. H., 662
Wicks, C. E., 226, 540, 545
Wiebelt, J. A., 235, 239, 248, 265
Wilke, C. R., 193, 226
Wilkes, G. B., 30
Wilson, R. E., 226, 540, 545
Wittke, D. D., 640, 663
Worthing, A. G., 30

Young, G., 515
Yuan, S. W., 377

Zenz, F. A., 737, 746, 748
Zerkle, R. D., 181, 185
Zuber, N., 640, 644, 662, 663
Zwick, S. A., 640, 663

SUBJECT INDEX

Absorption band, figure, 256
Absorption coefficient, monochromatic, 257
Absorptivity, 228
Acceleration:
 absolute, 298
 convective, 272
 linear, uniform, 280
 local, 272
Acoustic wave, 547
Adhesion, 76
Adiabatic flow with friction, 573
Aerostatics, 73
Alternate depths, 703
Angle factor, 238
Apparent stress, 484
Archimedes' principle, 85
Arnold diffusion cell, 200
Aspect ratio, 793
Atmosphere:
 iso-lapse, 74
 isothermal, 73
 model, 73
 standard, 74
Atomic volume, table, 190

Barometer, 69
Bernoulli function, 316
Bernoulli's equation, 315, 328
Bessel equation, 122
Bessel functions, 123
 table, 765
Binary diffusion equation, 198
Binary mixture, 188
Biot modulus, 153
Blackbody, 230
Blake-Kozeny equation, 728
Blasius solution, 397
Blowing parameter, boundary layer, 520
Bluff bodies, flow over, 481
Body force, 65
Boiling:
 flow, 636
 pool, 636
Boiling crisis, 638
Boiling curve, 636
Borda mouthpiece, 464
Boundary, convective, 115, 142
Boundary condition, non-homogeneous, 132
Boundary layer, 348
 hydrodynamic: laminar, 398
 turbulent, 473
 mass transfer: laminar, 518
 turbulent, 525

Boundary layer:
 thermal: laminar, 416
 turbulent, 485
Bubble dynamics, 619
Buckingham π-theorem, 788
Bulk flow, 195
Bulk fluid boiling, 639
Bulk modulus of elasticity, isothermal, 10
Bulk temperature, 429, 490
Buoyancy, 84
Buoyant force, 85
Burke-Plummer equation, 730
Burnout point, 638

Capacitance, thermal, 156
Capillarity, 78
Catalytic surface, 209
Cauchy-Riemann equations, 358
Cavitation, 624
Cavitation number, 627
Centigrade scale, 35
Centipoise, 21
Centistoke, 21
Centroids, table, 762
Chemical reactions:
 heterogeneous, 196, 209
 homogeneous, 196, 213
Chezy equation, 700
Chilton-Colburn analogy, 531
Choked flow, 563
Choking velocity, 742
Circulation, 351
Clausius-Clapeyron equation, 640
Clausius inequality, 54
Coefficient:
 absorption, monochromatic, 257
 of compressibility, 9
 convective-mass-transfer, 200, 517
 diffusion, 189
 discharge, 571
 drag, 484
 linear expansion, 9
 overall heat-transfer, 116
 performance (COP), 54
 pressure, 793
 skin-friction: average, 402, 477
 local, 402, 477
 tension, 9
 volume expansion, 9, 596
Cohesion, 76
Colburn equation, 487, 490
Colburn factor, 486
Compressibility, 10

Compressibility:
 coefficient of, 9
 isothermal, 9
Concentration, 7
Condensation:
 dropwise, 649
 film, 649
Conductive heat transfer:
 multidimensional, 127
 one-dimensional, 107
 transient, chart, 164
Conductive shape factor, table, 137
Conductivity, thermal, 22
Configuration factor, 238
Conservation, principle of, 27
Conservation equations:
 energy, 28
 momentum, 28
 species, 28
Continuity equation, 29, 290
 differential form, 295
 integral form, 293
Continuum, 5
Contractions, 462
Control surface, 32
Control volume, 32, 289
 noninertial, 307
Convective boundary, 115, 142
Convective mass transfer:
 coefficient of, 200
 forced, 186
 interphase, 186
 natural, 186
 turbulent, 186
Convective transport, 115
Counterdiffusion, equimolal, 192
Critical depth, 703
Critical heat flux, 638
Critical point, 38
Critical state, 557
Curl operator, 349, 785
Curvilinear squares, 134
Cycle, 43
Cylinder:
 rotating: flow about, 371
 force on, 373
 stationary: flow about, 367
 force on, 369

D'Alembert's paradox, 370
Darcy-Weisbach equation, 389, 700
Darcy's law, 724
Density, 6, 24
Departure from nucleate boiling (DNB), 638

SUBJECT INDEX

Derivative, fluid (total, substantial), 272, 785
Difference equation, 178
Diffuse, 229
Diffuser, 465
Diffusion, 25
  binary, 198
  coefficient of, 189
  forced, 186
  heterogeneous chemical reaction, 209
  homogeneous chemical reaction, 213
  molecular, 186
  pressure, 186
  thermal, 186
  transient: multidimensional systems, 217
    one-dimensional systems, 216
    table, 218
    semi-infinite medium, 215
Diffusivity:
  mass, 27, 188
    binary, 26
    in gas, 189, 763
    in liquids, 193, 763
    in solids, 194, 764
  momentum, 27
  thermal, 24, 27, 105
Dimensional analysis, 786
Dimensions, 12, 786
Discharge coefficient, 571
Displacement thickness, 405, 477
Dittus-Boelter equation, 491
Doublet, 366
Drag:
  on bluff bodies, 481
  form, 479
  rotating cylinder, 373
  stationary cylinder, 370
Drag coefficient, 484
  for single particle, figure, 720
Dropwise condensation, 655
Dufour effect, 187
Dynamic pressure, 279

Eddy transport theory, 531
Eddy viscosity, 277, 455
Eigenvalues, 164
Electrical analog, 108, 136
Electromagnetic radiation, thermal, 227
Electromagnetic spectrum, 227
Emissive power, 231
  monochromatic, 233
  monochromatic hemispherical, 247
Emissivity, 231
  carbon dioxide: figure, 258
    table, 768

Emissivity:
  water vapor, figure, 259
Emittance, hemispherical, 236
Energy:
  conservation of, 322
  internal, 45
  kinetic, 45
  potential, 45
Energy equation, 411
  compressible flow, 552
  integral boundary-layer, 421
Energy generation, 105
Entropy, 55
Entry length, 386, 393
  thermal, 428
  velocity, 428
Equation of state, 8
Equilibrium, 32
  mechanical, 86
  neutral, 86
  relative, 280
  stable, 86
  thermodynamic, 40
  unstable, 86
Equipotential line, 358
Equipotential surfaces, 359
Equivalent length, 461
Ergun equation, 731
Error function, Gauss, 160
Euler number, 279, 793
Eulerian viewpoint, 15, 271
Euler's equation of motion, 314
Excess temperature, 639
Expansion:
  abrupt, 464
  linear, coefficient of, 9
  volume, coefficient of, 9

Falling-rate period, 217
Fanno flow, 576
  table, 774
Fick's diffusion equation, 25
Fick's law, 188
Fick's second law of diffusion, 198
Fictive film, 204
Field intensity, 17
Film boiling, 645
Film temperature, 426, 490
Film theory, 204
Filter cake, 733
Filtration, 733
Fins, 118
  effectiveness, 123
  efficiency, 123
First law of thermodynamics, 43, 322, 328
Fixed bed, 718
Flat plate:
  flow over: laminar, 395
    turbulent, 473
  heat transfer from: laminar, 416

Flat plate:
  heat transfer from: turbulent, 485
  mass transfer from: laminar, 518
    turbulent, 525
Flotation, 84
Flow boiling, 636, 646
Fluid derivative, 272, 785
Fluidization, 736
Fluidized bed, 718
  mass-transfer, 536
Flux, 16
  geometric, 244
  mass, 25, 188
    binary mixtures, 26
    molal, 189
Flux density, 16
Fouling factors, 689
Fourier equation, 105
Fourier heat-conduction equation, 22, 103
Fourier modulus, 156
Free-convection heat transfer:
  enclosed layers, 605
  horizontal cylinders, 605
  horizontal plates, 605
  inclined plates, 605
  miscellaneous shapes, 605
  vertical cylinder, 605
  vertical plates, 594, 605
Free vortex, 370
Frequency, 227
Friction factors, 388
  in commercial piping, 459
  in concentric annuli, figure, 392
  in ducts, figure, 392
  in packed beds, 727
  in pipe flow, 455
Froude number, 279, 698, 794
Fully-developed flow, 386

Gas constants, table, 767
Gas dynamics, 10
Gauss error function, 160
Geometric flux, 244
Geopotential altitude, 75
Gradient, 17
  energy, figure, 316
  hydraulic, figure, 316
  potential, 17
Graetz number, 429, 614
Graetz solution, 427
Grashof number, 594, 796
Gravitation constant, 10
Gray body, 247
Grober chart, 167

Hagen-Poiseuille equation, 389, 728
Hausen equation:
  heat transfer, 429
  mass transfer, 524

Head:
elevation, 317
piezometric, 317
pressure, 317
velocity, 317

Heat, 42
Heat engine, 51
Heat exchanger:
compact, 691
counter-flow, 669
cross-flow tube-bank, 669
effectiveness, 682
flat-plate, 669
parallel-flow, 669
shell-and-tube, 669

Heat flux, 104
Heat generation, 111
Heisler chart, 165
Helmholtz instability theory, 644
Hemispherical emittance, 236
Hohlraum, 230
Hydraulic diameter, 392
Hydraulic jump, 703
Hydraulic radius, 727
Hydrodynamics, 10
Hydrometer, 86
Hydrostatic equation, 69
Hydrostatic paradox, 69
Hydrostatics, 68
Hypersonic flow, 550

Ideal gas, 50
Integral method:
frictional drag, 403
heat transfer, 476
mass transfer, 527
Intensity:
radiation, 236
of turbulence, 474

Interface, fluid, 76
Interphase mass transfer, 540
Irrotationality, 350
Isentropic flow, 553
parameters, table, 770
Isothermal bulk modulus of elasticity, 10
Isothermal compressibility, 9

J factor, 486, 536

Kirchhoff's law, 233, 248
Kopp's law, 190
Kutta-Joukowski theorem, 374

Lagrangian viewpoint, 15, 271
Laminar flow, 275
Laplace equation, 105
Laplacian, 106
Lapse rate, 74
Leidenfrost phenomenon, 638
Lewis number, 27, 521
Lift, 370
rotating cylinder, 373

Lift:
stationary cylinder, 370
Line integral, 351
Linear-momentum equation:
differential form, 312
integral form, 300
noninertial control volume, 310

Liquid metals, 501
Log-mean mass fraction, 202
Log mean temperature difference, 674
Loss coefficient, table, 462
Lumped thermal system, 154

Mach angle, 550
Mach cone, 548
Mach number, 279, 548, 794
Manning equation, 700
Manometer, 72
Manometry, 71
Mass density, 188
Mass diffusion equation, 196
Mass diffusivity, 27, 188
Mass fraction, 188
Mean free path, 6
Mean-value theorem, 291
Membrane, diffusion through, 207
Meniscus, 71, 77
Minor losses, 461
Mixing-cup temperature, 429
Mixing-length theories, 445
Modified Bessel functions, table, 766
Modulus of elasticity, isothermal bulk, 10
Molal density, 189
Molar concentration, 189
Mole fraction, 189
Molecular volume, table, 190
Molecular weight, 189
Moment of inertia, 81
table, 762
Moment of momentum, 318
Momentum, 298
Momentum diffusivity, 27
Momentum thickness, 406, 477
Moody diagram, 459

Navier-Stokes equations, 380
general, 384
incompressible, 385
Networks, piping, 472
Newton, Sir Isaac, 11
Newtonian cooling, 157
Newtonian fluids, 21
Newton's law of cooling, 115
Newton's viscosity equation, 18
Nodal point, 139
Noncircular ducts, 392
Nonuniform flow, 275

Normal-shock parameters, table, 772
Normal-shock waves, 566
Nozzles:
compressible flow, 561
convergent, 562
convergent-divergent, 564

Nucleate boiling, 640
Number of transfer units (NTU), 685
Numerical solution, 138
Nusselt modulus, 419
Nusselt number, 796

Open-channel flow:
classification of, 697
optimum cross-section, 701
specific energy, 702
transition in, 702

Order-of-magnitude analysis, 395
Overall heat-transfer coefficient, 116
approximate, 668

Packed beds, mass transfer in, 536
Packing area, 727
Packings, 731
Parallel-axis theorem, 81
Parallel flows, 393
Partial pressure, 188
Pascal's law, 66
Path lines, 275, 292
Peclet number, 429, 502
Percolation, 726
Perfect fluid, 347
Perfect gas, 8, 49
Performance, coefficient of (COP), 54
Permeability, 207
table, 725
Phase, 36
Phase change, conduction with, 655
Planck radiation functions, table, 235
Planck's constant, 228
Plotting, flux, 135
Pohlhausen solution, 418
Poise, 21
Poisson equation, 105
Pool boiling, 636
Porosity, 722
Potential flow, 350
Power law, one-seventh-, 451, 527
Prandtl number, 27, 418, 519, 796
Prandtl-Taylor analogy, 534
Prandtl's boundary-layer equations, 397
Prandtl's mixing length, 446, 532
Pressure, 6, 65

SUBJECT INDEX

Pressure:
 absolute, 69
 barometric, 70
 center of, 80
 coefficient, 279
 gauge, 69
 local atmospheric, 69
 standard atmospheric, 69
 vacuum, 69
Pressure prism, 80
Process:
 irreversible, 51
 isentropic, 42
 isovolumetric, 42
 polytropic, 41
 quasi-steady, 40
 reversible, 40
Properties:
 of alloys, table, 761
 extensive, 36
 of gases, table, 756
 intensive, 36
 of liquid metals, table, 756
 of liquids, table, 756
 of metals, table, 761
 of nonmetals, table, 754
 of water, table, 779

Quality, 36, 46
 flow, 631

Radiation:
 intensity, 236
 thermal, 227
Radiosity, 249
Ray tracing, 250
Rayleigh equation, 623
Rayleigh flow, 582
 parameters, table, 776
Rayleigh line, 568
Rayleigh number, 595n.
Reaction rates:
 finite, 210
 instantaneous, 209
Reciprocity theorem, 243
Rectilinear flow, 362
Recuperator, 667
Reference temperature, free-
  convection, 604
Reflectivity, 228
Regenerator, 691
Relative roughness, 460
Relaxation, 138
 block, 140
 over, 140
Reradiating surfaces, 252
Reservoir, thermal, 51
Residual equation, 141
 table, 143
Resistance, thermal, 108, 156
Reynolds analogy:
 heat transfer, 485
 mass transfer, 530
Reynolds number, 277, 279,
  794

Reynolds stress, 444, 484
Rheology, 22
Rotation, 348
 uniform, 283
Roughness, relative, 460
Roughness factors, open-
  channel flow, table, 778
Roughness regimes, pipe
  flow, 449

Saltation velocity, 742
Schmidt number, 27, 519, 797
Second law of thermody-
  namics, 50
 Clausius, statement of, 53
 Kelvin-Planck, statement
  of, 53
Separation, 479
 of variables, 128, 162
Shape-factor:
 conduction, 134
 radiation, 238
Shear stress, 4, 18
Shear velocity, 448
Shell mass balance, 200
Sherwood number, 520, 797
Shielding, radiation, 254
Shock waves, 566
Sieder-Tate equation, 492
Similar profiles, 597
Similarity, 397, 519
 dynamic, 278
 geometric, 278
 kinematic, 278
Similitude, 277
Singular point, 353
Sink, 51, 364
Skin-friction coefficient, 402
Slip ratio, 630
Sluice gate, 703
Solar constant, 261
Solids transport, 718, 741
 dense phase, 742
 dilute phase, 744
 intermediate phase, 743
Sonic velocity, 547
Soret effect, 187
Sound, speed of, 546
Source (thermal reservoir), 51
Source flow, 363
Source-sink flow, 364
Specific gravity, 7, 72
Specific heat, 24, 49
Specific volume, 7
Specific weight, 7
Spectroradiometric curves,
  233
Specular reflectors, 229
Speed of sound, 546
Sphericity, 721
Stability, 86
Stagnation state, 554
Stanton number, 486
State, thermodynamic, 36
Steady flow, 274

Stefan-Boltzmann constant,
  231
Stefan's problem, 655
Stoichiometry, 209
Stoke, 21
Stokes' hypothesis, 382
Stokes' law, 719
Stokes' theorem, 353
Strain:
 normal, 383
 rate of, 382
 shearing, 383
Stream filament, 269
Stream function, 354
Stream tube, 269
Streamline, 269, 292
 equation for, 354
Stress:
 apparent, 484
 normal, 323
 Reynolds, 484
 shear, 323
Subcooled boiling, 639
Submerged surfaces:
 curved, 82
 plane, 79
Subsonic flow, 550
Substantial derivative, 272,
  785
Suction parameter, boundary
  layer, 520
Supercavitation, 627
Superficial fluid velocity, 723
Superposition, principle of,
  130
Supersonic flow, 550
Surface force, 65
Surface tension, 77
 of liquids, table, 621
System, thermodynamic,
  32, 289

Taylor's vorticity transport
  theory, 447
Teledeltos paper, 136
Temperature, 7, 33
 absolute zero, 34, 53
 bulk, 429, 490
 excess, 639
 film, 426, 490
 mixing-cup, 429
 reference, free-convection,
  604
Tension, coefficient of, 9
Tensor, 16
Terminal settling velocity,
  719
Thermal capacitance, 156
Thermal conductivity, 22–24
Thermal diffusivity, 24,
  27, 105
Thermal efficiency, 51
Thermal resistance, 108
Thermodynamics, 31
Thermometer, 34

Thermometer:
  Celsius, 34
  Fahrenheit, 34
Third law of thermodynamics, 56
Throat, 559
Time average, 276
Total derivative, 272, 785
Total head, 316
Transient conduction, 162–180
Transmissivity, 228
Transonic flow, 550
Triple point, 38
Troposphere, 74
Turbulence intensity, 474
Turbulent flow, 275
Two-component flow, 717
Two-phase flow, 628, 717

Uniform flow, 275, 362
Units, 12
Universal velocity, 449, 532
Unsteady flow, 274

Vapor pressure, 38
Velocity potential, 357
Vena contracta, 462
View factor, 238
Viscometer:
  Engler (German), 21
  Redwood Admiralty (English), 21
  Redwood No. 1 (English), 21
  Saybolt Furol, 21
  Saybolt Universal, 21
Viscosity:
  absolute, 5, 18, 752
  dynamic, 18, 752
  eddy, 277, 445
  kinematic, 20, 27
  kinematic conversion, table, 751
  second coefficient of, 383
Viscous-dissipation function, 415
Void fraction, 630, 722

Volume, atomic, table, 190
Von Kármán analogy, 535
Von Kármán integral technique, 527
Vorticity, 348

Wake, 479
Water table, 705
Wave:
  acoustic, 547
  surface, 696
Wavelength, 227
Weber number, 280, 794
Weight, 10
Wien's displacement law, 234
Work, 38
  flow, 323
  shaft, 323
  surface, 323

Zeroth law of thermodynamics, 33